Hydrological Processes and Water Security in a Changing World

8th Global FRIEND-Water Conference 2018

Proceedings of the International Association of Hydrological Sciences (PIAHS) Volume 383, 2020

Beijing, China
6 - 9 November 2018

Editors:

Z. Yu
C. Lu
D. Yu
J. Cai
G. Mahe
A. Mishra
C. Cudennec
H.A.J. Van Lanen
D. Orange
A. Amani

ISBN: 978-1-7138-2058-1

Printed from e-media with permission by:

Curran Associates, Inc.
57 Morehouse Lane
Red Hook, NY 12571

Printed with permission by Curran Associates, Inc. (2021)

International Association of Hydrological Sciences
Centre for Ecology and Hydrology
Wallingford, Oxfordshire OX10 8BB
United Kingdom

Phone: 441 491 692 442
Fax: 441 491 692 448

http://iahs.info/

Additional copies of this publication are available from:

Curran Associates, Inc.
57 Morehouse Lane
Red Hook, NY 12571 USA
Phone: 845-758-0400
Fax: 845-758-2633
Email: curran@proceedings.com
Web: www.proceedings.com

TABLE OF CONTENTS

Proc. IAHS, 383, 1–1, 2020
https://doi.org/10.5194/piahs-383-1-2020

Open Access
PIAHS

Foreword

Shamila Nair-Bedouelle

Natural Sciences, UNESCO, Paris, France

Correspondence: Shamila Nair-Bedouelle (s.nair-bedouelle@unesco.org)

Published: 15 September 2020

The Flow Regimes from International Experimental and Network Data (FRIEND–Water) is an international collaborative network of experts that aims to generate new understanding about regional hydrology and multi scale water cycle processes. FRIEND Water is investigating long term variations and changes in hydrological variables to better understand the climate, river basin and human interventions on the spatial and temporal distribution of water. The research and training supported by the FRIEND–WATER programme are critical for evidence based decision-making and ensuring capacity for Water Resources Management, Socio-Economic development, safeguarding the environment and assessing the impact of global change, including climate change and human activities.

As a cross cutting theme of UNESCO's Intergovernmental Hydrological Programme (IHP), FRIEND–Water contributes to research on (1) regional water resources, (2) water related disasters such as floods and droughts, (3) global change and water cycle, and water education and capacity building. The FRIEND Water programme complements and interacts with many national projects and international initiatives.

FRIEND–Water interacts with and is relevant to all six-core IHP-VIII themes. It was launched during the third phase of IHP from (1984–1988) with the initial focus on the data-rich basins of northwest Europe and progressively expanded with the different phases to all the regions. The programme has evolved in more than 35 years of its existence. Since then it kept growing interest among the scientific community with up to eight regional components around the world, that is: SA FRIEND–Water (Southern Africa), CONGO FRIEND–Water, FRIEND–Water NILE, AMIGO FRIEND–Water (Latin America), EURO-FRIEND–Water, MEDFRIEND-Water (Mediterranean countries), AOC FRIEND–Water (West and Central Africa), ASIAN PACIFIC FRIEND–Water.

The 8th series of the Global FRIEND–Water conference held in 6–9 November 2018, in Beijing China was co-organized by UNESCO with Government of China through the Department of Hydrology, Ministry of Water Resources of the People's Republic of China within the framework of the UNESCO Intergovernmental Hydrological Programme (IHP) in Beijing China.

The Conference focused on the theme of hydrological processes and water security in a changing world and was a contribution to the United Nations 2030 Agenda for Sustainable Development, and provided the opportunity to reflect on the interactions between society and the hydrological system in a changing environment. Furthermore, participants of the conference also stressed the relevance of adopting multidisciplinary approaches to understand better climatic and anthropogenic impacts on the hydrological cycle ranging from local to global scales.

I would like to thank the Government of China for all the support and the conference organizing committee for bringing participants from more than 40 countries to present scientific and research findings during the oral and postal sessions. This proceeding constitutes the main outcome of the conference, and covers the following areas: hydrological processes and climatic impact; data monitoring, observation, and analysis extremes such as floods and droughts and water resources and environmental impacts and management.

I would like to thank the conference organizing committee, editorial members and International Association of Hydrological Sciences for compiling papers in the Proceedings of the International Association of Hydrological Sciences (PIAHS) publication. UNESCO will continue to support countries to address the local, national, regional and global challenges towards achieving water security by promoting and leading international hydrological research, facilitating education and capacity development and enhancing governance for sustainable and peaceful water resources management.

Proc. IAHS, 383, 3–4, 2020
https://doi.org/10.5194/piahs-383-3-2020

Open Access
PIAHS

Preface: Hydrological processes and water security in a changing world

Zhongbo Yu[1], Chunhui Lu[1], Jianyuan Cai[2], Dazheng Yu[2], Gil Mahe[3], Anil Mishra[4], Christophe Cudennec[5], Henny A. J. Van Lanen[6], Didier Orange[7], and Abou Amani[4]

[1]State Key Laboratory of Hydrology–Water Resources and Hydraulic Engineering, Hohai University, Nanjing, China
[2]Department of Hydrology, Ministry of Water Resources, Beijing, China
[3]HSM, Hydrosciences, IRD, UM, CNRS, MSE, 34090 Montpellier, France
[4]UNESCO, Division of Water Sciences, 7 place de Fontenoy, 75352 Paris CEDEX 07, France
[5]UMR SAS, Institut Agro, INRAe, 35000 Rennes, France
[6]Hydrology and Quantitative Water Management Group, Environmental Sciences Department, Wageningen University, Droevendaalsesteeg 3a, 6708 PB Wageningen, the Netherlands
[7]Eco&Sols, UM, IRD, CIRAD, INRAe, SupAgro Montpellier, 34090 Montpellier, France

Correspondence: Zhongbo Yu (zyu@hhu.edu.cn, zhongbo.yu@gmail.com)

Published: 15 September 2020

The world's freshwater environment is changing rapidly and facing unprecedented pressures. A growing global population and economic development have led to increasing demands on water resources and water management. The impact of climate change and intensive anthropogenic activities is now posing a great threat on the hydrological cycle leading to various water issues, such as floods, droughts, over-exploitation of water resources, pollution, and ecosystem degradation in many countries. On a global scale, more than two billion people live in countries experiencing high water stress. Consequently, understanding hydrological responses to stresses at local, regional and global scales is critical for addressing these water issues, including water resources planning and management for everywhere on the planet.

In order to advance the knowledge of hydrological processes and flow regimes in different regions, UNESCO initiated in the 1980s the Flow Regimes from International Experimental and Network Data (FRIEND) programme (Van Lanen et al., 2014), within the framework of Intergovernmental Hydrological Program (IHP). The FRIEND programme is an international research programme that helps to set up regional networks for analyzing hydrological data. It aims to develop better understanding of hydrological variability and similarity across time and space, through the mutual exchange of data, knowledge and techniques at the regional level. FRIEND's track record shows its strengths as a global network supporting a science-based network of high-level hydrological research at the river basin level and larger scales. These activities undertaken by FRIEND regional scientific groups of researchers, resulted in several topics related to regional hydrology and scientific publications, which is a contribution on water security within the framework of the IHP VIII and beyond to water related sustainable development goals (SDGs) of the United Nations Agenda 2030 for sustainable development, Paris Agreement and Sendai Framework.

The 8th Global FRIEND conference was co-organized by UNESCO and Government of China from 6–9 November 2018 in Beijing, with the focus on the theme of hydrological processes and water security in a changing world, under which eight thematic topics were discussed during the conference: (1) hydrological observations under the changing environment and scarcity; (2) river regimes and hydrological extremes under the changing environment; (3) simulation and prediction of surface water and groundwater processes under the impact of human activities; (4) urban hydrology and sponge city; (5) multi-objective water resources allocation and operation; (6) integrated watershed management including eco-hydrology and socio-hydrology; (7) water quality and sediment transport; and (8) river health and ecological baseflow under changing environment. The conference explored how society, water systems and the environment interact under changing conditions. It highlighted the advance in hydrological science and innovation in water management, as well as the solutions to improve this relationship, particularly in view of contributing to the 2030 Agenda and the Paris Agreement. For the first time the two IHP programs FRIEND and Ecohydrology jointly organized a side event about the importance to bring ecological solutions to hydrological problems.

The series of FRIEND conferences, workshops and scientific publications since 1989 are proven to be essential for fostering cooperation among FRIEND network and beyond. More than 200 participants from 44 countries attended the 8th Global FRIEND conference. This conference reminded that knowledge from all parts of the world need to be shared to improve our capacity to manage water as a common resource. More than 100 papers were presented at the conference. Eighty-nine papers were submitted for publication in this PIAHS volume, from which eventually 50 papers were accepted. These papers cover study areas in precipitation and climate impact; observation, analysis and simulations of hydrologic processes; floods in the changing environments; drought monitoring and analysis; water resources and environmental impacts.

Quantifying the natural and anthropogenic impacts on the hydrological processes and water resources are very important in current research efforts to understand how hydrological processes vary among climates and through time, as regional and global climates undergo change, and how modified hydrological processes affect water resources at different scales. Despite that many advances in these research areas have been achieved, the comprehensively integrated data and multidisciplinary approaches are required to address solutions for the implementation of SDGs and particularly water related targets. Furthermore, the outcome of the conference presented in the proceedings will be shared and discussed widely among IHP networks which would lead to recommend inputs for sustainable water resources management at different scale, but also provide feedbacks for the preparation of strategic plan for the ninth phase of IHP (IHP IX, 2022–2029).

The hydrological mechanisms, including flood and drought events, pollutions and water scarcity, underlie the climate-soil-vegetation dynamics and control the most basic ecological patterns and processes, both for biota and materials. Then the regional hydrology and flow regime characterization can emphasize the comprehension and role of the hydrological template that regulates ecohydrological functioning of the nature. Hydrological processes and water security are intrinsically linked to ecological and societal based approaches for watershed management. The future effort is at the nexus of water-soil-energy-food; and the related fields of geoinformatics, cyberinfrastructure in watersheds and remote measurements for scientific studies and water resources assessment should be helpful.

References

Van Lanen, H. A. J., Demuth, S., Daniell, T., Hannah, D.M., Laaha, G., Mahe, G., and Tallaksen, L. N.: Over 25 years of FRIEND-Water: an overview, in: Hydrology in a Changing World: Environmental and Human Dimensions, Proceedings of FRIEND-Water 2014, Montpellier, France, October 2014, edited by: Daniell, T. M., Van Lanen, H. A. J., Demuth, S., Laaha, G., Servat, E., Mahe, G., Boyer, J.-F., Paturel, J.-E., Dezetter, A., and Ruelland, D., Sci. Ed., IAHS Publ., 363, 1–7, 2014.

Proc. IAHS, 383, 5–12, 2020
https://doi.org/10.5194/piahs-383-5-2020

Variations of Seasonal Precipitation in the Yellow River Basin and Its Relationship to General Circulation and SST

Jing Liu[1], **Lijun Jin**[1], **Changxing Huang**[2], **and Lina Zhang**[1]

[1]Hydrology Bureau, Yellow River Conservancy Commission, Zhengzhou 450004, China
[2]Information Center (Hydrology Monitor and Forecast Center), Ministry of Water Resources, Beijing 100053, China

Correspondence: Jing Liu (liujing0208@163.com)

Published: 16 September 2020

Abstract. Based on the precipitation data from 1961 to 2017 in the Yellow River Basin and NCEP/NCAR reanalysis data, the variation characteristics of seasonal precipitation in the Yellow River Basin and its relationship to the circulation patterns were analyzed. Furthermore, effect of sea surface temperature (SST) on the seasonal precipitation was discussed. The results were as follows: (1) The precipitation in spring, summer and autumn all presented non-significant decreasing trends, while that in winter presented a non-significant upward trend. (2) The inter-annual and inter-decadal fluctuations of the precipitation of each season were evident. The fluctuation in winter was the most obvious, followed by that in spring and autumn, with that in summer most stable. Moreover, the precipitation in summer, autumn and winter in 1990s was the least. (3) When the precipitation in spring, autumn and winter was significantly above (below) normal, the 500 hPa height anomaly field over Eurasia mid-high latitude of the corresponding period presented "positive (negative) in east and negative (positive) in west". While, when the precipitation in summer was significantly above (below) normal, the height anomaly field presented "+(−), −(+), +(−)", and the West Pacific subtropical high was stronger (weaker) and lying more northward (southward). Additionally, when the precipitation in each season was significantly above (below) normal, the 850 hPa wind anomalies showed abnormal southerly (northerly) winds in Eastern China, which was favorable (unfavorable) to transport water vapor from the South China Sea and the Western North Pacific to the Yellow River Basin. (4) The spring precipitation was above (below) normal during the period of El Niño (La Niña), while the summer, autumn and winter precipitation presented the opposite characteristics.

1 Introduction

The Yellow River Basin with a vast territory is the second largest river in China. Precipitation is the main source of water resources, having significant effect on industrial and agricultural production as well as people's livelihood (Shao et al., 2015). Hence, it is a very meaningful work to study the precipitation variations in the Yellow River Basin. The annual mean precipitation increases from the northwest to the southeast. Generally, the spatial distribution of the seasonal precipitation is identical to that of the annual mean precipitation, but there are differences in some local area as the season changes (Chang et al., 2014; Zhu et al., 2005; Xu and Zhang, 2006). In recent 50 years, the annual, spring, summer and autumn precipitation all presented decreasing trends except for that in winter (Chang et al., 2014; Yao and Wu, 2014). Wang et al. (2012) proposed that there was significant inter-annual fluctuation for the summer precipitation with rainy years alternating with dry years. Chen et al. (2016) considered that the seasonal precipitation distribution changed in 1990s, namely, if the proportion of the seasonal precipitation gradually increased before 1990s, it would decrease after 1990s.

There had been many researches on the causes of the summer precipitation anomaly in the Yellow River Basin. Wang et al. (2012) proposed that monsoon, the West Pacific subtropical high, extratropical cyclone and the depression in Lake Baikal were the main circulation systems affecting the summer precipitation. Shao et al. (2015) and Song et al. (2013) considered that when the upper-level jet stream in East Asian and the West Pacific subtropical high were stronger (weaker) and lying more northward (southward), and simultaneously the trough in Lake Baikal was deeper (higher), there would be more (less) precipitation in the middle reaches of the Yellow River. Chang et al. (2014) discovered that the break-point of the precipitation in the Yellow River Basin was around 1986, and the change was mainly closely related to the monsoon variation. Lan et al. (2007) considered that Ural blocking highs existed mostly when the summer precipitation was plentiful in the upper reaches of the Yellow River, and vice versa. Weather and climate is directly affected by atmospheric circulation, and is also obviously influenced by the sea surface temperature (SST) via acting on atmospheric circulation. Many studies found that El Niño and La Niña had a significant relationship with precipitation distribution in China. In El Niño years, precipitation was more (less) than normal in the south (north) of China, and that was the opposite in La Niña years (Gong et al., 1998; Xu et al., 2005). Zhao et al. (2018) considered that the annual mean precipitation was decreased in the El Niño years and increased in La Niña years. Chang et al. (2013) defined the NINO3 SST difference index from the previous October to current April and indicated that the index had a significant negative correlation with summer precipitation in the Yellow River Basin.

Predominantly, the data used in the previous studies was mostly of the period before 2010, and the objects were about the summer precipitation in the Yellow River Basin. While in this paper, the data was prolonged to 2017 and the evolution characteristics of the spring, summer, autumn and winter precipitation were analyzed. Furthermore, the effects of atmospheric circulation as well El Niño and La Niña on the seasonal precipitation were discussed to provide a basis for seasonal precipitation forecast in the Yellow River Basin.

2 Data and methods

2.1 Data

The monthly area rainfall data of 1961–2017 applied in this paper was from the Information Center (Hydrology Monitor and Forecast Center), Ministry of Water Resources; The monthly reanalysis data of geopotential height and wind fields were from the U.S National Center for Environmental Prediction and National Center for Atmospheric Research (NCEP/NCAR), and the horizontal resolution was $2.5° \times 2.5°$. The monthly SST data was from the Met Office Hadley Centre for Climate Prediction and Research, and the horizontal resolution was $1.0° \times 1.0°$ (Rayner et al., 2003).

2.2 Methods

The year was divided into four seasons: spring (March, April and May), summer (June, July and August), autumn (September, October, November) and winter (December, January and February of the next year). The least square method was used to calculate the linear variation trends and the significance was estimated by T test (Wei, 1999). The seasonal precipitation was standardized and the obtained value greater (less) than 1 (-1) was defined as significantly above (below) normal. Furthermore, composite analysis of the general circulation anomaly fields for the periods with extremely excessive or deficient seasonal precipitation was applied, and then the differences between the two situations were estimated and tested by T test (Wei, 1999). Specially, it should be noted that the climate fields were of the period from 1961 to 2017.

3 Variation characteristics of seasonal precipitation in the Yellow River Basin

3.1 Season distribution of annual precipitation

The mean spring, summer, autumn and winter precipitation in the Yellow River Basin was respectively 89.4, 275.2, 123.8 and 16.0 mm, accounting for 17.7 %, 54.5 %, 24.5 % and 3.2 % of the mean annual precipitation (see Table 1), which was consistent with the previous studies (Chang et al., 2014; Chen et al., 2016). It was thus clear that the precipitation in summer was the most, followed by that in autumn and spring, with that in winter the least. The precipitation amount in summer was 16 times more than that in winter. Correlations between different period precipitation were measured and the results showed that the correlation coefficients between the annual precipitation and the spring/summer/autumn precipitation were respectively 0.39/0.71/0.61, which were all at 0.01 significance level, while the correlation coefficient between the annual precipitation and the winter precipitation and those between each seasonal precipitation were not significantly correlative. Overall, the annual precipitation amount was largely determined by the summer and autumn precipitation.

3.2 Inter-annual variation of the seasonal precipitation

The inter-annual fluctuation of each seasonal precipitation in the Yellow River Basin was evident (see Fig. 1). During the period from 1961 to 2017, the spring precipitation was extremely excessive in 1963, 1964, 1967, 1983, 1990, 1991 and 1998, while extremely deficient in 1962, 1981, 1995, 2000 and 2001. Specifically, the maximum/minimum spring precipitation appeared in 1964/1962, and accordingly the amount was 174.4/36.1 mm. The summer precipitation was extremely excessive in 1964, 1966, 1973, 1976, 1988, 1996, 2003, 2013 and 2016, while extremely deficient in 1965, 1969, 1972, 1974, 1986, 1991, 1997, 1999, 2001 and

Table 1. Seasonal precipitation eigenvalues in the Yellow River Basin.

Season	Mean precipitation (mm)	Proportion to the annual mean precipitation (%)	Maximum precipitation Value (mm)	Maximum precipitation Year	Minimum precipitation Value (mm)	Minimum precipitation Year	Variable coefficients (Cv)
Spring	89.4	17.7	174.4	1964	36.1	1962	0.32
Summer	275.2	54.5	388.3	1988	160.9	1997	0.18
Autumn	123.8	24.5	234.2	2011	54.6	1998	0.33
Winter	16.0	3.2	43.4	1989	2.2	1998	0.48

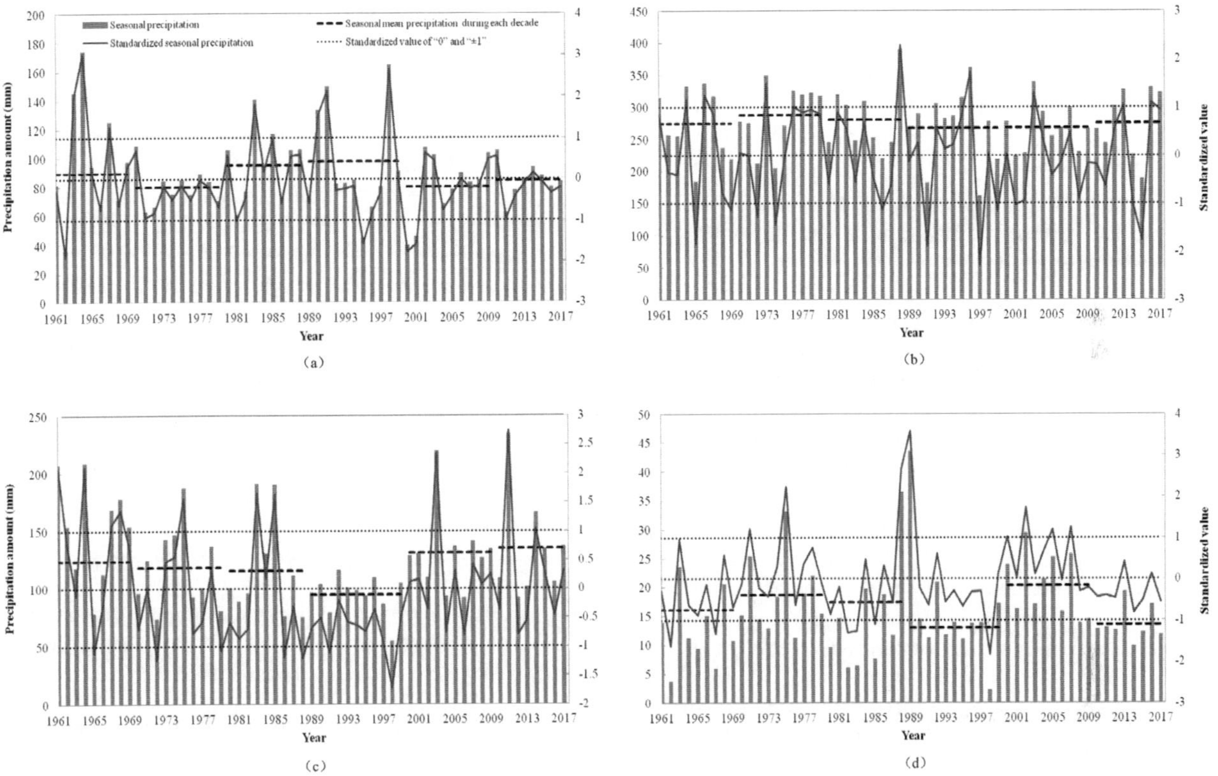

Figure 1. Time series of the spring **(a)**, summer **(b)**, autumn **(c)** and winter **(d)** precipitation in the Yellow River Basin.

2015, which were slightly different from the previous studies due to the using of data of different periods (Shao et al., 2015; Song et al., 2013; Zhang, 2011). The maximum/minimum summer precipitation appeared in 1988/1997, and accordingly the amount was 388.3/160.9 mm. The autumn precipitation was extremely excessive in 1961, 1964, 1967, 1968, 1975, 1983, 1985, 2003, 2011 and 2014, while extremely deficient in 1965, 1972, 1979, 1986, 1988, 1991 and 1998. Specifically, the maximum/minimum autumn precipitation appeared in 2011/1998, and accordingly the amount was 234.2/54.6 mm. The winter precipitation was extremely excessive in 1971, 1975, 1988, 1989, 2000, 2001, 2005 and 2007, while extremely deficient in 1962, 1967, 1982, 1983, 1985 and 1998. The maximum/minimum winter precipitation appeared in 1989/1998, and accordingly the amount was 43.4/2.2 mm. Variable coefficients (Cv) of each seasonal precipitation during the period from 1961 to 2017 were analyzed, and the results showed that the Cv of the summer precipitation was the minimum, followed by that of the spring and autumn precipitation, with that of the winter precipitation the maximum (see Table 1). Namely, the precipitation series distribution in summer was the most concentrated, while that in winter was the most dispersed.

3.3 Inter-decadal variation of the seasonal precipitation

The variation trends of the spring, summer, autumn and winter precipitation in the Yellow River Basin were respectively −1.9, −1.6, −2.2 and 0.3 mm per 10 years, which all failed the significance test. The inter-decadal variation for each seasonal precipitation was analyzed (see Fig. 1), and the results showed that the spring precipitation with 97.8/79.9 mm during the 1990s/1970s was the most/least. Conversely, the summer precipitation with 287.4/267.5 mm during the 1970s/1990s turned out to be the most/least. As for the autumn precipitation, the average amount with 130 mm since 2010 was the most, while the amount with 94.8 mm during the 1990s was the least. For the winter precipitation, the average amount with 20.3/13 mm during the 2000s/1990s was the most/least. In summary, the precipitation in summer, autumn and winter uniformly presented the least during the 1990s. Besides, the autumn precipitation successively showed decline and increase tendency before and after 1990s.

4 Atmospheric circulation characteristics for the years with abnormal seasonal precipitation in the Yellow River Basin

There were mainly two reasons that caused the abnormal precipitation: abnormal moisture condition; abnormal ascending movement. Moisture was transported mainly at 850 hPa and ascending movement was crucially provided by the background fields at 500 hPa (Hao et al., 2016). For the years with abnormal precipitation in the Yellow River Basin, the wind field at 850 hPa and the geopotential height field at 500 hPa definitely presented abnormal. Thus in order to find the relationship between the abnormal precipitation and the general atmospheric circulation, composite analyses of the two element fields were conducted respectively for the years with extremely excessive and deficient seasonal precipitation.

4.1 Characteristics of the geopotential height field at 500 hPa

When the precipitation was abnormal in spring, autumn or winter in the Yellow River Basin, the geopotential height anomaly fields of the corresponding period presented similar spatial distributions. For the years with extremely excessive precipitation, the anomaly field over Eurasia mid-high latitude presented "positive in east and negative in west", namely the anomaly field presented positive in coastal region of East Asia and negative in Lake Balkhash and the area north to it. While the anomaly field presented opposite distribution for the years with extremely deficient seasonal precipitation. Ulteriorly the differences between the height anomaly fields for the years with extremely excessive precipitation and those for the years with extremely deficient precipitation were calculated and the results were analyzed by T test. It could be seen clearly that the spatial distribution of the difference field presented a positive centre in the Sea of Japan and a negative centre in the area to the north of Lake Balkhash, passing the significance test at 0.05 level, which was in accordance with the anomaly field for the years with extremely excessive precipitation. Especially, the difference field in winter was the greatest with the positive centre of "60 pm" and the negative centre of "−70 gpm" (see Fig. 2a, c and d).

When the precipitation was abnormal in summer, the spatial distribution of the height anomaly field at 500 hPa showed distinctly different. When the precipitation in summer was extremely excessive, the anomaly filed presented a pattern of "+, −, +" over Eurasia mid-high latitude with positive anomaly around the Ural Mountains and the Sea of Okhotsk and simultaneously negative anomaly around the Lake Baikal. Over mid-low latitude, there showed positive anomaly around the Yangtze-Huaihe Rivers and the south of the Yangtze River, which indicated that the West Pacific subtropical high was stronger and lying more northward and the warm moist southerly flow in the west of it met with the dry cold air diffused from the bottom of Lake Baikal Trough around the Yellow River Basin, leading to abundant precipitation. On the contrary, the anomaly field presented opposite distribution when the precipitation in summer was extremely deficient: both the trough located in the east of the Ural Mountains and that in coastal region of East Asia were stronger; there was a ridge around the Lake Baikal; the West Pacific subtropical high was weaker and lying more southward. Under this background, the most area of the Yellow River Basin was controlled by the north air stream behind the trough, which was unfavourable to bring about precipitation. Ulteriorly the differences between the height anomaly fields for the years with extremely excessive precipitation and those for the years with extremely deficient precipitation in summer were calculated and the results were analyzed by T test. It could be seen that the differences around the Ural Mountains, the Sea of Okhotsk and the Yangtze-Huaihe Rivers were positive, which all passed the significance test at 0.1 level (see Fig. 2b).

4.2 Characteristics of the wind field at 850 hPa

When the precipitation was significantly above normal in spring, summer, autumn and winter in the Yellow River Basin, the height anomaly fields at 500 hPa of the corresponding period were all beneficial to the formation of the abnormal southerly winds. Accordingly, the wind anomaly fields at 850 hPa indicated that there were uniformly abnormal southerly or southeasterly winds in Eastern China when the precipitation was extremely excessive, which was favourable to transport water vapor from the South China Sea and the Western North Pacific to the Yellow River Basin. Conversely, there were uniformly abnormal northerly winds in Eastern China when the precipitation was extremely deficient, which was not favourable to transport the water vapour to the north. The differences of the wind anomaly fields at 850 hPa were analyzed, and the results showed that the abnormal southerly winds in most of Eastern China in spring, autumn and winter passed the significance test at 0.05 level,

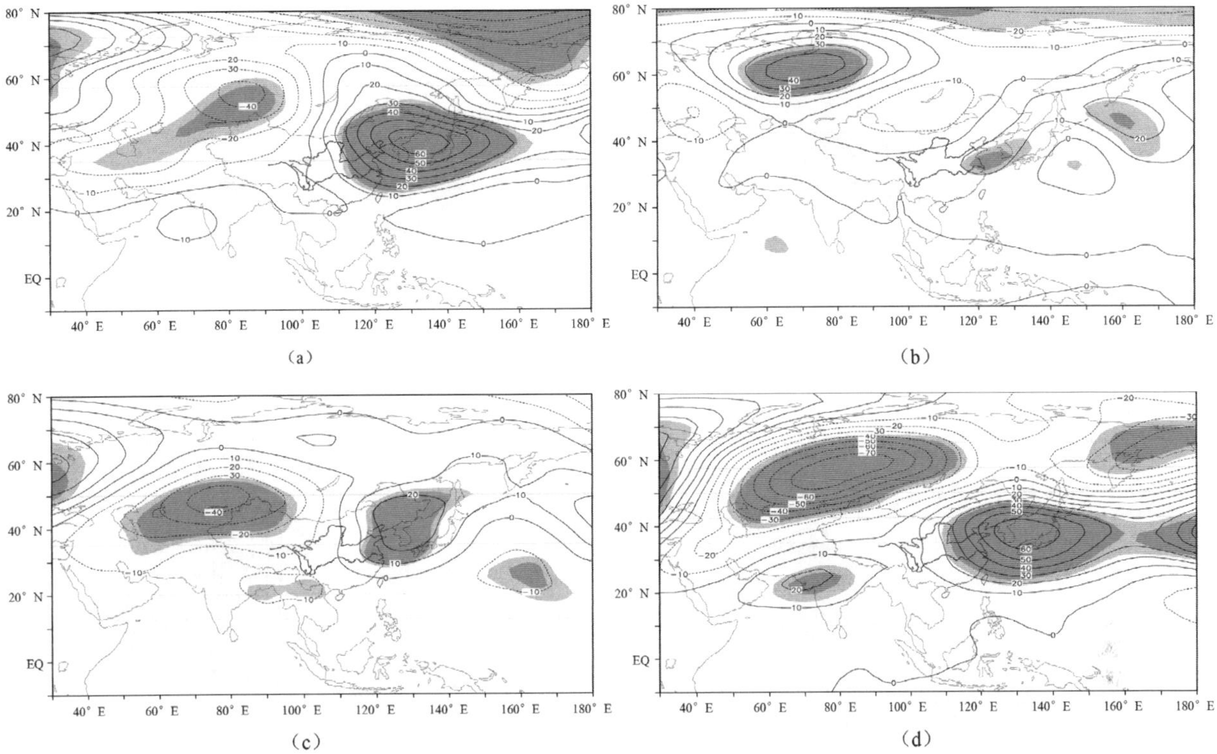

Figure 2. The differences between the height anomaly fields at 500 hPa for the years with extremely excessive precipitation and those for the years with extremely deficient precipitation in spring **(a)**, summer **(b)**, autumn **(c)** and winter **(d)**. (The yellow and orange parts respectively indicated the area passing the significance test at 0.1 and 0.05 levels.)

while only those in areas from southern china to the lower reaches of the Yellow River in summer passed the significance test (see Fig. 3).

5 Effects of El Niño and La Niña on seasonal precipitation in the Yellow River Basin

The correlation analysis between seasonal precipitation in the Yellow River Basin and the corresponding SST (see Fig. 4) showed that spring precipitation was positively correlated with SST over the equatorial central eastern Pacific Ocean, while summer, autumn and winter precipitation were negatively correlated. Moreover, the correlation in spring, summer and autumn was stronger than that in winter. The region with high correlation in each season was uniformly located in the eastern Pacific, where was exactly similar to the NINO3 district (150–90° W, 5° S–5° N). In consideration of this factor, the NINO3 index (the regional average of the SST anomaly in NINO3 district) was adopted to discuss the influence of El Niño and La Niña on seasonal precipitation in the Yellow River Basin.

The El Niño (La Niña) phenomenon was characterized by high (low) value of NINO3 index above 0.5° (below −0.5°). Based on this criterion, the years for each season under the conditions of El Niño and La Niña were selected. Furthermore, the average precipitation anomaly percentage for each season under the two conditions were respectively calculated (see Fig. 5). It was indicated that the spring precipitation was above (below) normal during the period of El Niño (La Niña) , while the summer, autumn and winter precipitation presented the opposite characteristics. The average NINO3 indices for each season when the corresponding precipitation was significantly above (below) normal were calculated. The results showed that the average NINO3 indices in spring, summer, autumn and winter were respectively 0.38, −0.47, −0.42 and −0.63° (−0.28, 0.6, 0.4, −0.08°) when the corresponding precipitation was significantly above (below) normal. Thus it could be seen that the relationship between precipitation in spring, summer and autumn and SST in NINO3 district was consistent with what was illustrated in Fig. 4, while precipitation in winter was more closely related to El Niño than to La Niña, further indicating that the correlation between winter precipitation and SST in NINO3 district was relatively weaker.

On the whole, the spring precipitation was increased (decreased) during the period of El Niño (La Niña) , while the summer, autumn and winter precipitation was decreased (increased) during the period of El Niño (La Niña).

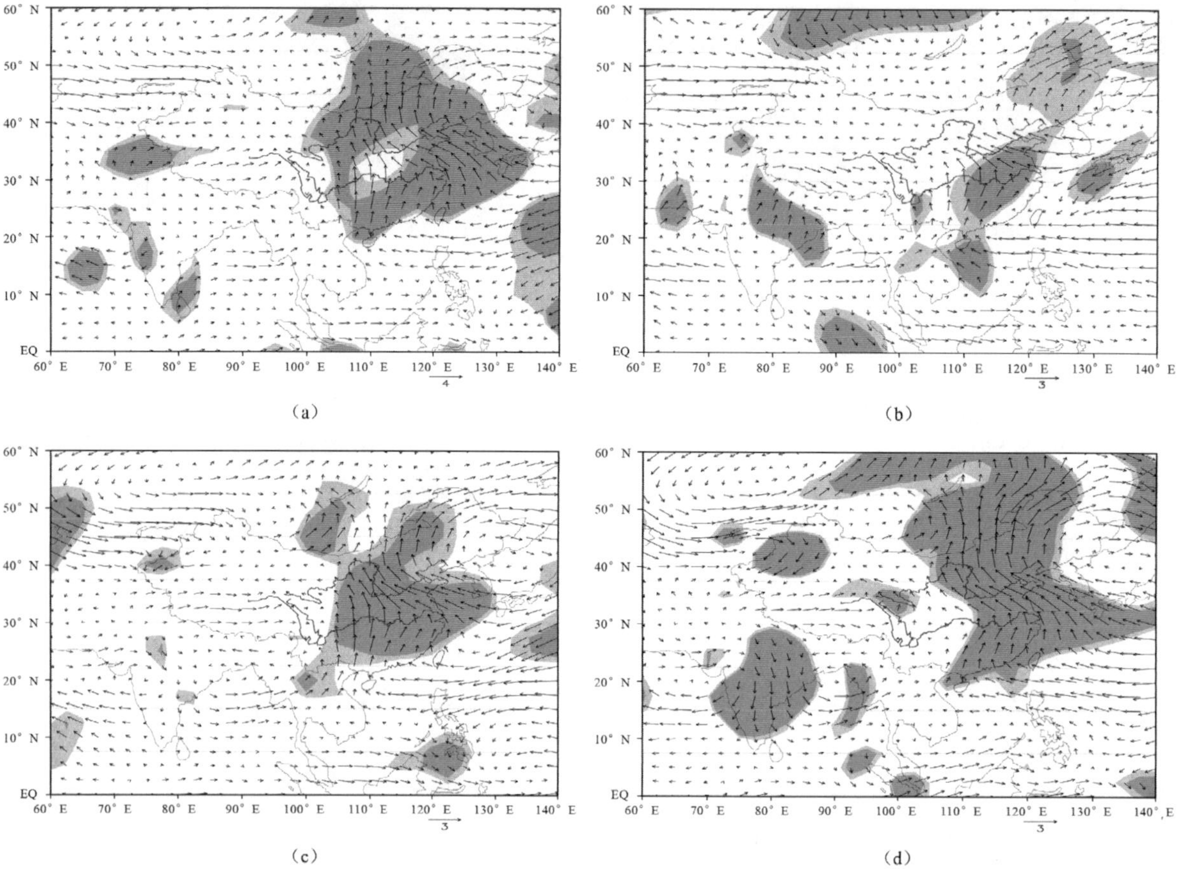

Figure 3. The differences between the wind anomaly fields at 850 hPa for the years with extremely excessive precipitation and those for the years with extremely deficient precipitation in spring **(a)**, summer **(b)**, autumn **(c)** and winter **(d)**. (The yellow and orange parts respectively indicated the area passing the significance test at 0.1 and 0.05 levels.)

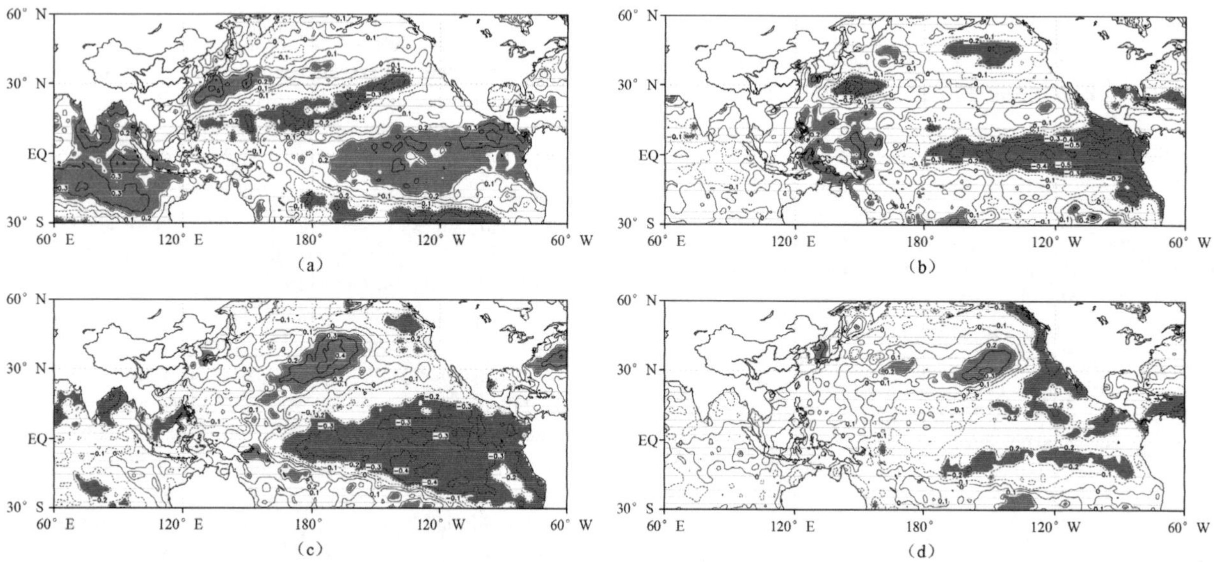

Figure 4. The correlation between precipitation in spring **(a)**, summer **(b)**, autumn **(c)** and winter **(d)** in the Yellow River Basin and the corresponding SST. (The shadow indicated the area passing the significance test at 0.1 level.)

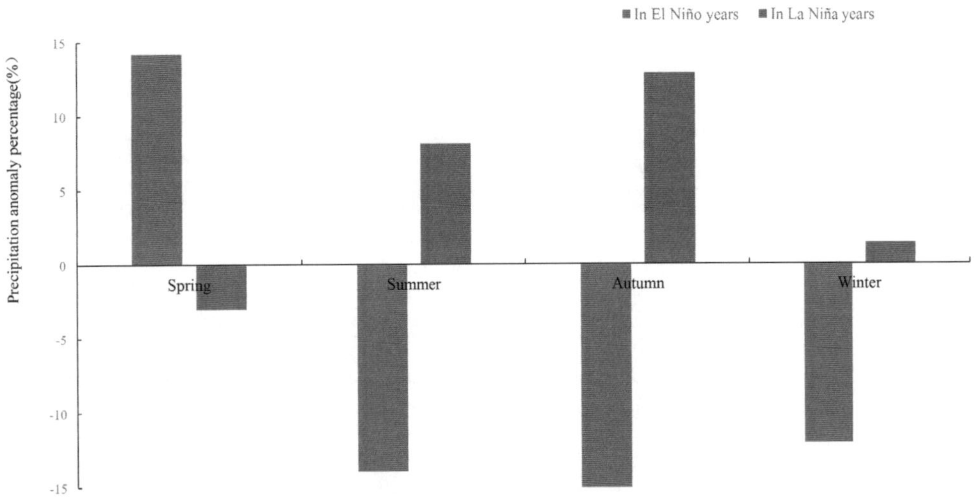

Figure 5. The average precipitation anomaly percentage for each season under the conditions of El Niño and La Niña in the Yellow River Basin.

6 Conclusion

The change tendency as well as the inter-annual and inter-decadal variations of the spring, summer, autumn and winter precipitation in the Yellow River Basin were analyzed. On the basis, the circulation characteristics for the periods when the seasonal precipitation was abnormal were discussed and the results were as follows.

1. Variation trends of the precipitation in spring, summer, autumn, and winter of the Yellow River Basin were respectively −1.9, −1.6, −2.2 and 0.3 mm per 10 years, which all failed the test of significance.

2. The inter-annual and inter-decadal fluctuations of the precipitation of each season were evident. The fluctuation in winter was the most obvious, followed by that in spring and autumn, with that in summer most stable. In terms of the inter-decadal variation, precipitation in spring, summer, autumn, and winter reached the largest in respectively 1990s, 1970s, 2010–2017 and 2000s, while recorded the smallest in respectively 1970s, 1990s, 1990s and 1990s.

3. When the precipitation in spring, autumn and winter was significantly above (below) normal, the 500 hPa height anomaly field over Eurasia mid-high latitude of the corresponding period presented "positive (negative) in east and negative (positive) in west". While, when the precipitation in summer was significantly above (below) normal, the height anomaly field presented "+ (−), − (+),+ (−)", and the West Pacific subtropical high was stronger(weaker) and lying more northward (southward). Additionally, when the precipitation in each season was significantly above (below) normal, the 850 hPa wind anomalies showed abnormal southerly (northerly) winds in Eastern China, which was favorable (unfavorable) to transport water vapor from the South China Sea and the Western North Pacific to the Yellow River Basin.

4. The spring precipitation was above (below) normal during the period of El Niño (La Niña) , while the summer, autumn and winter precipitation was below (above) normal during the period of El Niño (La Niña).

Data availability. The precipitation data is not publicly available due to research institution policy. The reanalysis data of geopotential height and wind fields was from the U.S National Center for Environmental Prediction and National Center for Atmospheric Research at: https://psl.noaa.gov/data/gridded/data.ncep.reanalysis.pressure.html (last access: 8 August 2020). The SST data was from the Met Office Hadley Centre for Climate Prediction and Research at: https://www.metoffice.gov.uk/hadobs/hadisst/ (last access: 8 August 2020).

Author contributions. JL developed the main idea and wrote the manuscripts. LJ and LZ collected and analyzed the data. CH participated in the translating of parts of the paper.

Competing interests. The authors declare that they have no conflict of interest.

Special issue statement. This article is part of the special issue "Hydrological processes and water security in a changing world". It is a result of the 8th Global FRIEND–Water Conference: Hydrological Processes and Water Security in a Changing World, Beijing, China, 6–9 November 2018.

References

Chang, J., Wang, Y. G., and Zhao, Y.: Impact of variation of sea surface temperature in Nino3 district on summer precipitation over Yellow River Basin, Meteorol. Monthly, 39, 1133–1138, 2013.

Chang, J., Wang, Y. G., Zhao, Y., and Li, F. X.: Characteristics of climate change of precipitation and rain days in the Yellow River Basin during recent 50 years, Plateau Meteorology, 33, 43–54, 2014.

Chen, L., Wang, Y. M., Chang, J. X., and Wei, J.: Application of weighted Markow Chain in forecasting rainfall in Jinan City, Yellow River, 38, 8–12, 2016.

Gong, D. Y. and Wang, S. W.: Impact of ENSO on the seasonal rainfall in China, Journal of Natural Disasters, 7, 44–52, 1998.

Hao, L. S., Ding, Y. H., and Min, J. Z.: Relationship between summer monsoon changes in East Asia and abnormal summer rainfall in North China, Plateau Meteorology, 35, 1280–1289, 2016.

Lan, Y. C., Lin, S., Wen, J., and Chang, J. J.: Studies on atmospheric circulation anomaly of flood period and prefreshet period in the highest and the lowest flow years in the upper reach of Yellow River, Plateau Meteorology, 26, 1652–1658, 2007.

Rayner, N. A., Parker, D. E., Horton, E. B., Folland, C. K., Alexander, L. V., Rowell, D. P., Kent, E. C., and Kaplan, A.: Global analyses of sea surface temperature, sea ice, and night marine air temperature since the late nineteenth century, J. Geophys. Res., 108, 4407, https://doi.org/10.1029/2002JD002670, 2003.

Shao, P. C., Li, D. L., and Wang, C. X.: Spatial and temporal changes of summer rain in the Yellow River Basin and its relation to the East Asia subtropical westerly jet in last 50 years, Plateau Meteorology, 34, 347–356, 2015.

Song, W. L., Gu, W., Liu, Y. J., and Liu, C. Z.: Characteristics of anomalous sign and prediction of summer precipitation in the middle reaches of Yellow River, Meteorological Monthly, 39, 1204–1209, 2013.

Wang, N., Li, D. L., and Z, J.: Variations and circulation pattern of drought and flood in summer at middle and upper reaches of Yellow River, Arid Land Geography, 35, 754–763, 2012.

Wei, F. Y.: Statistical diagnosis and prediction technology of modern climate, Meteorological, China, 1999.

Xu, W., Cheng, M. J., and Song, W. W.: A review of studies on the influence of ENSO events on the climate in China, Scientia Meteorologica Sinica, 25, 212–220, 2005.

Xu, Z. X. and Zhang, N.: Long-term trend of precipitation in the Yellow River basin during the past 50 years, Geogr. Res., 25, 27–34, 2006.

Yao, W. Y. and Wu, D.: Characteristics and variation trends of temperature and precipitation in the Yellow River Basin during 1961–2010, China Rural Water and Hydropower, 8, 104–109, 2014.

Zhang, S. Q.: Characteristics of changes of summer drought and flood and impact of climatic physical factors on these changes in the Yellow River valley, J. Meteorol. Environ., 27, 40–44, 2011.

Zhao, P. P., Lv, H. S., and Wang, C. Y.: Impact of ENSO event on meteorological drought and hyrological drought in Weihe River Basin, Water Resources and Power, 36, 9–13, 2018.

Zhu, Q. L., Zhang, L. Z., Yu, G. R., Dai, D., Cai, F., and Liu, X. A.: The spatial and temporal variability characteristics of precipitation in the Yellow River Basin of recent 30 years, J. Natural Resour., 20, 477–482, 2005.

Proc. IAHS, 383, 13–24, 2020
https://doi.org/10.5194/piahs-383-13-2020

Open Access
PIAHS

Precipitation and water stage variability under rapid developments of urbanization in Taihu Basin

Juan Wu[1], Hejuan Lin[1], Zhiyong Wu[2], Song Jin[1], Jian Wu[3], Haiping Ji[1], Min Liu[1], and Yueyun Gan[1]

[1]Bureau of Hydrology Information Center of Taihu Basin Authority, Shanghai 200434, China
[2]College of Hydrology and Water Resources, Hohai University, Nanjing 210098, China
[3]Urban and Rural Development Bureau of Suzhou New District, Suzhou 215163, China

Correspondence: Zhiyong Wu (wzyhhu@gmail.com)

Published: 16 September 2020

Abstract. As one of the most developed regions in China, hydrological process of Taihu basin was influenced under rapid development of urbanization. Based on original gauge data and water stage correction data from Bureau of Hydrology, variability in precipitation and water stage was investigated by Mann–Kendall test. The results demonstrated that: (1) The highest water stage (HWS), lowest water stage (LWS), average water stage (AWS) of all months had increasing trends, especially significantly in non-flood season. The LWS also increased significantly in May, July, September, and November, while the AWS increased significantly in September and November. (2) Water diversion from Yangtze River to Taihu Basin and supplying to the downstream, had strong effects on water stage increase of Taihu Lake. Finally, recommendations based on the correlation between precipitation and water stage was proposed to protect effective utilization of water resources from the impacts of rapid urbanization.

1 Introduction

The influences of urbanization on river systems have been widely recognized as the most significant among all human activities (Dudgeon, 2006; Shi et al., 2010). Under the circumstance of urban occupation, tributaries have been narrowed gradually and even disappeared, while the impervious surface area has been increased significantly, leading to the increase in runoff volume of main rivers (Gregory, 2006; Elmore and Kaushal, 2008; Yuan et al., 2006). Most rivers of Taihu basin were interrupted by sluices, pump stations and floodgates (Cheng et al., 2007; Zhao et al., 2011). There are 68 sluices along the Yangtze River to defend flood and meet water demand of water resources, 230 rivers with 186 sluices around the Taihu Lake, and 9 sluices along the Hangzhou Bay to solve the drainage of the flood (Yin et al., 2009). The flow regimes in Taihu basin are not only facing the effects of the changing climate, but also intense urbanization process (Zhou et al., 2013; Liu and Xu, 2015). By the end of 2015, the population of Taihu Basin reached to 59.97 million and the GDP is 6688 billion Yuan, representing about 4.4 % and 9.9 % of the nation's totals, respectively (Taihu Basin Authority of Ministry of Water Resources, 2015). Due to its rapid socio-economic development, Taihu basin suffers high vulnerability to natural disasters (Zhao and Wen, 2012). As a result, detecting and comparing the trends of precipitation and water stage of Taihu Basin is important to realize the evolution of water resources. In this study, urbanization is defined as the rural lifestyle changing into city lifestyle, according to G. F. Li et al. (2013), after the Reform and Opening-up in 1978, urbanization in China was in the state of acceleration.

Most research on Taihu Basin were focused on hydrological simulation (Liu et al., 2013), flood risk management (Xie et al., 2013; Gong and Lin, 2008), water quality (Zhao et al., 2011; Yu et al., 2007; W. Li et al., 2013; Wang et al., 2014) and ecological problems. Research on hydrological response to urbanization has received wide concerns for recent years (Zeng and Wang, 2012). Zhu et al. (2015) found that the river connectivity of Shanghai had been significantly degraded from 1965 to 2006, and the land use changed significantly with more building areas and fewer water regions and agricultural land. Yang et al. (2014) analyzed that precipitation differences between cities and suburbs of sub-region in Taihu Basin from 1961 to 2006 and revealed that the amount of precipitation increased with the level of urbanization at spatial and temporal scales, while maximum daily precipitation and the number of rainstorms in urban areas also increased. Deng et al. (2015) investigated the temporal and spatial change of river system in the Taihu region in conjunction with urbanization, and the results demonstrated that the changes in river density, the water surface ratio and main river area length ratio in rapid urbanization period were much greater than those in the slow urbanization period. Wang et al. (2016) assessed the contributions of precipitation and human activity on water stage increase in the plain river network region of Taihu Basin, which demonstrated that human activities played more and more important roles in the extreme water stage changes since the late 1980s.

In this article, the Mann-Kendall test was used to detect precipitation change of Taihu basin and water stage of Taihu lake from 1954 to 2015. Secondly, based on observed data from annual Water Year Book and water-level correction data from Hydrology Bureau of Taihu basin, the precipitation–water stage relationship was investigated. Thirdly, we discussed possible causes of water stage trends variation, including the variation of precipitation, regulation of sluices and other influencing factors. This research can provide a scientific foundation to further researches, such as water projects regulation impacts on flow regime, flood control and water resources management in Taihu basin.

2 Research area

Taihu Basin (30°28′ ∼ 32°15′ N, 119°11′ ∼ 121°53′ E) locates in the core area of Yangtze River Delta with Yangtze River in the north, Qiantangjiang in the south, East Sea in the east and the mountainous areas such as Mount Tianmu and Mount Maoshan as the boundary in the west. In terms of administrative division, Taihu Basin, the total area is 36 895 km^2, belongs to Jiangsu Province with area 19 399 km^2, Zhejiang Province 12 095 km^2, Shanghai Municipality 5176 km^2 and Anhui Province 225 km^2. The hydrographic net of Taihu Basin is complicated (Deng et al., 2016), with rivers and lakes accounting for 17 % of the total area of the basin. As the largest lake of the basin with area 2338 km^2 (Hu and Wang, 2009), Taihu Lake is the center of floodwater storage and regulation, whose average water depth is 1.89 m and water content is 4.428 billion m^3.

The Taihu basin lies in the subtropical zone (Wang et al., 2011) and has a monsoon climate with an average annual precipitation of 1218.1 mm, mainly concentrated in flood season ranging from May to September. There are three rainy seasons annually in the Taihu Basin: Spring rainy period, Meiyu period and Typhoon period (Chen and Wang, 1999). The spring rainy period from March to May accounts for about 24.2 % of the annual precipitation. The Meiyu period with

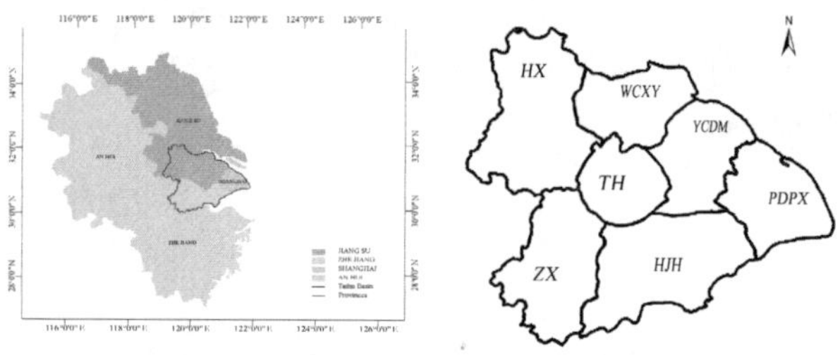

Figure 1. Taihu Basin and its sub-regions in China.

abundant precipitation from June to July accounts for about 19.8 % of the annual precipitation. However, there exist large changes in precipitation days and amount of the Meiyu period, for example, the Meiyu period did not appear in 1958 and 1978, whereas the Meiyu period lasted for 43 d with 681 mm in 1999 and 55 d with 645 mm in 1991, resulting in two catastrophic floods. The Typhoon period from August to October has high precipitation intensity and often leads to severe regional floods, for instance, the No.14 typhoon brought 150.1 mm for Taihu Basin, which occurred on 5 September 1962.

After witnessed basin-wide catastrophic flood in 1991 (Wu and Guan, 2000), in accordance with the planning of water conservancy, the first round of comprehensive regulation started intermediately. The 11 key projects have been built in the basin: Wangyu River, Taipu River, Lake Levee, Hangjiahu South Discharge, Huxi Cited Discharge, Wuchengxi Cited Discharge, Tiaoxi Flood Protection in East and West, Hangjiahu North Exhaust Passage, Red Flag Pond, expanding Lanlugang and Xietang, Huangpu River Upstream Flood Control. Combined with the existing water projects in the basin, the Taihu Basin has formed one project system, through which the flood flows northwards into the Yangtze River, eastwards into the Huangpu River and southwards into the Hangzhou Bay. The first four projects were completed in 1999, while the last seven projects were accomplished in 2002. By the end of 1999, the problems of quality-induced water shortage and water environment deterioration over Taihu Basin were very serious, where 80 % of river network was subject to pollution of varying degrees, and more than 70 % of the water in Taihu was in a eutrophic state. In order to improve the water environment, the Taihu Basin Authority (TBA) began to carry out Water Diversion from Yangtze River to Taihu Basin through Wangyu River and supplying water to the downstream through Taipu River since 2002. Therefore, 2002 can be considered as an important demarcation point to study Taihu water stage variation. The flood control capacity is further enhanced after 2002, since the complex system of flood control infrastructure has formed, including dikes, numerous large sluice gates, and pumping stations.

Based on spatial and temporal distribution of precipitation and characteristics of the watershed runoff and confluence, the Taihu basin was divided into seven sub-regions, Zhexi sub-region (ZX) with 15 stations to calculate average precipitation, Huxi sub-region (HX) with 17 stations, and Taihu sub-region (TH) with 8 stations, Wuchengxiyu sub-region (WCXY) with 10 stations, Yangchengdianmao sub-region (YCDM) with 11 stations, Hangjiahu sub-region (HJH) with 11 stations and Pudongpuxi sub-region (PDPX) with 9 stations. The first three sub-regions are called the upper area, while the last four sub-regions are in the downstream, which was shown in Fig. 1. The arithmetic mean method was applied in calculating average precipitation of each sub-region, while the weighting mean method was applied to calculate average precipitation of Taihu Basin based on sub-region area.

The water stage of Taihu Lake is an important index for the Taihu Basin, which has significant effects on regional flood defense, water supply safety, and water resources management. In this study, the water stage of Taihu Lake is determined by the mean value of the water stages from the five stations surrounding the lake: Dapukou (DPK), Wangting (WT), Xishan (XS), Xiaomeikou (XMK), and Jiapu (JP). Location of the research area, hydrometric stations were shown in Fig. 2.

Suffering from severe land subsidence caused by extensive groundwater exploitation since 1980s, the maximum cumulative subsidence was over 1 m for some stations in Taihu Basin (Wu et al., 2009). However, water stage in Water Year Book was observed based on stationary datum for most hydrometric stations. The occurrence of land subsidence leads to height datum distortion, thus, the water stage should be corrected. Four kinds water stage correction: average distribution method, segment correction method, datum reference method, and region approximation method were applied in our study (Zhao et al., 2015). The water stage of five stations surrounding the lake was modified before tendency analysis.

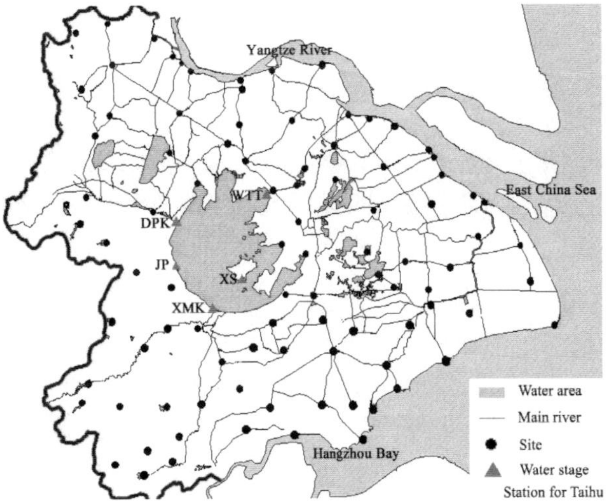

Figure 2. Location of study area and hydrometric stations selected.

3 Methodology

3.1 Mann-Kendall test

Mann-Kendall test is a simple but useful method to identify changes for long-term hydrological series (Mann, 1945; Kendall, 1975), especially for complicated basins where mathematics modeling was very difficult to provide satisfactory results (Yin et al., 2012). In the Mann-Kendall test (Xu et al., 2003; Zhao et al., 2007), the null hypothesis H_0 is that the data $(X_1, \ldots, X_n)$ are a sample of n independent and identically distributed random variables. The alternative hypothesis (Peng et al., 2016) H_1 of a two-sided test is that the distribution of x_k and x_j is not identical, and $j \geq k$. The test statistic S is computed by:

$$S = \sum_{k=1}^{n-1} \sum_{j=k+1}^{n} \operatorname{sgn}(x_j - x_k) \tag{1}$$

Where

$$\operatorname{sgn}(x_j - x_k) = \begin{cases} 1 & x_j - x_k > 0 \\ 0 & x_j - x_k = 0 \\ -1 & x_j - x_k < 0 \end{cases} \tag{2}$$

When $n \geq 8$, the statistics S is approximately normally distributed with the mean and the variance as follows:

$$E(S) = 0 \tag{3}$$

$$\operatorname{Var}(S) = \frac{n(n-1)(2n+5) - \sum_{i=1}^{m} t_i (i-1)(2i+5)}{18} \tag{4}$$

Where t_i is the number of ties of extent i. The standardized statistics (Z) is formulated as:

$$Z = \begin{cases} \frac{S-1}{\sqrt{\operatorname{Var}(S)}} & S > 0 \\ 0 & S = 0 \\ \frac{S+1}{\sqrt{\operatorname{Var}(S)}} & S < 0 \end{cases} \tag{5}$$

A positive and negative Z values indicate the upward and downward trends, respectively. The existing trend is considered to be statistically significant if $|Z| > Z_{1-\alpha/2}$ at significance level of α. In this study, α is set to be 0.05 and 0.10 with $Z_{1-\alpha/2}$ equal to 1.96 and 1.65, respectively.

3.2 API (Antecedent Precipitation Index) hydrological model

API (Antecedent Precipitation Index) hydrological model is commonly used in streamflow analysis. In order to build correlation between precipitation and runoff, the first step is to select precipitation and runoff event with single peak, secondly, antecedent influence precipitation (P_a) of each flood was computed, thirdly, $P - P_a - R$ curve can be completed,

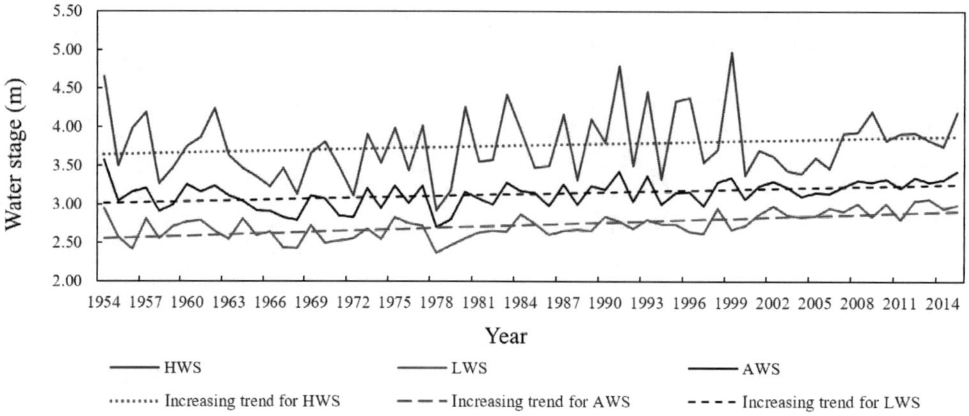

Figure 3. Increasing trend for HWS, LWS, and AWS from 1954 to 2015.

which was made up of several lines characterizing the relation between precipitation and runoff under equal value intervals of antecedent influence precipitation. P_a is calculated as:

$$P_{a,t+1} = K \times \left(P_{a,t} + P_t\right) \tag{6}$$

$$K = 1 - \frac{\overline{E}_m}{W_m} \tag{7}$$

Where P_a, $t+1$ is the antecedent influence precipitation at time $t+1$, $P_{a,t}$ is the antecedent influence precipitation at time t, P_t is the precipitation at time t, K is the regression coefficient of soil, $\overline{E}_m$ is monthly average potential evaporation, and W_m is maximum water storage. P_a, $t+1$ is no more than W_m, W_m is 100 mm in Taihu Basin, the initial P_a, t is 40 mm according to artificial experience in this study.

4 Results

4.1 Tendencies of precipitation and Taihu water stage variations

The increasing trend of annual precipitation and precipitation in flood season (from May to September) over Taihu basin were not statistically significant, with Z value of 0.72 and 0.02. The monthly precipitation decreased significantly at the significance level of 0.05 in April, May, and September, with Z value of -2.54, -2.33, and -2.82. The precipitation in non-flood season (January to March, October to December) demonstrated increasing trends, but only the precipitation in January increased significantly at significance level of 0.05. For the upper area, the annual precipitation and precipitation of flood season over ZX had an insignificant decreasing trend, while the annual precipitation and precipitation in flood season over HX had an insignificant increasing trend. For the downstream areas, the annual precipitation and precipitation in flood season over WCXY, YCDM, PDPX had an insignificant increasing trend. The monthly precipitation in January increased significantly at the significance level of 0.05 for each sub-region, whereas monthly precipitation in April, May and September decreased significantly at the significance level of 0.1 except HX. The precipitation in non-flood season (February, March, November and December) and precipitation in flood season (July and August) demonstrated increasing trends for each sub-region. The monthly precipitation over Taihu basin was shown in Table 1.

The increasing tendency of annual highest Taihu water stage (HWS) was not significant, with Z value 1.45. However, The Z value of annual lowest Taihu water stage (LWS) reached to 5.08, while the annual average Taihu water stage (AWS) reached to 2.57, which indicated that LWS and AWS increased significantly at a significance level of 0.05 (Fig. 3). The HWS, LWS, AWS of all months had increasing trends, especially significantly in January, February, March, April, August, and December, which was shown in Table 2. Apart from this, the LWS in May, July, September, November also increased significantly, while the AWS in September and November increased significantly.

Considering the 11 key projects completed and Water Diversion started in 2002, the HWS, LWS, and AWS variation were compared before 2001 and after 2002, which was showed in Table 3. From December to April, the HWS increased more than 0.18 m, the LWS increased more than 0.19 m and the AWS increased more than 0.17 m. The largest increasing extent of the HWS, LWS, and AWS reached more than 0.29 m in March, whereas the least increasing extent reached more than 0.04 m in July.

4.2 Tendencies of net water diversion, inflow and outflow variations

The HWS, LWS, and AWS of April, May and September increased, although monthly precipitation of the same period decreased. Therefore, monthly net water diversion and drainage amount of sluices along the Yangtze River, as well

Table 1. Tendencies analysis of precipitation over Taihu basin from 1954 to 2015.

Z value	year	Flood season	Jan.	Feb.	Mar.	Apr.	May	June	July	Aug.	Sept.	Oct.	Nov.	Dec.
Basin	0.72	0.02	**2.30**	1.45	0.46	**−2.54**	**−2.33**	0.21	1.22	1.49	**−2.82**	0.49	0.94	0.71
ZX	−0.21	−0.98	*1.92*	1.30	0.12	**−2.88**	**−2.62**	−0.41	1.49	0.92	**−2.69**	0.67	1.00	0.26
HX	1.03	0.57	**2.19**	*1.73*	0.74	−1.45	−0.75	0.62	1.25	0.67	−1.43	0.18	0.57	0.55
TH	0.00	−0.19	**2.55**	1.46	0.35	**−2.48**	*−1.73*	−0.06	0.60	1.22	**−2.53**	0.58	0.60	0.61
WCXY	0.09	0.04	*1.93*	1.17	0.16	*−1.92*	−1.48	−0.12	1.29	0.96	*−1.71*	−0.21	0.29	0.48
YCDM	1.00	0.38	**2.44**	1.64	0.58	**−1.98**	**−2.28**	0.01	1.09	**2.21**	**−2.35**	0.90	1.18	0.72
HJH	0.72	−0.29	**2.19**	1.13	0.54	**−2.76**	**−2.72**	0.22	0.90	1.43	**−2.59**	0.24	1.12	0.79
PDPX	1.48	0.68	**2.55**	1.25	0.58	**−2.36**	**−2.42**	0.85	0.79	**2.60**	**−2.36**	1.20	1.17	0.98

Texts in Bold indicate the results were significant at a significance level of 0.05; texts in italic indicate the results were significant at a significance level of 0.1; negative value means decreasing and positive value means increasing.

Table 2. Tendencies analysis of HWS, LWS, and AWS from 1954 to 2015.

Z value	Year	Jan.	Feb.	Mar.	Apr.	May	June	July	Aug.	Sept.	Oct.	Nov.	Dec.
HWS	1.45	**3.67**	**4.02**	**4.11**	**3.10**	0.27	0.83	1.59	**2.14**	1.59	0.88	1.51	**2.80**
LWS	**5.08**	**4.57**	**4.28**	**4.26**	**3.84**	**2.02**	0.95	**1.99**	**2.98**	**2.96**	1.36	**2.05**	**4.29**
AWS	**2.57**	**4.13**	**4.05**	**4.18**	**3.41**	0.70	0.91	1.60	**2.50**	**2.33**	0.88	*1.76*	**3.55**

Texts in Bold indicate the results were significant at a significance level of 0.05; texts in italic indicate the results were significant at a significance level of 0.1; negative value means decreasing and positive value means increasing.

as net inflow into Taihu Lake, should be analyzed, both of which were defined as daily accumulative amount of water diversion minus drainage in this study. The net water diversion and drainage amount were dominant by 4 sluices of HX sub-region (Jianbi Sluice, Jiuquhe Sluice, Xiaohe Sluice and Weicun Sluice), 3 sluices of WCXY sub-region (Zhangjiagang Sluice, Dingbo Sluice, and Shiyiwei Sluice) and Wangyu Sluice. The net inflow into Taihu Lake was dominant by HX, Wangting Sluice, and ZX, and the net outflow from Taihu Lake was dominated by Taipu Sluice, YCDM, WCXY, and HJH.

The net water diversion and drainage amount along the Yangtze River were demonstrated in Table 4, while the annual net water diversion amount for Wangyu Sluice was demonstrated in Fig. 4. The net water diversion amount of Wangyu Sluice was mainly negative before 2001, which changed into positive after water diversion from Yangtze River to Taihu Basin through Wangyu River after 2002. Wangyu Sluice had a significant increasing trend from January to May, as well as September to December. HX had a significant increasing trend in September while decreasing trends in January, February, April, and November. All months of the non-flood season (January to April, October to December) in WCXY had increasing trends, which was significant in January, April, October, November, and December.

The net water diversion and drainage variation were compared before 2001 and after 2002, which was showed in Table 5. Wangyu Shuice increased 239 million m^3 in September, with an increase of 52 to 179 million m^3 in non-flood season. HX increased by 120 million m^3 in September, with an increase of 17 to 57 million m^3 from June to October. WCXY increased 27 million m^3 in April, with 5 to 14 million m^3 increasing in other months of the non-flood season.

As shown in Table 6, the net inflow of HX into Taihu Lake had significant increasing trends for all months, and a net inflow of Wangting Sluice into Taihu Lake had insignificant increasing trends in non-flood season except April. The net outflow of Taipu Sluice from Taihu Lake had decreasing trends in non-flood season except March, with significant decreasing trends in October. The net outflow of YCDM from Taihu Lake in April, May, and September had insignificant decreasing trends, whereas HJH and WCXY from Taihu Lake had increasing trends.

The net inflow and outflow variation into Taihu Lake were compared before 2001 and after 2002, which was showed in Table 7. The net inflow from ZX into Taihu Lake decreased 36 to 202 million m^3 in flood season, and decreased 1 to 76 million m^3 in non-flood season, however, HX increased 120 to 402 million m^3 for all months, while Wangyu Shuice increased 5 to 185 million m^3 for all months. The net outflow of Taipu Sluice decreased 49 to 72 million m^3 in non-flood season, while the net outflow of YCDM decreased 11 to 121 million m^3 from April to November, and increased 1 to 43 million m^3 from January to March. The net outflow of HJH increased 32 to 155 million m^3 for all months, whereas the net outflow of WCXY increased 11 to 34 million m^3 for non-flood season.

The average annual inflow and outflow variation into Taihu Lake were compared before 2001 and after 2002, which was

Table 3. The highest, lowest and average variation of Taihu water stage before 2001 and after 2002.

	Jan.	Feb.	Mar.	Apr.	May	June	July	Aug.	Sept.	Oct.	Nov.	Dec.
HWS (m)	0.22	0.28	0.29	0.18	0.08	0.08	0.04	0.14	0.07	0.08	0.09	0.18
LWS (m)	0.22	0.25	0.31	0.19	0.14	0.10	0.12	0.22	0.20	0.11	0.12	0.22
AWS (m)	0.22	0.26	0.33	0.17	0.10	0.09	0.06	0.19	0.13	0.09	0.10	0.20

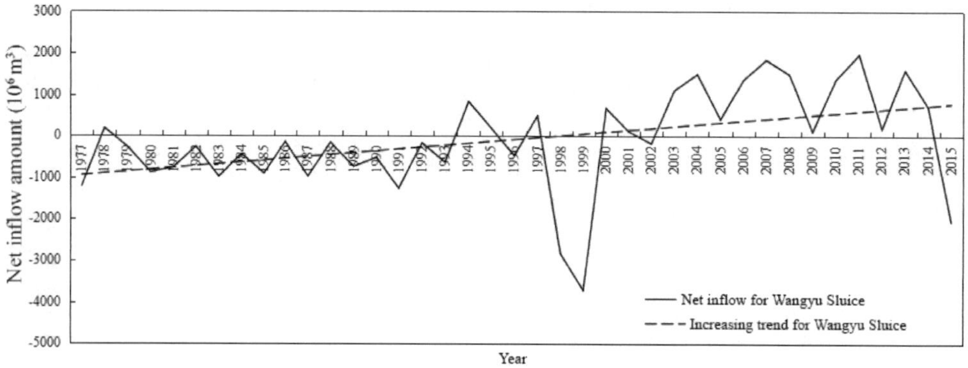

Figure 4. Increasing trend of annual net inflow amount for Wangyu Sluice.

showed in Table 8. Compared with inflow into Taihu Lake before 2001, ZX decreased 0.23, HJH decreased 0.73 and WCXY decreased 0.30 billion m^3. However, HX increased 2.36 while Wangting Sluice increased 0.83 billion m^3. Compared with the outflow from Taihu Lake before 2001, ZX increased 0.48, and HJH increased 0.41 billion m^3. Nevertheless, Taipu Sluice decreased 0.31 and YCDM decreased 0.21 billion m^3. Compared with the net flow (differences between inflow and outflow) before 2001, the net inflow of HX increased 2.49, Wangting Slucie increased 0.71 while ZX decreased 0.71 billion m^3; on the other hand, net outflow of Taipu Sluice decreased 0.30, YCDM decreased 0.35, while net outflow of HJH increased 1.14 and WCXY increased 0.19 billion m^3.

The annual inflow and outflow percentage of sub-regions were compared before 2001 and after 2002, which was showed in Fig. 5. The flood movement pattern changed a lot, with bigger differences existing in inflow percentage of sub-regions. The inflow percentage of ZX, HJH sub-region decreased, whereas HX and Wangting Sluice increased. Total inflow percentage of ZX and HX sub-region were 79 %, and total inflow percentage of Wangting and HX sub-region reached to 74 %, which indicated inflow from Wangting sluice and HX sub-region contribution became more and more important. On the other hand, outflow percentage of ZX and HJH increased, while Taipu Sluice and YCDM decreased. Different from inflow concentrating in HX and Wangting sluice, outflow distribution was relatively uniform.

In order to solve problems in pollution induced water shortage areas, the water diversion from Yangtze River to Taihu Basin through Wangyu River and supplying water to the downstream through Taipu River was initialized by Taihu Basin Authority of Ministry of Water Resources in 2002. From 2002 to 2015, total water diversion from Yangtze River through Wangyu River reached to 27.24 billion m^3, with net inflow into to Taihu Lake 12.42 billion m^3, and water supplying to downstream through Taipu River reached to 18.28 billion m^3. Therefore, the increase of net water diversion along the Yangtze River leads to significant net inflow increase of HX, Wangting Sluice into Taihu Lake. Under this condition, the net outflow from Taihu Lake converted into net inflow into Taihu Lake after 2002.

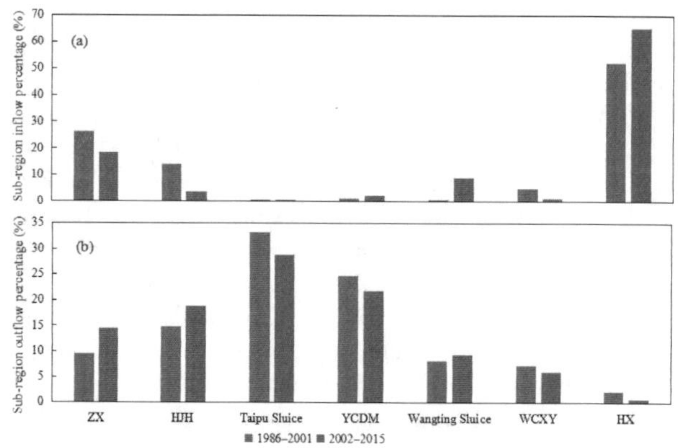

Figure 5. Multi-year average annual inflow **(a)** and outflow **(b)** percentage of sub-regions.

Table 4. Tendencies analysis of net water diversion and drainage amount along the Yangtze River.

Z value	Jan.	Feb.	Mar.	Apr.	May	June	July	Aug.	Sept.	Oct.	Nov.	Dec.
Wangyu Sluice	**4.52**	**4.36**	**1.97**	**1.11**	**2.13**	−0.64	−0.77	0.10	**1.98**	**3.58**	**5.36**	**5.48**
HX	**−2.13**	**−2.06**	−0.92	**−2.61**	−0.63	0.48	0.82	0.53	**2.49**	0.65	**−2.31**	−1.50
WCXY	**3.47**	0.99	0.47	**3.18**	−0.45	**−2.54**	**−1.74**	**−2.21**	0.23	**2.08**	**2.34**	**2.14**

Texts in Bold indicate the results were significant at significance level of 0.05; texts in italic indicate the results were significant at significance level of 0.1; negative value means decreasing and positive value means increasing.

Table 5. Net water diversion and drainage variation along the Yangtze River before 2001 and after 2002.

	Jan.	Feb.	Mar.	Apr.	May	June	July	Aug.	Sept.	Oct.	Nov.	Dec.
Wangyu Sluice	165	173	125	52	70	-30	−9	99	239	179	165	179
HX ($\times 10^6$ m^3)	−4	−6	−5	−12	−1	27	51	57	120	17	−18	−1
WCXY ($\times 10^6$ m^3)	14	9	5	27	−4	−41	−24	−34	5	14	13	9

Under the circumstance of rapid urbanization developments, the HWS, LWS, and AWS of April and May increased, although monthly precipitation of the same period decreased, which may be caused by the water diversion of Wangyu Sluice from Yangtze River. The HWS, LWS, and AWS in August and September demonstrated significant increasing trends , although the precipitation of the same period decreased significantly, which may be caused by the water diversion of Wangyu Sluice and HX sub-region from Yangtze River for storing water resources of non-flood season. The HWS, LWS, and AWS in November and December increased significantly, which may due to precipitation increase as well as water diversion of Wangyu Sluice and WCXY sub-region from Yangtze River in order to satisfy water requirements and improve regional water environment.

5 Discussion

During flood fluctuation process, the antecedent influence precipitation had great influence on Taihu water stage increasing, which can be considered as main reflection of land surface condition and soil moisture. The Taihu lake is located in plain river network region with indefinite flow direction, which increased difficulty in runoff depth computation. Therefore, instead of runoff depth, the water stage increase of Taihu Lake was analyzed based on precipitation event (Yan et al., 2015). However, few researches concentrated on the correlation between average precipitation of Taihu Basin and water stage of Taihu Lake under the circumstance of rapid urbanization developments. In this study, the relationship between water stage increase and heavy precipitation event was compared before 2001 and after 2002. Based on flood control experience, a significant increase of Taihu water stage was caused by daily average precipitation of Taihu Basin more than 20.0 mm. In order to eliminate effects of previous precipitation or compound flood, the first and last four days of the selected event were less than 1.0 mm, while the starting point to water stage peak lasted for no more than 10 d. Total precipitation amount was defined as daily average precipitation accumulation, and water stage increase was defined as differences from starting point to water stage peak.

For the period before 2001, we selected 116 events with precipitation 57.8 mm, while the water stage increased from 3.10 to 3.22 m averagely. For the period after 2002, we selected 39 events with precipitation 55.9 mm, while the water stage increased from 3.27 to 3.38 m averagely. Compared with water stage increase for the same precipitation before 2001 shown in Table 9, water stage increase after 2002 did not change too much when precipitation was less than 100.0 mm, while the water stage increased more significantly when precipitation was more than 120.0 mm. The more precipitation was, the more quickly water stage increased. The water stage increased 0.10 m for 50.0 mm precipitation, namely 5.0 mm precipitation increased 0.01 m water stage averagely. The water stage increased 0.25 m for 100.0 mm precipitation, namely 4.0 mm precipitation increased 0.01 m averagely. The water stage increased almost 0.40 m for 150.0 mm precipitation, namely 3.8 mm precipitation increased 0.01m averagely.

Monthly water stage variation increased 0.01 to 0.02 m in non-flood season for the same monthly multi-year average precipitation. Monthly water stage variation increased more when precipitation is abundant, while monthly water stage variation decreased less when precipitation is scare. However, For the same monthly multi-year average precipitation in flood season, monthly water stage variation after 2002 increased 0.04 m in May, 0.01 m in June, 0.08 m in September, which was caused by the water diversion of Wangyu

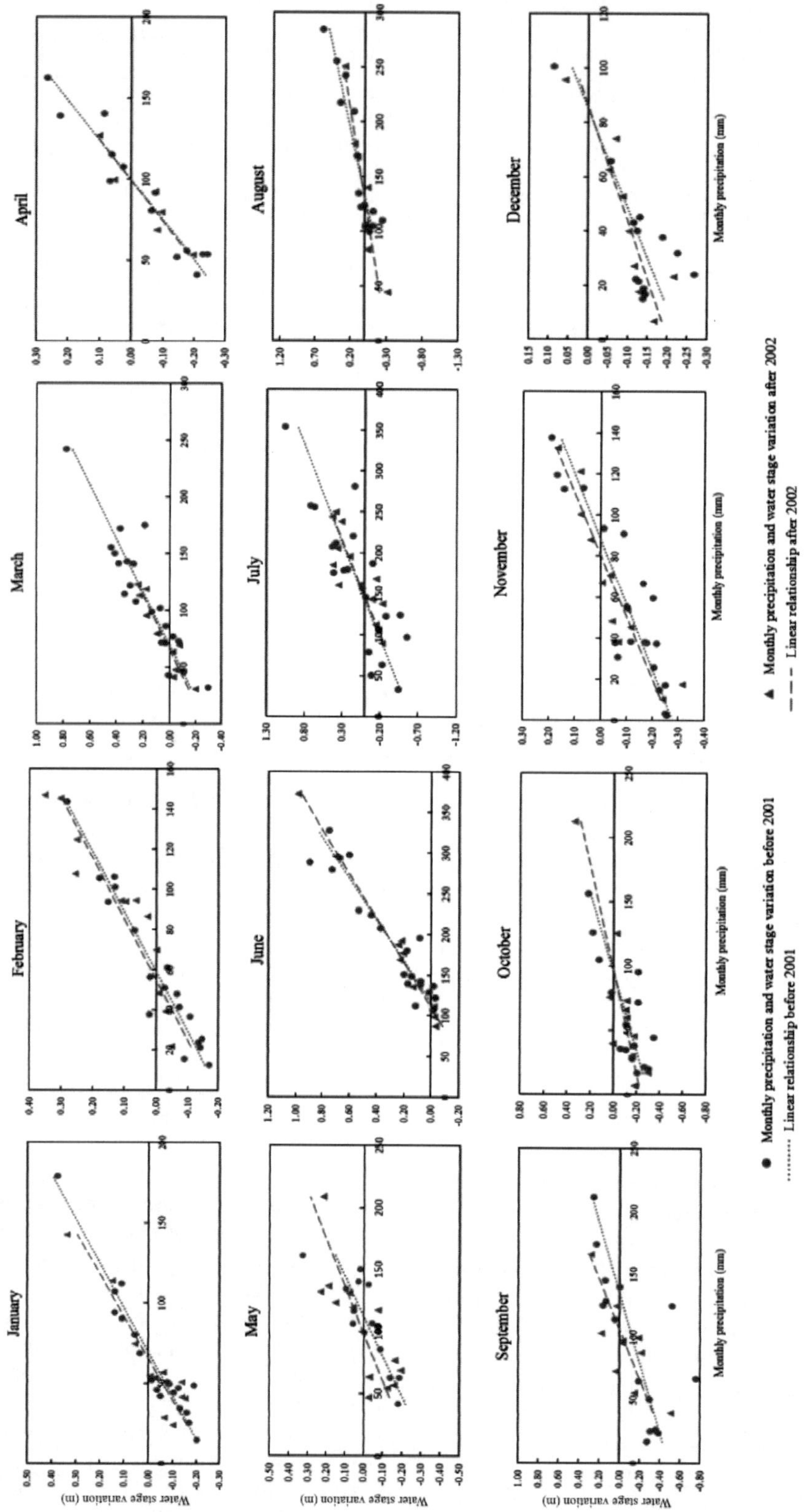

Figure 6. Relationships between monthly precipitation and water stage variation before 2001 and after 2002.

Table 6. Tendencies analysis for net inflow and outflow of Taihu Lake.

Z value	Jan.	Feb.	Mar.	Apr.	May	June	July	Aug.	Sept.	Oct.	Nov.	Dec.
ZX	**−2.69**	−0.38	−1.09	−1.46	−1.45	−0.57	−0.52	−0.07	−1.43	0.39	−0.46	*−1.70*
HX	**3.68**	**3.85**	**1.98**	**2.89**	**4.00**	**3.59**	**2.00**	**3.93**	**3.09**	**3.28**	**3.03**	**4.18**
Wangting Sluice	1.21	1.12	1.18	1.61	0.20	1.45	1.05	0.41	0.09	1.21	1.14	0.66
Taipu Sluice	−1.64	−0.96	0.95	0.32	0.41	**1.78**	−0.21	−0.16	−0.57	**−2.32**	−1.46	−0.66
YCDM	**2.19**	0.93	1.87	−0.73	**−3.14**	**−3.27**	−0.20	0.00	−0.46	−0.57	−0.45	1.52
HJH	**2.43**	**2.25**	**2.69**	**3.85**	**2.77**	*1.87*	*1.86*	*1.70*	**3.19**	0.43	1.12	*1.77*
WCXY	1.07	**2.16**	**3.19**	0.93	0.57	0.73	1.14	0.00	−0.55	0.13	1.23	0.50

Texts in Bold indicate the results were significant at significance level of 0.05; texts in italic indicate the results were significant at significance level of 0.1; negative value means decreasing and positive value means increasing.

Table 7. Net inflow and outflow of Taihu Lake variation before 2001 and after 2002.

	Jan.	Feb.	Mar.	Apr.	May	June	July	Aug.	Sept.	Oct.	Nov.	Dec.
ZX ($\times 10^6\,m^3$)	−76	−7	−48	−32	−36	−121	−202	−36	−180	57	−1	−23
HX ($\times 10^6\,m^3$)	120	156	132	134	275	283	402	332	241	121	126	178
Wangting Sluice ($\times 10^6\,m^3$)	65	88	46	16	63	14	5	62	158	62	55	77
Taipu Sluice ($\times 10^6\,m^3$)	−65	−49	47	28	57	55	−68	−53	−60	−72	−64	−56
YCDM ($\times 10^6\,m^3$)	33	1	43	−35	−60	−86	−121	−76	−65	−25	−11	37
HJH ($\times 10^6\,m^3$)	106	77	138	116	78	155	146	91	87	32	35	75
WCXY ($\times 10^6\,m^3$)	23	11	34	17	12	28	3	−1	−9	22	33	18

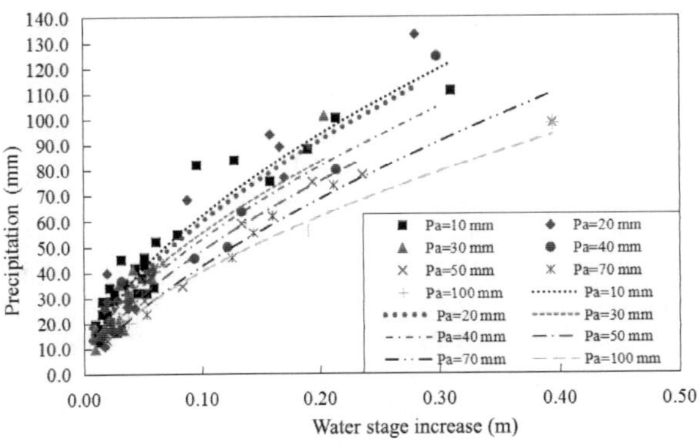

Figure 7. $P - P_a - R$ curve.

Sluice into Taihu lake from Yangtze River for storing water resources of non-flood season after 2002. However, monthly water stage variation decreased 0.01 m in August for the same monthly multi-year average precipitation after 2002, monthly water stage variation decreased more in August in abundant precipitation. Relationships between monthly precipitation and water stage variation were demonstrated in Fig. 6.

Accumulated precipitation amount more than 20.0 mm concentrated in flood season from 2002 to 2015, was selected in the $P - P_a - R_{\text{curve making}}$, which was demonstrated in Fig. 7. With the value of P_a increased from 10 to 100 mm, the $P - P_a - R$ curve got closer to abscissa axis, and the upper part of curves were approximately parallel. Under the same precipitation amount with 50 mm, water stage increased 0.07 m when P_a equals 10 mm water stage increased 0.08 m when P_a equals 20 mm, water stage increased 0.09 m when P_a equals 30 mm, water stage increased 0.095 m when P_a equals 40 mm, water stage increased 0.11 m when P_a equals 50 mm, water stage increased 0.13 m when P_a equals 70 mm, and water stage increased 0.14 m when P_a equals 100 mm.

Based on forecasts application requirement and technical level, the permissible error of Taihu water stage peak was set as 0.10 m according to Chinese national standards for hydro-

Table 8. Average annual inflow and outflow of Taihu Lake variation before 2001 and after 2002.

	ZX	HX	Wangting Sluice	Taipu Sluice	YCDM	HJH	WCXY	Total
Inflow ($\times 10^9$ m^3)	−0.23	2.36	0.83	−0.01	0.14	−0.73	−0.30	2.06
Outflow ($\times 10^9$ m^3)	0.48	−0.13	0.12	−0.31	−0.21	0.41	−0.11	0.25
Net flow ($\times 10^9$ m^3)	−0.71	2.49	0.71	−0.30	−0.35	1.14	0.19	1.81

Negative value means decreasing and positive value means increasing.

Table 9. Relationships between precipitation and water stage increase before 2001 and after 2002.

Precipitation (mm)	20	30	40	50	60	70	80	90	100	120	150
water stage increase before 2001 (m)	0.01	0.04	0.07	0.10	0.13	0.15	0.18	0.21	0.24	0.29	0.38
water stage increase after 2002 (m)	0.00	0.03	0.06	0.10	0.13	0.16	0.19	0.22	0.25	0.31	0.41

logical forecasting (Chinese National Standards, 2008). The success rate was defined as ratio between Taihu water stage peak forecasts number within permissible error not more than 0.10 m and total forecasts number. 30 precipitation events in 2016 and 2017 were selected for parameters validation with success rate 93 %, which can achieve first grade in real-time flood forecasts.

6 Conclusions

In this paper, precipitation, water stage, net water diversion along the Yangtze river, as well as net inflow and outflow of Taihu Lake using long time data, were analyzed under rapid urbanization developments. Moreover, the correlation between precipitation and water stage were discussed. The following conclusion is gained:

The monthly precipitation of Taihu Basin and its sub-regions decreased significantly in April, May, and September, while increased in January, February, March, October, November and December. The HWS, LWS, AWS of all months had increasing trends, and HWS increased more than 0.18 m, LWS 0.19 m and AWS 0.17 m in the non-flood season after 2002.

The monthly net water diversion and drainage amount along the Yangtze River in HX sub-region increased 27 to 120 million m^3 in flood season, while Wangyu Shuice increased 239 million m^3 in September, with 52 to 179 million m^3 increase in non-flood season after 2002. Total net inflow into Taihu increased 1.81 billion m^3, of which HX increased 2.49 billion m^3 and Wangting Sluice increased 0.71 billion m^3 after 2002.

The correlation between precipitation and water stage increase after 2002 did not change too much when precipitation was less than 100.0 mm. The water stage peak had strong relationships with starting point.The higher starting point was, the higher water stage peak was. The more precipitation for the same period was, the more quickly water stage increased. Thus, meeting the demands of water resources utilization, reducing initial water level of Taihu Lake by applying pre-discharging method could take an active role of flood control.

Monthly water stage variation increased 0.01 m to 0.02 in non-flood season for the same monthly multi-year average precipitation. For the same monthly multi-year average precipitation in flood season, monthly water stage variation after 2002 increased 0.04 m in May, 0.01 m in June, 0.08 m in September, which was caused by the water diversion of Wangyu Sluice into Taihu lake from Yangtze River for storing water resources of non-flood season after 2002. With the value of P_a increased from 10 to 100 mm, the $P-P_a-R$ curve got closer to abscissa axis, and the upper part of curves were approximately parallel.

In conclusion, urbanization has an evident influence on the increase of Taihu water stage. Nevertheless, urbanization on water stage increase is the interaction of precipitation and underlying surface variation. Thus, in order to evaluate the impacts of urbanization on Taihu Basin in detail, each characteristic of urbanization indicator contributed to water stage rise should be quantified and assessed in the following study.

Data availability. The data use in this paper are not publicly accessible.

Author contributions. JW and ZW initiated the concept, structured and wrote the article. SJ, HL reviewed the manuscript and provided valuable suggestions. JW, HJ, ML, YG processed precipitation, water stage, net inflow of Taihu Basin, net water diversion data along the Yangtze River.

Competing interests. The authors declare that they have no conflict of interest.

Special issue statement. This article is part of the special issue "Hydrological processes and water security in a changing world". It is a result of the 8th Global FRIEND-Water Conference: Hydrological Processes and Water Security in a Changing World, Beijing, China, 6–9 November 2018.

Financial support. This research has been supported by the National Key Research and Development Program (grant no. 2018YFC0407900), the National Science Foundation of China (grant no. 51779071), the National Science Foundation of China (grant no. 51579065), the National Key Research and Development Project (grant no. 2017YFC1502403), and the Special Public Sector Research Program of Ministry of Water Resources of China (grant no. 201501014).

References

Chen, Z. Y. and Wang, Z. H.: Yangtze Delta, China: Taihu lake-level variation since the 1950s, response to sea-level rise and human impact, Environ. Geol., 37, 333–339, 1999.

Cheng, J., Yang, K., Zhao, J., Yuan, W., and Wu, J. P.: Variation of river system in center district of Shanghai and its impact factors during the last one hundred years, Scientia Geographica Sinica, 27, 85–91, https://doi.org/10.13249/j.cnki.sgs.2007.01.85, 2007.

Chinese National Standards: Standard for hydrological information and hydrological forecasting, Chinese Ministry of Water Resources-GB/T 22482-2008, 5–6, 2008.

Deng, X. J., Xu, Y. P., Han, L. F., and Wang, Y. F.: Impacts of urbanization on river systems in the Taihu Region, China, Water, 7, 1340–1358, https://doi.org/10.3390/w7041340, 2015.

Deng, P. X., Xu, Y. P., Han, L.F., Yang, M. N., Yang, L., Song, S., Li, G., and Wang, Y. F.: Spatial-temporal evolution of the distribution pattern of river systems in the plain river network region of the Taihu Basin, China, Quaternary Int., 392, 178–186, https://doi.org/10.1016/j.quaint.2015.04.010, 2016.

Dudgeon, D.: The impacts of human disturbance on stream benthic invertebrates and their drift in North Sulawesi, Indonesia, Fresh Water Biol., 51, 1710–1729, https://doi.org/10.1111/j.1365-2427.2006.01596.x, 2006.

Elmore, A. J. and Kaushal, S. S.: Disappearing headwaters: Patterns of stream burial due to urbanization, Front. Ecol. Environ., 6, 308–312, https://doi.org/10.1890/070101, 2008.

Gong, Z. and Lin, Z. X.: Strategy of flood control in Taihu Basin, Advances in Water Resources and Hydraulic Engineering-Proceedings of 16th IAHR-APD Congress and 3rd Symposium of IAHR-ISHS, 3, 1011–1016, 2008.

Gregory, K. J.: The human role in changing river channels, Geomorphology, 79, 172–191, https://doi.org/10.1016/j.geomorph.2006.06.018, 2006.

Hu, Q. F. and Wang, Y. T.: Impact assessment of climate change and human activities on annual highest water level of Taihu Lake, Water Sci. Eng., 2, 1–15, https://doi.org/10.3882/j.issn.1674-2370.2009.01.001, 2009.

Li, G. F., Xiang, X. Y., Tong Y. Y., and Wang, H. M.: Impacts assessments of urbanization on flood risk in the Yangzte river Delta, Stoch. Environ. Res. Risk Assess., 27, 1683–1693, https://doi.org/10.1007/s00477-013-0706-1, 2013.

Li, W., Yang, M. Y., Liang, Z. W., Zhu, Y., Mao, W., Shi, J. Y., and Chen, Y. X.: Assessment for surface water quality in Lake Taihu Tiaoxi River Basin China based on support vector machine, Stoch. Environ. Res. Risk Assess., 27, 1861–1870, https://doi.org/10.1007/s12665-016-6051-1, 2013.

Liu, L. and Xu, Z. X.: Hydrological projections based on the coupled hydrological-hydraulic modeling in the complex river network region: a case study in the Taihu basin, China, J. Water Clim. Change, 6, 386–399, https://doi.org/10.2166/wcc.2014.156, 2015.

Liu, L., Xu, Z. X., Reynard, N. S., Hu, C. W., and Jones, R. G.: Hydrological analysis for water stage projections in Taihu Lake, China, J. Flood Risk Manage., 6, 14–22, https://doi.org/10.1111/jfr3.12015, 2013.

Mann, H. B.: Nonparametric tests against trend, Econometrica, 13, 245–259, https://doi.org/10.2307/1907187, 1945.

Peng, D. Z., Qiu, L. H., Fang, J., and Zhang, Z. Y.: Quantification of Climate Changes and Human Activities That Impact Runoff in the Taihu Lake Basin, China, Math. Probl. Eng., 2016, 1–7, https://doi.org/10.1155/2016/2194196, 2016.

Shi, Y. L., Wang, R.S., Fan L. Y., Li, J. S., and Yang, D. F.: Analysis on land-use change and its demographic factors in the original-stream watershed of Tarim river based on GIS and statistic, Procedia Environ. Sci., 2, 175–184, https://doi.org/10.1016/j.proenv.2010.10.021, 2010.

Taihu Basin Authority of Ministry of Water Resources: Taihu basin and southeast rivers water resources bulletin, available at: http://www.tba.gov.cn/slbthlyglj/szygb/content/slth1_8931163d6fc84971b633faea493e0a40.html (last access: 7 August 2020), 2015.

Wang, L., Cai, Y. L., Chen, H. Q., Dag, D., Zhao, J. M., and Yang, J.: Flood disaster in Taihu Basin, China: causal chain and policy option analyses, Environ. Earth Sci., 63, 1119–1124, https://doi.org/10.1007/s12665-010-0786-x, 2011.

Wang, Y. F., Xu, Y. P., Xu. Y., Song, S., Li, G., and Wu, L.: Changing patterns of extreme water stages in urbanizing plain river network region of Taihu Basin, China: characteristics and causes, Hydrol. Earth Syst. Sci., 1–24, https://doi.org/10.5194/hess-2016-184, 2016.

Wang, Y. K., Sheng, D., Wang, D., Ma, H. Q., Wu, J. C., and Xu, F.: Variable Fuzzy Set Theory to Assess Water Quality of the Meiliang Bay in Taihu Lake Basin, Water Resour. Manage., 28, 867–880, https://doi.org/10.1007/s11269-014-0521-6, 2014.

Wu, J. C., Shi, X. Q., Ye, S. J., Xue, Y. Q., Zhang, Y., and Yu, J.: Numerical simulation of land subsidence induced by groundwater overexploitation in Su-Xi-Chang area, China, Environ. Geol., 57, 1409–1421, https://doi.org/10.1007/s00254-008-1419-5, 2009.

Wu, H. Y. and Guan, W. Q.: 1991's flood in Taihu basin, Beijing, China Water Power Press, 170, 2000.

Xie, J. B., Sayers, P., Sun, D. Y., and Zhang, H. H.: Broad-scale reliability analysis of the flood defence infrastructure within the Taihu Basin, J. Flood Risk Manag., 6, 42–56, https://doi.org/10.1111/jfr3.12034, 2013.

Xu, Z. X., Takeuchi, K., and Ishidaira, H.: Monotonic trend and step changes in Japanese precipitation, J. Hydrol., 279, 144–150, https://doi.org/10.1016/S0022-1694(03)00178-1, 2003.

Yan, Q. H., Lei, T. W., Yuan, C. P., Lei, Q. X., Yang, X.S., Zhang, M. L., Su, G. X., and An, L. P.: Effects of watershed management practices on the relationships among rainfall, runoff, and sediment delivery in the hilly-gully region of the Loess Plateau in China, Geomorphology, 228, 735–745, https://doi.org/10.1016/j.geomorph.2014.10.015, 2015.

Yang, M. N., Xu, Y. P., Pan, G. B., and Han, L. F.: Impacts of urbanization on precipitation in Taihu Lake Basin, China, J. Hydrol. Eng., 19, 739–746, https://doi.org/10.1061/(ASCE)HE.1943-5584.0000852, 2014.

Yin, Y. X., Xu, Y. P., and Chen, Y.: Relationship between flood/drought disasters and ENSO from 1857 to 2003 in the Taihu Lake basin, China, Quat. Int., 208, 93–101, https://doi.org/10.1016/j.quaint.2008.12.016, 2009.

Yin, Y. X., Xu, Y. P., and Chen, Y.: Relationship between changes of river-lake networks and water stages in typical regions of Taihu Lake Basin, China, Chinese Geographical Science, 22, 673–682, https://doi.org/10.1007/s11769-012-0570-9, 2012.

Yu, G., Xue, B., Lai, G. Y., and Liu, X. M.: A 200-year historical modeling of catchment nutrient changes in Taihu basin, China, Hydrobiologia, 581, 79–87, https://doi.org/10.1007/978-1-4020-6158-5_9, 2007.

Yuan, W., Philip, J., and Yang, K.: Impact of urbanization on structure and function of river system: case study in China, Chinese Geographical Science, 16, 102–108, 2006.

Zeng, C. F. and Wang, L. C.: A study of hydrologic responses to land use and cover change in Taihu Lake Basin, southeast China, J. Food Agr. Environ., 10, 1404–1408, 2012.

Zhao, Q. J. and Wen, Z. M.: Integrative networks of the complex social-ecological systems, Proc. Environ. Sci., 1383–1394, https://doi.org/10.1016/j.proenv.2012.01.131, 2012.

Zhao, D. Y., Ge, L. B., and Huang, G. Y.: Research on water stage correction for depressed area, Journal of China Hydrology, 35, 72–76, 2015.

Zhao, F. F., Xu, Z. X., and Huang J. X.: Long-term trend and abrupt change for major climate variables in the upper Yellow River basin, Acta Meterol. Sin., 21, 204–214, https://doi.org/10.1029/2006JD007471, 2007.

Zhao, G. J., Gao, J. F., Tian, P., and Ni, G. H.: Spatial-temporal characteristics of surface water quality in the Taihu Basin, China, Environm. Earth Sci., 64, 809–821, https://doi.org/10.1007/s12665-011-0902-6, 2011.

Zhao, J., Shan, F. Z., Yang, K., and Wu, E. K.: River sinuosity in tidal plain and its response to rapid urbanization, Adv. Water Sci., 22, 631–637, 2011.

Zhou, F., Xu, Y. P., Chen, Y., and Du, J. K.: Hydrological response to urbanization at different spatio-temporal scales simulated by coupling of CLUE-S and the SWAT model in the Yangtze River Delta region, J. Hydrol., 485, 113–125, https://doi.org/10.1016/j.jhydrol.2012.12.040, 2013.

Zhu, H. F., Ren, X. Y., Jin, Y., Yang, K., and Che, Y.: Multilevel analysis of a riverscape under rapid urbanization in the Yangtze delta plain, China: 1965–2006, Environ. Monitor. Assess., 187, 711–724, https://doi.org/10.1007/s10661-015-4931-6, 2015.

Proc. IAHS, 383, 25–30, 2020
https://doi.org/10.5194/piahs-383-25-2020

Analysis of water level variations in the Huangzhuang reach (Hanjiang River, China) during recent years

Xiao Xiao, Beiping Mao, and Bao Qian

Middle Changjiang River Bureau of Hydrology and Water Resources Survey, Wuhan, 430012, China

Correspondence: Bao Qian (jacber@163.com)

Published: 16 September 2020

Abstract. The 2017 Hanjiang autumn flood appeared to have higher surface water levels at the same discharge rate than previously at the Huangzhuang gauging station. It is seriously impacted the flood control. This paper analyses the 2000–2017 data from the Huangzhuang reach in the Hanjiang River, including surface water levels, the surface water fall, erosion-deposition variation and the engineering activities. A comparison of the 2016–2017 data with the 2000–2015 data in the Huangzhuang reach, revealed that the surface water levels were about 1.47 and 0.13 m higher in the high and intermediate discharge ranges, respectively. In the low discharge range, the 2016–2017 and 2010–2015 surface water levels were similar. However, a comparison of the 2016–2017 with the 2000–2009 data in the low discharge range, shows that the surface water levels were 0.28–0.4 m lower. The surface water fall in the Zhuandouwan–Huangzhuang's reach decreased in 2016–2017 for discharges above 5000 $m^3 s^{-1}$, whereas in the Huangzhuang–Datong reach the opposite was found. For other time periods and discharge ranges the fall decreased. In the Datong-Shayang reach the fall reduced in 2014–2015 relative to other periods in 2000–2017 in the discharge range above 5000 $m^3 s^{-1}$. The erosion-silting analysis for the Huangzhuang reach showed scouring of local river deposits due to bridge engineering. However, near the bridge the substantial silting was found. The collated surface water level and scouring-silting data suggest that reduced surface water levels during low discharge are mainly due to the overall river erosion. The increased surface levels in the high discharge range are due to local river siltation, especially because of the increased water resistance as response to bridge engineering.

1 Introduction

The Hanjiang River is a large tributary of the Yangtze River and has its confluence near Wuhan (China). The Hanjiang River is divided into the upper, middle and lower reaches (Zhang et al., 2004). The Huangzhuang reach is at the middle and lower reaches of the Hanjiang River, which is a critical part to prevent floods. Hence, its flood characteristics are important. The climate, including precipitation characteristics of the Hanjiang River basin lead to a distinct difference in flooding hazard between summer and autumn (Sun et al., 2014; Dai et al., 2012). Li et al. (2017) concluded that the autumn flood of the Hanjiang River in 2017 was mainly characterized by a long duration and frequent occurrence of floods. The flood peak was not so high, but the flood volume was big. The river water level rose significantly at the middle and lower reaches. Xia et al. (2006) analyzed the flood characteristics of the autumn flood of the Hanjiang River in 2005, which was also was characterized by a large flood volume, but this had a high flood peak. Wu and Guan (2017) and Duan et al. (2018) have studied the flood control and regulation scheme for the middle and lower reaches of the Hanjiang River. Besides analyzing floods in the Hanjiang River, many scholars have also researched the water level characteristics of the Hanjiang River. For example, Lu et al. (2009) and Shi et al. (2013) have studied the impact of the building of the Danjiangkou Reservoir, which is in the upper reach, on the discharge characteristics of the Hanjiang River. Wang et al. (2017) have studied the hydrological characteristics of the lower reach, and Feng et al. (2016) have analyzed the impact of discharge variation on the state of the middle and lower reaches of the Hanjiang River.

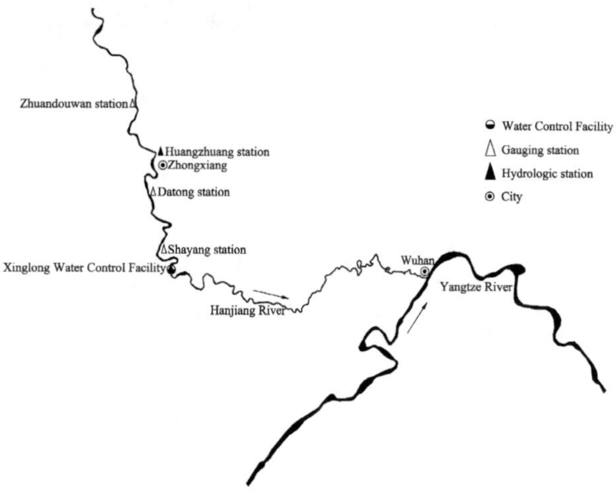

Figure 1. Sketch map of the research river.

The Huangzhuang hydrologic station is the main control station at the Huangzhuang reach of the Hanjiang River. About 32 km downstream the Datong gauging station can be found. During the autumn flood of the Han River in 2017, the actual hydrological measurements indicate that the observed water level of the Huangzhuang station was a bit high for the discharge at that time. To further understand the change in surface water level pattern of the Huangzhuang reach and the underlying reasons, this paper will study the relationship of the surface water level and discharge, and find out the changing pattern of water level at the Huangzhuang station. We will use data from the period 2000–2017, such as the highest water level, lowest level and corresponding discharges in these years. Data from four stations are analyzed, namely, from upstream to downstream, the Zhuandouwan station, Huangzhuang station, Datong station and Shayang station. Three important time moments are considered, namely, 2009 (when the Xinglong Water Control Project commenced to be constructed), 2013 (when the Xinglong Water Control Project was completed), and 2016 (when the Highway No. 2 bridge over the Hanjiang River at Zhongxiang commenced to be constructed). The following time sections were identified: 2005–2008, 2009–2013, 2014–2015 and 2016–2017, which were associated with average discharges of below 1000, 1000–5000, 5000–10 000 and 10 000 $m^3 s^{-1}$. The scour and silting changes within the Nianpanshan-Maliang reach (68.9 km) were analyzed based on data from 31 fixed large cross sections in the period between 2012 and 2016. Finally, the impact of the Highway No. 2 bridge project, which included damming near the bridge, on the water level of the Huangzhuang reach was analyzed according to the changing relationship of water level and the discharge Han.

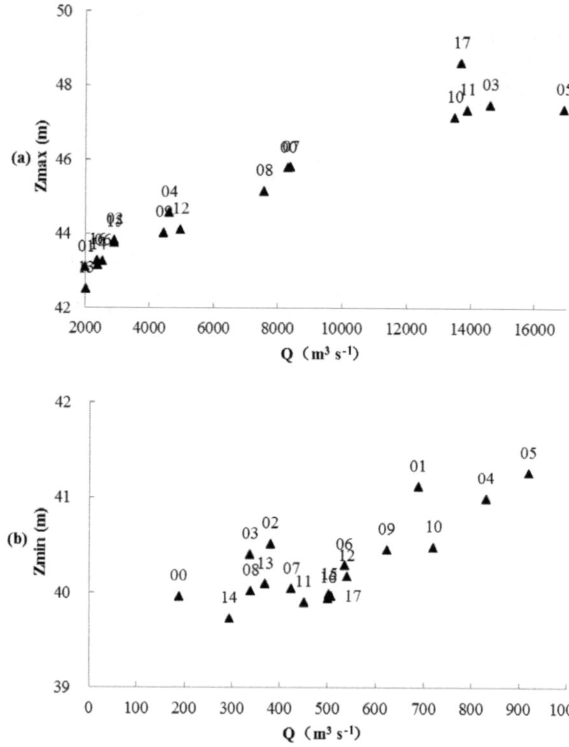

Figure 2. Relationship between annual average daily water levels and corresponding discharge at the Huangzhuang station for the period 2000–2017, **(a)** maximum water level, and **(b)** minimum water level. The numbers refer to the years, e.g. 00 means 2000.

2 Analysis of Flood Characteristics of the Huangzhuang Reach

2.1 Water level changes at the Huangzhuang station

Relationships between the annual average daily highest water level (Fig. 2a), annual average daily lowest water level (Fig. 2b) and corresponding discharge at the Huangzhuang station between 2000 and 2017, revealed that at high discharges during 2016–2017 (discharge: 13 000–17 000 $m^3 s^{-1}$), the water level was about 1.47 m higher than that during 2000–2015. At intermediate discharges (2000–10 000 $m^3 s^{-1}$), the water level was about 0.13 m higher. At low discharges (less than 3000 $m^3 s^{-1}$), the water level was similar to the level during 2010–2015, whereas the water level was 0.28–0.41 m lower than during 2000–2009. So, for the same discharge, the water level at the Huangzhuang station witnessed a significant rise in the high flows range in recent years. The change mainly happened after 2016. For the low flow range the opposite was found, although change is smaller.

 https://doi.org/10.5194/piahs-383-25-2020

2.2 Analysis of Huangzhuang reach surface fall

The distances between the Zhuandouwan and Huangzhuang stations, Huangzhuang and Datong stations, and Datong station and Shayang stations is 22, 32 and 39 km, respectively. The surface water fall (decrease of surface water level over reach) for different discharge ranges and the defined time periods is shown in Fig. 3. The surface water fall (decrease of surface water level over reach) was calculated for each time period and discharge range as the average of all daily falls. Figure 3 indicates that the surface water fall in the Zhuandouwan-Huangzhuang reach shows a decreased fall in the two most recent periods (Fig. 3a). The surface water fall in the Huangzhuang-Datong reach during 2016–2017 witnessed a big increase when the discharge was above 5000 $m^3 s^{-1}$, while the opposite was found for the time sections 2005–2008 and 2009–2013 (Fig. 3b). The surface water fall of Datong-Shayang reach during 2014–2015 witnessed a big drop (only available when the discharge was lower than 5000 $m^3 s^{-1}$). The Datong-Shayang reach was affected by the Xinglong Water Control facility, while the Huangzhuang-Datong reach was affected by related the project of bridges in the reach.

3 Analysis of Influencing Factors

3.1 Analysis of scour and silting changes in the Huangzhuang reach

Scour and silting data were obtained from three measurements in the Nianpanshan-Maliang reach that were conducted in December 2012, December 2015 and December 2016. Channel storage and scour-silting in the reach was calculated with a cross-section method (Duan et al., 2001; Zhang et al., 1989; Wang et al., 2018). The scour-silting conditions of the reach since 2012 are shown in Table 1. During 2012–2016, the Nianpanshan-Maliang reach was a scour as a whole under different calculated water levels. From 2012 to 2015, 15.36 million m^3 was scoured from the channel at high water levels. From 2015 to 2016, this number was 8.26 million m^3 as a whole. When the channel was at a low water level, the reach scoured 2.57 million m^3. during 2012–2015. Scouring is assumed to be an influencing factor for dropping of the water level.

For further analysis, we investigated scouring and silting in the Hanjiang River in the reach from Huangzhuang to the Road No. 2 bridge Hanjiang Zhongxiang. This reach has a length of 8.1 km, in which 4 fixed large cross-sections occur. The reach showed scouring between 2012 and 2015 (Table 2). The opposite holds for water levels of 42–46 m in the period 2015–2016. In this period the reach showed silting. Partial silting of this reach may be the cause for the rising of the Hanjiang River at Huangzhuang at a high water levels. However, the influence was not distinct.

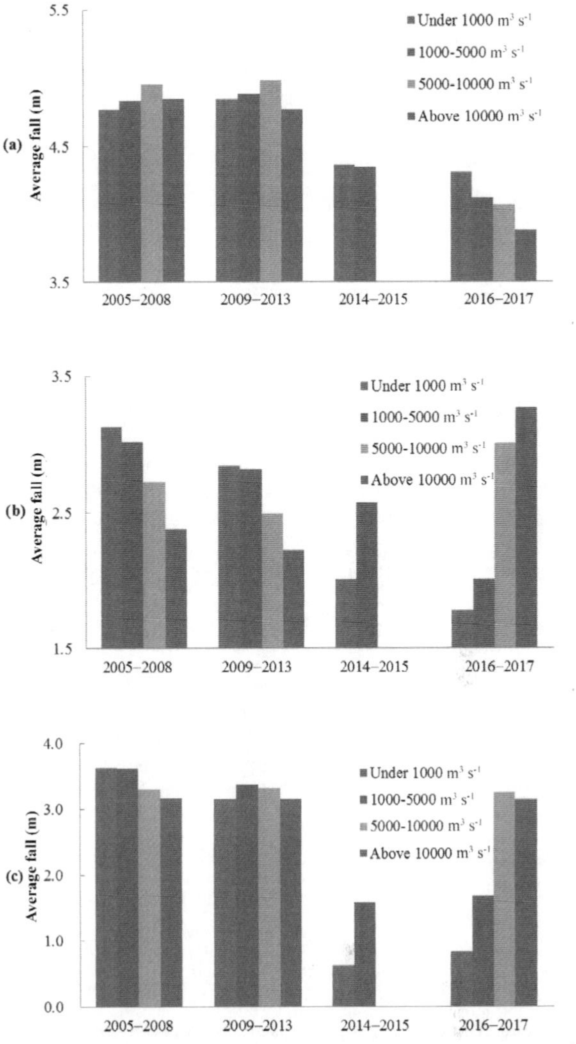

Figure 3. Surface water fall in different reaches of the Hanjiang River for different discharge ranges and time periods in 2005–2017, **(a)** Zhuandouwan-Huangzhuang reach, **(b)** Huangzhuang-Datong reach, and **(c)** Datong-Shayang reach.

The Xinglong Water Control facility lies about 100 km downstream of the Huangzhuang station, which commenced to be constructed in 2009 and was completed in 2013. The water of the reservoir returned to the Hanjiang River about 70 km upstream of the Maliang station. The analysis of surface water levels and surface fall indicate that after the building of the Xinglong Water Control facility, the water level was higher when the discharge was less than 5000 $m^3 s^{-1}$ at the Shayang station. Meanwhile, the surface water fall in the Datong-Shayang dropped. Compared to other stations, impact on water level and surface fall here was small.

From 2013 to 2016, many spur dikes, revetments, bottom protection measures, channel filling and beach protec-

Table 1. Statistics of scour and silting amount in the Nianpanshan–Maliang reach riverbed (+silting, −scour).

River reach	Calculated water	Scour and silting amount of each time section (* 10 000 m^3)		
	level (m)	2012–2015	2015–2016	2012–2016
Nianpanshan-Maliang 68.9 km	42	−257	−23	−280
	43	−467	−125	−592
	44	−606	−229	−835
	45	740	−350	−1090
	46	−868	−512	−1380
	47	−1122	−672	−1795
	48	−1360	−778	−2138
	49	−1536	−826	−2362

The calculated water level uses the national 85 elevation.

Table 2. Statistics of scour and silting amount in the Huangzhuang–Zhongxiang reach riverbed (+silting, −scour).

River reach	Calculated water	Scour and silting amount of each time section (* 10 000 m^3)		
	level (m)	2012–2015	2015–2016	2012–2016
Huangzhuang-to the Road No. 2 bridge in the Hanjiang River at Zhongxiang (8.1 km)	42	−80	+34	−46
	43	−174	+37	−136
	44	−182	+22	−160
	45	−183	+23	−160
	46	−193	+22	−171
	47	−205	0	−205
	48	−214	-6	−221
	49	−220	−8	−228

The calculated water level uses the national 85 elevation base.

tion belts have been built within the reach from Nianpanshan to the Xinglong Water Control facility. They were intended to regulate the channel, with the aim to deepen the main river channel. The elevation of the jetty head of the spur dikes was about 41 m and the dike body was 2–3 m high, and around 5 m wide. They were densely placed in the upper and lower reaches of the Hanjiang River near Huangzhuang. Due to the water blocking effect of the dikes and the building of other projects, such as spur dikes and supporting revetments and bottom protection, the bottom roughness factor of river channels close to the structures was assumed to increase.

The analysis of surface water levels and surface water falls, showed that the channel regulation projects influenced the entire reach, but did not caused distinct changes.

During 2016–2017, the Road No. 2 bridge over the Hanjiang River at Zhongxiang was under construction. The bridge is located 9.7 km downstream the Huangzhuang station. The relationship between the surface water level and discharge, and the surface water fall near the bridge was measured. The changes in the relationship between the surface water level and discharge relative to the data from 2012 to 2016, shows that the road bridge caused a big impact(see also Sect. 3.2).

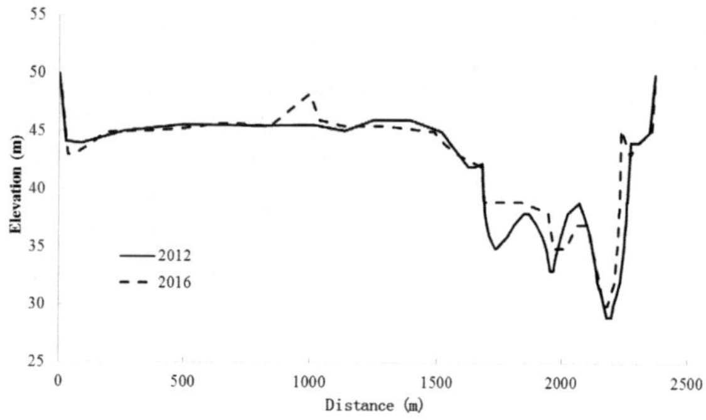

Figure 4. The cross section at the bridge in the Hanjiang River.

3.2 Changes of cross section at the bridge

Figure 4 shows the cross section at the bridge in 2012 and 2016. The figure shows that from 2012 to 2016, the silting on the left bank of the main river channel stretched over about 150 m, with a maximum silting thickness of 4 m. The silting line is about 20 m from the right bank of the main river channel, and the elevation of the lowest point of the riverbed rose from 29 to 30 m. The blocking area of the cofferdam during the construction period in 2016 was about 600 m^2. The dam had an elevation of 39 m and a length of about 150 m on the left bank of the cross section. We estimate that the water blocking area of the three main piers was about 243 m^2, and that the one of the pier on the bottomland was about 50 m^2. It suggests that the water blocking during the construction was the main factor for the rising of the surface water level.

The Manning formula (Zhao et al., 2005) was used to calculate the stage-discharge relationship at for the conditions in 2012 and 2016 (Fig. 5). The following assumptions were made: (i) roughness factor of 0.03 based on the stage-discharge relationship of the Huangzhuang station, and (ii) gradient of 0.1 ‰ applicable to the Huangzhuang-Datong reach on 5 October 2017. The stage-discharge relationship indicates that when the water level was lower than 38 m, the curves of 2012 and 2016 basically coincide. When the water level was higher than 38 m, the surface water level in 2016 was higher than in 2012 for the same discharge, i.e. 0–1.3 m higher.

3.3 Analysis of damming at the Road No. 2 bridge

The damming at Road No. 2 bridge over of the Hanjiang River in Zhongxiang is was analyzed using water level observations the upstream and downstream of the cross section of the bridge. Surface water levels were observed at 17:00–19:00 h on 6 October 2017. The positions of the observations points are shown in Figure 6 and the measuring results in Table 6. The measuring data show that the surface water fall is big, that is, 0.085, 0.143 m and 1296 m at 994 m, at upstream of the bridge, at the bridge, and at 1000 m downstream of the bridge, respectively. A small negative gradient has occurred

Table 3. Observed surface water fall and gradient at the Road No. 2 bridge.

Point x	Actual measured, water level H (m)	Distance, D (m), $D_{x-1}-D_x$	Surface fall between two points (m), $H_{x-1}-H_x$	Surface water gradient $H_{x-1}-H_x$ / $(D_{x-1}-D_x)$ (‰)	Position relative to bridge
1#	46 243				2031 m upstream
2#	46 160	1037	0.083	0.08	994 m upstream
3#	46 163	482	−0.003	−0.006	512 m upstream
4#	46 078	293	0.085	0.29	209 m upstream
5#	46 071	104	0.007	0.067	105 m upstream
6#	45 928	105	0.143	1.36	Bridge
7#	45 923	107	0.005	0.047	107 m downstream
8#	45 926	96	−0.003	−0.031	203 m downstream
9#	45 918	304	0.008	0.026	507 m downstream
10#	44 622	493	1296	2.63	1000 m downstream position

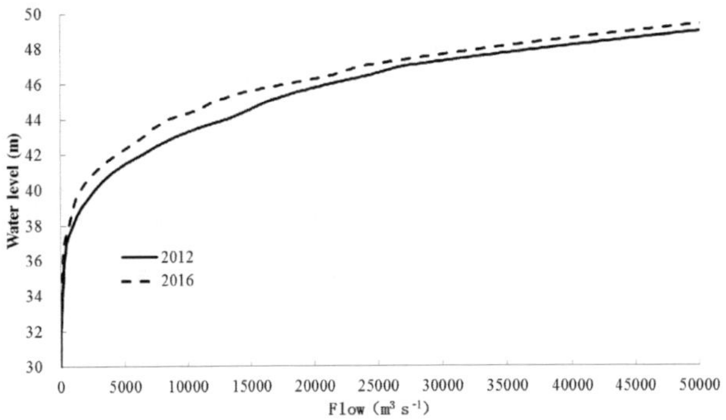

Figure 5. Impact of construction work in 2016 on the stage-discharge relationship at the Road No. 2 road bridge over of the Hanjiang River in Zhongxiang.

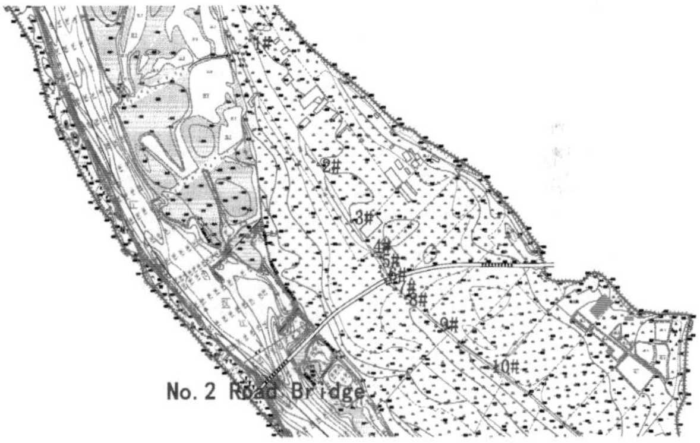

Figure 6. The measuring points at the Road No. 2 bridge.

at two places, both upstream and downstream of the bridge. On 6 October 2017, the average surface water fall over the Huangzhuang-Datong reach (32 km) was 2.94 m. The surface water fall over the reach from observation points 1# to 10# (3031 km) was 1621 m, accounting for 55.1 % of the entire surface fall from Huangzhuang to Datong. It indicates that the damming at the bridge is the main reason for the bit high surface water level at the Huangzhuang station.

4 The study led to the following conclusions

1. Compared with 2000–2015, the surface water level at the Huangzhuang station was about 1.47 m higher in 2016–2017 in the high discharge range. In the middle discharge range, the water level was 0.13 m higher. The surface water level during 2016–2017 was basically similar with that during 2010–2015 in the low discharge range, but 0.28–0.41 m lower than in 2000–2009. The rise of the surface water level at the Huangzhuang station mainly occurred in the high discharge range predominantly after 2016. At the lower water level, the water level showed a trend of dropping.

2. The analysis of the changes in the surface water level and surface water fall, revealed that the water level of the Huangzhuang station during 2016–2017 mainly changed due to projects built in parts of the Huangzhuang-Datong reach. The building of the Xinglong Water Control facility mainly made the water levels of the Shayang station rise and the surface water fall of the Datong-Shayang station drop.

3. The Nianpanshan-Maliang reach of the Hanjiang River showed a trend of scouring as a whole. During 2012–2016, 2.80 million m^3 was scoured in total from the reach. The overall scouring in the reach was the main reason for the drop of the surface water level in the low discharge range. Silting in the reach from Huangzhuang to the Road No. 2 bridge over Hanjiang River at Zhongxiang might have been a cause for the rising water level at Huangzhuang.

4. During 2012–2016, silting was observed in the cross section at the Road No. 2 bridge. When the water level exceeded 38 m, it was 0–1.3 m higher in 2016 than in 2012 under the same discharge. The surface water fall in the rather short subreach near the bridge was 1621 m, accounting for 55.1 % of the entire fall in the Huangzhuang-Datong reach (subreach is about 10 % of the whole reach) . The analysis reveals that the damming at the bridge was the main reason that the water level was a bit high in the high discharge range at the Huangzhuang station.

5. In summary, the main cause for the higher water levels at the Huangzhuang station were the impacts from projects in parts of the Huangzhuang-Datong reach, particularly the Road No. 2 bridge at Zhongxiang.

Data availability. The data file will be sent separately to the editorial department.

Supplement. The supplement related to this article is available online at: https://doi.org/10.5194/piahs-383-25-2020-supplement.

Author contributions. XX is responsible for data analysis and manuscript writing. Other authors participate in data collection and analysis.

Competing interests. The authors declare that they have no conflict of interest.

Special issue statement. This article is part of the special issue "Hydrological processes and water security in a changing world". It is a result of the 8th Global FRIEND-Water Conference: Hydrological Processes and Water Security in a Changing World, Beijing, China, 6–9 November 2018.

Acknowledgements. We would like to thank our colleagues from Middle Changjiang River Bureau of Hydrology and Water Resources Survey for their hard work.

Financial support. This work is supported by the Hubei Provincial Natural Science Foundation of China under the contract no. 2017CFB312, the Fundamental Research Funds for the Central Universities (no. 2017B20514) and PAPD and the Open Foundation of State Key Laboratory of Hydrology-Water Resources and Hydraulic Engineering (no. 20174911111).

References

Changhong, W., Xiaobin, C., and Wangsheng, L.: Study on the hydrological characteristics of the lower reaches of Hanjiang River[J], China Water Transport, 17, 116–119, 2017.

Guangyin, F., Wenjie, L., and Lihua, Z., Dong, C.: Influence of flow process variation on river health in middle and lower reaches of Hanjiang River[J], Journal of China Hydrology, 36, 46–50, 2016.

Guobin, L., Yi, L., Xianglin, Z., Zhenghua, Z., and Tao, C.: Impact of the Danjiangkou reservoir on the flow regime in the middle and lower reaches of Hanjiang river[J], Resources and Environment in the Yangtze Basin, 18, 959–963, 2009.

Hongmei, M.: Analysis of river bed evolution and river regime control in the MANhekou to Shayang of Hanjiang River[D], The sixth session of the national river management and Ecological Restoration Technology Seminar, 114–119, 2014.

Jintai, Z.: Analysis of scouring and silting in Chenglingji to Hankou reach of Changjiang River[R], Bureau of Hydrology, Changjiang Water Resources Commission, 124 pp., 1989.

Linbo, W.: Han Jia Zhang Zi of raoyanhe hydrological characteristic water level for flood control (flow) study[J], Yangtze River, 49, 68–71, 2018.

Minglong, D., Lily, Y., and Yuanyuan, L.: Study on encounter rules of floods from Hanjiang River and upper Yangtze River[J], Yangtze River, 43, 48–51, 2012.

Wangchun, C., Jing, Z., Tao, Z., and Yubing, C.: Analysis of the storage capacity in the Shayang to Hankou reach of the downstream Hanjiang River[J], Express Water Resources & Hydropower Information, 38, 41–44, 2017.

Wei, X., Junya, M., Xianqun, Y., and Beiping, M.: Characteristics of flood peak discharge in the lower reaches of Hanjiang River in 2005[J], Yangtze River, 37, 28–34, 2006.

Weixin, D., Shenglian, G., Jun, Z., Wenhui, X., and Huanhuan, B.: Study on dynamic control scheme of water level of Danjiangkou Reservoir during flood season[J], Yangtze River, 49, 7–12 , 2018.

Wenzhong, D., Yahui, Z., and Jianjun, L.: Causes of water level rising in Chenglingji-Luoshan reach of Changjiang River[J], Journal of Hydraulic Engineering, 2, 29–34, 2001.

Xia, W. and Honglin, G.: Study on the flood control plan for Yicheng to Shayang section of middle and lower reaches of Hanjiang River[J], China Rural Water and Hydropower, 4, 35–40, 2017.

Xiaojun, Z., Jiaqing, H., and Huangyin, F.: Analysis of flood characteristics of middle reach of Hanjiang River[J], Journal of Changjiang Engineering Vocational College, 21, 6–7, 2004.

Xiurui, S., Xianlin, K., and Bin, Y.: Changes in the relationship between water level and discharge after Danjiangkou reservoir construction and its impact on downstream flood control[J], Express Water Resources and Hydropower Information, 20, 8–10, 1999.

Yunfa, L.: Recent scouring in the middle stream of Hanjiang River[J], Journal of Yangtze River Scientific Research Institute, 31, 1–5, 2015.

Yurong, L., Jun, Z., and Xiao, Z.: Autumnal flood analysis and forecast operation of Hanjiang River in 2017[J], Yangtze River, 48, 1–10, 2017.

Zhaohua, S., Qiufen, F., Jianqiao, H., and Qixin, C.: Reasons for heavy deposition of sandbar at confluence of Yangtze River and Han river during autumn flood season of 2011[J], Journal of Hydroelectric Engineering, 33, 111–117, 2014.

Zhenxing, Z. and Jianjing, H.: Hydraulic[D], Tsinghua University Press: Beijing, China, 458 pp., 2005.

Proc. IAHS, 383, 31–41, 2020
https://doi.org/10.5194/piahs-383-31-2020

Open Access
PIAHS

Climate change impact on hydrological characteristics and water availability of the Mountain Pamir Rivers

Inom Normatov[1] **and Parviz Normatov**[2]

[1]Meteorology and Climatology Department, Tajik National University, Dushanbe, Tajikistan
[2]Institute of Water Problems, Hydropower and Ecology Academy of Sciences Republic of Tajikistan, Dushanbe, Tajikistan

Correspondence: Inom Normatov (inomnor@gmail.com)

Published: 16 September 2020

Abstract. Results of monitoring accumulated snow cover in upstream areas of the Transboundary Pyanj River (Central Asia) are presented. It is found that the formation of the snow cover and the spatial distribution of atmospheric precipitation in the Mountain Pamir is determined by the orography of the terrain. Orography influences air mass movement in mountain areas, which contributes in different climatic zones to a shift in periods when the maximum amount of snow is falling. Completely different scenarios for the development of meteorological variables in the western and eastern parts of the Gunt River Basin were found, due, firstly, to the influence of the mountainous area orography and, secondly, to the penetration of various air masses. It is observed that in the western part of the basin the average annual precipitation remained almost unchanged over the period 1944–2014, whereas there is a decreasing trend in the eastern part. Assessment of the climate change impact on the formation of the Gunt River water flow was made by comparing the trend in the change of discharge using data from two observation periods 1940–1970 and 1986–2016. Calculations show a decrease of the Gunt River discharge by 5 % over a period of more than 70 years. The influence of climate warming on the river flow is indicated by comparison of river hydrograph in two periods 1940–1970 and 1986–2016. The hydrograph of the Vanch River in the earlier-mentioned periods shows a shift in the maximum of the monthly discharge towards the left, indicating an earlier melting of snow and glaciers in the upstream regions of the river and a significant increase in discharge in the period 1986–2016.

1 Introduction

Mountain watersheds serve as important water sources by providing fresh water for downstream human activities (Viviroli et al., 2003; Langston et al., 2011). As a result of snow and glacier melt, the magnitude and timing of runoff from these watersheds tend to be very sensitive to changes in the climate (Immerzeel et. al., 2009, 2010, 2012, 2013; Jeelani et al., 2012; Lutz et al., 2014, 2016; Gain et al., 2011). Changes of melt runoff may even affect the sustainable development of downstream cities in the long run (Verbunt et al., 2003; Zhang et al., 2012). Modeling snow and glacier melt runoff processes is therefore quite important for local water supply, hydropower management and flood forecasting (Klok et al., 2001). However, melt runoff modeling in such regions faces two challenges: scarcity of meteorological data and uncertainty in parameters due to limited understanding of the complex hydrological processes.

Growing evidence indicates mountain glacier and snow covered area on average have declined on both hemispheres (IPCC, 2007, 2018). Geographic areas where the water cycle is dominated by snow and glacier melt hydrology in particular expected to be susceptible to climate change as it affects the seasonality of runoff (Adam et al., 2009). Changes in seasonal snow covered and glaciated regions may alter the variability of stream flow and hence water availability that sustains a large population downstream. The reason for the uncertainty is not that people do not understand the regional importance, but it is the complexity of the processes and the terrain that is hard to access to carry out measurements/observations. These climate driven responses of mountainous river hydrology may pose significant challenges for this region. Regional climate projections by IPCC (2018) indicate that in Central Asia by the end of the 21st century the median temperature will rise by of 3.7 °C with largest warming over higher altitudes particularly in the Tibetan Plateau and the Himalayas (Bhattarail and Regmi, 2011).

As global temperatures rise, the world's snow resources are predicted to change significantly (Hosaka et al., 2005; Christensen et al., 2007; Deser et al., 2010). Long-term changes in global, regional, and local snow depth, snow water equivalent (SWE), and extent will ultimately have major ramifications for ecosystem functions, human utilization of snow resources, and the climate itself through feedback mechanisms like snow albedo (Barry, 1996). Unfortunately, only extent snow cover area (SCA) of the three snow metrics listed above is easily monitored using satellites. This monitoring, which is under way for several decades (Robinson, 1993; Frei anbd Gong, 2005) has shown that global SCA has been decreasing for the past 30 years (Lemke et al., 2007; Matthew et al., 2010).

Snow accumulation generally increases with elevation because of the combined effect of the prevailing lower temperatures and the increased frequency of precipitation events caused by orographic effects.

Distributed data on snow depth, density and snow water equivalent (SWE) with a high spatial and temporal resolution are essential for validation of and/or as input to snow drift models (Prokop et al., 2008) and snow melt

runoff models (Lindström et al., 2010). Thus, there is a great need for distributed snow data, mainly for SWE data (i.e. snow depth and density). Spatially distributed SWE data is important for many stakeholders, for example, it can be used as an input to the new generation of hydrological models predicting snowmelt run-off (Kolberg and Gottschalk, 2006; Udnæs et al., 2007; Arheimer et al., 2008).

Alford and Armstrong (2010) and Alford et al. (2009) examined glaciers and glacier discharge in Nepal zooming in on nine subcatchments of the Ganges left tributaries with a total glaciated area of 3644 km^2 and concluded that glaciers contribute by 2 %–3 % to the discharge of all rivers, flowing from Nepal, i.e., 5.38 km^3 in total, which indicates that specific glacier runoff from this area is about 1500 mm yr^{-1}. In another research, conducted recently in the Nepal Himalaya, the glacier and seasonal snow melt contribution to the Mean Annual Flow (MAF) is estimated as 14 km^3, i.e., about 10 % of MAF from Nepal (Andermann et al., 2012).

In the Amu Darya and Syr Darya basins, meltwater contributes by 69 % and 79 %, respectively, to mean annual streamflow, and the share of seasonal snowmelt by far outweighs that of glaciers.

The study area includes the climatic zones of the Pamir Mountains. Precipitation and snow depth data from meteorological stations in the relevant climatic zones of the Pamir Mountains were obtained from the Agency for Hydrometeorology of Tajikistan. Mean monthly temperatures and precipitation of the Gunt River Basin were obtained from meteorological stations in the Khorog and Bulunkul regions for the period 1944–2014.

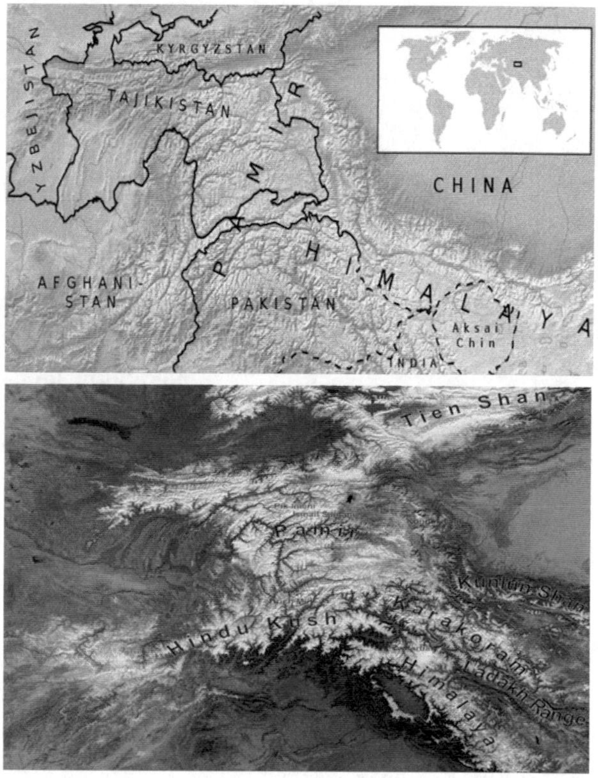

Figure 1. Maps of Mountain Pamir. Sources: (**a**) https://upload.wikimedia.org/wikipedia/commons/thumb/7/79/Location_map_Pamir_mhn.svg/1200px-Location_map_Pamir_mhn.svg.png, last access: 12 May 2020, (**b**) https://upload.wikimedia.org/wikipedia/commons/thumb/9/9e/High_Asia_Mountain_Ranges.jpg/540px-High_Asia_Mountain_Ranges.jpg, last access: 18 June 2020.

2 Data and Methodology

The roof of the World – so often translate the name of the Pamir mountain system. Pamir is located on the territory of Tajikistan, China, Afghanistan and Pakistan and closely connected with other highest mountain systems of the world – Tien Shan, Karakorum, Hindu Kush. Himalayas. In the Pamirs, there are several peaks with a height of more than 7000 m a.s.l. The highest point of the Pamirs – Kongur peak (7649 m) located in China (Fig. 1).

In Tajikistan, there are such peaks as Ismoili Somoni (7495 m), Lenin (7134 m) and Korzhenevskaya (7105 m). Mountains Pamir is divided into Eastern and Western parts. The Eastern Pamirs in physical and geographical terms is a high plateau located above 3500 m a.s.l. The Western Pamirs is occupied by high mountain ranges separated from each other by deep valleys. The valleys are located at altitudes of 1700–2500 m, and the mountain ranges exceed 5000 m. Glaciers and eternal snow characterize the Western Pamirs. The Pamirs has the largest glaciation with the largest valley glacier Fedchenko (about 70 km).

The diversity of climatic conditions in Central Asia, the patterns of changes in meteorological processes, depending on the geographic and geoecological features of the region leads to the need for climatic zoning. The Pamir Mountains is considered to be the area where the influence of moist, cold Mediterranean precipitation decreases and dry Central Asian commences. The main zone of generation of the Pyanj River and its main tributaries Gunt and Vanch rivers water flow is located in the Pamir Mountains (Fig. 2).

The Gunt River is a large tributary of the Transboundary Pyanj River by length of 246 km and 14 840 km^2 of the basin area. Its source (Alichur River) originates in the Pamir Highlands, where the river flows through a marshy plain bearing traces of ancient glaciation (Fig. 3). After leaving the lake Yashilkul River is called Gunt. On the stretch from the lake to the mouth, the Gunt flows with a violent rapid current, forming numerous rapids on its way. Before flowing into the Pyanj, Gunt takes on the left a large tributary of the Shakhdarya River (Knowledge Base, 2003).

The Pamir is at the front of the contact between two air masses, the Western (in the winter half-year) and the Indian summer monsoons (in the summer half-year), which dominate in the two climatic zones of the Gorno-Badakhshan Autonomous Region of the Republic of Tajikistan. The Pamir differs from the neighboring Himalayas and Hindu Kush by the fact that the water flow of the formed rivers is mainly determined by the melting of snow and glaciers (Pohl et al., 2015; Fuchs et al., 2013; Aizen et al., 2009; Syed et al., 2006; Palazzi et al., 2013). The Yashilkul Lake on the middle of the Gunt River divides the river basin to the western and eastern parts. The orography of the western and eastern parts of the basin makes a significant contribution to the penetration of air masses and the formation of atmospheric precipitation. A pronounced relief with steep erosional valleys characterizes the western part but the eastern territory is a high plateau with smaller heights and wider valleys. Lake Yashilkul to some extent smooths the average annual value of water consumption and leads to the fact that in low water and in winter the water flow remains almost constant (Meier et al., 2016).

The Gunt River Basin (Fig. 3) is generally located in the transit zone of various air masses, although the formation of a certain amount of atmospheric precipitation in river basin significantly affected by western Indian monsoons and southern cyclones (Meier et al., 2013).

The penetration of air masses from the Mediterranean and Caspian seas is also characteristic of the Vanch River basin. The Vanch River is one of the tributaries of the transboundary Pyanj River that is formed after the confluence of equivalent rivers Kasholyakh and Abdukahor. The river Vanch can be considered a representative of river arteries, in the formation of water flow of which an important place played groundwater, dictated by the geological structure and distribution in the basin of permeable rocks. Measurements carried out in 1965 showed that the annual runoff module in the basin of the Geographical society glacier on the upstream of Vanch River (basin area 206 km, the share of glacial runoff 62 %) is 15–60 $L\,km^2\,s^{-1}$. In the Abdukahor River Basin on the upstream of Vanch river (the basin area is 329 km, the share of glacial runoff is 42 %) 28.4 $L\,km^2\,s$ (Adam et al., 2009). The general tendency reduction of glaciers in Central Asia in the last century also covered the glaciers of the Vanch River Basin. The processing of satellite images LANDSAT ETM+ and TERRA (ASTER) allowed the authors (Prokop et al., 2008; Lindström at al., 2010) to establish that the glacier area of the Vanch river basin for the period 1961–2000 decreased by 23.4 % and throughout the basin of the transboundary Pyanj River by 32.7 %.

The territory of the Republic of Tajikistan is characterized by four climatic zones. In turn, the Gorno-Badakhshan Autonomous Region (GBAO) that covers almost the entire mountain Pamir and is a zone of formation of the Transboundary Pyanj River is characterized by three climatic conditions (Fig. 4).

3 Results and Discussion

3.1 Spatial distribution of precipitation and snow cover on the Mountain Pamir climatic zones

The dependence of the depth of the snow cover and the amount of atmospheric precipitation on the altitude of the mountainous terrain as seen from Fig. 5 does not described by the known regularities.

For example, the meteorological station Shaimak located at an altitude more 3800 m a.s.l. by the precipitation means differs little from the values typical to the Irkht of meteorological station at the height of 3276 m a.s.l. The same can be observed at comparison of the snow depth and precipitation values at other meteorological stations.

The spatial heterogeneity of atmospheric precipitation in mountain areas is primarily due to the influence of the terrain orography on the distribution of moist air masses at more moderate altitudes (Darvaz), the average annual temperature increase occurs at a more perceptible rate than at the headwaters (Normatov et al., 2017).

On terraces of the Darvaz meteorological station (1288 m a.s.l.) location the precipitation mostly drops out in the liquid state and seldom observed precipitation in the solid state (under a favorable temperature condition) characterized by short lifetime.

The vertical gradient at the Southern and Central zones borders is about 40 mm $(100\,m)^{-1}$ of elevation that indicates the more humid foothills and the existence of hallway to pen-

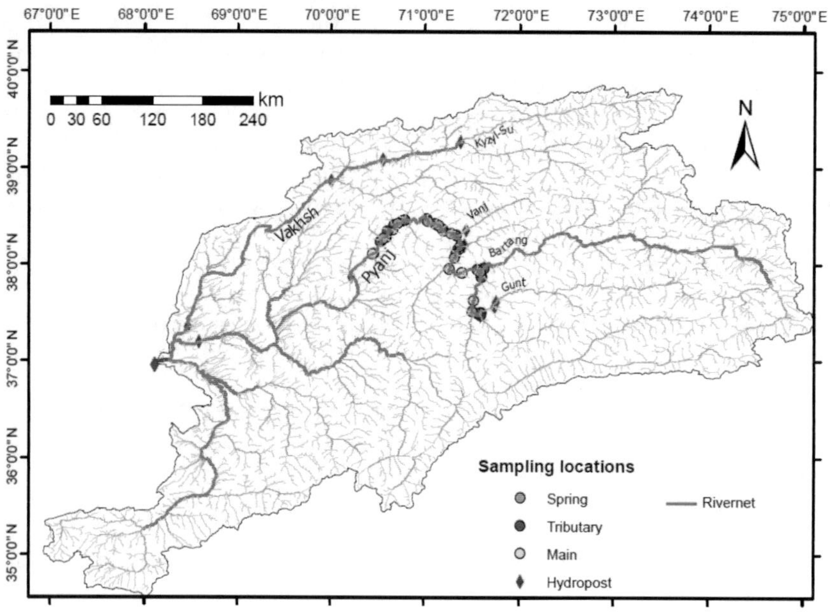

Figure 2. Maps of Pyanj River and tributaries basins.

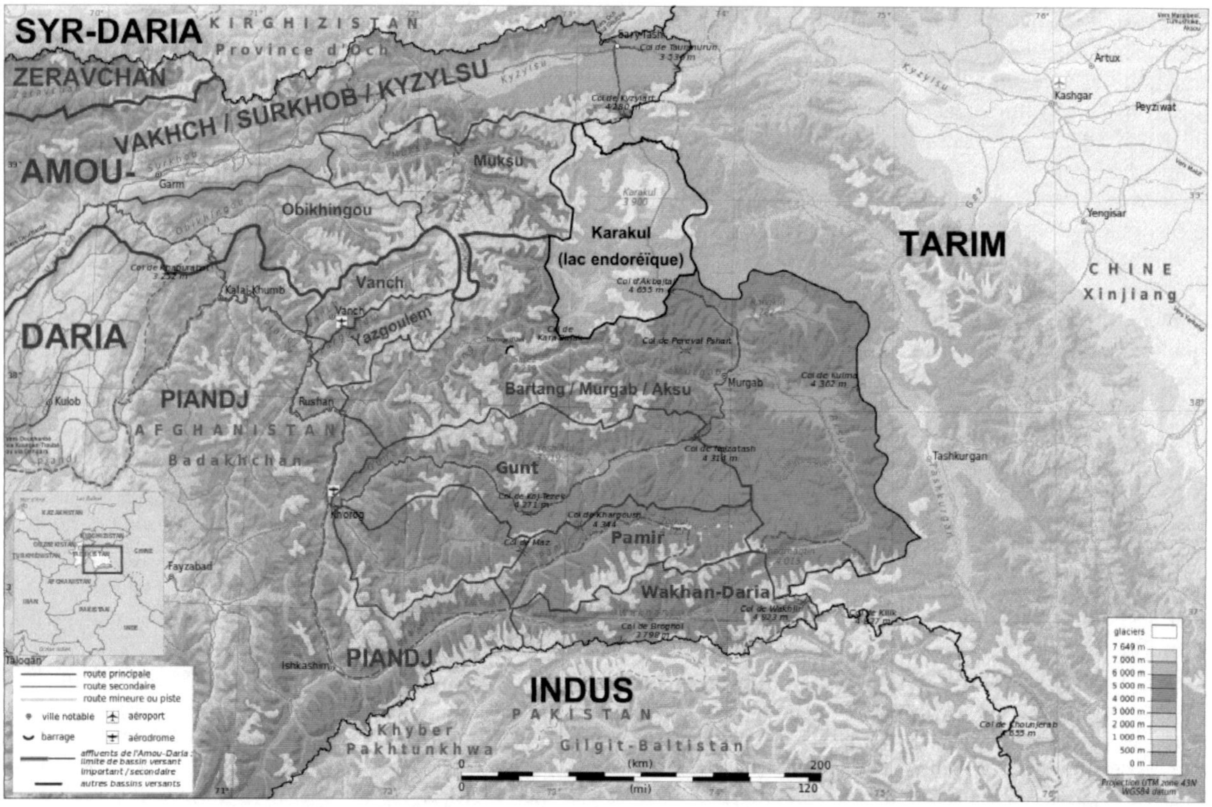

Figure 3. Gunt River Basin on Hydrological Map of the Republic of Tajikistan. Source: https://commons.wikimedia.org/wiki/File:Pamir_drainage_basin_map-fr.svg, last access: 7 July 2020.

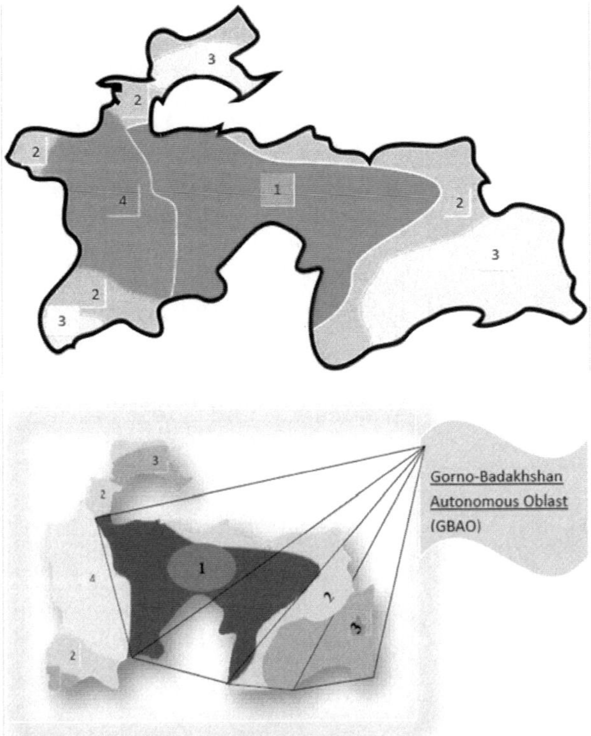

Figure 4. Climatic zones of the Republic of Tajikistan and Gorno Badakhshan Autonomous Oblast: 1 – warm continental climate; 2 – cold semi-arid climate; 3 – dry cold climate; 4 – warm Mediterranean climate.

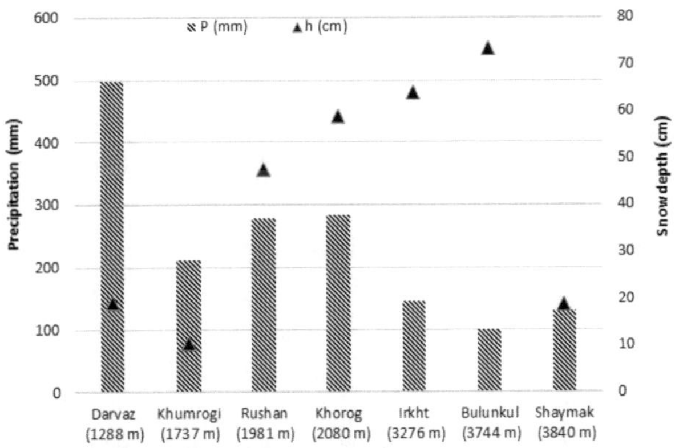

Figure 5. Average annual precipitation and snow depth for seven meteorologica stations in the Pyanj River Basin depending on altitude for the period 1944–2014.

etration of the west humid airflows. As the airflow moves deeper into the mountain range and passes through ridges moist air converts to precipitation and becomes dry.

The average annual precipitation in the Eastern Pamirs is insignificant (40–140 mm at average long-term value of about 76 mm). Precipitation deficit in the Eastern Pamir is

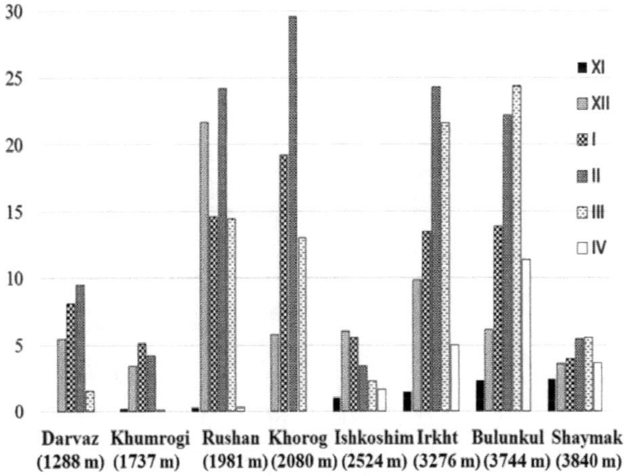

Figure 6. The average long-term monthly snow covers at the meteorological stations of the Pyanj River Basin for the period 1944–2014.

explained by the fact that in the Western Pamir, characterized by high mountain ranges (5000–6000 m a.s.l.), there is a discharge of moist air with heavy precipitation and the air passing through its ridges becomes dry (Normatov et al., 2017).

In principle, there should be a time interval between the periods of maximum snowfall in the western, central climatic zone of the Pamir characterized by precipitation that is more abundant and the eastern dry climatic zone. This is primarily because the mass of dry air that crossed the mountain ranges must be saturated by water vapor. The average long-term monthly snow cover at the meteorological stations of the Pyanj river basin as can be seen from the Fig. 6 shows that the maximum value of the snow covers in the eastern meteorological stations Bulunkul and Shaimak corresponds to the March.

The appearance of the maximum value of the snow cover at the Ishkoshim meteorological station in December (Fig. 7) is connect probably by the penetration of air masses from Iran and Afghanistan. At comparing the histograms on the Fig. 6, it becomes clear that the Western and Central climatic zones (Khorog, Irkht, Rushan and Bulunkul) are characterized by a sufficient amount of precipitation and the result of the existence of a temperature regime with a sufficient height of the snow cover.

3.2 Gunt River Basin Meteorology and Hydrology

The Gunt river runoff changes and meteorological conditions of the Western and Eastern parts of the Gunt river basin was considered for the period 1944–2016.

The trend of the atmospheric precipitation changes relative to the average long-term for the period 1944–2016 according to Khorog meteostation it shown on the Fig. 8.

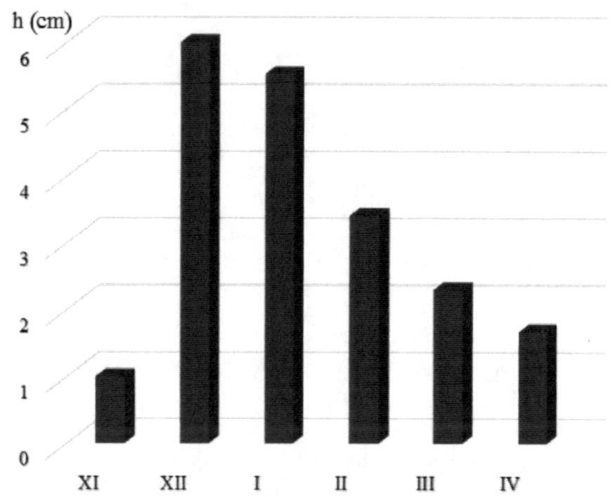

Figure 7. The maximum snow cover at the Ishkoshim meteorological station in the Pyanj river basin for the period 1944–2014.

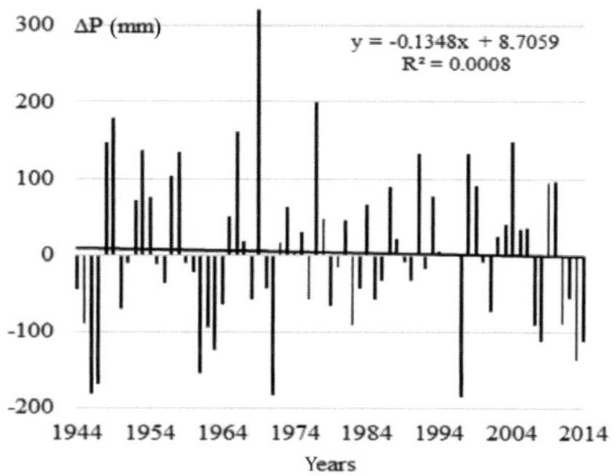

Figure 8. Deviation of annual precipitation from the long-term average (1944–2016) and trend for the Khorog meteorological station.

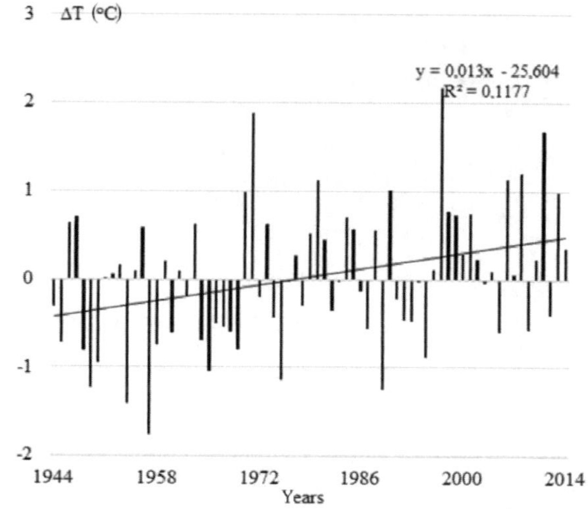

Figure 9. Deviation of annual temperature from the long-term average (1944–2016) and trend for the Khorog meteorological station.

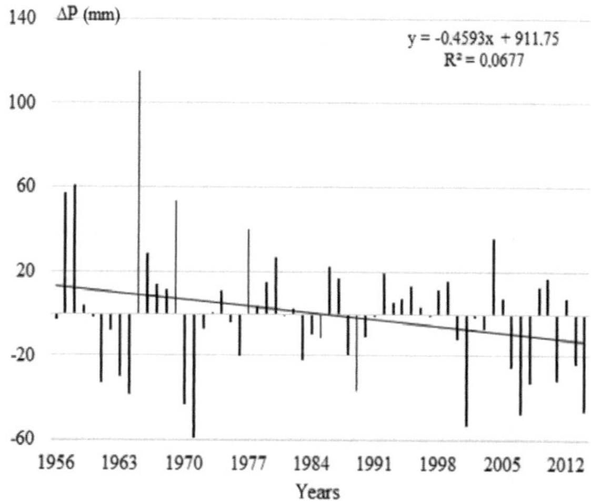

Figure 10. Deviation of annual precipitation from the long-term average (1956–2012) and trend for the Bulunkul meteorological station.

For more than 70 years (1944–2016), as shown in Fig. 8, precipitation in the Western part of the Gunt river basin was almost constant. However, the temperature increased at a rate of $0.007\,°C\,yr^{-1}$ (Fig. 9). The Middle, North and Caspian Seas air masses penetration to the Gunt River Basin is confirmed by the results of isotope analyzes of the waters of the northern and southern tributaries of the Gunt River (Meier et al., 2016). Despite the small difference in the height of the ridges in the eastern and western parts of the Gunt River Basin rivers is enriched by heavy isotopes. This is primarily due to atmospheric circulation and penetration of the Mediterranean and Caspian cyclones.

An opposite picture of the change in temperature and the amount of atmospheric precipitation in the eastern part of the Gunt river basin was observed (Fig. 10).

The decreasing trend of the atmospheric precipitation and almost constant value of the temperature according on the Bulunkul meteostation data for the period 1956–2012 indicates about the influence of the terrain orography on the formation of climatic conditions (Fig. 10). The Western part of the Gunt river basin due to its location, experiences greater and longer solar radiation. Ridges, thereby quickly getting rid of seasonal snow due to the reflection of radiation create an additional thermal effect. Conversely, the processes taking place in the eastern part of the basin, due to the inadequacy of solar radiation, are more conservative. This phenomenon contributes to the fact that the steady-state temperature retains its long-term value (Fig. 11).

The precipitation reducing trend in the eastern part is probably due to the fact that in this part of basin inflow of dry

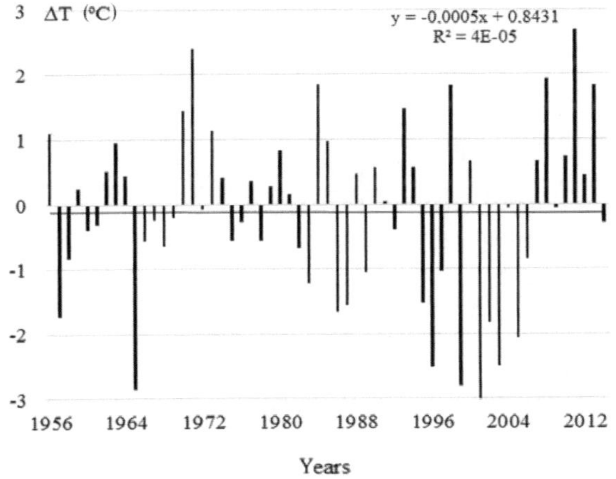

Figure 11. Deviation of annual temperature from the long-term average (1956–2012) and trend for the Bulunkul meteorological station.

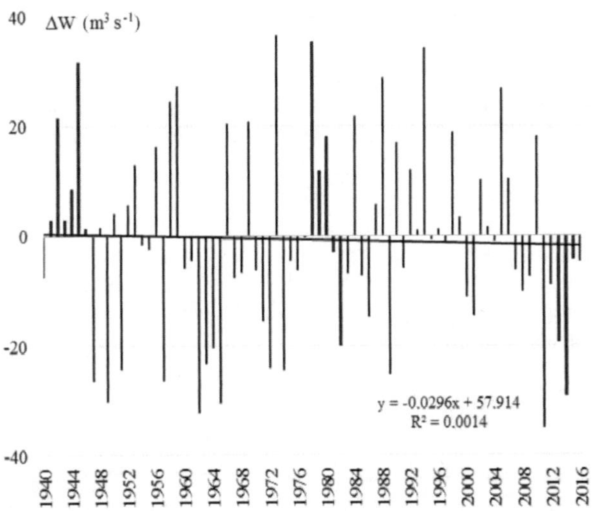

Figure 12. Deviation of annual discharge of the Gunt River from the long-term average (1940–2016) and trend.

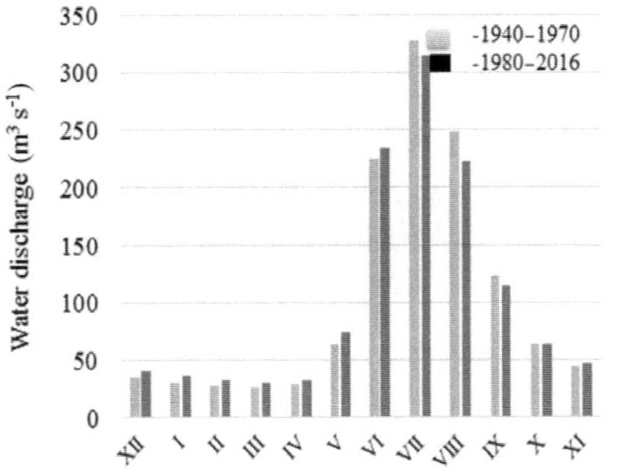

Figure 13. Hydrological regime of the Gunt River for the periods 1940–1970 and 1986–2016.

mass. The Indian monsoon air mass is unloaded during the passage of high ridges and only the dry residue comes in the eastern part of the Gunt river basin. Naturally, the fluctuation of meteorological parameters cannot but affect the hydrological characteristics of the river system of the basin as well as the Gunt River. Calculations carried out have shown that for more than 70 years the water discharge in the Gunt River has decreased by 5 % (Fig. 12).

To assess the impact of climate change on the Gunt River discharge from the entire spectrum of observations two periods were selected: 1940–1970 and 1986–2016. The river hydrograph for the respective periods is shown on the Fig. 13. The hydrograph of the Gunt River shows that the assumption of the authors (Pohl et al., 2015) on the role of Lake Yashilkul in smoothing the water level of the river is erroneous since a seasonal distribution of the river runoff is observed.

Calculations to determine the river runoff in the two-selected observation periods show that the volume of runoff in the first and second quarter of the period 1986–2016 is more than their values in the respective quarters of the period 1940–1970. For example, water volume in the first quarter of the period 1986–2016 is more than 19 %, in the second quarter about 12 % more than in the corresponding quarters of the period 1940–1970. Such a picture of the runoff change could be attributed to the climate-warming factor. However, in the third and fourth quarters of the 1940–1970 period water volumes exceed 4 % and more than 2 % of the respective quarters of the period 1986–2016.

The observed phenomena indicate that in the basin of the Gunt river and its tributary (Alichur River) the water supply sources are degraded, which is to reduce the size of glaciers, as well as the contribution of underground sources to the formation of the hydrograph of the winter period of the river.

3.3 Runoff Formation of the Vanch River Tributary of the Transboundary Pyanj River

The average annual runoff of the Vanch River in relation to the long-term for the period 1940–2016 is present on the Fig. 14a.

The period is characterized by an increasing trend of river flow. In order to determine the influence of the climatic factor on the hydrological regime of the Vanch River a comparison of the dynamics of changes in the water flow of the river was made for the periods 1940–1970 and 1986–2016 (Fig. 14b). The hydrograph of the river for two periods (1940–1970, 1986–2016) presented on the Fig. 14b. It can see that the average monthly value of the water flow of the Vanch River for the period 1986–2016 exceeds the analogous values of the period 1940–1970. The nature change of river runoff in two periods indicates about the impact of climate change on

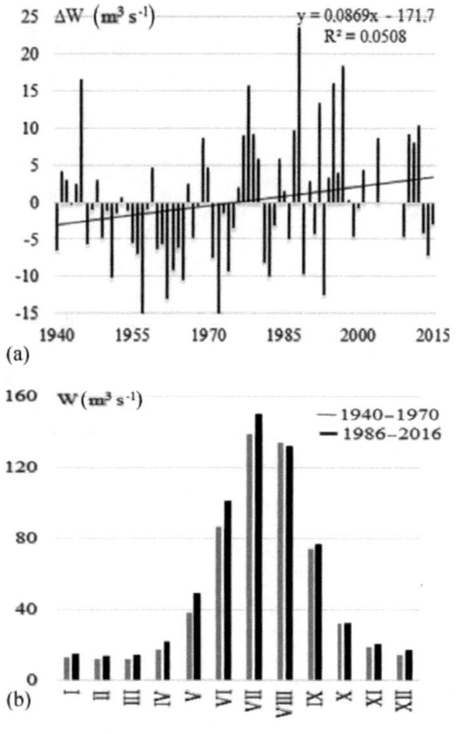

Figure 14. Deviation of annual discharge of the Vanch River from the long-term average (1940–2016) and trend **(a)**, and hydrological regime for the periods 1940–1970 and 1986–2016 **(b)**.

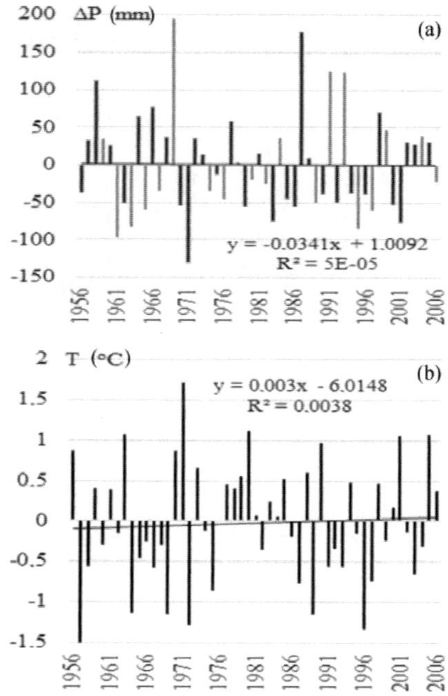

Figure 15. Deviation of annual temperature **(a)** and atmospheric precipitation **(b)** from the long-term average of the Vanch River basin for the period 1956–2006.

the state of the ice-snow reserves of the basin on the river upstream. If take into account the low temperature changes (3×10^{-3} °C yr^{-1}) for considered period, as can be seen from Fig. 15b and the almost constant value of atmospheric precipitation (Fig. 15a) then what factors contribute to the positive development trend of the river runoff.

To receive an answer to this question the Vanch River hydrograph make up in the form shown on the Fig. 16a. As can see from the Fig. 16a the maximum of the hydrograph of the Vanch river for the period 1986–2016 is shifted to the left. Therefore, the maximum value of the water runoff is observed earlier than the corresponding value of the period 1940–1970. The observed phenomenon can be explained in the framework of the assumption that the source of the river (glacier) is subject to degradation as a result of warming and is not a dense but loosened structure. For such structures, a small impact is sufficient to effect an aggregate transformation. Based on the obtained results it can be concluded that the glacial area of the Vanch River upstream is subject to a significant reduction. In order to establish the type of supply of the Vanch River it is sufficient to look to the Fig. 14b.

As can be seen from Fig. 14b, the maximum amount of precipitation in the Vanch river basin is March–April but the maximum runoff of the river is July. This means that the Vanch River is characterized by glacier feeding. Naturally, the contribution of seasonal snows to the formation of the river's water flow is not excluded. The value of water flow observing on the Fig. 16b indicates the predominance of the glacial feeding of the river.

4 Conclusion

The spatial distribution of atmospheric precipitation and snow depth in the upstream regions of the Pyanj River, including the Gunt and Vanch Rivers, covering different climatic zones of the Pamir is caused by the orography of the mountainous terrain and irregularities in air mass movements. The western climate zone is characterized by more abundant precipitation than the eastern zone. It is assumed that the duration of snow cover is determined by the temperature regime of the area. Penetration of air masses from the Mediterranean and Caspian Sea to the Gunt River Basin kept the atmospheric precipitation almost constant for more than 70 years in the western part of the river basin. The opposite was found in the eastern part of the Gunt River Basin. It is assumed that the decreasing trend in atmospheric precipitation in the eastern part is because of inflow of dry air mass. The air mass of the Indian monsoon is unloaded during the passage of the high mountain ranges and only dry air enters the eastern part of the Gunt River Basin. Comparison of the hydrological regimes of the Gunt River for the periods 1940–1970 and 1986–2016 shows a change in the seasonal

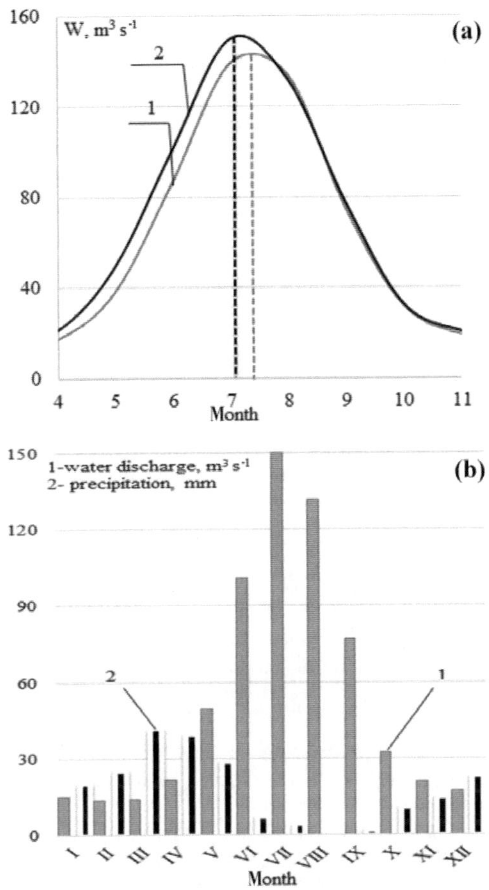

Figure 16. Hydrograph of the Vanch River for the period 1940–1970 (1), 1986–2016 (2) **(a)** and the average monthly discharge values of the Vanch River and atmospheric precipitation for the period 1940–2016 **(b)**.

distribution of the river discharge. The change indicates that the Yashilkul Lake lost its influence that earlier smoothed the water level of the Gunt River. In the Vanch River Basin precipitation remained almost constant during the period 1956–2016, whereas temperature shows a slightly increasing trend. A comparison of the river discharge for the periods 1940–1970 and 1986–2016 revealed an increase, which is associated with the degradation of the ice sheet in the headwater of the Vanch River Basin.

Data availability. Meteorological, hydrological and other data are supplied by the Agency for Hydrometeorology of the Republic of Tajikistan (http://www.meteo.tj, last access: 18 June 2020).

Author contributions. The contribution of PN to the work consists in the collection, systematization and processing of meteorological data, design of graphic materials and discussion of the results. The contribution of IN is the formulation of the problem, planning of fieldwork, preparation of the manuscript, analysis of literature sources and discussion of the results.

Competing interests. The authors declare that they have no conflict of interest.

Disclaimer. Authors' point of view may differ from that of USAID.

Special issue statement. This article is part of the special issue "Hydrological processes and water security in a changing world". It is a result of the 8th Global FRIEND–Water Conference: Hydrological Processes and Water Security in a Changing World, Beijing, China, 6–9 November 2018.

Financial support. This research has been supported by USAID (PEER project).

References

Adam, J. C., Hamlet, A. F., and Lettenmaier, D. P.: Implications of global climate change for snowmelt hydrology in the twenty-first century, Hydrol. Process., 23, 962–972, 2009.

Aizen, V. B., Mayewski, P. A., Aizen, E. M., Joswiak, D. R., Surazakov, A. B., Kaspari, S., Grigholm, B., Krachler, M., Handley, M., and Finaev, A.: Stable-isotope and trace element time series from Fedchenko glacier (Pamirs) snow/firn cores, J. Glaciol., 55, 275–291, 2009.

Alford, D. and Armstrong, R.: The role of glaciers in stream flow from the Nepal Himalaya, The Cryosphere Discuss., 4, 469–494, https://doi.org/10.5194/tcd-4-469-2010, 2010.

Alford, D., Armstrong, R., and Racoviteanu, A.: Glacier retreat in the Nepal Himalaya. The role of glaciers in the hydrologic regime of the Nepal Himalaya, Technical Report for the World Bank, South Asia Sustainable Development (SASDN), Environment and Water Resources, Unit, World Bank, Kathmandu, Nepal, p. 86, 2009.

Andermann, C., Longuevergne, L., Bonnet, S., Crave, A., Davy, P., and Gloaguen, R.: Impact of transient groundwater storage on the discharge of Himalayan Rivers, Nat. Geosci., 5, 127–131, 2012.

Arheimer, B., Lindström, G., Pers, C., Rosberg, J., and Strömqvist, J.: Development and test of a new Swedish water quality model for small-scale and large-scale applications, Proc. XXV Nordic Hydrol. Conf., 11–13 August 2008, Reykjavik, Iceland, 2008.

Barry, R. G.: The parameterization of surface albedo for sea ice and its snow cover, Prog. Phys. Geogr., 20, 63–79, 1996.

Bhattarail, B. C. and Regmi, D.: Impact of Climate Change on Water Resources in View of Contribution of Runoff Components in Stream Flow: A Case Study from Langtang Basin, Nepal, Hydrol & Met., 9, 75–84, 2011.

Christensen, J. H., Hewitson, B., Busuioc, A., Chen, X., Gao, I., Held, R., Jones, R. K., Kolli, W. T., Kwon, R., Laprise, V., Magaña Rueda, L., Mearns, C. G., Menéndez, J., Räisänen, A., Rinke,

A., Sarr, A., and Whetton, P.: Regional Climate Projections. In: Climate Change 2007: Chapter 11. The Physical Science Basis. Contribution of Working Group I to the Fourth Assessment Report of the Intergovernmental Panel on Climate Change, edited by: Solomon, S., Qin, D., Manning, M., Chen, Z., Marquis, M., Averyt, K. B., Tignor, M., and Miller, H. L., Cambridge University Press, Cambridge, UK and New York, NY, USA, 847–940, 2007.

Deser, C., Thomas, R., Alexander, M., and Lawrence, D.: The seasonal atmospheric response to projected Arctic sea ice loss in the late twenty-first century, J. Climate, 23, 333–351, 2010.

Frei, A. and Gong, G.: Decadal to century scale trends in North American snow extent in coupled atmosphere-ocean general circulation models, Geophys. Res. Lett., 32, 235–239, 2005.

Fuchs, M. C., Gloaguen, R., and Pohl, E.: Tectonic and climatic forcing on the Pyanj river system during the Quaternary, Int. J. Earth Sci., 102, 1985–2003, 2013.

Gain, A. K., Immerzeel, W. W., Sperna Weiland, F. C., and Bierkens, M. F. P.: Impact of climate change on the stream flow of the lower Brahmaputra: trends in high and low flows based on discharge-weighted ensemble modelling, Hydrol. Earth Syst. Sci., 15, 1537–1545, https://doi.org/10.5194/hess-15-1537-2011, 2011.

Hosaka, M., Nohara, D., and Kitoh, A.: Changes in snow cover and snow water equivalent due to global warming simulated by a 20 km-mesh global atmospheric model, SOLA, 1, 93–96, 2005.

Immerzeel, W. W., Droogers, P., De Jong, S. M., and Bierkens, M. F. P.: Large-scale monitoring of snow cover and runoff simulation in Himalayan river basins using remote sensing, Remote Sens. Environ., 113, 40–49, 2009.

Immerzeel, W. W., Van Beek, L. P. H., and Bierkens, M. F. P.: Climate change will affect the Asian water towers, Science, 328, 1382–1385, 2010.

Immerzeel, W. W., Van Beek, L. P., Konz, M., Shrestha, A. B., and Bierkens, M. F. P.: Hydrological response to climate change in a glacierized catchment in the Himalayas, Climatic Change, 110, 721–736, https://doi.org/10.1007/s10584-011-0143-4, 2012.

Immerzeel, W. W., Pellicciotti, F., and Bierkens, M. F. P.: Rising river flows throughout the twenty-first century in two Himalayan glacierized watersheds, Nat. Geosci., 6, 742–745, https://doi.org/10.1038/ngeo1896, 2013.

IPCC: Climate change 2007, Synthesis Report, edited by: Pachauri, R. K. and Reisinger A., Geneva, Switzerland, 234 pp., 2007.

IPCC: Summary for Policymakers of IPCC Special Report on Global Warming of 1.5 °C approved by governments, available at: http://www.ipcc.ch (last access: 24 April 2020), 2018.

Jeelani, G., Feddema, J., Van der Veen, C. J., and Stearns, L.: Role of snow and glacier melt in controlling river hydrology in Liddar watershed (western Himalaya) under current and future climate, Water Res. Res., 1, 48–56, 2012.

Klok, E. J., Jasper, K., Roelofsma, K. P., Gurtz, J., and Badoux, A.: Distributed hydrological modelling of a heavily glaciated Alpine river basin, Hydrolog. Sci., 46, 553–570, 2001.

Knowledge Base: Land and Water Resources Use in the Aral Sea Basin: Rivers, River Systems, available at: http://www.cawater-info.net (last access: 2 March 2019), 2003.

Kolberg, S. A. and Gottschalk, L.: Updating of snow depletion curve with remote sensing data, Hydrol. Process., 20, 2363–2380, 2006.

Langston, G., Bentley, L. R., Hayashi, M., McClymont, A., and Pidlisecky, A.: Internal structure and hydrological functions of an alpine proglacial moraine, Hydrol. Process., 25, 2967–2982, 2011.

Lemke, P., Ren, J., Alley, R. B., Allison, I., Carrasco, J., Flato, G., Fujii, Y., Kaser, G., Mote, P. W., Thomas, R. H., and Zhang, T.: Observations: Changes in snow, ice and frozen ground, Climate Change 2007: The Physical Sciences Basis, edited by: Solomon, S., Qin, D., Manning, M., Chen, Z., Marquis, M., Averyt, K. B., Tignor, M., and Miller, H. L., Cambridge University Press, Cambridge, UK and New York, NY, USA, 337–383, 2007.

Lindström, G., Pers, C., Rosberg, J., Strömqvist, J., and Arheimer, B.: Development and test of the HYPE (Hydrological Predictions for the Environment) model – a water quality model for different spatial scales, Hydrol. Res., 41, 295–319, 2010.

Lutz, A. F., Immerzeel, W. W., Shrestha, A. B., and Bierkens, M. F. P.: Consistent increase in High Asia's runoff due to increasing glacier melt and precipitation, Nat. Clim. Chang., 4, 587–592, https://doi.org/10.1038/nclimate2237, 2014.

Lutz, A. F., Immerzeel, W. W., Kraaijenbrink, P. D. A., Shrestha, A. B., and Bierkens, M. F. P.: Climate change impacts on the upper Indus hydrology: Sources, shifts and extremes, PLoS One, 9, 1–33, https://doi.org/10.1371/journal.pone.0165630, 2016.

Matthew, S. M., Taras, B., Liston, G. E., Derksen, C., Jonas, T., and Lea, J.: Estimating Snow Water Equivalent Using Snow Depth Data and Climate, Classes, J. Hydrometeorol., 11, 1380–1394, 2010.

Meier, K., Knoche, M., Merz, R., and Weise, S.: Stable isotopes in river waters in the Tajik Pamirs: regional and temporal characteristics, Isotope Envir. Health Study, 49, 542–554, 2013.

Meier, K., Knoche, M., Brehme, M., Merz, R., and Weise, S. M.: Discharge pattern in the Gunt valley, western Pamir, with respect to δ^2H, δ^{18}O and hydrochemistry, in: Chapter 43. Geostatistical and Geospatial approaches for characterization of Natural resources in the Environment, edited by: Janardhana Raju, N., Springer, New York, USA, Dordrecht, the Netherlands, London, UK, 281–284, 2016.

Normatov, P. I., Markaev, B. A., and Normatov, I. S.: Meteorological Features of Climatic Zones in the Basin of the Transboundary River Pyanj, Bull. Irkutsk St. Univ., 21, 106–113, 2017.

Palazzi, E., Hardenberg, J. V., and Provenzale, A.: Precipitation in the Hindu-Kush Karakoram Himalaya: observations and future scenarios, J. Geophys. Res.-Atmos., 118, 85–100, 2013.

Pohl, E., Knoche, M., Gloaguen, R., Andermann, C., and Krause, P.: Sensitivity analysis and implications for surface processes from a hydrological modelling approach in the Gunt catchment, high Pamir Mountains, Earth Surf. Dynam., 3, 333–362, https://doi.org/10.5194/esurf-3-333-2015, 2015.

Prokop, A., Schirmer, M., Rub, M., and Lehning, M.: A comparison of measurement methods: terrestrial laser scanning, tachymetry and snow probing for the determination of the spatial snow depth distribution on slopes, Ann. Glaciol., 49, 210–216, 2008.

Robinson, D. A.: Hemispheric snow cover from satellites, Ann. Glaciol., 17, 367–371, 1993.

Syed, F. S., Giorgi, F., Pal, J. S., and King, M. P.: Effect of remote forcings on the winter precipitation of central southwest Asia, part 1: observations, Theor. Appl. Climatol., 86, 147–160, 2006.

Udnæs, H. C., Alfnes, E., and Andreassen, L. M.: Improving runoff modelling using satellite-derived snow covered area, Nord. Hydrol., 38, 21–32, 2007.

Verbunt, M., Gurtz, J., Jasper, K., Lang, H., Warmerdam, P. M. M., and Zappa, M.: The hydrological role of snow and glaciers in alpine river basins and their distributed modeling, J. Hydrol., 282, 36–55, 2003.

Viviroli, D., Weingartner, R., and Messerli, B.: Assessing the hydrological significance of the world's mountains, Mt. Res. Dev., 23, 32–40, 2003.

Zhang, S. Q., Gao, X., Ye, B. S., Zhang, X., and Hagemann, S.: A modified monthly degree-day model for evaluating glacier runoff changes in China Part II: application, Hydrol. Process., 26, 1697–1706, 2012.

Proc. IAHS, 383, 43–50, 2020
https://doi.org/10.5194/piahs-383-43-2020

Open Access
PIAHS

Stress testing as complement to climate scenarios: recharge scenarios to quantify streamflow drought sensitivity

Michael Stoelzle[1], **Maria Staudinger**[2], **Kerstin Stahl**[1], **and Markus Weiler**[1]

[1]Faculty of Environment and Natural Resources, University Freiburg, Freiburg, Germany
[2]Department of Geography, University of Zurich, Zurich, Switzerland

Correspondence: Michael Stoelzle (michael.stoelzle@hydro.uni-freiburg.de)

Published: 16 September 2020

Abstract. Precipitation deficits and temperature anomalies are often the main cause for low flows and summer streamflow droughts. However, where groundwater is the main contribution to sustain water availability and ecological integrity during dry spells, the role of recharge and catchment storage is crucial to understand streamflow drought sensitivity. Here we introduce recharge stress tests as complement to climate scenarios to characterize and quantify the streamflow drought sensitivities of catchments. The stress tests are presented by applying them to six headwater catchments in Switzerland with various catchment and streamflow characteristics. The stress tests drive the bucket-type hydrological model HBV in a framework, in which pre-drought recharge conditions can be decreased to test how catchments respond to and recover from drought. We identified an upper limit of stress test durations around 12 months as indicator of maximum recharge- and storage-memory for the study catchments. Varying response on stress testing across the catchments suggests different storage properties and thus different recovery times from drought. From the stress test simulations, we found up to 200 d longer summer streamflow droughts with additional streamflow deficits which account for up to 40 d of median flow. Using a worst-case pre-drought recharge in stress test simulation leads to minimum flow reductions of 50 %–80 % compared with the reference simulation. Based on the results we conclude with recommendations for further stress test research in drought hydrology.

1 Introduction

To assess the sensitivity of catchments to drought it is important to understand how hydrological systems might respond to changes in future climate. Unfortunately, climate change studies often fail to distinguish between inherent climate variability and projected climate change signal. For instance, different temporal structures of future climate input can affect low flows, but the sequencing of simulated wet and dry spells is not altered (Vormoor et al., 2017). Climate change scenarios introduce large uncertainties to hydrological assessment of future streamflow droughts and low flow events (Addor et al., 2014). A way forward to reduce these uncertainties is a generic model framework in which climate input of a reference simulation from a hydrological model can be altered to test the response of catchments regarding periods with low streamflow (Stoelzle et al., 2014; Staudinger et al., 2015). Here we introduce a novel stress testing framework based on historical data to quantify how sensitive catchments are to decreased recharge before a drought. In the stress test (ST) framework pre-drought recharge is reduced while the catchment-specific climate variability is preserved. Through alteration of pre-drought recharge and thus water availability, a new sequencing in drought propagation is tested.

Besides uncertainties of climate scenarios another reason to apply drought stress testing is the increasing demand of complementary information for water agencies to develop reliable water management plans to make provision for future streamflow droughts. In many regions worldwide regional climate models project decreasing summer precipitation and lower summer water availability. Those projections often provide regional and seasonal patterns of hydrological change but miss out on catchment- and event-specific drought propagation characteristics. By using stress tests water managers could answer typical "what-if"-questions to better understand how extreme historical streamflow droughts were and what could have happened if drought preconditions had been more severe.

2 General stress test design

Magnitude, frequency and duration of low flow periods and streamflow droughts are often primarily controlled by groundwater contribution. This groundwater contribution depends on a varying connectivity between groundwater and rivers and the recharge rates that (re-)fill groundwater storages. Recharge compared with precipitation and temperature is hence considered as a dominant control of low flow magnitude and eventually streamflow drought severity. Recharge conditions in pre-drought periods can influence deficit, duration and intensity of subsequent drought periods. As streamflow regimes are more or less groundwater-dominated the duration of these periods should vary to identify and account for different relevant time scales of drought propagation.

In the presented STs recharge before summer low flow periods is altered. This alteration is implemented in the bucket type hydrological model HBV as model experiment, which allows to obtain ST simulations and compare them all to a reference streamflow simulation. As a direct permutation of recharge in HBV is not supported we realize the stress testing by replacing the original precipitation (P) and temperature (T) data of a certain period with P and T data from other years that have less simulated reference recharge. In stress testing "stress" is defined as systematic decrease of recharge, and "test" is the quantification of change in streamflow response in a systematic framework related to this decrease. STs are not about hydrological prediction but identifying the sensitivity of a catchment to drought conditions. Similar to climate change model chain experiments, all streamflow simulations from STs are compared with the HBV reference simulation and never against observed streamflow to account for model biases. In the HBV model, recharge is defined as the percolation from the soil box into the groundwater box. The percolation rate can be decreased by evapotranspiration and increased by snow melt.

We present two types of stress testing approaches. The first stress test is based on annual natural variability (ST_{NAT}) with a fixed one-year permutation scheme in which the original pre-drought year is altered. The second stress test (ST_{WORST}) embeds different stress durations, extracts worst-case pre-drought recharge periods from data and starts at the event-specific onset of the drought period. An overview of the stress testing is shown in Fig. 1 and technical procedures are described in Sect. 3.3 and 3.4.

3 Methods

3.1 Study sites and data

We perform STs in six mostly rainfall-dominated catchments in Switzerland that can be classified as headwater catchments with minor anthropogenic influences, small urbanization and near-natural streamflow. However, influences due to hydropeaking and wastewater treatment are possible, but considered to have a negligible influence with respect to ST design. Table 1 gives an overview on catchment characteristics and some hydrometeorological metrics. Annual precipitation in the catchments is relatively high (1100–2000 mm a^{-1}) and they differ in catchment area (32–416 km^2), runoff ratios (0.40–0.76), low flow stability (Q_{85}/Q_{50}, Q_{95}/Q_{50}) and flashiness. Gridded daily precipitation P (MeteoSwiss RHiresD, 2 km interpolated observations data set) and mean daily air temperature T, (MeteoSwiss, TabsD, 2 km interpolated observations data set) – are aggregated to daily averages for the catchments. Observed streamflow is provided by the Swiss Federal Office for the Environment (FOEN).

3.2 HBV modelling

For the HBV reference simulation, the model parameters were derived from a GAP-calibration (Seibert, 2000). We used the HBV model in the version HBV-light 4.0 with 100 m elevation zones and snow-, soil and groundwater routines, where groundwater is separated into a faster and a slower box (Seibert and Vis, 2012). The calibration period was set to 1980–2000, but P- and T-data were available for the period 1971–2018. All HBV runs (calibration and ST simulation) have the same model warm-up, i.e. 10 years of data (1961–1970) that are not included for ST period selection. Gradients of P and T were derived based on input data. For the calibration we used an objective function Eff_w, that combines the adapted Kling-Gupta-Efficiency (70 % weight) from Pool et al. (2018) and the Mean Absolute Relative Error MARE (30 % weight) to account for low flows and volumetric bias reduction in the calibration.

3.3 Natural variability stress test (ST_{nat})

The overriding question of the ST on natural variability (ST_{NAT}) is: what would have happened in a drought year if the pre-drought year with its specific recharge rates was changed? To answer this question, we permute the preconditions of each drought year (e.g. 2003) by replacing the pre-drought year (e.g. 2002) with all other 39 years from the 40-year time series (1976–2015). P- and T-input for HBV is hence replaced between 31 May of the drought year (e.g. 31 May 2003) and the 1 June of the pre-drought year (e.g. 1 June 2002) to obtain 39 different ST simulations for each drought year (i.e. 6 drought years $\times 39$ permutations $= 234$ ST_{NATs} per catchment). For this ST recovery always starts

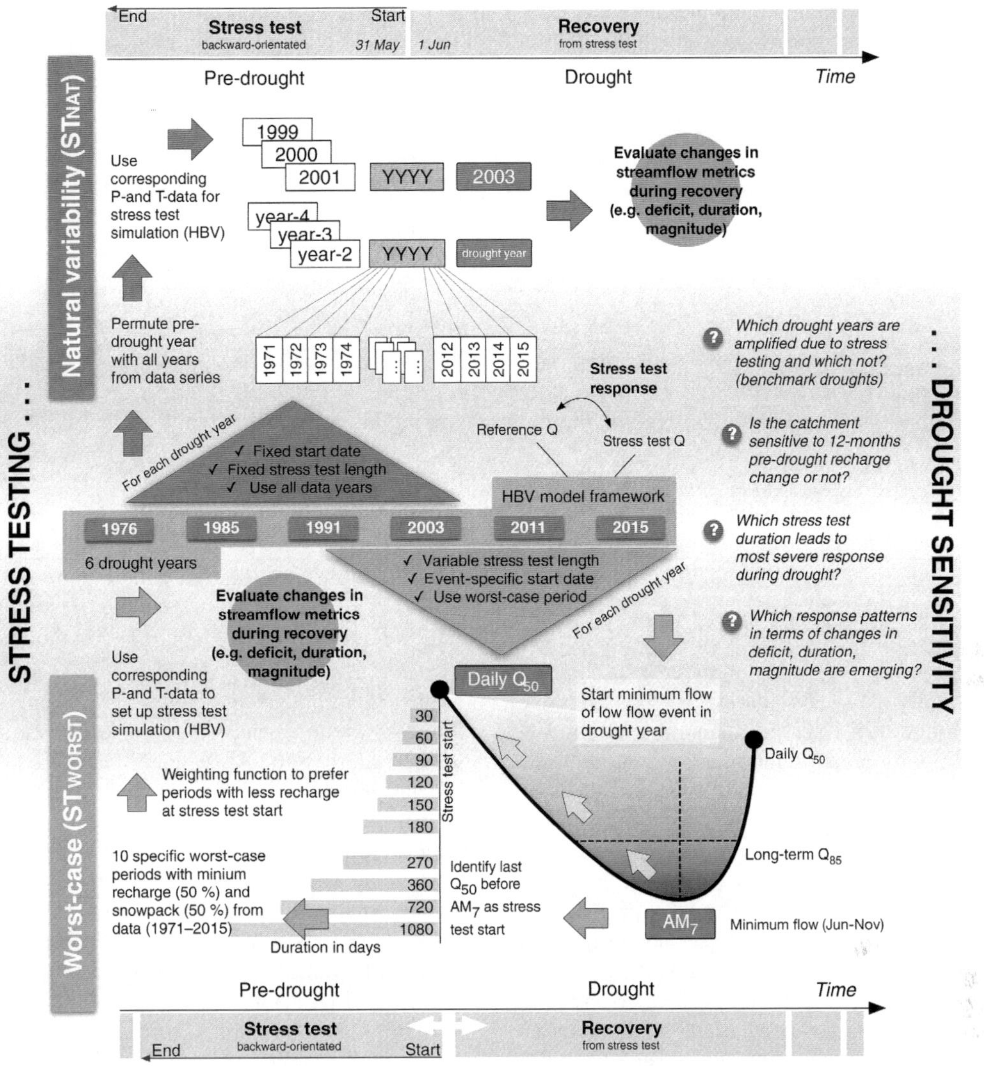

Figure 1. Overview of general stress testing approach, research questions and the procedures for stress tests ST_{NAT} and ST_{WORST}.

from 1 June of the drought year. After the 31 May the HBV-model input is the same for reference and ST simulation.

3.4 Worst-case stress test (ST_{worst})

As the yearly time scale of ST_{nat} might be too long to test a catchments' sensitivity to decreased recharge, ST_{worst} is based on drought year- and catchment-specific low flow events and variable ST durations. First the minimum flow is identified for each catchment by the day of the 7 d minimum flow, AM_7, from June through November for all six drought years. The last exceedance of daily Q_{50} before AM_7 is defined as the onset of the low flow period and the first exceedance of daily Q_{50} after the occurrence of AM_7 is defined as termination of the low flow period. The date of the onset of low flow is also the end date for all ST_{WORST} (Fig. 1). The ST end date is the starting point for ten different simulations with durations of 30, 60, 90, 120, 150, 180, 270 d and 1, 2 and 3 years. For each duration the worst-case period with minimum recharge is identified based on the reference simulation, i.e. the worst-case period is selected from the period 1971–2015 with 45 data years. Occasionally the original period is the worst-case period, then no ST can be performed. Here more data years compared to ST_{NAT} are used to allow for prolonged STs before the drought year 1976.

To prefer periods with drier conditions near the end of the ST (i.e. 31 May) a weighted recharge sum is used. A triangular function weights the daily recharge at the beginning of the ST with factor 0, the last day of ST with factor 1 and uses a linear increase to derive daily weights in between. Zappa and Kan (2007) have shown that snow- and glacier-melt substan-

Table 1. Catchments characteristics, climate and streamflow metrics. Metrics are based on observed precipitation (P) and simulated streamflow (Q, Q_{50}, Q_{85}, Q_{95}) time series from 1976 to 2015. Ratios in brackets are related to P and Q_{50}, respectively. Flashiness is a metric to quantify and compare the path length of the hydrograph (Baker et al., 2004), the higher the value, the flashier the hydrograph.

ID	Catchment name Gauging station	Area km^2	Mean elev. Max elev. m a.s.l.	Mean slope %	P Q (Q/P) mm a^{-1}	Q_{50} mm d^{-1}	Q_{85} (Q_{85}/Q_{50}) Q_{95} (Q_{95}/Q_{50}) mm d^{-1}	Flashiness (R-B-Index) (–)
BRO	Broye, Pay. Caserne	416	715 1495	6.6	1229 649 (0.53)	0.79	0.38 (0.48) 0.23 (0.29)	0.20
LAN	Langeten, Huttwil	60	760 1085	8.3	1328 649 (0.49)	1.28	0.91 (0.71) 0.67 (0.52)	0.08
MEN	Mentue, Yvo. La Maugue.	105	675 923	5.3	1109 481 (0.43)	0.55	0.26 (0.47) 0.15 (0.27)	0.18
WIG	Wigger, Zofingen	366	656 1395	10.8	1215 481 (0.40)	0.83	0.54 (0.65) 0.38 (0.45)	0.11
BIB	Biber, Biberbrugg	32	1002 1483	11.2	1874 1040 (0.56)	1.30	0.59 (0.45) 0.34 (0.26)	0.43
ALP	Alp, Einsiedeln	47	1157 1885	15.3	2010 1525 (0.76)	2.08	1.08 (0.52) 0.66 (0.32)	0.43

tially buffered summer low flows during the 2003 heatwave. This underpins that both recharge and snow pack must be considered to identify reliable worst-case periods for stress testing. Hence, snow water equivalent (SWE) from the reference simulation at the end date of ST is added to the weighted recharge sums. In this study both components are equally weighted, but also another weighting between recharge and snow pack might be plausible to select worst-case periods. By adding up recharge and snowpack it is possible to rank all potential worst-case periods according to their potential "water availability" during the pre-drought period. From the ranking of 44 potential worst-case periods the one with the lowest combination of weighted recharge sum and snow pack is set to be the "worst-case period". This period must have a smaller amount of unweighted recharge sum combined with snow pack than the reference simulation. If not, the next period in ranking with slightly higher amount of weighted recharge and snow pack is used. A new ST simulation is prepared where P- and T-data as model input from the reference simulation is replaced by P- and T-data from the selected worst-case period.

3.5 Evaluation of stress test response

Recovery duration (in days) is used to quantify how long the system requires to return to "normal" state, i.e. recovery is terminated on the day when ST streamflow converges with references streamflow (with 2 % tolerance) after the ST has ended. Recovery in this study does not imply total recovery from drought, but is used as a metric to quantify the persistence of the ST signal.

Minimum streamflow (change in %) during streamflow droughts is important for ecological integrity of rivers, various water users, water quality and water temperature. We thus test the effect of pre-drought recharge stress on the relative change in minimum flow (AM_7) during the original drought period.

Streamflow deficit (mm and normalized days) is an integrative measure of streamflow drought severity, because some events show rather low minimum flow but others show prolonged durations. We calculate streamflow deficit below the seasonal Q_{85} (June–November) of the reference streamflow to ensure that the additional ST streamflow deficit is meaningful regarding drought intensification. Subsequently streamflow deficit is normalized based on Van Loon et al. (2014) by Q_{50}-days to compare ST effects across catchments and relate the deficit to an average water availability in a catchment (i.e. the recovery potential of a catchment).

4 Results and Discussion

4.1 Performance of HBV model

From the GAP-calibration the model parameterization from the simulation with the highest Eff_w out of 100 simulations is used for further analysis. Eff_w ranges between 0.81 and 0.87 for all catchments. Since all STs are compared to the reference runs the differences between model reference and observation are not further considered in this study. However, the model efficiency illustrates that in general hydrograph representation is suitable for stress testing.

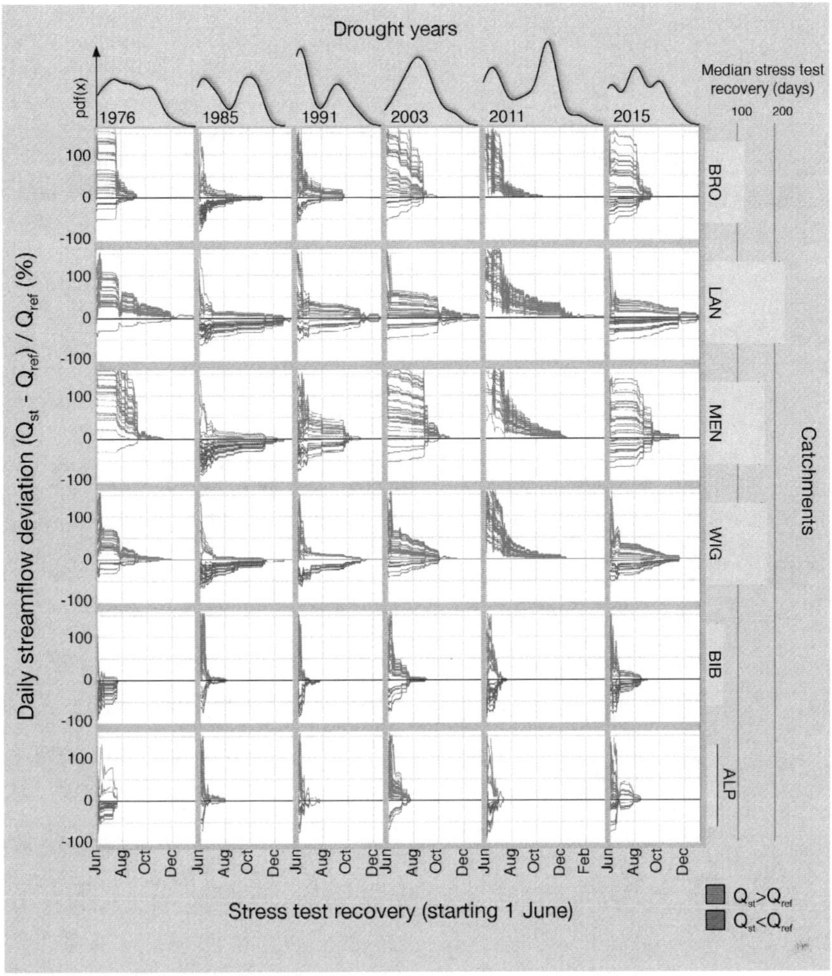

Figure 2. Evaluation of recovery time after ST_{NAT} for different drought years (columns) and catchments (rows) where Q_{st} is streamflow from stress test simulation and Q_{ref} is streamflow from reference simulation. Density curves (top) and barplots with median values (right) show the distributions of recovery times across drought years or catchments. y-axis is truncated at +150 % to improve visualization.

4.2 Evaluation of ST_{NAT}

The permutation of pre-drought years leads to drier and wetter preconditions and causes changes of the original drought periods (after 1 June) with more and less streamflow (Fig. 2). However, response patterns vary markedly across the catchments and drought years. Median recovery times are ∼ 200 d (LAN), ∼ 150 d (MEN, WIG), 100 d (BRO) and 25–45 d (BIB, ALP). For the last group the altered pre-drought conditions lead only occasionally to long recovery times (< 100 d) indicating that those catchments are less sensitive to alteration of complete pre-drought years. One reason for that could be the relatively high flashiness indicating short turnover times between precipitation input and streamflow response. In other catchments (LAN, MEN, WIG) we found longer recovery times where the ST signal persists until the end of the drought year. Focusing on differences between the drought years it is apparent that altered pre-drought years could have led to more severe droughts in 1985 and 2015. On the contrary, original pre-conditions in 1976 and 2011 must have been very dry as the analysis shows only recovery lines with positive streamflow deviations compared to the reference (turquoise-coloured lines in Fig. 2). Indeed, climate records show that April and May 2011 were exceptionally warm and dry in Switzerland. However, negative deviations of ST streamflow compared to reference streamflow for ALP and BIB in 2011 give some evidence that altered pre-drought years could have decreased low streamflow in those flashier catchments until end of July.

Comparing prominent recent drought years (i.e. 2003 and 2015) we found that in 2003 across all catchments 72 % of the ST_{NAT} pre-drought years were drier than the original pre-drought year, but only in 14 % the recovery periods

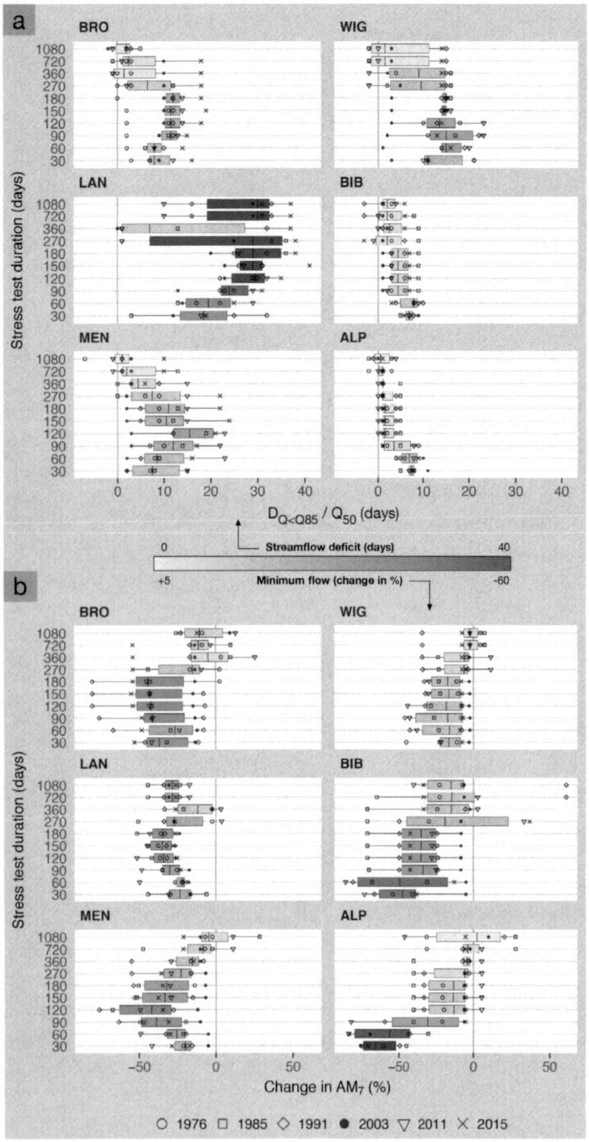

Figure 3. Streamflow response due to ST_{WORST} with different durations (30–1080 d) analysed by **(a)** streamflow deficit and **(b)** change in minimum streamflow. Higher colour saturation of boxplots indicates more severe droughts due to stress testing.

are drier (i.e. less streamflow) than the reference. For the 2003 drought this suggests that the year between May 2003 and June 2002 was relatively wet. For 2015 around 50 % of pre-drought years were drier and in 41 % the recovery periods have less streamflow than the reference. This unveils that drought propagation is controlled by a combination of catchment characteristics (e.g. hydrogeology) and drought event characteristics (e.g. pre-conditions, seasonal timing of drought, precipitation and temperature anomalies). The 2003 drought appears to be more controlled by the extreme heatwave during the actual streamflow drought event than by pre-drought recharge deficits (i.e. the winter 2002/2003). In contrast, stress testing the drought year 2015 suggests that this event could be amplified and prolonged by pre-drought recharge decrease. In both cases the early parts of the streamflow droughts might be classified as flash droughts due to their rapid rate of event intensification (Otkin et al., 2018). Additionally, the drivers and characteristics of those flash droughts appear to be different and hence might be classified as "heatwave" flash drought (2003) and "precipitation" flash drought (2015) following the definition of Mo and Lettenmaier (2016) interpreting the drivers of hydrological drought. Unimodal distributions of ST recovery times (years 1976 and 2003 in Fig. 2) might be an indicator of heatwave-induced flash droughts when high temperature anomalies superimpose the differences in pre-drought recharge across the permuted years. However, more data analysis is needed to justify this assumption, although evaluation of ST_{WORST} in the next section supports these findings.

4.3 Evaluation of ST_{WORST}

The major difference in stress test design between ST_{NAT} and ST_{WORST} is that the latter allows for event-specific starting dates and different ST durations. In 5 % of all ST_{WORST}-simulations the original pre-drought recharge was already the worst case (i.e. absolute minimum recharge in data). Only for 30 and 60 d ST_{WORST} new worst-case periods could be found for all catchments and drought years. We used two different drought metrics (i.e. streamflow deficit and change in minimum flow) to evaluate a catchments' sensitivity on reduced recharge (Fig. 3). Results show that up to 40 d of median flow are needed to refill the deficit induced by stress testing (Fig. 3a). In general, the deficit curves showed their maximum at ST durations around 120–270 d. However, larger deficits for shorter durations can be found in the WIG, BIB and ALP catchments. Here stress testing with longest ST durations leads occasionally to negative deficits (i.e. a streamflow surplus). For BRO, LAN and MEN the largest streamflow deficit occurs in the drought year 2015, for BIB and ALP the drought year 1985 is prominent if the response of very short STs is neglected. Maxima in streamflow deficit for STs with durations smaller than 360 d indicate that our study catchments are rather sensitive to intra-annual recharge deficits than to multi-year reduction in pre-drought water availability. Focusing on ST durations a negative correlation between ST duration and pre-drought recharge reduction is revealed. STs with 30, 60–150 and $\geq$ 360 d duration are based on 90 %, 60 %–70 % and 20 %–25 % combined recharge- and snowpack-reduction, respectively. However, the snowpack/recharge-ratio at the end of the STs is in 95 % of all simulations were below 1 %, only in 1976 we found considerable snowpack in one catchment (5 %–60 %). The most severe response in the ALP catchment is generated with a 30 d ST indicating that this catchment is more exposed to "flash droughts" (Otkin et al., 2018) due to assumingly smaller catchment storages (Staudinger et al., 2017) and/or smaller water retention ability (i.e. higher Q/P-ratios). In comparison the LAN catchment shows in some drought years increasing streamflow deficit when longer ST durations are applied (> 120 d).

Analogous to streamflow deficit the patterns of change in minimum flow (AM_7) also show a tendency for larger decrease in AM_7 when shorter STs are used (Fig. 3b). Maximum AM_7 reduction is $\sim 50\,\%$ (LAN, WIG) and $\sim 80\,\%$ (BRO, MEN, BIB, ALP) which is caused by ST durations $\leq$ 150 d. Longer ST durations ($\geq$ 360 d) lead to increased AM_7 in some cases. The different response patterns depending on drought years and ST durations highlight that drought assessment is a complex task and that multiple drought events should be analysed during stress testing to gain a comprehensive picture on streamflow drought sensitivity of a catchment. Without multiple drought years the most extreme changes in streamflow deficit or minimum flow might be missed. However, ST durations of 1–2 years as upper limit appear to be sufficient for the analysis of sensitivity.

5 Conclusions and outlook

In this study we present a hydrological model experiment to stress pre-drought recharge in order to test drought sensitivity across several headwater catchments in Switzerland. The concept of stress testing is complementary to climate change scenarios and is hence of great interest for hydrological research and water management. Event-specific stress tests with variable durations (ST_{WORST}) appear to be superior to stress tests with fixed start dates and fixed durations (ST_{NAT}). However, with ST_{NAT} a direct assessment of the severity of historical drought years and their potential intensification is possible. In general, varying ST durations show that the six study catchments have a recharge/storage-memory of around 6 to 12 months when focusing on summer low flows. From ST analysis it is apparent that recovery last 25–200 d, minimum flows during summer droughts are 50 %–80 % smaller and additional ST-induced streamflow deficits account for 10–40 d of median flow (Q_{50}). These metrics are often-requested hydrological estimates for improved water management. A catchment classification by such hydrological metrics can help better predict future low flow hazards. It is, for example, important to note that more groundwater-dominated catchments tend to have larger streamflow deficits during drought, and more flashy catchments are more sensitive to larger decrease in minimum flow (Fig. 3). Here streamflow metrics like flashiness or the Q_{95}/Q_{50}-ratio help to classify different catchments. Our results suggest that future research on stress testing should put more emphasis on event analysis and different drought (propagation) types (e.g. precipitation vs. heatwave flash droughts). However, worst-case periods from data are limited to test extreme pre-conditions with duration longer than 90–120 d as the historical climate in Switzerland provides no longer lasting "zero recharge" periods. To cope with this limitation the stress test design may include estimated recharge droughts based on return periods (e.g. 100-year recharge drought) for different stress test durations. An extension of the stress tests to more catchments, other streamflow regimes (e.g. snowmelt-dominated) and other climatic regions (i.e. varying P- and T-patterns as drivers of summer streamflow droughts) appears useful to further explain future low flow sensitivity.

Data availability. Data are not freely available, but streamflow and meteorological data can be requested through the agencies (Swiss Federal Office for the Environment (FOEN) and Federal Office of Meteorology and Climatology MeteoSwiss). The hydrological model HBV light is freely available (Seibert and Vis, 2012).

Author contributions. All authors developed the main research idea. MiS and MaS performed the analyses and evaluated the stress tests. MiS wrote the manuscript with contributions and revisions from all authors.

Competing interests. The authors declare that they have no conflict of interest.

Special issue statement. This article is part of the special issue "Hydrological processes and water security in a changing world". It is a result of the 8th Global FRIEND–Water Conference: Hydrological Processes and Water Security in a Changing World, Beijing, China, 6–9 November 2018.

Acknowledgements. Streamflow and catchment metadata were provided by the Hydrology Division of the Swiss Federal Office for the Environment (FOEN; BAFU). We thank Irene Kohn and Marit van Tiel for their comments on the manuscript. This study contributes to the projects "Low Flows and Groundwater" (Swiss Federal Office of the Environment FOEN) and TrenDHy (German Research Foundation DFG).

References

Addor, N., Rössler, O., Köplin, N., Huss, M., Weingartner, R., and Seibert, J.: Robust changes and sources of uncertainty in the projected hydrological regimes of Swiss catchments, Water Resour. Res., 50, 7541–7562, https://doi.org/10.1002/2014WR015549, 2014.

Baker, D. B., Richards, R. P., Loftus, T. T., and Kramer, J. W.: A new flashiness index: Characteristics and applications to midwestern rivers and streams, J. Am. Water Resour. Assoc., 40, 503–522, https://doi.org/10.1111/j.1752-1688.2004.tb01046.x, 2004.

Mo, K. C. and Lettenmaier, D. P.: Precipitation Deficit Flash Droughts Over the United States, J. Hydrometeorol., 17, 1169–1184, https://doi.org/10.1175/JHM-D-15-0158.1, 2016.

Otkin, J. A., Svoboda, M., Hunt, E. D., Ford, T. W., Anderson, M. C., Hain, C., and Basara, J. B.: Flash Droughts: a Review and Assessment of the Challenges Imposed by Rapid-Onset Droughts in the United States, B. Am. Meteorol. Soc., 99, 911–919, https://doi.org/10.1175/BAMS-D-17-0149.1, 2018.

Pool, S., Vis, M. J. P., and Seibert, J.: Evaluating model performance: towards a non-parametric variant of the Kling-Gupta efficiency, Hydrolog. Sci. J., 63, 1941–1953, https://doi.org/10.1080/02626667.2018.1552002, 2018.

Seibert, J.: Multi-criteria calibration of a conceptual runoff model using a genetic algorithm, Hydrol. Earth Syst. Sci., 4, 215–224, https://doi.org/10.5194/hess-4-215-2000, 2000.

Seibert, J. and Vis, M. J. P.: Teaching hydrological modeling with a user-friendly catchment-runoff-model software package, Hydrol. Earth Syst. Sci., 16, 3315–3325, https://doi.org/10.5194/hess-16-3315-2012, 2012.

Staudinger, M., Weiler, M., and Seibert, J.: Quantifying sensitivity to droughts – an experimental modeling approach, Hydrol. Earth Syst. Sci., 19, 1371–1384, https://doi.org/10.5194/hess-19-1371-2015, 2015.

Staudinger, M., Stoelzle, M., Seeger, S., Seibert, J., Weiler, M., and Stahl, K.: Catchment water storage variation with elevation, Hydrol. Process., 31, 2000–2015, https://doi.org/10.1002/hyp.11158, 2017.

Stoelzle, M., Stahl, K., Morhard, A., and Weiler, M.: Streamflow sensitivity to drought scenarios in catchments with different geology, Geophys. Res. Lett., 41, 6174–6183, https://doi.org/10.1002/2014GL061344, 2014.

Van Loon, A. F., Tijdeman, E., Wanders, N., Van Lanen, H. J., Teuling, A. J., and Uijlenhoet, R.: How climate seasonality modifies drought duration and deficit, J. Geophys. Res.-Atmos., 119, 4640–4656, https://doi.org/10.1002/2013JD020383, 2014.

Vormoor, K., Rössler, O., Bürger, G., Bronstert, A., and Weingartner, R.: When timing matters-considering changing temporal structures in runoff response surfaces, Climatic change, 142, 213–226, https://doi.org/10.1007/s10584-017-1940-1, 2017.

Zappa, M. and Kan, C.: Extreme heat and runoff extremes in the Swiss Alps, Nat. Hazards Earth Syst. Sci., 7, 375–389, https://doi.org/10.5194/nhess-7-375-2007, 2007.

Proc. IAHS, 383, 51–59, 2020
https://doi.org/10.5194/piahs-383-51-2020

Changing climate and implications for water use in the Hetao Basin, Yellow River, China

Ian White[1], Tingbao Xu[1], Jicai Zeng[2], Jian Yu[3], Xin Ma[3], Jinzhong Yang[2], Zailin Huo[4], and Hang Chen[4]

[1]Fenner School of Environment and Society, Australian National University, Canberra, ACT, 0200, Australia
[2]State Key Laboratory of Water Resources and Hydropower Engineering Science, Wuhan University, Wuhan, 430068, China
[3]Water Resources Research Institute of Inner Mongolia, No. 11, Genghis Khan East Road, New Town, Hohhot, Inner Mongolia, 010020, China
[4]Centre for Agricultural Water Research in China, China Agricultural University, No. 17, East Rd, Haidian, Beijing, 100083, China

Correspondence: Ian White (ian.white@anu.edu.au)

Published: 16 September 2020

Abstract. Balancing water allocations in river basins between upstream irrigated agriculture and downstream cities, industry and environments is a global challenge. The effects of changing allocations are exemplified in the arid Hetao Irrigation District on the Yellow River, one of China's three largest irrigation districts. Amongst the many challenges there, the impact of changing climate on future irrigation water demand is an underlying concern. In this paper we analyse trends in local climate data from the late 1950s and consider the implications for irrigation in the Basin. Since 1958, daily minimum temperatures, T_{min} in the Basin have increased at three times the rate of daily maximum temperatures, T_{max}. Despite this, there has been no significant increases in annual precipitation, P or pan evaporation, E_{pan}. The difference between the increasing trends in T_{max} and T_{min} means that the average annual diurnal temperature range, DTR, has decreased very significantly, part of a global phenomenon. Hargreaves empirical approach is used to estimate changes in both incoming solar radiation, R_{s}, and potential evaporation, ET_0. Changes in estimated ET_0 correlated well with changes in measured pan evaporation, E_{pan}. Paradoxically, the estimated decreasing trend in R_{s} does not correspond to a significant decreasing trend in E_{pan}. Implications of changing climate on water use and soil salinity in the Basin are discussed.

1 Introduction

The world-wide challenge of allocating water fairly between competing upstream and downstream users and the environment is exemplified in the Yellow River, the so-called "cradle of civilization" in China (Xu et al., 2005). The challenge is compounded when upstream uses include vital irrigated food production systems. The 10 000 km^2 Hetao Irrigation District on the northern side of the Ordos loop of the Yellow River is one of China's three largest irrigation areas. This arid, cold region (Peel et al., 2007) in Inner Mongolia has been farmed for 2300 years, since the Qin dynasty. It is an important source of grains, oil seeds, sugar, fruits, vegetables and poultry.

Increasing water abstractions from the Yellow River for irrigation, downstream demands from rapidly expanding industrial and urban developments, obligatory environmental flows, coupled to channel elevation changes due to sediment deposition have all severely stressed the river system (Xu et al., 2005; Tang et al., 2008; Lehane, 2015). In response the Yellow River Water Conservancy Commission mandated a 15 % reduction in irrigation supply to Hetao to be phased in over time with a reduced supply target of 4.0 km^3 yr^{-1} for the Hetao District.

Recent research in Hetao has focussed on reducing water use while maintaining or increasing productivity (e.g. IWC-IM, 1999; Xu et al., 2010; Xue and Ren, 2017). Air temperatures have increased at a significant rate in northern and

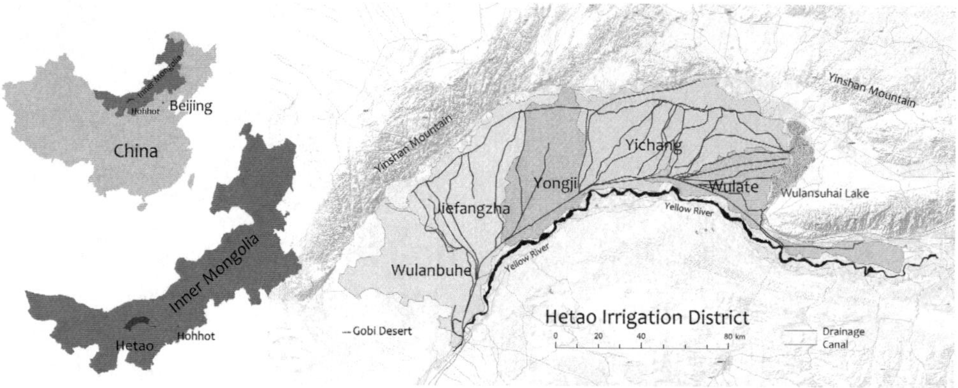

Figure 1. The Hetao Irrigation District Inner Mongolia, showing the five irrigation sub-districts.

western China (Tang et al., 2010) and the impacts of climate change on irrigation water demand is a continuing concern for both regulators and irrigators. In this work we analyse trends in local measured climate parameters across Hetao, relate them to physical mechanisms and consider the implications of the already observed climate changes on irrigation water demand and salinity management.

2 Materials and methods

2.1 Hetao irrigation district

The general features of the flat, bow-shaped, approximately 200 km long by 50 km wide Hetao Irrigation District are in Fig. 1 and listed in Table 1. The District is bounded to the west by the Gobi Desert, to the north by the western portion of the Yinshan Mountains and by the Yellow River in the south. This region of northern China has experienced significant increasing trends in maximum and temperatures over the past 60 years (Fig. 2). From April until early November, water is diverted from the Yellow River at Dengkou on the southwestern edge of the region and distributed throughout the District by a series of multi-level canals. Drainage discharges into Wulansuhai Lake in the eastern edge of the region.

The Basin is divided into five sub-districts: Yigan (previously called Wulanbuhe), Jiefangzha, Yongji, Yichang and Wulate comprising 19.2 %, 20.7 %, 16.2 %, 30 % and 13.9 % of the total area of the District. Annual pan evaporation, E_{pan} across the District, far exceeds annual precipitation and the variability of annual rainfall decreases with increasing rainfall across the District from west to east as does E_{pan} (Table 1).

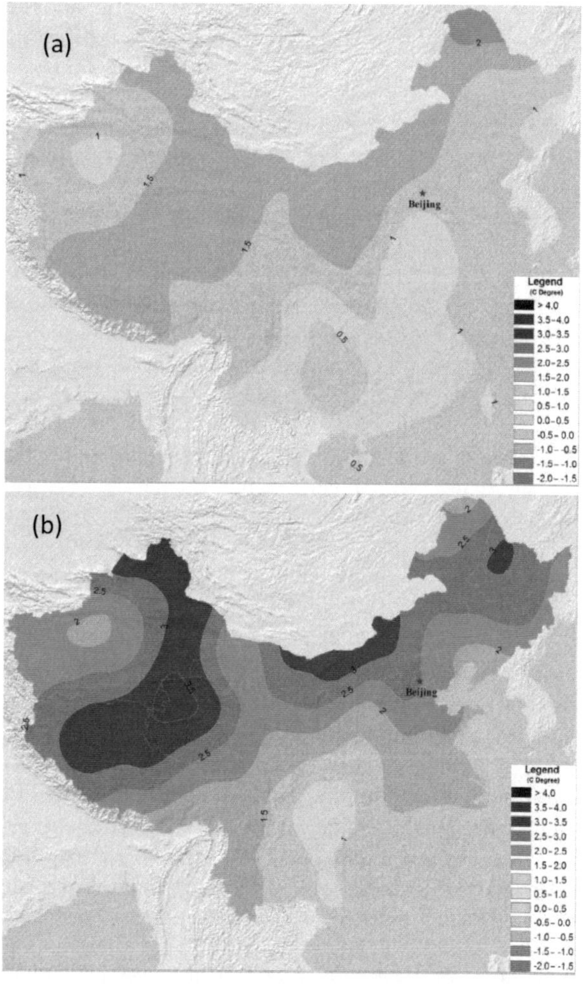

Figure 2. Trends in spatially interpolated **(a)** mean daily maximum and **(b)** minimum temperatures (°C/100 years) in China between 1956 and 2015.

2.2 Sources of data

This study used data drawn from a range of disparate and scattered sources collected by different organisations over diverse time periods (Table 2).

Table 1. Characteristics of the Hetao Irrigation District.

Characteristic	Value	Units
Elevation range west to east	1052–1018	m
Area of Irrigation District*	10 677	km^2
Area of Irrigated Land*	5744	km^2
Growing Season	May to October	
Mean annual air temperature	7	°C
Mean January air temperature	−10.1	°C
Mean July air temperature	23.8	°C
Range of spatial mean annual precipitation west to east (1956–2015)	64–266	$mm\,yr^{-1}$
Range Variability index annual precipitation east to west (1956–2015)	0.74 to 1.03	
Range mean annual pan evaporation east to west (1957–2013)	2016–2466	$mm\,yr^{-1}$
Number of Farmers	~ 1 000 000	

* Yang (2005).

Table 2. Data sources used in this work.

Data	Years	Source
Precipitation	1954–2013	China Meteorological Data Centre, CMDC
Pan Evaporation	1957–2013	China Meteorological Data Centre, CMDC
Evapotranspiration	1990–2010	Yang (2005); Hao and Yang (2016)
	2000–2010	Yang et al. (2012)
Irrigated area	1958-2013	Mongolia, 1983–2013; Yang (2005)

Table 3. The five climate stations used to assess precipitation across Hetao (west to east) for the period 1957 to 2013.

Sub-district	Station	Latitude N	Longitude E	Elevation (m)
Yigan	53419	40°20′	107°00′	1055
Jiefangzha	53420	40°54′	107°08′	1038
Yongii	53513	40°46′	107°24′	1042
Yichang	53337	41°06′	108°17′	1027
Wulate	53433	40°44′	108°39′	1025

2.2.1 Basin climate

Monthly precipitation data was available from five available stations, each within one of the five irrigation sub-districts (Table 3). Monthly E_{pan} data was available for this study from the Yongii and Wulate stations in Table 3 for the period 1957 to 2013 together with daily maximum and minimum temperatures, T_{max} and T_{min}, from 1957 and 1954, respectively, to 2013. Mean monthly precipitation peaks in July and August. Because of this, mean pan evaporation peaks earlier in May but mean temperature peaks in July. On average mean atmospheric temperatures are at or below zero for 5 months of the year.

2.2.2 Temporal trends in climate data

Trends in annual temperature, DTR, precipitation, and pan evaporation, irrigation supply, specific irrigation drainage, and groundwater and drainage salinity were determined using linear regression (LR). Trends were considered not significant when the probability, p, was less than 0.9. Because of the sensitivity of linear regression trends to start and end values of a series, weaker trends ($p < 0.95$) were also tested using the nonparametric (NP) Theil–Sen slope estimator (Theil, 1950). The significance of nonparametric trends was determined using the Mann–Kendall test (Kendall, 1970).

2.2.3 DTR and incoming solar radiation, R_s

The daily DTR (°C) is defined as:

$$\mathrm{DTR} = T_{max} - T_{min}. \quad (1)$$

FAO-56 (Allen et al., 1998) pointed out that DTR can be used as an indicator of the fraction of extra-terrestrial radiation, R_a ($MJ\,m^{-2}\,d^{-1}$), that reaches the earth's surface. Hargreaves and Samani (1982) (see also Hargreaves and Allen, 2003) proposed that the ratio of incoming solar radiation incident on the earth's surface, R_s ($MJ\,m^{-2}\,d^{-1}$), to R_a can be approximated by the empirical relation:

$$R_s/R_a = k_{R_s}\mathrm{DTR}^{1/2} \quad (2)$$

where k_{R_s} (°C$^{-0.5}$) is an empirical adjustment coefficient (0.16 for interior regions, 0.19 for coastal regions).

Using estimates of mid-monthly R_a from FAO-56 and the latitude of the stations in Table 3, together with $k_{R_s} = 0.16$ (°C$^{-0.5}$) in Eq. (2) and DTR data, temporal changes in the annual incoming solar radiation to Hetao can be estimated.

Good daily T_{max} and T_{min} data are available from 1958 to 2013 for two sites, Yongii and Wulate in the Hetao Basin listed in Table 3.

2.2.4 DTR and potential evaporation, ET_0

Temperature is not the main driver of ET_0. In the Penman–Monteith approach, ET_0 can be expressed as (FAO-56, Allen et al., 1998):

$$ET_0 = C_1\left[(1-\alpha)\cdot R_S - R_{nl} - G\right] + C_2 f(u, \Delta e). \qquad (3)$$

In Eq. (3), C_1, C_2 are factors which depend weakly on temperature, pressure and crop characteristics, α the surface albedo, R_{nl} the net outgoing long wave radiation, G the soil heat flux, u the wind speed and Δe the water vapour pressure deficit. The first term in Eq. (3) represents the key contribution of R_s to ET_0, the second term represents the "drying power of the air". For arid locations, such as Hetao, the "drying power of the air" is expected to contribute significantly to ET_0, although evaporation of surrounding irrigation water may confound this locally.

For Hetao, available long-term data necessary to fully evaluate trends in ET_0 via the Penman–Monteith approach is not readily available. FAO-56 suggest that the approach developed by Hargreaves (Hargreaves and Samani, 1982, 1985; Hargreaves, 1994), based on daily T_{max} and T_{min} give useful estimates of ET_0, particularly over periods longer than 5 d:

$$ET_0 = 0.0023 \cdot R_a(\overline{T} + 17.8)\mathrm{DTR}^{0.5} \qquad (4)$$

with $\overline{T} = (T_{max} - T_{min})2$.

Using Eq. (2), Eq. (4) can be written in terms of R_s as:

$$ET_0 = 0.0023 \cdot (\overline{T} + 17.8)R_s/k_{R_s}. \qquad (5)$$

FAO-56 noted that the 1985 Hargreaves' method, Eq. (4), "has shown reasonable ET_0 results with global validity". Hargreaves and Allen (2003) did an extensive evaluation of Eq. (4) over a very wide range of global locations against the FAO-56 Penman Monteith method and concluded that: "Where equipment cost is a consideration, where data quality is questionable, or where historical data are missing, both the reduced set FAO-PM or the 1985 Hargreaves are recommended, since the two methods are surprisingly equivalent over a wide range of climates." Use over 5 d or longer is recommended because of the influence of moving weather fronts, and large variations of wind speed and cloud cover at daily time steps.

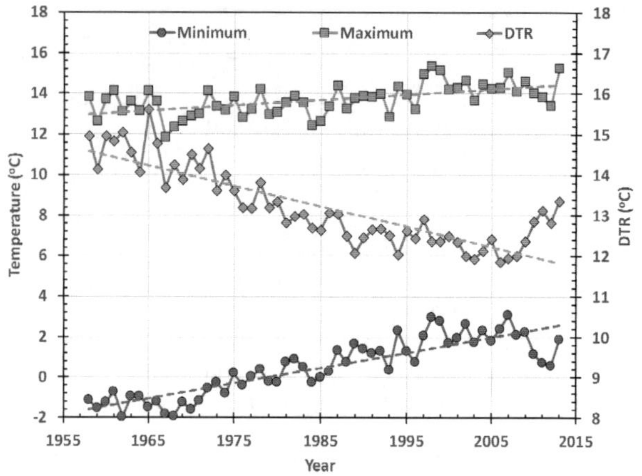

Figure 3. Increasing significant trends in the area-averaged annual mean daily maximum and minimum temperatures combined from the two stations, Yongii and Wulate in Table 3 and the decreasing significant trend in mean DTR for the period 1958 to 2013.

Equation (4) appears to compensate for the lack of R_n, humidity and wind speed data required by the Penman–Monteith method. The diurnal temperature range in the equation implicitly accounts for the influence of cloudiness because DTR decreases with increasing cloudiness. As well, it has been shown that DTR correlates with relative humidity and vapor pressure deficit and is inversely influenced by wind run (Hargreaves and Allen, 2003). Hargreaves and Allen (2003) recommend that Eq. (4) should be calibrated against Penman–Monteith combination equations to produce a regionally calibrated relationship.

The temporal trend in ET_0 with time follows from Eq. (5),

$$\frac{\partial ET_0}{\partial t} = \frac{0.0023}{k_{R_s}}\left(R_s \cdot \frac{\partial \overline{T}}{\partial t} + [\overline{T} + 17.8].\frac{\partial R_s}{\partial t}\right). \qquad (6)$$

Whether estimated ET_0 will increase or decrease with time will depend on the relative magnitude of the two terms $R_s \frac{\partial \overline{T}}{\partial t}$ and $[\overline{T} + 17.8]\frac{\partial R_s}{\partial t}$ in Eq. (6) which may be of opposite sign.

3 Climate trends in the basin

3.1 Temperature

The trends in annual mean daily T_{max} and T_{min} at Yongii and Wulate over the period 1958 to 2013 were identical. As a result, the area-averaged mean temperature is used here as representative of the Hetao Basin. Figure 3 plots the average annual mean daily T_{max} and T_{min} average and annual mean daily precipitation, pan evaporation and temperature as a function of time. Table 4 shows that both increasing temperature trends were highly significant with T_{min} increasing almost three times the rate of T_{max}.

Table 4. Increasing trends in the average annual daily maximum, mean and minimum temperatures and DTR in Hetao for the period 1958 to 2013.

Temperature	Trend (°C per decade)	Probability
Maximum	0.26 ± 0.05	> 0.9999
Mean	0.50 ± 0.05	> 0.9999
Minimum	0.75 ± 0.06	> 0.9999
DTR	-0.49 ± 0.04	> 0.9999

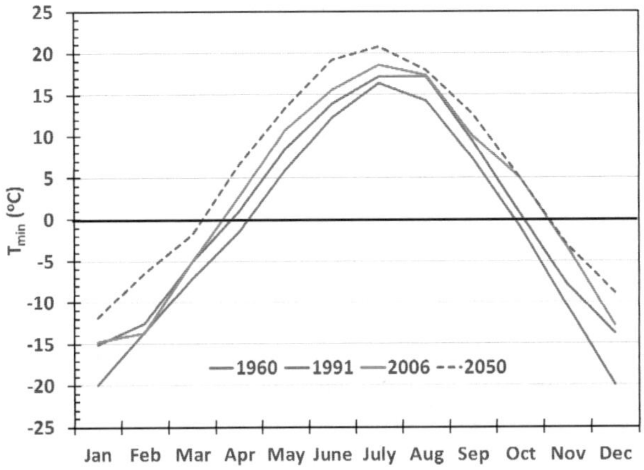

Figure 4. Change in the distribution of area-averaged monthly minimum temperatures in Hetao. Values for 1960, 1991 and 2006 are actual measured distributions from years with similar annual precipitation and that for 2050 is projected from monthly trends.

Analysis of monthly mean daily temperature data in Hetao shows that T_{min} in all months of the year have increased very significantly and are also increasing at a greater rate than monthly T_{max}. Figure 4 shows the change in the distribution of measured monthly minimum temperatures with time in Hetao as well as the projected temperature distribution for 2050 using the highly significant monthly increasing trends. Because the distribution of monthly temperatures depends on precipitation, the years chosen in Fig. 4 were selected because they had similar annual precipitations, (1960: $P = 155$ mm; 1991: $P = 151$ mm; 2006; $P = 156$ mm; with the projected precipitation for 2050 of 157 mm).

Figure 4 demonstrates that by 2006, the period over which the minimum temperature in Hetao was below zero had decreased by almost 2 months relative to that for 1960 so that the length of the growing season in Hetao is expanding due to increasing minimum temperatures. This also suggests that the depth to which the soil is frozen and the length of time it is frozen for has decreased. This may have implications for salt flushing from topsoil.

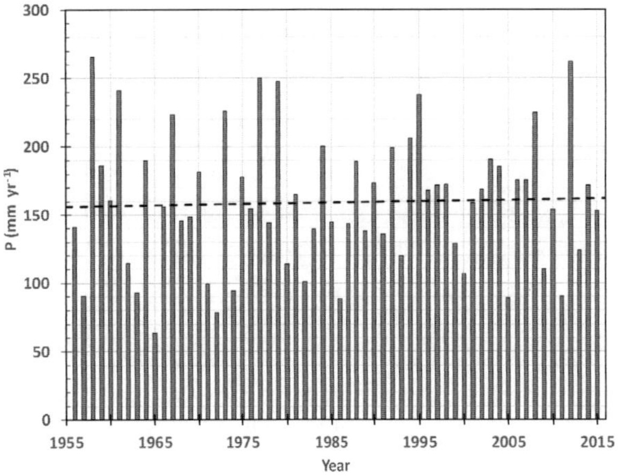

Figure 5. Areal-averaged annual precipitation of the Hetao Basin, 1957 to 2013. The linear trend is not significant.

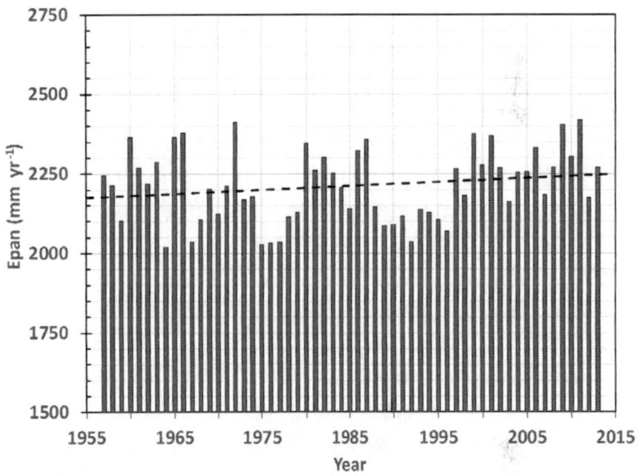

Figure 6. Area-averaged pan evaporation for the Hetao Basin, 1957 to 2013. The linear trend is not significant.

3.2 Annual precipitation and E_{pan}

Area-weighted average annual precipitation for the Basin between 1957 and 2013 is plotted in Fig. 5. The large variations in annual rainfall over the period 1957 to 2013, from 64 to 288 mm yr^{-1}, are apparent. The small trend, 0.3 ± 4.3 mm per decade, however, is not significant ($p < 0.9$).

The area-averaged E_{pan} for Hetao, between 1957 and 2013 is plotted in Fig. 6. An increasing linear regression trend of $+13 \pm 9$ mm per decade can be seen. NP analysis gave a similar trend of $+15$ mm/ per decade. For both analyses, however, the trends were not significant ($p < 0.9$). So, the highly significant increasing annual maximum and minimum temperatures in Hetao are not associated with any significant increases in annual precipitation or annual pan evaporation between 1957 and 2013.

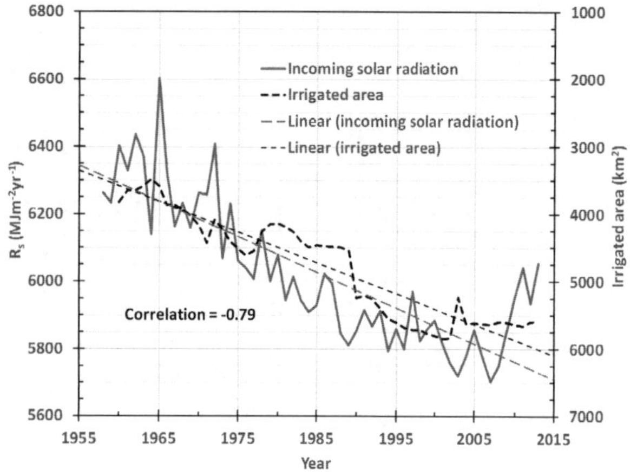

Figure 7. Estimated temporal change of incoming solar radiation to Hetao from 1958 to 2013 using Eq. (2). Also shown is the change in area of land irrigated (plotted in reverse order) and significant linear trends.

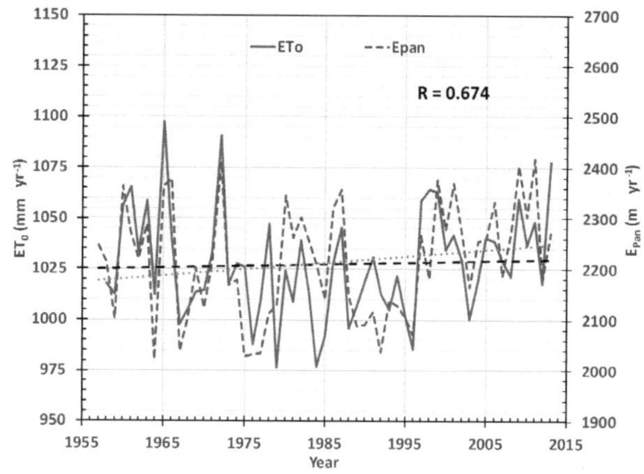

Figure 8. Comparison between annual potential evaporation estimated from Eq. (5) and the area weighted annual pan evaporation for Hetao over the period 1958 to 2013. The linear temporal trends are not significant.

3.3 Trends in DTR

Because $T_{\min}$ increased at three times the rate of $T_{\max}$, DTR decreased significantly (Fig. 3) at $-0.49\,°\mathrm{C}$ per decade over the period 1958–2013. This is the same absolute rate as annual mean daily temperature is increasing (Table 4).

Analyse of the data revealed that between 1958 and 1985, DTR decreased at faster rate, $-0.85 \pm 0.10\,°\mathrm{C}$ per decade ($p > 0.9999$) than the full period 1958 to 2013. Between 1985 to 2013, however, the trend $-0.03 \pm 0.09\,°\mathrm{C}$ per decade was not significant ($p < 0.9$). This change in trend is consistent with regionally-averaged northern hemisphere data where DTR was found to decrease significantly from 1950 to the late 1970's. Between 1979 and 2004, however there was no significant trend in DTR (Vose et al., 2005). This appears to coincide with a fundamental shift in the El Niño–Southern Oscillation phenomenon (Easterling et al., 1997).

3.3.1 Decreasing DTR and changes in R_s

Equation (2) was used to estimate temporal changes in R_s using the DTR data using the method of Hargreaves and Samani (1982). The results are given in Fig. 7 which suggests the annual incoming solar radiation to Hetao decreased at a rate of $-105.6 \pm 10.2\,\mathrm{MJ\,m^{-2}}$ per decade ($p > 0.9999$) between 1958 and 2014, or 9 % over 50 years. This is close to the 10 % to 20 % reduction reported in many places over the same period (Stanhill and Cohen, 2001). This global phenomenon is attributed by some to global "dimming" (a decrease in net radiation due to increased cloudiness or aerosols) (Easterling et al., 1997; Vose et al., 2005). It is noted in Fig. 7, however, that from about 2009 onwards, there appears to have been an increase in annual R_s.

The overall decrease in estimated R_s cannot be attributed to increasing cloudiness due to increasing annual precipitation, since, as shown above, annual precipitation in Hetao has not increased significantly between 1957 and 2013. The correlation between estimated annual R_s and annual precipitation, P is weakly significant ($p > 0.9$) with $\partial R_S/\partial P = -1.0 \pm 0.6\,\mathrm{MJ\,m^{-2}\,mm^{-1}}$. This correlation between R_s and P appears to be mainly due to a correlation with fluctuations around the decreasing trend in R_s Some of the fluctuations of R_s about the trend line in Fig. 7 may be caused by annual changes in cloud cover associated with annual variations in precipitation.

From Eq. (3) it is expected that the decrease in incoming solar radiation in Fig. 7 associated with a decrease in potential evaporation in Hetao. We next examine that proposition.

3.3.2 DTR and ET_0

In discussing Eq. (6) above it was noted that any trend in ET_0 will depend on the terms $R_S\frac{\partial \overline{T}}{\partial t}$ and $[\overline{T} + 17.8]\frac{\partial R_s}{\partial t}$. In Hetao, $\frac{\partial \overline{T}}{\partial t} > 0$ (Table 4) but $\frac{\partial R_s}{\partial t} < 0$ (Fig. 7), so depending on their relative magnitudes, ET_0 could increase, decrease or remain constant with time.

Monthly values of ET_0 were estimated from Eq. (5) which were summed to give an estimate of annual ET_0. The estimated change of ET_0 between 1958 and 2013 is plotted in Fig. 8.

The temporal trend in annual ET_0 estimated from Eq. (5), shown in Fig. 8, using linear regression, is $\partial ET_0/\partial t = +0.8 \pm 2.2\,\mathrm{mm\,yr^{-1}}$ per decade. Nonparametric analysis gave a trend of $+1.7\,\mathrm{mm\,yr^{-1}}$ per decade. Both trend estimates were not significant ($p < 0.9$) over the period 1958 to 2013.

Over this period, the mean annual potential evaporation estimated from Eq. (5) was 1027 ± 26. This is considerably larger than the actual evapotranspiration evaluated by Hao and Yang (2016) using two approaches based on a water balance approach and using NDVI and the FAO-56 Penman–Monteith equation. Their water balance for the period 1960 to 2013 gave actual annual evaporation of 601 $mm\,yr^{-1}$ and the NDVI Penman–Monteith approach gave 605 $mm\,yr^{-1}$ for the period 2006–2010. Yang et al. (2012) used remote sensing to estimate evapotranspiration over the cropping season and found a mean value of 605 $mm\,yr^{-1}$ between the years 2000 and 2010. Over the May to September growing season Eq. (5) predicts that ET_0 was 742 ± 21 mm. Hargreaves and Allen (2003) recommended that Eq. (6) should be calibrated against Penman–Monteith combination equations to produce a regionally calibrated relationship.

3.3.3 Estimated ET_0 and measured E_{pan}

Figure 8 shows the comparison between annual ET_0 estimated from Eq. (8) and area weighted annual E_{pan} for Hetao. Between 1958 and 2013, the correlation between ET_0 and E_{pan} was very significant ($p > 0.99999$) with correlation, $R = 0.67$. As discussed previously, the linear regression trend in pan evaporation, $\partial E_{pan}/\partial t = +13 \pm 9$ mm per decade was not significant ($p < 0.9$) as was the trend in estimated ET_0. From 1958 to 1975 the correlation between ET_0 and E_{pan} was stronger, $R = 0.83$ ($p > 0.9999$) but between 1976 and 2013, the correlation was lower, $R = 0.59$ but still very significant ($p > 0.9999$).

This comparison has demonstrated a reasonably good correlation between annual ET_0estimated from Eq. (8) and area weighted mean pan evaporation, E_{pan}, for Hetao. It also reveals that despite the significant increasing maximum and minimum temperatures and the significant decreases in DTR and in estimated R_s, there are no significant trends in either estimated ET_0 or in E_{pan} over the period 1958 and 2013.

4 Discussion

The dramatic trends found in annual mean daily T_{max} and T_{min} between 1958 and 2013 in the Hetao Basin are not accompanied by any significant trends in annual precipitation or pan evaporation. The significant disparity between the rates of change T_{max} and T_{min} mean that DTR is decreasing. This is accord with world-wide observations (e.g. Stanhill and Cohen, 2001). In Hetao, however, the rate of decline of DTR seems more intense and has persisted longer than elsewhere. Decreases in DTR and in E_{pan} (not found here) have been variously attributed to global "dimming" (a decrease in net radiation due to increased cloudiness or aerosols) or global "stilling" (a decrease in wind speed close to the surface) (Easterling et al., 1997; Stanhill and Cohen, 2001; Roderick and Farquhar, 2002; Vose et al., 2005).

Numerous hypotheses have been advanced for the changes occurring globally in DTR and in E_{pan} including changes in cloud cover, aerosol concentrations, precipitation, soil moisture, atmospheric circulation, land use and decreasing wind speeds (Peterson et al., 1999; Roderick and Farquhar, 2004; Vose et al. 2005; Roderick et al., 2007; McVicar et al., 2012). The consensus is that decreasing DTR is a consequence of firstly a general decrease in net radiation, and secondly a decrease in wind speed close to the surface.

It was found here that that between 1958 and 1985, DTR decreased at faster rate than over the full period 1958 to 2013. Between 1985 to 2013, however, the was not significant. This change in trend is consistent with regionally-averaged northern hemisphere data (Vose et al. 2005). This change in trend appears to coincide with a fundamental shift in the El Niño–Southern Oscillation phenomenon (Easterling et al., 1997).

In Hetao, one explanation for the changing trend in DTR could be the impact of evaporation of surrounding irrigation water. Increase in water vapour concentration in the atmosphere decreases R_s but increases in coming long-wave radiation at night, raising T_{min}. If this is the case, one might expect that the area of land being irrigated might impact on estimated R_s. Figure 8 compares the decreasing trend in estimated R_s with the increasing trend in irrigated land area. There is a strong, highly significant negative correlation, $R = -0.79$. The decline in trend in R_s after 1985 also corresponds approximately to a period of almost no growth in irrigated area and to the use of plastic film on the soil surface to minimise soil evaporation. While correlation does not imply causality, the impact of the extent of irrigation on incoming short wave and long-wave radiation is worth researching.

Despite the estimated decline in R_s, there was no significant trend in E_{pan} or estimated ET_0 between 1958 and 2013. This absence of a trend appears to imply that the declining radiative component is balanced by an increasing drying power of the air. This is counter intuitive since the 46 % increase in area irrigated in Hetao between 1960 and 2013 should have decreased the water saturation deficit. In terms of Eq. (5), the absence of a significant annual trend in ET_0 over the full period indicates $R_s \frac{\partial \overline{T}}{\partial t} \approx -[\overline{T} + 17.8] \frac{\partial R_s}{\partial t}$.

The annual ET_0 estimated here using the empirical approach of Hargreaves were higher than other available estimates but correlated well with measured E_{pan}. It has been recommended that despite widespread testing of the Hargreaves approach, its estimates of ET_0 should be calibrated locally.

Implications of changing climate

The available data in the Hetao Basin shows that between 1958 and 2013, annual mean daily minimum temperatures increased at three times the rate of maximum temperatures. Between 1960 and 2010, the period over which minimum temperatures were below zero decreased by over one month. If these increases continue, by 2050 that period could decrease by two months. That may imply a longer growing season and increased water demand in Hetao.

These increasing temperatures also mean that topsoil in the Basin will be frozen for a shorter period and to a shallower depth than in the past. Soil freezing has at least two important roles. Firstly, it appears to promote formation of soil structure and improves early season infiltration and root growth. Secondly, the propagation of the freezing front downwards may expel salt that accumulated over the growing season deeper into the soil profile and away from the root zone. A decrease in the time and depth the soil is frozen may lead to a lower efficiency of salt removal.

5 Conclusions

Annual daily temperatures in the Inner Mongolia region of northern China have increased dramatically over the past 60 years. The limited available records in Hetao show that the period that temperatures are below zero has shortened significantly suggesting that the length of the growing season is expanding. Because minimum temperatures are increasing at three times the rate of maximum temperatures, DTR is decreasing. This implies that either incoming solar radiation or wind run is decreasing or both.

Despite the dramatic change in temperatures, annual rainfall, pan evaporation and estimated potential evaporation, have no significant trends over this period. This means that only an expansion of the length of the growing season could cause an increase in future water demand.

The rapid increase in T_{min} means that if the trend continues, the soil will be frozen for a shorter period and to a shallower depth. This has implications for soil structure formation, infiltration and root growth as well as the efficiency of salt flushing from the soil. All require further research.

Data availability. All data sources used in this work are given in Table 2.

Author contributions. JYu and JYa initiated, co-hosted and supported the research, IW drafted the paper and analysed trends, TX carried out the spatial interpolation of mean daily maximum and minimum temperatures throughout China, JZ provided critical insights particularly on all aspects of water in the Hetao Basin as well as invaluable translations, XM provided details of irrigation practices in the basin, ZH and HC shared their research results on the Hetao Basin, especially on evapotranspiration.

Competing interests. The authors declare that they have no conflict of interest.

Special issue statement. This article is part of the special issue "Hydrological processes and water security in a changing world". It is a result of the 8th Global FRIEND–Water Conference: Hydrological Processes and Water Security in a Changing World, Beijing, China, 6–9 November 2018.

Acknowledgements. This work was supported by the Major Special Project of Inner Mongolia Science and Technology: Research and Demonstration of the Key Technologies of Drip Irrigation Using the Multi-source of Water in Yellow River Irrigation Areas of Inner Mongolia (2013–2017).

Financial support. This research has been supported by the Major Special Project of Inner Mongolia Science and Technology: Research and Demonstration of the Key Technologies of Drip Irrigation Using the Multi-source of Water in Yellow River Irrigation Areas of Inner Mongolia (grant no. 2013-2017).

References

Allen, R. G., Pereira, L. S., Raes, D., and Smith, M.: Crop Evapotranspiration, FAO Irrigation and Drainage Paper 56, FAO, Rome, 300 pp., 1998.

Easterling, D. R., Horton, B., Jones, P. D., Peterson, T. C. Karl, T. R., Parker, D. E., Salinger, M. J., Razuvayev, V., Plummer, N., Jamason, P., and Folland, C. K.: Maximum and minimum temperature trends for the globe, Science, 277, 364–367, 1997.

Hao, P. and Yang, J.: Spatial distribution of crop evapotranspiration in Hetao Irrigation District based on NDVI and FAO56 Penman–Monteith, J. Irrig. Drain., 35, 20–25, 2016.

Hargreaves, G. H.: Simplified coefficients for estimating monthly solar radiation in North America and Europe, Departmental Paper, Dept. of Biol. And Irrig. Engrg., Utah State University, Logan, Utah, 1994.

Hargreaves, G. H. and Allen, E. G.: History and evaluation of Hargreaves evapotranspiration equation, J. Irrig. Drain. Eng., 129, 53–63, https://doi.org/10.1061/(ASCE)0733-9437(2003)129:1(53), 2003.

Hargreaves, G. H. and Samani, Z. A.: Estimating potential evapotranspiration, J. Irrig. Drain. Eng.-ASCE, 108, 223–230, 1982.

Hargreaves, G. H. and Samani, Z. A.: Reference crop evapotranspiration from temperature, T. ASAE, 1, 96–99, 1985.

IWC-IM: Construction and rehabilitation planning project for water-saving in Hetao Irrigation District of the Yellow River basin, Inner Mongolia, Institute of Water Conservancy and Hydropower of Inner Mongolia, Hohhot, 1999.

Kendall, M. G.: Rank Correlation Methods, 4th Edn., Griffin, London, 1970.

Lehane, S. Water Management in the Yellow River Basin, Future Directions International Pty Ltd., Dalkeith, WA, 9 pp., available at: http://www.futuredirections.org.au/publication/water-management-in-the-yellow-river-basin/ (last access: 23 July 2020), 2015.

McVicar, T. R., Roderick, M. L., Donohue, R. J., Li, L. T., Van Niel, T. G., Thomas, A., Grieser, J., Jhajharia, D., Himri, Y., Mahowald, N. M., Mescherskaya, A. V., Kruger, A. C., Rehman, S., and Dinpashoh, Y.: Global review and synthesis of trends in observed terrestrial near-surface wind speeds:

Implications for evaporation, J. Hydrol., 416–417, 182–205, https://doi.org/10.1016/j.jhydrol.2011.10.024, 2012., 2012.

Peel, M. C., Finlayson, B. L., and McMahon, T. A.: Updated world map of the Köppen–Geiger climate classification, Hydrol. Earth Syst. Sci., 11, 1633–1644, https://doi.org/10.5194/hess-11-1633-2007, 2007.

Peterson, T. C., Gallo K., Lawrimore, J., Owen, T., Huang, A., and McKittrick, D.: Global rural temperature trends, Geophys. Res. Lett., 26, 329–333, 1999.

Roderick, M. L. and Farquar, G. D.: Changes in Australian Pan Evaporation from 1970 To 2002, Int. J. Climatol., 24, 1077–1090, 2004.

Roderick, M. L., Rotstayn, L. D., Farquhar, G. D., and Hobbins, M. T.: On the attribution of changing pan evaporation, Geophys. Res. Lett., 34, L17403, https://doi.org/10.1029/2007GL031166, 2007.

Stanhill, G. and Cohen, S.: Global Dimming: A Review of the Evidence for a Widespread and Significant Reduction in Global Radiation with Discussion of Its Probable Causes and Possible Agricultural Consequences, Agr. Forest Meteorol., 107, 255–278, https://doi.org/10.1016/S0168-1923(00)00241-0, 2001.

Tang, G., Ding, Y., Wang, S., Ren, G., Liu, H., and Zhang, L.: Comparative Analysis of China Surface Air Temperature Series for the Past 100 Years, Adv. Clim. Change Res., 1, 11–19, https://doi.org/10.3724/SP.J.1248.2010.00011, 2010.

Tang, Q., Oki, Y., Kanae, S., and Hu, H.: Hydrological cycles change in the Yellow River Basin during the last half of the Twentieth Century, J. Climate, 21, 1790–1806, https://doi.org/10.1175/2007JCLI1854.1, 2008.

Theil, H.: A rank-invariant method of linear and polynomial regression analysis, I, II, III, in: Proceedings of the Royal Netherlands Academy of Arts and Sciences, Koninklijke Nederlandse Akademie van Wetenschappen, Amsterdam, 386–392, 521–525, 1397–1412, 1950.

Vose, R. S., Easterling, D. R., and Gleason, B.: Maximum and minimum temperature trends for the globe: An update through 2004, Geophys. Res. Lett., 32, L23822, https://doi.org/10.1029/2005GL024379, 2005.

Xu, X., Huang, G., Qu, Z., and Pereira, L. S.: Assessing the groundwater dynamics and impacts of water saving in the Hetao Irrigation District, Yellow River basin, Agr. Water Manage., 98, 301–313, 2010.

Xu, Z., Takeuchi, K., Ishidaira, H., and Liu, C.: An Overview of Water Resources in the Yellow River Basin, Water Int., 30, 225–238, 2005.

Xue, J. and Ren, L.: Assessing water productivity in the Hetao Irrigation District in Inner Mongolia by an agro-hydrological model, Irrig. Sci., 35, 357–382, https://doi.org/10.1007/s00271-017-0542-z, 2017.

Yang, J.: Water Resources Planning Report [R], Water Supplies Bureau of Bayan Nur, 2005.7, Wuhan University, Wuhan, 2005.

Yang, Y., Shang, S., and Jiang, L.: Remote sensing temporal and spatial pattern of evapotranspiration and response to water management in a large irrigation district in North China, Agr. Forest Meteorol., 164, 112–122, 2012.

Proc. IAHS, 383, 61–68, 2020
https://doi.org/10.5194/piahs-383-61-2020

Open Access
PIAHS

Study of the spatial and temporal variability of rainfall in the Middle and Lower Cheliff (Algeria)

Ali Hadour[1,2], **Gil Mahé**[2], **and Mohamed Meddi**[1]

[1]Ecole Nationale Supérieure de l'Hydraulique, 29 route de Soumaâ, Blida, Algeria
[2]UMR HydroSiences Montpellier/IRD, Montpellier, France

Correspondence: Ali Hadour (hydroali15@gmail.com)

Published: 16 September 2020

Abstract. The purpose of this study is to identify homogeneous rainfall regions and to study the spatial and temporal variability of rainfall in the Cheliff basin using the regional vector method and the statistical approach (Pettitt test, Lee Heghinian test and Hubert segmentation) and the geostatistic approach (inverse distance weighting method). In terms of results, the regional vector method highlighted six (6) homogeneous rainfall regions. The downward trend occurred in the study area in 1972, affecting a few coastal stations. In 1976, this decline extended to the South West and throughout the coastal region. In 1980, the drop covered the entire basin. This decline has resulted in an estimated deficit of 30 % on average in the eastern region, the coastal region and the Mina. However, the central part of the basin experienced a 20 % decrease compared to the period before the break (1968–1980). The same spatial irregularity in rainfall was observed during the pre-break and post-break periods (1981–2010). On the other hand, throughout the basin, the areas corresponding to the rainfall ranges identified during the 1968–1980 period experienced an average decrease of 100 mm during the post-break period, except in the eastern region, where the decrease exceeded 200 mm.

1 Introduction

Water demand has never been so high. This is due to the population growth, the expansion of irrigated areas or the increasing needs of industry (WWAP, 2018). In the Mediterranean area, liquid precipitation is the main source of water for rivers and underground reservoirs. In addition, rainfall is the third stage of the water cycle; a cycle influenced by different natural meteorological fluctuations, and its intensification and acceleration is expected because of global warming (IPCC, 2007, 2014).

Located on the Southern shore of the Mediterranean Sea, Algeria is characterized by a contrasting and diversified climate; its evolution shows an alternation between wet and dry periods. Besides, the great variability of the 1970s allowed the extension of semi-arid zones and the restriction of wetlands.

In such a context, the study of the spatial and temporal evolution of climate parameters, including rainfall, is essential to understand this variability, to quantify it and map it, as well as to improve management approaches, to put in place adaptation plans and to develop new orientations if necessary.

This study examined the spatial and temporal variability of rainfall in the Cheliff basin. In this regard, the regional vector method on monthly rainfall data was used for two purposes: to criticize and synthesize rainfall information; and to build and validate homogeneous rainfall regions. In addition, the statistical approach was used to study the temporal variability of rainfall. For this purpose, three tests integrated into the Khronostat software were selected (Pettitt test, Lee Heghinian test and Hubert segmentation). Subsequently, the spatialization of rainfall was done by the inverse distance weighting method over two different periods: the pre-break period and the post-break period.

Published by Copernicus Publications on behalf of the International Association of Hydrological Sciences.

2 Materials and methods

2.1 Presentation of the study area

The Cheliff basin is the largest watershed in Algeria. Its surface area is about 44 900 km^2. It includes three different regions: the Boughezoul upstream; the High and Middle Cheliff; the Lower Cheliff and the Mina. It extends between 0 and 3°30′ E and 34 and 36° N. It is surrounded to the North by the Dahra Mountains and to the South by the Saharan Atlas, to the West by the Oranese basin and to the East by Algerian basin. The region experiences a semi-arid Mediterranean climate. The rainy season extends from November to March, and the rest of the year is characterized by dry and sunny periods.

2.2 Origin of the data

The monthly rainfall data come from the National Hydraulic Resources Agency (ANRH). The spatial distribution of the 50 stations included in this study is heterogonous. Indeed, there is a high concentration of measurement sites on the plains and a significant insufficiency qualifies the relief areas. The measure covers the period from the hydrologic year 1968 to the hydrologic year 2010; a hydrologic year starts on 1 September and ends on 31 August. 24 stations present complete series. On the other hand, the remaining 26 contain deficiencies; the number and the date differ from one station to another.

2.3 The regional vector method

The regional vector method was developed by Hiez (1977) at ORSTOM/IRD (Institute of Research for Development). It is based on the pseudo-proportionality hypothesis.

As a result, the method gives a regional vector corresponding to a time series of monthly rainfall indices, coming from the extraction of the information most probably contained in a set of observation sites previously grouped by homogeneous region. The most probable information is extracted using the principle of pseudo-proportionality of total rainfall between stations and the principle of likelihood. This method has been applied by several authors in West Africa and Mediterranean area (Mahé and Olivry, 1995; Bodian, 2014; Kotti et al., 2016; Hallouz et al., 2013).

This technique is used in this study to validate the previously delineated homogeneous regions based on the results of inter-station correlation studies, and to review and synthesize information of each homogeneous region. The corrections proposed by the method are intended to minimize the error matrix; they are not taken into account since a good analysis of extremes in the study area is not available.

The study area was divided into 6 homogeneous regions on which the regional vector method was applied to the monthly time step, using IRD's Hydraccess software (Vauchel, 2005).

2.4 Detection of breaks in time series

Statistical analysis was chosen to study and identify rainfall variability in our study area. For this purpose, a classical statistical test, a Bayesian method and a segmentation technique were selected. Our choice was based on the ease of implementation; they work even when the position of the jumps is unknown and require fewer application conditions.

The classical statistical test is a non-parametric approach to identify a change in mean or variance in a data series. It is developed by Pettitt (1979); it does not require any particular conditions, except continuity.

Lee and Heghinian's test (Lee and Heghinian, 1977) based on the univariate Bayesian method is adapted to identify a change in mean in the data series. It imposes normality, non-autocorrelation and constancy of variance of the time series studied.

The segmentation technique chosen is based on optimal segmentation in the least-squares sense and uses the simple step-by-step technique as combinatory. It is developed and presented by Hubert and Carbonnel (1987). It is used to search several jumps in the average. It requires the normality of the studied distribution.

The implementation of these three tests on all observed rainfall time series is performed by the Khronostat software (Boyer, 1998).

2.5 Spatialization of rainfall data

The objective is to determine and map the spatial distribution of rainfall in our study area before and after the break period. The geostatic method chosen is a deterministic approach, called inverse distance weighting (Davis, 1986; Franke, 1982). This method assigns a weight inversely proportional to the distance between the site and the point to be estimated.

Unlike stochastic methods (kriging) that require a large number of application assumptions and allow the consideration of the spatial structure of the data and their positions, the deterministic approach does not assume any hypothesis on the data and only takes into account the positioning of the stations in relation to each other.

3 Results and discussion

3.1 Data review and validation of homogeneous regions

Homogenization consists in grouping rainfall stations that have a similar behavior, based on a proportionality test integrated into the regional vector method (Mahé and L'Hôte, 1992). The close behavior of the rainfall series is validated if and only if the proportionality coefficient for each station is below 0.20. Above this value, the risk that the evolution of the station will differ from that of the other stations in the group increases (Kotti et al., 2016). The results of the regression approach are used as the first filter for selecting homogeneous regions. In addition, the regional vector method validated six homogeneous regions with a proportionality coefficient below the threshold (0.20) in each homogeneous group.

Stations showing close rainfall variability are grouped together in the same region. The spatial distribution of the regions is oriented along two axes: (1) from West to East, (2) from North West to South West. The first region includes 6 stations located on the Atlas, a mountainous range with a relief varying from 800 to 1400 m. The second region is based in the Northern valleys of the Atlas Mountains and reaches some plains in the Eastern part of the basin; it concerns 15 stations. The fourth region is locate on the plains in the middle of the basin and consists of 11 stations. The sixth region is concentrated in the coastal zone and contains 4 stations. The third region gathers 3 stations located on the South Eastern mountainous area of the Atlas with altitudes exceeding 1400 m. 5 stations exhibit an isolated behavior. In other words, they do not represent any similarity with the validated regions; they are situated on the periphery of the basin, in high relief areas Fig. 1.

The regional vector also allows us to synthesize the rainfall signal in each region as a vector. This makes it easily accessible and simplified to visualize and understand the evolution of rainfall in each region as well as to compare regions. Figure 2 shows that the six regions have the same fluctuation in the rainfall signal. In contrast, the amplitude of the rainfall signal is varying according to the regions. It is relatively the same in regions 1, 2 and 3. In addition, region 4 shows a variation in amplitude close to that of region 6. On the other hand, region 5 located in the South Western part of the basin is characterized by the lowest amplitudes.

3.2 Study and identification of the rainfall temporal variability

The observation period for all stations extends from 1968 to 2010, except for the stations 013002 and 012503 that have observation periods from 1930 to 2007 and from 1936 to 2010, respectively.

Figure 3 shows the rainfall evolution in these two stations. It highlights the alternation of dry and wet years along the observation period and also shows a gradual decrease in rainfall in the mid-1970s. It is noted that rainfall from 1930 to 1970 fluctuated around a relatively higher average than that of the 1970–2010 period.

The break tests were applied to the 50 rainfall time series considered in this study. The first homogeneous region experienced a break (a decrease in annual rainfall) in 1979; this jump was observed in 4 stations among 6 stations included in this group. Rainfall in the second homogeneous region began to decline in 1979 and 1980; among the 15 stations in this group, 8 had a break in 1980 and 4 in 1979. 2 out of 3 stations in the third homogeneous group showed no downward trend. From the 11 stations belonging to region 4, 3 stations exhibited a decreasing trend from 1978, 4 experienced a break in 1979 and 3 stations reported a jump in 1980. Rainfall in the fifth homogeneous region began to decrease in 1976, a date detected at 4 of the 5 stations belonging to this region. In the coastal region, the downward trend was introduced in 1972 and affected all the stations by the end of 1977. In short, this downward trend appeared in the Cheliff basin in 1972, where it occurred in some coastal stations. In 1976, the decrease in rainfall extended to both South West and coastal stations. This decline began to reach the central part of the basin in 1978, and then it spread to the East and South East of the Cheliff in 1980. These results are consistent with the work of other researchers on Algeria (Hallouz et al., 2013; Meddi and Hubert, 2003), and even at the regional level (Singla et al., 2010; Kingumbi et al., 2005).

It is known that climatic factors strongly influence the variability of rainfall at different time scales. However, the real reasons for this sudden decline in the 1980s are now unknown. On the one hand, reliable conclusions cannot be made about the cyclical pattern of rainfall in the study area, given the limited number of observation years. On the other hand, conclusions on the type of oscillation index influencing rainfall in the study area are contradictory. For example, Meddi et al. (2010) and Taibi et al. (2017) assert that the variability of rainfall in the study area is positively correlated with the Southern Oscillation Index (SOI) while Salameh (2008) and Xoplaki et al. (2004) argue that climate variability in the Mediterranean area is influenced by the North Atlantic and Mediterranean Oscillations. In the current context, such a decrease can also be attributed to climate change and its possible impacts on rainfall.

Figure 4 illustrates the deviation between the average of the pre-break period (1968–1980) and the average of the post-break period (1981–2010) for each station. In terms of deficits, the first homogeneous region experienced an estimated average decline by 31.5 %. In addition, the 15 stations in the second homogeneous region showed an average deficit of 22.8 %. The 11 stations belonging to the homogeneous group 4 also reported an estimated average decrease of 18.73 %, while those in the 5th homogeneous region dropped by an average of 31.8 %. The coastal stations belonging to the 6th region recorded an estimated decrease of 30.75 %.

It can be seen that the stations located in high relief areas and in the coastal region are more affected by the decline. In contrast, the stations in the central part of the basin, in plains with moderate relief, experienced a relatively small decrease. Meddi and Meddi (2007), Ghenim and Megnounif (2013), Meddi et al. (2010) and Taibi et al. (2013) showed the same deficit magnitudes over the study area and surrounding regions.

3.3 Study of the spatial variability of rainfall

The spatialization was performed over 2 different periods, the first period is from 1968 to 1980 and the second one is from 1980 to 2010. This distribution was chosen in order to

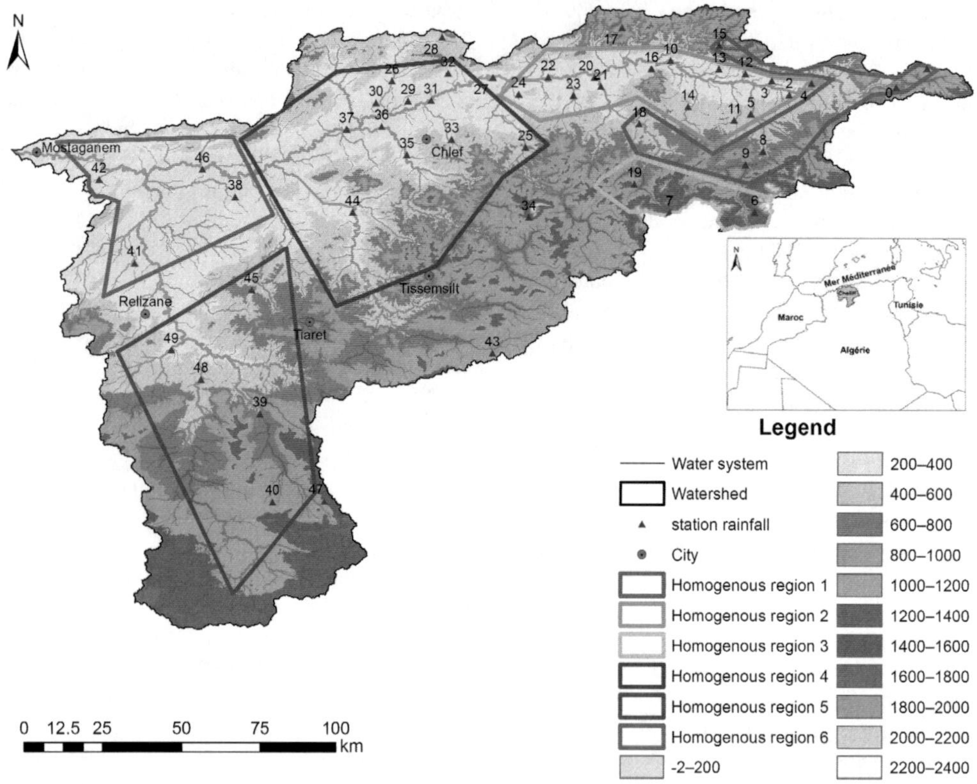

Figure 1. Study area and delineation of homogeneous regions.

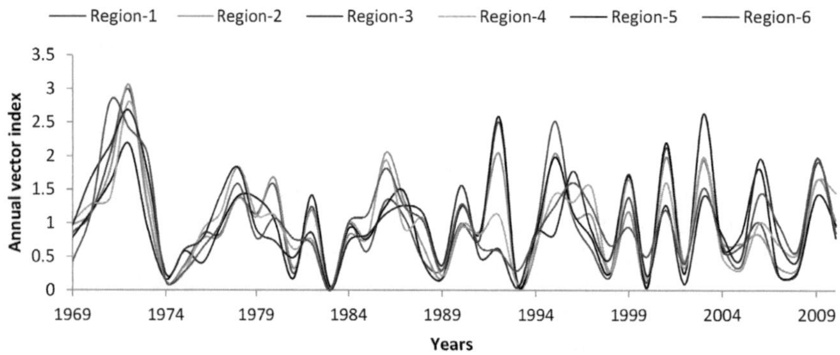

Figure 2. Evolution of the annual vectors in the six region.

understand the spatial structure of rainfall after and before the downward trend over the entire basin.

Figure 5 shows the distribution of inter-annual average rainfall over the study area from 1968 to 1980, the year during which the decreasing trend in annual totals spread to the entire basin. This map displays the increase in rainfall from West to East.

The coastal region of the Cheliff wadi received an average ranging from 311 to 400 mm during the 22 years of observations. In addition, the central plains of the basin and the upstream part of the Mina River experienced an average rainfall varying from 400 to 450 mm, and the valleys connecting the central plains and the mountain ranges of the Atlas were watered with an average from 450 to 550 mm. Besides, the Eastern and North Eastern mountain areas were provided with rainfall between 550 and 800 mm.

Following the propagation of the break through the whole basin in 1980, the average inter-annual rainfall calculated over a period of 30 years (1981–2010) showed the same spatial irregularity with an increase along a West-to-East ori-

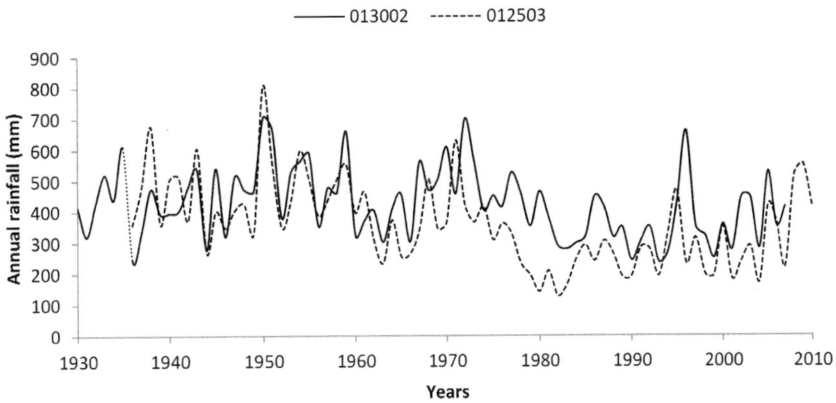

Figure 3. Evolution of rainfall in the stations 013002 and 012503.

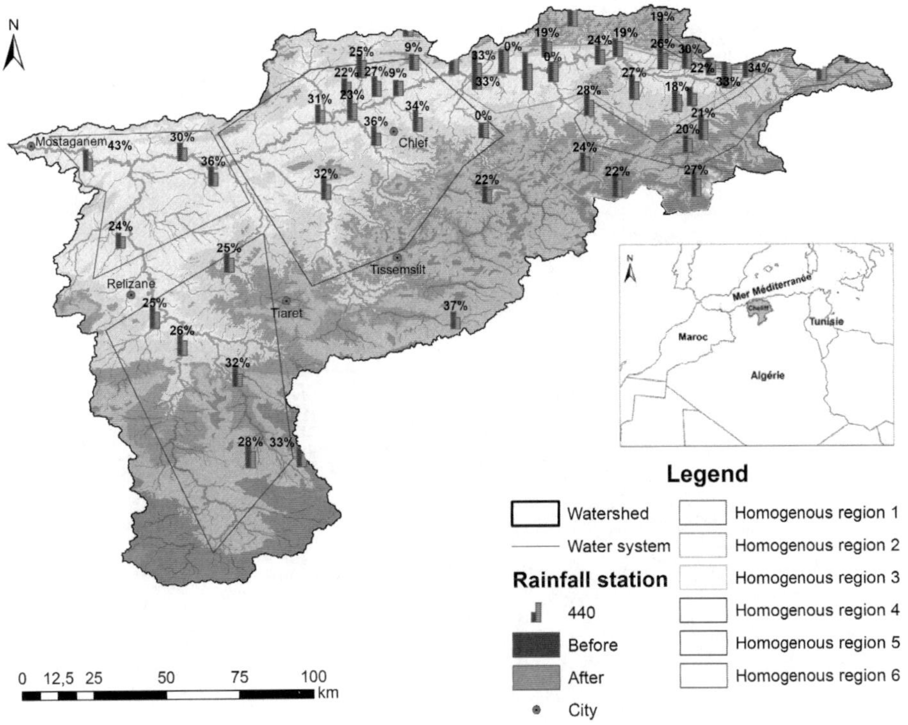

Figure 4. Rainfall deficit in the Cheliff Basin, Algeria.

ented axis. On the other hand, the rainfall amounts received by the various regions identified above have decreased relatively in the entire basin Fig. 6. Thus, the coastal region experienced rainfall ranging from 240 to 300 mm. In addition, rainfall dropped from (400, 450 mm) to (300, 350 mm) after 1980 in the central plains region and upstream the Mina. Besides, the valley area received rainfall varying from 400 to 450 mm while the Eastern and North Eastern mountains experienced rainfall ranging from 450 to 580 mm. In sum, it is noted that the annual totals decreased on average by 100 mm in the study area, except for the high relief region located in the Eastern and North Eastern part, which had a rainfall decrease of over 200 mm.

4 Conclusion

A study of the spatial and temporal variability of rainfall, as well as the determination of homogeneous rainfall units in the Cheliff basin, was conducted based on data from 50 rainfall stations over a 42-year observation period (1968–2010).

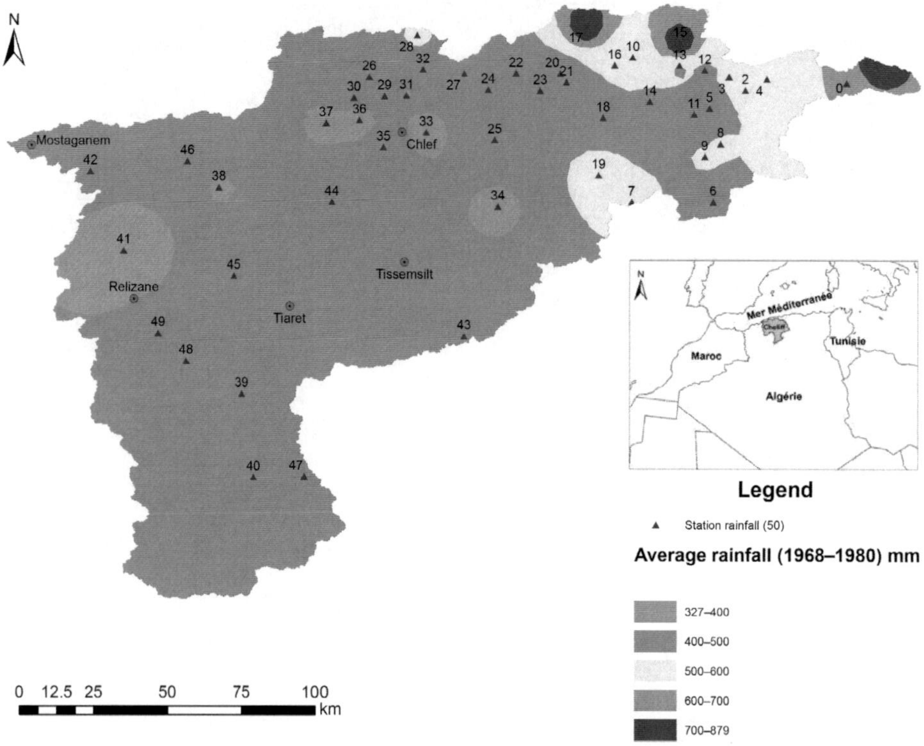

Figure 5. Inter-annual rainfall map in the Cheliff Basin (1968–1980).

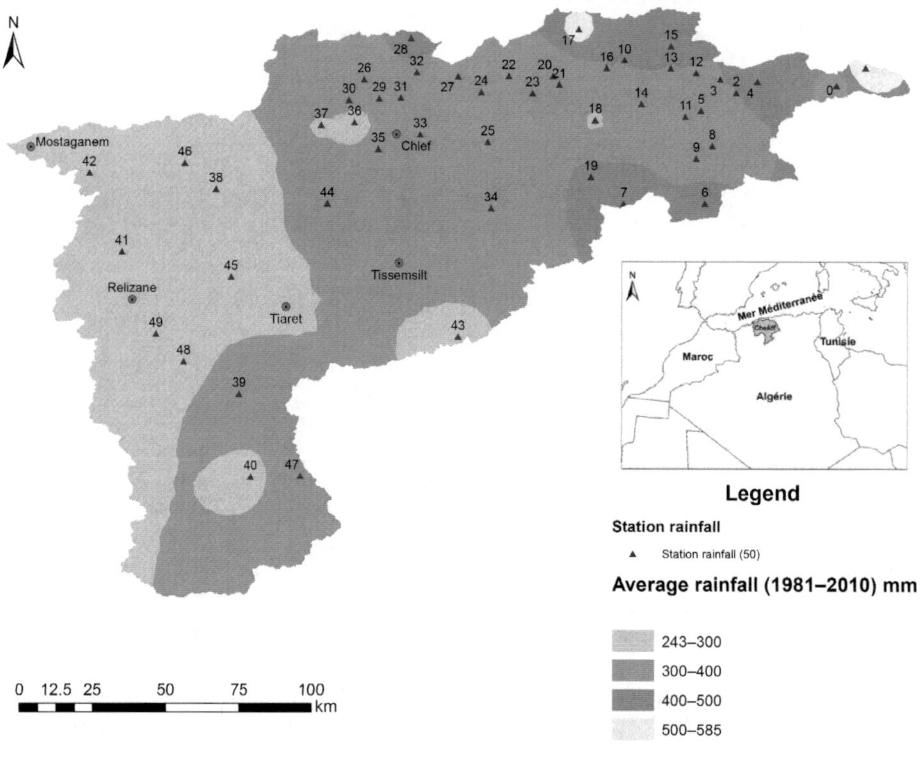

Figure 6. Inter-annual rainfall map in the Cheliff Basin (1981–2010).

https://doi.org/10.5194/piahs-383-61-2020

The regional vector method validated 6 homogeneous regions. Subsequently, the rainfall signal was synthesized in each region. The comparison between the six regional vectors developed showed a similar evolution of rainfall in the East and the central part of the basin. However, the rainfall pattern in the coastal region and the South Eastern part was different from the rest of the basin.

The 3 statistical tests used in the study of the temporal variability of rainfall showed a general downward trend over the entire basin in 1980. The rainfall deficit recorded over the period 1981–2010 compared to the period 1968–1980 is estimated at an average of 30 % in the Eastern part of the basin, the coastal region and the upstream part of the Mina. In addition, the plain domain located in the central part of the basin has experienced an average estimated decrease of 20 %.

The study area has a high spatial variability in rainfall. This variability is directly related to the topographical variety that characterizes it (plain, valley, plateaus, isolated or chain mountains). The same spatial distribution of rainfall during the periods 1968–1980 and 1981–2010 is observed, with an increase in rainfall from West to East. On the other hand, rainfall decreased by an average of 100 mm during the post-break period over the entire basin, except in the Eastern part where the decrease exceeded 200 mm.

The reduction in rainfall since 1980 has resulted in a considerable rainfall deficit. Such a decline caused a dramatic impact on the availability of water resources. To this end, a good understanding and estimation of rainfall variability is of crucial importance in any approach to preserve and manage water resources.

Data availability. The data are not publicly accessible.

Author contributions. GM and MM provided the work supervision and AH performed the calculation and analysis.

Competing interests. The authors declare that they have no conflict of interest.

Special issue statement. This article is part of the special issue "Hydrological processes and water security in a changing world". It is a result of the 8th Global FRIEND–Water Conference: Hydrological Processes and Water Security in a Changing World, Beijing, China, 6–9 November 2018.

Acknowledgements. This study was carried out as part of the referenced TASSILI research project (40010RJ/18MDU109). We would like to thank the Directorate-General of Scientific Research and Technological Development (Algeria) and the Joint Research Unit HydroSciences Montpellier (France).

Financial support. This research has been supported by the Partenariat Hubert Curien (PHC) TASSILI program (grant no. 18MDU109).

References

Bodian, A.: Caractérisation de la variabilité temporelle récente des précipitations annuelles au Sénégal (Afrique de l'Ouest), Physio-Géo. Géographie physique et environnement, 8, 297–312, 2014.

Boyer, J.-F.: Logiciel Khronostat d'analyse statistique de séries chronologiques. IRD UR2 Programme 21 FRIEND AOC, Equipe Hydrologie UMRGBE, Université de Montpellier II, Ecole des Mines de Paris, France, 1998.

IPCC: Climate Change 2007: The Physical Science Basis. Contribution of Working Group I to the Fourth Assessment Report of the Intergovernmental Panel on Climate Change, edited by: Solomon, S., Qin, D., Manning, M., Chen, Z., Marquis, M., Averyt, K. B., Tignor, M., and Miller, H. L., Cambridge University Press, Cambridge, UK and New York, NY, USA, 996 pp., 2007.

IPCC: Climate Change 2014: Mitigation of Climate Change. Contribution of Working Group III to the Fifth Assessment Report of the Intergovernmental Panel on Climate Change, edited by: Edenhofer, O., Pichs-Madruga, R., Sokona, Y., Farahani, E., Kadner, S., Seyboth, K., Adler, A., Baum, I., Brunner, S., Eickemeier, P., Kriemann, B., Savolainen, J., Schlömer, S., von Stechow, C., Zwickel, T., and Minx, J. C., Cambridge University Press, Cambridge, UK and New York, NY, USA, 161 pp., 2014.

Davis, J. C.: Statistics and Data Analysis in Geology, John Wiley and Sons, New York, USA, 656 pp., 1986.

Franke, R.: Scattered data interpolation: tests of some methods, Math. Comput., 38, 181–200, 1982.

Ghenim, A. N. and Megnounif, A.: Analyse des précipitations dans le Nord-Ouest algérien, Science et changements planétaires/Sécheresse, 24, 107–114, 2013.

Hallouz, F., Meddi, M., and Mahe, G.: Modification du régime hydroclimatique dans le bassin de l'Oued Mina (nord-ouest d'Algérie), Revue des sciences de l'eau/Journal of Water Science, 26, 33–38, 2013.

Hiez, G.: L'homogénéité des données pluviométriques, Cahiers ORSTOM, série Hydrologie, 14, 29–173, 1977.

Hubert, P. and Carbonnel, J. P.: Approche statistique de l'aridification de l'Afrique de l'Ouest, J. Hydrol., 95, 165–183, 1987.

Kingumbi, A., Bargaoui, Z., and Hubert, P.: Investigation of the rainfall variability in central Tunisia/Investigations sur la variabilité pluviométrique en Tunisie centrale, Hydrolog. Sci. J., 50, p. 508, https://doi.org/10.1623/hysj.50.3.493.65027, 2005.

Kotti, F. C., Mahe, G., Habaieb, H., Dieulin, C., Calvez, R., and Ali, H. B.: Etude des pluies et des débits sur le bassin versant de la Medjerda, Tunisie Study of rainfall and discharges in the Medjerda watershed, Tunisia, Bulletin de l'Institut Scientifique, Rabat, 38, 19–28, 2016.

Lee, A. F. and Heghinian, S. M.: A Shift Of The Mean Level In A Sequence Of Independent Normal Random Variables – A Bayesian Approach, Technometrics, 19, 503–506, 1977.

Mahé, G. and L'Hôte, Y.: Utilisation de la Méthode du Vecteur Régional pour la description des variations pluviométriques interannuelles en Afrique de l'Ouest et centrale de 1951 à 1989,

Régionalisation en hydrologie: application au développement. Huitièmes Journées Hydrologiques de l'ORSTOM, Montpellier, France, 22–23, 1992.

Mahé, G. and Olivry, J. C.: Variations des précipitations et des écoulements en Afrique de l'Ouest et centrale de 1951 à 1989, Science et changements planétaires/Sécheresse, 6, 109–117, 1995.

Meddi, M. and Hubert, P.: Impact de la modification du régime pluviométrique sur les ressources en eau du nord-ouest de l'Algérie, in: Hydrology of the Mediterranean and Semiarid Regions, IAHS Publ, 278, 229–235, 2003.

Meddi, H. and Meddi, M.: Variabilité spatiale et temporelle des précipitations du Nord-Ouest de l'Algérie, Géographia technica, 2, 49–55, 2007.

Meddi, M., Assani, A., and Meddi, H.: Temporal variability of annual rainfall in the Macta and Tafna catchments, Northwestern Algeria, Water Resour. Manag., 24, 3817–3833, 2010.

Pettitt, A.: A non-parametric approach to the change-point problem, J. Roy. Stat. Soc. C-App., 28.2, 126–135, 1979.

Salameh, T.: Modelisation multi-echelles de la circulation atmospherique sur le bassin mediterraneen, Doctoral dissertation, Ecole Polytechnique X, Paris, France, 2008.

Singla, S., Mahé, G., Dieulin, C., Driouech, F., Milano, M., El Guelai, F. Z., and Ardoin-Bardin, S.: Evolution des relations pluie-débit sur des bassins versants du Maroc, Global Change: Facing Risks and Threats to Water Resources, Proc. Of the Sixth World FRIEND Conference, Fez, Morocco, IAHS. 340, 679–687, 2010.

Taibi, S., Meddi, M., Souag, D., and Mahé, G.: Évolution et régionalisation des précipitations au nord de l'Algérie (1936–2009), Climate and land surface changes in hydrology, IAHS Publ, 359, 191–197, 2013.

Taibi, S., Meddi, M., Mahé, G., and Assani, A.: Relationships between atmospheric circulation indices and rainfall in Northern Algeria and comparison of observed and RCM-generated rainfall, Theor. Appl. Climatol., 127, 241–257, 2017.

Vauchel, P.: Hydraccess: Software for Management and processing of Hydro–meteorological data software, Version 2.1.4, available at: https://www.documentation.ird.fr/hor/fdi:010026171 (last access: 5 August 2020), 2005.

WWAP (United Nations World Water Assessment Programme)/UN-Water: The United Nations World Water Development Report 2018: Nature-Based Solutions for Water, UNESCO, Paris, France, 154 pp., 2018.

Xoplaki, E., González-Rouco, J. F., Luterbacher, J., and Wanner, H.: Wet season Mediterranean precipitation variability: influence of large-scale dynamics and trends, Clim. Dynam., 23, 63–78, 2004.

Proc. IAHS, 383, 69–78, 2020
https://doi.org/10.5194/piahs-383-69-2020

Extreme Rainfall and Temperature Changes Assessment on late-twenty-first-century over the Mono River Basin, Togo (West-Africa)

Batablinlè Lamboni[1], **Agnidé Emmanuel Lawin**[2], **Celestin Manirakiza**[1], **and Dagbégnon Luc Olabodé**[1]

[1]Institute of Mathematics and Physical Sciences, University of Abomey-Calavi, BP 613 Porto-Novo, Benin
[2]Applied Hydrology Laboratory, University of Abomey-Calavi, BP 4521 Cotonou, Benin

Correspondence: Batablinlè Lamboni (lbata1982@yahoo.fr)

Published: 16 September 2020

Abstract. This study assessed the extreme rainfall and temperature changes over Mono river basin by the end of 21st century under the highest greenhouse gas emission scenario RCP8.5. Simulations of eight regional climate models (RCMs) provided by Africa-CORDEX program were considered. The maximum and minimum temperatures and rainfall were analyzed. Two groups of indices were analyzed. The first group consists of frequency indices: maximum number of consecutive dry days and wet days. The second group is intensity indices: five-day maximum rainfall and simple daily intensity index. These variables were calculated at annual and seasonal scales. Changes from the baseline period 1971–2000 were computed for far future 2071–2100. As result, almost all the RCMs considered predicted a decrease of the rainfall and increase of maximum and minimum temperatures over most parts of the Mono basin, particularly in the south. Declining mean monthly precipitation and irregular rainy seasons at all scales were exhibited by most of the models. Simple daily intensity and five-day maximum rainfall are projected to decrease by the majority of the used models. As for dry and wet sequences, the RCMs showed an increase of the consecutive dry days and a decrease of wet days.

1 Introduction

A global view of climate change impacts is important to quantify the aggregated effects in a given region since large scale processes can influence driving forces at sub regional or mesoscale. Indeed, the detection of the planetary warming trend, attributed to anthropogenic activities, have been presented with greater certainty in the 4th report of the IPCC. According to this report, the average global temperature will increase and could reach 4.5 °C by 2100, the precipitation will change and the extreme events will be increasingly recurring. Several studies (Aguilar et al., 2013; IPCC, 2013; Masson et al., 1999) showed that impacts of climate extremes are mostly depending on the spatial scale considered and vary from a region to another, from country to country and even from one local area to another within the same country. For Africa, many previous studies have investigated extreme events and have showed that changes in extreme temperature and precipitation are contrasting (Yèkambèssoun et al., 2017.). Current and future trends suggested that regions: West Africa, including the Sahara, the Horn of Africa, part of Central Africa and southern hemispheres, could undergo a warming of order of 3 to 6 °C by 2100. Pluviometric regimes will also be affected and may decrease of more than 20 % to 30 % compared to the reference 1961–1990 (Bigot, 2003). Other studies indicated an increase in extreme rainfall events, particularly in Sahel (Nicholson and Webster, 2007) but also a decrease in Nigeria (Oguntundé et al., 2011) and (Easterling et al., 2000), Guinea Conakry (Aguilar et al., 2009), in eastern Niger (Ozer et al., 2009), in Ivory-Coast (Soro et al., 2014), in Benin (Yèkambèssoun et al., 2017) and in South Africa (Mason et al., 1999). In the northern part of Africa, there is a tendency towards wetter conditions; in contrast with the eastern part, experienced with more drying trends, although, these trends are of low significance. The understanding and characterizing climate variability at reduced scales are so fundamental to understanding the impacts of cli-

Published by Copernicus Publications on behalf of the International Association of Hydrological Sciences.

mate change, local vulnerability and for future development projects (Emmanuel et al., 2019).

For the specific case of Togo, according to the report of republic of Togo (MERF, 2009), floods, drought, late rains, high temperatures and high winds, are the major risks in the country. The drought covers the entire territory except for the coastal region which is however facing sea level rise risk. It should be noted that among the latest climatic events, flooding takes precedence over other risks with its share of material damage and loss of life. It is becoming more and more dangerous to the whole country. Activities which are most exposed to these risks are farming, livestock farming, marketing of agricultural products and market gardening (MERF, 2009). Therefore, it compels scientists to conduct studies on climate change projection in Togo in order to take adaptation measures to enhance these communities in their development. However, studies of the future impact of climate change on precipitation and temperature are very few. Nelson et al. (2010) assessed impacts of precipitation and temperature changes on agricultural productivity by 2050 in Togo. This study used global climate models (GCMs) with spatial resolution of 200 to 300 km that cannot give details at less scales. So these models are only important when it comes to representing climate on a very large scale of space. But at regional level, they present many limitations (Hulme et al., 2001). In order to fill this resolution issue, Regional Climate Models (RCMs) from CORDEX (COordinated Regional Climate Downscaling EXperiment) are now proposed in our study because of their high spatial resolution (10 to 100 km). Furthermore Batablinlè et al. (2018) showed, at the Mono river basin scale, there was very few study focuses on future evolution of the extreme temperatures, number of extremely heavy rainfall days, consecutive dry days, consecutive wet days, five-day maximum rainfall, simple daily intensity index. We have therefore found it necessary to evaluate the future changes of climate indices using the World Meteorological Organization (WMO) reference period 1971–2000 computed from Regional Climate Models (RCMs). In this study, rain and temperature simulated by 8 RCMs from the Africa-CORDEX program have been analyzed through maximum and minimum temperatures, average rainfall and extreme precipitation events by 2100 on the Mono river basin which is the major economical part of Togo in terms of agricultural and energy outputs. The objective of this work is therefore to provide knowledge and more details about extreme rainfall and temperature for the population of Mono river basin.

2 Data and the methods

2.1 Study area

This study area has already been described by certain authors (Batablinlè et al., 2018; Lamboni et al., 2019b).

Mono basin is located in the Gulf of Guinea region, more precisely Between 06°16′ and 09°20′ N and 0°42′ and 20°25′ E. The Mono basin is almost entirely in Togo although its lower course is between Togo and Benin. Our study considered the part in Togo (Fig. 1). It houses a dam of hydroelectric power plant called Nangbéto. Elongated, covering 25 400 km^2, the Mono basin is the major river of Togo. In general, West Africa's climate is controlled by the interaction of two air masses; the inuence of which varies throughout the year with the north-south movement of the Intertropical Convergence Zone (ITCZ). Hot and dry continental air masses originating from the high pressure system above the Sahara desert give rise to dusty Harmattan winds over most of West Africa from November to February. In summer, moist equatorial air masses originating over the Atlantic Ocean bring annual monsoon rains. Within this West African context, rainfall in the study area is characterized by two types of rainfall regimes. In southern basin there are two rainy seasons which extend from mid-March to mid-July and from mid-August to October. In northern basin, there is one rainy season which extends from April to October. Indeed, according to the report of PANA-Togo (MERF, 2009), the country is inuenced by two large climate systems: The tropical Soudanian system in the north with a wet season running from May to October and a dry season from November to April. In this area pluviometry varies from 900 to 1100 mm per year with an average temperature of 28 °C; the tropical Guinean system in the south is characterized by two dry seasons and two wet seasons of unequal duration with pluviometry located between 1000 and 1600 mm yr^{-1} and an average temperature of 27 °C. According to the report of WAEMU in 2006, the population of the basin is approximately two million, with an annual increase of 2.9 %. This population also distributed at high densities in the south of the basin has as main activities, agriculture (mainly rainfed); in the lower valleys, fishing and salt-farming are the major activities (Ernest et al., 2015).

2.2 Data and Sources

The observed daily rainfall and temperature data used in this study, are provided by the Togo National Meteorology Agency (Météo Togo) for the period 1970–2000 for eight stations spatially located as shown in Fig. 1 and complemented by other sources such as the archives of ORTM Togo and the Nangbeto dam. The missing data values were replaced statistically by similar day averages for the previous years where there were existing data values. Table 1 shows the Cordex models used. In order to develop future climate change projections on Mono river basin, the most recent Regional Climate Models (RCMs) provided by the COordinated Regional Downscaling EXperiment (CORDEX) Africa were analyzed. These models were run using different emission scenarios, so-called Representative Concentration Pathways (RCPs) which provide different assumptions of greenhouse gas emission pathways. One of RCPs was selected

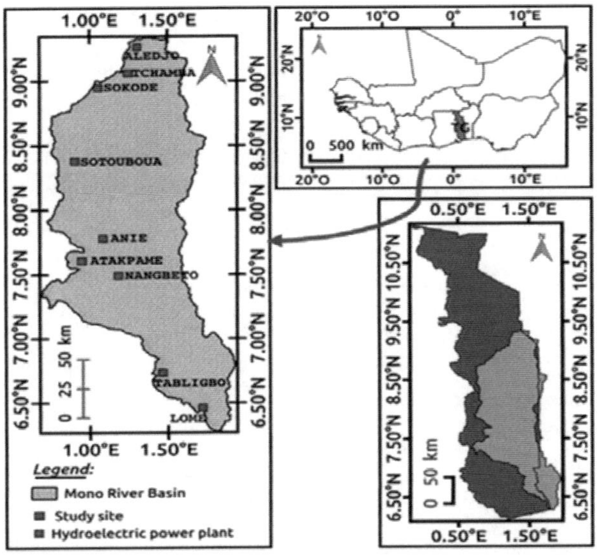

Figure 1. The geographical location (sources Batablinlè et al., 2018).

in this study to cover a wide high (RCP8.5) global anthropogenic radiative forcing. We made the choice to download the values of several climate variables of eight models available in a spatial resolution 0.44° or 50 km. It has already been used over Mono basin by certain authors as (Batablinlè et al., 2018; Emmanuel et al., 2019; Lamboni et al., 2019a).

2.3 Methods

The seasonal averages as well as the annual averages were calculated using the Climate Data Operator Commands (Batablinlè et al., 2018; Emmanuel et al., 2019). Then, the free software R is used to compute all the statistical parameters and all the plots presented in the following. In this study, four climatic indices associated to the precipitations and temperatures were considered as potentially impacting on the availability of water resources, population and agricultural productivity in this zone (Emmanuel et al., 2019). These indices, computed with Climate Data Operator Commands, are defined in Table 2 as follow. To bias-correct the model future projection, the cumulative distribution function matching (CDFm) of the model in the reference period is used to identify the corresponding percentile values of the future period (Batablinlè et al., 2018). Later the observation CDFm is used to find the climate variable value for the same future cumulative probability, which represents the bias corrected future value. The drawback of the above procedure of bias correction is the assumption of constant bias between the reference and future periods . However, we will accept this assumption as most of the recent studies did (Sarr and Camara, 2017) as mentioned earlier, bias correction and spatial disaggregation relies on finer-scale observation data and consists of the following steps: (1) aggregate finer-scale observed data to a RCM-specific scale; (2) correct the RCM's systematic biases through quantile mapping; and (3) disaggregate bias-corrected outputs to a finer scale (Batablinlè et al., 2018). An equiratio cumulative distribution function matching (equiratio CDFm): The statistical transformation can be written as:

$$X_{\mathrm{m-padjst}} = X_{\mathrm{m-p}} \frac{h_1\left(X_{\mathrm{m-p}}\right)}{h_2\left(X_{\mathrm{m-p}}\right)} \tag{1}$$

and

$$X_{\mathrm{m-padjst}} = X_{\mathrm{m-p}} + h_1\left(X_{\mathrm{m-p}}\right) - h_2\left(X_{\mathrm{m-p}}\right) \tag{2}$$

with (Eq. 1) for precipitation and (Eq. 2) for temperature, where $F_{\mathrm{m-p}}$ is the cumulative distribution function (CDF) of the model for a future projection period, and $F_{\mathrm{o-c}}^{-1}$ and $F_{\mathrm{m-c}}^{-1}$ are quantile functions for observations and model in the reference period, respectively. h_1 and h_2 were defined by:

$$h_1 = F_{\mathrm{o-c}}^{-1} F_{\mathrm{m-p}} \tag{3}$$

and

$$h_2 = F_{\mathrm{m-c}}^{-1} F_{\mathrm{m-p}} \tag{4}$$

A complete description of this method is described by Gudmundsson et al. (2012).

Before this procedure, statistical measures such as the relative error or mean absolute error were used to describe and compare the observed and simulated variables. The relative error (E_{P} and E_{T}), respectively for precipitations and temperatures of the estimated values of the rainfall may be calculated from the following Eqs. (1) and (2):

$$E_{\mathrm{P}} = \frac{(X_{\mathrm{o}} - X_{\mathrm{s}})}{X_{\mathrm{o}}} \tag{5}$$

$$E_{\mathrm{T}} = (X_{\mathrm{o}} - X_{\mathrm{s}}) \tag{6}$$

and finally, the changes from the reference period are assessed as shown by Eqs. (7) and (8) respectively for precipitations and indices ($\Delta X_{\mathrm{P,I}}$) and temperatures (ΔX_{T})

$$\Delta X_{\mathrm{P,I}} = \left(\frac{\mathrm{FF} - \mathrm{HIST}}{\mathrm{HIST}}\right) \times 100 \tag{7}$$

or

$$\Delta X_{\mathrm{T}} = \mathrm{FF} - \mathrm{HIST} \tag{8}$$

where FUT and HIST represent respectively for the future (by 2100) and the historical or reference period (1971–2000).

3 Results and discussions

3.1 Results

Before assessing the climate change scenario we evaluated the considered models' performances to represent the climate

Table 1. Used climatic models, global model under which they are run (column 1), their Institute of origin (column 2), their short name (column 3) and RCM model Name (column 4).

Global Model Name	Institute ID	Model Short Name	RCM Model Name
GFDL-ESM2M	NOAA-GFDL	NOAA	SMHI-RCA4
NorESM1-M	NCC	NCC	SMHI-RCA4
MPI-ESM-LR	MPI-M	MPI	SMHI-RCA4
MIROC5	MIROC	MIROC	SMHI-RCA4
IPSL-CMA5-MR	IPSL	IPSL	SMHI-RCA4
EC-EARTH	ICHEC	ICHEC	KNMI-RCAMO22T
CNRM-CM5	CRNM-CERFACS	CNRM	SMHI-RCA4
CanESM2	CCCma	CCCMA	SMHI-RCA4

Table 2. Selected indices and their signification.

ID	Name	Definition	Units
Frequency indices (adapted from WMO 2009)			
CDD	Consecutive dry days	Maximum number of consecutive days with PRCP* <1 mm	d
CWD	Consecutive wet days	Maximum number of consecutive days with PRCP* >1 mm	d
Intensity indices (adapted from WMO 2009)			
RX5day	Max5-day precipitation amount Annual	Maximum consecutive 5 d precipitation	Mm
SDII	Simple daily intensity Index	Annual total precipitation divided by the number of wet day (defined as PRCP* >1 mm) in the year	mm d^{-1}

* PRCP – daily total precipitation

of the study region. Table 3 showed that the spatial mean rainfall MAE (Mean absolute error) of the selected RCMs are weak and not greater than 3 mm d^{-1} in mean during the main part of the rainy season which is June to September (the agricultural period). NOAA model is the most accurate model while MIROC model showed the biggest MRE (Mean relative error). The higher bias between simulated and observed data, may be caused by errors in measuring data and errors associated with the model. The statistical analysis show that there is no enough an agreement between simulated and observed variables which such as mean temperatures (minimum and maximum) and precipitations (P_m). Obviously, current RCMs still have substantial deficiencies in simulating the regional climate over Mono. To produce the reliable projections required for regional climate assessment, we need to bias-correct the raw RCM outputs.

3.1.1 Changes of Minimum and Maximum temperature

Predicted changes of maximum temperature and minimum temperature between the far future (2071–2100) and the reference period (1971–2000) are shown by Fig. 2. For the maximum temperature, all the models predicted an increase for the whole basin except model canESM2 which showed a slight decrease for a small part of the southern basin. The lowest increase is 0.42 °C and is projected by the model CNRM-CM5. The highest increase is 3.3 °C and is projected by the rest of models. The north of the basin is forecasted to experience the greatest increase according to almost all the models. The projections exhibited clearly an increase in days of high heat in the north as shown by the models CNRM-CM5 (1), GFDL-ESM2M (6) , EC-EARTH (3), IPSL-CMA5-MR (4), MPI-ESM-LR (6), NorESM1-M (7) and Models-Mean (9). However, for models CCCMA, the maximum temperature will be relatively cooler on some area in the southward of the basin (mean value −0.29 °C) and a strong increase on the rest of this area. In the north and the center, the maximum temperature will increase in average by 1 °C for model CNRM-CM5 (1) and model MPI-ESM-LR (5), 0.22 °C for model GFDL-ESM2M (6), 1.45 °C for models EC- EARTH (3), IPSL-CMA5-MR (4) and NorESM1-M (7). Globally, the Models-Mean (9) predicted 1 °C as rate of increase.

Table 3. Spatial mean Rainfall MAE (in mm) and MRE (in %) between observations and bias corrected simulations over Mono river basin.

						Main part of the rainy season							
Models	Biases	J	F	M	A	M	J	J	A	S	O	N	D
NOAA	MAE	0	0	1.1	1.9	3.7	0.8	1.3	1.6	3.7	1.9	0.3	0
	MRE	0	7	6	33	29	5	7	8	18	15	5	0
NCC	MAE	1.6	1.4	1.1	3.7	3.4	2.3	2.6	2.2	3.9	0.8	0.9	1.3
	MRE	6	9	6	51	34	14	14	10	15	12	37	6
MPI	MAE	0.4	1.0	1.1	0.3	1.2	2.5	3.8	2.7	2.6	1.1	2.1	1.3
	MRE	4	12	7	4	11	16	20	12	10	5	11	3
MIROC	MAE	1.2	2.2	1.3	0.5	0.5	3.1	2.1	3.2	4.5	1.4	0	1.1
	MRE	6	6	15	15	47	20	12	12	16	6	0	2
IPSL	MAE	1.7	1.7	1.2	3.0	4.8	3.9	3.3	1.0	1.3	0.8	0.7	1.8
	MRE	8	8	4	32	37	24	22	35	17	4	3	2
ICHEC	MAE	1.8	0.8	1.3	2.7	3.8	4.0	0.5	1.3	1.8	2.2	0.9	3.1
	MRE	9	3	4	28	31	26	31	27	19	11	15	4
CNRM	MAE	2.1	0.2	1.4	2.0	2.9	2.2	1.3	2.0	3.3	3.2	0.4	0.5
	MRE	2	26	5	33	28	13	18	22	15	13	11	6
CCCMA	MAE	1.2	0.8	1.3	3.5	3.6	3.1	2.5	1.4	1.2	0.3	1.8	0.1
	MRE	4	2	3	34	25	16	28	12	9	14	8	22
Models Mean	MAE	0	0.1	0	1.9	1.3	2.8	3.1	2.5	2.3	2.7	0.2	0.3
	MRE	0	9	0	11	13	8	9	18	10	7	2	3

The analysis of projected minimum temperature for far future revealed that there is a relative uniform evolution over the entire study period for all models except models CNRM-CM5 (1) and EC-EARTH (3). Indeed, in the northward, the minimum temperature were projected and will increase for (mean value 1.55 °C), (mean value 1.45 °C), (mean value 2 °C), (mean value 1.4 °C), (mean value 2 °C), (mean value 2.4 °C) and (mean value 1.2 °C), respectively, for models MIROC5 (2), IPSL-CMA5-MR (4), MPI-ESM-LR (5), GFDL-ESM2M (6), NorESM1-M (7), canESM2 (8) and Models-Mean (9) . For the models CNRM-CM5 (1) and EC- EARTH (3), the results exhibit an increasing average for the minimum temperature in the north. When considering the southward and the center, models used predict an increasing average for, (mean value 2.5 °C), (mean value 0.51 °C), (mean value 1.82 °C), (mean value 2 °C), (mean value 1.22 °C), (mean value 1.55 °C) and (mean value 1.83 °C), respectively, for models MIROC5 (2), IPSL-CMA5-MR (4), MPI-ESM-LR (5), GFDL-ESM2M (6), NorESM1-M (7), canESM2 (8) and Models-Mean (9). While for models CNRM-CM5 (1) and (C) EC-EARTH (3), the results show a relative rise of the minimum temperature but more than 1 °C.

3.1.2 Precipitation frequency indices

We have analyzed also the maximum duration of the number of consecutive wet days (CWD) and the maximum number of consecutive dry days (CDD) in Fig. 3. The changes relating to the maximum duration of the number of consecutive wet days (CWD) between the far future and the historical period are shown in Fig. 3. The lowest decrease is −40 % and is projected by the model (M8) CanESM2. The highest increase is 32 % and is projected by model (M5) MPI-ESM- LR. Indeed, the projection models used showed a variability of the consecutive wet days on the northern part of the basin from 0 % to −14.5 %, −15 % to −21 %, −10 % to −20 %, −5 % to 15 %, −10 % to 11 %, −10 % to 17 %, −10 % to 5 %, −10 % to 20 % and −10 % to −15 % respectively for models CNRM-CM5 (M1), MIROC5 (M2), EC-EARTH (M3), IPSL-CMA5-MR (M4), MPI-ESM- LR (M5), GFDL-ESM2M (M6), NorESM1-M(M7), CanESM2 (M8) and Mean-models (M9). The southward and center (except models MPI-ESM-LR (M5) and GFDL-ESM2M (M6)), there is a decreasing average in general for the maximum duration of the number of consecutive wet days (CWD), from 0 % to −15 %, 0 % to −10 %, −5 % to −15 %, −15 % to 5 %, 0 % to −10 %, −10 % to −30 % and −10 % to −5 % respectively for models CNRM-CM5 (M1), MIROC5 (M2), EC-EARTH (M3), IPSL-CMA5-MR (M4), NorESM1-M (M7), CanESM2 (M8) and Mean-models (M9). In summary, sev-

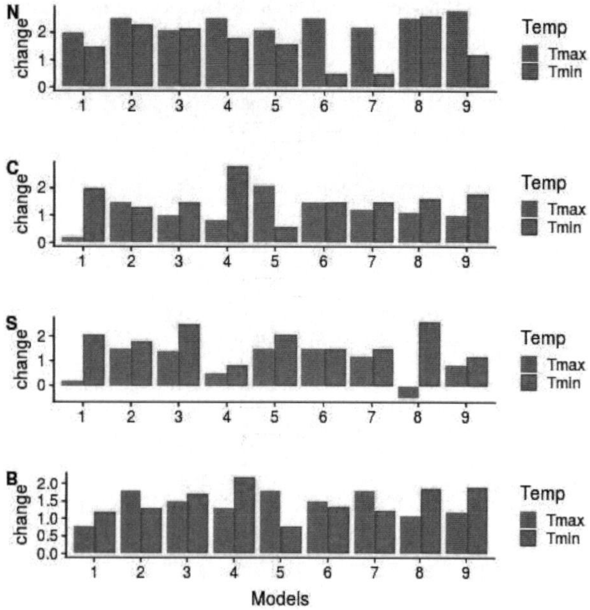

Figure 2. Maximum and minimum temperature changes (°C) between far future (2071–2100) and historical period (1971–2000) during May to September for models: (1) CNRM-CM5, (2) MIROC5, (3) EC-EARTH, (4) IPSL-CMA5-MR, (5) MPI-ESM-LR, (6) GFDL-ESM2M, (7) NorESM1-M, (8) CanESM2 and (9) Models-Mean (ensemble mean of Model) in the North (N), Center (C), South (S) and over mono basin (B).

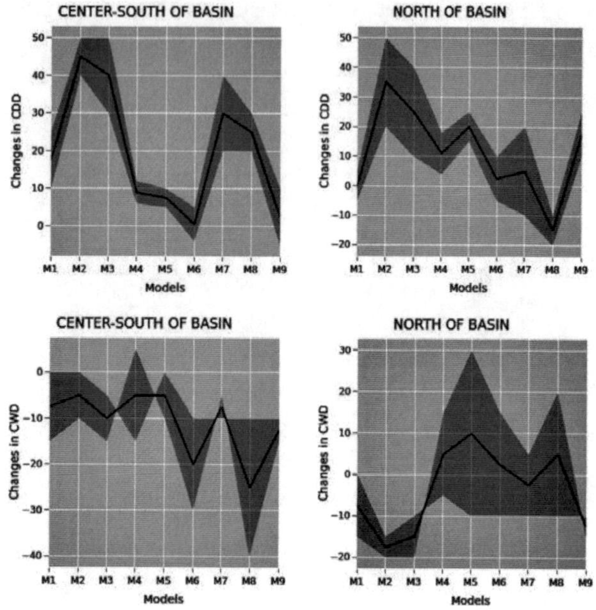

Figure 3. Mean consecutive wet days (CWD) and Mean consecutive dry days (CDD) changes between far future (2071–2100) and historical period (1971–2000) during May to September for models: (M1) CNRM-CM5, (M2) MIROC5, (M3) EC-EARTH, (M4) IPSL-CMA5-MR, (M5) MPI-ESM-LR, (M6) GFDL-ESM2M, (M7) NorESM1-M, (M8) CanESM2 and (M9) Models-Mean (ensemble mean of Model).

eral models used have announced clearly a decrease the northward of the maximum duration of the number of consecutive wet days (CWD) and over the south and center of the Mono basin. To quantify the potential impact of climate change on future drought events, the change in the number of dry periods was analyzed. A dry period is defined as at least five consecutive days for which the daily precipitation is less than 1 mm. The relative deviations of the maximum duration of dry sequences between the far future and the historical period are represented in the Fig. 3. Contrary to the case of the consecutive wet days (in the same figure), the projection models used show a variability of the consecutive dry days on the northern part of the basin from −5 % to 5 %, 20 % to 50 %, 10 % to 40 %, 4 % to 18 %, 15 % to 25 %, −5 % to 10 %, −10 % to 20 %, −20 % to −10 % and 10 % to 25 %, respectively, for models CNRM-CM5 (M1), MIROC5 (M2), EC-EARTH (M3), IPSL-CMA5-MR (M4), MPI-ESM-LR (M5), GFDL-ESM2M (M6), NorESM1-M (M7), CanESM2 (M8) and Mean-models (M9). The southward and center, models project an increase of the maximum duration of the number of consecutive dry days (CDD), from 10 % to 25 %, 40 % to 50 %, 30 % to 50 %, 6 % to 12 %, 5 % to 10 %, −4 % to 5 %, 20 % to 40 %, 20 % to 30 % and −5 % to 10 %, respectively, for models CNRM-CM5 (M1), MIROC5 (M2), EC-EARTH (M3), IPSL-CMA5-MR (M4), MPI-ESM-LR (M5), GFDL-ESM2M (M6), NorESM1-M (M7), CanESM2 (M8) and Mean-models (M9). The RCMs and their mean in general showed that the maximum number of consecutive rainy days (CWD) will decrease over the Mono basin while the maximum duration of dry sequences will go up.

Other important aspects of daily precipitation characteristics for evaluating climate variability and change impacts is the Max 5 d precipitation and the simple daily intensity index (intensity indices) within the rainy season. The section following is reserved to the analysis of the intensity indices.

3.1.3 Precipitation intensity indices

The relative differences in the number of rainy days during the rainy season (May–June–July–August–September) between the far future and historical period are shown in the Fig. 4. The model MPI-ESM-LR (M5) indicated lowest decrease (mean value −9 %) against the model NorESM1-M (M7) which indicates an increase of the max 5 d precipitation the northward (mean value 20 %). In general, the projection models used showed a variability of the max 5 d precipitation in the north of the basin varying between, −7.5 % and −5 %, −4 % and 4 %, 5 % and 15 %, 5 % and 20 %, −30 % and 10 %, −5 % and 5 %, 0 % and 20 %, 10 % to 15 % and −10 % and 20 %, respectively, for models CNRM-CM5 (M1),

MIROC5 (M2), EC-EARTH (M3), IPSL-CMA5-MR (M4), MPI-ESM-LR (M5), GFDL-ESM2M (M6), NorESM1-M (M7), CanESM2 (M8) and Mean-models (M9). The southward models projected also a fluctuation of the max 5 d, between −8.5 % and −7.5 %), −6 % and −2 %, −5 % and 0 %, −5 % and 0 %, −20 % and 10 %, −15 % and −5 %, 0 % and 20 %, 10 % and 15 %, and −10 % and 20 % , respectively, for models CNRM-CM5 (M1), MIROC5 (M2), EC-EARTH (M3), IPSL-CMA5-MR (M4), MPI-ESM-LR (M5), GFDL-ESM2M (M6), NorESM1-M (M7), CanESM2 (M8) and Mean-models (M9). In the center and south, models NorESM1-M (M7) and CanESM2 (M8) show clearly a decrease of the max 5 d, respectively, for, −5 % and −10 %. The RCMs and their mean showed in general that the max 5 d will decrease over the south and center of the Mono basin.

The relative differences in the intensity of rainy days between the far future and the historical period are shown in the Fig. 4. For the results of the variability of the simple daily intensity index, the models have shown dipole structures, as the frequency indices : The projection models used show a variability of the simple daily intensity index in the north of the basin varying between, −6 % and −1 %, −4 % and 3 %, −2 % and 8 %, −16 % to 8 %, −10 % and 10 %, −12 % and 4 %, −15 % and 5 %, −15 % and 5 % and −5 % and 0 % , respectively, for models CNRM-CM5 (M1), MIROC5 (M2), EC-EARTH (M3), IPSL-CMA5-MR (M4), MPI-ESM-LR (M5), GFDL-ESM2M (M6), NorESM1-M (M7), CanESM2 (M8) and Mean-models (M9). The southward models project an increase and decrease of the simple daily intensity index (Fig. 4), from −3 % to −1 %, 0 % to 3 %, −6 % to −2 %, −16 % to −10 %, −2.5 % to 5 %, −12 % to −3 %, −10 % to 5 %, −10 % to 5 % and −15 % to −5 %, respectively, for models CNRM-CM5 (M1), MIROC5 (M2), EC-EARTH (M3), IPSL-CMA5-MR (M4), MPI-ESM-LR (M5), GFDL-ESM2M (M6), NorESM1-M (M7), CanESM2 (M8) and Mean-models (M9). The RCMs and their mean in general showed that the simple daily intensity index will decrease over a big part of the Mono basin during the rainy season (May–June–July–August–September) particularly the southward and in the center.

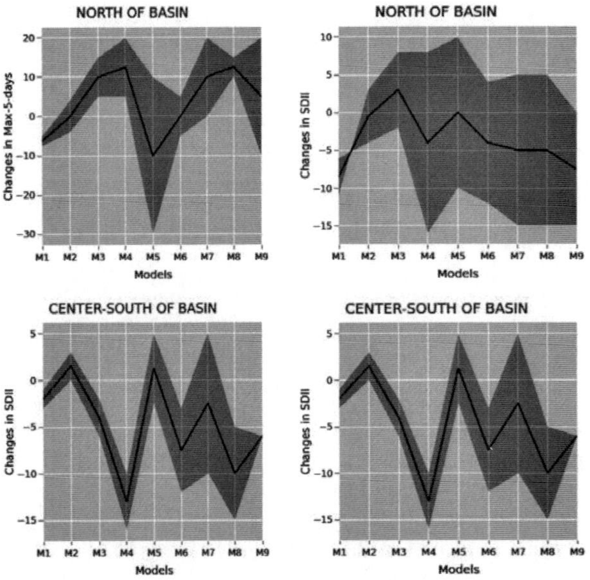

Figure 4. Max 5 d precipitation andsimple daily intensity index changes (in %) between far future (2071–2100) and historical period (1971–2000) during May to September for models: (M1) CNRM-CM5, (M2) MIROC5, (M3) EC-EARTH, (M4) IPSL-CMA5-MR, (M5) MPI-ESM-LR, (M6) GFDL-ESM2M, (M7) NorESM1-M, (M8) CanESM2 and (M9) Models-Mean (ensemble mean of Model).

3.1.4 Analysis of inter-annual irregularity

The analysis of the results presented in Table 4 shows that K2 tends to decrease in two stations (Tchamba and Lomé) from 1971–2000, indicating a tendency towards regularity. On the other hand, in the six (6) others, K2 varies by increasing, showing a tendency of increasing the irregularity of the precipitations. For the period 2071–2100, excepted stations Alédjo, Anié, Soutouboua and Lomé, which will be marked by increases but less significant, the other resorts will have to increase to reach 9.41 at Sokodé 7.12 Tabligbo, 5.18 at Atakpamé and 5.52 at Tchamba. Declining mean precipitation heights at all scales and irregular rainy seasons may be associated with a palpable increase in temperatures in the Mono river watershed by 2100.

3.2 Discussion

The analysis of extreme weather indices shows that all climate projections indicate an overall increase in extreme temperatures in the Mono catchment. The variations of the rainfall and their indices will depend strongly on emission scenarios and models. Indeed, some models suggested an increase in these indices while others indicated a decrease. To have better comprehension of results of in this paper it is important to look at the results of other regions and countries. The large disparity observed in the evolution of precipitation by 2100 could be due to the type of forcing models or the convection pattern used in West Africa (Sylla et al., 2016). This divergence of models on climate projections on precipitation in West Africa is still uncertain. Yèkambèssoun et al. (2017) estimated that the spatially inconsistent trends of climate indices noticed are likely linked to the high heterogeneousness of the rainfall. Indeed, the Mono basin is characterized by two types of rainfall regimes. In southern basin, there are two rainy seasons which extend from mid-March to mid-July and from mid-August to October. In northern basin, there is one rainy season which extends from April to October. For Nelson et al. (2010), in the future (specifically

Table 4. Results obtained by calculation of the K_2 coefficient of inter-annual rainfall irregularity in the form of table of 1971–2000 and 2071–2100 for the Tabligbo, Atakpamé, Anié, Soutouboua, Sokodé, Tchamba , Alédjo and Lomé stations.

Station	1971/1980	1981/1990	1991/2000	2071/2080	2081/2090	2091/2100
Atakpamé	1.09	2.92	2.51	5.18	2.19	3.54
Tabligbo	1.53	1.22	2.04	7.12	4.12	2.01
Anié	1.86	2.38	2.17	1.51	1.38	2.01
Lomé	2.23	4.96	4.11	6.51	1.85	2.12
Sokodé	3.04	7.12	9.41	8.56	2.85	3.52
Tchamba	3.08	4.74	3.50	5.36	5.52	3.52
Alédjo	1.48	2.14	1.19	1.95	1.98	4.9
Soutouboua	2.54	1.13	7.38	1.56	3.80	8.02

by 2050), precipitation may rise or fall on average depending on which model we look to. Despite everything, in the Mono basin, several models used agree on an increase in temperature (maximum and minimum) and a decrease of precipitation on a large part of the basin. Nelson et al. (2010) using models GCM (general circulation models), showed that, in the future (specifically by 2050) the temperature is expected to increase between 1 and 2.5 °C in Togo. Diallo et al. (2013) pointed out that the decrease in mean rainfall in this area of West Africa could probably be induced in part by a weakening of the moisture from local sources, slowing down the hydrological cycle. Over the Mono basin, our work is also consistent with many works (Tall et al., 2016; Batablinlè et al, 2018; Lamboni et al., 2019a, b), which found a strong decrease in precipitation by 2100 under the RCP8.5 scenario. It is also in line with the recent special IPCC report which

states that West Africa will likely experience longer and more intense droughts in the near future (IPCC, 2013). In general, when considering the number of rainy days and the intensity of rainy days (mostly declining) were coherent with that of rain simulated average on the Mono basin. By consequence, Extreme rains in some areas could lead to some extreme events such as floods and droughts. The analysis of the maximum duration of wet sequences shows a sharp decline in the Mono basin (Batablinlè et al., 2018). Ly et al. (2013) showed a gradual decrease in the maximum duration of the number of wet sequences in West Africa since 1960. Our work shows that such an evolution diagnosed on the present climate is likely to persist in the future. This decrease of the maximum duration of wet sequences in Togo and particularly on the Mono basin appears to be consistent with the evolution of the maximum duration of dry sequences because a sharp decrease in the maximum duration of wet sequences could favor an increase in the maximum length of the dry sequences. This decrease (respectively increase) of the wet sequences (respectively dry sequences) in the Mono basin is a response to global warming (Sylla et al., 2016). In Benin, the work on climate indices made. Yèkambèssoun et al., 2017 showed that only the annual total precipitation, the annual total of wet days and the annual maximum rainfall recorded during 30 d present a significant decreasing trend while the other rainfall indicators appear to remain stable for the period 1960–2000. Likewise rainfall during the months of June to September appears to have witnessed declining trends over 1961–1993 in Nigeria (Oguntundé et al., 2011; Easterling, et al., 2000).

The contribution of our study in the area is the use of several variables (analyzed with several models) based exclusively on precipitation and other parameters (e.g. temperature) to analyze the rainfall season patterns and determine the factors of their temporal variability. The results of this study could enable policy makers to adapt strategies needed for better management of water and natural disasters and the development of a resilient agricultural system in the face of climate change. Since assessments of the impacts of climate change in West-Africa are one of the issues most addressed by the IPCC, the use of multi-models for future extreme temperature and precipitation assessment under the greenhouse gas emission scenario RCP4.5 and RCP8.5 over Mono basin is a promising prospect that would contribute to a better understanding of the potential effects of climate change

4 Conclusions

A new multi-source dataset is used in this paper to analyze changes in extreme temperature and rainfall events over the Mono basin on the far future (2071–2100) and the historical period (1971–2000). The analysis of these climatic data has contributed to the knowledge of extreme phenomena and their evolution in the future over the Mono basin. These extreme climatic phenomena are very random and highly variable in space and time. Their knowledge is therefore fundamental to protect against natural disasters. The RCMs as well as their average showed that rain will decrease over the Mono basin while temperatures (Maximum and minimum) will increase. Moreover, others climatic indices have also been analyzed in order to characterize these events. In this way, for the simple daily intensity index and five-day maximum rainfall, the models have presented variable structures but the average of the models showed a moderate diminution on almost whole Mono basin except on some parts. The analysis of wet and dry sequences showed that the most of RCMs point out a fairly significant decrease in maximum duration of the number of consecutive rainy days and an increase of the maximum duration of the number of consecutive dry days almost all over the Mono basin. Projections of the RCMs showed that these extreme rain events do not evolve in the same direction as precipitation and this could lead to flooding intense in some parts of Mono basin particularly in the north. Seasonal evolutions were investigated to provide more detail about intra-annual behavior in extreme parameters for the Mono river basin. The results show that declining mean precipitation heights at all scales and irregular rainy seasons may be associated with a palpable increase in temperatures in the Mono river watershed by 2100.

Data availability. All data can be accessed via the references: The Table 1 presents the data used in this study. It is a set of eight climatic models outputs with daily time step and spatial resolution of 0.44° and available at: https://esgf-data.dkrz.de/projects/esgf-dkrz/ (last access: 3 August 2017). The observed data used, are provided by the Togo National Meteorology Agency (Météo Togo). Please, contact the authors if you are interested in this research data.

Author contributions. BL and AEL conceived and designed the research; CM and DLO contributed in providing data; BL performed the research, analyzed the data and wrote the paper.

Competing interests. The authors declare that they have no conflict of interest.

Special issue statement. This article is part of the special issue "Hydrological processes and water security in a changing world". It is a result of the 8th Global FRIEND–Water Conference: Hydrological Processes and Water Security in a Changing World, Beijing, China, 6–9 November 2018.

Acknowledgements. I acknowledge the financial support from the Deutscher Akademischer Austausch Dienst (DAAD).

All my gratitude to UNESCO-IHP and local organizers who gave me financial support and took charge of my participation in the 8th World Conference on Large River Basins in China. We are also grateful to the anonymous reviewers whose thoughtful and constructive comments have improved the quality of the article.

Financial support. This research has been supported by UNESCO-IHP.

References

Aguilar, E., Barry, A. A., Brunet, M., Ekang, L., Fernandes, A., Massoukina, M., and Zhang, X.: Changes in temperature and precipitation extremes in western central Africa, Guinea Conakry, and Zimbabwe, 1955–2006, J. Geophys. Res.-Atmos., 114, 1–11, https://doi.org/10.1029/2008JD011010, 2009.

Batablinlè, L., Lawin, E., and Agnide, C. M.: Africa Cordex Simulations Projection of Future Temperature, Precipitation, Frequency and Intensity Indices Over Mono Basin in West Africa, J. Earth Sci. Clim. Change, 9, 490, https://doi.org/10.4172/2157-7617.1000490, 2018.

Bazyomo, S. D., Ouedraogo, A., and Lawin, E. A.: Seasonal trends solar radiation available at the earth's surface and implication of future annual power outputs changes on the photovoltaic systems with one and two tracking axes, J. Clim. Weather Forecast., 5, 1–9, 2017.

Bigot, P.: Variability climatic, interaction and modification environnement, Variabilite climatique, Interaction et Modification, Environ. J. Hydrol., 4, 8141, 2003.

Diallo, I. and Camara, M.: Comparaison du climat et de la variabilite interannuelle de la pluie simulée au Sahel par les modèles climatiques regionaux (Comparison of climate and interannual variability of rainfall simulated in the Sahel by regional climate models), 24, 96–106, 2013.

Easterling, D., Evans, J., Groisman, P., Karl T., Kunkel, K., and Ambenye, P.: Observed variability and trends in extreme climate events: A brief review, B. Am. Meteorol. Soc., 81, 417–425, 200.

Emmanuel, L. A., Lamboni, B., Célestin, M., and Hodabalo, K.: Future ExtremesTemperature:Trends and Changes Assessment over the Mono River Basin, Togo (West Africa), J. Water Resour. Protect., 11, 82–98, 2019.

Ernest, A., Yves, T., Totin, H. S. V., Gil, M., and Pierre, C.:Dynamique et modélisation des crues dans le bassin du Mono à Nangbéto (Togo/Bénin), Hydrol. Sci. J., 59, 2060–2071, https://doi.org/10.1080/02626667.2013.871015, 2015.

Gudmundsson, L., Bremnes, J. B., Haugen, J. E., and Engen-Skaugen, T.: Technical Note: Downscaling RCM precipitation to the station scale using statistical transformations – a comparison of methods, Hydrol. Earth Syst. Sci., 16, 3383–3390, https://doi.org/10.5194/hess-16-3383-2012, 2012.

Hulme, M., Doherty, R. M., Ngara, T., New, M. G., and David, L.: African climate change: 1900–2100, Clim. Res, 17, 145–168, https://doi.org/10.3354/cr017145, 2001.

IPCC: Climate Change 2013: The Physical Science Basis, Contribution of WorkingGroup I to the Fifth Assessment Report of the Intergovernmental Panel on Climate Change, edited by: Stocker, T. F., Qin, D., Plattner, G.-K., Tignor, M., Allen, S. K., Boschung, J., Nauels, A., Xia, Y., Bex, V., and Midgley, P. M., Cambridge University Press, Cambridge, United Kingdomand New York, NY, USA, 2013.

Lamboni, B., Emmanuel, L. A., and Manirakiza, C.: Future extremes Temperature and Rainfall: Trends and Changes Assessment over Mono Basin in WEST AFRICA, XXXII Colloque l'Association Internationale de Climatologie; Climatic Change, Variability and Climate Risks, Thessaloniki Greece, 9–14, 2019a.

Lamboni, B., Emmanuel, L. A., Manirakiza, C., and Djibib, Z. M.: Variability ofFuture Rainfall over the Mono River

Basin of West-Africa, Am. J. Clim. Change , 8, 137–155, https://doi.org/10.4236/ajcc.2019.81008, 2019b.

Ly, M., Traoré, S. B., Alhassane, A., and Sarr, B.: Evolution of Some Observed Climate Extremes in the West African Sahel, Weather Clim. Ext., 1, 19–25, 2013.

Mason, S., Waylen, P., Mimmack, G., Rajaratnam, B., and Harrison, J.: Changes in extreme rainfall events in South Africa, Climatic Change, 41, 249–257, 1999.

MERF: Plan d'action national d'adaptation aux changements climatiques (PANA), (National Action Plan for Adaptation to Climate Change), Lomé, Republic of Togo, 27–134, 2009.

Nelson, G. C., Rosegrant, M., Palazzo, A., Gray, I., Ingersoll, C., and Robertson, R.: Food Security, Farming, and Climate Change to 2050: Scenarios, Results, Policy Options, Washington, 2010.

Nicholson, S. E. and Webster, P. J.: Physical basis for the interannual variability of rainfall in the Sahel, Q. J. R. Meteorol. Soc., 133, 2065–2084, 2007.

Oguntundé, P., Abiodunb, B., and Lischeid, G.: Rainfall trends in Nigeria, 1901–2000, J. Hydrol, 411, 207–218, 2011.

Ozer, P., Hountondji, Y., and Laminou Manzo, O.: Evolution des caractéristiques pluviométriques dans l'est du Niger de 1940 à 2007, Evolution of rainfall characteristics in eastern Niger from 1940 to 2007, GEO-ECO-Trop, 33, 11–30, 2009.

Sarr, A. B. and Camara, M.: Projections Climatiques Sur Le Sénégal, Europ. Sci. J., 13, 206–222, https://doi.org/10.19044/esj.2017.v13n17p206, 2017.

Soro, G. E., Anouman, D., Goula Bi, T. A., and Srohorou, B.: Characterization of sequences of meteorological dryness at various time scales in Sudanese climate: case of the extreme northwest of the Ivory Coast, Larhyss J., 18, 107–124, 2014.

Sylla, M. B., Elguindi, N., and Giorgi, F. W. D.: Projected robust shift of climate zones over West Africa in response to anthropogenic climate change for the late 21st century, Climatic Change, 134, 241–253, https://doi.org/10.1007/s10584-015-1522-z, 2016.

Tall, M., Sylla, M. B., Diallo, I., Pal, J. S., Mbaye, M. L., and Amadou, T. G.: Projected impact of climate change in the hydroclimatology of Senegal with a focus over the Lake of Guiers for the 21st century, Theor. Appl. Climatol., 129, 655–665, https://doi.org/10.1007/s00704-016-1805-y, 2016.

Yèkambèssoun, N. O., Emmanuel, A. L., André, A., Benjamin, K. Y., Ganiyu, T. O., and Abel, A. A.: Decreasing past and mid-century rainfall indices over the ouémé river basin, benin (West Africa), Climate, 5, 1–28, 2017

Proc. IAHS, 383, 79–84, 2020
https://doi.org/10.5194/piahs-383-79-2020

Hydroclimatic variability in Tomi at Sibut, Gribingui at Kaga-Bandoro and Fafa at Bouca basins, in the Central African Republic

Cyriaque Rufin Nguimalet[1] **and Didier Orange**[2]

[1]Département de Géographie, Faculté des Lettres et Sciences Humaines, Université de Bangui, BP 1037, Bangui, République Centrafricaine
[2]IRD, UMR210-Eco&Sols, University Montpellier, CIRAD, INRA, Supagro Montpellier, Montpellier, France

Correspondence: Cyriaque R. Nguimalet (cyrunguimalet@gmail.com) and Didier Orange (didier.orange@ird.fr)

Published: 16 September 2020

Abstract. The research of ruptures on rainfall and discharge serial data from 1950 to 1995 of three small catchments (from 2000 to 6000 km^2) of the Central African Republic, at the boundary between Chad and Congo basins, has shown a high spatial variability. The rupture observed in 1970 at the subcontinental scale, which started the drought period in West and Central Africa, is observed only on the Northward basin, the driest. Then it was difficult to compare the hydroclimatic periods from a basin to another one. However, all the studied basins have shown a degradation of the hydrological regime from the end of the 1980s onward, with a severe level since the end of 1980s. The depletion coefficients have the same range for the 3 studied basins than for the Ubangi River basin, widening the drought impact.

1 Introduction

In Western and Central Africa, the major climatic disruption in 1970 due to a weak annual pluviometry is well documented (Olivry et al., 1998; Wesselink et al., 1996; Nguimalet and Orange, 2013). But very few studies have analyzed this effect on small river basins (below 10 000 km^2). Goula et al. (2006) and Fadika et al. (2008) revealed a latest impact of the drought rupture on small river basins towards 1980 in Western Africa, compared to the large rivers. Nevertheless, Goula et al. (2006) have shown in Ivory Coast an earliest rainfall reduction of −14 % (1967–1993) on the N'Zo basin (4310 km^2) against a greater impact late over the annual mean discharges (−32 % of reduction over 1981–1993 period). Thus, this paper deals with the effect of hydroclimatic variabilities on 3 close small basins, located at the boundary between Chad River (Gribingui and Fafa) and Congo River basins (Tomi). The aim is to confirm and to date the periods of hydrological variabilities of the targeted small river basins and to compare with the hydroclimatic periods defined in the Ubangi and Congo basins (Orange et al., 1997; Laraque et al., 1997, 2001). We'll discuss the impact on the groundwater's regimes per basin.

2 Data and methods

2.1 Overview of the studied basins

Tomi at Sibut (2380 km^2), Gribingui at Kaga-Bandoro (5680 km^2) and Fafa at Bouca (4380 km^2) are located in the Central-South and Central-North of the CAR (Fig. 1), located at the boundary between the upper Chad basin (Fafa and Gribingui) and and the Congo basin (Tomi). The elevation, the geology, pedology, vegetation, land-use and demographic pressure are quite similar: the elevation ranges from around 700 to 400 m (Nguimalet, 2013). The studied river basins are located in the Sudano-guinean subtype of the wet tropical climate, with two contrasted seasons: a rainy season (from April to October) and a dry season (from November to March). The 3 rivers are mostly perennial but some periods of discharge stop may occur due to the ongoing drought severity. The people density of 6.2 inhabitants km^{-2} is weak, without

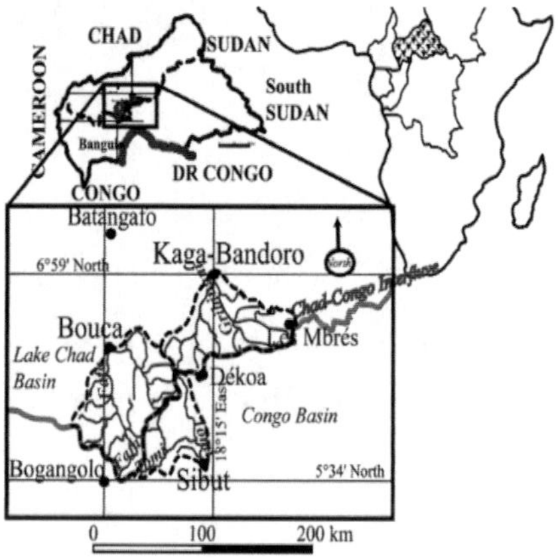

Figure 1. Location of the 3 studied catchments (Fafa at Bouca, Tomi at Sibut and Gribingui at Kaga-Bandoro).

any pressure on natural resources in these basins (Nguimalet, 2018).

2.2 Data and methods used

Annual rainfall (R) and annual discharge (Q) have been gathered over the 1950–1995 period from the ASECNA at Bangui and was supplemented by the Institut de Recherche pour le Développement (IRD). Raingauge data were recorded in six gauging stations distributed over all basins-slopes such as Bogangolo, Sibut, Dékoa, Bouca, Kaga-Bandoro and Les Mbrés (Fig. 1). The gaps of a few months were filled by the averages framing method among the monthly totals or average. Because of the lack of measurements over the 1950–1957 period, annual mean discharges of the Fafa river at Bouca have been estimated by linear regression from those of the Ouham river at Batangafo. These data (R and Q) were disposed according to hydrological year that goes from 1 April to 31 March and analyzed with different statistical tests to search ruptures, with Khronostat 1.01 software (Lubes-Niel et al., 1998). The rainfall index and the flow index (the rate of annual amount to the interannual average of the series over the targeted period) was calculated per basin. Daily mean Q data were used to calculate the depletion coefficients and the potential groundwater volume (Goula et al., 2006) in order to appreciate their pluriannual evolution.

3 Results

3.1 Rainfall and rupture points in the time series for the 3 basins (1950–1995)

Yearly extremes are 1729 mm (1980–1981) and 918 mm (1951–1952) in the Tomi basin, 1628 mm (1975–1976) and 694 mm (1990–1991) in the Gribingui basin, and 1638 mm (1961–1962) and 1101 mm (1972–1973) in the Fafa basin. Based on the rainfall index (Fig. 2), only the Gribingui basin seems to be marked by a drought period, starting from 1968. The results show a weak rainfall interannual variability (from $-1\,\%$ to $-3\,\%$) at the boundary between the upper Chad basin (Fafa and Gribingui) and the Congo basin (Tomi), compared to the highest observed in the Lake Chad basin by Ardoin-Bardin et al. (2009). That would be explained by (i) the basins' small size, (ii) their altitudinal position on the top of both Chad and Congo upper basins, and (iii) also the major role of vegetation cover and drained altered nodular soils.

The analysis of ruptures confirms that the major climatic rupture measured on the larger basin such as Ubangi is not high on these 3 basins, even absent on the Tomi basin (Table 1). The break of the large drought recorded in West and Central Africa, marked in 1968 on the Ubangui Basin, is only recorded on the Fafa basin. The Pettitt's test and the Bayesian method have given identical breaks dates in the Gribingui (in 1982) and the Fafa (in 1969) basins, which diverge with those obtained by Hubert's segmentation. According to the Hubert's segmentation, the Gribingui basin has known two ruptures (1989, 1990) with two rainfall homogeneous periods: 1314 mm in 1951–1989, 694 mm in 1990–1990, and 1367 mm in 1991–1995. In fact, the 1990 break shows an unique year dropped. To resume, only the Fafa basin, located between those of Gribingui in the North and Tomi in the South, has recorded a real rainfall rupture in 1968, with homogeneous periods, $+6\,\%$ over the wet period (1951–1968) and $-3\,\%$ over the dry period (1969–1994). But on Ubangi basin, there are two homogeneous dry periods since 1968: less dry on 1969–1980 ($-1\,\%$) and dry (1981–1994) over this 1950–1995 period (Table 1). Although a comparison is done between the 3 basins' trend and the Ubangi's one, its largest size basin (499 000 km^2) subjected to the high spatial heterogeneity leading to its rainfall distribution and reduction. Due to the 3 basins small size (from 2000 to 6000 km^2), the diversity of observed ruptures is probably directly linked to their basin characteristics. The particularity of Tomi basin which does not record any rupture could be justified by the geographical features (smallest and steeper basin, density of vegetation, drained nodular soils) showing alternated short wet and dry rainfall trend over time.

3.2 Discharges and ruptures of stationarity of the studied basins

The interannual mean discharge on 1950–1995 period is, from North to South, 22 $m^3\,s^{-1}$ for Gribingui, 37 $m^3\,s^{-1}$ for Fafa and 18 $m^3\,s^{-1}$ for Tomi, with respectively specific discharge 3.9, 8.4, and 7.6 $L\,s^{-1}\,km^{-2}$. The Gribingui basin is twice drier than Fafa and Tomi basins. Such as the rainfall pattern, the discharge evolution of Gribingui is quite equivalent to that recorded on the Ubangi at Bangui. At the opposite, the discharge evolution of Tomi didn't show any marked drought period: just a weak humid period from 1961 to 1970, then a high peak of discharges corresponding to heavy rains

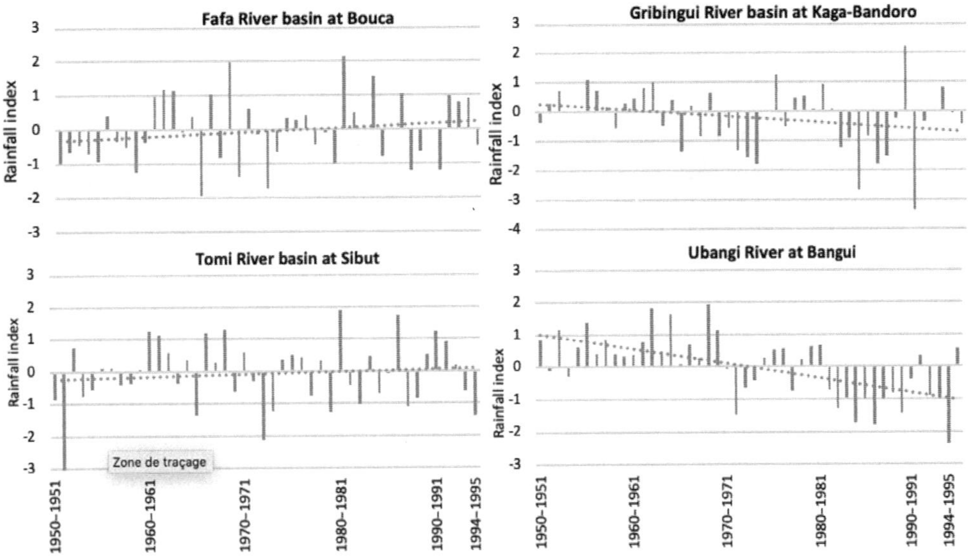

Figure 2. Rainfall index from 1950 to 1995 for 3 studied basins and the Ubangi basin at Bangui.

Table 1. Years of ruptures detected in rainfall time series per basin.

River Basins	Interannual rainfall (mm)	Pettitt	Bayesian	Hubert's Segmentation (mm)	Period characteristic	Period ratio (%)
Gribingui	1315	1982	1982	1951–1989: 1314	Dry	−1
				1990–1990: 694	Very dry	−53
				1991–1995: 1367	Wet	+4
Fafa	1389	1969	1969	1951–1968: 1476	Wet	+6.3
				1969–1994: 1340	Dry	−3
Tomi	1403	–	1994	1951–1995: 1,417	Normal	+1
Ubangi	1437	1969	1980	1950–1968: 1519	Wet	+6
				1969–1980: 1426	Less Dry	−1
				1981–1994: 1328	Dry	−8

The period ratio is the rainfall amount of one period divided by the interannual rainfall average.

during 3 consecutive years (1991, 1992, 1993) (Fig. 2), with important flood events. Indeed, the 1990s is marked by a rainfall recovery observed on Fafa and Tomi basins, which was not recorded in the Ubangi River basin. The discharge evolution in the Fafa basin is a balance between the Gribingui and Tomi basins. There is a humid period before the 1970s and a peak discharge from 1991 to 1993 due to short rainfall recovery at this time (Fig. 3).

The rupture tests over the hydrological series (Table 2) are varied to results found on rainfall series: one rupture for Tomi in 1990/1991, one break only for Gribingui in 1984 (2 years after the rainfall rupture) and a severe rupture for Fafa in 1968, and mainly three ruptures (1958, 1969 and 1980) for Ubangi (Table 1).

3.3 Dynamics of the depletion coefficient

In the 3 studied basins, there are the same range of the depletion coefficient than for the Ubangi at Bangui (Nguimalet and Orange, 2019), from 0.010 to 0.035 d^{-1} over the studied period. Maxima are of 0.035 d^{-1} in 1987 and 1989 over Tomi, 0.030 d^{-1} in 1989 over Gribingui, 0.032 d^{-1} in 1988 over Fafa and 0.025 d^{-1} in 2000 over Ubangi (Fig. 4). These maxima are all recorded in the 1990's period. And all the minima are situated around the year 1970: 0.08 d^{-1} in 1973 on Gribingui, 0.010 d^{-1} in 1968 on Fafa, and 0.015 d^{-1} in 1970 on Tomi. On the Ubangi, the minimum was of 0.013 d^{-1} in 1953. These figures show a critical evolution of the groundwater contribution to the discharge: from 0.054 km^3 in 1987 to 0.4 km^3 in 1991 on Tomi, from 0.095 km^3 in 1987 to 0.24 km^3 in 1992 on Gribingui, from 0.071 km^3 in 1987 to 0.5 km^3 in 1994 on Fafa, and from 55 km^3 in 1974 to 21 km^3

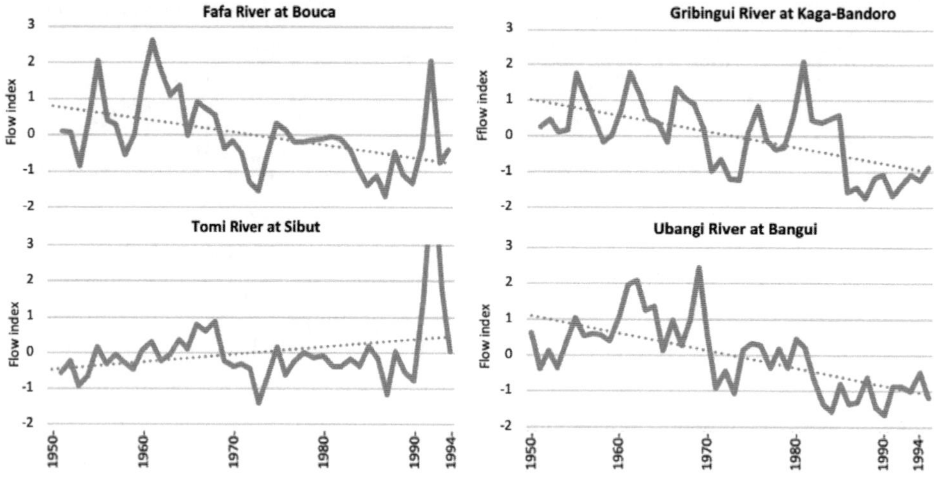

Figure 3. Interannual evolution of flow index over 1951–1995.

Table 2. Years of ruptures detected in hydrological time series per basin.

River basins	Interannual discharge ($m^3 s^{-1}$)	Pettitt	Bayesian	Hubert's Segmentation ($m^3 s^{-1}$)	Period	Period ratio (%)
Gribingui	22	1984	1984	1951–1983: 26 1984–1993: 8.4	Wet Very dry	+19 −61
Fafa	37	1968	1968	1951–1967: 57 1968–1993: 29	Wet Dry	+43 −22
Tomi	18	–	–	1951–1990: 17 1991–1991: 66 1992–1993: 26	Dry More Mean wet	−6 +67 +44
Ubangi	3748	1970	1970	1950–1958: 4044 1959–1969: 4877 1970–1980: 3618 1981–1994: 2729	Mean wet Wet Dry Very dry	+8 +30 −3 −27

The period ratio is the discharge amount of one period divided by the interannual discharge average.

in 1990 on Ubangi. Though the 1970 rupture is not so severe, and even not visible over the Tomi basin, the total potential groundwater per catchment seems to diminish significantly everywhere during the most recent studied years. Fafa and Gribingui described the same image as Ubangi for both the evolution of depletion coefficient and total potential groundwater, each time around of 1970 (Fig. 4). Due to this global trend, groundwater level increases since the 1988 severe drought benefiting of the short rainfall recovery for Fafa, Tomi and also for Ubangi while it decreases for Gribingui.

4 Discussion: groundwater's responses to these hydroclimatic ruptures

Nowadays, the anthropization of the 3 studied basins is still weak, not only because of weak human densities (6.24 inhabitants km^{-2}), but also because of the extensive agriculture practices. Then this weak human impact on the environment cannot drive any pressure on the hydrological regimes of these basins. The evolution of the rainfall-discharge relationship from 1950 to 1995 is quite similar inside each small studied basin, but they are not comparable between each basin. First it is noticeable that the great drought recorded within all the large rivers from West and Central Africa has been recorded only on Fafa basin that flows towards Lake Chad. On this basin, the drought period started in 1968 as recorded on the Ubangi river at Bangui, with successive drops (−3 % and −27 % respectively). It is the case of the Pipi basin at Ouadda in the NE of the Ubangi basin, where the drought begins in 1968 (Nguimalet and Ndjendolé, 2008). Besides, there is an absence of rainfall rupture in the Tomi basin meanwhile the discharge rupture occurred in 1990. That would be due both to its smallest size (2380 km^2) and to its nodular soils and forested vegetation

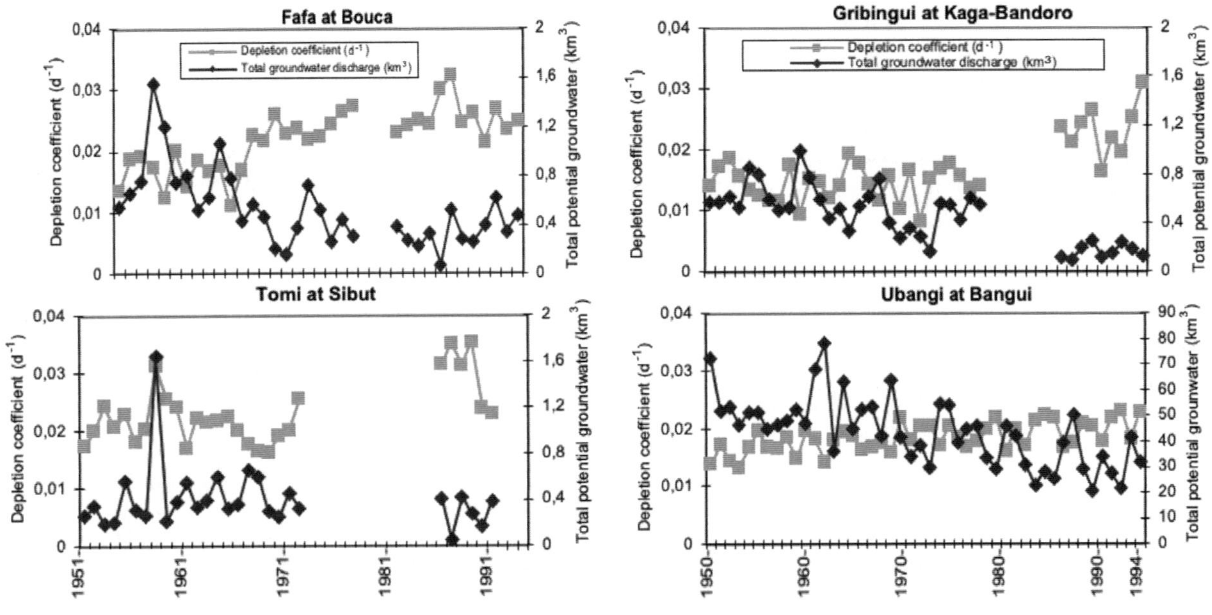

Figure 4. Interannual evolution of depletion coefficient and total potential groundwater discharge (1950–1995).

towards its common sources with Fafa. For both, the importance of vegetation cover and respective soils play a main role into rainy processes. In addition, the singular behavior of Tomi would come from its high elevation's difference ($0.0024\,\mathrm{m\,m^{-1}}$), compared to Gribingui ($0.0014\,\mathrm{m\,m^{-1}}$) and Fafa ($0.0011\,\mathrm{m\,m^{-1}}$). A reverse situation is observed in the Gribingui basin with both a rupture year for rainfall in 1989 against only one for the annual mean discharge serial data. The diversity of these results is probably linked to the hydrological abundance of each basin. On the driest northwards, Gribingui with a specific discharge of half the two other studied basins, the river discharge is not really impacted by the drought of the 1970s but only the recent drought on 1988 recorded on the 3 rivers is marked, even on Ubangi river. The regime of Tomi is quite stable, consequence of a homogeneous rainfall pattern from 1950 and probably due also to a good groundwater contribution, marked by a specific discharge over $7\,\mathrm{L\,s^{-1}\,km^{-2}}$ and a smooth evolution of the depletion coefficient due to its basin' geographical features (Fig. 4). The Fafa basin shows an intermediate behavior between Gribingui in the North and Tomi in the South. Then it seems that a drought effect into these catchments is more marked according to a South-North gradient through ruptures on rainfall and discharge regimes at different dates. Conway et al. (2009) also noticed that Eastern Africa's rivers have known the hydrological degradation later during the 1980s, when the drought was extensively recorded in many areas over subSaharan Africa. As for rainfall regimes modifications, Kouassi et al. (2018) talked about "late break " observed in 1980s, compared to "earliest" that occurred in the late 1960s. Thus one supposes that variable "size of basin" would intervene according to brutality/ spontaneity (in case of Tomi) of flow regime under the effect of slope (basins of small size) or of delayed and differed response of basin to rainy events.

However, the groundwater time series are more impacted than the discharge ones. On the Tomi and the Fafa basins, the depletion coefficient and the total potential groundwater discharge have a similar evolution than that of the Ubangi basin at Bangui since the end of 1980s, mainly in 1988s. Only the basin of Gribingui at Kaga-Bandoro shows a weak decrease of the groundwater potential from early 1990s.

5 Conclusion

This research of ruptures on the rainfall and discharge long serial data on 3 small close basins (from 2000 to $6000\,\mathrm{km^2}$) of Central Africa north of the Equator, has shown a high variability over the period 1950 to 1995. The continental rupture of 1970 marking the beginning of the drought in West and Central Africa is not clearly observed. Each basin displays the specificity of its hydroclimatic functioning. Then it is difficult to compare the hydroclimatic periods from one basin to another one. However, all the studied basins have shown a degradation of their hydrological regimes since the end of the 1980s due to the drought severity. That is confirmed by Mahé (1995) that the 1981–1989 period is dry, mainly in the North (Ubangi) at the interface of Chad and Congo basins where located the 3 studied basins. The potential of groundwater discharge has decreased on the 3 studied basins, with a minimum level around 1988 for each basin. Nevertheless, it seems to increase in the 3 basins with a rainfall recovery noted at the 1990s starting.

This study has demonstrated that even if the discharge evolution is not really marked by the drought of 1970, all the studied basins are nowadays under water stress with their potential groundwater discharge at the weakest state.

Data availability. The original data are available online at https://www.researchgate.net/publication/343111160_RCA_Tomi_Gribingui_Fafa_Data#fullTextFileContent (Orange, 2020).

Author contributions. Both authors contributed equally to this work.

Competing interests. The authors declare that they have no conflict of interest.

Special issue statement. This article is part of the special issue "Hydrological processes and water security in a changing world". It is a result of the 8th Global FRIEND–Water Conference: Hydrological Processes and Water Security in a Changing World, Beijing, China, 6–9 November 2018.

Acknowledgements. The authors acknowledge the FRIEND–Water network for the opportunity to publish these original data. Special thanks for the anonymous reviewers.

References

Ardoin-Bardin, S., Dezetter, A., Servat, E., Paturel, J. E., Mahé, G., Niel, H., and Dieulin, C.: Using general circulation model outputs to assess impacts of climate change on runoff for large hydrological catchments in West Africa, Hydrolog. Sci. J., 54, 77–89, 2009.

Conway, D. P., Persechino, A., Ardoin-Bardin, S., Hamandawana, H., Dieulin, C., and Mahé, G.: Rainfall and river flow variability in sub-Saharan Africa during the 20th century, J. Hydrometeorol., 10, 41–59, 2009.

Fadika, V., Goula, B. T. A., Kouassi, F. W., Doumouya, I., Koffi, K., Kagamate, B., Savane, I., and Srohorou, B.: Variabilité interannuelle et saisonnière de l'écoulement de quatre cours d'eau de l'Ouest côtier de la Côte d'Ivoire (Tabo, Dodo, Néro, San Pédro) dans un contexte de baisse de la pluviométrie en Afrique de l'Ouest, Eur. J. Sci. Res., 21, 406–418, 2008.

Goula, B. T. A., Savané, I., Konan, B., Fadika, V., and Kouadio, G. B.: Impact de la variabilité climatique sur les ressources hydriques des bassins de N'Zo et N'Zi en C. I. (Afrique tropicale humide), VertigO, 7, 1–12, 2006.

Kouassi, K. M., Kouassi, K. L., Lorougnon, G. D. J., Yao, K. B., Mélédjé, N. H., Biemi, J., Lasm, T., and Rouché, N.: Variabilité des extrêmes pluviométriques sur le bassin versant de la rivière Bia (Sud-Est, Cote d'Ivoire), Eur. J. Sci. Res., 14, 134–147, 2018.

Laraque, A., Olivry, J. C., Orange, D., and Marieu, B.: Variations spatio-temporelles des régimes pluviométriques et hydrologiques en Afrique Centrale du début du siècle à nos jours, FRIEND '97 – Regional Hydrology: Concepts and Models for Sustainable Water Resource Management, IAHS Publ., 246, 257–263, 1997.

Laraque, A., Mahé, G., Orange, D., and Marieu, B.: Spatiotemporal variations in hydrological regimes within Central Africa during the twentieth century, J. Hydrol., 245, 104–117, 2001.

Lubes-Niel, H., Masson, J. M., Paturel, J. E., and Servat, E.: Variabilité climatique et statistiques. Étude par simulation de la puissance et de la robustesse de quelques tests utilisés pour vérifier l'homogénéité de chroniques, Rev. Sci. Eau, 11, 383–408, 1998.

Mahé, G.: Modulation annuelle et fluctuations interannuelles des précipitations sur le bassin-versant du Congo, in: Actes du Colloque PEGI Grands Bassins Fluviaux Péri-Atlantiques: Congo, Niger, Amazonem edited by: Boulègue, J. and Olivry, J. C., INSU, CNRS, ORSTOM, Paris, France, 13–26, 1995.

Nguimalet, C. R.: Climate change analysis across rainfall-discharge variability in selected river catchments of Kenya and Central African Republic, Palaeoecol, Afr,, 32, 195–220, 2013.

Nguimalet, C. R.: Comparison of community-based adaptation strategies for droughts and floods in Kenya and the CAR, Water International, 43, 183–204, 2018.

Nguimalet, C. R. and Ndjendolé, S.: Les extrêmes hydrologiques: des indicateurs d'hydrodynamisme ou d'hydraulicité du plateau gréseux de Mouka-Ouadda sur la rivière Pipi à Ouadda (République Centrafricaine), Z. Geomorphol., 52, 125–141, 2008.

Nguimalet, C. R. and Orange, D.: Dynamique hydrologique récente de l'Oubangui à Bangui (Centrafrique): impacts anthropiques ou climatiques?, Geo-Eco-Trop, 37, 101–112, 2013.

Nguimalet, C. R. and Orange, D.: Caractérisation de la baisse hydrologique actuelle de la rivière Oubangui à Bangui, République Centrafricaine, Houille Blanche, 1, 78–84, 2019.

Olivry, J. C., Briquet, J. P., and Mahé, G.: Variabilité de la puissance des crues des grands cours d'eau d'Afrique intertropicale et incidence de la baisse des écoulements de base au cours des deux dernières décennies, Water Resources Variability in Africa during the XXth Century, IAHS Publ, 252, 189–197, 1998.

Orange, D.: RCA_Tomi Gribingui Fafa Data, ResearchGate, available at: https://www.researchgate.net/publication/343111160_RCA_Tomi_Gribingui_Fafa_Data#fullTextFileContent, last access: 21 July 2020.

Orange, D., Wesselink, A. J., Mahé, G., and Feizouré, C. T.: The effects of climate changes on river baseflow and aquifer storage in Central Africa. In: Sustainability of water resources under increasing uncertainty, IAHS Publ, 240, 113–123, 1997.

Wesselink, A., Orange, D., Feizoure, C., and Randriamiarisoa: Les régimes hydroclimatiques et hydrologiques d'un bassin-versant de type tropical humide: l'Oubangui (République Centrafricaine), in: L'Hydrologie Tropicale: géoscience et outil pour le développement, IAHS Publ, 238, 179–194, 1996.

Proc. IAHS, 383, 85–91, 2020
https://doi.org/10.5194/piahs-383-85-2020

The Runoff Observation and Composition Analysis in the Niyang River Basin on the Tibetan Plateau

Hongwei Liu[1,2,3], Jiufu Liu[1,2,3], Wenzhong Wang[1,2,3], Xing Min[1,2,3], Hao Zheng[1,2,3], Xiyuan Deng[1,2,3], and Niu Wang[1,2,3]

[1]Hydrology and Water Resources Department, Nanjing Hydraulic Research Institute, Nanjing, Jiangsu, 210029, China
[2]The State Key Laboratory of Hydrology-Water Resources and Hydraulic Engineering, Nanjing, Jiangsu, 210029, China
[3]Research Center for Climate Change, The Ministry of Water Resources of the People's Republic of China, Nanjing, Jiangsu, 210029, China

Correspondence: Hongwei Liu (hwliu@nhri.cn)

Published: 16 September 2020

Abstract. In order to study on the hydrometric methods in the high-altitude area, and learn more about the hydrological mechanism in the Qinghai-Tibet Plateau, a series of observation research were carried out in the Niyang River watershed, a tributary of the Yarlung Zangbo River. The glacier recession and the runoff variation which conducted by the climate change asked people to study more on the plateau hydrology, where is extremely lack of the hydrological monitoring and dataset in the past due to the high altitude and the cold environment. In this study, 4 basins with the areas of 182, 216, 243, 213 km^2 was equipped with different types of hydrometric instruments, so that the precipitation (rainfall and snow), water level, flow speed and the meteorological elements could be measured. In addition, the soil and water in the watershed were sampled at the altitude from 3667 to 6140 m, which could be used to analyze the runoff composition. With the equipment, the applicability of the hydrometric methods and instruments were discussed according to the monitoring situation. With the samples' analysis results (including the δD, δ^{18}O, and the chemical components), the runoff composition characteristics were analyzed. This research is funded by the NSFC project 91647111 and 91647203, which are included in the *Runoff Change and its Adaptive Management in the Major Rivers in Southwestern China* Major Research Plan.

1 Introduction

The Qinghai-Tibet plateau, so-called *Asian Water Tower* (Immerzeel et al., 2010; Viviroli et al., 2007), origins many large rivers in Asia, including the Yangtze River, Yellow River, Yarlung Zangbo River (Brahmaputra River), Lancang River, Nu River, etc (Fig. 1). Water resources and hydro energy are quite rich in this area, and more than 500 billion m^3 fresh water flows out of China through the Yarlung Zangbo river, Lancang river, and Nu river (Yao et al., 2000) The region is a climate change sensitive area, and its water resources situation is obviously affected by climate change. In the past 50 years, the glaciers in the southwestern river source area have been retreated, and the snow line has risen. Accordingly, the runoff in the Southwest China increased significantly (Immerzeel et al., 2010), but whether it has encountered the inflection point is still unclear (Piao et al., 2010). In a changing climate with the water resources demands growth, the role that Asian Water Tower plays for the regional water river supply have to be identified more closely, which is based on understanding of the runoff variation trends and mechanism in the plateau area.

Although there is great need for the knowledge of the runoff evolution and biogenic substances migration and transformation for developing strategies for adaptive utilization of runoff, progress in study on these hydrological questions lags behind in the Qinghai-Tibet plateau than the other regions because of lack of observation dataset (Liu et al.,

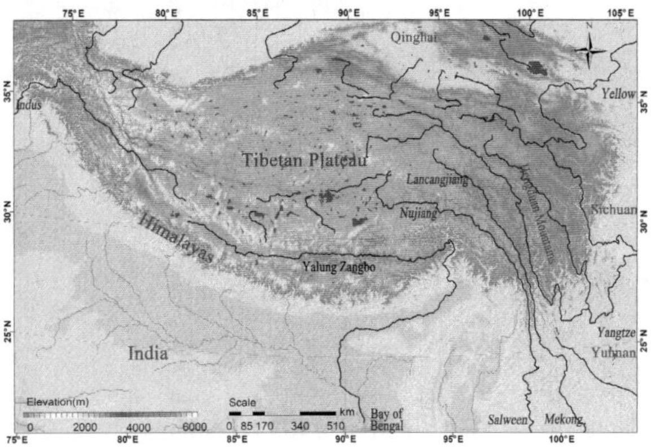

Figure 1. Large rivers flow out of Tibet Plateau.

2016; Yin et al., 2016). Scarce hydrometric network was built in the extremely topographically biased alpine region, for what accumulated observing records were far from adequate for researching the highly spatial heterogeneity of hydrological, meteorological and climatological patterns (Briggs and Cogley, 1996; Liu et al., 2016; Viviroli et al., 2007). A consensus is that the research in data scarce areas would be a long-term mission for the hydrologists, due to that the IAHS (International Association Of Hydrological Sciences) started an PUB plan (Predictions in Ungauged Basins) in the 23rd IUGG conference in 2003 (Sivapalan et al., 2003). The PUB methods also depend on the regional hydrological and meteorological knowledge, or the new remote sense satellites and the data fusion (Blöschl et al., 2013; Liu et al., 2010). Consequently, sound monitoring dataset is an irreplaceable foundation for the hydrological research and development basically. Fortunately, new methods development, especially the sensing and information technology, provide more possibilities for the high altitude, extremely cold, and mountainous plateau, with which hydrological research will focus on much wider range of spatial and temporal scales and interactions between water, earth, and biological systems (Montanari1 et al., 2015).

Benefit from the information technology development and internet of things, traditional sensors are redesigned to monitor, record, and transformation hydrometric data automatically (Shu, 2005). The monitoring data could break away from the spatial distance and physical network, which make building the hydrometric network in the depopulated alpine zones (Meyer and Huey, 2006), such as the discharge measurement based on the intelligent control of hydrological cable (Yuan et al., 2004), the automatic hydrometric boat (Wu and Kong, 2017), ADCP (Liu, 2016), and so on. Generally, the online monitoring hydrometric technologies with the contact-type sensors are able to use practically in the hydrometric networks. And more non-contact-type sensing technologies are developing, including grand-based, airborne-based, and satellite based remote sensors. The ground-based

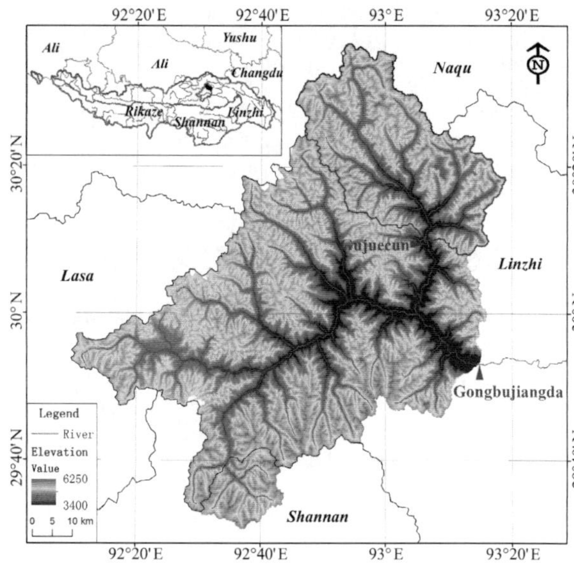

Figure 2. Map of the study area (1. Yarlung Zangbo, 2. Gongbujiangda, 3. Gujuecun Basin).

water level and discharge radar, and the ultrasonic water level meter had been widely used in the hydrometric stations (Shu, 2005). Airborne lidar was used for the terrain elevation scan and modelling, with which the hydraulic model could be develop to calculate the river discharge (Roub et al., 2012). Air-based flow velocity measurement technique is also developing. The satellite-based radar and multispectral camera data are widely used for the precipitation monitoring and forecasting (Wandishin et al., 2005), total water storage (Long et al., 2015), variation of the snow cover and glacier (Deems et al., 2013; Nolan et al., 2015).

Since 2015, the NSFC launched a Major Research Plan named *Runoff Change and its Adaptive Management in the Major Rivers in Southwestern China*, which focus on the research of the analysis of the water and its solute sources, runoff variation, and the adaptive management mainly in the Qinghai-Tibet plateau region. Observation of the hydrological information is a critical portion of the plan, which focus on the effective monitoring methods of the water quantity, water quality and other relevant elements in the alpine environment.

The high-altitude region with low air pressure, low temperature, and strong ultraviolet ray bring a lot of problems on using the hydrological and meteorological monitoring instruments, for what A series of hydrological monitoring and water components analysis field experiments were carried out in a sub-basin of the Yarlung Zangbo River, named Niyang River Basin. It is expected to develop the hydrometric network construction principles and understand the runoff components better.

Table 1. Area and land-use of the interested basins.

Basin name	Area	Land-use (%)		
	km^2	woodland	Grassland	Snow/Ice
Yarlung Zangbo	253 000	13.21	61.98	9.81
Gongbujiangda	6400	24.48	69.35	5.95
Gujuecun	1547	37.11	51.65	11.18
Burulangqu	182	22.42	26.18	51.40
Yimenlangqu	216	57.35	18.25	24.34
Chuqu	243	61.54	36.46	1.95
Jinnongqu	213	11.54	85.17	1.02

2 Study Area

2.1 Site Locations

The study area locates in the Niyang River Watershed, which is a tributary of the Yarlung Zangbu River in the Qinghai-Tibet Plateau. The interest basin named Gujuecun Basin is in the area of Gongbujiangda Country belongs to the Tibet autonomous region (Fig. 2), which is a sub-basin of the Niyang River Basin.

The Yarlung Zangbo River as one of the highest rivers in the world origins from Jiemayangzong Glacier in western Tibet and flows more than 2000 km in China before entering India, in which the name changes to Brahmaputra River. The Niyang River is one of the 5 biggest tributaries with 307.5 km long and 2.2×10^{10} m^3 / a volume of runoff. The Gujuecun basin is a rugged mountainous sub-basin with the elevation between 3667–6140 m.

2.2 Land-use

Land-use types in the experimental area are mainly the forest, grassland, snow and ice (including temperate glacier), as shown in the Table 1 and Fig. 3.

Four sub-basins could be divided from the Gujuecun Basin, and named I-Burulangqu Creek, II-Yimenlangqu Creek, III-Chuqu Creek, and IV-Jinnongqu Creek. The Glacier spreads in the east part of the Gujuecun Basin, so that more focus is paid to Burulangqu Creek and Yimenlangqu Creek with more comprehensive land-use combinations.

2.3 Cross-sections Gauged

Gujuecun Basin is an ungauged basin because of the mountainous topography and lack of electricity and communication lines. In order to get the basic hydrological dataset, 5 monitoring gauges were built at 4 outlets of the sub-basins and the general outlet of Gujuecun Basin (Fig. 3).

3 Survey and Observation

Besides the hydrological data, the dataset needed for the runoff component analysis includes the geochemical and the

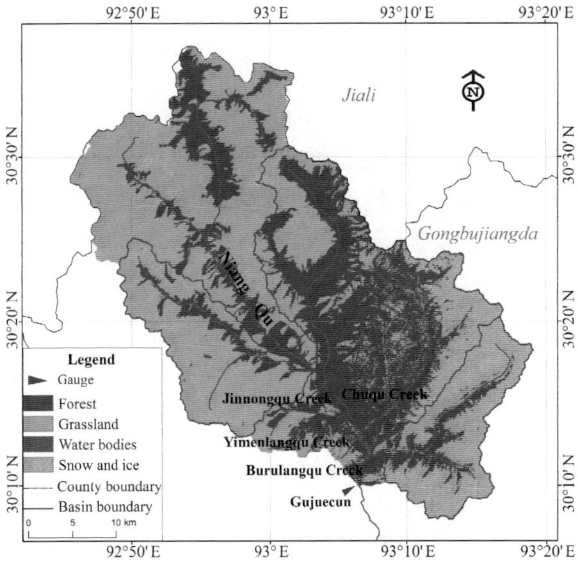

Figure 3. Map of the land-use distribution of gauges in the outlets of the sub-basins.

stable isotopic data in the river water, soil water, and vegetation water. A possible effort to make runoff components clear in this ungauged basin is the End Member Analysis (EMMA) finally, but the first issue should be solved is the fundamental dataset.

The measures for achieving the dataset generally through three kinds of work, observing in site, gauging, sampling and analysis.

3.1 Observation in site

The necessary terrain information, gauging cross-section, and some chemical concentrations of the water were measured in site.

Although the overall topographic data could use the satellite DEM dataset, the detailed terrain data necessary around the 5 gauges must be surveyed in site. A 3-D laser scanner (Rigel VZ1000 type) was used to scan the ground surface, and the points cloud was record to establish the terrain model (Fig. 4). The underwater elevation was measured by the dip sticks. The gauging sites' cross-section were able to be drawn by combining the surface and underwater survey results. The pH, dissolved oxygen, water temperature, water conductivity, and measured with Hash HQ40d Portable Meter with sensors in every sampling site (will be introduce in Sect. 3.3).

3.2 Gauging

As shown in Figs. 3, 5 gauges was established in 2017 to launch the hydrometric monitoring. Radar water level meter was installed on a iron frame hanging over the river channel. Data was stored in a local memory and transmitted by

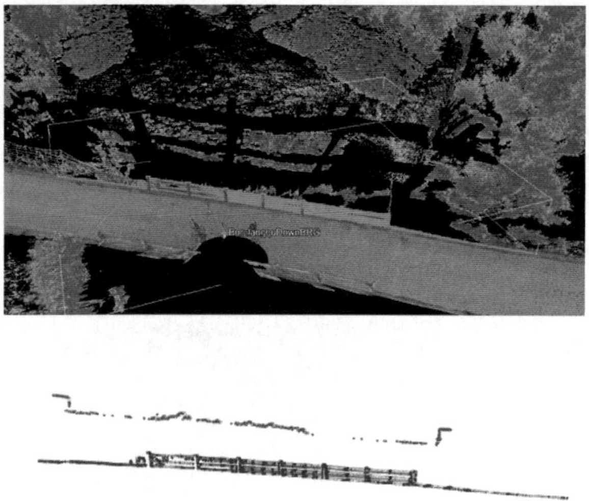

Figure 4. Elevation points cloud scanned by Rigel VZ1000.

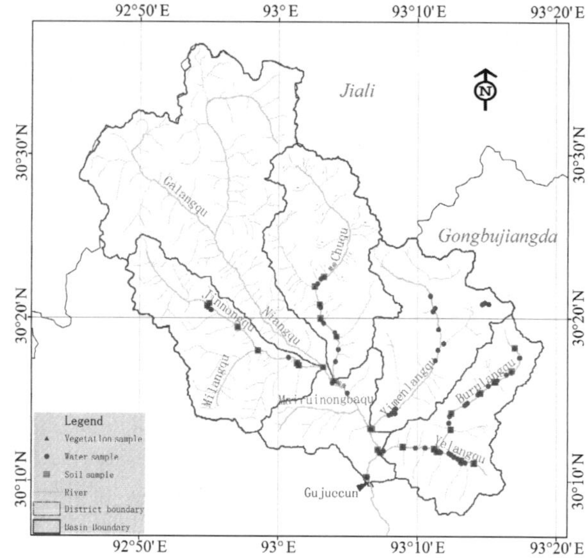

Figure 5. Sampling points in the field trip.

Beidou satellite, and the satellite receiver was installed in an experimental field of Nanjing Hydraulic Research Institute, 2300 km away.

3.3 Field Trip and Sampling

More data necessary was not able to monitor by the 5 gauges, consequently, field trips and sampling were launched in the area. Focus on the comprehensive land-use area, sampling was concentrated in the Burulangqu Creek and Yimenlangqu Creek, which have glacier and snow. River water, sprint water, snow, glacier ice, soil, and vegetation samples were taken ever 5 m × 5 m grid, if the types of samples exist (Fig. 5). Three times sampling trips were arranged in one year, winter (before snowmelt), summer (wet season), autumn (before the snow accumulated).

3.4 Sample Analyzing

The water samples were examined by the LGR isotopic water analyser, DIONEX ICS-2100 ion chromatographic system, and PE ICP-OES Optima 7000DV, with which δ^{18}O, δ^{2}H, K^+, Na^+, Ca^{2+}, Mg^{2+}, Sr, Si, F^-, Cl^-, NO_3^-, SO_4^{2-}, etc. were analysed.

The soil and vegetation samples were treated with the cryogenic vacuum distillation system, with which the water could be extracted, and δ^{18}O, δ^{2}H were analysed.

4 Results and Discussion

4.1 Precipitation and river water isotopes

The steady isotopes show regularities in the stable isotopic composition of the precipitation water and surface water along the river according to the elevation change in the Gujuecun Basin obviously, that the higher elevation samples show lower δ^{2}H and δ^{18}O content, and the δ^{2}H vs δ^{18}O scatter plots spread around a line generally. Three Nanjing rainfall and snow samples isotopic values were included to extend the comparation. According to the rainfall and the snow isotope result in Fig. 6, most of the data point spread over the World Meteoric Water Line (WMWL, or GMWL), the Lasa Local Meteoric Waterline (LMWL), and the Namucuo LMWL, while 3 points falls quite near the 3 lines. It shows that the different sources of the atmospheric water which formed the snow and rain, some precipitation comes from the similar sources to the Namucuo Lake and Lasa City, but there're also some different sources. Combine the surface water isotopes, it seems most of the runoff trends to fit the Lasa and Namucuo LMWLs, so that more observation needed before a sound Gujuecun LMWL could be drawn. But we still could speculate that the Gujuecun LMWL should be a little bit higher than the other 3 lines.

The δ^{2}H and δ^{18}O isotope values of the local river water in Fig. 7 are not meet the Lasa LMWL very well nor the Namucuo LMWL generally. In the mid and -value area, the points spread between the two LMWLs, while in the negative-value area dataset deviated higher from the lines.

The points in the relative higher value area in Fig. 7 trends closer to the Lasa LMWL, which generally lower than the Namucuo line and higher than the global precipitation line WMWL. With the decreasing of the δ^{2}H and δ^{18}O isotope values, the points spread higher and higher gradually, reaching the Namucuo LMWL and across then. The samples with poorer value points in the left side of the graph are all from the higher altitude area, which could accord with the precipitation attributes better shown in Fig. 6. Considering the com-

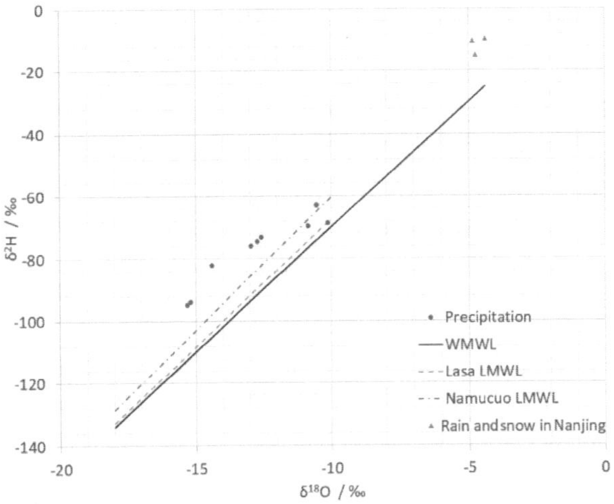

Figure 6. Rainfall and snow δ^2H and δ^{18}O isotopes.

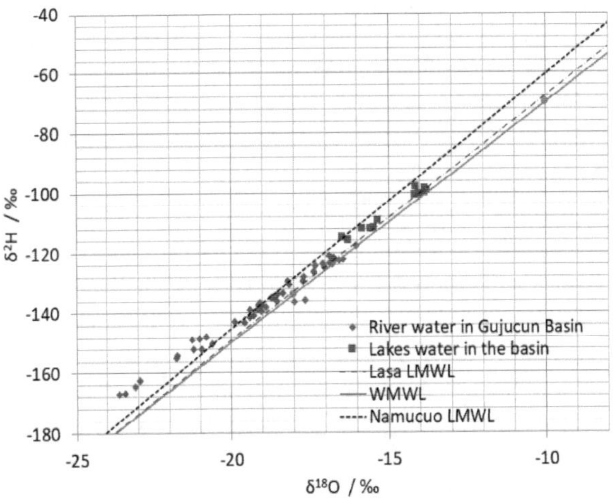

Figure 7. δ^2H vs δ^{18}O of the surface runoff water.

plexity of the surface water sources, the river and lake may contain runoff from rain, snow, glacier, underground runoff, or other sources formed at different time, which may undergo different degrees of fractionation. Because of sampling happen only in some particular months rather than the monthly series, the mixed surface water is regarded as better representative of the local isotope attribute, which imply that the vapor sources of the precipitation in Gujuecun basin are similar to Lasa and Namucuo, but different sources or source combinations may also exist. More observations and sampling are needed to make it clear.

4.2 Spatial variation of the geochemical indicators

The geochemical elements could also help to identify the sources of the runoff. According to the Fig. 8, K, Ca, Na, Mg,

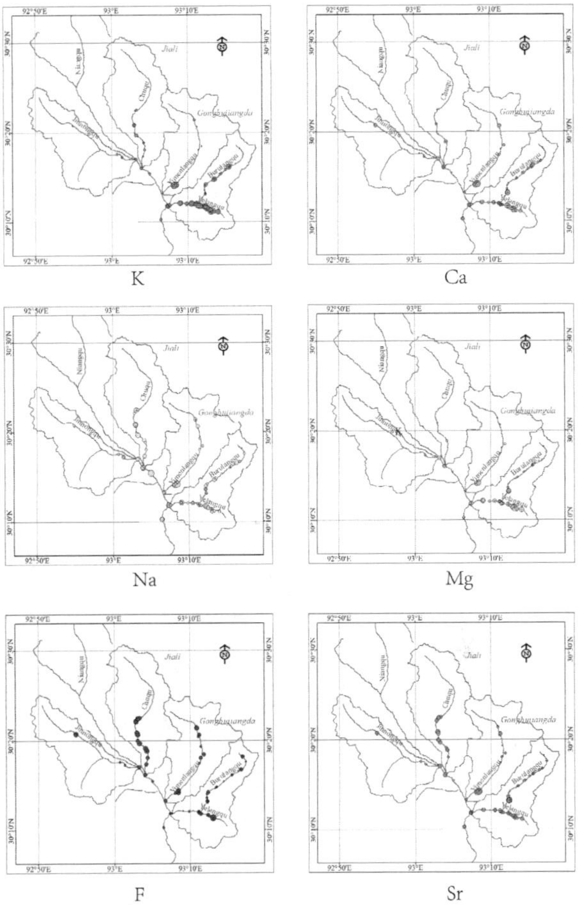

Figure 8. Geochemical elements distribution map.

F, Sr elements shows different spatial distribution characteristics. The K, Ca, Mg elements shows a significant source of the Yelangqu Creek, which is a branch of the Burulangqu Creek. The Na and Sr elements

contribute more by the Chuqu Creek and Yelangqu Creek. The F spread similar in different branches. High value points could be found in Yimenlangqu downstream and the Burulangqu downstream, that may relate to the grazing points. With the concentration of different geochemical elements at Gujuecun Gauge, distribution of the elements could help us to estimate the general sources of the mixed river water. An EMMA method is possible to be launched with the isotopes and geochemical elements, and the runoff composition could be analylized quantitively in the coming further research.

4.3 Isotopes and Ions in the Soil and Vegetation

The Fig. 9 is for the trees' stems isotope characteristics. The stems' values trend to spread under the WMWL and the LMWL, which shows that the water in the stems must be mainly from the precipitation event most recently. The trees

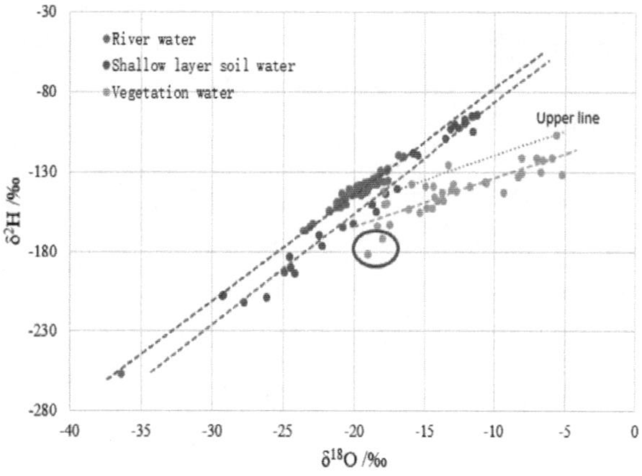

Figure 9. Isotopes of the soil and vegetation water.

seem take use of not only the local water in the soil around the root, but also the fractured water formed by the multiple precipitation events or supplied by the higher place. Certainly, there may also be some of the soil pore water used. According to the points which are closer to the LMWL may use more water from the river.

The vegetation water isotopes spread scattered. It may be caused by different reasons, but the water source should be an essential factor. Most of the vegetation seems comes from the similar water source according to the Fig. 9, for the points trend spread along a line significantly. But the points near the upper line (the trendline of the top 6 vegetation water) and in the lower circle should not from the same source. The vegetation may utilize not only the soil water supplied by the surface runoff, but also the deeper water sources with different isotopic characteristics. It could be speculated that the water in the stems were mainly from the precipitation event most recently. The trees seem take use of not only the local water in the soil around the root, but also the fractured water formed by the multiple precipitation events or supplied by the higher place.

5 Conclusions

This study was aimed at finding a series of feasible methods on the hydrometric monitoring and runoff composition analysis in the alpine environment. According to the high mountainous and low air pressure environment, automatically experimental gauges were established. Although the automatically instrument could monitor the water level data, but lack of public transition signal and the low guarantee degree of Beidou satellite channel, the received data are not quite complete. The equipment is still adjusting and improving.

Survey and sampling are still necessary for the study of the hydrological scheme of the alpine basin lack of observation data. Hydrogen and oxygen isotopes and geochemical elements were examined in the interest basin. Results show that δ^2H and δ^{18}O stable isotopes of the precipitation, river water, lake water was analyzed, and there are obvious regularities according to the elevation change. The steady isotopes show regularities in the stable isotopic composition of the precipitation water and surface water along the river according to the elevation. The river water δ^2H and δ^{18}O generally meet the local meteoric water line nearby, which shows the main source of the runoff is the precipitation and snow. Different geochemical elements indicate the different spatial sources of the runoff water, which could further analyzed with the end member algorithm methods later.

Vegetation water in the stems must be mainly from the river water and precipitation event most recently. Although the trees seem mainly take use of the similar water sources, but a few observation data also show different sources existence, such as fractured water formed by the multiple precipitation events or supplied by the higher place.

More observation and analysis will be done in the coming 3–4 years to collect more data, which should help to understand the hydrological and the ecohydrological processes characteristics better in the plateau area. More quantitative analysis and modeling method will be used when the dataset can support it.

Data availability. Topographic, geochemical and isotopic data are measured by authors' research group, which would be collected, summarized and published by the NSFC KEY RESEARCH PLAN Runoff Change and its Adaptive Management In the Major Rivers in Southwestern China.

Author contributions. HL and JL contribute to the concept of the study; HL generally organized the measuring, sampling, and analyzing; WW supported the GIS analysis and survey; XM contributed to the sampling and survey; HZ, XD, and NW contributed to the samples' geochemichal and isotopical analysis.

Competing interests. The authors declare that they have no conflict of interest.

Special issue statement. This article is part of the special issue "Hydrological processes and water security in a changing world". It is a result of the 8th Global FRIEND-Water Conference: Hydrological Processes and Water Security in a Changing World, Beijing, China, 6–9 November 2018.

Acknowledgements. This study is funded by NSFC projects, (1) the Experiment Study on Runoff Components Analysis in Niyang River Watershed Based on the Geochemical Indicative Tracing Factors (No. 91647111) and (2) Multi-scale monitoring of hydrological processes in the Yarlung Zangbo River Basin (No. 91647203), which are included in the Major Research Plan – National Natural Science Foundation of China.

Financial support. This research has been supported by the National Natural Science Foundation of China (grant nos. 91647111 and 91647203).

References

Blöschl, G., Sivapalan, M., Wagener, T., Viglione, A., Venije, H. S., and Processes, S.: Runoff Prediction in Ungauged Basins, ISBN 9781139235761, https://doi.org/10.1017/CBO9781139235761, 2013.

Briggs, P. R. and Cogley, J. G.: Topographic Bias in Mesoscale Precipitation Networks, J. Climate, 9, 205–218, 1996.

Deems, J. S., Painter, T. H., and Finnegan, D. C.: Lidar measurement of snow depth: A review, J. Glaciol., 59, 467–479, https://doi.org/10.3189/2013JoG12J154, 2013.

Immerzeel, W. W., Van Beek, L. P. H., and Bierkens, M. F. P.: Climate change will affect the asian water towers, Science, 328, 1382–1385, https://doi.org/10.1126/science.1183188, 2010.

Liu, C., Bai, P., Wang, Z., Liu, S., and Liu, X.: Study on prediction of ungaged basins: A case study on the Tibetan Plateau, Shuili Xuebao/Journal Hydraul. Eng., 47, 272–282, https://doi.org/10.13243/j.cnki.slxb.20150925, 2016.

Liu, S., Liu, C., and Zhao, W.: Towards the Methodology for Predictions in Ungauged Basins, available at: http://www.progressingeography.com/EN/10.11820/dlkxjz.2010.11.014, Prog. Geogr., 29, 1333–1339, 2010.

Liu, Y.: Review on Development of ADCP Technology and Its Application, Hydrogr. Surv. Charting, 36, 45–49, https://doi.org/10.3969/j.issn.1671-3044.2016.02.011, 2016.

Long, D., Longuevergne, L., and Scanlon, B. R.: Global analysis of approaches for deriving total water storage changes from GRACE satellites, Water Resour. Res., 51, 2574–2594, https://doi.org/10.1002/2014WR016853, 2015.

Meyer, M. L. and Huey, G. M.: Telemetric system for hydrology and water quality monitoring in watersheds of northern New Mexico, USA, Environ. Monit. Assess., 116, 9–19, https://doi.org/10.1007/s10661-006-7242-0, 2006.

Montanari1, A., Bahr, J., Bloschl, G., Cai, X., Mackay, D. S., Michalak, A. M., Rajaram, H., and Sander, G.: Fifty years of Water Resources Research:Legacy and perspectives for the science of hydrology, Water Resour. Res., 51, 6797–6803, https://doi.org/10.1002/2015WR017998, 2015.

Nolan, M., Larsen, C., and Sturm, M.: Mapping snow depth from manned aircraft on landscape scales at centimeter resolution using structure-from-motion photogrammetry, The Cryosphere, 9, 1445–1463, https://doi.org/10.5194/tc-9-1445-2015, 2015.

Piao, S., Ciais, P., Huang, Y., Shen, Z., Peng, S., Li, J., Zhou, L., Liu, H., Ma, Y., Ding, Y., Friedlingstein, P., Liu, C., Tan, K., Yu, Y., Zhang, T., and Fang, J.: The impacts of climate change on water resources and agriculture in China, Nature, 467, 43–51, https://doi.org/10.1038/nature09364, 2010.

Roub, R., Hejduk, T., and Novák, P.: Automating the creation of channel cross section data from aerial laser scanning and hydrological surveying for modeling flood events, J. Hydrol. Hydromechanics, 60, 227–241, https://doi.org/10.2478/v10098-012-0020-5, 2012.

Shu, D.: On Modernization of Hydrological Information System, Hohai University, (phD thesis), 2005.

Sivapalan, M., Takeuchi, K., Franks, S. W., Gupta, V. K., Karambiri, H., Lakshmi, V., Liang, X., McDonnell, J. J., Mendiondo, E. M., O'Connell, P. E., Oki, T., Pomeroy, J. W., Schertzer, D., Uhlenbrook, S., and Zehe, E.: IAHS Decade on Predictions in Ungauged Basins (PUB), 2003–2012: Shaping an exciting future for the hydrological sciences, Hydrol. Sci. J., 48, 857–880, https://doi.org/10.1623/hysj.48.6.857.51421, 2003.

Viviroli, D., Dürr, H. H., Messerli, B., Meybeck, M., and Weingartner, R.: Mountains of the world, water towers for humanity: Typology, mapping, and global significance, Water Resour. Res., 43, W07447, https://doi.org/10.1029/2006WR005653, 2007.

Wandishin, M. S., Baldwin, M. E., Mullen, S. L., and Cortinas, J. V: Short-Range Ensemble Forecasts of Precipitation Type, Society, 20, 609–626, https://doi.org/10.1175/WAF871.1, 2005.

Wu, C. and Kong, F.: Scheme Research on IoT Hydrological Measuring and Reporting System Based on Unmanned Ship, Sh. Electron. Eng., 37, 136–139, https://doi.org/10.3969/j.issn.1672-97302017.08.032, 2017.

Yao, T., Liu, X., Wang, N., and Shi, Y.: Amplitude of climatic changes in Qinghai-Tibetan Plateau, Sci. Bull., 45, 1236–1243, 2000.

Yin, Z., Feng, Q., Liu, S., and Zou, S.: The application progress of hydrological model in quantifying the contribution of glacier runoff to total watershed runoff, available at: http://www.bcdt.ac.cn/CN/10.7522/j.isnn.1000-0240.2016.0028, J. Glaciol. Geocryol., 38, 248–258, 2016

Yuan, D., Li, D., Zhang, L., Tian, Z., and Zhang, F.: Study and Development on Hydrometric Cableway of Automation Observation and Control System, Yellow River, 26, 11–12, 2004.

Proc. IAHS, 383, 93–97, 2020
https://doi.org/10.5194/piahs-383-93-2020

Study on Hydrologic Forecasting at the Construction Period of Datengxia Hydroproject

Kangming Lu

Bureau of Hydrology of Pearl River Water Resources Commission, Guangzhou, 510611, China

Correspondence: Kangming Lu (kangminglu@126.com)

Published: 16 September 2020

Abstract. To provide better service of hydrologic forecasting in the construction period of Datengxia Hydroproject that is the largest one under construction on the Pearl River, this paper focuses in analysing the characteristics of the regional floods, and reviewing the hydrologic forecasting work since starting construction, and discussing the hydrologic forecasting methods of the period not only the river closure but also the initial filling. The paper presents some practical work experiences and suitable forecasting methods on the basis of the construction schedule and the river diversion schemes. The study will conducive to improve our service ability of hydrologic forecasting and provide the important technical support of safe construction during the flood season.

1 Introduction

Datengxia hydroproject is the largest one under construction on the Pearl River, southern of China. After completion, the hydroproject will play an important role in flood control, shipping, power generation, Pearl River Delta water supply, irrigation, etc. (Yue, 2014).

Hydrologic forecasting is an important non-engineering measure to prevent from flood at the construction period. It is necessary to do some research on forecasting methods on the basis of the construction schedule and the river diversion schemes. These forecasting methods have provided important technical support for reasonable construction arrangement.

1.1 Brief description of the hydroproject

The hydroproject is located in the main stream of Pearl River which is also called West River (Wu and Zhang, 2017), about 6.6 km upstream from the Rainbow Bridge of Guiping City (Cheng et al., 2016). The hydrologic network of Datengxia Basin as shown in Fig. 1. The area of Datengxia catchment is 198 612 km^2, which accounts for 56 % of the area of West River Basin. The storage capacity of the hydroproject for flood control is about 1.50 billion m^3 (Li and Zhan, 2018).

1.2 Schedule of construction diversion

The construction period of the hydroprojet is divided into two stages. The diversion standard of the longitudinal cofferdam is a 20-year return flood (39 300 $m^3 s^{-1}$) (Zeng et al., 2013), so that it will work at both first-stage river diversion of right bank and second-stage river diversion of left bank.

The first-stage river diversion of right bank has began at September 2015, and it will last for 51 months. The construction includes the lock, the power plant of left bank, the sluices of left bank, the auxiliary dams of West River and Nanmu River. The intended completion date will be before early May 2019. The date of the river closure will be planned in November 2019.

The second-stage river diversion of left bank will begin at September 2019, and it will last for 49 months. The storage period of the first-stage project will be from February to April 2020.The intended completion date will be in December 2023. The construction will include the power plant of right bank, the sluices of right bank and the fishway.

2 Flood characteristic and composition of Qian Reach

The hydroproject is located in the Qian Reach of the West River. The Wuxuan Hydrologic Station has long term measured data of the reach. According to the analysis of Wux-

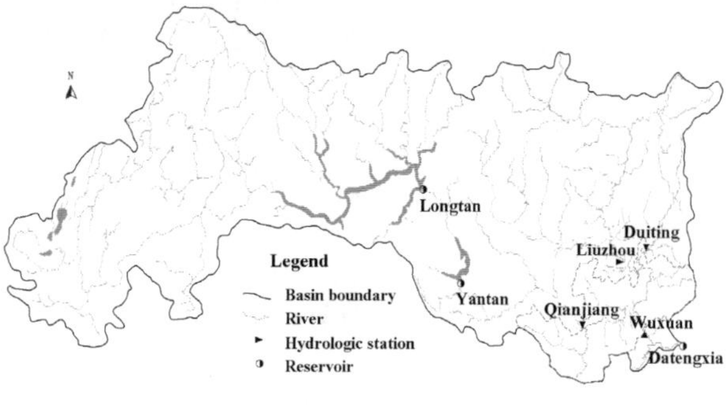

Figure 1. The hydrologic network of Datengxia Basin.

uan's data, the flood characteristic of Qian Reach is found that the flood occur last a long time with high peak and huge volume, and the occurrence probability of double flood peak can reach about 50 %.

The flood of Qian reach mainly comes from Hongshui River and Liu River. The stream gauging staion of Hongshui River is called Qianjiang Station. The stream gauging staions of Liu river are called Liuzhou Station and Duiting Station.

Because of the upstream reservoirs' operation (such as Longtan and Yantan) and topographic features (such as karst funnel and cenotes), the flood of Hongshui River spreads slowly and the peak is gentle. The influence of Hongshui River on flood peak in Wuxuan station is far less than that of Liu River. According to the maximum three days volume analysis of 110 floods from 2006 to 2016, the proportion of Wuxuan's floods that mainly come from Hongshui River is only 3 %, the proportion of Wuxuan's floods that mainly come from Liu River is 47 %, and the other Wuxuan's floods mainly come from both the uncontrolled area (see the dotted line area in Fig. 1) and Liu River.

The Liu River is located on the windward slope of the mountains in the northern of the Pearl River Basin. This terrain is conducive to emergence of heavy rain (Ma and Li, 2013). And Liu River Basin is fan-shaped, so that the flood is rapidly gathering. That is why that Liu River is prone to multimodal floods with high peak and huge volume. The flood peak time of Wuxuan station corresponds to that of Liu River, and the proportion of flood volume of Liuzhou Station is positively correlated with that of Wuxuan Station. According to statistics, the maximum three days volume of 31 Wuxuan's floods are over 3.00 billion m^3 in the past ten years. And among those floods, 74 % obey the law of average travel time of the peak in the reach from Liuzhou to Wuxuan, and the maximum three days flood volume of Liuzhou accounts for 70 %–80 % of that of Wuxuan.

3 Review of hydrologic forecasting work since starting construction

The construction of the first-stage project has started since 2015. And during the period from September 2015 to May 2016, it was the construction period of the first-stage diversion cofferdam. The first-stage project were protected by sub cofferdams before the longitudinal and transverse cofferdams had been built to the design elevation. So we do the hydrologic work for the flood control standard of the sub cofferdams.

After the construction of cofferdams completed, we set up two temporary gauging stations on the upstream and downstream transverse cofferdams. Based on the observed data of three floods between June and August 2017, the correlation diagram had been used to work out the water level forecasting scheme of the temporary gauging stations. The forecasting scheme has been applied to the practical work since March 2018, and the result shows that the accuracy of the water level forecasting can give important technical support for the construction of the hydroproject.

3.1 Hydrologic forecasting in the construction period of cofferdams

During the construction period of cofferdams, we continuously worked on the flow forecasting of Wuxuan Station based on the observed data and the latest numerical forecasting of rainfall. The results indicated that both accuracy and reliability of the flood forecasting work had been satisfactory.

There was a obvious rainfall process happen in Datengxia Basin on 10–13 November 2015. A rare winter flood occurred in Qian Reach, which was the second largest flood to the previous record for that period. On the morning of 11 November, we forecasted that the flow of Wuxuan would exceed the design standard of the sub cofferdams in the drought period (6720 $m^3\,s^{-1}$). And we informed Datengxia Company of the forecasting result and reported to the Office of Pearl River Flood Control and Drought Relief Headquarters immediately. On the morning of 12 November, we forecasted that the flow of Wuxuan would exceed 6000 $m^3\,s^{-1}$ in the evening of the same day, and then issued a flood warning. The forecasting results were correct, the flow of Wuxuan reached 6000 $m^3\,s^{-1}$ at 18:00 LT (BJT) of 12 November, and it exceeded 6720 $m^3\,s^{-1}$ at two hours later.

From April to May 2016, there were several heavy rainfall occurred in the basin. The flood volume of Qian Reach in mid Apr. was the largest since 1936. The construction was hit by two consecutive floods that exceeded the design standard of the sub cofferdams in the flood season (10 500 $m^3\,s^{-1}$). And the extraordinary flood was happen once again in early May. The accurate forecasts were made in the morning of 10 April, 13 April and 8 May, which give important technical support for the completion of cofferdams on schedule.

Table 1. Correlation coefficient of the fitting synthetic curves at different time intervals n.

time intervals n/h	upstream transverse cofferdam	downstream transverse cofferdam
0	0.915	0.907
1	0.924	0.917
2	0.925	0.929
3	0.944	0.939
4	0.953	0.948
5	0.958	0.953
6	0.991	0.990
7	0.965	0.962
8	0.968	0.965
9	0.967	0.965

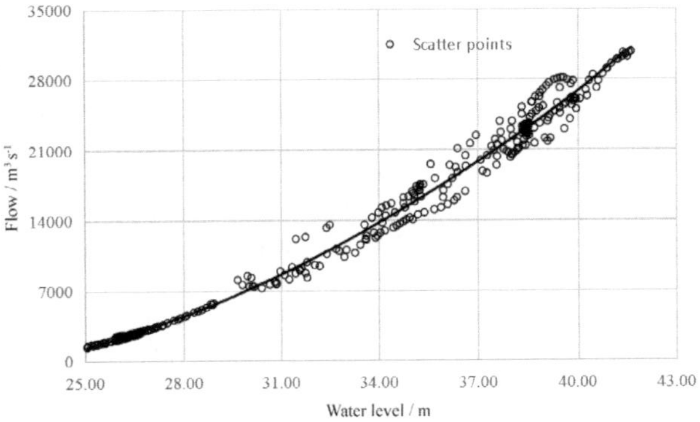

Figure 2. The scatter diagram with $T-6$ h time flow of Wuxuan and T time water level of upstream transverse cofferdam.

3.2 Water level forecasting of upstream and downstream transverse cofferdams

According to observed data of the temporary gauging stations from June to August 2017, we worked out a water level forecasting scheme by variable relation scatter graph. It is easily to draw the scatter diagrams with $T-n$ ($n=1,2,\ldots,9$ time flow of Wuxuan and T time water level of upstream and downstream transverse cofferdams. Because the scattered points showed parabolic trend, then we could set up the quadratic equation with one unknown, which was fitted by the least square method. The correlation coefficient of the fitting synthetic curves were separately calculated in Table 1. The result showed that $T-6$ h flow of Wuxuan has the highest correlation with T time water level of upstream and downstream transverse cofferdams, as shown in Figs. 2 and 3.

Considering pre-forecasting error, we should amend the forecasting result of the fitting synthetic curve. In view of the three floods in 2017, there were still some differences in the fitting curve of each flood and the fitting synthetic curve. For example, when the flow of Wuxuan exceeded 15 000 $m^3\,s^{-1}$, the water level of upstream transverse cofferdam in Jul. is

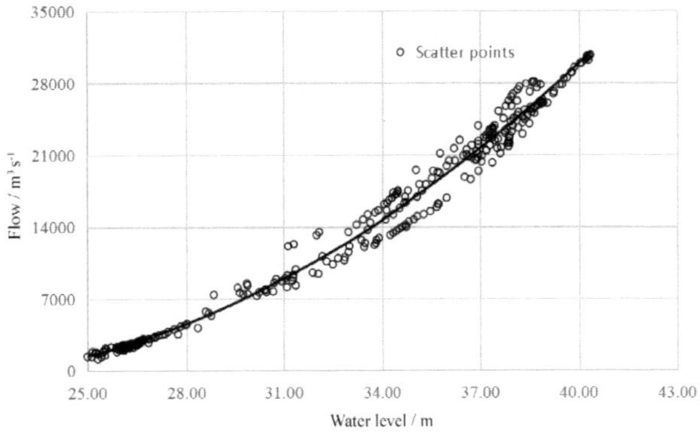

Figure 3. The scatter diagram with $T-6$ h time flow of Wuxuan and T time water level of downstream transverse cofferdam.

higher than that in Aug. in the situation of the same flow. According to analysis, the water level was related to not only current flow but also early inflow of the reach. So the forecasting result should be corrected by pre-error before issued.

The forecasting scheme has been applied to the practial work since March 2018. The absolute average forecasting error of water level of upstream and downstream transverse cofferdams were only 0.15 m at the foreseeable period of 24 h. The result showed that the accuracy of water level forecast could give important technical support for the construction of the hydroproject.

4 Hydrologic forecasting of the period not only the river closure but also the initial filling

The flow of Qian Reach will become more complex in the period of the river closure and the initial filling, so that the forecasting scheme of Sect. 3.2 will not work any longer after demolishing the upstream and downstream transverse cofferdams. By combining hydrology method and computational hydraulics method, we attempt to find a new forecasting scheme which is suitable for the period not only the river closure but also the initial filling.

4.1 Construction of forecasting scheme

Because of the diversion of the longitudinal cofferdam, the water will flow though the sluices of left bank and the closure gap of right bank in the period of the river closure. According to the algorithm of ring river network (Li and Wang, 2005), we can construct a computational hydraulics model of Qian Reach. The study area is generalized as upstream reach, first-stage diversion reach, second-stage diversion reach and downstream reach of the hydroproject. The sketch map of river network in the study area as shown in Fig. 4. The forecasting scheme will be applicable for different construction stages only if topographic conditions of the reaches are changed.

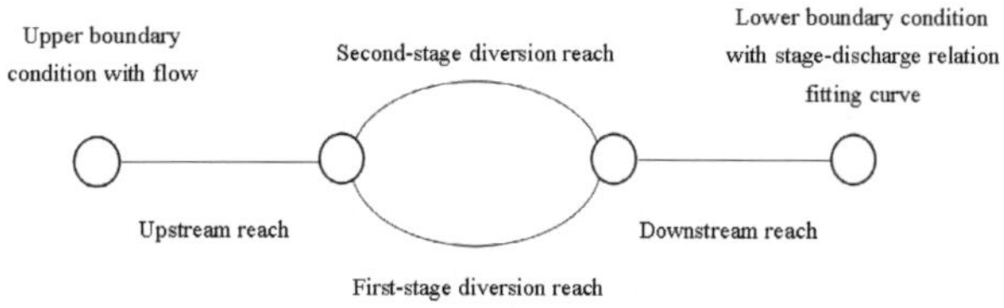

Figure 4. Sketch map of river network in the study area.

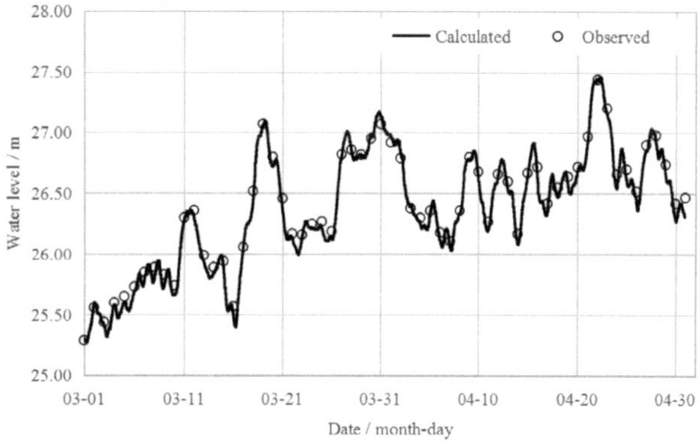

Figure 5. Water level process of forecasting and observed at upstream transverse cofferdam in March and April 2018.

4.2 Calibration of parameters

The parameter of the forecasting model is the roughness of each reach. According to the design flow value of the river closure, the parameters are calibrated by the observed flow of Wuxuan between 2500 and 3500 $m^3\,s^{-1}$. The second-stage diversion reach should be generalized as a closed section which suppose that all the sluices of left bank are closed during the construction period of the first-stage project. The upper boundary of the model is the forecasting flow that made by hydrologic model, and the lower boundary condition is stage-discharge relation fitting curve. Through the tentative calculation, the correlation coefficients of forecasting and observed water level process should be as large as possible by constantly adjusting the roughness of each reach.

As the forecasting results in March and April 2018, the correlation coefficients of the water level of upstream and downstream of the cofferdams are all over 0.98. It is shown that the model can accurately simulate the water level process of Qian Reach, and it can be applied to water level forecasting in drought period. The water level process of forecasting and observed at upstream and downstream transverse cofferdams as shown in Figs. 5 and 6.

4.3 Prospect of application

The model can be applied to forecast the water level of Qian Reach in the period not only the river closure but also the initial filling by changing the generalization structure of river network. According to the schedule, the construction of left bank project will be completed before May 2019. The sluices of left bank will work after demolishing the upstream and downstream transverse cofferdams, and therefore this reach should be generalized as weir flow with sluice control. The closure gap of first-stage diversion reach should be generalized as free weir flow, and the width of closure gap should be changed in real time. The model can be applied in the period of the river closure without changing neither topographic conditions of the other reaches nor the upper and lower boundary conditions. And after February 2020, the model can be applied in the period of the initial filling only if the first-stage diversion reach is generalized as weir flow with sluice control.

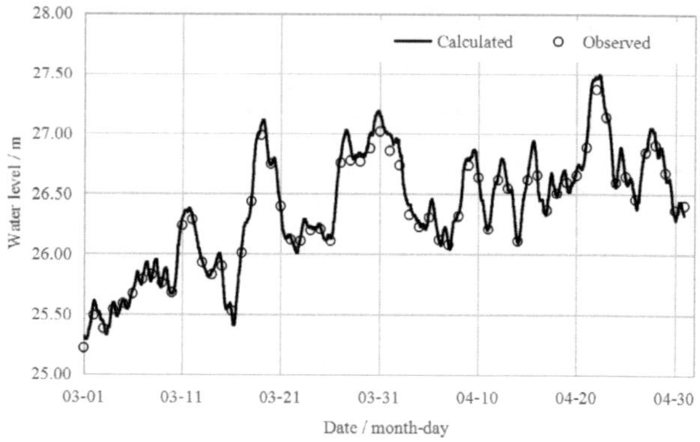

Figure 6. Water level process of forecasting and observed at downstream transverse cofferdam in March and April 2018.

It noteworthy that two work must be actualized to ensure the hydrologic forecasting work continues. The first is that the river cross-section measurement of left bank side of longitudinal cofferdam needs to be carried out immediately after demolishing the upstream and downstream transverse cofferdams. The second is that the topographic survey of upstream reach of the hydroproject must be done when the river closure is completed. Thus the generalization of river network can match the river network topography as far as possible until the above information available.

5 Conclusions

This paper analyses the flood characteristic and the flood composition on Qian Reach, and reviews the hydrologic forecasting work since starting construction, and presents the water level forecasting scheme of the upstream and downstream transverse cofferdam by variable relation scatter graph. The practical application shows that the accuracy of water level forecasting can give important technical support for the construction of the Datengxia Hydroproject currently. In the end, a forecasting model suitable for the river closure and the initial filling is discussed, which will lay the foundation of further forecasting work.

Data availability. The majority of the data presented here are from an ongoing and yet incomplete project. The datasets generated and analyzed during the current study are available from the corresponding author on reasonable request.

Competing interests. The author declares that there is no conflict of interest.

Special issue statement. This article is part of the special issue "Hydrological processes and water security in a changing world". It is a result of the 8th Global FRIEND–Water Conference: Hydrological Processes and Water Security in a Changing World, Beijing, China, 6–9 November 2018.

Acknowledgements. I would like to extend my great gratitude to my graduate tutor Guangchi Li for passing on the algorithm of ring river network.

References

Cheng, Y., Li, Y., Chang, W., and Zhao, X.: Location Selection of ship lock in Datengxia hydroproject, Water Resources & Hydropower of Northeast China, 7, 3–4, https://doi.org/10.14124/j.cnki.dbslsd22-1097.2016.07.002, 2016.

Li, G. and Wang, C.: Universal algorithm for river basin flood routing model, Journal of Hohai University (Natural Sciences), 6, 624–628, https://doi.org/10.3321/j.issn:1000-1980.2005.06.005, 2005.

Li, X. and Zhan, X.: Review on quality management of Datengxia water conservancy control project, Water Conservancy Construction and Management, 7, 4–6, https://doi.org/10.16616/j.cnki.11-4446/TV.2018.07.02, 2018.

Ma, X. and Li, Y.: Hydrologic analysis and calculation of Datengxia hydroproject, Water Resources & Hydropower of Northeast China, 5, 38–40, https://doi.org/10.14124/j.cnki.dbslsd22-1097.2013.05.025, 2013.

Wu, Y. and Zhang, X.: Impact of Datengxia gorge water conservancy project on navigation condition, Guangxi Water Resources & Hydropower Engineering, 4, 93–95, https://doi.org/10.16014/j.cnki.1003-1510.2017.04.029, 2017.

Yue, Z.: Promote the construction of green Pearl River by improving basin management capacity, China Water Resources, 24, 73–74, https://doi.org/10.3969/j.issn.1000-1123.2014.24.022, 2014.

Zeng, X., Ma, C., Li, Y., and Yang, S.: Discuss on construction layout and diversion design of Datengxia water conservancy control project, Pearl River, 1, 46–49, https://doi.org/10.3969/j.issn.1001-9235.2013.01.012, 2013.

Proc. IAHS, 383, 99–110, 2020
https://doi.org/10.5194/piahs-383-99-2020

Open Access PIAHS

Analysis of Event-based Hydrological Processes at the Hydrohill Catchment Using Hydrochemical and Isotopic Methods

Na Yang[1,2], Jianyun Zhang[1,3], Jiufu Liu[1], Guodong Liu[2], Aimin Liao[1], and Guoqing Wang[1,3]

[1]State Key Laboratory of Hydrology-Water Resources and Hydraulic Engineering, Nanjing Hydraulic Research Institute, Nanjing 210029, China
[2]College of Water Resource & Hydropower, Sichuan University, Chengdu 610065, China
[3]Yangtze Institute for Conservation and Development, Nanjing 210098, China

Correspondence: Guoqing Wang (gqwang@nhri.cn)

Published: 16 September 2020

Abstract. Hydrochemical and isotopic techniques have been widely applied in hydrological sciences because isotopic tracers can identify water sources and hydrochemical tracers can discern runoff flow paths. To better understand the hydrological process, we combined hydrochemical and isotopic techniques under controlled experimental conditions to investigate hydrological process from rainfall to runoff in the Hydrohill experiment catchment, a typical artificial catchment in Chuzhou, China. Hydrochemical and isotopic data, i.e., pH, electric conductivity (EC), total dissolved solids (TDS), anions (Cl^-, NO_3^-, SO_4^{2-} and HCO_3^-), cations (K^+, Na^+, Ca^{2+} and Mg^{2+}) and dissolved Si, ^{18}O and D in water samples were collected during a rainfall event in 2016, and used to determine the hydrochemical and isotopic characteristics of rainfall and runoff components. We applied EC, TDS, SO_4^{2-}, Ca^{2+}, Mg^{2+}, ^{18}O and D as tracers to investigate rainfall-runoff processes in the experimental catchment. Runoff flow paths could be well identified by the relationship between ^{18}O and EC, TDS, SO_4^{2-}, Ca^{2+} and Mg^{2+}. The quantity of flow flux and mass fluxes of main hydrochemical and isotopic tracers gauged at the catchment outlet shows applicable tracers (Ca^{2+}, Mg^{2+}, SO_4^{2-}, and ^{18}O) are mainly from deep groundwater runoff (from soil layer of 60–100 cm beneath ground surface). Contributions of the event water and pre-event water to the total runoff during the rainfall-runoff process are different. The quantitative results were very encouraging as a basis to develop hydrological models for further study.

1 Introduction

Runoff sources and their pathways play a significant part in runoff generation and flow concentration, and thus have been widely studied during the past decades (Zhao, 1989; Gu et al., 2018). Variations of temporal and spatial dynamics inside the catchment indicate different flow paths and water sources contributing to the total runoff outlet. Hydrochemical and isotopic processes of total runoff outlet are mainly affected by the solutes sources, hydrological pathways and the flow components contributions (Yang et al., 2012). Hydrochemical tracers (Uhlenbrook et al., 2002; Tardy et al., 2004) and stable isotopes (Carey et al., 2004) widely applied in hydrological science have enhanced our understanding of runoff generation processes. Environmental tracers are commonly used on a catchment scale because it is possible to determine the source areas of runoff, flow pathways, residence times and the hydraulic characteristics of flow systems. Hydrochemical indexes such as pH, electrical conductivity (EC), total dissolved solids (TDS) or the concentration of dissolved silicon and various anions and cations (including K^+, Na^+, Ca^{2+}, Mg^{2+}, Cl^-, SO_4^{2-} and HCO_3^-) have been used to separate and quantify runoff components, so that the contributions of different runoff components can be determined and the hydrological pathways and the solutes sources can be investigated (Pilla et al., 2006; Mul et al., 2008).

Several methods, such as chemical/isotope (tracer) hydrograph separation, rainfall-runoff models and base flow filters, have often been applied to quantify runoff components (Gonzales et al., 2009; Cartwright et al., 2014). Hydrograph separation combining hydrochemical and isotopic tracers is a method commonly used to identify the origin and pathways of surface and subsurface runoff and thus reveal the mechanisms of runoff generation (Hooper et al., 1986; Klaus et al., 2013). For example, quick flow, which contributes to the total outlet immediately after the rainfall event, can include water from different sources (Hrachowitz et al., 2011; Cartwright et al., 2014). More studies have used two-component mixing models, three-component mixing models and multi-component mixing models to distinguish pre-event water with event water (Huth et al., 2004; Hugenschmidt et al., 2014) and identify water sources and flow paths, in which isotope tracers (^{18}O and ^{2}H) are applied in conjunction with geochemical tracers (Vincent et al., 2001; Brian et al., 2010; Stadler et al., 2012). Mixing models can enhance the accuracy of research about hydrological processes. The results obtained by separation techniques based on the mass balance approach suggest a significant contribution of pre-event water to catchment runoff, as documented in the majority of studies performed worldwide (Klaus et al., 2013). For example, streamflow components were determined by analyzing ^{18}O, ^{2}H, SO_4^{2-}, NO_3^-, Cl^-, Na^+, K^+, Ca^{2+}, Mg^{2+}, EC and other parameters mainly focusing on the temporal and spatial variations of the sources which come from precipitation, stream waters, soil solution and spring water. ^{18}O and Si were finally used to evaluate contributing sources by using mass balance equations and end-member mixing diagrams (Ladouche et al., 2001). Based on hydrochemical tracers, a three-component hydrograph separation was carried out in a steep, remote and monsoon-dominated study site ($7\,km^2$) in northern Thailand (Hugenschmidt et al., 2014). In the study, silica and EC, K^+ and Ca^{2+} were used as indicators for fractions of runoff components, surface runoff dynamics, and to give a better understanding into groundwater behavior. In the study, groundwater accounted for the largest contribution to stormflow (62 %–80 %) throughout all events, followed by shallow subsurface flow (17 %–36 %) and surface runoff (2 %–13 %). Thus, calculating the contributions of different runoff components to the total runoff is quite important for accurate predictions of water supply from mountain watersheds (Zhang et al., 2018).

Water delivery to the catchment outlet can be affected by variations of flow paths connecting the residence time variations of water inside the catchment (Hrachowitz et al., 2016). Surface water and subsurface water contributions to the whole catchment and the inherent flow paths indicate the details of hydrological processes. Hydrochemical data and stable water isotopes can be used in mass balance to gain an understanding of source water and flow paths in complex terrain (Robinson et al., 2009; Williams et al., 2011). End-Member Mixing Analysis (EMMA) is a hydrological mixing model that can be used to conceptualize runoff generation processes and has been successfully applied to many catchments around the world (Maurya et al., 2011; Baraer et al., 2015). For example, Zhang et al. (2018) utilized hydrochemical tracers from 2009 to 2011 to statistically model streamflow sources within a mesoscale ($264\,km^2$) watershed. Snowmelt and rainwater from the subalpine zone and groundwater sampled from the upper montane zone contributed 54 %, 22 %, and 24 % of the annual streamflow, respectively, which contribute to a better understanding of streamflow source waters in complex mountain terrain.

More controlled experiments are needed to seek new fundamental understanding through new mechanistic experimentations of how watershed systems work. Field hydrological experiments at the Hydrohill experimental catchment at the Chuzhou hydrological experiment base have improved the basic theory of hydrology. For example, it has been confirmed that the runoff components include surface runoff, interflow in the unsaturated zone and groundwater runoff in the saturated zone and their runoff generation mechanisms (Liu et al., 2018a, b). However, a further study examining the contributions and water residence time of different runoff components to the total outlet of the catchment is still required. Moreover, understanding of characteristics of tracers in rainfall-runoff processes is also quite necessary and important. Kendall et al. (2001) investigated the relations between Cl^-, Si, and $\delta^{18}O$ of subsurface waters and rainfall at Hydrohill to identify event water and pre-event water contributions to the outlet of the catchment, and found the Cl^- and Si in quick-flow runoff originated from event water were not regarded as suitable conservative tracers for either water sources or flow-paths in this catchment. However, due to the advanced instruments in this research for testing more tracers, such as conventional indexes, cations and anions, we have more choices.

Using data from a rainstorm in October 2016 at the Hydrohill experimentation area in Eastern China, the paper aims to use the tracer analysis approach to investigate the hydrological processes and their relationship with both hydro-chemical and isotopic processes to get a better understanding of the interaction mechanism between different fluxes in the critical zone. The main objectives of this work are (i) to analyse the characteristics of tracers and select the applicable ones that could be more effective in describing hydrological processes, (ii) to test the effectiveness of combining isotopic and hydrochemical indicators in identifying runoff pathways, (iii) to investigate the mass fluxes of selected tracers in different runoff components to the total outlet.

2 Materials and methods

2.1 Study site

The study was performed in the $512\,m^2$ Hydrohill artificial experimental catchment, located in Chuzhou, southeastern

China. The bottom consists of two intersecting slopes that are inclined at 10° and have an overall downslope gradient of 14° (Fig. 1c). Impermeable concrete walls were set up on this aquiclude with the final objective to enclose the catchment to prevent any lateral exchanges of underground flow. The silt-loam soil removed before excavation was put back to a depth of approximately 1 m. The surface of the basin was covered by grass and a central drainage trench was constructed at the intersection of the two slopes after three years of soil subsidence and a water-sampling instrument was installed (Gu and Freer, 1995). The rainfall is unevenly distributed during the year, with June, July and August accounting for 49 % of the annual rainfall. The mean annual precipitation is 1060 mm (period 1952–2016), and the average annual temperature is 14.9°.

This experimental catchment is equipped with a separate runoff observation system, which can monitor surface runoff and subsurface runoff from soil layers with depths of 0–30, 30–60, 60–100 cm (inferred as SSR30, SSR60, and SSR100) (Fig. 1b). Each trough has a 20 cm aluminum lip that extends horizontally into the soil layer to prevent leakage between layers. Hydrological cycling factors observation devices including tipping bucket-type micro pluviometers, rain cylinders, standard water collectors, soil moisture sensors, pressure water level gauges were installed at the Hydrohill catchment (Fig. 1).

2.2 Data collection

Rainfall data was measured by the four rain gauges located in the lower and upper parts of the catchment. The runoff data for surface water (SR), SSR30, SSR60, and SSR100 and the total runoff (TR) can be measured by the five 90° sharp crested V-notch weirs and logarithmic weirs at the outlet of Hydrohill (Fig. 1d). Continuous measurements were taken by using pressure-type stage recorders.

2.3 Sample collection and analysis

During 25–29 October 2016, 125 samples were collected at the catchment. These included rainwater, surface water and subsurface water. Rainfall samples were collected through a rainfall water collector with a diameter of 40 cm located on the observation room roof. The rain immediately arrived at the observation room through a pipeline. Rainfall samples were collected every hour during the rainfall event. Section negative pressure sampling was used to collect surface runoff and subsurface runoff samples. When there were typical points like peak points and valley points on the hydrograph, we started to collect water samples. Samples were stored in 50 mL high-density polyethylene (HDPE) bottles. HDPE bottles were washed by deionized water and rinsed with sample water three times at the time of collection. All samples were filtered through a 0.45 µm filter and then sealed before laboratory analysis. All the samples were transported to the laboratory in Nanjing and preserved at 4° until analysis.

All samples were analyzed for general parameters (pH, EC), major ions (K^+, Na^+, Ca^{2+}, Mg^{2+}, Si, Cl^-, NO_3^-, SO_4^{2-} and HCO_3^-), and stable isotopes (^{18}O and ^2H). pH and EC were measured in the field using a portable EC digital analyzer (HQ14d, Hach, USA) and a multi-parameter digital analyzer (HQ40d, Hach, USA). The hydrochemistry of all water samples was analyzed within two weeks of collection. Cations (K^+, Na^+, Ca^{2+} and Mg^{2+}) and dissolved Si were analyzed by Inductively Coupled Plasma Optical Emission Spectrometry (ICP-OEC), and the anions (Cl^-, NO_3^- and SO_4^{2-}) were analyzed by Dionex ICS-2100 Ion Chromatography. All samples were filtered through a 0.45 µm filter before laboratory analysis. The concentration of HCO_3^- was determined by a titration assay on-site. ^{18}O and ^2H analyses were carried out by the 4th generation Los Gatos Research Liquid Water Isotope Analyser and expressed according to the Vienna Standard Mean Ocean Water (VSMOW).

3 Results and discussion

3.1 Hydrological response in rainfall events

This rainfall event happened from 23:00 on 25 October to 12:00 on 28 October and can be divided into three major rainfall periods (Fig. 2). Because the whole depth of Hydrohill is 1m and the groundwater level is higher than 40 cm during this rainfall, SSR100 is regarded as groundwater runoff. This event was one of the biggest storms following a low-intensity rainfall event. The general hydrological characteristics of these floods, including rain volume, the maximum intensity in 1 h, mean intensity and peak discharge in separate runoff and total runoff, are concluded in Table 1.

The first rainfall period occurred from 23:00 on 25 October to 23:00 on 26 October and the daily rainfall was 54 mm. Because there had been no recent storms, and the soil before the event was dry, there was a minor flow response. The discharge of the surface flow was nearly the same as the discharge of SSR30 and SSR60, all of which were smaller than SSR100. This indicates that the rainfall water infiltrated quickly and there was probably 25 % existing pre-event water. The second period was the largest and was a complex process among the three periods which occurred between 07:00 and 17:00 on 27 October . The runoff response was so quick due to the high rain volume. SR shows the quickest response and the volume of SSR100 was also the highest in all runoff components, with a little longer residence time than others. The third period which occurred during the whole day of October 28 was the shortest period. Though its process was also complex, the runoff response was slow and the SSR100 remaining was the largest runoff volume. The total runoff efficient (V/R ratio) in Table1 showed that more than 60 % of the rainfall had infiltrated the soil reaching the total outlet of the catchment and the V/R ratio of total runoff in the third

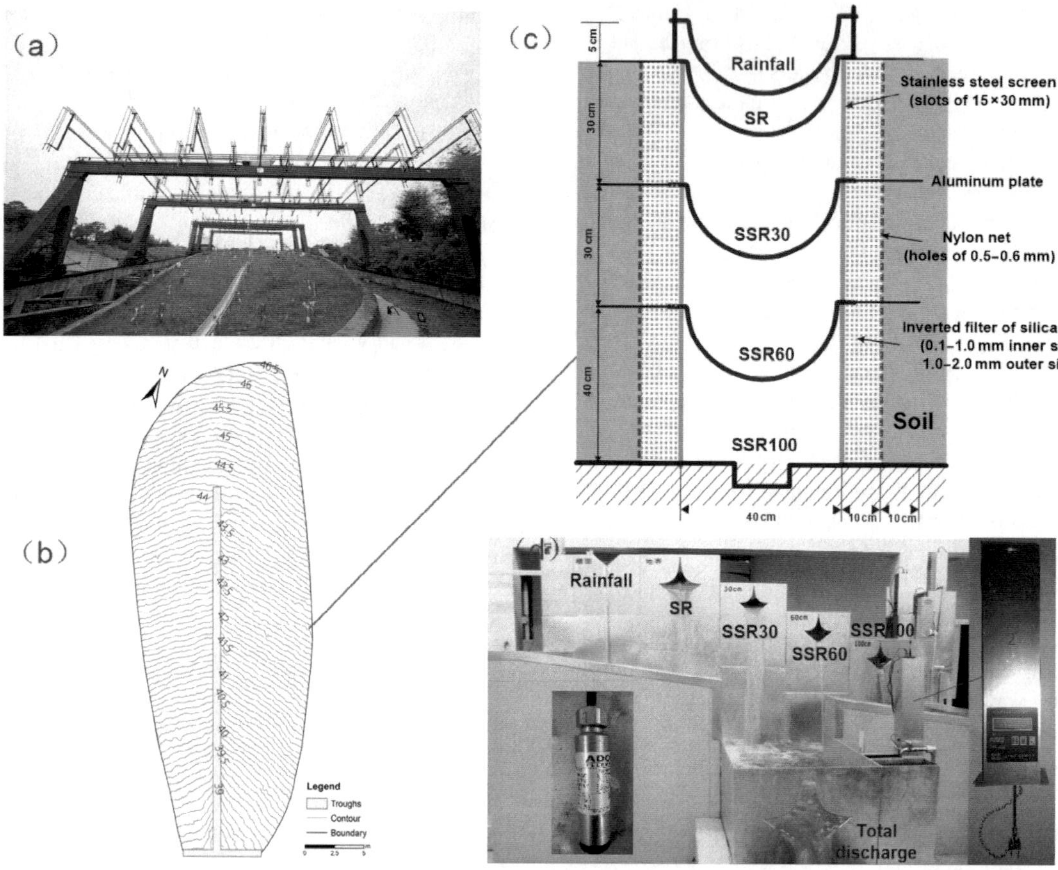

Figure 1. Aerial view of the Hydrohill experimental catchment and measurements of runoff components, **(a)** aerial view of the Hydrohill catchment with observation devices, **(b)** contour line on the Hydrohill catchment, **(c)** and **(d)**, runoff components measuring system at the outlet of the Hydrohill catchment. SR: surface runoff; SSR30, SSR60, and SSR100: subsurface runoff from soil layers of 0–30, 30–60, and 60–100 cm in depth.

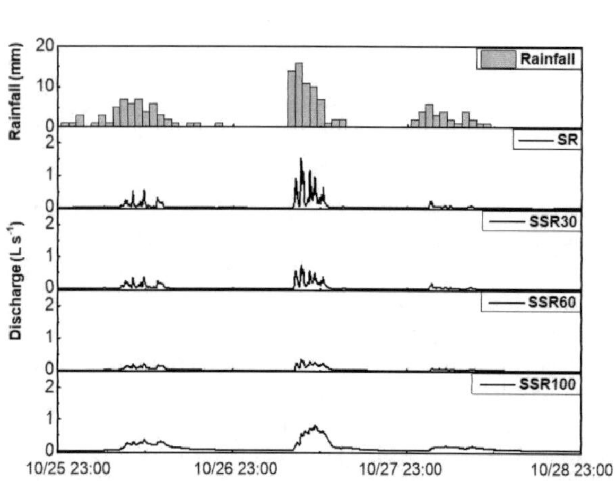

Figure 2. Hydrographs of the separate runoff in the 2016 autumn rainfall event.

period was the biggest (nearly 1.0) in all three rainfall periods, indicating that soil had almost been saturated. In the three successive periods, the V/R ratio of SR, SSR30 and SSR60 in the second period was bigger than that in the other two periods. However, the V/R ratio of SSR100 in the three periods was higher than that of SR, SSR30 and SSR60, indicating that other than the infiltrated water, existing water before this rainfall also contributed to the total runoff.

3.2 Hydrochemical Process

In this research, we analysed many more chemical parameters than previous studies to find more suitable tracers. The statistical results, including maximum values, minimum values and mean values of all indicators in this rainfall event, are shown in Table 2.

All three rainfall periods in this study showed vertical features of chemical components. The pH values of most samples ranged from 7.06 to 8.92, with an average value

Table 1. Principal hydrological characteristics of the flood event in 2016.

		25 Oct	27 Oct	28 Oct
Rainfall (mm)		54.00	63.00	33.00
Duration (h)		24	10	12
Maximum intensity in 1 h ($mm\,h^{-1}$)		7.00	16.00	6.00
Mean intensity ($mm\,h^{-1}$)		2.25	6.30	2.75
Total flow volume (mm)	SR	5.60	14.60	3.10
	SSR30	5.30	9.50	2.00
	SSR60	5.00	6.70	2.90
	SSR100	18.30	24.20	24.20
	TR	34.10	55.10	32.10
Peak discharge ($L\,s^{-1}$)	SR	0.54	1.55	0.21
	SSR30	0.35	0.71	0.17
	SSR60	0.20	0.33	0.09
	SSR100	0.36	0.81	0.18
	TR	1.42	2.93	0.538
Runoff coefficient	SR	0.10	0.23	0.09
	SSR30	0.10	0.15	0.06
	SSR60	0.09	0.11	0.09
	SSR100	0.34	0.38	0.73
	TR	0.63	0.88	0.97

of 7.82, indicating an alkaline nature. The average pH values were 8.44 in rainfall, 7.58 in surface runoff, 7.72 in SSR30, 8.07 in SSR60, 7.95 in SSR100. The values of EC in water samples ranged from 9.81 to 394.00 $\mu S\,cm^{-1}$ in this rainfall. Average EC values were 17.86 $\mu S\,cm^{-1}$ in rainfall, 37.28, 51.42, 118.11 $\mu S\,cm^{-1}$ in SSR30, SSR60 and SSR100, respectively. TDS is often calculated by adding together all the major cations: $Na^+ + K^+ + Ca^{2+} + Mg^{2+}$, and major anions: $HCO_3^- + SO_4^{2-} + Cl^- + NO_3^-$ in $mg\,L^{-1}$ (Walton, 1989). The lowest TDS value was recorded in the rainfall and the highest ranged between 76.65 and 197.77 $mg\,L^{-1}$ found in SSR100. The TDS was 13.63–28.64 $mg\,L^{-1}$ for surface water, 19.72–41.22 $mg\,L^{-1}$ for SSR30 and 27.22–117.51 $mg\,L^{-1}$ for SSR60. Average TDS values were 6.72 $mg\,L^{-1}$ in rainfall, 20.40 $mg\,L^{-1}$ in surface runoff, 28.03, 53.59 and 123.10 $mg\,L^{-1}$ in SSR30, SSR60 and SSR100, respectively. So we can find that when the flow path is longer, both the TDS and EC values will show an increasing trend from rainfall to SSR100.

During the observation periods, the main conventional physical and chemical indicators (EC, TDS), and main ions (Mg^{2+}, Ca^{2+} and SO_4^{2-}) showed similar characteristic patterns, i.e., with the depth of the water flow path increasing, hydrochemical tracers increased gradually (Fig. 3). Even though there are similar characteristics with Mg^{2+}, Ca^{2+} and SO_4^{2-} at some stage, it is still not certain whether K^+, Na^+, Cl^-, and HCO_3^- are applicable to hydrological process research. These indicators should be studied further. In addition, there are no apparent rules in values of pH, NO_3^- and dissolved Si, which cannot be regarded as applicable tracers.

According to the characteristics of all the chemical tracers, three groups of the chemical behavior of the catchment discharge can be identified at the catchment outlet:

- Group I: EC, TDS, Ca^{2+}, Mg^{2+} and SO_4^{2-}, which can be used as good tracers to identify runoff paths;
- Group II: K^+, Na^+, Cl^-, and HCO_3^-, which are less applicable as tracers;
- Group III: pH, NO_3^-, and dissolved Si, which are not suitable as tracers.

The patterns of concentrations in all the runoff components reflect the variable contributions of flow components to the catchment outlet according to the main origin of the chemical elements. The measurement of conductivity reflects the ion concentration in water (Hendrickx et al., 2002). In reverse, the increase of conductivity in each separate runoff explained the process of runoff components. During the migration process, the TDS of four runoff components originating from gas precipitation, dissolved solid precipitation, dissolved salts and ion exchange gradually rose (Gibbs, 1970).

SO_4^{2-} originates essentially from the atmosphere but is concentrated within the catchment by evaporation processes and storage in soils (Liu, 1970). Ca^{2+} and Mg^{2+} are more affected by atmospheric sources and mineral weathering within the soil (Mcdonnell et al., 1990). Cl^- would not form an insoluble salt with the main cations, so it is a stable component and is suitable to be a tracer (Shen, 2010). During the runoff process, Cl^- will accumulate gradually with the growth of runoff path (Ladouche et al., 2001); however, in this study,

Table 2. Results of the hydrological and isotopic analysis for various samples.

Samples type	statistics	pH	EC ($\mu S\ cm^{-1}$)	TDS ($mg\ L^{-1}$)	$\delta^{18}O$ (‰)	δD (‰)	Cl^- ($mg\ L^{-1}$)	NO_3^- ($mg\ L^{-1}$)	SO_4^{2-} ($mg\ L^{-1}$)	HCO_3^- ($mg\ L^{-1}$)	K^+ ($mg\ L^{-1}$)	Na^+ ($mg\ L^{-1}$)	Ca^{2+} ($mg\ L^{-1}$)	Mg^{2+} ($mg\ L^{-1}$)	Si ($mg\ L^{-1}$)
Rainfall	Mean	7.97	17.86	6.72	−11.65	−75.53	0.22	0.94	1.15	6.59	0.09	0.41	1.48	0.18	1.45
	SD	0.92	9.84	3.12	2.24	19.21	0.12	0.97	0.79	3.66	0.01	0.56	1.06	0.05	1.17
SR	Mean	7.58	37.28	20.40	−12.24	−82.07	0.59	0.02	1.20	22.18	1.27	0.29	5.39	0.55	3.57
	SD	0.29	9.78	4.14	1.69	15.518	0.25	0.03	0.59	7.48	0.38	0.42	1.21	0.16	1.43
SSR30	Mean	7.72	51.42	28.03	−12.29	−81.47	0.72	0.07	1.42	28.77	1.16	0.26	9.22	0.80	4.30
	SD	0.27	14.49	6.58	1.59	13.06	0.43	0.05	0.60	9.72	0.29	0.19	3.49	0.28	1.85
SSR60	Mean	8.07	118.11	53.59	−11.83	−79.46	0.62	0.08	2.50	53.34	0.72	0.44	21.19	1.37	4.50
	SD	0.21	39.39	22.24	1.13	10.141	0.13	0.06	1.19	28.43	0.15	0.21	8.49	0.53	2.18
SSR100	Mean	7.95	297.93	123.10	−10.85	−69.99	0.43	0.23	6.81	109.10	0.51	1.41	54.83	4.33	6.87
	SD	0.20	61.77	35.92	1.27	6.56	0.12	0.27	2.14	7.01	0.12	0.39	8.80	0.97	1.21

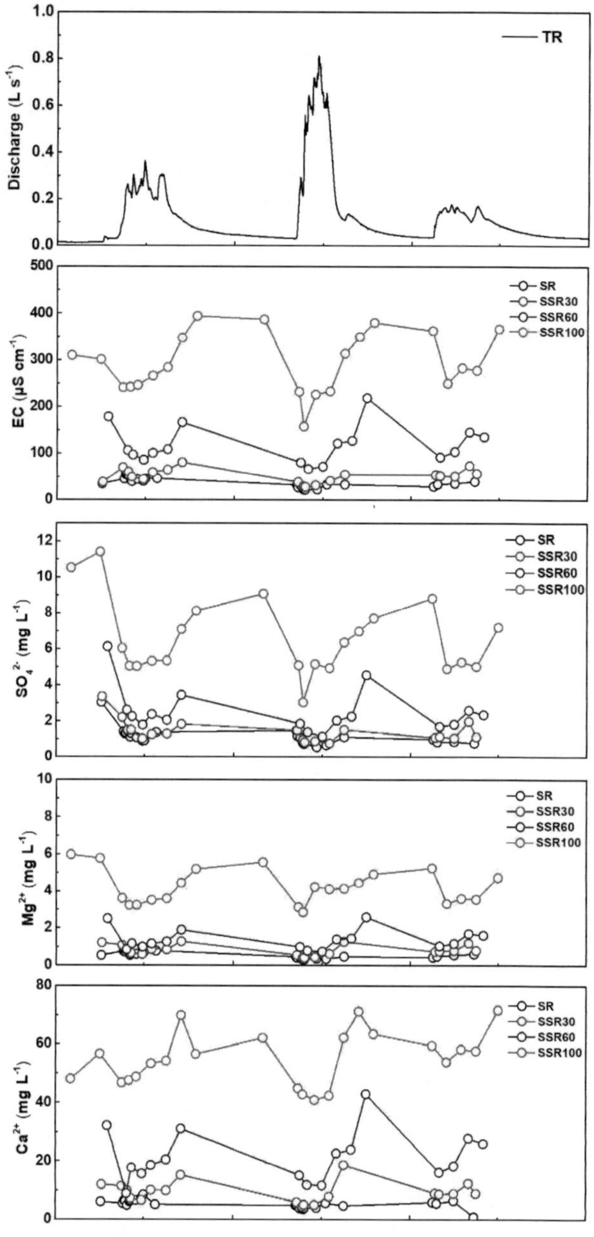

Figure 3. Processes of EC, SO_4^{2-}, Mg^{2+} and Ca^{2+} in different runoff components. TR: total runoff.

Cl^- didn't reflect the same pattern as other studies, and the change of Cl^- is a complex mixing process. This finding is consistent with early work at the site by Kendall et al. (2001). The time-process changes of Ca^{2+}, Mg^{2+} and SO_4^{2-} were consistent, indicating that the controls on hydrochemical ions were uniform. Gu et al. (2007) previously found that the time-process changes of hydrochemical ions in runoff were similar to that in rainfall. This led to the conclusion that rainfall plays a leading role in this change; however, it is not

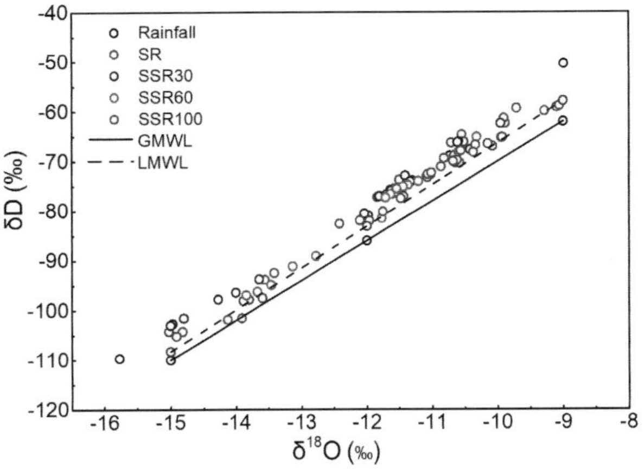

Figure 4. Relationship between δ^{18}O and δD in rainfall, SR, SSR30, SSR60 and SSR100. GMWL: Global Meteoric Water Line (Craig, 1961); LMWL: Local Meteoric Water Line (Tan et al., 2009).

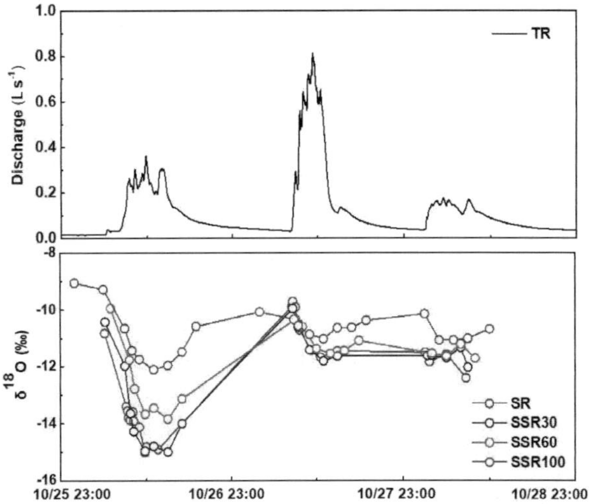

Figure 5. Processes of total discharge and δ^{18}O in different runoff components.

the only source because the ion concentration gradually increases, which can also be affected by pre-event water. In addition, hydrochemical ions can be the result of biological processes and water-rock interaction in the soil, contributing to the increase of hydrochemical ions. So, it is possible that there is a mixture of pre-event water and event water in this rainfall, and a mixture of upper layer runoff infiltration and lower layer runoff in space.

3.3 Isotopic compositions of waters

In this study area, there was a lack of hydrogen and oxygen data in multi-year rainfall, but the data for Nanjing was adequate, allowing the local meteoric water line in Nanjing to be used in this area. Rainfall, SR, SSR30, SSR60 and SSR100 of this storm samples are distributed along the local meteoric water line (Fig. 4) in a δ^{18}O-δD relation diagram (Tan et al., 2009):

$$\delta D = 8.4\,\delta^{18}O + 17.71 \quad (1)$$

which means that interflow water and groundwater are not influenced by the evaporation process (McDonnell et al., 1990). The rain seeps through the soil and the unsaturated zone and mixes with non-evaporated water from pre-event water (Ladouche et al., 2001).

During the above three rainfall periods, the sampling of SR, SSR30, SSR60 and groundwater runoff showed a common isotopic characteristic (Fig. 5): a negative correlation between discharge and δ^{18}O. The temporal isotopic content of δ^{18}O in all the runoff components had tended to return to the initial value during the first recession stage, during which the δ^{18}O sampling of groundwater was the richest and that of SR and SSR30 the most depleted, indicating that SR, SSR30, SSR60 may have been partially updated and diluted by the event water while there was still a large part of the pre-event water in the groundwater runoff. At the start of the rainfall, there was infiltration of depleted rainfall into the soil layer with enriched ^{18}O, and rainfall and soil water mixed. However, due to the combination of event water and the water in the soil, the values of δ^{18}O in SR, SSR30 and SSR60 became similar, which was apparent in the third period. As a result, the proportion of pre-event water became smaller as event water infiltrated. From Fig. 5, we can find that even though the rainfall volume of the second period was the largest, the difference between all the runoff components was smaller than that in the first stage, which indicated that the combination of rainfall and the soil water was almost over. The contributions of event water (rainfall) and pre-event water (soil water) during the rainfall event will continue to be studied in the following research. In this study, due to the linear relationship between δ^{18}O and δD, δD shows consistent feature with δ^{18}O.

3.4 Identification of rainfall-runoff flow paths

From the relationship between δ^{18}O and different hydrochemical indexes (Fig. 6), we can find a linear relationship between δ^{18}O and EC, TDS, SO_4^{2-}, Ca^{2+} in different runoff and can identify the runoff path, including surface runoff and subsurface runoff. In detail, δ^{18}O and EC, TDS, SO_4^{2-}, Ca^{2+} had a negative relationship in surface runoff and SSR30 and showed a positive correlation in SSR60 and SSR100, and the values of hydrochemical indicators increased at depth.

In previous studies, Gu (1992), Gu and Freer (1995), and Gu et al. (2018) concluded that there are three runoff components including surface runoff, interflow in the unsaturated zone and groundwater runoff in the saturated zone at the Hydrohill catchment and obtained runoff production meth-

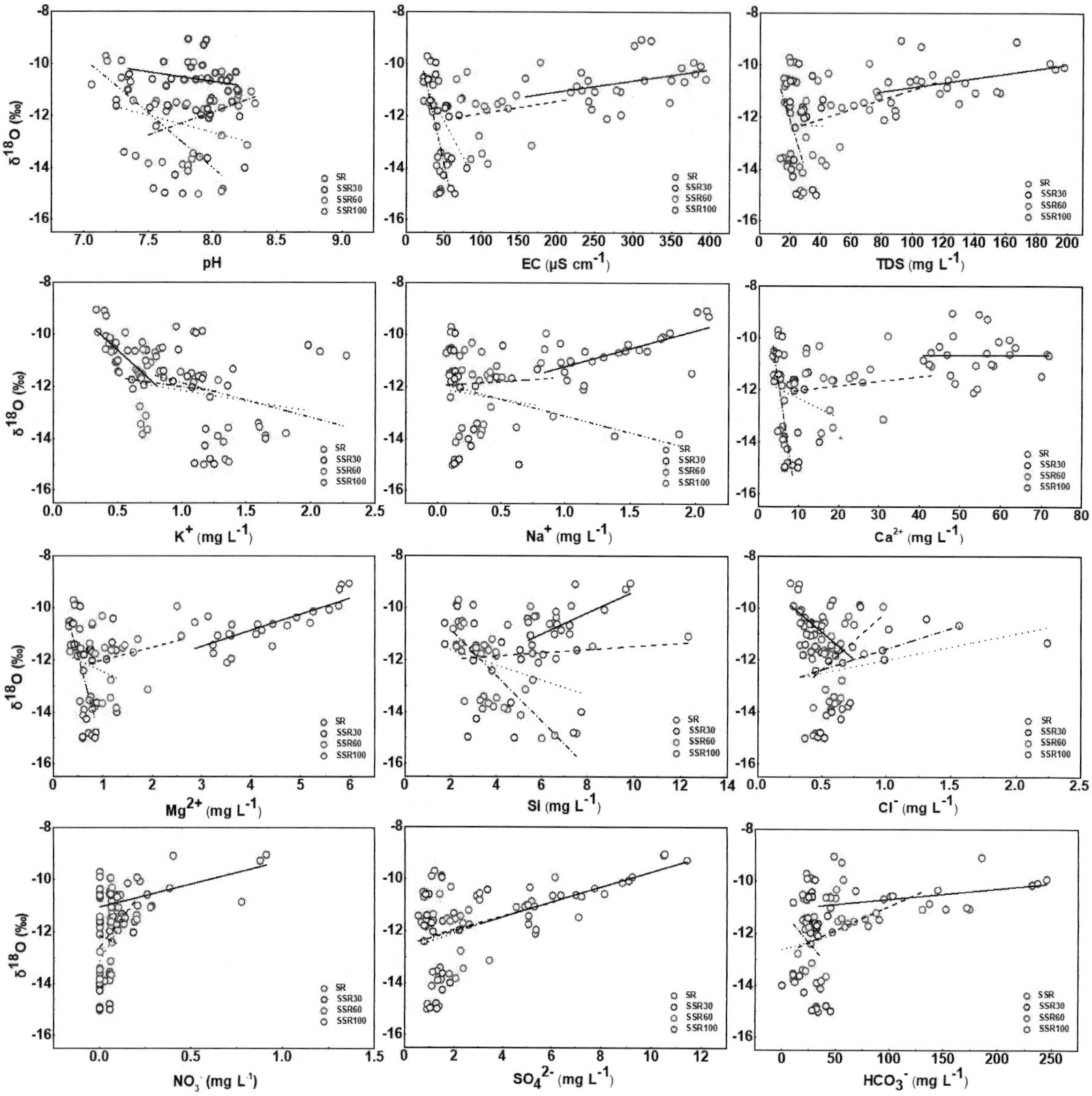

Figure 6. Relationship between δ^{18}O and different hydrochemical indexes.

ods combined with hydrochemistry and isotope tracing. The processes of hydrochemitry and ^{18}O in rainfall and different runoff components are related to its cause, flow paths, rainfall volume and intensity, etc. Hydrochemical ions and isotopes can enter the catchment with precipitation and flow through surface runoff and subsurface runoff to the total outlet. During this process, biological interaction and water-rock interaction take place. The bottom of Hydrohill is closed, and the final water source for runoff generation is rainfall. Though rainfall plays a leading role in the generation of runoff, it does not play a dominant role in hydro-geochemical components (Gu et al., 2007). The composition of hydrochemical and isotopic tracers in Hydrohill can vary due to the interaction between water and soil. Thus, tracers can be effective tools to identify flow pathways in rainfall-runoff processes.

3.5 Hydrological and isotopic flux

During the three consecutive rainfall periods, surface runoff and subsurface runoff meet at the total outlet section. Meanwhile, the hydrochemical and isotopic indicators in the surface runoff and subsurface runoff make separate contributions to the total outlet output. Thus, the quantity of flow flux and mass fluxes of hydrochemical and isotopic indica-

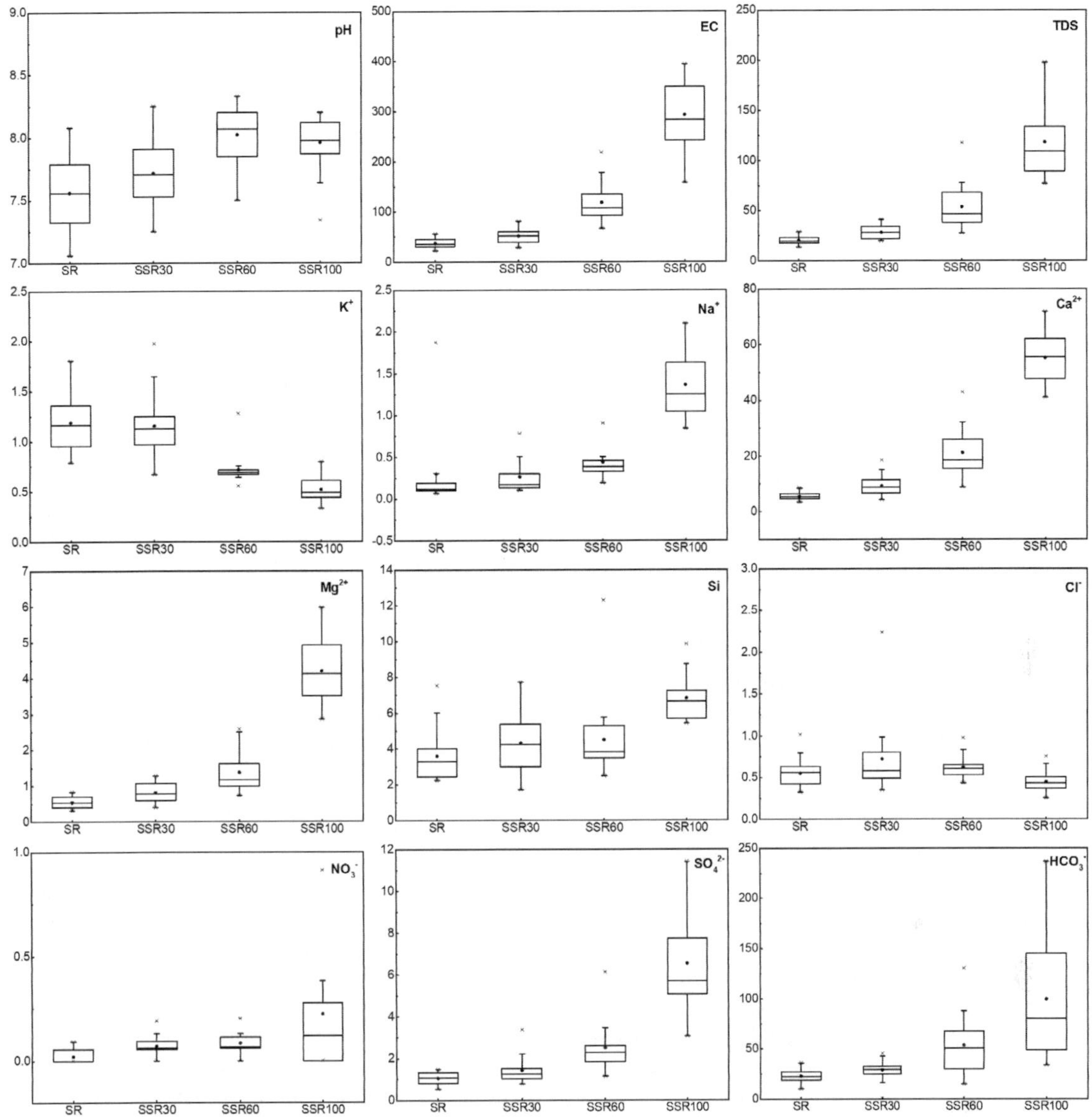

Figure 7. Concentrations of K^+, Na^+, Cl^-, HCO_3^- in different runoff components for the rainfall-runoff events (The asterisk represents an outlier).

tors outputted from surface runoff, SSR30, SSR60, SSR100 and total runoff can be calculated by Eqs. (2) and (3),

$$R=\int_{t_1}^{t_2} Q\mathrm{d}t \tag{2}$$

$$M=\int_{t_1}^{t_2} Qc\mathrm{d}t \tag{3}$$

where R is the quantity of flow flux from runoff components and M is the mass fluxes of hydrochemical and isotopic indicators outputted in the runoff; t_1 and t_2 represent two moments, respectively, Q is the quantity of flow at one sampling moment; c is the concentration of the solutes. The results of their contribution ratios for the total runoff in different runoff components in three rainfall events are shown in Table 3.

In this rainfall event, runoff can mainly generate from SSR100 (groundwater runoff), contributing much more than SR, SSR30 and SSR60. Meanwhile, the mass fluxes of Ca^{2+}, Mg^{2+}, SO_4^{2-}, and $\delta^{18}O$ indicated that subsurface runoff contributed more to the total outlet, especially from groundwater runoff. As a result, the quantity of flow flux and mass fluxes of hydrochemical and isotopic indicators in groundwater runoff showed a positive relationship. The SR, SSR30 and SSR60 also showed a positive relationship, except that Ca^{2+}

Table 3. The contribution rate of the selected indexes during the hydrological periods.

Runoff components	Runoff		Ca^{2+}		Mg^{2+}		SO_4^{2-}		$\delta^{18}O$	
	R (m^3)	CR (%)	M (g)	CR (%)	M (g)	CR (%)	M (g)	CR (%)	M (g)	CR (%)
SR	19.94	32.14	91.79	6.03	8.71	7.42	17.28	9.75	235.37	32.94
SSR30	11.15	17.97	72.40	4.76	6.43	5.47	12.96	7.31	135.09	18.90
SSR60	5.33	8.58	81.43	5.35	5.35	4.56	10.09	5.69	64.28	8.99
SSR100	25.63	41.31	1276.08	83.86	96.92	82.55	136.91	77.25	279.84	39.16
TR	62.05		1521.70		117.41		177.23		714.58	

in SSR30 was smaller than SSR60, having an opposite relationship with the quantity of flow flux in SSR30 and SSR60. Even though different flow pathways existed in different rainfall events, in recent years, the hydrological community has come to a consensus that subsurface flow is generally dominated by the preferential flow of various kinds (Jones, 2010; Lin, 2010; Uhlenbrook, 2010). Subsurface preferential flow can route water through the subsurface, rather vertically and laterally. In addition, the contributions of Ca^{2+}, Mg^{2+}, SO_4^{2-} and $\delta^{18}O$ from SR, SSR30, SSR60 and SSR100 to the outlet in the three events indicated that runoff components originate from the mixing of event water and pre-event water, which is consistent with results of Kendall's study (Kendall et al., 2001). During the rainfall-runoff processes, different proportions of pre-event water have been updated by a different percentage of event water, but the detailed percentage of updated water was not calculated in this research.

4 Discussion

In this study, the characteristics of hydrochemical and isotopic factors in one typical rainfall event were analyzed. Through the analysis, flow pathways and main water sources reaching the total outlet could be identified. In Fig. 7, there was an outlier among the values of almost all the indicators, and the outlier of K^+ in SSR60, Na^+ in SR, dissolved Si in SSR60 and Cl^- in SSR30 exceeded much more than the upper limit. The uncertainty of the values may be mainly due to a measurement error or sampling error. Thus it is a little difficult to identify whether they can be good tracers to analyze the hydrological processes and they are classified in the second group and the third group. There is a need to collect more water samples from different rainfall events to determine this.

Studying hydrological processes enables us to know more about the flow paths, residence time distribution, and spatial distribution patterns of soil (Keith et al., 2010; Spence et al., 2010). This also means that hydrological processes are closely related to soil properties. For example, soil pores can influence flow pathways and soil moisture can affect hydrological processes, including solute transport and land-atmosphere interactions, as well as a range of geographic and pedogenic processes (Western et al., 2004). However, due to the lack of soil water data in this study, such as the soil water content and the hydrochemical and isotopic data of soil water, some further issues cannot be analyzed. Also, hydrograph separation for the event water and pre-event water and residence time distribution of runoff components in the Hydrohill catchment need to be studied.

5 Conclusions

Hydrochemical and isotopic methods are effective in tracing water movement from rainfall to runoff components at outlets of an artificial experimental catchment with separate runoff processes. EC, TDS, SO_4^{2-}, Ca^{2+}, Mg^{2+}, ^{18}O and D are more applicable to investigating rainfall and runoff processes at the Hydrohill experimental catchment. $\delta^{18}O$ and EC, TDS, SO_4^{2-}, Ca^{2+}, Mg^{2+} had a linear relationship and their correlations can be used to identify runoff flow paths including surface runoff, interflow and groundwater runoff.

Flow flux and mass fluxes of main hydrochemical and isotopic tracers gauged at the Hydrohill catchment outlet, Ca^{2+}, Mg^{2+}, SO_4^{2-}, and ^{18}O in the total runoff are mainly from groundwater runoff (SSR100). However, the quantitative results were very encouraging as a basis to develop hydrological models for further study, including examination of hydrograph separation for event water and pre-event water and residence time of different runoff components.

Data availability. Data is available based on request to the corresponding authors.

Author contributions. NY contributed results analysis and drafting the manuscript, JZ and GW structured the manuscript and contributed results discussion, JL contributed methodology of the work, GL contributed analysis on results reasonability, and AL contributed data collection and water sampling analysis.

Competing interests. The authors declare that they have no conflict of interest.

Special issue statement. This article is part of the special issue "Hydrological processes and water security in a changing world". It is a result of the 8th Global FRIEND–Water Conference: Hydrological Processes and Water Security in a Changing World, Beijing, China, 6–9 November 2018.

Acknowledgements. We appreciate Weizu Gu for his valuable comments and suggestions, Hao Zheng, Niu Wang for their assistance of data collection and laboratory analysis and Chaoyu Zheng, Tongping Liu for their water sampling.

Financial support. This research has been supported by the National Natural Science Foundation of China (grant nos. 41830863, 51879162, 51609145, and 91647203), the Second Qinghai-Tibet Plateau Comprehensive Scientific Research Project (grant no. 2019QZKK0203), and the National Key Research and Development Programs of China (grant no. 2016YFA0601501).

References

Brian, A. P., Wilfred, M. W., Feng, X. H., and Charles, J. V.: The application of electrical conductivity as a tracer for hydrograph separation in urban catchments, Hydrol. Process., 22, 1810–1818, https://doi.org/10.1002/hyp.6786, 2010.

Baraer, M., McKenzie, J., Mark, B. G., Gordon, R., Bury, J., Condom, T., Gomez, J., Knox, S., and Fortner, S. K.: Contribution of groundwater to the outflow from ungauged glacierized catchments: A multi-site study in the tropical Cordillera Blanca, Peru, Hydrol. Process., 29, 2561–2581, https://doi.org/10.1002/hyp.10386, 2015.

Carey, G. and Feng, X. H.: A stable isotope study of soil water: evidence for mixing and preferential flow paths, Geoderma, 119, 97–111, https://doi.org/10.1016/S0016-7061(03)00243-X, 2004.

Cartwright, I., Gilfedder, B., and Hofmann, H.: Contrasts between estimates of baseflow help discern multiple sources of water contributing to rivers, Hydrol. Earth Syst. Sci., 18, 15–30, https://doi.org/10.5194/hess-18-15-2014, 2014.

Craig, H.: Isotopic variations in meteoric waters, Science, 133, 1702–1703, https://doi.org/10.1029/93wr01684, 1961.

Gibbs, R. J.: Mechanisms controlling world water chemistry, Science, 170, 1088–1090, https://doi.org/10.1126/science.172.3985.870, 1970.

Gonzales, A. L., Nonner, J., Heijkers, J., and Uhlenbrook, S.: Comparison of different base flow separation methods in a lowland catchment, Hydrol. Earth Syst. Sci., 13, 2055–2068, https://doi.org/10.5194/hess-13-2055-2009, 2009.

Gu, W. Z.: Experimental Research on Catchment Runoff Responses Traced by Environmental Isotopes, Advances in Water Science, 3, 246–254, 1992 (in Chinese).

Gu, W. Z. and Freer, J.: Patterns of surface and subsurface runoff generation, in: Tracer Technologies for Hydrological Systems, Proceedings of Symposium Boulder, 229, 265–273, 1995.

Gu, W. Z., Lu, J. J., Zhao, X., and Norman, E. P.: Responses of hydrochemical inorganic ions in the rainfall-runoff processes of the experimental catchments and its significance for tracing, Adv. Water Res., 18, 1–7, 2007 (in Chinese).

Gu, W. Z., Liu J. F., Lin, H., Lin, J., Liu, H. W., Liao, A. M., Wang, N., Wang, W. Z., Ma, T., Yang, N., Li, X. G., Zhuo, P., and Cai, Z.: Why Hydrological Maze: The Hydropedological Trigger? Review of Experiments at Chuzhou Hydrology Laboratory. Frontiers in Hydropedology: Interdisciplinary Research from Soil Architecture to the Critical Zone, Vadose Zone J., 17, 1–27, https://doi.org/10.2136/vzj2017.09.0174, 2018.

Hendrickx, J. M. H., Das, B., Corwin, D. L., Wraith, J. M., and Kachanoski, R. G.: Relationship between soil water solute concentration and apparent soil electrical conductivity, Anal. Methods, 4, 1275–1282, 2002.

Hooper, R. P. and Shoemaker, C. A.: A comparison of chemical and isotopic hydrograph separation, Water Resour. Res., 22, 1444–1454, https://doi.org/10.1029/WR022i010p01444, 1986.

Hrachowitz, M., Bohte, R., Mul, M. L., Bogaard, T. A., Savenije, H. H. G., and Uhlenbrook, S.: On the value of combined event runoff and tracer analysis to improve understanding of catchment functioning in a data-scarce semi-arid area, Hydrol. Earth Syst. Sci., 15, 2007–2024, https://doi.org/10.5194/hess-15-2007-2011, 2011.

Hrachowitz, M., Benettin, P., Van Breukelen, B. M., Fovet, O., Howden, N. J., Ruiz, L., and Wade, A. J.: Transit times-the link between hydrology and water quality at the catchment scale, Wiley Interdisciplinary Reviews, Water, 3, 629–657, https://doi.org/10.1002/wat2.1155, 2016.

Hugenschmidt, C., Ingwersen, J., Sangchan, W., Sukvanachaikul, Y., Duffner, A., Uhlenbrook, S., and Streck, T.: A three-component hydrograph separation based on geochemical tracers in a tropical mountainous headwater catchment in northern Thailand, Hydrol. Earth Syst. Sci., 18, 525–537, https://doi.org/10.5194/hess-18-525-2014, 2014.

Huth, A. K., Leydecker, A., and Sickman, J. O.: A two-component hydrograph separation for three high-elevation catchments in the Sierra Nevada, California, Hydrol. Process., 18, 1721–1733, https://doi.org/10.1002/hyp.1414, 2004.

Jones, J. A.: Soil piping and catchment response, Hydrol. Process., 24, 1548–1566, https://doi.org/10.1002/hyp.7634, 2010.

Keith, D. M., Johnson, E. A., and Valeo, C.: A hillslope forest floor (duff) water budget and the transition to local control, Hydrol. Process., 24, 2738–2751, https://doi.org/10.1002/hyp.7697, 2010.

Kendall, C., McDonnell, J. J., and Gu, W. Z.: A look inside "black box" hydrograph separation models: a study at the hydrohill catchment, Hydrol. Process., 15, 1877–1902, https://doi.org/10.1002/hyp.245, 2001.

Klaus, J. and Mcdonnell, J. J.: Hydrograph separation using stable isotopes: review and evaluation, J. Hydrol., 505, 47–64, https://doi.org/10.1016/j.jhydrol.2013.09.006, 2013.

Ladouche, B., Probst, A., Viville, D., Idir, S., Baqué, D., Loubet, M., Probst, J. L., and Bariac, T.: Hydrograph separation using isotopic, chemical and hydrological approaches (Strengbach catchment, France), J. Hydrol., 242, 255–274, https://doi.org/10.1016/S0022-1694(00)00391-7, 2001.

Lin, H.: Linking principles of soil formation and flow regimes, J. Hydrol., 393, 3–19, https://doi.org/10.1016/j.jhydrol.2010.02.013, 2010.

Liu, C. Q.: Biogeochemical processes and cycling of nutrients in the earth's surface: cycling of nutrients in soil–plant systems of karstic environments, Southwest China, Science Press, Beijing, China, 1970 (in Chinese).

Liu, J. F., Gu, W. Z., Liao, A. M., Wang, N., Lu, J. J., Lin, J., Liu, H. W., Wang, W. Z., Ma, T., Cai, Z., Liao, M. H., Li, X. G., Zhuo, P., and Yang, N.: Hydrology of artificial and controlled experiments, Practice on the Watershed Hydrological Experimental System Reconciling Deterministic and Stochastic Subjects Based on the System Complexity: 1, Theoretical Study (chap. 11), 227–251, https://doi.org/10.5772/intechopen.78721, 2018a.

Liu, J. F., Liao, A. M., Wang, N., Lin, J., Liu, H. W., Wang, W. Z., Ma, T., Cai, Z., Liao, M. H., Li, X. G., Zhuo, P., Yang, N., Lu, J. J., and Gu, W. Z.: Hydrology of artificial and controlled experiments, Practice on the Watershed Hydrological Experimental System Reconciling Deterministic and Stochastic Subjects Based on the System Complexity: 2, Practice and Test (chap. 12), 253–281, https://doi.org/10.5772/intechopen.79357, 2018b.

Maurya, A. S., Shah, M., Deshpande, R. D., Bhardwaj, R. M., Prasad, A., and Gupta, S. K.: Hydrograph separation and precipitation source identification using stable water isotopes and conductivity: River Ganga at Himalayan foothills, Hydrol. Process., 25, 1521–1530, https://doi.org/10.1002/hyp.7912, 2011.

Mcdonnell, J. J., Bonell, M., Stewart, M. K., and Pearce, A.: Implications for Stream Hydrograph Separation, Water Resour. Res., 26, 455–458, 1990.

Mul, M. L., Mutiibwa, R. K., Uhlenbrook, S., and Savenije, H. H.: Hydrograph separation using hydrochemical tracers in the Makanya Catchment, Tanzania, Phys. Chem. Earth, 33, 151–156, https://doi.org/10.1016/j.pce.2007.04.015, 2008.

Pilla, G., Sacchi, E., Zuppi, G., Braga, G., and Ciancetti, G.: Hydrochemistry and isotope geochemistry as tools for groundwater hydrodynamic investigation in multilayer aquifers: a case study from Lomellina, Po plain, South-Western Lombardy, Italy, Hydrogeol. J., 14, 795–808, https://doi.org/10.1007/s10040-005-0465-2, 2006.

Robinson, Z. P., Fairchild, I. J., and Arrowsmith, C.: Stable isotope tracers of shallow groundwater recharge dynamics and mixing within an Icelandic sandur, Skeiđarársandur, Hydrology in mountain regions: Observations, processes and dynamics, 119–125, 2009.

Shen, X.: Natural Hydrochemistry, China Environmental Science Press, 102 pp., 2010 (in Chinese).

Spence, C., Guan, X. J., Phillips, R., Hedstrom, N., Granger, R., and Reid, B.: Storage dynamics and streamflow in a catchment with a variable contributing area, Hydrol. Process., 24, 2209–2221, https://doi.org/10.1002/hyp.7492, 2010.

Stadler, S., Sültenfuß, J., Holländer, H. M., Bohn, A., Jahnke, C., and Suckow, A.: Isotopic and geochemical indicators for groundwater flow and multi-component mixing near disturbed salt anticlines, Chem. Geol., 294, 226–242, https://doi.org/10.1016/j.chemgeo.2011.12.006, 2012.

Tan Z. C., Lu, B. H., Wang, J., and Sun, Y. Y.: Characteristics of stable hydrogen and oxygen isotopes of precipitation and runoff in Wudaogou Hydrological Experimental Catchment, J. Hohai Univ., 37, 650–654, 2009 (in Chinese).

Tardy, Y., Bustillo, V., and Boeglin, J. L.: Geochemistry applied to the watershed survey: hydrograph separation, erosion and soil dynamics. A case study: the basin of the Niger River, Africa, Appl. Geochem., 19, 469–518, https://doi.org/10.1016/j.apgeochem.2003.07.003, 2004.

Uhlenbrook, S.: Catchment hydrology-a science in which all processes are preferential, Hydrol. Process., 20, 3581–3585, https://doi.org/10.1002/hyp.6564, 2010.

Uhlenbrook, S., Frey, M., Leibundgut, C., and Maloszewski, P.: Hydrograph separations in a mesoscale mountainous basin at event and seasonal timescales, Water Resour. Res., 38, 1–14, https://doi.org/10.1029/2001WR000938, 2002.

Vincent, M., Jean-Francois, D. L., and Couren, M.: Investigation of the hydrological processes using chemical and isotopic tracers in a small Mediterranean forested catchment during autumn recharge, J. Hydrol., 247, 215–229, https://doi.org/10.1016/s0022-1694(01)00386-9, 2001.

Walton, N. R. G.: Electrical Conductivity and Total Dissolved Solids-What is Their Precise Relationship?, Desalination, 72, 275–292, https://doi.org/10.1016/0011-9164(89)80012-8, 1989.

Western, A. W., Zhou, S. L., Grayson, R. B., McMahon, T. A., Blöschl, G., and Wilson, D. J.: Spatial correlation of soil moisture in small catchments and its relationship to dominant spatial hydrological processes, J. Hydrol., 286, 113–134, https://doi.org/10.1016/j.jhydrol.2003.09.014, 2004.

Williams, M. W., Barnes, R. T., Parman, J. N., Freppaz, M., and Hood, E.: Stream Water Chemistry along an Elevational Gradient from the Continental Divide to the Foothills of the Rocky Mountains, Vadose Zone J., 10, 900–914, https://doi.org/10.2136/vzj2010.0131, 2011.

Yang, Y. G., Xiao, H. L., Zou, S. B., Zhao, L. J., Zhou, M. X., Hou, L. G., and Wang, F.: Hydrochemical and hydrological processes in the different landscape zones of alpine cold region in China, Environ. Earth Sci., 65, 609–620, https://doi.org/10.1007/s12665-011-1108-7, 2012.

Zhang, Q., Knowles, J. F., Barnes, R. T., Cowie, R. M., Rock, N., and Williams, M. W.: Surface and subsurface water contributions to streamflow from a mesoscale watershed in complex mountain terrain, Hydrol. Process., 32, 954–967, https://doi.org/10.1002/hyp.11469, 2018.

Zhao, R. J.: Catchment Hydrological Modelling, Water Conservancy Press, Beijing, 76–79, 1989 (in Chinese).

Proc. IAHS, 383, 111–119, 2020
https://doi.org/10.5194/piahs-383-111-2020

Analysis of the influence of environmental change on river runoff in Loess Plateau of China

Yong Jing[1,2] **and Zuhao Zhou**[1]

[1]China Institute of Water Resources and Hydropower Research, Beijing, 100038, China
[2]Hydrology and Water Resources Survey Bureau of Shaanxi Province, Xi'an, 710068, China

Correspondence: Yong Jing (yong.j@126.com) and Zuhao Zhou (zhzh@iwhr.com)

Published: 16 September 2020

Abstract. The double interaction between climate change and human activity affects the changes in the environmental conditions of catchment runoff and confluence. Using 1956–2012 57 years of river runoff series data of 27 rivers in hilly gully area of Loess Plateau in Shaanxi province 39 hydrological station data, reduction of social and economic water consumption, plotting the annual precipitation and annual runoff double cumulative curve of annual runoff flow series consistency test, the consistency processing of annual runoff data and the quantitative analysis of its influence were also made. The results show that: the consistency of annual runoff data sequence of 21 hydrological stations is affected by the change of environmental conditions of runoff generation and confluence, and the turning point (year) and the degree of impact can be divided into three situations or three periods. One is the 12 station in 2000 after the annual runoff system is small; Two, there are 4 stations before and after in 1970, the annual precipitation and annual runoff double cumulative curve is obviously turning point. Before the turning point, a series of systems has a large high production period. After turning, it shows that a series of systems with low runoff yield caused by the Changes in environmental conditions of runoff and confluence in the underlying surface and climate and so on; the three is to have 5 stations occurred both before and after the 1970 high low flow period of the transition period in 2000, and after the annual runoff series of small low again.

1 Foreword

In China, the observed annual runoff of the Yellow River, Haihe River and the Pearl River has decreased sharply, the trend of drought and flood has intensified (Zhang et al., 2012), and the research upsurge of climate change and human activities affecting the runoff variation in the river basin has lasted for more than 10 years (Hui et al., 1993; Jing et al., 2008a–d; Jing, 2016; Peng, 2016; Wang et al., 2001, 2012; Zhang et al., 2012; Zhu et al., 2008).The quantitative study of the effects of climate factors and human activities, in particular various human activities, on the runoff of rivers undoubtedly contributes to the scientific evaluation of increasingly strained water resources and their exploitation and utilization, It is of great significance to support social economic and political development and the construction of ecological civilization.

The present quantitative analysis of climate factors and the changes of the environmental conditions of the catchment production and confluence made up of human activities are summarized. The methods of thinking which affect the river runoff (Hui et al., 1993; Jing et al., 2008a, b, d; Jing, 2016; Peng, 2016; Wang et al., 2001, 2012; Zhang et al., 2012; Zhu et al., 2008) are more consistent : In the long series of hydrological data, the short series of "natural runoff"in the period of high yield flow on the underlying surface of the original ecology with slight influence on early human activities were classified. This requires finding a turning point (year) in the impact of strong human activities; and then using the "underlying surface of the primordial" high-yielding "natural"runoff data as a "baseline". The difference between the recent runoff data of "current underlying surface" and that of low-yield runoff is the comprehensive influence value of the environmental change of watershed runoff. Finally, on the basis of data and methods, the climate factors and human activities can be separated from this comprehensive ef-

Published by Copernicus Publications on behalf of the International Association of Hydrological Sciences.

fect value, and all kinds of factors of the effect of human activity can also be analyzed.

Taking the study area of Loess Plateau in the northern part of Shaanxi Province as an example, the time-series cumulative value correlation curve method is used in this paper. This paper analyzes and proves the great influence of the great change of environmental conditions on the abortion and confluence of river path: the "present condition underlying surface"of low yield water in the basin has been irreversible to the "primordial underlying surface" of its early high yield water. The hydrological data monitored under the condition of high yield water in the early stage of "original ecological underlying surface" no longer have the basis of "consistency of underlying surface" of "low yield water" in the near future, and three kinds of influence conditions are analyzed and summarized. In this paper, three kinds of correction methods of the "consistency" of the current underlying surface are proposed for the source of the data. On this basis, the comprehensive influence of environmental change on river runoff is analyzed quantitatively.

The research area is the central part of China's Loess Plateau and an important national energy and chemical base. In this area, the gullies are developed, the topography is broken, the precipitation is less, and the water resources are poor, so it is the most serious and important area of soil and water loss in history. The whole area is about 93 000 km^2, the population is 5.635 million people, the annual GDP scale is about CNY 385.6 billion, the annual output value of agriculture, forestry, animal husbandry and fishery industry is about CNY 48.67 billion. Since the 1960s and 1970s, the massive soil and water loss control has continued to this day, with the rapid development of modern energy and chemical industry park and urbanization, the development and utilization of regional soil and water mineral energy resources, as well as the geomorphological and hydrogeological conditions of the underlying surface of natural production and confluence in the basin are deeply affected.

2 Research methods and data sources

2.1 Research methods

Covering the hilly and gully region of the Loess Plateau in Shaanxi, a 57-year series from 1956 to 2012 of the annual runoff data of 39 hydrologic stations on 27 rivers were selected. The series of runoff data are calculated by "reduction" of the amount of water consumed directly by human society year by year, and the method and steps of 57-year series consistency check and correction are adopted, To comprehensively study and verify the effects of long-term large-scale human activities on the gradual accumulation of runoff production and confluence on the underlying surface of the basin, and the changes of climatic factors such as precipitation, which have a dual interaction with each other, And finally, the environmental conditions of runoff production and confluence in the river basin will result in a turning point change of low yield flow under the "current underlying surface", and make the river runoff decrease.

Data preprocessing: under the condition that the condition of the underlying surface of the river basin remains consistent, the annual river runoff series should be kept basically consistent after the amount of river water consumed directly by human society is "reduced" to the measured amount of water measured by the hydrologic station. Before analyzing and judging whether the consistency of annual river runoff series is destroyed by the change of underlying surface, we should carry out the "reduction" of the direct consumption of river water by human society by the measured data of hydrologic station. As far as possible, the actual annual runoff of the hydrological section can reflect the "true value" of the current year's runoff yield and confluence of the watershed, that is, the "reduction calculation" of the "true value" of the measured annual runoff in the current year. The "reduction calculation" adopts the method of investigation and analysis of agricultural irrigation, industry, urban and rural life, reservoir water storage variables and the amount of water diversion across the river basin by means of the itemized investigation and reduction method.

By using the method of subdivision investigation and reduction: $W_{\text{natural}} = W_{\text{measured}} + W_{\text{irrigation}} + W_{\text{industrial}} + W_{\text{urban life}} + W_{\text{rural life}} + W_{\text{reservoir storage variable}} \pm W_{\text{diversion of water (drainage area elicited or introduced)}}$, the amount of water consumed by human society is added to the measured value of hydrology, that is, the natural river runoff is reduced to "natural river runoff". Year by month, agricultural irrigation, industry, urban and rural life, reservoir water storage variables and cross-basin water diversion are investigated and calculated month by month, forming the series of "natural river runoff" after the reduction of human society's consumption of river water by every hydrological station from 1956 to 2012. In order to further analyze whether the underlying surface and other environmental factors have an impact on its series.

2.2 Data sources

39 hydrological station, 95 precipitation station from 1956 to 2012, 57 years monitoring data, analysis and calculation and involves a large number of agricultural irrigation, industrial and urban life, reservoir storage and inter-basin water transfer variable water volume, water and soil conservation engineering construction status and the related data, such as social economic indicators from the author of nearly 30 years in the handbook of shaanxi province hydrology, the shaanxi province water resources investigation and water resources in Shanxi Province and other preparation work, the relevant government water conservancy, statistics, meteorological departments to collect, textual research and practical investigation data accumulation.

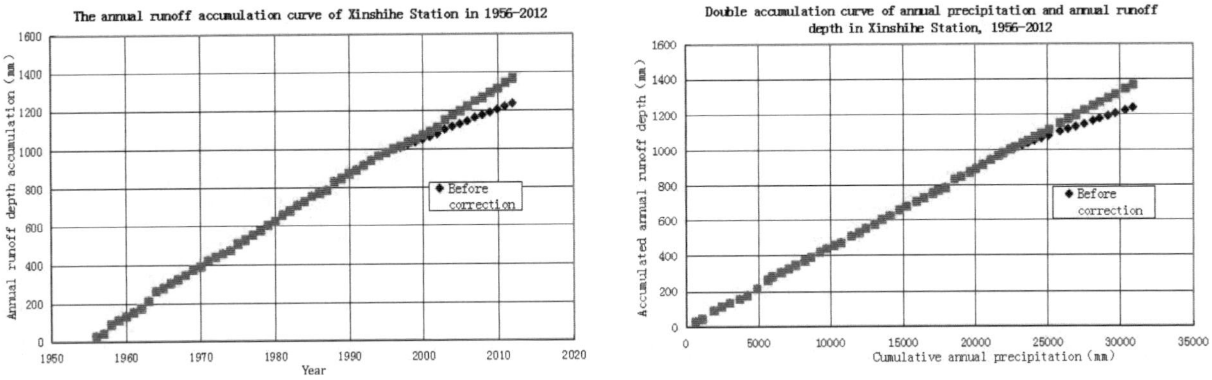

Figure 1. Annual runoff accumulation and double accumulation curve of annual precipitation runoff at Shenmu station.

Table 1. Statistical tables of small hydrological stations in the annual runoff series system around 2000 (Data Series 1956–2012).

Hydrologic station	Series of damage turning points (year)	Series of damage turning points (years) before the series: Annual runoff depth mean (mm)	Series of damage turning points (years) before the series: Annual mean precipitation (mm)	Series of damage turning points (years) after the series: Annual runoff depth mean (mm)	Series of damage turning points (years) after the series: Annual mean precipitation (mm)	Correction of the annual mean runoff in the same period after the break year (mm)	After the destruction turning point (year) series: Average annual mean value of uncorrected runoff is less than that before the revised series (%)	After the destruction turning point (year) series: Annual precipitation increased or decreased (%)	Annual precipitation increased or decreased (%)
HuangPu	1999	49.6	385.8	11.5	379.6	46.5	−76.8	−1.6	−6.3
WangDaoHengTa	1996	56.7	378.9	19.3	350.7	56.8	−66.0	−7.4	0.2
ShenMu	1998	70.4	399.2	24.3	381.3	67.1	−65.5	−4.5	−4.7
WenJiaChuan	1996	77.2	406.7	26.7	383.1	68.7	−65.4	−5.8	−11.0
XinShiHe	1995	24.7	554.2	15.0	516.2	22.3	−39.3	−6.9	−9.7
CaoPing	1996	48.0	456.7	24.9	451.7	46.3	−48.1	−1.1	−3.5
ZiChang	2003	47.6	493.3	28.3	519.4	47.5	−40.5	5.3	−0.2
YanChuan	2003	44.1	448.3	27.8	468.4	45.7	−37.0	4.5	3.6
YanAn	2003	42.7	484.8	23.8	441.0	38.8	−44.3	−9.0	−9.1
GanGuYi	2003	38.4	506.6	24.9	471.4	36.1	−35.2	−6.9	−6.0
DaCun	1996	37.8	563.0	23.6	523.9	33.6	−37.6	−6.9	−11.1
DanShi	2000	41.9	375.8	25.3	415.6	43.1	−39.6	10.6	2.9

3 Result analysis and discussion

3.1 Consistency analysis of annual river runoff series based on environmental change

After the amount of water consumed by human society is "reduced" to the measured amount of water, in general, the annual river runoff series should be kept consistent basically, but the mechanism that the underlying surface condition affects the consistency of runoff production and confluence is complicated and the process is long. It is very difficult to judge and quantitatively calculate its influence. Through the upstream and downstream balance check of the river runoff after the "actual measurement and reduction of the amount of water consumed by human society", and the comparative analysis of the trunk and tributaries, the double and single accumulative curves of the annual precipitation and the annual runoff of 39 hydrologic stations are drawn. The results show that the consistency of the annual runoff data series at 21 stations is broken by the change of the environmental conditions of runoff production and confluence, the damage turning point (year) and the influence degree can be divided into three situations.

1. the slope of the cumulative curves of the 12 stations, such as Shenmu River, slowed significantly after 2000, showing that the annual runoff sequence system was smaller. The details are Fig. 1 and Table 1.

2. the double accumulation curve of annual precipitation runoff of 4 stations had a significant turning point around 1970, and the series before the turning point showed a large and systematic high-yield flow period. After turning, a series of low yield periods are shown as a result of the change of environment conditions such as underlying surface and climate. See Fig. 2 and Table 2 for details.

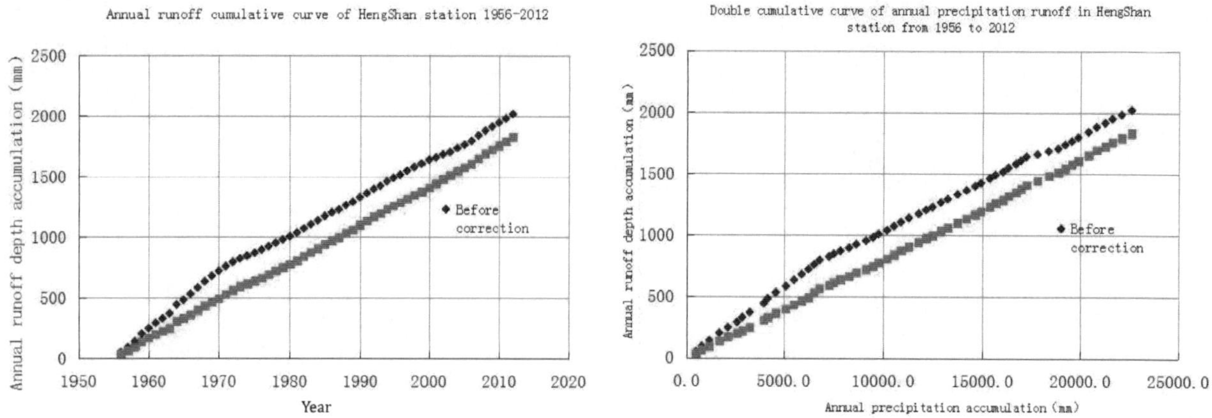

Figure 2. Annual runoff accumulation and double accumulation curve of annual precipitation runoff at Hengshan station.

Table 2. Statistical tables of small hydrological stations in the annual runoff series system around 1970 (Data Series 1956–2012).

Hydrologic(al) station	Series of damage turning points (year)	Series of damage turning points (years) before the series		After the destruction turning point (year) series		Correction of the mean annual runoff depth after the high-yield current period before the breakage turning point (mm)	Series of damage turning points (years) ago		Difference of the average annual runoff of the revised series ratio of high-yield current period "after the destruction turning point" (%)
		Annual runoff depth mean (mm)	Annual mean precipitation (mm)	Annual runoff depth mean (mm)	Annual mean precipitation (mm)		Average annual runoff depth of the modified high-yield flow period series is lower than that of the pre-correction period (%)	Annual precipitation over the same period (mm)	
HengShan	1970	48.2	415.4	30.9	390.3	32.8	−32.0	415.4	6.1
JingJiaGou	1970	52.1	445.3	40.1	391.0	44.2	−15.2	445.3	10.2
QingYangCha	1970	55.8	459.6	40.2	432.1	43.1	−22.8	459.6	7.2
BaiJiaCha	1970	52.8	484.4	36.9	447.8	38.3	−27.5	484.4	3.8

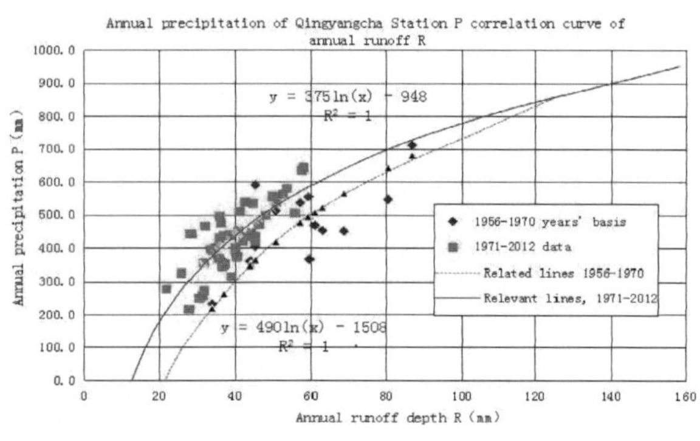

Figure 3. Relationship between annual precipitation and runoff at Qingyangcha station.

Figure 3 of the relationship between annual precipitation and annual runoff at Qingyangcha station also shows that in the same order of magnitude of annual precipitation, the annual runoff of the 1956–1970 series before 1970 is higher than that of the later series 1971–2012. It is necessary to revise the annual runoff of the high-yield flow period of "primordial underlying surface" with slight disturbance of human activities to the period of low-yield flow under the environment of "current underlying surface" after the change of strong human activities and climate factors, in order to maintain the consistency of environmental conditions of production and confluence (see Sect. 3.2).

3. In five stations, a turning point occurred in the period of high and low runoff of "primordial underlying surface" around 1970, and after 2000, the annual runoff series system in the environment of "current underlying surface" was again smaller than the low yield period, as shown in Fig. 4 and Table 3, respectively.

Table 3. Statistical tables of small hydrological stations in the annual runoff series system around 1nd 2000 (Data series 1956–2012).

Hydrologic(al) station	(1) Turning point of high yield flow period of "underlying surface of primordial ecosystem" (year)	(2) Turning point of secondary failure in the environment of "current underlying surface" during the period of low yield and current flow (year)	The series before the turning point of high-yield flow period in "underlying surface of primordial ecology" (1)			Turning point (1) and (2) "current underlying surface" environmental low-yield flow period series		"current underlying surface" environmental low-yield flow period after the turning point (2) series			The modified series of high yield flow period compared with turning point (1) and (2): "current underlying surface" environmental low yield flow period series		After the correction of the low yield flow period turning point (2), the series is higher than that between turning point (1) and (2). The "current underlying surface" environmental low yield flow period series	
			Annual runoff depth mean (mm)		Annual mean precipitation (mm)	Annual runoff depth mean (mm)	Annual mean precipitation (mm)	Annual runoff depth mean (mm)		Annual mean precipitation (mm)				
			Before correction	Behind correction				Before correction	Behind correction		Annual runoff mean difference (%)	Annual precipitation mean difference (%)	Annual runoff mean difference (%)	Annual precipitation mean difference (%)
MaHuYu	1970	1998	64.5	59.8	455.0	39.3	342.0	32.0	48.7	396.0	52.2	33.0	23.9	15.8
GaoShiYa	1978	1998	83.6	68.1	443.1	51.6	375.4	17.9	50.6	385.3	32.0	18.0	−1.9	2.6
GaoJiaBao	1971	1999	176.3	153.3	445.4	128.2	355.7	88.0	130.6	421.1	19.6	25.2	1.9	18.4
GaoJiaChuan	1971	1999	134.6	116.4	449.7	101.5	355.2	70.6	104.0	430.7	14.7	26.6	2.5	21.3
ShenJiaWan	1971	1998	96.1	72.1	394.1	48.5	326.1	28.7	56.1	360.6	48.7	20.9	15.7	10.6

3.2 Consistency correction of "underlying surface" of annual runoff series

3.2.1 Correction method

Different ways of thinking and methods were used to correct the environmental changes of the three kinds of production and confluence mentioned above.

For the series which only changed around 2000, it was considered that the environmental conditions of the "current underlying surface" and "primordial underlying surface" were basically consistent with those of the "original ecological underlying surface" during the high-yield flow period. And the runoff series is relatively small due to the lack of "natural" amount of the measured annual runoff in that year, which is corrected by the relationship of precipitation and runoff before the turn.

For the series that only turned after 1970, Table 2 shows that although the series precipitation decreases after turning, Fig. 3 the precipitation runoff relationship curve shows that the two series are obviously larger and smaller before and after the transition. The investigation confirmed the physical causes of the disturbance of strong human activities such as persistent large-scale soil and water conservation in these areas before and after the turning year, and considered that the high-yield flow period of the "underlying surface of the original ecology" has been transformed into the period of low yield flow under the environmental condition of "current underlying surface". The correction coefficient method is used to correct the high yield flow before turning to the low yield flow condition on the "current underlying surface". Operation steps of the modified coefficient method (Li, 2003): plot the relationship between precipitation and runoff before and after turning year (Fig. 3). Select an annual precipitation value, Two annual runoff depth values R_1 and R_2 were found on two curves. The annual runoff attenuation rate and correction coefficient were calculated by formulas $\alpha = (R_1 - R_2)/R_1 \cdot 100\,\%$ and $\beta = R_2/R_1$. The depth of runoff in the period of low yield flow on "current underlying surface"; Make annual precipitation correction coefficient Table 4; draw P–β relation curve diagram 5; check the correction coefficient of P–β relation curve, multiply by the annual runoff (depth) of the year to be revised, we can get the corresponding corrected natural annual runoff.

Table 4. Statistical table of annual precipitation and correction coefficient of Qingyangcha station.

P (mm)	200	300	400	500	600	700	800
A (%)	34.6	30.4	25.9	21.1	16.0	10.5	4.8
β	0.65	0.70	0.74	0.79	0.84	0.89	0.95

For the long series with two transitions around 1970 and 2000, it is the combination of the two situations. Table 3 shows that the two short series of precipitation before the turn of 1970 and after the turn of 2000 are greater than the mean of the series between the two turning points. It is considered that it is more reasonable to unify the high yield flow before 1970 and the low yield flow after 2000 to the "current underlying surface" low yield flow period between the two turning points. The correction coefficient method is used for high yield flow correction and the relationship of precipitation and runoff between the two turning points is corrected for low yield flow after 2000.

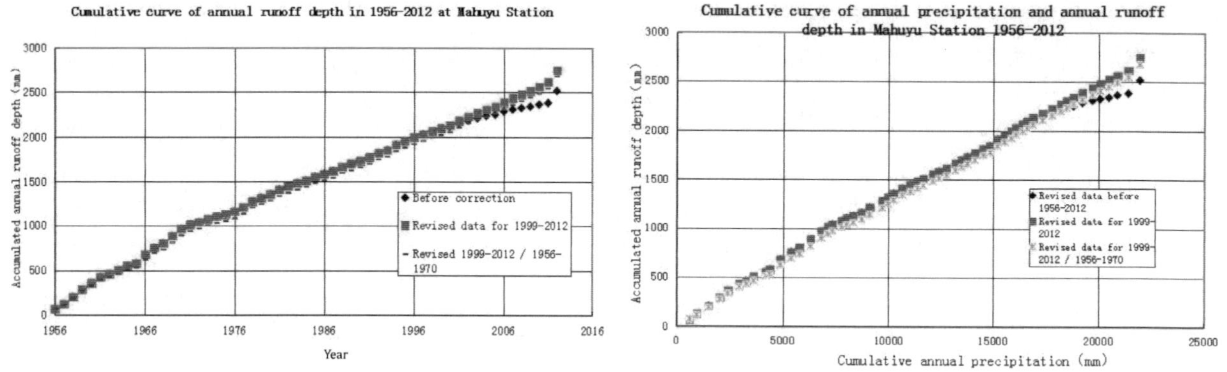

Figure 4. Annual runoff accumulation and double accumulation curve of annual precipitation runoff at Ma Huyu station.

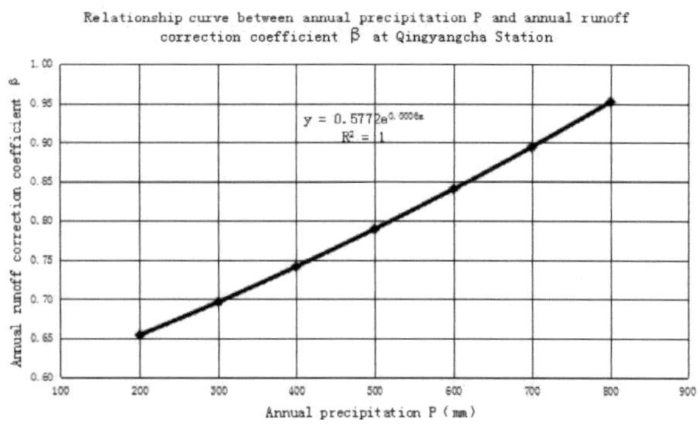

Figure 5. P–β relationship curve of Qingyangcha station.

3.2.2 Consistency Test of modified sequences

Plotting the double accumulation curve of the annual precipitation runoff and annual runoff single accumulation curve from 1956 to 2012 after correction, and Compared with the series of annual precipitation runoff relation curves before and after revision. From the cumulative line to the near straight line, the relative concentration degree and the correlation coefficient of the data scatter point are greatly improved, which shows that the modified sequence keeps a good consistency. The correlation coefficient of Gaoshiya station has been raised from 0.38 to 0.76, which is shown in Figs. 1, 2, 4, 6.

3.3 Analysis of annual runoff affected by environmental change

3.3.1 Analysis on Variation of annual Precipitation Series from 1956 to 2012

The accumulated annual precipitation curve of 21 hydrologic stations, whose annual runoff series consistency has been destroyed due to the great change of underlying surfaces, shows that 17 stations have maintained a good stability of 57 years from 1956 to 2012. It is further proved that after excluding the influence of no mutation of precipitation factors and direct consumption of river water by human activities, only the consistency of the underlying surface is greatly changed and the abrupt turning system of annual runoff series is reduced. There are 4 hydrological stations, such as Qingyangcha, where the annual precipitation series is larger after 2000 than in 2000–2012 and 1956–1999,the mean value of the series is increased from 9 % to 18.6 %, which indicates that the rainfall runoff fitting relationship before the previous year 2000 runoff series is reasonable. See Fig. 7 and Table 5 for details.

3.3.2 Quantitative Analysis of the impact of Environmental changes such as precipitation and Human activities on Annual runoff

Based on the average annual runoff of the series of high yield runoff in the "underlying surface of the original ecology", the series mean after the consistency correction is compared with it to judge the influence degree of the environmental change. Tables 1 and 2 show that the average effect of the corresponding two cases was 49.6 % and 24.3 %, the highest was 76.8 % and 32 %, and the lowest was 35.2 % and 15.2 %. The mean influence degree of the five stations with two turns around 1970 and 2000 was 15 % and 30 %, respectively, as shown in Table 6.

The special Wudinghe River and Beiluohe River Basin environmental changes on regional effects of the runoff are quantitatively analyzed from the author (Jing et al., 2008a, b; Jing, 2016): The effect of soil and water conservation measures on runoff decreased by more than 50 % in the middle reaches of Wudinghe River, while the decrease of rainfall in the whole basin affected the runoff decrease by 76.4 %, and the influence of human activity accounted for 23.6 %,among them, the water consumption in human society accounted for 18.8 %, and the effect of soil and water conservation measures was 4.8 %.

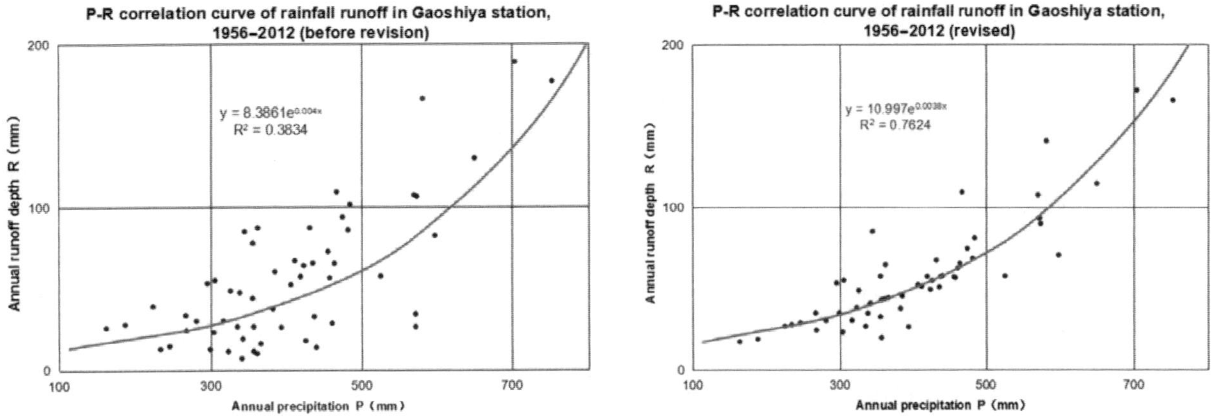

Figure 6. Comparison of precipitation runoff curve before and after correction at Gaoshiya station.

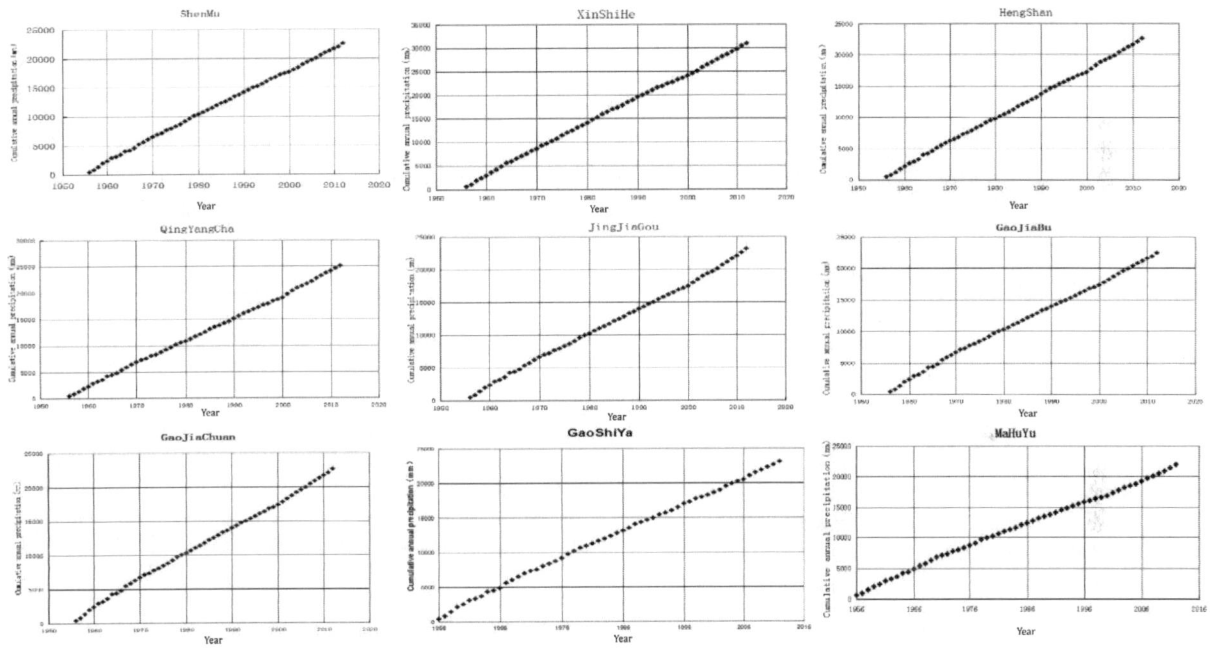

Figure 7. Accumulated curves of annual precipitation in 1956–2012 at 9 hydrologic stations such as Shenmu, Hengshan and Qingyangcha ect.

4 Conclusion

The underlying surface of the watershed in the hilly and gully region of the Loess Plateau in Shaanxi Province is gradually accumulated due to the long-term interactive and comprehensive disturbance of climate factors and strong human activities. The "primordial underlying surface" of most rivers in the high-yield flow period can be extended to about 2000 or so. Some rivers changed from "original ecological underlying surface" high yield flow period to low yield flow period under "current underlying surface" environment condition around 1970, and some rivers again appeared lower "current underlying surface" environment low yield flow period after 2000 or so. The consistency of the environmental conditions of the abortion and confluence of most of the rivers, and the consistency of the hydrological monitoring data sequence of the river runoff have been destroyed simultaneously. With the same order of precipitation, the yield and confluence of the watershed tend to decrease.

In the near future, the "current underlying surface" of the basin has a low yield, which is irreversible to the environmental condition of high yield in the early stage of the "underlying surface of the original ecology". Based on the principle of maintaining the consistency of the environmental

Table 5. Statistical unit of large hydrologic station in the annual precipitation series after 2000 Units: mm.

Hydrologic(al) Station	River location	Average annual precipitation 1956–1999	Average annual precipitation 2000–2012	After 2000, the series increases compared with the previous series	Increase in the average value of the series after 2000 compared with the previous series by %
Qingyangcha	Wuding River tributary Dali River (Middle reaches)	424.2	490.4	66.2	15.6
Dingjiagou	Wuding River main stream (middle reaches)	388.8	461.1	72.3	18.6
Gaojiabu	Tu Wei River main stream (middle and upper reaches)	386.4	421.1	34.7	9.0
Gaojiachuan	Tu Wei River main stream (downstream near estuary)	387.4	430.7	43.3	11.2

Table 6. 1970 and 2000 annual runoff series affected by environmental changes.

Hydrologic (al) station	High yield flow period of "underlying surface of primordial ecology"							Low production flow period of "current underlying surface" environment after 2000				
		Annual runoff mean	Human society consumes rivers		Fixed to "current underlying surface" low yield period between 1970 and 2000				Human society consumes rivers		Fixed to "current underlying surface" low yield period between 1970 and 2000	
	Period	(mm)	Wastage (mm)	Influence degree (%)	Wastage (mm)	Influence degree (%)	Period	Period	Wastage (mm)	Influence degree (%)	Wastage (mm)	Influence degree (%)
MaHuYu	1956–1970	64.5			−4.7	−7.3	1971–1997	1998–2012			−15.8	−24.5
GaoShiYa	1956–1978	83.6	2.07	2.48	−15.5	−18.5	1979–1997	1998–2012	3.39	4.06	−33.0	−39.5
GaoJiaBao	1956–1971	176.3	0.28	0.16	−23.0	−13.0	1972–1997	1999–2012	5.05	2.86	−45.7	−25.9
GaoJiaChuan	1956–1971	134.6	0.56	0.42	−18.2	−13.5	1972–1997	1999–2012	6.03	4.48	−30.6	−22.7
ShenJiaWan	1956–1971	96.1	1.19	1.24	−24.0	−25.0	1972–1997	1998–2012	4.1	4.27	−40.0	−41.6

conditions of the production and confluence of the current underlying surface in the near future, the hydrological observation data of the river should be revised to a unified basic standard, so that the hydrological sequence data can keep the consistency of the source "underlying surface" of the hydrological sequence. It is more meaningful in practical application.

In the region controlled by most rivers in the hilly and gully of the Loess Plateau in northern Shaanxi, the annual precipitation has maintained the stability of the 57 years series of 1956–2012 without systematic abrupt changes. After 2000, the annual precipitation in the whole basin of TuWeihe river was about 10 % larger a year, over the middle reaches of Wudinghe river, there are about 17 %, while the annual precipitation in the lower reaches of Wudinghe river keeps the stability of 57 years series.

The correction method in this paper can only meet the accuracy of the consistency correction of annual runoff series when the regional water resources evaluation is satisfied. The difficulty is how to discretize the revised data as random events occur, which should be concretely analyzed, selected and used.

Data availability. The meteorological data used in the previous study came from the China Meteorological Network (http://data.cma.cn, last access: 26 June 2018), and the agricultural irrigation, industrial and urban domestic water use data came from the Shaanxi Water Resources Bulletin (http://slt.shaanxi.gov.cn/sxmwr-zxfw-dfkj-3-show-93956, last access: 26 June 2018). The hydrological data were collected from the the Yellow River Water Conservancy Commission (YRWCC), Ministry of Water Resources of China and the Hydrology and Water Resources Survey Bureau of Shaanxi Province. These data are proprietary or confidential in nature and may only be provided with restrictions.

Author contributions. JY conceived the idea, data analysis and wrote the manuscript; ZZ supervised the data analysis and writing process. Both authors have read and agreed to the published version of the manuscript.

Competing interests. The authors declare that they have no conflict of interest.

Special issue statement. This article is part of the special issue "Hydrological processes and water security in a changing world".

It is a result of the 8th Global FRIEND–Water Conference: Hydrological Processes and Water Security in a Changing World, Beijing, China, 6–9 November 2018.

Financial support. This research has been supported by the the National Key Research and Development Program of China (grant no. 2016YFC0402405).

References

Hui, Y. Y., Liu, T. H., and Tong, X. Q.: Water and sediment changes in Wuding River, Yanhe River and Qingjian River and their development trend forecast, Yellow River Water and Sediment Change Research Foundation, China, Dissertation, 1993 (in Chinese).

Jing, Y., Chen, F. L., and Ge, F. L.: Study on Idiosyncrasy of Water and Sediment on Area headstream in Beiluohe Watershed, Hydrology and Water Resources Survey of Shaanxi Province, Shaanxi, China, Report, 2008a (in Chinese).

Jing, Y., Chen, F. L., Ge, F. L., and Gu, M. X.: Analysis of characteristics of Water and sediment in Source area of North Luohe River, Hydrology, 030, 92–96, https://doi.org/10.3969/j.issn.1000-0852.2010.04.023, 2010b (in Chinese).

Jing, Y., Ge, F. L., and Tian, H. W.: Department of Water Resources of Shaanxi Province, Water Resources Bulletin (2000–2016), Shaanxi, China, Report, 2016c (in Chinese).

Jing, Y., Li, Z. Y., and Pang, L.: Technical methods of Shaanxi Provincial Hydrological calculation Manual, Shaanxi Provincial Bureau of Hydrology and Water Resources Survey, Shaanxi, China, Report, 2017d (in Chinese).

Jing, Y.: Analysis on flow consistency and effect of human being in Wu Dinghe River , in: 2008 Academic Forum on Water Ecology Monitoring and Analysis, Shenyang, China, 2 December 2008, 272–276, 2008 (in Chinese).

Li, Y. Y.: Work document for the preparation of the national water resources comprehensive planning, The office of the leading group of the national water resources comprehensive planning work, Beijing, China, Report, 2003 (in Chinese).

Peng, R. S.: A preliminary study on the correction of water and sediment data, Study on water resources, 5, 368–378, https://doi.org/10.12677/JWRR.2016.54043, 2016 (in Chinese).

Wang, G. Q., Jia, X. A., Chen, J. N., Zhang, Y., and Li, H. B.: Interference point analysis of the significant influence of human activities on Hydrologic sequence, Northwest Water Resources and Water Engineering, 12, 13–15, https://doi.org/10.3969/j.issn.1672-643X.2001.03.004, 2001 (in Chinese).

Wang, G. Q., Zhang, J. Y., Liu, J. F., and He, R. M.: Quantitative Analysis of the effects of Climate change and Human activities on runoff, China Water, 2, 55–58, https://doi.org/10.3969/j.issn.1000-1123.2008.02.016, 2012 (in Chinese).

Zhang, A. J., Wang, B. D., and Cao, L.: Summary of studies on the contribution of climate change and human activities to runoff, Northeast China Water Conservancy and Hydropower, 1, 7–8, https://doi.org/10.14124/j.cnki.dbslsd22-1097.2012.01.001, 2012 (in Chinese).

Zhu, H. F., Zhao, W. W., Kang, M. Y., and Guo, W. W.: Analysis on Obvious Disturbing Point in Time Variation Process of Sediment Discharge of the Yanhe River, Yellow River, 8, 80–81, https://doi.org/10.3969/j.issn.1000-1379.2008.08.036, 2008 (in Chinese).

Proc. IAHS, 383, 121–127, 2020
https://doi.org/10.5194/piahs-383-121-2020

The New Situation of Diatom Bloom in the Middle and Lower Reaches of Hanjiang River under River Chief System

Bao Qian[1,2], **Xiao Xiao**[2,3], **Hefan Wang**[2], **and Jincheng Wang**[2]

[1]Key Laboratory of Integrated Regulation and Resources Development of Shallow Lakes of Ministry of Education, Hohai University, Nanjing, 210098, China
[2]Bureau of Hydrology, Changjiang River Water Resources Commission, Wuhan, 430010, China
[3]Engineering Research Center of Eco-environment in Three Gorges Reservoir Region, Ministry of Education, China Three Gorges University, Yichang, 443002, China

Correspondence: Bao Qian (jacber@163.com)

Published: 16 September 2020

Abstract. In the middle and lower reaches of the Hanjiang River in February 2018, another serious diatom bloom occurred. The Changjiang Water Resources Commission promptly initiated an emergency plan to use the cascade reservoirs at the middle and lower reaches of the Hanjiang River to carry out water scheduling. At the same time, six monitoring sections were set up to daily monitor in the middle and lower reaches of the Hanjiang River. From 9 February to 18 March, this bloom had lasted more than 30 d. It is the longest since the record was recorded. The emergency dispatch measures adopted during the outbreak of the diatom bloom have stopped the further development of diatom bloom and played a good role. Analysing the causes of diatom bloom can provide a more comprehensive basis for further optimizing emergency dispatch. This paper analyzes the key hydrological factors that occur in this diatom bloom and proposes a water dispatching optimization plan and specific countermeasures. However, to permanently solve the diatom bloom problem, it is necessary for local environmental protection departments to work actively under the current requirements of the river chief system. Comprehensive protection of the Hanjiang River water ecological environment, completely solve the watershed problem of non-point source pollution.

1 Introduction

Under the current background where the river chief system is fully promoted, local governments at each level pay a lot of attention to the water environment issue in the regions under their jurisdiction. Particularly at the boundaries of administrative regions, the environment issue of water systems, rivers and lakes has become a focus of attention from each concerned party. Hanjiang is the largest tributary of the Yangtze River. Danjiangkou reservoir at the middle reach is a water source for the middle line of the south-to-north water transfer project, shouldering the heavy responsibility for easing the water consumption pressure of the north of China. The mainstream of the middle and lower reaches flow past 14 cities, which is also an important source of water consumption for industry, agricultural and the life of residents in Shaanxi and Hubei. However, in recent years, eutrophication has escalated in the water bodies of the middle and lower reaches of Hanjiang River, and diatom bloom has occurred frequently (Dou et al., 2002). Since the first large-scale outbreak of diatom bloom in the spring of 1992, it would cause a heavy impact to the ecological environment of water bodies along the river every other year. Particularly after 2010, diatom bloom has broken out more frequently and showed a trend of gradual expansion, which has seriously threatened the ecological safety of the region and caused a big challenge to the implementation of river chief system (Zhang et al., 2016).

As a typical example of river algae bloom, the algae bloom of Hanjiang River is in substance diatom bloom. Unlike blue and green algae bloom in still-water ecological systems such as lakes and reservoirs, most algae bloom is highly resistant to a low temperature and prefers a water body with a certain flowing speed. They have a certain demand for nutritive salt such as nitrogen and phosphorus, they synthesize cell walls with the silicate in the environment. Therefore, the content of silicon is a critical factor influencing the growth of diatom. All these are the preconditions for the occurrence of diatom bloom in Hanjiang River under specific seasonal (spring) and hydrological conditions (dry season). Relevant research findings have been achieved in China and abroad on the cause and biological mechanism of diatom bloom. Most researchers argue that the diatom bloom is caused by a high level of nutritional salt and appropriate weather and hydrological conditions. However, diatom bloom is diversified and occurs in a wide spectrum of ecological environment, diatom bloom may be caused by different factors in different ecological systems. For example, most researchers the river diatom bloom can be attributed to a low flowing speed, a high level of nutritional salt and an appropriate temperature (Pan et al., 2014). However, some scholars point out, the building of large water conservancy projects, particularly dams, has a significant impact to the structure of plankton in the rivers (Xie et al., 2005; Li et al., 2016). Xie et al. point out that water quality factors such as nitrogen and phosphorus and climatic factors such as water temperature are not the critical factors to restrict the occurrence of diatom bloom in Hanjiang River. Hydrological factors such as flow and speed under the influence of water conservancy projects are the very critical factors (Xie et al., 2004).

An analysis of latest research fruits reveals that a consensus has been reached on diatom bloom, namely, physical factors such as precipitation and temperature are usually more important than factors such as the level of nutritional salt in the water body. Hydrological factors are very probably the limiting factors in the outbreak of diatom bloom (Ren et al., 2017). Based on the theory, Changjiang Water Resources Commission, an administration body of the river, has ever tackled the diatom bloom of Hanjiang River by regulating the cascaded reservoir in the middle and lower reaches of Hanjiang River, which used to achieve a good effect. However, viewed from the latest outbreak of diatom bloom in February 2018, the situation of diatom bloom in Hanjiang River was recorded with some new changes. As a result, the effect of tackling diatom bloom with the same method was not so good. This paper will make a detailed analysis of the new situation around the outbreak of diatom bloom in Hanjiang River, put forward new solutions to the new problem and provide a scientific reference for actively promoting the river chief system and the prevention and control of diatom bloom in Hanjiang River.

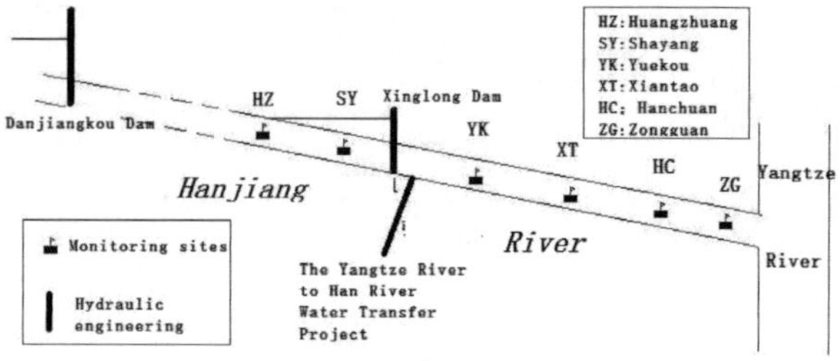

Figure 1. Layout of Emergency Monitoring Cross Sections.

2 The Outbreak Process of Diatom bloom in Hanjiang River in 2018

In the first third of February 2018, the staff of hydrometric stations along the river found abnormality in the water body in the middle and lower reaches of Hanjiang River, which seemed diatom bloom. It was verified after the environment protection staff monitored the site that starting from 9 February, a slight diatom bloom occurred in the section of Hanjiang River below Huangzhuang station. The water color of the river section showed abnormality and became increasingly dark. By 11 February, the water color became darker and turned brown, accompanied by a strong stench. Based on these phenomena, they judged that Hanjiang River had been hit by another outbreak of diatom bloom. To guarantee the water safety for urban and rural residents at the middle and lower reaches of Hanjiang River, Changjiang Water Resources Commission launched a contingency plan, increased the discharged volume of cascaded reservoirs to regulate the flow and carried out a daily monitoring on the algae in the section ranging from Huangzhuang to Wuhan (about 400 km) in the middle and lower reaches of Hanjiang River. By around 9 March, the diatom bloom of the entire river section faded away significantly, the water color changed from brown to light green. By far, the diatom bloom had lasted for more than 30 d which was recorded as the longest one in the history.

The monitoring set six sections, including Huangzhuang, Shayang, Yuekou, Xiantao, Hanchuan and Zongguan (the positions of the sections are shown in Fig. 1), of which Huangzhuang represents the inlet section of Xinglong reservoir, Shayang is the cross section of Xinglong reservoir, Yuekou, Xiantao, Hanchuan are respectively the cross sections of the river at the lower reach of Xinglong dam, while Zongguan represents the cross sections of the estuary at the lowest reach of Hanjiang River. The monitoring parameters of the cross sections include water temperature, pH, dissolved oxygen, saturation degree, chlorophyll a, algal density, dominant species of algae and corresponding hydrological parameters.

According to the monitoring results, after the outbreak of diatom bloom, Changjiang Water Resources Commission immediately launched a contingency plan to increase the discharged volume of water control projects such as Danjiangkou, Cuijiaying and Xinglong, increase the flow of the lower reaches and eliminate the key hydrological factors causing the diatom bloom. From 12 to 22 February, the flow of Huangzhuang station fluctuated at 834–962 $m^3 s^{-1}$, the flow of Xinglong station fluctuated at 648–890 $m^3 s^{-1}$, the flow of Xiantao station fluctuated at 834–962 $m^3 s^{-1}$ and the average flowing speed at the cross section of Huangzhuang and Shayang fluctuated at 0.4–0.5 $m s^{-1}$, and the flow speed of Xinglong and Xiantao stations fluctuated at 0.6–0.8 $m s^{-1}$. The minimal water temperature of each monitored cross section during the outbreak of diatom bloom was 8.5°, and the maximum temperature was 17.7°. The average temperature of Shayang cross section was 10°, the average temperature of Xiantao cross section was 11°, the average temperature of Zongguan cross section was 12.6°. The pH value rose from 8.1 to 8.9, the changing range of dissolved oxygen was 9.71–15.7 $mg L^{-1}$. The saturation degree had once risen to 151 %. The above parameters had surpassed the warning values for the outbreak of diatom bloom (water temperature $\geq 10°$, pH ≥ 8.0, DO $\geq 12 mg L^{-1}$) (Lu et al., 2000). This indicates that the outbreak of diatom bloom was serious in Hanjiang River. According to the monitoring of algal density in each cross section (Fig. 2), the extreme point of the outbreak of diatom bloom appeared around 21 February, near 10 d after its outbreak. The algal density of Zongguan cross section at the lower reach estuary of Hanjiang River reached 32 million cells per litre, which later began to decline, but the decline was slight. By 5 March, the algal density began to decline significantly. By 7 March, the algal density in each cross section had declined to around 5 million cells per litre. Meanwhile, the dominant species of algae monitored at Huangzhuang cross section changed from diatom to diatom and green algae, which indicated that diatom began to die and the new algal species took the dominant position. Diatom bloom began to fade away. By 9 March, the water color at each cross section under monitoring became increasingly normal, and the algal density at each cross section was around 1.40–3.20 million cells per litre.

3 Causal Analysis of Diatom bloom

In the earlier causal analysis of the diatom bloom in Hanjiang River, most researchers mainly focus on three areas. The first one is the water quality factors. The discharge of pollutants became increasingly serious in the middle and lower reaches of Hanjiang River. There were excessive nutritional salts such as nitrogen and phosphorus needed for the growth of algae. The Second one is the climatic factors. The warmth of spring and appropriate sunshine facilitated the mass propagation of algae. The third one is hydrological factors. Due to the high-level backwater effect of the Yangtze River and the decrease of the flow volume of Hanjiang River, the flowing speed of Hanjiang has slowed down. As a result, the water

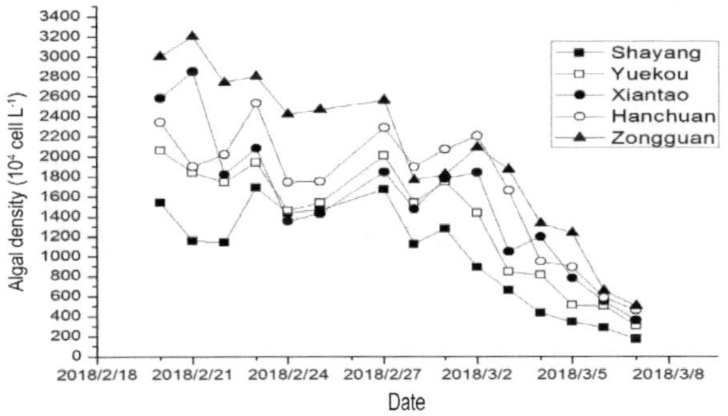

Figure 2. The tendency chart of algal density in each cross section during the outbreak of diatom bloom.

body of Hanjiang River shows the characteristics of a lake, which provides an environment for the occurring of diatom bloom (Wang et al., 2004). However, compared with past formation conditions, new situations had appeared this time.

1. *The scope of the outbreak of diatom bloom had expanded.* The previous outbreaks of diatom bloom were mainly concentrated in the section ranging from Xiantao to Wuhan. The scope of influence mainly covered the areas 200 km from the estuary. The outbreak of diatom bloom this time had extended to Huangzhuang cross section, which was about 400 km from the estuary of the river. During the outbreak of diatom bloom, the algal density monitored at Shayang cross section in Xinglong reservoir reached 17 million cells per litre. This indicated that the outbreak of diatom bloom had extended to Xinglong reservoir.

2. *Washed by mass flows, the diatom bloom had not faded away for a long period.* Many previous researches of diatom bloom indicate that when the flow at the cross section of Xiantao is above 500 $m^3 s^{-1}$, the diatom bloom will fade away gradually. Yin et al. have found by analyzing historical data that the diatom bloom could be solved when the flow volume of Xiantao station was higher than 800 $m^3 s^{-1}$ and the corresponding assurance rate reached 90 % by regulating the water volumes of reservoirs (Yin et al., 2017). However, from the early stage of the outbreak of diatom bloom this time, the discharged volume of reservoirs like Danjiangkou and Xinglong was increased, the flow volume of Xiantao station remained above 800 $m^3 s^{-1}$ and even reached 1230 $m^3 s^{-1}$ at the maximum. However, the diatom bloom still did not fade away.

In the face of such a situation, we have concluded the following causes based on the analysis of monitoring data gained during the outbreak of diatom bloom and our historical experience.

1. *The issues concerning the dominant species of diatom.* All the time, cyclotella has been studied as the dominant species of diatom in the diatom bloom of Hanjiang River (Kuang et al., 2000). However, some scholars like Zheng Lingling and Yang Qiang have discovered by scanning with an electronic microscope that the dominant species in the diatom bloom of Hanjiang River is *Stephanodiscus* (Yang et al., 2011; Zheng, 2005). Though the two kinds of algae both belong to diatom, their living habits are different to a certain extent. For example, the former prefers an environment of still water and abundant phosphorus, while it is easier for the latter to dominate in an environment with a great disturbance to water body, with a certain flowing speed and a big fluctuation of the concentration of nutritional salt. Therefore, inferred from the facts such as algae still dominated as water from the upstream kept increasing, the dominant algal species was most likely to be Stephanodiscus rather than cyclotella.

2. *The issues concerning the control of pollution source.* Diatom bloom was identified at the Huangzhuang cross section at the inlet end of Xinglong reservoir, which indicated that the diatom bloom this year has extended to Xinglong reservoir. This means that the eutrophication was increasingly intensified at Xinglong reservoir. The water body flew out of Danjiangkou reservoir and past Xiangyang. At the cross section of Yujiahu, it showed a rise to a certain extent in the level of ammonia nitrogen and nitrate nitrogen. This indicated that the water body of Xinglong reservoir took in part of pollutants from the upstream, which included the pollutants discharged from upstream cities like Xiangyang. The level of nitrogen and phosphorus was also found to rise gradually at Xiantao and Jijiazui (estuary). This indicated that the water bodies at the middle and lower reaches of Hanjiang River were being attacked by the pollution along the river. The pollution control on both banks along the river was a daunting task. By comparing the N / P ratio at the monitored cross sections, we find that the N / P ratio at Xiantao and Jijiazui was closer to the threshold value suitable for the propagation of algae (N / P ratio = 13) (Wang, 2010), while the N / P ratio at Yujiahu cross section reached 86, far higher than the range suitable for algae. This also explained why no diatom bloom had occurred near the river section of Xiangyang.

3. *Special hydrological and weather situations.* By analyzing the changing trend of algal density at each monitored cross section in Fig. 2, we know that the algal density of each monitored showed a trend of gradual rising from the upstream Shayang to Zongguan near the estuary, showing a phenomenon of "time lag" in the spatial and temporal distribution, which was mainly because other than enduring its own algal propagation, the downstream river section also needed to receive the algae washed down from the upstream and showed a state of accumulation. Meanwhile, affected by the weather, the propagation capacity of algae also fluctuated, which is another aspect of the "time lag". For example, after the rainfall on 20 February, the algal density at each cross section declined to different extents on 21 and 22 February. Later, as the weather became fine, the algal density was rising again. This indicates that the weather, particularly sunshine, plays an important role in the eruption of diatom bloom. The changing trend of diatom bloom at each monitored cross section shown in Fig. 2 indicates that from 20 February to around 2 March, the algal density at the cross sections below Xinglong was declining, which may be because the result of water volume regulation had checked the further eruption of algae to a certain extent. However, in the same period the algal density at Shayang cross section had risen. Therefore, we need to analyze why the water volume regulation effect was not distinct at its upstream Danjiangkou. By analyzing relevant hydrology monitoring data, we find that taking 3 March as an example, the cross-section flow volume of Huangzhuang and Shayang stations in the same period was both $1100\,m^3\,s^{-1}$. However, the existence of Xinglong reservoir had significantly "weakened" the scour effect of water volume from upstream, which had also provided a buffer space for the propagation of algae in the reservoir. Therefore, we can infer that the main reason for the lasting of diatom bloom this time was mainly that the diatom bloom in Xinglong reservoir had not been effectively controlled. As Xinglong water control project increased discharged volume, it constantly sent algae to the downstream of the dam. As a result, at the macro level, the water volume regulation effect of Danjiangkou was not distinct.

4. *Other reasons.* The above three aspects may be the main reasons for the outbreak of diatom bloom this time surpassed the expectation. In addition, the backwater effect of the Yangtze River and the significant drop of silt level at the middle and lower reaches of Hanjiang River may have also stimulated the outbreak of the diatom bloom. By looking up water level and flow volume data of Hankou hydrology station during the outbreak of diatom bloom, we find that the water level of Wuhan section of the Yangtze River was higher than the same period of past years. Taking the hydrology condition of Hankou station on 10 February as an example, the average flow volume of cross section on the day was $15\,100\,m^3\,s^{-1}$, which increased near 50 % than the average flow volume of February of past years (about $10\,000\,m^3\,s^{-1}$). This indicates that the backwater effect of Yangtze River had reduced the water surface drop of the middle and lower reaches of Hanjiang River, which reduced the water surface flow speed to a certain extent and was also a

Table 1. Monitoring results of nutritional salts of some cross sections in the middle and lower reaches of Hanjiang River during February.

Cross section	Ammonia nitrogen ($mg\,L^{-1}$)	Nitrate nitrogen ($mg\,L^{-1}$)	Total nitrogen ($mg\,L^{-1}$)	Total phosphorus ($mg\,L^{-1}$)	Nitrogen / phosphorus ratio ($mg\,L^{-1}$)
Below Danjiangkou dam	0.09	1.24	1.44	0.03	48
Yujiahu	0.12	1.60	1.72	0.02	86
Xiantao	0.01	1.76	1.77	0.09	19
Jijiazui	0.21	1.68	1.89	0.10	18

factor affecting the outbreak of diatom bloom. In addition, the issue of silts in Hanjiang River. With the establishment of water surface flow speed of cascaded reservoirs, the silt level in the water bodies in the lower reach became increasingly low. Analyzed from the physiological characteristics of algae, this would amplify the role played by sunshine in the propagation of algae. The silicon shell at the surface of diatom is relatively heavy, they are generally suspending at the middle and lower levels of the water body, which hinder their photosynthesis and growth. However, with the decreasing of silt, sunshine is easier to penetrate the water body, providing more sunshine for the growth of algae. As a result, it was easier to result in an outbreak of diatom bloom under the same weather conditions.

4 Countermeasures under River Chief System

By analyzing the main causes of the outbreak of diatom bloom this time, the paper will put forward the following countermeasures.

4.1 Strict control of pollution input

The serious eutrophication in the middle and lower reaches of Hanjiang River is the fundamental cause for the outbreak of diatom bloom. At present, the nitrogen and phosphorus level in the water body at the middle and lower reaches of Hanjiang River is high, which fully satisfies the necessary conditions for the formation of diatom bloom. The further extension of diatom bloom scope this year indicates that the pollution prevention and control needs to be further intensified for the middle and lower reaches of Hanjiang River. Particularly under the promotion of river chief system, each prefecture city along the river should pay full attention to the pollution control within the reach, prevent industrial and agricultural waste water and sanitary sewage, etc. that fail to reach the standard from being discharged into the river and gradually enhance the prevention and control capacity against the pollution source. Only by improving the overall water ecology environment of Hanjiang River is the fundamental method to permanently prevent the occurrence of diatom bloom in Hanjiang River.

4.2 Optimizing water volume regulation scheme

Through co-regulation of reservoirs, the regulation of ecological flow volume is an emergency measure for preventing and control of diatom bloom in Hanjiang River. The purpose of ecological regulation is to eliminate the hydrological conditions for formation of diatom bloom and prevent the occurrence of diatom bloom. The emergency regulation this time tells us that in the future we should further optimize the regulating scheme, particularly during the sensitive period before outbreak of diatom bloom. If there is any sign of diatom bloom in Xinglong reservoir, first of all we should increase the discharged volume of Xinglong to reduce the water level of the reservoir, increase the water surface flow speed of the reservoir and the reach downstream the dam, to control the overall hydrological environment for propagation of algae. Then we should increase the discharged volume of Danjiangkou to further increase the water surface flow speed of the reservoir, replenish the water volume of the reservoir and dilute the algal density. Meanwhile, on the basis of guaranteeing other water demands, at the mainstream of Yangtze River we can reduce the discharged volume of the Three Gorges reservoirs, reduce the water level of the mainstream of Yangtze River at the estuary of Hanjiang River, to increase the flow speed at the middle and lower reaches of Hanjiang River and eliminate the influence of backwater from Yangtze River. Viewed from the experience of water volume regulation this time, the too early increase of discharged flow volume of Danjiangkou will do limited help to the prevention and control of diatom bloom and even waste the precious water source for South-to-north water diversion. In addition, considering that the eutrophication of Xinglong reservoir is increasingly serious, the daily frequency of water renewal of the reservoir needs to be increased. Calculated by a designed capacity of Xinglong reservoir at 485 million m^3, if the water were discharged by a flow volume of $800\,m^3\,s^{-1}$ at the time of emergent regulation, it will at least take around 1 week, which is extremely adverse to the control of algae during the outbreak of diatom bloom.

4.3 Improve management mechanism

Considering that the water environment protection of Hanjiang River is a long-term task, and the ecological environment in Hanjiang River concerns the pursuit of a good life by about 40 million people of 78 counties and cities in 20 regions of six provinces along the river, we should actively improve related management mechanism, strengthen unified management of water volume and quality of Hanjiang River, and intensify the coordination, guidance, supervision and monitoring of organizations along the river, so as to build an ecological good environment as soon as possible.

5 Conclusion

1. The diatom bloom of Hanjiang River had lasted more than 30 d, which was the longest one in the history of records. The influencing scope of diatom bloom had extended from the original section from Wuhan to Xiantao to Xinglong reservoir and the situation of diatom bloom has witnessed new changes. By tracking and monitoring the diatom bloom this time, we can infer that the advantaged species of the diatom bloom may be Stephonidiscus. The maximum algal density during the outbreak of diatom bloom reached 32 million cells per litre. By comparing with the conventional threshold value of diatom bloom occurrence (algal density at 10^5 cells per litre), the water condition had basically restored normal when the algal density in the diatom bloom this time was 10^6cells per litre.

2. The emergency regulation measure taken by Changjiang Water Resources Commission during the outbreak of diatom bloom has obviously checked the further escalation of diatom bloom to a certain extent. However, with the changing of situation, the regulation effect this time was not ideal compared with previous years, which brought new inspirations and new requirements for the emergency regulation work.

3. The fundamental cause of the diatom bloom was the eutrophication of the water body at the middle and lower reaches of Hanjiang River. Therefore, under the present situation, we should take the full implementation of river chief system as a chance to strictly control the pollution input, strengthen the prevention and control capacity against the pollution, improve the water ecology environment. As an emergency measure during the outbreak of diatom bloom, the water volume regulation needs to be gradually optimized and improved along with the changing environmental issues, so as to enhance the coordination, guidance, supervision and monitoring of the organizations in the river chief system.

Data availability. Data used to produce Fig. 2 is available in the Supplement.

Supplement. The supplement related to this article is available online at: https://doi.org/10.5194/piahs-383-121-2020-supplement.

Author contributions. BQ is responsible for data analysis and manuscript writing. Other authors participate in data collection and analysis

Competing interests. The authors declare that they have no conflict of interest.

Special issue statement. This article is part of the special issue "Hydrological processes and water security in a changing world". It is a result of the 8th Global FRIEND–Water Conference: Hydrological Processes and Water Security in a Changing World, Beijing, China, 6–9 November 2018.

Acknowledgements. We would like to thank our colleagues from Bureau of Hydrology, Changjiang River Water Resources Commission for their hard work in the emergency monitoring process.

Financial support. This work is supported by the Hubei Provincial Natural Science Foundation of China (contract no. 2017CFB312), the Fundamental Research Funds for the Central Universities (no. 2017B20514) and PAPD. Project research center of the ministry of education of the three gorges reservoir area open fund (no. KF2018-05).

References

Dou, M., Xie, P., Xia, J., Shen, X. L., and Fang, F.: Study on algalbloom in Hanjiang River, Adv. Water Sci., 5, 557–561, https://doi.org/10.14042/j.cnki.32.1309, 2002.

Kuang, Q. J., Tan, Y. Y., Wan, D. B., and Zhang, J. Y.: On the phytoplankton in the middle and lower reaches of the Hanjiang River and the Prevention of water-blooms, Resource and environment in the Yangtze Basin, 9, 63–70, 2000.

Li, B. S., Li, H. Y., and Zhou, P. J.: Study on eco-environmental effect assessment for cascade hydropower development in Hanjiang River basin, Yangtze River, 47, 16–22, https://doi.org/10.16232/j.cnki.1001-4179.2016.23.004, 2016.

Lu, D. Y., Liu, P. G., and Fan, T. Y.: The investigation of water bloom in the downstream of the Han River, Res. Environ. Sci., 13, 28–31, https://doi.org/10.13198/j.res.2000.02.31.ludy.009, 2000.

Pan, X. J., Zhu, A. M., and Zheng, Z. W.: Structural characteristics and influencing factors of phytoplankton community in the middle and lower reaches of Hanjiang River during spring season, Chinese J. Ecol., 33, 33–40, https://doi.org/10.13292/j.1000-4890.20131220.0012, 2014.

Ren, J., Zhu, G. W., and Jin, Y. W.: Combined effects of water exchange rate and nutrient on diatom proliferation in Hengshan Reservoir, Taihu Basin, J. Lake Sci., 29, 604–616, https://doi.org/10.18307/2017.0309, 2017.

Wang, H. P., Xia, J., Xie, P., and Dou, M.: Mechanisms for hydrological factors causing algal blooms in Hanjiang River, Resource and environment in the Yangtze Basin, 13, 282–285, 2004.

Wang, P. L.: Studies on the formation mechanism of diatom blooms in Hanjiang River from hydrodynamics and nutrition, Master's Degree dissertation, Huazhong Agricultural university, Wuhan, 2010.

Xie, P., Xia, J., and Dou, M.: Research into the effects of the middle route of China's south-to-north water transfer project on water bloom in the middle-down stream of Hanjiang River and the countermeasures – Part 1: an analysis of the key factors generating water bloom in Hanjiang River, J. Natural Resour., 19, 418–423, 2004.

Xie, P., Dou, M., and Xia, J.: Water bloom occurrence probability calculation model in Hanjiang River under different water transfer schemes of the middle route of South to North water transfer project, Shuili Xuebao, 36, 727–732, https://doi.org/10.13243/j.cnki.slxb.2005.06.015, 2005.

Yang, Q., Xie, P., Xu, J., Shen, H., Zhang, M., Wang, S. B., and Wang, P. L.: Reseach advances of diatom blooms in rivers, Resource and environment in the Yangtze Basin, 20, 159–165, 2011.

Yin, D. C., Yin, Z. J., and Yang, C. H.: Key hydrological thresholds related to algae bloom in middle and lower reaches of Hanjiang River and studies on mitigation measures, China Water Resour., 09, 31–34, 2017.

Zhang, D., Qian, B., and Zhang, Y. D.: Bloom characteristic parameters of water quality analysis in Hanjiang River, J. Water Resour. Res., 5, 402–408, https://doi.org/10.12677/jwrr.2016.54046, 2016.

Zheng, L. L.: The physiological and ecological research about the dominant species in Hanjiang River diatom bloom, Master's Degree dissertation, Fujian Normal University, Fujian, 2005.

Proc. IAHS, 383, 129–134, 2020
https://doi.org/10.5194/piahs-383-129-2020

Geomorphometry-based modelling of discharge series in ungauged basins – Robustness regarding DEM sources

Stéphane Ecrepont and Christophe Cudennec

UMR SAS, Institut Agro, INRAe, 35000 Rennes, France

Correspondence: Christophe Cudennec (cudennec@agrocampus-ouest.fr)

Published: 16 September 2020

Abstract. The sensitivity of a geomorphology-based hydrological modelling is evaluated according to four DEMs from 5 to 50 m resolution in Brittany, France. A set of 8 basins (5–565.7 km^2) is used in a pseudo-ungauged context to explore the potential of Prediction in Ungauged Basin (PUB). The results show that despite slight differences on the stream networks extracted from DEMs and associated transfer functions, a coarse-worldwide DEM such as SRTM (25 m) supported similar performances than the finer French DEM (5 m) based on three validation indices. Finer DEMs may be useful only on headwater basins to gain a marginal performance.

1 Introduction

Prediction in Ungauged Basin remains an ambitious task, especially when dealing with discharge time series, i.e. hydrographs, and derived statistics of regimes (Blöschl et al., 2013; Hrachowitz et al., 2013). Since the beginning of DEM availability, geomorphology-based (or geomorphometry-based) approaches have proposed parsimonious, robust and flexible models with few calibration needs (Cudennec, 2007; Cudennec and de Lavenne, 2015; Rigon et al., 2015). The new availability of quasi-worldwide DEMs open new avenues for applications to simulate hydrographs with short time steps at ungauged sites, which can then be used to explore regime statistics. Cudennec (2000) proposed the inversion of a geomorphology-based transfer function to assess the space averaged net rainfall series (defined as the basin-average water signal produced by hillslopes to the river network) from a measured discharge series. The simulated discharge series at the outlet of a similar target basin is then obtained by transposition from the donor basin, i.e. generalisation of the net rainfall series, then the convolution with the corresponding transfer function. The latter is available for the target basin, thanks to the geomorphometric observation of its relief and stream network, and a robust assumption on the water velocity. This approach has been tested in three different well described contexts: semi-arid Tunisia (Boudhraâ et al., 2006, 2009, 2018), temperate oceanic France (de Lavenne et al., 2015, 2016), and pluvio-nival Québec (Ecrepont et al., 2019). A comparison of the approach with Top-Kriging (Skøien et al., 2006) has been performed and showed results of similar quality (de Lavenne et al., 2016). The robustness of the method to configuration calibration has been demonstrated on several occasions (Boudhraâ et al., 2018; de Lavenne et al., 2016), and simulations in truly ungauged basins have been performed beyond the pseudo-ungauged test configurations (de Lavenne and Cudennec, 2014, 2019; Dallery et al., 2020). But the sensitivity of the transfer function to the quality and resolution of the input DEM remains to be assessed in order to apply to less-described basins and regions, as it is often the case in the FRIEND approaches. In this study, a set of 8 basins is used to compare geomorphology-based transposition at a 15 min time step, with four DEMs spanning from coarser to accurate resolutions, including the quasi-worldwide SRTM one.

2 Material and methods

2.1 Study area

The study is performed in the central Brittany peninsula (France), in the Blavet and Oust river basins, around the Coët Dan basin at Kervidy, of the long-term hydrological observatory AgrHys (named Co in Table 1; https://www6.inra.fr/ore_agrhys_eng/, last access: 6 July 2020, Fovet et al., 2018; as in de Lavenne et al., 2015). This area of the Palaeozoic Armorican mountain range rises up to 385 m and is domi-

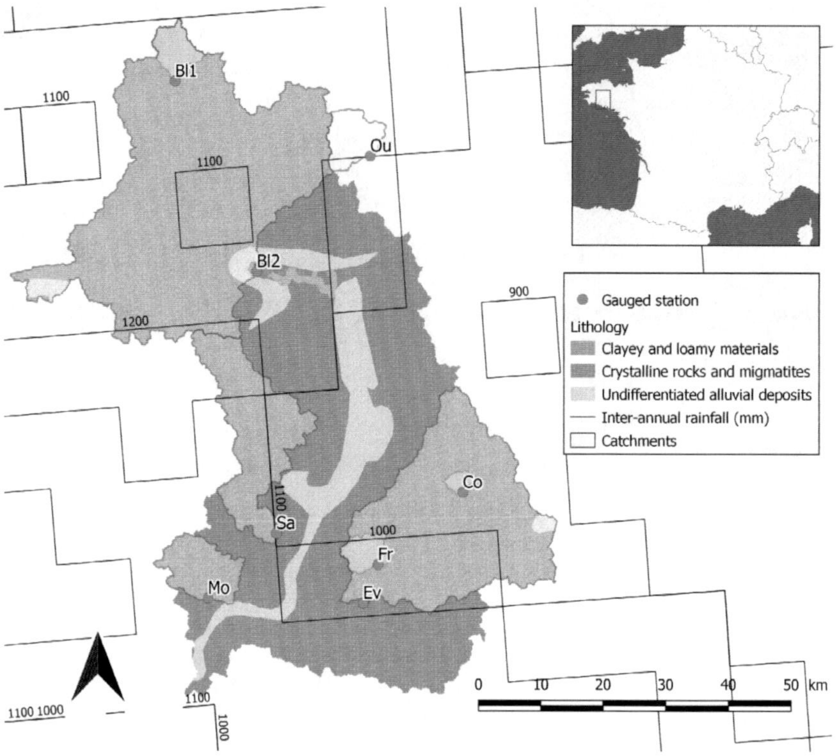

Figure 1. Geology and interannual rainfall (estimated on SAFRAN spatial grid – Habets et al., 2008) and location of the 8 basins.

Table 1. Geomorphological characteristics of the 8 basins (DEM ALTI25).

Abbreviated name	Surface (km^2)	Strahler order
Co	5	2
Fr	15.6	3
Bl1	19.2	3
Ou	28.3	4
Mo	45.4	4
Sa	125	4
Ev	314.8	5
Bl2	565.7	6

nated by schist and sandstone, while the south relates to hard rocks such as granite and gneiss. The location under prevailing wind from the Atlantic Ocean on the West induces a high rainfall climatology with a mean interannual precipitation up to 1200 mm. An overview of the geology, interannual isohyets, and location of the 8 basins is available in Fig. 1. It also displays the geographic neighbouring and partly nesting configuration of the 8 basins used for the systematic pseudo-ungauged transposition analysis, while Table 1 details their sizes.

Table 2. DEMs resolution and provider.

Name	Resolution (m)	Provider
RGE	5	IGN
ALTI25	25	IGN
SRTM	25	NASA
ALTI50	50	IGN

2.2 Data

Observed runoff discharge series were extracted from the French national database "banque HYDRO" (http://www.hydro.eaufrance.fr/, last access: 6 July 2020), from January 2012 to December 2016. Four different DEMs with increasing resolution are used, three from the French IGN national agency covering France, and the SRTM one from the US NASA federal agency displaying a quasi-worldwide coverage (Table 2).

2.3 Methods

Simulated discharge $\boldsymbol{Q}$ is the convolution of net rainfall $\boldsymbol{R}_{\mathrm{n}}$ (water flux equivalent to a height, flowing out of hillslopes to the river network), and a transfer function $\boldsymbol{TF}$, rescaled by

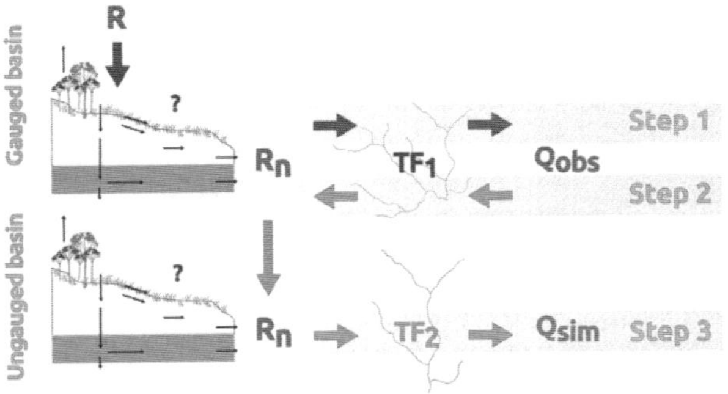

Figure 2. Deconvolution of the observed discharge series Q_{obs} of the donor gauged basin through the inversion of its geomorphology-based transfer function TF_1, generalization of net rainfall to a similar target basin, and simulation of the discharge series Q_{sim} at the outlet of the target (pseudo)-ungauged basin through the convolution with its transfer function TF_2 (from de Lavenne et al., 2015).

the basin surface S (Eq. 1).

$$\boldsymbol{Q} = S\,\boldsymbol{TF} * \boldsymbol{R}_{\text{n}} \tag{1}$$

For each basin and DEM, the river network was extracted with a minimum drainage threshold of $0.25\,\text{km}^2$; then the corresponding probability density function of hydraulic length L, pdf(L), was assessed, based on a morphometric analysis with GRASS 7.4.0 (Aouissi et al., 2013). Considering a water drop falling inside a basin and moving towards the outlet, the hydraulic length L corresponds to the travelling distance within the river network. Following the assumption of a linear transfer function (Blöschl and Sivapalan, 1995; Robinson et al., 1995; Cudennec, 2007), pdf$(t) = \boldsymbol{TF}$ is then deduced from pdf(L) thanks to the estimation of the average river flow velocity v. This characteristic velocity has been regionalized for the whole Brittany by de Lavenne (2014) as in Eq. (2), where $\overline{L}$ is the mean hydraulic length of the considered basin.

$$v = 8.5947 \times 10^{-4} \times \overline{L}^{0.6146} \tag{2}$$

According to the theory of inverse problem (Tarantola and Valette, 1982), $\boldsymbol{R}_{\text{n}}$ can be estimated by the inversion of Eq. (1) (Boudhraâ et al., 2006; Cudennec, 2000). A solution of this inversion is approached by maximum likelihood on a priori net rainfall $\boldsymbol{R}_{\text{n}}^{\text{ap}}$ (Boudhraâ et al., 2018) thanks to Eq. (3).

$$\boldsymbol{R}_{\text{n}} = \boldsymbol{R}_{\text{n}}^{\text{ap}} + \mathbf{C}_R \boldsymbol{TF}^t (\boldsymbol{TF}^t \mathbf{C}_R \boldsymbol{TF} + \mathbf{C}_Q)^{-1} \boldsymbol{Q}_{\text{m}} - \boldsymbol{TF}\,\boldsymbol{R}_{\text{n}}^{\text{ap}} \tag{3}$$

where $\mathbf{C}_R$ and $\mathbf{C}_Q$ are the matrices of covariance for vectors $\boldsymbol{R}_{\text{n}}$ and $\boldsymbol{Q}$, and t the transposed matrix. Contrary to $\boldsymbol{Q}$, $\boldsymbol{R}_{\text{n}}$ is independent from scale and can be transferred in another (eventually ungauged) basin, then convoluted into the discharge series in the target basin, according to its own $\boldsymbol{TF}$ (Fig. 2).

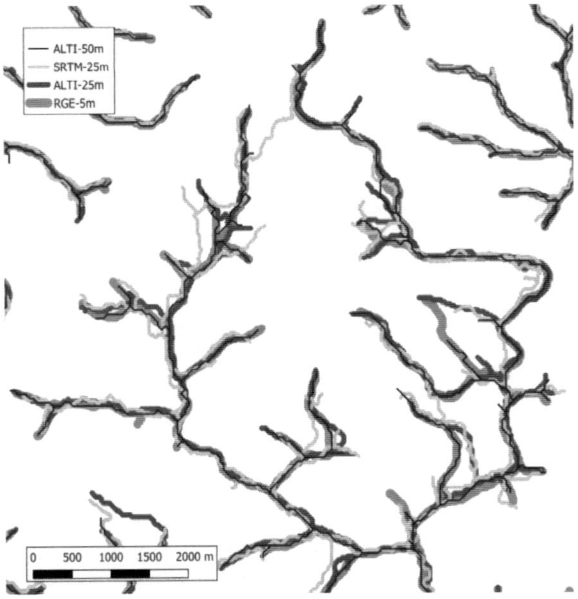

Figure 3. Comparison of the stream network extraction for all DEMs on a random zone of Bl2.

The evaluation of the transposition in pseudo-ungauged basins is performed through 3 criteria given by Eqs. (4), (5) and (6); the classical "NSE" (Nash and Sutcliffe, 1970), the NSE on square rooted discharge "NSE_{sqrt}" (Oudin et al., 2006; Seiller et al., 2017) and the volumetric efficiency VE (Criss and Winston, 2008).

$$\text{NSE} = 1 - \frac{\sum \left(\boldsymbol{Q}_{\text{obs},i} - \boldsymbol{Q}_{\text{sim},i}\right)^2}{\sum \left(\boldsymbol{Q}_{\text{obs},i} - \overline{\boldsymbol{Q}}_{\text{obs}}\right)^2} \tag{4}$$

$$\text{NSE}_{\text{sqrt}} = 1 - \frac{\sum \left(\sqrt{\boldsymbol{Q}_{\text{obs},i}} - \sqrt{\boldsymbol{Q}_{\text{sim},i}}\right)^2}{\sum \left(\sqrt{\boldsymbol{Q}_{\text{obs},i}} - \overline{\sqrt{\boldsymbol{Q}}_{\text{obs}}}\right)^2} \tag{5}$$

$$\text{VE} = 1 - \frac{\sum \left|\boldsymbol{Q}_{\text{obs}} - \boldsymbol{Q}_{\text{sim}}\right|}{\sum \boldsymbol{Q}_{\text{obs}}} \tag{6}$$

3 Results

3.1 Stream network

Structural differences of stream networks extracted from the different DEMs are illustrated on Fig. 3. The shapes of the networks are similar between DEMs, but some differences may appear, especially in headwaters. ALTI25, ALTI50 and RGE give similar stream networks despite their strongly different resolutions. The stream network extracted from SRTM is in the same range of differences.

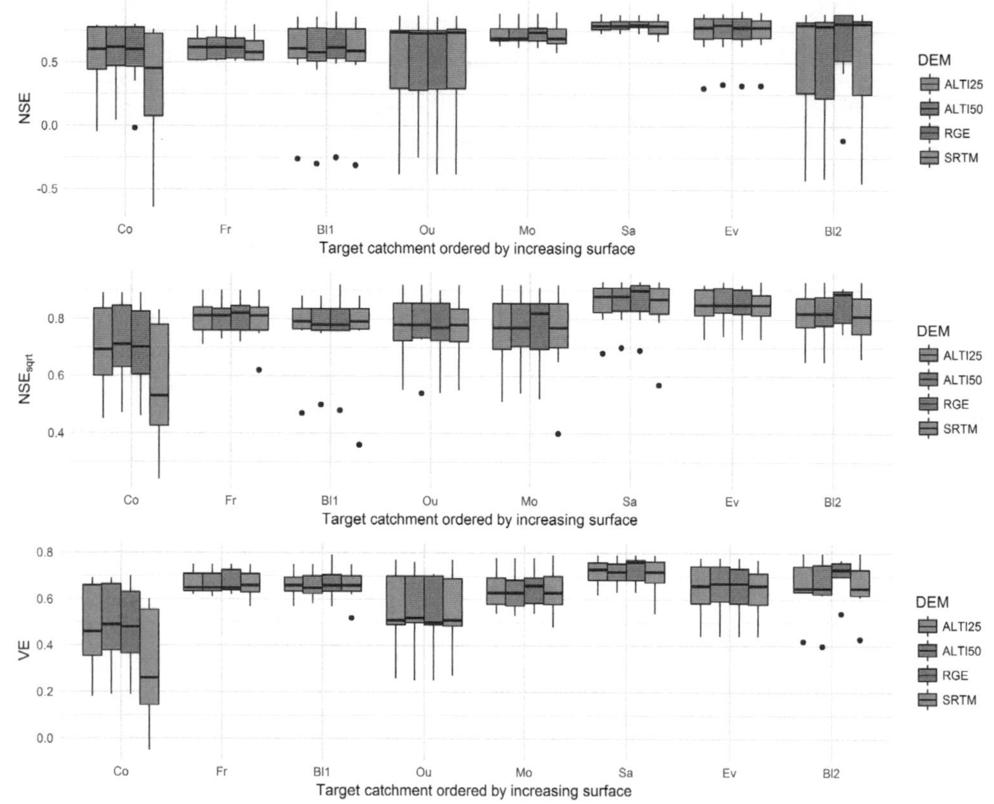

Figure 4. Simulation performance for 3 criteria and 4 DEMs, for all basins, and from all donor basins.

3.2 Performance in simulation

An overview of the results of simulation is displayed in Fig. 4 as a boxplot. For each basin and each DEM, all transpositions from other basins are gathered, for the continuous period 2012–2016, which allows a global comparison of the sensitivity to the DEM data source. Close and good results are obtained for all criteria and DEMs (median of NSE over 0.5). Lower values were assessed for Co, the smallest basin of the dataset ($5\,km^2$) using SRTM, which illustrates the importance of an accurate DEM for geomorphology-based modelling in headwater basins. However, a coarse DEM such as SRTM looks sufficient to perform geomorphology-based modelling on larger basins. The use of a high quality DEM such as RGE-5 m does not improve significantly the results. This could be explained by the local geomorphology characterized by low drainage densities and longer hillslopes, sufficiently captured by a regular DEM.

3.3 Hydrographs

Figure 5 displays an example (from Fr to Sa) of a discharge time series obtained by geomorphology-based transposition. Figure 5a and b is obtained from the high resolution national DEM (RGE, 5 m) and the medium resolution quasi-worldwide DEM (SRTM, 25 m), respectively. Identical scores are obtained for the 3 criteria, for these two neighbouring basins.

4 Conclusion

This study aimed to show the sensitivity of a robust geomorphology-based transposition framework: deconvolution of discharge time series in a gauged donor basin, generalisation of net rainfall towards another target basin and simulation of discharge time series in the (pseudo)-ungauged target basin, to the input DEM used to assess transfer functions. Simulations were performed in a systematic way between 8 basins selected from the center part of the Brittany region in the West of France, quite similar in hydroclimate and with slight differences in geology, and organized in a geographic neighbouring and partially nesting configuration. Regarding 3 criteria, NSE, NSE_{sqrt} and VE, good (median NSE ranging from 0.5 to 0.8) and consistent results were obtained for a 5-year simulation. The experiment shows that a coarse DEM such as SRTM, available internationally, gives similar results than a high-resolution DEM. The smallest basin Co (order 2) showed however the poorest results when using SRTM, which is explained by the limitation of this DEM to accu-

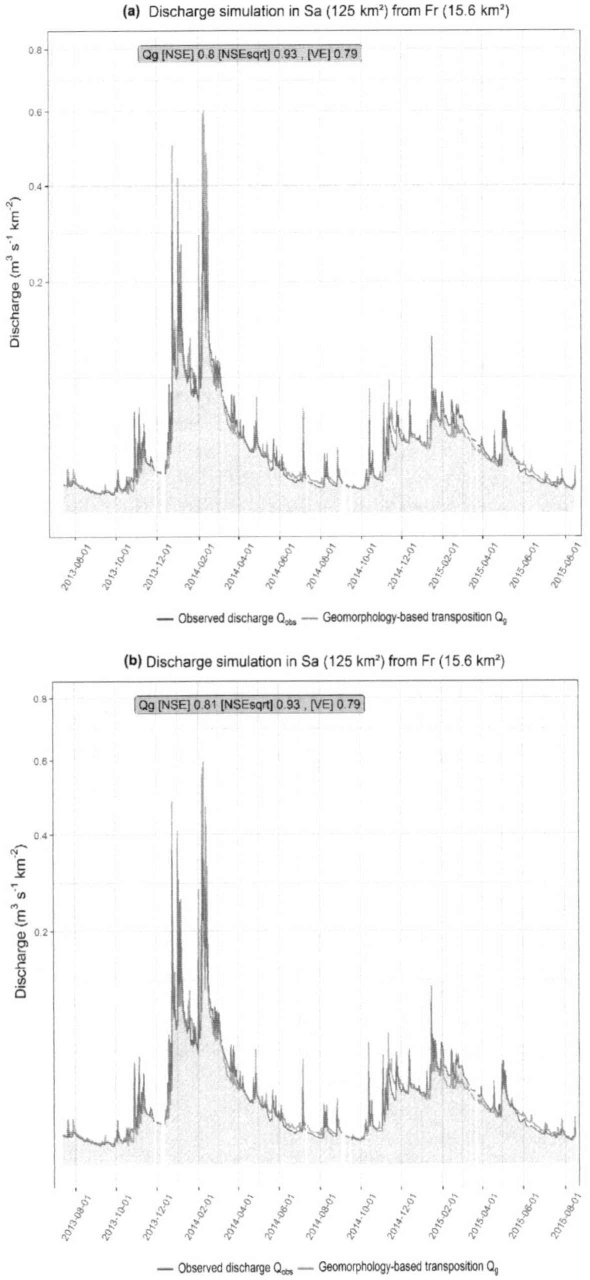

Figure 5. Discharge simulated series in pseudo-ungauged Sa ($125\,\mathrm{km}^2$) from Fr ($15.6\,\mathrm{km}^2$) between 8 January 2013 and 8 January 2015, using transfer functions extracted from **(a)** RGE (5 m) and **(b)** SRTM (25 m) – and observed series as a reference.

rately describe the stream network geomorphology. The systematically good results obtained on larger basins (order 3 and more) in a data-rich region are a proof of concept which opens great perspectives for a PUB application in any location of the world, eventually with a poor discharge gauging density (Dixon et al., 2020), and with a high robustness regarding available geomorphometric data.

Data availability. All data are available open access through the SIMFEN Web Service at http://bit.do/simfen (last access: 6 July 2020, Dallery et al., 2020).

Author contributions. CC and SE contributed equally to the design of the study and writing of the article. SE performed the analysis. CC fundraised the study.

Competing interests. The authors declare that they have no conflict of interest.

Special issue statement. This article is part of the special issue "Hydrological processes and water security in a changing world". It is a result of the 8th Global FRIEND–Water Conference: Hydrological Processes and Water Security in a Changing World, Beijing, China, 6–9 November 2018.

References

Aouissi, J., Pouget, J. C., Boudhraâ, H., Storer, G., and Cudennec, C.: Joint spatial, topological and scaling analysis of river network geomorphometry, Géomorphologie, 1, 7–16, https://doi.org/10.4000/geomorphologie.10082, 2013.

Blöschl, G. and Sivapalan, M.: Scale issues in hydrological modelling: A review, Hydrol. Process., 9, 251–290, https://doi.org/10.1002/hyp.3360090305, 1995.

Blöschl, G., Sivapalan, M., Wagener, T., Viglione, A., and Savenije, H.: Runoff Prediction in Ungauged Basins: Synthesis across Processes, Places and Scales, Cambridge University Press, Cambridge, 2013.

Boudhraâ, H., Cudennec, C., Slimani, M., and Andrieu, H.: Inversion d'une modélisation de type hydrogramme unitaire à base géomorphologique: interprétation physique et première mise en œuvre, Proc. IAHS, 303, 391–397, 2006.

Boudhraâ H., Cudennec C., Slimani M., and Andrieu H.: Hydrograph transposition between basins through a geomorphology-based deconvolution-reconvolution approach, Proc. IAHS, 333, 76–83, 2009.

Boudhraâ, H., Cudennec, C., Andrieu, H., and Slimani, M.: Net rainfall estimation by the inversion of a geomorphology-based transfer function and discharge deconvolution, Hydrolog. Sci. J., 63, 285–301, https://doi.org/10.1080/02626667.2018.1425801, 2018.

Criss, R. E. and Winston, W. E.: Do Nash values have value? Discussion and alternate proposals, Hydrol. Process., 22, 2723–2725, https://doi.org/10.1002/hyp.7072, 2008.

Cudennec, C.: Description mathématique de l'organisation du réseau hydrographique et modélisation hydrologique, PhD thesis, ENSA Rennes, 2000.

Cudennec, C.: On width function-based unit hydrographs deduced from separately random self-similar river networks and rainfall variability. Hydrolog. Sci. J., 52, 230–237, https://doi.org/10.1623/hysj.52.1.230, 2007.

Cudennec, C. and de Lavenne, A.: Hydrogeomorphology – a long term scientific interface, Hydrol. Res., 46, 175–179, https://doi.org/10.2166/nh.2015.100, 2015.

Dallery, D., Squividant, H., de Lavenne, A., Launay, J., and Cudennec, C.: An end user friendly hydrological Web Service for hydrograph Prediction in Ungauged Basins, Hydrol. Sci. J., https://doi.org/10.1080/02626667.2020.1797045, online first, 2020.

de Lavenne, A.: Modélisation hydrologique à base géomorphologique de bassins versants non jaugés par régionalisation et transposition d'hydrogramme, PhD thesis, Agrocampus Ouest, 226 pp., 2014.

de Lavenne, A. and Cudennec, C.: Prediction of streamflow from the set of basins flowing into acoastal bay, IAHS-AISH publication, International Association of Hydrological Sciences-Association Internationale des Sciences Hydrologiques publication, 365, 55–60, 2014.

de Lavenne, A. and Cudennec, C.: Assessment of freshwater discharge into a coastal bay through multi-basin ensemble hydrological modelling, Sci. Total Environ., 669, 812–820, https://doi.org/10.1016/j.scitotenv.2019.02.387, 2019.

de Lavenne, A., Boudhraâ, H., and Cudennec, C.: Streamflow prediction in ungauged basins through geomorphology-based hydrograph transposition, Hydrol. Res., 46, 291–302, https://doi.org/10.2166/nh.2013.099, 2015.

de Lavenne, A., Skøien, J. O., Cudennec, C., Curie, F., and Moatar, F.: Transferring measured discharge time series: Large-scale comparison of Top-kriging to geomorphology-based inverse modeling, Water Resour. Res., 52, 5555–5576, https://doi.org/10.1002/2016WR018716, 2016.

Dixon H., Sandström S., Cudennec C., Lins H., Abrate T., Bérod D., Chernov I., Ravalitera N., Sighomnou D., and Teichert F.: Intergovernmental cooperation for hydrometry – what, why, how?, Hydrol. Sci. J., https://doi.org/10.1080/02626667.2020.1764569, online first, 2020.

Ecrepont, S., Cudennec, C., Anctil, F., and Jaffrézic, A.: PUB in Québec: A robust geomorphology-based deconvolution-reconvolution framework for the spatial transposition of hydrographs, J. Hydrol., 570, 378–392, https://doi.org/10.1016/j.jhydrol.2018.12.052, 2019.

Fovet, O., Ruiz, L., Gruau, G., Akkal, N., Aquilina, L., Busnot, S., Dupas, R., Durand, P., Faucheux, M., Fauvel, Y., Fléchard, C., Gilliet, N., Grimaldi, C., Hamon, Y., Jaffrézic, A., Jeanneau, L., Labasque, T., Le Henaff, G., Mérot, P., Molénat, J., Petitjean, P., Pierson-Wickmann, A.-C., Squividant, H., Viaud, V., Walter, C., and Gascuel-Odoux, C.: AgrHyS: An observatory of response times in agrohydrosystems, Vadose Zone J., 17, 180066, https://doi.org/10.2136/vzj2018.04.0066, 2018.

Habets, F., Boone, A., Champeaux, J. L., Etchevers, P., Franchistéguy, L., Leblois, E., Ledoux, E., Le Moigne, P., Martin, E., Morel, S., Noilhan, J., Quintana Seguí, P., Rousset-Regimbeau, F., and Viennot, P.: The SAFRAN-ISBA-MODCOU hydrometeorological model applied over France, J. Geophys. Res.-Atmos. 113, D06113, https://doi.org/10.1029/2007JD008548, 2008.

Hrachowitz, M., Savenije, H. H. G., Blöschl, G., McDonnell, J. J., Sivapalan, M., Pomeroy, J. W., Arheimer, B., Blume, T., Clark, M. P., Ehret, U., Fenicia, F., Freer, J. E., Gelfan, A., Gupta, H. V., Hughes, D. A., Hut, R. W., Montanari, A., Pande, S., Tetzlaff, D., Troch, P. A., Uhlenbrook, S., Wagener, T., Winsemius, H. C., Woods, R. A., Zehe, E., and Cudennec, C.: A decade of Predictions in Ungauged Basins (PUB) – a review, Hydrol. Sci. J., 58, 1198–1255, https://doi.org/10.1080/02626667.2013_803183, 2013.

Nash, J. E. and Sutcliffe, J. V.: River flow forecasting through conceptual models part I – A discussion of principles, J. Hydrol., 10, 282–290, https://doi.org/10.1016/0022-1694(70)90255-6, 1970.

Oudin, L., Andréassian, V., Mathevet, T., Perrin, C., and Michel, C.: Dynamic averaging of rainfall-runoff model simulations from complementary model parameterizations, Water Resour. Res., 42, W07410, https://doi.org/10.1029/2005WR004636, 2006.

Rigon, R., Bancheri, M., Formetta, G., and de Lavenne, A.: The geomorphological unit hydrograph from a historical-critical perspective, Earth Surf. Proc. Land., 41, 27–37, https://doi.org/10.1002/esp.3855, 2015.

Robinson, J. S., Sivapalan, M., and Snell, J. D.: On the relative roles of hillslope processes, channel routing, and network geomorphology in the hydrologic response of natural catchments, Water Resour. Res., 31, 3089–3101, https://doi.org/10.1029/95WR01948, 1995.

Seiller, G., Roy, R., and Anctil, F.: Influence of three common calibration metrics on the diagnosis of climate change impacts on water resources, J. Hydrol., 547, 280–295, https://doi.org/10.1016/j.jhydrol.2017.02.004, 2017.

Skøien, J. O., Merz, R., and Blöschl, G.: Top-kriging – geostatistics on stream networks, Hydrol. Earth Syst. Sci., 10, 277–287, https://doi.org/10.5194/hess-10-277-2006, 2006.

Tarantola, A. and Valette, B.: Inverse problems = quest for information, J. Geophys., 50, 150–170, 1982.

Proc. IAHS, 383, 135–140, 2020
https://doi.org/10.5194/piahs-383-135-2020

Long-term persistence in discharge time series of mountainous catchments in the Elbe River basin

Ondrej Ledvinka[1] **and Thomas Recknagel**[2]

[1]Hydrology Database and Water Budget Department, Czech Hydrometeorological Institute, Na Sabatce 2050/17, Prague 412, 14306, Czechia
[2]Department M4, Federal Institute of Hydrology, Am Mainzer Tor 1, Koblenz, 56068, Germany

Correspondence: Ondrej Ledvinka (ondrej.ledvinka@chmi.cz)

Published: 16 September 2020

Abstract. Statistical detection of trends in hydrometeorological time series is a crucial task when revealing how river systems react to environmental and human-induced changes. It was shown that the autocorrelation structure of a series influences the power of parametric and nonparametric trend tests. While the order of short-memory processes can be sufficiently captured by AR(I)MA models, the determination of the Hurst exponent, which describes the long memory, is still challenging, considering that the available methods partially give different results. In the Elbe River basin, Europe, several studies focusing on the detection (or description) of long-term persistence were performed. However, different lengths of series and different methods were used. The aim of the present work is to gather the results gained in various parts of the Elbe basin in Central Europe and to compare them with our estimation of the Hurst exponent using six discharge series observed in selected subbasins. Instead of the dependence of the exponent on the catchment area suggested by the theory of aggregated short-memory processes, we rather found a relationship between this parameter and the series length. As the theory is not supported by our findings, we suppose that the Hurst phenomenon is caused by a complex interplay of low-frequency climate variability and catchment processes. Experiments based on distributed water balance models should be the further research objective, ideally under the umbrella of mutual international projects.

1 Introduction

The question of the existence and cause for trends in hydrometeorological (HM) time series has already been posed since the 19th century (e.g. v. Berg, 1867). In the course of the 20th century, predominantly nonparametric trend tests have been employed to analyze HM series. Theoretical works have shown (often through Monte Carlo experiments; e.g. Yue et al., 2002a) that these classical tests, although powerful in many cases (e.g. Yue et al., 2002b), are in fact sensitive to the violation of the assumption of independence in time series. Because HM series, as well as many other geophysical series, often reveal autocorrelation which is incompatible with that assumption, many modifications of trend tests came into being. This especially applies to the assumption of short-term persistence (STP) in HM time series that can be addressed by AR(I)MA models (or their variants) before utilizing trend tests themselves. However, building on others, Hamed (2008) showed that also long-term persistence (LTP) can adversely affect trend analyses through influencing the variance of test statistics.

Bearing the afore-mentioned facts in mind, we focus here specifically on the Elbe River basin located in Europe for which the decision on the presence of LTP in various HM series is very important for predicting the future development of discharge response to climate change, especially in connection with better estimating the uncertainty limits, which should be beneficial for future activities related to the Elbe and its tributaries, such as river navigation, water supply, hydropower production, irrigation, and so on. As there has been neither a closing result about the causes of LTP in HM series nor a single superior approach to the estimation of its parameter H, we offer a concise overview of the state of the discussion and put our own results into the context outlined by the outcomes of previous studies.

Published by Copernicus Publications on behalf of the International Association of Hydrological Sciences.

2 Theoretical background

The Hurst phenomenon (along with LTP) is well known among today's hydrologists. Therefore, we only stress some important points and possible misconceptions. An overview of the history of investigating LTP can be found in Beran (1994), Graves et al. (2017) or Montanari (2003). Hurst analysed the rescaled range R (see Sect. 3.1) as a measure of reservoir capacity and found that for many runoff series Eq. (1) applies:

$$R/S \propto N^{H}, \tag{1}$$

with standard deviation S, sample size $N$and the parameter H taking a mean value of 0.73 which differs clearly from the theoretical value 0.5 (Montanari, 2003). This is equivalent to the fact that the series autocorrelation function does not decay exponentially, but follows a power law and diverges. A large part of the amazement, these findings have caused, originates from the assumptions that runoff can be characterized by a stationary process for which H about 0.5 is expected. Mandelbrot and Van Ness (1968) and Mandelbrot and Wallis (1968) proposed a statistical model based on stationarity that should account for LTP: the fractional Brownian motion (fBm) model, which brought two other terms: infinity and self-similarity. Unwittingly, this model maybe pushed the search for the comprehension of the Hurst effect towards an unfavourable direction. The Gaussian distributed random process underlying runoff was already questioned by Louis M. Laushely in the discussion part of Hurst's seminal paper (Hurst, 1951). Even Hurst himself mentioned shifting mean as a possible reason for the phenomenon (Montanari, 2003). Eventually, Mandelbrot and Wallis (1968) stated that models underestimate the complexity of hydrological fluctuations and considered fluctuations on all timescales: macrometeorological, anthropological, climatological and paleoclimatological (Mandelbrot and Wallis, 1969). The fBm model was therefore never supposed to provide an explanation for LTP discovered in runoff series, but rather to be a tool able to simulate these series in a satisfactory manner. Although Mandelbrot and Wallis (1969) explained how the term "infinity" should be understood, a number of hydrologists refused to use the fBm approach together with its spooky "infinite memory" and considered LTP a mysterious intrinsic property of runoff series whose stochastic behaviour should be distinguished from deterministic trends (e.g. Kantelhardt et al., 2001). The concept was already questioned early by others. For example, Klemeš (1974) wondered "by what sort of physical mechanism" a measured value should be dependent on all previous values. Later, the operational school drew a lot of criticism. Klemeš (1978) stated that it failed "to see that the statistical and stochastic properties of hydrologic processes have definite physical causes". Mesa and Poveda (1993) wrote that "the so-called Hurst effect and other related anomalities are probably the result of a mixture of scales more than infinite memory. Knowledge of those scales is a more fundamental issue from a physical viewpoint". Beran (1994) demanded models "that can be explained by the physical mechanism that generates the data". One example of such a model is the concept of storage cascade mechanisms, which was proposed by Klemeš (1974), and in which Mudelsee (2007) saw the explanation for the Hurst phenomenon. However, Klemeš (1974) also postulated that LTP observed in hydrological series is caused by a complex interplay of different processes on different scales. This postulate, as Montanari (2003) stated, can nowadays have a broad acceptation within the hydrological community.

There are different approaches to quantify the strength of LTP through the parameter H called the Hurst exponent (or coefficient). In the following section, we briefly describe the two methods that have been applied in the Elbe River basin and also in our case study.

3 Estimation methods

We used (1) rescaled range (R/S) analysis which is based on the works of Hurst (1951) and Mandelbrot and Wallis (1969), and (2) detrended fluctuation analysis (DFA) which was proposed by Peng et al. (1994). With regard to validity and performance, Willems and Min (2016) compared functions of the programming language R which have implemented R/S or DFA. Inspired by the review, we used the package "pracma" (Borchers, 2019) for R/S and the package "nonlinearTseries" (Garcia, 2020) for DFA.

3.1 Rescaled range analysis

The adjusted range (R) is defined as the difference between the most positive and the most negative departures from the mean cumulated (discharge) curve, which is normalized by the sample standard deviation (S) to get the rescaled adjusted range statistic $R/S(n)$. The procedure can be formalized as in Eq. (2):

$$R/S(n) = \frac{\max_{1\leq k\leq n}\{Y_k - \frac{k}{n}Y_n\} - \min_{1\leq k\leq n}\{Y_k - \frac{k}{n}Y_n\}}{\sqrt{\frac{1}{n}\sum_{k=1}^{n}\left(X_k - \frac{1}{n}Y_n\right)^2}}, \tag{2}$$

where X_k is a sequence of random variables and Y_n the nth partial sum $X_1 + \ldots + X_n$. The parameter H can then be estimated by the slope of a double logarithmic plot of $R/S(n)$ against $n/2$ (for details see e.g. Graves et al., 2017).

3.2 Detrended fluctuation analysis

The idea behind DFA is to remove trends prior estimating the scaling behaviour. It is also based on the sequence of partial sums Y_n which is divided into segments of equal length s. For each segment, a polynomial regression model is fitted to the part of the time series and subtracted from it. Subsequently, the variance σ is calculated for each segment. The fluctuation function is the square root of the average variance subject to the segment width: $F(s)=\sqrt{\frac{1}{2N_s}\sum_{t=1}^{2N_s}\sigma(t,s)}$. Similar to R/S analysis, H can be estimated by the slope of a double logarithmic plot of $F(s)$ against s.

4 Overview of studies performed in the Elbe basin

Mudelsee (2007) fitted an ARFIMA(1, d, 0) model to the data of six water-gauging stations in the Elbe basin and analyzed if there was a correlation between catchment area and LTP. Although Mudelsee (2007) pointed out that, using a permutation test, a positive relationship was found also in the case of the Elbe basin, looking at the figures presented in that paper, one could argue that such a finding might be questionable.

Marković and Koch (2005) analyzed LTP in monthly high flows at the gauge Dresden (Elbe) and compared six different estimation methods which were applied to the raw data and to four time series derived by various filtering techniques. The comparison of methods, such as R/S analysis or the level of zero crossings approach, resulted in a wide range of H estimates.

Kantelhardt et al. (2006) analyzed daily runoff series of the gauge Dresden, covering the period 1852–2002. Through DFA, they found that $H=0.80$.

Markovic and Koch (2014) determined H from discharge records of the gauges Neu Darchau (on the Elbe) and Niederstriegis (Mulde). Analyzing raw data, they found that for Neu Darchau $H=0.80$ and for Niederstriegis $H=0.63$. After filtering the series (i.e. subtracting annual and low-frequency components), they observed considerable decreases of H. They concluded that LTP in discharge was caused by seasonal and decadal variability.

Regarding the Czech part of the Elbe basin, mainly Ledvinka (2014, 2015a, b) carried out several calculations that incorporated also a number of water-gauging stations relevant to this study. Three series of mean daily discharge from the Ore Mts. (i.e. one of the headwater areas of the Elbe basin) were subjected to wavelet and R/S analyses, and to the aggregated variance method. Also, maximum likelihood estimation of the fractional differencing parameter for ARFIMA models was conducted. Unexpectedly, quite high values of H were detected. Furthermore, it was found that many series involving indices related to hydrological drought in Czechia reveal high values of H as well, which was confirmed by unit root tests. On the other hand, precipitation series in Czechia for the period 1961–2012 showed no LTP (Ledvinka, 2015c). The differences between discharge and climate variables from this point of view, using selected stations within the Czech part of the Elbe basin, were highlighted also in Ledvinka and Jedlicka (2018).

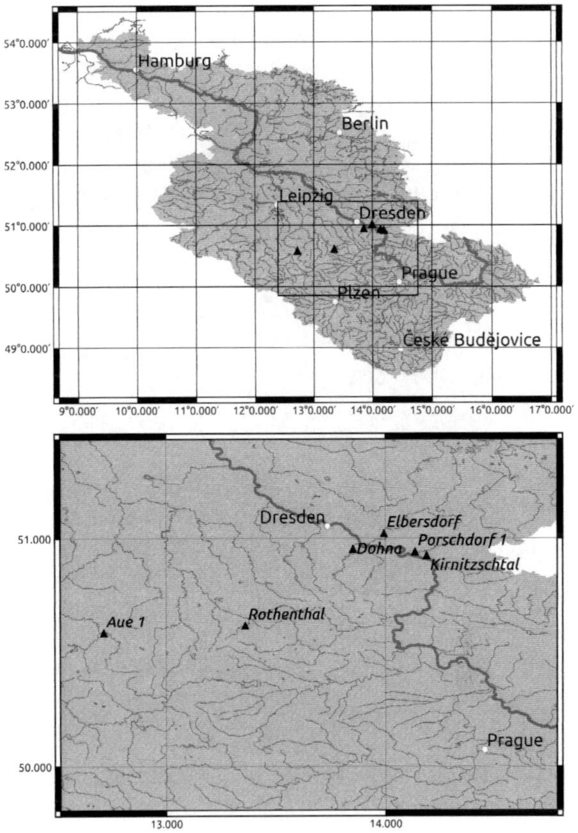

Figure 1. Location of six water-gauging stations selected for the case study within the Elbe River basin.

5 Case study

For our own analyses, we selected six discharge time series representing small tributaries in the central part of the Elbe basin close to the borders between Germany and Czechia. All gauges are located within a mountain range (see Figs. 1–2): Aue 1 (on the Schwarzwasser), Rothenthal (on the Natzschung) and Dohna (on the Müglitz) belong to the Ore Mts.; Elbersdorf (on the Wesenitz), Porschdorf 1 (on the Lachsbach) and Kirnitzschtal (on the Kirnitzsch) belong to the Elbe Sandstone Mts. The period, for which all gauges have complete mean daily discharge data, spans from 1929 to 2005. The gauges Kirnitzschtal and Porschdorf 1 had, at the time of writing, the longest time series (i.e. 1911–2017). The Hurst exponent was calculated using the R/S and DFA methods based on mean monthly discharge time series which were deseasonalized using the method described in Hipel and McLeod (1994) and implemented in the R package "deseasonalize" (McLeod and Gweon, 2013). The results are summarized in Table 1. For all gauges, we found that $H>0.5$, which means that, very likely, the series contain LTP. DFA gives higher values for H and shows a higher variability

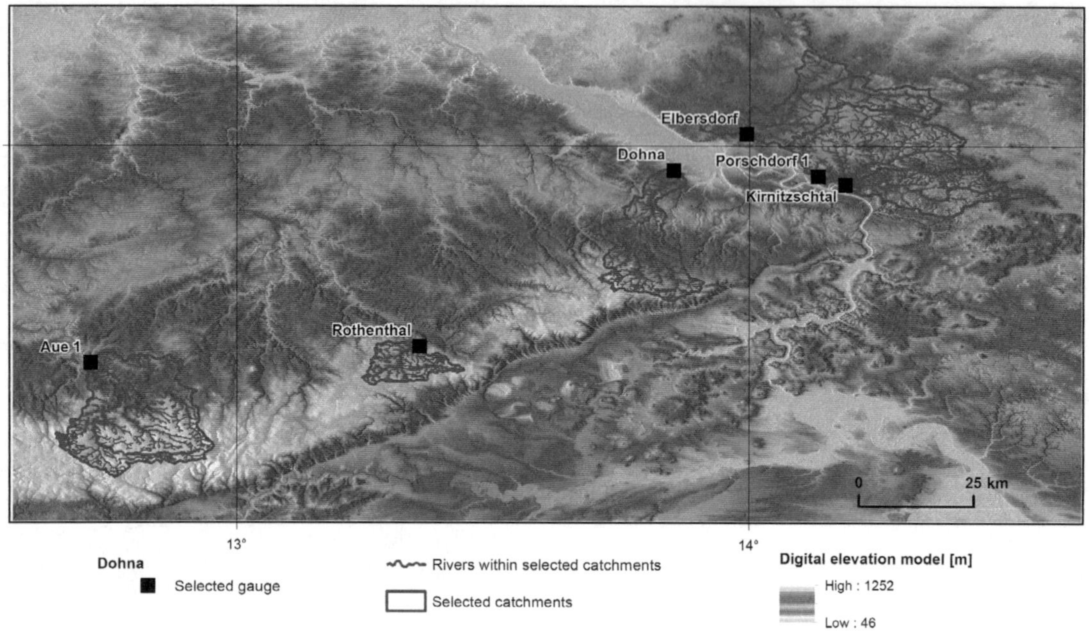

Figure 2. Closer look at the catchments represented by investigated gauges, river network within the catchments and elevation conditions in the case study area.

Table 1. Estimated H exponents for deseasonalized monthly mean discharge time series (1929–2005).

Gauge	R/S	DFA
Aue 1	0.60	0.64
Dohna	0.60	0.66
Elbersdorf	0.65	0.82
Kirnitzschtal	0.63	0.73
Porschdorf 1	0.60	0.76
Rothenthal	0.68	0.71

in comparison to R/S. Figure 3 shows a plot of H against catchment area. Based on the series of Porschdorf 1, Fig. 4 depicts how the estimates of H depend on varying sample size. It is obvious that for both applied methods, H decreases with increasing time series length.

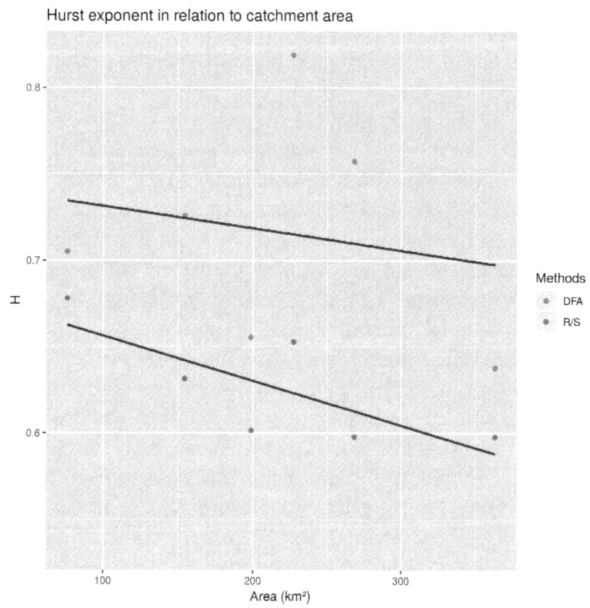

Figure 3. The values of the Hurst exponent H against catchment area.

6 Discussion and conclusions

The quantification of LTP in HM series is a difficult task, especially due to the fact that many estimation methods give different results regarding the magnitude of the LTP parameter H. This is evident also from our case study where only two methods were chosen (see Figs. 3–4). As shown in Fig. 4, also the length of available time series influences the estimation of H. Regarding the obvious tendency for H to decrease in Fig. 4, we must state that our series are still too short to give a clear answer to the question about the general behaviour of H for very long time series. Both convergence to the value of 0.5 as well as to the value signifying LTP might be possible. The issue of the time series length was demonstrated very clearly for precipitation by Markonis

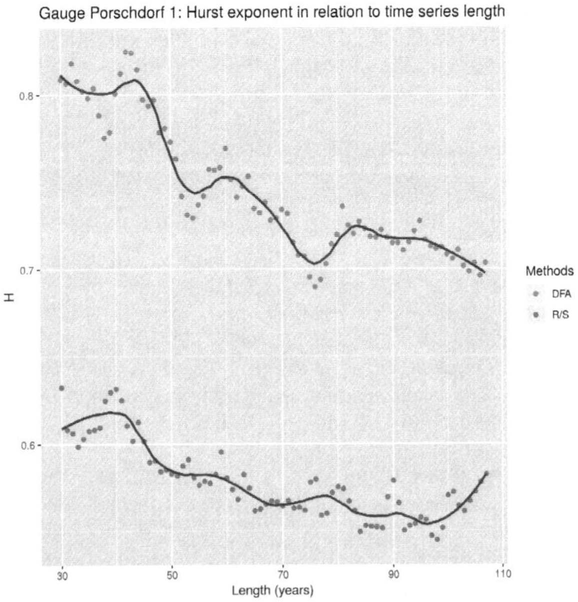

Figure 4. Dependence of the estimates of the Hurst exponent H on the length of analyzed series (shown for the Porschdorf 1 gauge).

and Koutsoyiannis (2016) who compared the behaviour of instrumental records with that of much longer time series derived from proxy data. However, when dealing with changes or trends in hydrological regimes, LTP is a key player since the outcomes of various tests strongly depend on the correct setting of null hypotheses. As the instrumental records are usually short, one should keep in mind the issues mentioned above before constructing models for practical purposes such as setting the lifetime of water works.

Overall, it is believed that LTP is stronger in discharge time series than in other HM series. Today, many hydrologists suggest that LTP is inherited by discharge from the low-frequency climate variability that may be reflected, for instance, in precipitation. The discrepancy between the behaviour of discharge and precipitation could be explained by physical processes similar to filtering (Milly and Wetherald, 2002). The theory that only aggregated short-memory processes are the cause for LTP is somewhat contradictory to our findings in the Elbe basin because the small catchments here reveal relatively high values of the parameter H as well. Moreover, Fig. 3 shows that there is no such dependence suggested by Mudelsee (2007). Further research into this topic should focus on the low-frequency variability of driving forces of runoff and their possible amplification by the catchment system. Distributed water balance models can be a tool for suitable experiments because studying time series of components, such as soil moisture, groundwater recharge, evapotranspiration and others, might give valuable answers regarding interactions and thresholds occurring in the Elbe River basin, many of which may still not be fully understood.

As already stated in several past papers (e.g. Koutsoyiannis, 2003, 2006; Koutsoyiannis and Montanari, 2007), not considering LTP properly may in practice lead to underestimation of process variability that, moreover, needs to be distinguished from change caused by external drivers. Especially in the fields of hydrology and water management, this is a very difficult but important task for designing water works. Therefore, the research into this topic should definitely continue.

Data availability. Demanded files with data and geodata from Saxony can be currently selected and downloaded using the iDA information system (https://www.umwelt.sachsen.de/umwelt/infosysteme/ida/, last access: 15 July 2020, LfULG, 2019). Individual tiles of digital elevation model for Fig. 2 were selected and retrieved using the USGS EarthExplorer Portal (https://earthexplorer.usgs.gov/, last access: 15 July 2020, USGS, 2020). Both sources require user's registration first, which is for free.

Author contributions. TR worked on the theoretical part, the case study (including the table and figures apart from Fig. 2) and the literature overview regarding Germany. OL worked on the rest of the manuscript.

Competing interests. The authors declare that they have no conflict of interest.

Special issue statement. This article is part of the special issue "Hydrological processes and water security in a changing world". It is a result of the 8th Global FRIEND–Water Conference: Hydrological Processes and Water Security in a Changing World, Beijing, China, 6–9 November 2018.

Acknowledgements. We would like to acknowledge Petra Walther (LfULG Sachsen) for providing suitable time series from the Saxon part of the Elbe basin.

References

Beran, J.: Statistics for Long-Memory Processes, Chapman & Hall, New York, 1994.

Borchers, H. W.: pracma: Practical Numerical Math Functions, R package version 2.2.9, available at: https://CRAN.R-project.org/package=pracma (last access: 21 April 2020), 2019.

Garcia, C. A.: nonlinearTseries: Nonlinear Time Series Analysis, R package version 0.2.8, available at: https://CRAN.R-project.org/package=nonlinearTseries (last access: 21 April 2020), 2020.

Graves, T., Gramacy, R., Watkins, N., and Franzke, C.: A brief history of long memory: Hurst, Mandelbrot and the road to ARFIMA, 1951–1980, Entropy, 19, 437, https://doi.org/10.3390/e19090437, 2017.

Hamed, K. H.: Trend detection in hydrologic data: the Mann–Kendall trend test under the scaling hypothesis, J. Hydrol., 349, 350–363, https://doi.org/10.1016/j.jhydrol.2007.11.009, 2008.

Hipel, K. W. and McLeod, A. I.: Time Series Modelling of Water Resources and Environmental Systems, Elsevier, Amsterdam, 1994.

Hurst, H. E.: Long term storage capacity of reservoirs (with discussions), Trans. Am. Soc. Civ. Eng., 116, 770–808, 1951.

Kantelhardt, J. W., Koscielny-Bunde, E., Rego, H. H., Havlin, S., and Bunde, A.: Detecting long-range correlations with detrended fluctuation analysis, Phys. Stat. Mech. Its Appl., 295, 441–454, https://doi.org/10.1016/S0378-4371(01)00144-3, 2001.

Kantelhardt, J. W., Koscielny-Bunde, E., Rybski, D., Braun, P., Bunde, A., and Havlin, S.: Long-term persistence and multifractality of precipitation and river runoff records, J. Geophys. Res.-Atmos., 111, D01106, https://doi.org/10.1029/2005JD005881, 2006.

Klemeš, V.: The Hurst phenomenon: a puzzle?, Water Resour. Res., 10, 675–688, https://doi.org/10.1029/WR010i004p00675, 1974.

Klemeš, V.: Physically based stochastic hydrologic analysis, in Advances in Hydroscience, vol. 11, edited by: Chow, V. T., Academic Press, New York, 285–356, 1978.

Koutsoyiannis, D.: Climate change, the Hurst phenomenon, and hydrological statistics, Hydrol. Sci. J., 48, 3–24, https://doi.org/10.1623/hysj.48.1.3.43481, 2003.

Koutsoyiannis, D.: Nonstationarity versus scaling in hydrology, J. Hydrol., 324, 239–254, https://doi.org/10.1016/j.jhydrol.2005.09.022, 2006.

Koutsoyiannis, D. and Montanari, A.: Statistical analysis of hydroclimatic time series: Uncertainty and insights, Water Resour. Res., 43, W05429, https://doi.org/10.1029/2006WR005592, 2007.

Ledvinka, O.: Are there nonstationarities and the Hurst phenomenon in discharge series within the Ore Mountains region?, in: Hydrology of a Small Basin 2014, vol. 1, edited by: Brych, K. and Tesař, M., Czech Hydrometeorological Institute, Praha, 287–295, 2014.

Ledvinka, O.: Evolution of low flows in Czechia revisited, Proc. IAHS, 369, 87–95, https://doi.org/10.5194/piahs-369-87-2015, 2015a.

Ledvinka, O.: Scaling of low flows in Czechia – an initial assessment, Proc. IAHS, 366, 188–189, https://doi.org/10.5194/piahs-366-188-2015, 2015b.

Ledvinka, O.: Nonstationarities in technical precipitation series in Czechia, Acta Hydrol. Slovaca, 16, 199–207, 2015c.

Ledvinka, O. and Jedlicka, M.: Freely available daily hydrometeorological data from Czechia: further insights, edited by: Kaźmierczak, B., Kutyłowska, M., Piekarska, K., and Jadwiszczak, P., E3S Web Conf., 44, 00093, https://doi.org/10.1051/e3sconf/20184400093, 2018.

LfULG: interdisziplinäre Daten und Auswertungen, available at: https://www.umwelt.sachsen.de/umwelt/infosysteme/ida/ (last access: 15 July 2020), 2019.

Mandelbrot, B. B. and Van Ness, J. W.: Fractional Brownian motions, fractional noises and applications, SIAM Rev., 10, 422–437, https://doi.org/10.1137/1010093, 1968.

Mandelbrot, B. B. and Wallis, J. R.: Noah, Joseph, and operational hydrology, Water Resour. Res., 4, 909–918, https://doi.org/10.1029/WR004i005p00909, 1968.

Mandelbrot, B. B. and Wallis, J. R.: Some long-run properties of geophysical records, Water Resour. Res., 5, 321–340, https://doi.org/10.1029/WR005i002p00321, 1969.

Markonis, Y. and Koutsoyiannis, D.: Scale-dependence of persistence in precipitation records, Nat. Clim. Change, 6, 399–401, https://doi.org/10.1038/nclimate2894, 2016.

Marković, D. and Koch, M.: Sensitivity of Hurst parameter estimation to periodic signals in time series and filtering approaches, Geophys. Res. Lett., 32, L17401, https://doi.org/10.1029/2005GL024069, 2005.

Markovic, D. and Koch, M.: Long-term variations and temporal scaling of hydroclimatic time series with focus on the German part of the Elbe River Basin, Hydrol. Process., 28, 2202–2211, https://doi.org/10.1002/hyp.9783, 2014.

McLeod, A. I. and Gweon, H.: Optimal deseasonalization for monthly and daily geophysical time series, J. Environ. Stat., 4, 1–11, 2013.

Mesa, O. J. and Poveda, G.: The Hurst effect: The scale of fluctuation approach, Water Resour. Res., 29, 3995–4002, https://doi.org/10.1029/93WR01686, 1993.

Milly, P. C. D. and Wetherald, R. T.: Macroscale water fluxes 3. Effects of land processes on variability of monthly river discharge, Water Resour. Res., 38, 1235, https://doi.org/10.1029/2001WR000761, 2002.

Montanari, A.: Long-range dependence in hydrology, in: Theory and Applications of Long-Range Dependence, edited by: Doukhan, P., Oppenheim, G., and Taqqu, M. S., Birkhäuser, Boston, Massachusetts, 461–472, 2003.

Mudelsee, M.: Long memory of rivers from spatial aggregation, Water Resour. Res., 43, W01202, https://doi.org/10.1029/2006WR005721, 2007.

Peng, C.-K., Buldyrev, S. V., Havlin, S., Simons, M., Stanley, H. E., and Goldberger, A. L.: Mosaic organization of DNA nucleotides, Phys. Rev. E, 49, 1685–1689, https://doi.org/10.1103/PhysRevE.49.1685, 1994.

USGS: EarthExplorer Portal, available at: https://earthexplorer.usgs.gov/, last access: 15 July 2020.

v. Berg: Wie kann man die Verminderung des Wasserstandes in der Elbe erklären?, Wiss. Beil. Leipz. Ztg., 13 June 1867.

Willems, W. and Min, A.: Praxisgerechte Detektion von Trends unter Berücksichtigung von Kurz- und Langzeit-Autokorrelationsstrukturen, IAWG, Ottobrunn, 2016.

Yue, S., Pilon, P., Phinney, B., and Cavadias, G.: The influence of autocorrelation on the ability to detect trend in hydrological series, Hydrol. Process., 16, 1807–1829, https://doi.org/10.1002/hyp.1095, 2002a.

Yue, S., Pilon, P., and Cavadias, G.: Power of the Mann–Kendall and Spearman's rho tests for detecting monotonic trends in hydrological series, J. Hydrol., 259, 254–271, https://doi.org/10.1016/S0022-1694(01)00594-7, 2002b.

Proc. IAHS, 383, 141–149, 2020
https://doi.org/10.5194/piahs-383-141-2020

Understanding and predicting large-scale hydrological variability in a changing environment

Nicolas Massei[1], Daniel G. Kingston[2], David M. Hannah[3], Jean-Philippe Vidal[4], Bastien Dieppois[3,5,6], Manuel Fossa[1], Andreas Hartmann[7], David A. Lavers[8], and Benoit Laignel[1]

[1]Normandie Univ, UNIROUEN, UNICAEN, CNRS, M2C, 76000 Rouen, France
[2]Department of Geography, University of Otago, Dunedin, New Zealand
[3]School of Geography, Earth & Environmental Sciences, University of Birmingham, Edgbaston, Birmingham, B15 2TT, UK
[4]INRAE, UR RiverLy, Centre de Lyon-Grenoble, 69625 Villeurbanne, France
[5]Centre for Agroecology, Water and Resilience (CAWR), Coventry University, Ryton-on-Dunsmore, Coventry, CV8 3LG, UK
[6]Department of Oceanography, University of Cape Town, Private Bag X3, Rondebosch, 7701, Republic of South Africa
[7]Chair of Hydrological Modeling and Water Resources, Freiburg University, Freiburg, Germany
[8]European Centre for Medium-Range Weather Forecasts (ECMWF), Shinfield Park, Reading, RG2 9AX, UK

Correspondence: Nicolas Massei (nicolas.massei@univ-rouen.fr)

Published: 16 September 2020

Abstract. In a context of climate, environmental, ecological and socio-economical changes, understanding and predicting the response of hydrological systems on regional to global spatial scales, and on infra-seasonal to multidecadal time-scales, are major topics that must be considered to tackle the challenge of water resource management sustainability. In this context, a number of strongly-linked key issues need to be addressed by the scientific community, including: (i) identifying climate drivers of hydrological variations, (ii) understanding the multi-frequency characteristics of hydroclimate variability, including evolution of extremes (meteorological/hydrological event scale to long-term natural/internal climate- or anthropogenic-driven variations and trends), (iii) assessing the influence of local- to regional-scale basin properties on hydrological system response to climate variability and change, (iv) identifying the evolving contribution of anthropogenic water use in observed hydrological variations. Based on pan-European collaborations, activities of the EURO-FRIEND "Large-scale variations in hydrological characteristics" group aim at generating new findings to improve our understanding of hydrological systems behavior sensu lato (i.e. surface and sub-surface) on large spatial and temporal scales (i.e continental – multidecadal). Through selected examples, this contribution emphasizes recent research developments in characterizing and modeling of climate-hydrology linkages at different temporal and spatial scales, as well as recent insights on climate-hydrology scaling characteristics (i.e. long-term persistence, dependance of processes, of hydrological behaviors, of large-scale climate/hydrology linkages on time-/spatial scales), long-term hydrometeorological reconstructions, and large-scale hydrological model refinement taking into account spatial heterogeneity of watershed physical characteristics.

1 Introduction

Understanding, characterizing and predicting changes in hydroclimatic conditions on land are essential to optimize water use and mitigate the impact of climate and environmental changes on water resources and related hazards (e.g. floods, droughts, erosive events, degradation of water quality, hydropower...). One of the main problems in predicting long-term hydroclimate variability is related to the limitations of climate models to adequately simulate internal modes of variability, whose impacts can sometimes worsen, sometimes attenuate the impact of climate change on regional hydrological systems, especially at decadal scales (Hawkins and Sutton, 2009; Deser et al., 2012, 2014; Roberts et al., 2015). For instance, natural phenomena such as the Interdecadal Pacific Oscillation and the Atlantic Multidecadal Oscillation substantially modify the rate of warming in many regions of the globe (Zhang et al., 2007; Meehl et al., 2016; Deser et al., 2012). It is then important to distinguish between trends and low-frequency variability inherited from climate system dynamics. Impacts usually affect large regional to continental spatial scales and highly depend on local physiographic conditions such as basin characteristics and land-use.

A number of key issues need to be addressed by the scientific community in order to understand and predict the large-scale hydrological response to climate variability and environmental changes, and to tackle the major challenge of adapting water resource management to climate change. This paper presents a reflexion on characterization, forcing factors, methodological approaches to hydrological variability, partly based on recent research of EURO-FRIEND Water Projet 3 "Large-scale variations in hydrological characteristics" members. It mainly focuses on the following particular points: (i) identification of the climate drivers of hydrological variations and changes, (ii) characterization of hydrological variability across time-scales, (iii) influence of basin characteristics on hydrological responses to climate variability and changes, (iv) the need for long-term (palaeo-)hydrological/climatic reconstructions to help understanding recent hydroclimatic variability and trends.

2 Identifying the climatic drivers of large-scale hydrology

Establishing links between hydrological variability and large-scale climate is crucial to understand how hydrology responds to climate changes and eventually to predict hydrological variations and develop climate-impact scenarios. Different complementary approaches can be undertaken. Some of them focus on describing statistical relationships between parts of the climate system and land surface water. However, understanding climate/hydrology links requires deeper analysis of mechanisms at play for developing a coherent and comprehensive understanding about how water flows from the ocean, through the atmosphere, to the land surface.

2.1 Atmospheric circulation-based analyses

At their most general and wide-ranging, studies of the role of atmospheric circulation as a control on large-scale hydrological variation involve correlation of hydrological metrics with gridded geopotential height data (e.g. Kingston et al., 2015). Alternatively, composite analysis of such gridded data sets may be used, based on a threshold of high or low values for some aspect of hydrological variation (e.g. river flow – Kingston et al., 2016a; precipitation – Lavers et al., 2013 and Massei et al., 2017). Such studies often focus on defined modes of coupled ocean-atmospheric circulation variation, such as the Atlantic Multidecadal Oscillation (AMO), North Atlantic Oscillation (NAO) and El Nino Southern Oscillation (ENSO), and at monthly or seasonal . At such spatial and temporal scales, the relatively high temporal and spatial autocorrelation of sea-surface temperatures (SSTs) can make this variable a particularly important focus for understanding large-scale variation in hydrological processes (e.g. Ionita et al., 2017; Kingston et al., 2015).

Rather than focusing on hemispheric or continental scale ocean-atmosphere variations, other studies of climate control on hydrological variation examine more closely the atmospheric circulation co-located with the immediate hydrological study. These typically make use of synoptic-scale weather classifications, such as the Lamb Weather Types (UK; Lamb, 1972), Kidson Weather Types (New Zealand; Kidson, 2000), or the European-scale Grosswetterlagen (James, 2007). Such classification schemes are typically calculated at daily or sub-daily scale, often making them suited to event-scale analyses (e.g. Fleig et al., 2011, in addition to longer-term characterization of the atmospheric circulation – land surface hydrology connections. Multi-scale analyses may also investigate relationships from SSTs to large-scale atmospheric circulation patterns and subsequently synoptic weather types (Hannah et al., 2014).

2.2 Atmospheric pathways to precipitation delivery

Analyses of large-scale geopotential height and wind fields can provide strong indications of delivery of warm/cold/wet/dry air-masses over a region based on corresponding changes in pressure gradient, orientation and presence of high and low pressure systems. However, such analyses do not ultimately provide quantitative evidence of changes in hydrologically relevant variables (i.e. precipitation supply). One approach to help filling the gap from large-scale atmospheric circulation to airmass/precipitation supply is to analyze more specifically the source and trajectory of air-masses through the atmosphere using reanalysis or climate model data. One method by which such an approach can be achieved is back trajectory analysis, whereby parcels of air spatially and temporally coincident with a hydrologically significant event (i.e. extreme precipitation, snowmelt or evaporation) are tracked backwards in time. The backwards trajectory thus reveals how parcels of air reach the hydrologically significant event and interact with larger-scale weather types and/or circulation patterns (e.g. Kingston and McMecking, 2015; Sahin et al., 2015). Alternatively to tracking the physical trajectory of air parcels, the geochemical characteristics of precipitation delivered to the land surface can also provide information on air-mass characteristics, precipitation source and moisture recycling from the surface to the atmosphere (e.g. Fiorella et al., 2018; Bass et al., 2014).

2.3 Moisture transport mechanisms: blocking, storm tracks and atmospheric rivers

A further means to provide more detail information on connections from large-scale circulation to land-surface hydrology is to focus on the role of specific atmospheric phenomena for delivering moist/dry/warm/cold air masses, particularly in relation to precipitation occurrence. For example, atmospheric blocking events are often associated with low precipitation totals, and outside of winter, with above average temperatures – such as during the 2015 summer drought in Europe (Ionita et al., 2017). Similarly, the presence of blocking typically results in diversion of prevailing winds and thus storm tracks – as was also observed during the 2015 European drought (Ionita et al., 2017). The relatively recent increased attention given to the role of atmospheric rivers (extremely intense and narrow "rivers" of atmospheric water vapour flux) has also led to new understanding of the processes leading to extreme rainfall and flooding events: for individual rivers (e.g. Lavers et al., 2015), regions, and globally (e.g. Kingston et al., 2016b; Lavers and Villarini, 2013; Paltan et al., 2017). Similarly, the absence of such features has been linked to drier than normal conditions, including drought (Paltan et al., 2017).

2.4 Hydrological prediction

Identification of concurrent climate-hydrology relationships such as those described in previous section provides important information on the series of processes that connect the climate system variation to large-scale hydrological variation. Whilst a critical step in developing understanding of the drivers of hydrology and in turn water resources, substantial additional benefit (particularly from an applied perspective) comes from the identification of lag-lead relationships in the ocean-atmosphere-land surface system, and so the potential for improved predictability of water resource availability. Seasonal lag-lead relationships typically focus on the role of slowly-varying components of the climate system, particularly SSTs (e.g. Kingston et al., 2013; Ionita et al., 2017), whereas on daily to weekly scales it has recently been shown that by focussing on water vapour transport instead of directly on precipitation has the potential to increase the forecast horizon by up to three days (Lavers et al., 2014).

3 Hydrological variability across time-scales

Understanding the long-term evolution of hydrological variations has been the subject of a large number of studies in recent decades (Hannaford et al., 2013; McCabe and Wolock, 2002; Pagano and Garen, 2005; Stahl et al., 2010). Most of them focused on the identification of trends in hydrological processes (precipitation, temperature, streamflow). However, characterization of high- to low-frequency hydrological variability, beyond trend identification only, now appears necessary for understanding the evolution and forcing factors of hydrological phenomena in a changing climate. In this framework, identifying intrinsic oscillations affecting both large-scale climate and hydrological variables over multiple time-scales provides new insights.

3.1 Trends and low-frequency variability

Assessing the low-frequency variability of hydrological and hydroclimatic variables received increasing attention for the past years. On the pan-European scale, for instance, Gudmusson et al. (2011) or Hannaford et al. (2013) addressed this issue. Gudmusson et al. (2011) investigated the low-frequency variability of runoff for 358 watersheds across Europe, and deduced that low-frequency behavior would follow well known continental-scale atmospheric features, whereas watershed properties would explain the proportion of variance of low-frequency fluctuations. Applying a so-called multi-temporal approach for trend analysis on a database comprising 132 european watersheds, Hannaford et al. (2013) highlighted that magnitude and direction of short-term trends were strongly influenced by interdecadal variability. Investigating river flow variations at 38 gauging stations in metropolitan France, Boé and Habets (2014) highlighted the large amplitude of the multi-decadal variations in French rivers. Similarly, using 500 years observational and reconstructed hydroclimatic records, Dieppois et al. (2016) highlighted increasing multidecadal variability in the Seine river watershed, whose contribution is greater than the general trend to wetter conditions, since in the mid-19th century. Both studies emphasized on the importance of quantifying such variability, which may modify the long-term effect of global climate change on hydroclimatic processes (amplification, attenuation or sometimes inversion). This is also of crucial importance for short-length observational time-series and classical trend detection approaches, which can be substantially affected by potential interactions with decadal to multidecadal fluctuations. In this sense, it also constitutes a significant source of uncertainty for near-future climate projections, as multidecadal fluctuations are largely underestimated in current state-of-the art models (e.g. Ault et al., 2012, 2013) and contributing to uncertainties associated with internal climate variability (Keenlyside et al., 2008).

3.2 Time-frequency approaches for characterizing hydrological variability

One consistent means in the aim of better characterizing low-frequency variability and trends relies on the full exploration of the total range of the time-series spectrum. Since knowledge about time-evolution of hydroclimate variability is important, multi-frequency or multiresolution approaches need to be used, as proposed by Anctil and Coulibaly (2004), Coulibaly and Burn (2004), Labat (2006), or more recently by Massei et al. (2010, 2017), Massei and Fournier (2012), Kingston et al. (2016a) and Dieppois et al. (2013, 2016). These authors used continuous wavelet transform or multiresolution techniques (e.care of mostg. wavelet-based, empirical mode decomposition) to explore the time-frequency spectral content of precipitation, streamflow or climate indices in order to identify the different oscillatory components explaining the variance of such signals and its change through time. Kingston et al. (2016) used such techniques in addition to correlation analysis, partial least-square regression and composite analysis for understanding of the circulation–climate–inflow relationships for three headwater lakes in the Waitaki basin (New-Zealand). The approach allowed them to demonstrate, not only the presence of statistical relationships, but also their temporal location and periodicity together with an identification of underlying physical mechanisms. Studying precipitation and streamflow on the Seine river watershed in France between 1950 and 2008, Massei et al. (2010, 2017) and Massei and Fournier (2012) found interannual ($\sim$ 7-years) and interdecadal ($\sim$ 17-years) oscillations, whose fluctuations are only significant, respectively, since the 1970s and 1990s, that would explain 23 % of daily Seine streamflow, and up to 35 % of daily deseasonalized flow. Using similar methodology, Dieppois et al. (2016) also identified multi-decadal fluctuations ($\sim$ 64 years) in the Seine river watershed, whose contributions significantly increase in the mid-19th century. Compared to low-pass filtering, spectral or multiresolution-based approaches allow for separating and extracting components from high-to low-frequencies, making it possible (i) to analyze separately the time evolution and the variability of all statistically significant frequency components, (ii) to look for possible different determinism associated to these components.

3.3 Linking large-scale climate and hydrology across temporal scales

As internal modes of climate variability and associated teleconnections oscillate over a number of different time-scales, so do hydrological variables, on regional or even local scales (e.g. at weather stations or small watershed outlets). This leads to a number of key-questions: do large-scale climate and local-scale variations co-vary similarly for all time-scales? Can we identify statistical and/or physical links between large- and local-scale variables that would be hidden when accounting for linear or nonlinear correlation analysis only? Can we define processes that are associated to characteristic oscillations? Which processes become eventually filtered out by local characteristics?

In northern France, Massei et al. (2007, 2010) and Massei and Fournier (2012) identified clear co-variation between the NAO index, precipitation and discharge in the Seine river watershed at specific time-scales: interannual ($\sim$ 7-years) and interdecadal ($\sim$ 17-years). Dieppois et al. (2013, 2016) then proposed that interdecadal to multidecadal hydroclimatic fluctuations in the Seine river watershed were associated with a south-eastward (southwestward) shift of the Icelandic Low (Azores High), which is particularly pronounced in Spring. This interdecadal shift in the NAO center action was consistent with Fritier et al. (2012), who showed that $\sim$ 17 years interdecadal fluctuations were significantly detected in the Icelandic low, only. According to Hilmer and Jung (2000), anomalous sea-ice and fresh through Fram Strait could have helped to drive such an eastward shift in the Icelandic Low over the recent decades, and at the interdecadal time-scale. Similar interactions between Arctic sea-ice, which would be linked to the AMO, could also occur at the multidecadal time-scale, as suggested by the recent work of Miles et al. (2014). In addition, combining composite/correlation analysis and multiresolution filtering technics, Massei et al. (2017) show that atmospheric patterns linked to rainfall and streamflow in the Seine river watershed, where progressively shifting from local/regional convective atmospheric patterns (i.e. rainfall associated with low-pressure system centered on the watershed) to large-scale zonal circulations (i.e. NAO-like patterns) from intraseasonal ($\sim$ 3 months) to interdecadal ($\sim$ 17 years) time-scales. While previous studies suggested climate-streamflow predictions were using raw data (e.g. Lavers et al., 2013), Massei et al. (2017) highlighted that prediction skills could thus be substantially improved by tracking the predictor at the different time-scales.

4 Research issues deserving increasing attention: assessing the role of large-scale basin characteristics and bridging the gap between historical period and palaeo-hydroclimatic time-scales

Amongst many scientific questions that still need to be addressed in the field of large-scale hydrology, investigating the role of basin properties on hydrological variability and the need for long-term hydrological reconstructions in particular constitute important issues to which more efforts should be focused. On the one hand, it appears crucial to gain new insights into understanding physical processes at play in watersheds, including for instance recharge processes or surface water/ground water interactions. In particular, improving upscaling procedures would be needed, that would allow for a more realistic consideration of local-scale processes in large-scale hydrological models. On the other hand, the need for a better understanding and more comprehensive characterization of long-term variability in a context of climate change requires "zooming out" hydrological time series; looking back at the past may constitute one important means to assess the significance of observed trends and low-frequency variations, and to disentangle natural and anthropogenic hydroclimatic variations.

4.1 The influence of basin characteristics on hydrological response to climate variability and change

Assessing the relationships between large-scale climate and hydrological variations from the watershed scale up to the regional or even continental scale very often involves comparison between hydrological output variations, e.g. streamflow (Kingston et al., 2006; Lavers et al., 2013; Massei et al., 2010), low-flow characteristics (Giuntoli et al., 2013), ground water level (Holman et al., 2011; El Janyani et al., 2012; Lavers et al., 2015) and large-scale atmospheric variables or circulation indices.

Most of these authors, however, highlighted the need for a better understanding of basins properties in order (i) to gain new insights into the large-scale climate and local-scale hydrological relationships, (ii) to improve projections of water resources availability under climate change.

This conclusion was reached for instance by Laizé and Hannah (2010), who attempted to assess the respective roles of large-scale atmospheric circulation, regional climate and basin properties on streamflow based on a database comprising 104 gaging stations in Great Britain. They underlined the important, but variable role of basin properties in modifying climate signals in river flow and the need to consider both sets of controls in evaluating hydrological sensitivity to climate change.

Basin properties operate different types of filtering of large-scale climate fluctuations according to time-scale. For shorter time-scales, the basins transfer functions smooth out the climate input. Beyond a certain higher time-scale, the input signal is no longer frequency-modulated (i.e., the climate oscillations are recovered at the exact same frequency in the output). This is can be easily related to the conclusion drawn by Gudmusson et al. (2011) for instance, who noticed that the low-frequency behavior of streamflow on the pan-European scale followed well known continental-scale atmospheric features, whereas watershed properties would explain the proportion of variance of such low-frequency fluctuations.

Since higher frequencies are being filtered out, the low-frequency variations corresponding to long time-scales are clearly expressed and explain very often a non-negligible part of variance in output signals (Slimani et al., 2009). The role of aquifers in this process, acting as potentially strong low-pass filters, is pre-eminent, especially where low-flows are strongly sustained by ground water or in places where ground water flooding may occur.

El Janyani et al. (2012) attempted to relate the proportion of variance explained by frequency components in chalk ground water levels in northwestern France according to geological heterogeneity (thichness of 2 types of surficial formations, vadose zone and aquifer). These authors conducted cluster analyses based on the frequency content of the data and detected a similar interannual variability in ground water level and winter-months NAO. They finally interpreted of the geological forcing factors of piezometric variability using ANOVA to define explanatory variables. The authors concluded that the location of the piezometer along the flowpath, as well as the size of the ground water sub-basin considered were determinant in the structure of the frequency content of piezometric time-series: the more downstream the piezometer, the higher the amplitude of low-frequencies.

Thiéry et al. (2018) used the results, obtained at an experimental site in northern France, of a 2.5-year monitoring of water content, soil temperature, matrix pressure in the chalk aquifer under various natural or forced infiltration conditions, to develop a vertical numerical flow model as input to a ground water flow model. The refinement of the modeling approach based on improved knowledge on infiltration processes eventually allowed the ground water model to reproduce very well the interannual variability of ground water level, which had been related to large floods in northern France and southern England in winter and spring 2000–2001.

As pointed out by Hartmann et al. (2017), accounting for local to regional scale heterogeneity can actually reveal crucial for more consistent modeling of hydrological response to climate variability and change and appropriate water resources management. In their study on carbonate rock regions of Europe, Northern Africa, and the Middle East, these authors showed that accounting for carbonate heterogeneity (presence of karstic terranes) results in actual recharge rates up to four times larger for present conditions and up to five times larger for potential future conditions than estimated when recharge is considered homogeneous.

4.2 Looking back at the past using hydrometeorological reconstructions

The release of global extended reanalyses like 20CR (Compo et al., 2011), ERA20C (Poli et al., 2016) or CERA-20C (Laloyaux et al., 2018) over the past few years now provide reconstructions of global atmospheric circulation back to the 19th century onwards. It thus opened the way for deriving local-scale meteorological reconstructions and resulting catchment-scale hydrological reconstructions based on statistical downscaling methods and hydrological modeling tools (Dayon et al., 2015; Minvielle et al., 2015; Kuentz et al., 2015; Caillouet et al., 2016, 2017). These new high-resolution datasets thus now allow studying the climate variability at the local scale, as well as the resulting hydrological variability for small catchments. Long-term analyses of these datasets for example highlighted the large multidecadal variability in precipitation and streamflow over France over the last 140 years, a variability already suggested by hydrometric studies for specific catchments (Kuentz et al., 2014) or from sparse streamflow measurements over metropolitan France (Boé and Habets, 2014).

Hydrological and climatological reconstructions also constitute a unique means to tackle the challenge of estimating long-term statistical properties of climate variability beyond the time-scale of observational records. These approaches attempt to bridge the gap between hydrological/climatic observations and palaeo-hydrological/palaeo-climatic records, and received growing attention in the last recent years. Studies seek to link the statistical information beared by historical records and that contained in climatic or hydrological proxies in order to gain additionnal insights into interpretation of low-frequency variability and trends in hydrological processes (Dieppois et al., 2016; Iliopoulou et al., 2018; Markonis and Koutsoyiannis, 2016; Markonis et al., 2018). Characterizing long-term persistence, for instance, may reveal particularly relevant for improving hydro-climatic extreme models or understand observed hydrological variations under climate change (Markonis and Koutsoyiannis, 2016). The conclusions drawn from such studies may have important implications in terms of water management, including floods and droughts, for instance. Among recent palaeo-hydrological studies, Wilhelm et al. (2015) could reproduce the main regional hydrological trends (periods of high flood-frequency, ranges of flood-frequency values and timing of the most intense events) and propose a palaeo-hydrological reconstruction back to the 14th century from lake sediments in the french mediterranean Alps. Studying a lake sequence in the same region, Wilhelm et al. (2016) highlighted a tendency towards higher frequencies of high-intensity flood events during the warm Medieval Climate Anomaly: consequently, they suggest a possible increase in the intensity of floods in this area under global warming.

Enhancing interactions and communication between hydrological, climatological and palaeo-climatological scientific communities appears more than ever essential to address such crucial issues as understanding the long-term variability of hydro-climatic processes and interpreting current variability and changes under natural climate and anthropogenic forcing.

5 Conclusion

This paper presented a summary of some recently used approaches to characterize and understand large-scale hydrological variability, partly based on research conducted by members of EURO-FRIEND Water project 3 "Large-scale-variations in hydrological characteristics". The paper highlighted some key-elements for hydrological variability assessment, understanding and prediction, including the analysis of the climate drivers and multi-time-scales characteristics of hydrological variations. As emphasized throughout this article, the concept of "large-scale variations" must be here understood from both spatial and temporal standpoints. Both dimensions are closely linked: low-frequency hydrological variations can be intuitively associated with homogeneous hydrological behaviors across large areas. In other words, long-term, slow and highly-inertial climate variability logically tends to affect vast regions. This constitutes the basis for hydrological classification, and a supporting principle of many approaches used in hydrology (e.g. double-mass analysis for data homogeneity assessment).

However, this article also underlines that from both spatial and temporal sides, the need appears for bridging the gap between the small and the large scales. Fine-scale hydrologic or hydrogeologic processes are generally well understood, but large-scale hydrological models very often suffer from too simplified representations of the physical characteristics of the basins, which may impair their ability to generate consistent enough hydrological responses to tackle water management challenges under climate change, for instance. Efforts should then continue on the assessment of the way physical properties modulate the climate signal to release hydrological variability on the basin scale. Similarly, from the temporal side, comparison of historical measurements of hydro-climatic variables with (palaeo-)hydro-climatic reconstructions back to the past centuries or millenia (e.g. as elaborated by Pages 2k groups) seems necessary to improve understanding and interpretation of the long-term variability (of both averages and extremes) as recorded in historical direct measurements of hydro-climatic variables. In this sense, enhancing interactions between hydrological and palaeo-hydrological/palaeo-climatic communities appears fundamental.

Data availability. No data sets were used in this article.

Author contributions. This article was based on a joint reflection/discussion of all authors. Each author was involved in the setting up of the outline, was more particularly in charge of specific sections according to their main field of expertise, and eventually reviewed the final version.

Competing interests. The authors declare that they have no conflict of interest.

Special issue statement. This article is part of the special issue "Hydrological processes and water security in a changing world".

It is a result of the 8th Global FRIEND–Water Conference: Hydrological Processes and Water Security in a Changing World, Beijing, China, 6–9 November 2018.

Acknowledgements. This paper forms a contribution to the UNESCO-IHP EURO FRIEND–Water programme.

Financial support. Andreas Hartmann has been supported by the Emmy Noether-Programme of the German Research Foundation (DFG; grant no. HA 8113/1-1).

References

Anctil, F. and Coulibaly, P.: Wavelet analysis of the interannual variability in Southern Québec Streamflow, J. Climate, 17, 163–173, 2004.

Ault, T. R., Cole, J. E., and St. George, S.: The amplitude of decadal to multidecadal variability in precipitation simulated by state-of-the-art climate models, Geophys. Res. Lett., 39, L21705, https://doi.org/10.1029/2012GL053424, 2012.

Ault, T. R., Deser, C., Newman, M., and Emile-Geay, J.: Characterizing decadal to centennial variability in the equatorial Pacific during the last millennium, Geophys. Res. Lett., 40, 3450–3456, 2013.

Bass, A. M., Munksgaard, N. C., O'Grady, D., Williams, M. J. M., Bostock, H. C., Rintoul, S. R., and Bird, M. I.: Continuous shipboard measurements of oceanic δ^{18}O, δD and δ^{13}CDIC along a transect from New Zealand to Antarctica using cavity ring-down isotope spectrometry, J. Marine Syst., 137, 21–27, 2014.

Boé, J. and Habets, F.: Multi-decadal river flow variations in France, Hydrol. Earth Syst. Sci., 18, 691–708, https://doi.org/10.5194/hess-18-691-2014, 2014.

Caillouet, L., Vidal, J.-P., Sauquet, E., and Graff, B.: Probabilistic precipitation and temperature downscaling of the Twentieth Century Reanalysis over France, Clim. Past, 12, 635–662, https://doi.org/10.5194/cp-12-635-2016, 2016.

Caillouet, L., Vidal, J.-P., Sauquet, E., Devers, A., and Graff, B.: Ensemble reconstruction of spatio-temporal extreme low-flow events in France since 1871, Hydrol. Earth Syst. Sci., 21, 2923–2951, https://doi.org/10.5194/hess-21-2923-2017, 2017.

Compo, G. P., Whitaker, J. S., Sardeshmukh, P. D., Matsui, N., Allan, R. J., Yin, X., Gleason, B. E., Vose, R. S., Rutledge, G., Bessemoulin, P., Brönnimann, S., Brunet, M., Crouthamel, R. I., Grant, A. N., Groisman, P. Y., Jones, P. D., Kruk, M. C., Kruger, A. C., Marshall, G. J., Maugeri, M., Mok, H. Y., Nordli, Ross, T. F., Trigo, R. M., Wang, X. L., Woodruff, S. D., and Worley, S. J.: The Twentieth Century Reanalysis Project, Q. J. Roy. Meteor. Soc., 137, 1–28, https://doi.org/10.1002/qj.776, 2011.

Coulibaly, P. and Burn, D.: Wavelet analysis of variability in annual Canadian streamflows, Water Resour. Res., 40, W03105, https://doi.org/10.1029/2003WR002667, 2004.

Dayon, G., Boé, J., and Martin, E.: Transferability in the future climate of a statistical downscaling method for precipitation in France, J. Geophys. Res., 120, 1023–1043, https://doi.org/10.1002/2014JD022236, 2015.

Deser, C., Phillips, A., Bourdette, V., and Teng, H.: Uncertainty in climate change projections: the role of internal variability, Clim. Dynam., 38, 527–546, https://doi.org/10.1007/s00382-010-0977-x, 2012.

Deser, C., Phillips, A. S., Alexander, M. A., and Smoliak, B. V.: Projecting North American climate over the next 50 years: uncertainty due to internal variability, J. Climate, 27, 2271–2296, https://doi.org/10.1175/JCLI-D-13-00451.1, 2014.

Dieppois, B., Durand, A., Fournier, M., and Massei, N.: Links between multidecadal and interdecadal climatic oscillations in the North Atlantic and regional climate variability of northern France and England since the 17th, J. Geophys. Res.-Atmos., 118, 4359–4372, 2013.

Dieppois, B., Lawler, D. M., Slonosky, V., Massei, N., Bigot, S., Fournier, M., and Durand, A.: Multidecadal climate variability over northern France during the past 500 years and its relation to large-scale atmospheric circulation, Int. J. Climatol., 36, 4679–4696, https://doi.org/10.1002/joc.4660, 2016.

El Janyani, S., Massei, N., Dupont, J.-P., Fournier, M., and Dörfliger, N.: Hydrological responses of the chalk aquifer to the regional climatic signal, J. Hydrol., 464, 485-493, 2012.

Fiorella, R. P., Poulsen, C. J., and Matheny, A. M.: Seasonal patterns of water cycling in a deep, continental mountain valley inferred from stable water vapor isotopes, J. Geophys. Res.-Atmos., 123, 7271–7291, https://doi.org/10.1029/2017JD028093, 2018.

Fleig, A. K., Tallaksen, L. M., Hisdal, H., and Hannah, D. M.: Regional hydrological drought in north-western Europe: linking a new Regional Drought Area Index with weather types, Hydrol. Process., 25, 1163–1179, https://doi.org/10.1002/hyp.7644, 2011.

Fritier, N., Massei, N., Laignel, B., Durand, A., Dieppois, B., and Deloffre, J.: Links between NAO fluctuations and inter-annual variability of winter-months precipitation in the Seine River watershed (north-western France), C. R. Geosci., 344, 396–405, 2012.

Giuntoli, I., Renard, B., Vidal, J.-P., and Bard, A.: Low flows in France and their relationship to large-scale climate indices, J. Hydrol., 482, 105–118, 2013.

Gudmundsson, L., Tallaksen, L. M., Stahl, K., and Fleig, A. K.: Low-frequency variability of European runoff, Hydrol. Earth Syst. Sci., 15, 2853–2869, https://doi.org/10.5194/hess-15-2853-2011, 2011.

Hannaford, J., Buys, G., Stahl, K., and Tallaksen, L. M.: The influence of decadal-scale variability on trends in long European streamflow records, Hydrol. Earth Syst. Sci., 17, 2717–2733, https://doi.org/10.5194/hess-17-2717-2013, 2013.

Hannah, D. M., Fleig, A. K., Kingston, D. G., Stagge, J. H., and Wilson, D.: Connecting streamflow and atmospheric conditions in Europe: state-of-the-art review and future direction, in: Hydrology in a Changing World: Environmental and Human Dimensions, Proceedings of FRIEND–Water 2014, 7–10 October 2014, Montpellier, France, IAHS Publ., 363, 401–406, 2014.

Hartmann, A., Gleeson, T., Wada, Y., and Wagener, T.: Enhanced groundwater recharge rates and altered recharge sensitivity to climate variability through subsurface heterogeneity, P. Natl. Acad. Sci. USA, 114, 201614941, 10.1073/pnas.1614941114, 2017.

Hawkins, E. and Sutton, R.: The potential to Narrow Uncertainty in Regional Climate Predictions, B. Am. Meteorol. Soc., 90, 1095–1108, 2009.

Hilmer, M. and Jung, T.: Evidence for a recent change in the link between the North Atlantic Oscillation and Arctic Sea ice export, Geophys. Res. Lett., 27, 989–992, https://doi.org/10.1029/1999GL010944, 2000.

Holman, I. P., Rivas-Casado, M., Bloomfield, J. P., and Gurdak, J. J.: Identifying nonstationary groundwater level response to north Atlantic ocean–atmosphere teleconnection patterns using wavelet coherence, Hydrogeol. J., 19, 1269–1278, 2011.

Iliopoulou, T., Papalexiou, S. M., Markonis, Y., and Koutsoyiannis, D.: Revisiting long-range dependence in annual precipitation, J. Hydrol., 556, 891–900, https://doi.org/10.1016/j.jhydrol.2016.04.015, 2018.

Ionita, M., Tallaksen, L. M., Kingston, D. G., Stagge, J. H., Laaha, G., Van Lanen, H. A. J., Scholz, P., Chelcea, S. M., and Haslinger, K.: The European 2015 drought from a climatological perspective, Hydrol. Earth Syst. Sci., 21, 1397–1419, https://doi.org/10.5194/hess-21-1397-2017, 2017.

James, P. M.: An objective classification method for Hess and Brezowsky Grosswetterlagen over Europe, Theor. Appl. Climatol., 88, 17–42, 2007.

Keenlyside, N. S., Latif, M., Jungclaus, J., Kornblueh, L., and Roeckner, E.: Advancing decadal-scale climate prediction in the North Atlantic sector, Nature, 453, 84–88, 2008.

Kidson, J. W.: An analysis of New Zealand synoptic types and their use in defining weather regimes, Int. J. Climatol., 20, 299–316, 2000.

Kingston, D. G. and McMecking, J.: Precipitation delivery trajectories associated with extreme river flow for the Waitaki River, New Zealand, Proc. IAHS, 369, 19–24, https://doi.org/10.5194/piahs-369-19-2015, 2015.

Kingston, D. G., Fleig, A. K., Tallaksen, L. M., and Hannah, D. M.: Ocean-atmosphere forcing of summer streamflow drought in Great Britain, J. Hydrometeorol., 14, 331–344, https://doi.org/10.1175/JHM-D-11-0100.1, 2013.

Kingston, D. G., Stagge, J. H., Tallaksen, L. M., and Hannah, D. M.: European-scale drought: Understanding connections between atmospheric circulation and meteorological drought indices, J. Climate, 28, 505–516, https://doi.org/10.1175/JCLI-D-14-00001.1, 2015.

Kingston, D. G., Webster, C. S., and Sirguey, P.: Atmospheric circulation drivers of lake inflow for the Waitaki River, New Zealand, Int. J. Climatol., 36, 1102–1113, https://doi.org/10.1002/joc.4405, 2016a.

Kingston, D. G, Lavers, D. A., and Hannah, D. M.: Floods in the Southern Alps of New Zealand: The importance of atmospheric rivers, Hydrol. Process., 30, 5063–5070, https://doi.org/10.1002/hyp.10982, 2016b.

Kuentz, A., Mathevet, T., Coeur, D., Perret, C., Gailhard, J., Guérin, L., Gash, Y., and Andréassian, V.: Historical hydrometry and hydrology of the Durance river watershed, Houille Blanche, 5, 57–63, https://doi.org/10.1051/lhb/2014039, 2014.

Kuentz, A., Mathevet, T., Gailhard, J., and Hingray, B.: Building long-term and high spatio-temporal resolution precipitation and air temperature reanalyses by mixing local observations and global atmospheric reanalyses: the ANATEM model, Hydrol. Earth Syst. Sci., 19, 2717–2736, https://doi.org/10.5194/hess-19-2717-2015, 2015.

Labat, D.: Oscillations in land surface hydrological cycle, Earth Planet. Sc. Lett., 242, 143–154, 2006.

Laizé, C. and Hannah, D. M.: Modification of climate–river flow associations by basin properties, J. Hydrol., 389, 186–204, https://doi.org/10.1016/j.jhydrol.2010.05.048, 2010.

Laloyaux, P., de Boisseson, E., Balmaseda, M., Bidlot, J.-R., Broennimann, S., Buizza, R., Dalhgren, P., Dee, D., Haimberger, L., Hersbach, H., Kosaka, Y., Martin, M., Poli, P., Rayner, N., Rustemeier, E., and Schepers, D.: CERA-20C: A Coupled Reanalysis of the Twentieth Century, J. Adv. Model. Earth Syst., 10, 1172–1195, https://doi.org/10.1029/2018MS001273, 2018.

Lamb, H. H.: British Isles Weather Types and a Register of Daily Sequence of Circulation Patterns, 1861–1971, Geophysical Memoire 116, HMSO, London, UK, 85 pp., 1972.

Lavers, D. A. and Villarini, G.: The nexus between atmospheric rivers and extreme precipitation across Europe, Geophys. Res. Lett., 40, 3259–3264, 2013.

Lavers, D. A., Prudhomme, C., and Hannah, D. M.: European precipitation connections with large-scale mean sea level pressure (MSLP) fields, Hydrolog. Sci. J., 58, 1–18, 2013.

Lavers, D. A., Pappenberger, F., and Zsoter, E.: Extending medium-range predictability of extreme hydrological events in Europe, Nat. Commun., 5, 5382, https://doi.org/10.1038/ncomms6382, 2014.

Lavers, D. A., Hannah, D. M., and Bradley, C.: Linking variations in groundwater level in southern England to large-scale atmospheric circulation, J. Hydrol., 523, 179–189, 2015.

Markonis, Y. and Koutsoyiannis, D.: Scale-dependence of persistence in precipitation records, Nat. Clim. Chang., 6, 399–401, 2016.

Markonis, Y., Hanel, M., Maca, P., Kysely, J., and Cook, E. R.: Persistent multi-scale fluctuations shift European hydroclimate to its millennial boundaries, Nat. Commun., 9, 1767, https://doi.org/10.1038/s41467-018-04207-7, 2018.

Massei, N. and Fournier, M.: Assessing the expression of large-scale climatic fluctuations in the hydrological variability of daily Seine river flow (France) between 1950 and 2008 using Hilbert-Huang Transform, J. Hydrol., 448–449, 119–128, 2012.

Massei, N., Laignel, B., Deloffre, J., Mesquita, J., Motelay, A., Lafite, R., and Durand, A.: Long-term hydrological changes of the Seine River flow (France) and their relation to the North Atlantic Oscillation over the period 1950–2008, Int. J. Climatol., 30, 2146–2154, 2010.

Massei, N., Dieppois, B., Hannah, D. M., Lavers, D. A., Fossa, M., Laignel, B., and Debret, M.: Multi-time-scale hydroclimate dynamics of a regional watershed and links to large-scale atmospheric circulation: Application to the Seine river catchment, France, J. Hydrol., 546, 262–275, https://doi.org/10.1016/j.jhydrol.2017.01.008, 2017.

McCabe, G. J. and Wolock, D. M.: A step increase in streamflow in the conterminous United States, Geophys. Res. Lett., 29, 2185, https://doi.org/10.1029/2002GL015999, 2002.

Meehl, G. A., Hu, A., Santer, B. D., and Xie, S.-P.: Contribution of the Interdecadal Pacific Oscillation to twentieth-century global surface temperature trends, Nat. Clim. Chang., 6, 1005–1008, 2016.

Miles, M. W., Divine, D. V., Furevik, T., Jansen, E., Moros, M., and Ogilvie, A. E. J.: A signal of persistent Atlantic multidecadal variability in Arctic sea ice, Geophys. Res. Lett., 41, 463–469, https://doi.org/10.1002/2013GL058084, 2014.

Minvielle, M., Pagé, C., Céron, J.-P., and Besson, F.: Extension of the SIM Reanalysis by combination of observations and statistical downscaling, in: Engineering Geology for Society and Territory, eidtd by: Lollino, G., Manconi, A., Clague, J., Shan, W., and Chiarle, M., 1, 189–192, https://doi.org/10.1007/978-3-319-09300-0_36, 2015.

Pagano, T. and Garen, D.: A recent increase in western U.S. streamflow variability and persistence, J. Hydrometeorol., 6, 173–179, 2005.

Paltan, H., Waliser, D., Lim, W. H., Guan, B., Yamazaki, D., Pant, R., and Dadson, S.: Global floods and water availability driven by atmospheric rivers, Geophys. Res. Lett., 44, 10387–10395, https://doi.org/10.1002/2017GL074882, 2017.

Poli, P., Hersbach, H., Dee, D. P., Berrisford, P., Simmons, A. J., Vitart, F., Laloyaux, P., Tan, D. G. H., Peubey, C., Thépaut, J.-N., Trémolet, Y., Hólm, E. V., Bonavita, M., Isaksen, L., and Fisher, M.: ERA-20C: An atmospheric reanalysis of the twentieth century, J. Climate, 29, 4083–4097, https://doi.org/10.1175/JCLI-D-15-0556.1, 2016.

Roberts, C. D., Palmer, M. D., McNeall, D., and Collins M.: Quantifying the likelihood of a continued hiatus in global warming, Nat. Clim. Chang., 5, 337–342, 2015.

Sahin, S., Turkes, M., Wang, S. H., Hannah, D. M., and Eastwood, W. J.: Large scale moisture flux characteristics of the Mediterranean basin and their relationships with drier and wetter climate conditions, Clim. Dynam., 45, 3381–3401, 2015.

Slimani, S., Massei, N., Mesquita, J., Valdés, D., Fournier, M., Laignel, B., and Dupont, J.-P.: Combined climatic and geological forcings on the spatio-temporal variability of piezometric levels in the chalk aquifer of Upper Normandy (France) at pluridecennal scale, Hydrogeol. J., 17, 1823–1832, 2009.

Stahl, K., Hisdal, H., Hannaford, J., Tallaksen, L. M., van Lanen, H. A. J., Sauquet, E., Demuth, S., Fendekova, M., and Jódar, J.: Streamflow trends in Europe: evidence from a dataset of near-natural catchments, Hydrol. Earth Syst. Sci., 14, 2367–2382, https://doi.org/10.5194/hess-14-2367-2010, 2010.

Thiéry, D., Amraoui, N., and Noyer, M.-L.: Modelling flow and heat transfer through unsaturated chalk – Validation with experimental data from the ground surface to the aquifer, J. Hydrol., 556, 660–673, 2018.

Wilhelm, B., Sabatier, P., and Arnaud, F.: Is a regional flood signal reproducible from lake sediments?, Sedimentology, 62, 1103–1117, https://doi.org/10.1111/sed.12180, 2015.

Wilhelm, B., Vogel, H., Crouzet, C., Etienne, D., and Anselmetti, F. S.: Frequency and intensity of palaeofloods at the interface of Atlantic and Mediterranean climate domains, Clim. Past, 12, 299–316, https://doi.org/10.5194/cp-12-299-2016, 2016.

Zhang, R., Delworth, T. L., and Held, I. M.: Can the Atlantic Ocean drive the observed multidecadal variability in Northern Hemisphere mean temperature?, Geophys. Res. Lett., 34, L02709, https://doi.org/10.1029/2006GL028683, 2007.

Proc. IAHS, 383, 151–158, 2020
https://doi.org/10.5194/piahs-383-151-2020

Modeling hydrologic processes and potential responses to climate change in an agro-silvo-pastoral watershed in the Mediterranean area

Youssef Brouziyne[1], Lahcen Benaabidate[2], Aziz Abouabdillah[3], Rachid Bouabid[3], and Abdelghani Chehbouni[1,4]

[1]Mohammed VI Polytechnic University (UM6P), International Water Research Institute, Ben Guerir, 43150, Morocco
[2]Laboratory of Functional Ecology and Environmental Engineering, Faculty of Sciences and Technology, University of Sidi Mohamed Ben Abdellah, Fez, 30 000, Morocco
[3]Ecole Nationale d'Agriculture de Meknèse, BP S/40, Meknès, 50000, Morocco
[4]CESBIO, Université de Toulouse, CNRS, CNES, IRD, Toulouse, France

Correspondence: Youssef Brouziyne (youssef.brouziyne@gmail.com)

Published: 16 September 2020

Abstract. Precipitation changes and water use patterns are two factors affecting the water quantity; obviously, hydrologic processes are always linked to many elements in the watershed scale, so to understand water management issues it is fundamental to analyze the different elements of hydrologic processes occurring in the watershed. In this study, the "SWAT" model (Soil and Water Assessment Tool) has been used to simulate the water balance for the present climate conditions on a semi arid watershed located in the central North of Morocco (R'dom). The study watershed covers an area of 1993 km^2, and is hosting farming, pasture and forestry related activities. The water stress situation in the R'dom watershed can be summarized as limited resource facing increasing water demand. SWAT model was first run and calibrated under current climate; and was driven with downscaled climate simulations to generate future hydrological projections for R'dom watershed in the 2031 to 2050 horizon under two Representative Concentration Pathways (RCPs): 4.5 and 8.5. The results of the study showed that the water balance in R'dom watershed is dominated by evapotranspiration and the water resources distribution within the watershed is uneven and follows a decreasing gradient matching the flow direction. The main results of climate change scenarios showed that R'dom watershed will undergo significant decrease of water resources availability with more economic impact under the scenario RCP8.5 as all areas hosting the economical activities will be affected and the highest changes of water yield should be under this scenario.

1 Introduction

The consequences of climate change on water cycle vary greatly from region to other. The IPCC (2019) estimates that streamflows should increase in high latitudes (such as in Northern Europe) and in certain tropical regions, but should decrease in lower latitudes (in particular in the Mediterranean region and in the subtropical zones).

Water resources in the Mediterranean region are particularly vulnerable to human activities; increase of populations, tourism and agriculture development, industrial growth are some examples of challenges faced by water resources in this area. In addition to these factors, climate change represents a further threat that can have a significant impact on hydrology in the Mediterranean area (Abouabdillah et al., 2010).

While the main purpose behind the use of models is to simulate a part of reality at the best possible approximation, many of conceptual models are useful for understanding a particular problem or for predicting the behaviors of a system in the future (Krysanova and Srinivasan, 2015). In addition to the use of models to interpret the complexity of a situation, results can sometimes be extrapolated to larger spatial or temporal scales or higher levels of organization with similar conditions (Abouabdillah et al., 2014; Moriasi et al., 2007).

Soil and Water Assessment Tool (SWAT) is a hydrological modeling program designed by the United States Department of Agriculture (Arnold et al., 1998). This model simulates the production of water and sediments in large-scale watersheds, complex basins, with different soil types, and uses and management conditions; it also simulates the effect of agriculture practices or any other business activity can have on water quality (Arnold et al., 2012; Moriasi et al., 2007).

SWAT model has never been applied in R'dom catchment, and its use on this watershed under climate change context is expected to reproduce the hydrological cycle components within different emission scenarios to simulate future conditions that would affect one or more of the physical processes involved in the water cycle.

In this case, the R'dom catchment is a natural unit in which forestry and agricultural activities are cohabitating and represent a typical basin for a big part of the Mediterranean area in terms of challenges faced by water resources from both quantity and quality aspects.

2 Objectives

The main purpose of this work is assessing water resources in R'dom watershed under two RCP scenarios downscaled from a regional circulation model in the near future (2031 to 2050) as a first millstone of 21st century investigations. The selected RCPs were picked to cover the realistic scenario of greenhouse gas concentration trajectory (RCP4.5) and the pessimistic one (RCP8.5).

SWAT model has been used to simulate different processes in the study area using the two RCP scenarios data with a focus on streamflow response at sub-basins levels. Results will help to understand potential climate change impacts on water resources in a typical agro-forestry watershed in the Mediterranean area, and will contribute to future water resources management strategies in any watershed similar to R'dom.

3 Materials and methods

3.1 Study area

Located in the north eastern of Morocco, R'dom watershed is covering an area of 1993 km^2 and includes two large plains (Saiss and Gharb). Being a part of one of the largest hydrologic watersheds in Morocco (Sebou basin), R'dom watershed is characterized by a semi-arid climate and has a big socio-economic importance due to the activities held inside it (farming and forestry activities) (Table 1).

Table 1. Occupation percentages of major land covers in R'dom watershed.

	Land use categories	Percentage (%)
Residential centre		3.69
Forest (oak)		7.98
Pasture and Barren		42.03
Wetlands/lakes		1.71
farming	Wheat	12.23
farming	Corn	9.83
farming	Olive trees	10.59
farming	Onion and potatoes	7.21
farming	Sunflower	4.73

3.2 Model presentation

SWAT model gives chance to simulate a large number of physical processes in the basin through the division of the basin to several sub-basins based on a threshold area value depending on the accuracy and the objectives of the study (Arnold et al., 2012). Sub-basins are then divided into hydrological response units (HRU) that represent a unique combination of soil type, land cover and slope. The subdivision of the watershed allows the model to reflect different hydrologic processes occurring at the level of each unit (evapotranspiration, percolation, runoff...) (Neitsch et al., 2002).

As all HRUs are connected and runoff drainage simulation is based on slope (Lu et al., 2015), the basin water balance was studied in the basin outlet where measured flow out data are available (Souk Elhad gauge: $x = 465943$, $y = 410043$).

3.3 Input data collection and preparation

Being a very important step of any modeling work, data collection and preparation for this study was performed with high precision in order to run SWAT with a start conditions reflecting the reality as much as possible.

For the topographic characterization of the study watershed, digital elevation model (DEM) of R'dom basin has been extracted from the Shuttle Radar Topography Mission (SRTM) with a resolution of 30 meters.

SWAT soil database was updated by soil parameters of R'dom watershed based on findings of the Agriculture Minister' studies on soil types of the area of Meknes-El Hajeb-Sidi Kacem (Soils study at 1 : 100 000 scale in the district of Meknes for farming development – 1989).

A couple of soil parameters (such as: Saturation hydraulic conductivity and Field capacity...) were meant to be calculated using an addition calculation program (Soil Water Characteristics software ©, Version 6.02.74).

Land cover of R'dom basin was built based on satellite image processing; after fields surveys in the main catchment area to have preliminary insight about field situation, two LANDSAT8 image scenes (taken on 2 April 2015) were pro-

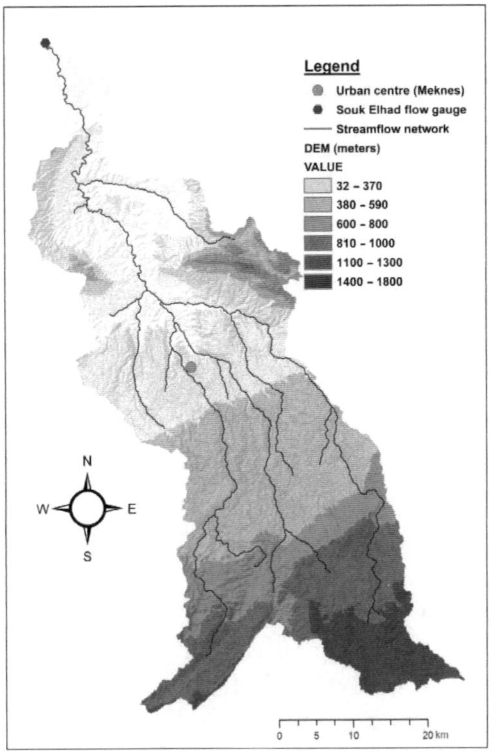

Figure 1. Digital elevation model of R'dom watershed.

cessed using ERDAS Imagine 2014 and ArcGis 10.2.2. Supervised classification method was used by creating land use classes based on their respective spectral signals.

Farming practices features were amended according to the real agronomic practices held in R'dom watershed, especially the operations that can have an effect on water cycle such as: start and end of season dates, irrigation amount, tillage machinery and frequency… etc.

Observed weather parameters were imported to the SWAT and that included daily temperature (max and min) records from 9 weather stations and daily rainfall from 4 recording stations in and around the watershed. Inputs data series were from January 2003 to December 2010.

3.4 SWAT simulation

3.4.1 SWAT parameterization on study catchment

Watershed delineation tool of ArcSWAT has been used to delineate sub-basins based on Digital Elevation Model (DEM) data. 27 connected sub-watersheds were identified and stream network of R'dom basin was defined. Watershed outlet was manually added to match the geographic location of Souk Elhad gauge (Fig. 1); the selected station where flow out process will be monitored to compare measured and predicted flows.

3.4.2 SWAT calibration and validation

In overall SWAT was run over the period of January 2004 to December 2009, the first two years have been allocated to warm up process, the period January 2006 to December 2007 was dedicated to model calibration. Validation was performed over the period January 2008 to December 2009.

For efficient calibration operation, sensitivity analysis was carried out based on relative sensitivity equation to identify the most influential SWAT parameters to focus on during the calibration process.

The used equation of relative sensitivity is (Haan, 2002):

$$S_r = \left[(O_{P+\Delta P} - O_{P-\Delta P})/O_P\right]/[(2\Delta P/P)], \tag{1}$$

where S_r is the relative sensitivity; $O_P + \Delta P$ are model outputs with the input parameter being studied, set at the base value (equal to the initial calibrated value) plus a specified percentage (often taken to be between 10 % and 25 %; $O_P - \Delta P$ are model outputs with the input parameter being studied, set at base minus a specified percentage (often taken to be between 10 % and 25 %); O_P is the model output with input parameter set at the base value; ΔP is the prescribed absolute change in the value of the input parameter and P is the initial calibrated value (base value) of the input parameter.

Following sensitivity analysis, Streamflow was examined during the calibration process. A maximal agreement between observed and predicted water budget components was the ultimate goal. This agreement was monitored by two parameters indicating the statistical goodness-of-fits and the optimization of the objective functions of this study.

The used statistics tests are:

$$\mathrm{NSE} = 1 - \frac{\sum_{i=1}^{n}(O_i - P_i)^2}{\sum_{i=1}^{n}\left(O_i - O_{\mathrm{avg}}\right)^2}, \tag{2}$$

and:

$$R^2 = \left(\frac{\sum_{i=1}^{n}\left(O_i - O_{\mathrm{avg}}\right)\left(P_i - P_{\mathrm{avg}}\right)}{\left[\sum_{i=1}^{n}\left(O_i - O_{\mathrm{avg}}\right)^2 \sum_{i=1}^{n}\left(P_i - P_{\mathrm{avg}}\right)^2\right]^{0.5}}\right)^2, \tag{3}$$

where O_i is the observed water yield for time period i; P_i is the simulated value for the same period; O_{avg} is the mean of observed water yield per time period and n is the number of time periods.

The Nash–Sutcliffe coefficient (NSE) represents the model's precision in simulating the flood streamflow; the maximum NSE value is 1.0 and occurs if predicted values are perfectly matching measured ones, The determination coefficient, R^2, provides an idea about how well is the correlation between and simulated values (Arnold et al., 2012; Viola et al., 2009).

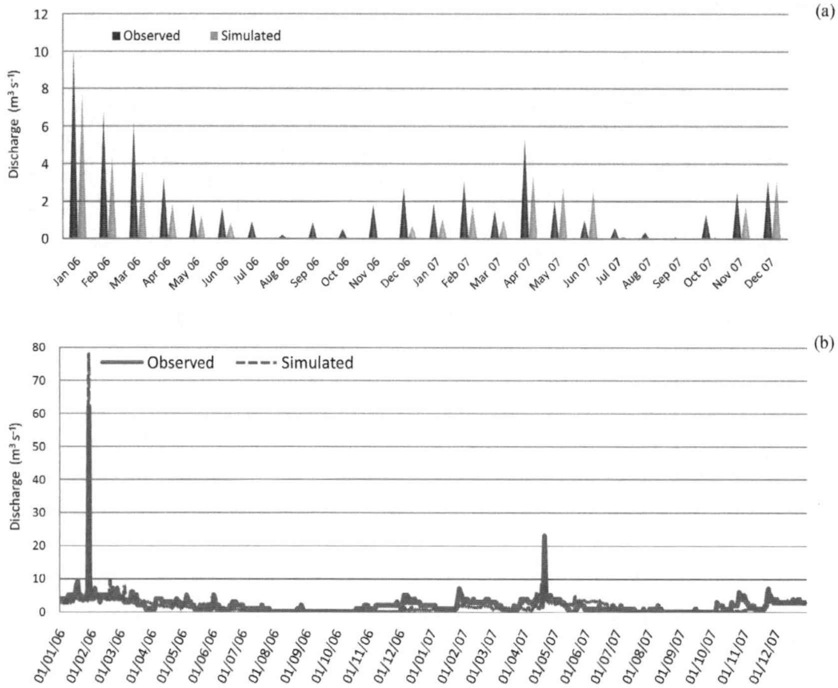

Figure 2. Observed Vs simulated monthly **(a)** and daily **(b)** discharge after manual calibration at the Souk Elhad flow gage.

3.4.3 Future climate data and model scenarios

The selected climate change scenarios for this study were generated from the output of the global climate model CNRM CM5 (Centre National de Recherches Meteorologique) based on IPCC Assessment Report (AR5). There are 4 different types of Representative Concentration Pathways (RCPs): RCP2.6, RCP4.5, RCP6.0 and RCP8.5 that represent different alternatives of global development (Kim et al., 2013). The CNRM participated in the Coordinated Regional Climate Downscaling Experiment (CORDEX) with different Regional Climate Models (RCMs) (Dosio, 2016).

In this study, the downscaled RCM CLMcom-CCLM and two RCPs (4.5 and 8.5) were used. Grid size was 12.5 km and the baseline was from January 1981 to December 2005. The future simulation period for both precipitations and temperature (max and min) was from January 2031 to December 2050. Linear scaling and variance scaling approaches were used during the bias correction in order to show some levels of bias against real measured precipitation and temperature data respectively.

Table 2. Relative sensitivity values for all tested SWAT input parameters.

Parameters	TWYD	Runoff	Baseflow
CN2	1.77	5.57	−0.55
SOL_AWC	−0.31	−1.25	−0.64
ESCO	0.06	0.10	0.06
GWQMN	−0.24	0.00	−0.38

With: CN2: initial SCS runoff curve number for moisture condition II; SOL_AWC: available water capacity of the soil layer (mm H_2O mm^{-1} soil); ESCO: soil evaporation compensation factor; GWQMN: threshold depth of water in shallow aquifer required for return flow to occur (mm H_2O).

4 Results

4.1 SWAT model application

4.1.1 Sensitivity analysis and calibration

After investigation of relative sensitivity of twelve different SWAT input parameters for flow, 4 parameters were revealed as the most influential (Table 2) with different sensitivity degrees.

During manual calibration process, the focus was on the 4 most influential parameters to parameterize SWAT to the local R'dom watershed condition by comparing model flow prediction with observed data at outlet level. The R^2 and NSE have been used to evaluate model prediction perfor-

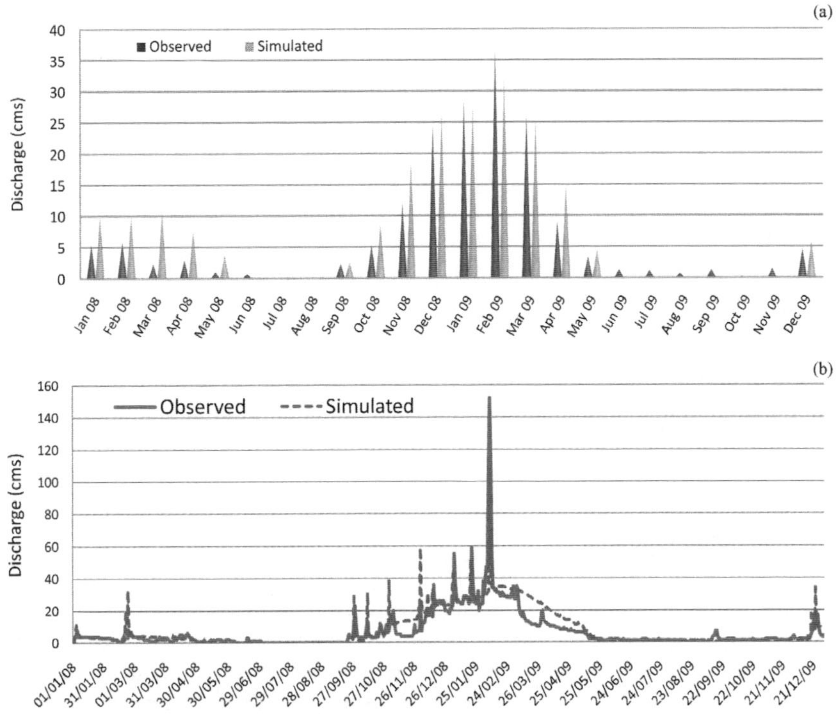

Figure 3. Observed Vs predicted monthly **(a)** and daily **(b)** discharge during the validation period at the Souk Elhad flow gage.

mance while bearing in mind that they should exceed 0.5 to be able to judge the calibration as satisfactory according to many papers (Arnold et al., 2012).

After several SWAT iterations, a satisfactory agreement between simulated and measured discharge in both monthly and daily time steps was found (Fig. 2); the NSE and R^2 values were respectively 0.68 and 0.85 for the monthly time step; 0.58 and 0.79 for the daily time step.

During SWAT model validation process, predicted discharge trend followed observed discharge pattern during all the validation period with NSE and R^2 values of 0.88 and 0.91 for the month time step respectively, 0.65 and 0.73 for the daily time step (Fig. 3).

4.1.2 Water balance

Simulated over the period January 2006 to December 2009, the R'dom water balance showed that actual evapotranspiration is very important and represents around 83 % of annual rainfall and irrigation, the surface runoff represents less than 2 % of the total annual rainfall (Table 3).

Total water yield is the sum surface runoff, lateral flow, groundwater contribution net transmission losses; its distribution within the R'dom watershed, Fig. 4 represents the average of simulated actual water yield by sub-basins. In overall the average total water yield is decreasing going from upstream (South) to upstream (North) taking the same trend of altitude which is high in the South and low in the North.

Table 3. Average of simulated monthly Water balance components of R'dom basin during the actual simulation period (2004–2009).

Month	Rain (mm)	Surf (mm)	Lat (mm)	TWYD (mm)	ET (mm)
1	71.15	7.96	3.17	19.28	11.43
2	37.75	4.28	1.13	13.92	16.15
3	29.5	3.98	1.14	9.44	15.68
4	20.35	1.78	0.1	7.9	32.5
5	15.79	0.5	0.1	1.27	33.86
6	5.38	0.14	0.08	0.73	25.52
7	4.75	0.04	0.04	0.43	14.47
8	2.19	0.38	0.06	0.65	14.15
9	20.32	0.7	0.13	0.98	16.31
10	33.47	0.97	0.13	4.28	12.62
11	49.28	1.74	0.09	13.06	12.7
12	56.33	6.85	3.1	17.22	10
Total	352.48	29.32	9.27	89.16	215.39
Percentage out of rainfall		8.31 %	2.62 %	25.29 %	61.1 %

With, Surf: surface runoff; Lat: lateral flow; TWYD: total water yield and ET: evapotranspiration.

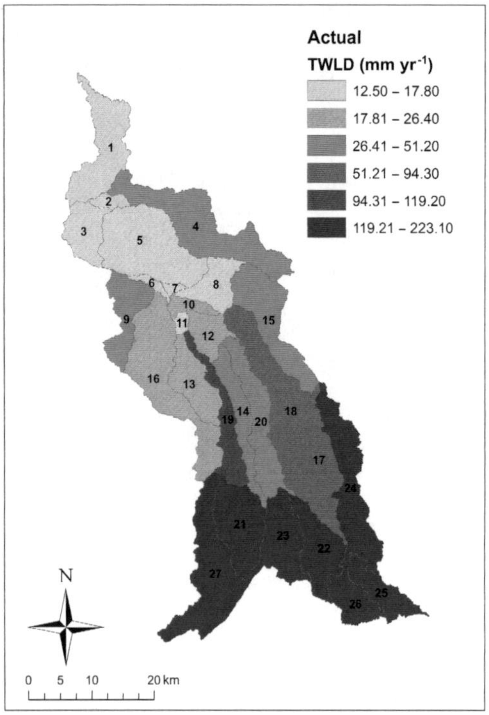

Figure 4. Average of simulated total water yield by sub-basins under for the period 2004 to 2009.

The main agricultural activities are held in the centre and northern part of the watershed, the southern part of the basin is covered mainly by forest and pasture lands.

4.2 Implication of climate change scenarios on streamflow

The climate change scenarios that have been generated and corrected statistically were used as inputs to the calibrated SWAT model over the R'dom watered conditions, in order to simulate future streamflow of the study watershed. The same HRUs and crop management setting used for SWAT calibration were kept for the future scenarios simulation with the future rainfall and temperature data as the only changes.

As the big focus of this study was the to assess the potential effect of two climate change scenarios on streamflow pattern in the R'dom watershed, the yearly average of total water yield (TWYL), mm yr^{-1}, was picked as the hydrologic variable to monitor by sub-basin over the watershed. Total water yield is considered as the sum of surface runoff, lateral flow, groundwater contribution net of transmission losses.

Figure 5 represents the percentage of change of simulated total water yield by SWAT at sub-basin scale under the future scenarios RCP4.5 and RCP8.5 during the period 2031 to 2050 comparing to the actual simulation period 2004 to 2009.

On watershed scale, both scenarios showed decrease on average water yield per sub-basin ranging from −3 % (RCP4.5) till −47 % (RCP8.5). The biggest decreases are more in the lower altitudes: the outlet and the watered centre.

For the scenario RCP4.5, it is in the northern and centre of the watershed where highest changes of TWYL average are expected; these areas are hosting most of the farming activities in the R'dom catchment. The highest percentage of average TWYL decrease is expected in the northern sub-basins and should range from −40 % to −49 %.

For the scenario RCP8.5, and in addition to almost all affected areas in the RCP4.5, nearly all western sub-basins of the catchment should be affected with percentages of average TWYL change ranging between −29 % and −40 %. It is worthy to mention that most of sub-basins affected exclusively in scenario 8.5 are home of forestry and pasture activities.

The analysis of the both climate change scenarios projected very significant impact on streamflow over the R'dom watershed, this impact have different geographic distribution inside the study catchment but the scenario 8.5 is likely to be more economically significant as the streamflow change is expected to be significant in areas where farming, pasture and forestry activities are held.

5 Conclusion

This study assesses the impact of climate change on water balance (represented by water yield variable) in the semi-arid R'dom watershed situated in Morocco. Two emission scenarios (RCP4.5 and RCP8.5) were used to evaluate the climate change impact. SWAT is used to simulate the changes in different components of water balance. The performance of SWAT sensitivity analysis, calibration and validation processes were satisfactory and confirmed the ability of this model to simulate hydrological cycle in the R'dom watershed under climate change effects in the near future (2031 to 2050).

Both emission scenarios projected a decrease of the total water yield in the entire R'dom watershed with more dispersed effect under the scenario 8.5 covering almost all the area where the main economical activities are carried out.

The decrease of water availability in R'dom watershed will definitely have negative impact on the availability of water resources for both rainfed and irrigated agriculture as well as on ecosystems in the forest within the watershed. This change would affect also the economical and social balances in the area as the main activities in the R'dom basin are pasture and agro-forestry. Appropriate adaptation strategies should then be implemented to face the climate change impacts.

This study tried to prove the ability of a hydrologic modeling tool to predict the situation of water resources under

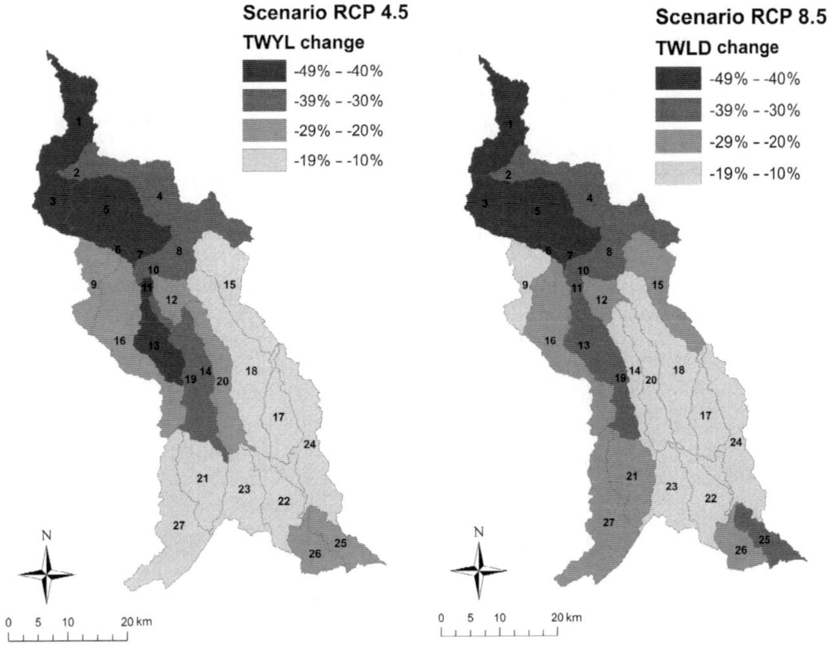

Figure 5. Changes of TWYL by sub-basins under future scenarios RCP4.5 and RCP8.5 for the period 2031 to 2050.

climate change effect in a Mediterranean basin. The main outcome is that, in such semi-arid area, water resources are very vulnerable. The results will be a starting point for further analysis of other water balance components response to climate change in the R'dom basin in order to provide more data that will be useful for future water resources management strategies in similar climatic conditions.

Data availability. The used data in this study are subject to intellectual property agreements with the study stakeholders and therefore, the data are not publicly accessible.

Author contributions. YB developed the methodology with RB and AA, he supervised the conceptualization with the help of LB, and he served as the primary writer. AA contributed in the writing of parts of the paper, and AC validated the used tools and methodology.

Competing interests. The authors declare that they have no conflict of interest.

Special issue statement. This article is part of the special issue "Hydrological processes and water security in a changing world". It is a result of the 8th Global FRIEND–Water Conference: Hydrological Processes and Water Security in a Changing World, Beijing, China, 6–9 November 2018.

References

Abouabdillah, A., Oueslati, O., De Gerolamo, A. M., and Lo Porto, A.: Modeling the impact of climate change in the mediterranean catchment (Merguellil, Tunisia), Fresenius Environ. Bull., 19, 2334–2348, 2010.

Abouabdillah, A., White, M., Arnold, J. G., De Girolamo, A. M., Oueslati, O., Maataoui, A., and Lo Porto, A.: Evaluation of soil and water conservation measures in a semi-arid river basin in tunisia using swat, Soil Use Manage., 30, 539–549, https://doi.org/10.1111/sum.12146, 2014.

Arnold, J. G., Srinivasan, R., Muttiah, R. S., and Williams, J. R.: Large area hydrologic modeling and assessment – part i, J. Am. Water. Resour. Assoc., 34, 73–89, https://doi.org/10.1111/j.1752-1688.1998.tb05961.x, 1998.

Arnold, J. G., Moriasi, D. N., Gassman, P., Abbaspour, K. C., White, M. J., Srinivasan, R., Santhi, C., Harmel, R. D., Van Griensven, A., Van Liew, M. W., Kannan, N., and Jha, M. K.: Swat: Model use, calibration, and validation, T. ASABE, 55, 1491–1508, 2012.

Dosio, A.: Projection of temperature and heat waves for Africa with an ensemble of CORDEX regional climate models, Clim. Dynam., 49, 493–519, https://doi.org/10.1007/s00382-016-3355-5, 2016.

Haan, C. T.: Statistical methods in hydrology, Iowa State Press, Ames, Iowa, 2002.

IPCC: IPCC special report on climate change, desertification, land degradation, sustainable land management, food security, and greenhouse gas fluxes in terrestrial ecosystems, C/O Imperial College, London, UK, 2019.

Kim, J., Choi, J., Choi, C., and Park, S.: Impacts of changes in climate and land use/land cover under ipcc rcp scenarios on streamflow in the hoeya river basin, Korea, Sci. Total. Environ., 452–453, 181–195, https://doi.org/10.1016/j.scitotenv.2013.02.005, 2013.

Krysanova, V. and Srinivasan, R.: Assessment of climate and land use change impacts with swat, Reg. Environ. Change., 15, 431–434, https://doi.org/10.1007/s10113-014-0742-5, 2015.

Lu, Z., Zou, S., Xiao, H., Zheng, C., Yin, Z., and Wang, W.: Comprehensive hydrologic calibration of swat and water balance analysis in mountainous watersheds in northwest china, in: Physics and chemistry of the earth, parts a/b/c, Water scarcity and integrated water resources management, edited by: Xiangsheng Deng, R. B. S., Liu, J., and Guneralp, B., Elsevier B.V., Amsterdam, 76–85, 2015.

Moriasi, D. N., Arnold, J. G., Van Liew, M. W., Binger, R. L., Harmel, R. D., and Veith, T.: Model evaluation guidelines for systematic quantification of accuracy in watershed simulations in watershed simulations, T. ASABE, 50, 885–900, 2007.

Neitsch, S. L., Arnold, J. G., Kiniry, J. R., Williams, J. R., and King, K. W.: Soil & water assessment tool-theoretical documentation – version 2000, Grassland, Soil and Water Research Laboratory – Agricultural Research Service, Temple, Texas, USA, 76502, 2002.

Viola, M. R., Mello, C. R., Acerbi, J., and Silva, A. M.: Modelagem hidrologica na bacia hidrografia do rio aiuruoca, Rev. Bras. Eng. Agric. Embient., 13, 581–590, https://doi.org/10.1590/S1415-43662009000500011, 2009.

Proc. IAHS, 383, 159–162, 2020
https://doi.org/10.5194/piahs-383-159-2020

Study of soil erosion risks using RUSLE Model and remote sensing: case of the Bouregreg watershed (Morocco)

Fatima Hara[1], Mohammed Achab[1], Anas Emran[1], and Gil Mahe[2]

[1]Geo-Biodiversity and Natural Patrimony Laboratory(GEOBIO,)Scientific Institute, Geophysics, Natural Patrimony and Green Chemistry Research Centre(GEOPAC), Mohammed V University in Rabat, 10106 Rabat, Morocco
[2]HydroSciences Montpellier/IRD, 34090 Montpellier, France

Correspondence: Fatima Hara (fatimahara1@gmail.com)

Published: 16 September 2020

Abstract. The Bouregreg watershed is located to the north-western center of Morocco, characterized by a semi-arid climate. It covers a total area of approximately 10 000 km^2. This basin is a very sensitive area to water erosion. This causes the degradation of its vegetation cover and its land. The most sensitive and poorly protected soils erode much more easily and lose their fertility.The objective of this work is to quantify soil losses by water erosion in the Bouregreg watershed using the Revised Universal Loss Equation (RUSLE) and Geographic Information Systems. The average annual rate of soil erosion in the Bouregreg watrershed are estimated at 20 t ha^{-1} yr^{-1}. The spatial distribution map of soil erosion show that 71 % of the total area has low risk of soil erosion (<3 t ha^{-1} yr^{-1}), while 28 % of the study area shows moderate to high risk of erosion (20–60 t ha^{-1} yr^{-1}). Areas of very high risk of erosion are also present in some sectors of the watershed covering 1 % of the total surface.

1 Introduction

Soil erosion has been recognized as the major cause of land degradation worldwide and becomes a serious environmental and economic problem when accentuated by human activities (Del Mar López et al., 1998). On the other hand, water erosion is the main factor of the deterioration of the soil resource (Wachal, 2007). The soil loss phenomena have accelerated around the world (De Graaf, 1996). Water erosion risk assessment involves mapping and analyzing many factors involved in the erosive process: rainfall erosivity, soil erodibility, slope length and vegetal cover factors (Wischmeier and Smith, 1978).

Morocco is one of the Mediterranean countries that have always suffered from land degradation and mismanagement. Soil erosion characterizes the majority of Morocco's reliefs. FAO study (1990) revealed that the situation continues to deteriorate and becomes more and more critical, and evaluate that almost 40 % of land in Morocco are affected by erosion.

The Bouregreg watershed (approximately 10 000 km^2) topic of this study is located between latitudes 32.8 and 34° N and longitudes 5.4 and 6.8° W. This watershed constitutes a very sensitive area to degradation of soil and vegetation by water erosion. the sediments resulting from this degradation contribute to the siltation of the SMBA dam, located to the north of the watershed and producing about 31 % of Morocco's drinking water specially those of two large Moroccan cities, Rabat and Casablanca.

The climate of Bouregreg watershed is Mediterranean. It ranks globally in the semi-arid bioclimatic stage. The average rainfall varies from 450 mm yr^{-1} in Rabat and 750 mm yr^{-1} in high altitude areas (Oulmès). The average temperatures are near Atlantic and in mountain areas. They vary between 15 and 25 °C. They are strongly influenced by warm currents in dry periods. The average high temperatures during (July–August) alternate between 30 and 34 °C in the Bouregreg and between 26 and 28 °C in the coastal zone (PDAIRE, 2011).

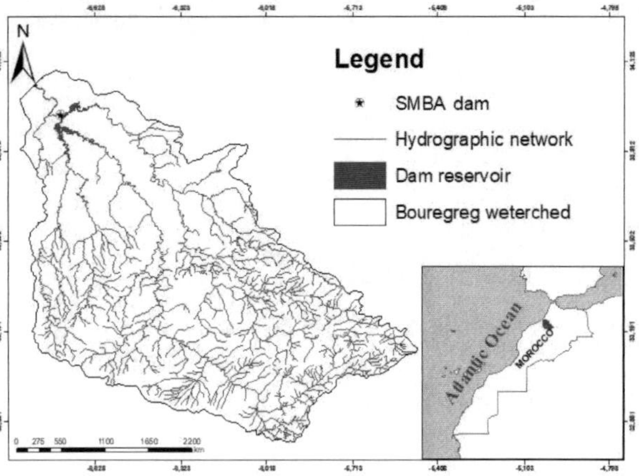

Figure 1. Location of the Bouregreg Watershed.

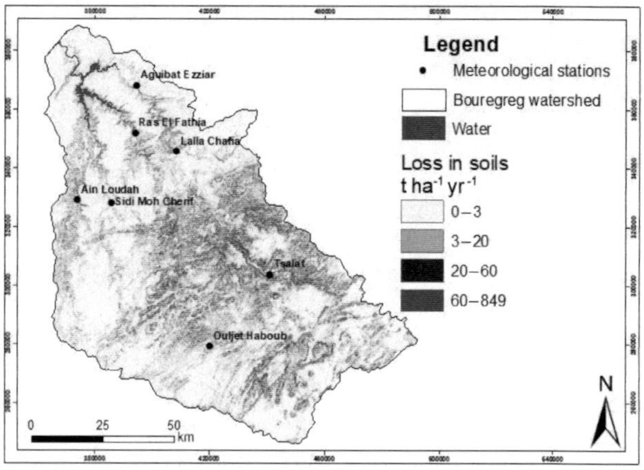

Figure 2. Spatial distribution map of soil losses in the Bouregreg watershed.

The proposed study attempts by using the Revised Universal Loss Equation (RUSLE) and Geographic Information Systems to quantify soil losses rate and the effect of vegetation and land cover/use changes on soil erosion in this watershed.

2 Methodology

The Quantification of soil losses rate in the Bouregreg watershed was performed by the application of the Revised Universal Soil Loss Equation (Wischmeier and Smith, 1978).

The USLE method of Wischmeier and Smith (1978) represented an important step in the history of water erosion modeling. It is the most used model for a global estimate of losses due to the process of surface erosion, with her revised RUSLE model, grace to the ease of its implementation and its compatibility with geographic information systems (Millward and Mersey, 1999). The RUSLE model has been recognized as the most widely used model for the prediction and spatial distribution of annual average soil loss by integrating different erosion factors (Renard et al., 1997). The equation is expressed by the relation:

$$A = R \cdot K \cdot LS \cdot C \cdot P \quad (1)$$

where A is the average annual soil loss ($\mathrm{t\,ha^{-1}\,yr^{-1}}$), R is The Rainfall and Runoff Factor ($\mathrm{MJ\,mm^{-1}\,ha^{-1}\,h^{-1}\,yr^{-1}}$), K is the Soil Erodibility Factor ($\mathrm{t\,ha^{-1}\,h^{-1}\,ha^{-1}\,MJ^{-1}\,mm^{-1}}$), LS is the Slope Factor, C is the Crop/Vegetation and Management Factor, P is the Support Practice Factor.

3 Results and discussion

3.1 Climate aggressiveness factor R

The rainfall erosivity is an indicator of precipitation aggression. This factor was estimated by the Arnoldus formula (Arnoldus, 1977) to eight stations in the Bouregreg watershed (Fig. 2) on 36-year period (1977–2013). The formula based on monthly mean precipitation (pi) and average annual rainfall (p). The interpolation values of these stations gave a distribution map of this factor ranging from 29.53 to 57.87. The map shows an increase in values from the northeast of the basin to the southwest, where there are the highest altitudes.

Table 1. Distribution of soil losses in the Bouregreg watershed.

Soil loss ($\mathrm{t\,ha^{-1}\,yr^{-1}}$)	Area (ha)(%)	Area	Erosion risk
0–3	705 300	71	Low
3–20	213 900	22	Moderate
20–60	61 000	6	High
60–849	6000	1	Very high
Total	989 800	100	

3.2 The erodibility factor K

The erodibility factor (K) is a function of soil texture, structure, organic matter content and permeability. In our case, due to the lack of this type of data, we opted to use FAO Digital SoilMap data from the World (http://www.fao.org/land-water/land/land-governance/land-resources-planning-toolbox/category/details/en/c/1026564/, last access: 12 August 2020). The results obtained show that the K factor varies from 0.0139 to 0.0214.

3.3 The topographic factor LS

The topographic factor (LS) was calculated from the inclination of the slopes and their length by the formula of Wischmeier and Smith (1978):

$$LS = (L/22.13)m(0.065 + 0.045 \cdot S + 0.065 \cdot S2) \quad (2)$$

where L is the slope length in m, S is the inclination of the slope in %, m is a parameter such that $m = 0.5$ if the slope is >5 %, $m = 0.4$ if the slope is 3.5 % to 4.5 %, $m = 0.3$ if the slope is 1 % to 3 % and $m = 0.2$ if the slope is <1 %.

LS factor values range from 0 to 1440. The length and degree of inclination of the slope are decisive in the process of erosion. LS values are very high in the southwestern part of the Bouregreg watershed.

3.4 The factor of vegetation covers C

The evaluation of this factor in Bouregreg watershed is made by adapting the equation of Van Der Knijff (2000).

$$C = \exp\left[-\alpha \times \mathrm{NDVI}/(\beta - \mathrm{NDVI})\right] \quad (3)$$

where α, β: parameters determining the shape of the NDVI-C curve with $\alpha = 2$ and $\beta = 1$ and NDVI is normalized difference vegetation index.

The NDVI was evaluated by operating a satellite image Landsat 8 dated 2018. The results obtained show that the C factor varies from 0.054 to 1.

3.5 The factor of erosion control practices P

In the case of our study, and in the absence of anti-erosion practices in the study area, the P factor is assigned a value of 1.

3.6 Evaluation of soil losses

Soil losses result from the combination of the RUSLE model factors, climatic aggressivity R, soil erodibility K, topographic factor LS, vegetation cover C and anti-erosion practices P. The combination of the maps of these The main factors in a GIS environment resulted in the map of soil losses in the watershed (Fig. 2).

Average soil losses in the Bouregreg watershed are estimated at $20\,\mathrm{t\,ha^{-1}\,yr^{-1}}$. Soil losses in the Bouregreg watershed have been grouped into 4 classes of values (Fig. 2 and Table 1).

4 Conclusions

The study of risk of erosion in the Bouregreg watershed was performed using the Revised Universal Equation of Sol Losses (RUSLE). The different factors of the model have been estimated (Eq. 1) and spatialised in a GIS environment. The combination of these different factors made possible the elaboration of the erosion risk map. The results obtained show that the study area is subject to climatic aggressiveness and show a moderate vulnerability of soils to water erosion. however, the risk of soil erosion is high in some sectors of the watershed.

Data availability. The availability and the nature of the numerical data (geological, climatological and pedological, land use) used in the calculation of the erosion and sedimentation processes in the study basin, is well prepared for the use of the universal equations of revised soil loss (RUSLE). These data have been integrated, listed and digitally processed in a Geographic Information System. Other image processing software were also used (QGis, Erdas, Envi) for the delimitation of the watershed and the extraction of their geometric properties (surface area, average altitude, etc.). For this we based ourselves on a digital terrain model (DEM) at 30 m resolution, which also served us as a basis for the extraction of the hydrographic network and the digital calculation of the different thematic maps of the basin.

Author contributions. FH write the paper, AM conducted review and editing, AE and GM project administration and resources.

Competing interests. The authors declare that they have no conflict of interest.

Special issue statement. This article is part of the special issue "Hydrological processes and water security in a changing world". It is a result of the 8th Global FRIEND–Water Conference: Hydrological Processes and Water Security in a Changing World, Beijing, China, 6–9 November 2018.

Acknowledgements. The authors would like to thank the Bouregreg and Chaouia watershed hydraulic agency for collaboration. They would also like to thank all of the reviewers for providing valuable input to earlier versions of this manuscript.

References

Arnoldus, H. M. J.: Methodologye used to determine the maximum potentialaverage annual soil loss due to shett and rill erosion in Morocco, FAO Soil Bull., 34, 39–51, 1977.

Del Mar López, T., Mitchel Aide, T., and Scatena, F. N.: The effect of land use on soil erosion in the Guadiana watershed in Puerto Rico, Carib. J. Sci., 34, 298–307, 1998.

De Graaf, J. (Ed.): Price of soil erosion: an economic evaluation of soil conservation and watershed development, Agricultural University Wageningen, 300 pp., 1996.

Millward, A. A. and Mersey, J. E.: Adapting the RUSLE to model soil erosion potential in a mountainous tropical watershed, CATENA, 38, 109–129, 1999.

PDAIRE: Project Study Plan Director Renovation An Integrated Water Resources hydraulic basin Bouregreg and Chaouia, available at: http://81.192.102.15/index.php/pdairleft (last access: 9 August 2020), 2011.

Renard, K. G., Foster, G. R., Weesies, G. A., McCool, D. K., and Yoder, D. C.: Predicting soil erosion by water – a guide to conservation planning with the Revised Universal Soil Loss Equation (RUSLE). United States Department of Agriculture, Agricultural Research Service (USDA-ARS) Handbook No. 703, 1997.

Van der Knijff, J. M., Jones, R. J. A., and Montanarella, L.: Soil Erosion Risk Assessment in Europe, EUR 19044 EN, 34 pp., 2000.

Wachal, D. J.: Integrating GIS and erosion modeling – A tool for watershed management, ESRI international user conference, Paper No. UC1038, 11 pp., 2007.

Wischmeier, V. H. and Smith, D. D.: Predicting rainfdl erosion losses – a guide to conservation planning, United States Department of Agriculture in cooperation with Purdue Agricultural Experiment Station. United States Department of Agriculture, Washington, Agriculture Handbook No. 282, 1978.

Proc. IAHS, 383, 163–169, 2020
https://doi.org/10.5194/piahs-383-163-2020

Calibration of the hydrological model GR4J from potential evapotranspiration estimates by the Penman-Monteith and Oudin methods in the Ouémé watershed (West Africa)

Domiho Japhet Kodja[1], **Arsène J. Sègla Akognongbé**[1], **Ernest Amoussou**[1,2], **Gil Mahé**[3], **E. Wilfrid Vissin**[1], **Jean-Emmanuel Paturel**[3], **and Constant Houndénou**[1,4]

[1]University of Abomey Calavi (Bénin)/Laboratory Pierre Pagney, Climate, Water, Ecosystems and Development (LACEEDE) 03 BP 1112 Cotonou, Republic of Benin
[2]University of Parakou (Bénin), Department of Geography and Spatial, Parakou, Republic of Benin
[3]University of Montpellier, IRD, UMR 5569 HSM HydroSciences Montpellier, France
[4]Institute of Geography, Spatial Planning and the Environment (IGATE), Abomey-Calavi, Republic of Benin

Correspondence: Domiho Japhet Kodja (japhdom@gmail.com)

Published: 16 September 2020

Abstract. The Ouémé watersheds at Bétérou and Bonou has been recently facing increased sensitivity to extreme hydroclimatic phenomena that occurred by flooding or drought events. In the same time, the population growth and the related socio-economic activities increased the pressure state on water resources. In this context, hydrological modeling is an important issue and this study aims at analyzing the calibration of the hydrological model GR4J based on PET Penman-Monteith and Oudin methods. Daily rainfall, Penman-Monteith and Oudin evapotranspiration and daily data flow from the Bétérou and Bonou hydrometric stations on the Ouémé Basin have been implemented in the GR4J model over the period 1971 to 2010. Oudin PET values are slightly higher than the Penman-Monteith PET ones. However, the difference between the two PET methods have only few impacts on the optimization and performance criteria of the GR4J model. The Nash values ranges from 0.83 to 0.91 in Bonou, and 0.52 to 0.70 in Bétérou for the calibration in dry period, while in validation, they are 0.59 to 0.78 in Bétérou, and 0.56 to 0.88 in Bonou in wet season. In view of these results, with the two PET methods used which do not result from the same climatic variables, it should be said that the formulation of PET has only few impacts on the results of GR4J for these tropical basins.

1 Introduction

The ongoing climatic changes are one of the factors responsible for the recurrence of extreme rainfall anomalies generating risks of flood and drought (Amoussou et al., 2015). Crétat et al. (2012) indicated that due to climate change, the intensity and frequency of heavy rainfall will increase in the future with impacts on groundwater and surface water resources. To predict the availability of these resources is essential for adapted solutions to development projects (Sighomnou, 2004). Rainfall-runoff models are good tools to define the use of water resources by reproducing flows at the basin scale from the measurements or the simulation of rainfall and PET (Afouda et al., 2015). For Paturel et al. (2003) a rainfall-runoff model is more interesting in developing countries because they can be used to estimate the resource available and also its evolution over time by combining it with climate scenarios. Choosing the efficient rainfall-runoff model to reproduce the flows depends on the available input data. Oudin (2004), indicated that sensitivity analysis is considered an important phase of the modeling process. Andréassian et al. (2004) studying the impact of the PET on the parameters and the performances of GR4J and TOPMO models, showed that the models' efficiency was very few influenced by the choice of the PET formula. In Benin, many hydrological modeling have been carried out with GR4J and GR2M (Vissin, 2007; Amoussou et al., 2015; Afouda et al., 2015). However, these authors did not focus their research on the type of PET methods relevant to the calibration of hydrological model. We thus question in this study the influence of the PET method on the performances of the GR4J model. We test two methods of PET estimation: Penman-Monteith and Oudin, in two sub-catchments of the Ouémé River, Bétérou in the upper catchment and Bonou at the outlet.

1.1 Geographic location of the study environment

The Ouémé watershed at Bonou outlet, covers an area of 46 200 km^2 (41 % of the Benin area); it spreads in its northern part on Dahomeen basement, and on sedimentary ones in the South. Some parts of the basin are in East of Nigeria, and West of Togo (Fig. 1), (Le Barbé et al., 1993).

The climatic structure of Ouémé watershed is presented as follows: in the south is under subequatorial climate with four-seasons alternatively wet and dry (at the outlet of Bonou), and in the north a Sudanian climate with a single-rain-season (at the outlet of Bétérou). In the central part, there is an intermediate climate of sub-sudanian type, transitional climate between the South and the North of the area. The Ouémé River is flowing from North to South along 510 km on the Beninese territory.

2 Data and methods

2.1 Data

We used daily rainfall recorded at 30 stations including 4 synoptic ones from the National Direction of Meteorology (NDM) database, related to the ASECNA, Agency for Aerial Navigation Safety in Africa and Madagascar. We use daily flows measured at the outlet of Bonou and collected from the Hydrology Department of the General Direction of Water (DGEau) of Benin. These data cover the years 1971 to 2010 and have been supplemented by the SIEREM/HSM database (Boyer et al., 2006).

2.2 Methods

We choose rainfall stations with less than 5 % lacks (Amoussou et al., 2015), with regionalization by the Thiessen method. We test the stationarity rupture probability using statistical tests of the Khronostat 1.01 software, developed by HydroSciences Montpellier laboratory (Boyer et al., 2006).

2.2.1 Calculation method for evapotranspiration

PET is not a measured variable, such as rain, but the result of a model combining various climatic variables (Oudin, 2004). In hydrology and especially in rain-flow modeling, the Penman (or Penman-Monteith) model is the most used. This approach nevertheless requires the collection of climate data. If the data collection does not allow a Penman PET calculation, several other formulas can be used to determine the PET. The choice of the formula is then based on the adequacy of the data needed to calculate the PET and the data available. In the case of this study, we highlighted the Penman Monteith PET and the Oudin PET.

Potential Evapotranspiration of Penman-Monteith

Several methods are used to calculate Potential Evapotranspiration (PET). In hydrological modeling, the Penman-Monteith evapotranspiration estimation method is used (Allen et al., 1998). It requires several climatic parameters including the air temperature (min and max), the average relative humidity, the insolation, the wind speed at 2 m s^{-1} of the ground, the latitude of the climatological station considered. The remaining data is generated by the software. The software Instat + is used to calculate the PET. It is obtained by the formula below Eq. (1):

$$ET_0 = \frac{0.408\Delta(R_n - G) + \gamma \frac{900}{T+273} U_2 (e_s - e_a)}{\Delta + \gamma(1 + 0.34U_2)} \quad (1)$$

with: ET_0 reference evapotranspiration (mm d^{-1}), R_n net radiation at the surface (MJ m^{-2} d^{-1}), G soil heat flux density (MJ m^{-2} d^{-1}), T mean temperature daily air at 2 m in height (°C), U_2 wind speed at 2 m in height (m s^{-1}), E_s saturation vapor pressure (kPa), and the actual vapor pressure (kPa) These are the saturation vapor pressure deficit (kPa), the slope of the vapor pressure curve (kPa °C^{-1}), the psychometric constant (kPa °C^{-1}).

Potential Evapotranspiration of Oudin

The method of Oudin has been used by other researchers (Kay and Davies, 2008; Amoussou et al., 2015). This method of PET calculating comes from the Jensen-Haise and McGuinness models (Amoussou et al., 2015). These models generally used in climatology take into account only the average daily temperature of the air and the solar radiation which depends on the latitude and the 365 d of the year. Potential Evapotranspiration is obtained by the formula Eq. (2):

$$PE = \frac{R_e T_a + K_2}{\gamma_\rho K_1}, \quad T_a + K_2 > 0, \quad PE = 0 \quad (2)$$

with PE: Potential Evapotranspiration (mm j^{-1}); R_e: Solar radiation (MJ m^{-2} d^{-1}); T_a: Average daily temperature (°C); γ: the latent heat flow (2.45 MJ kg^{-1}); ρ: the density of the water (kg m^{-3}) K_1 (°C) and K_2 (°C): are fixed parameters of the model.

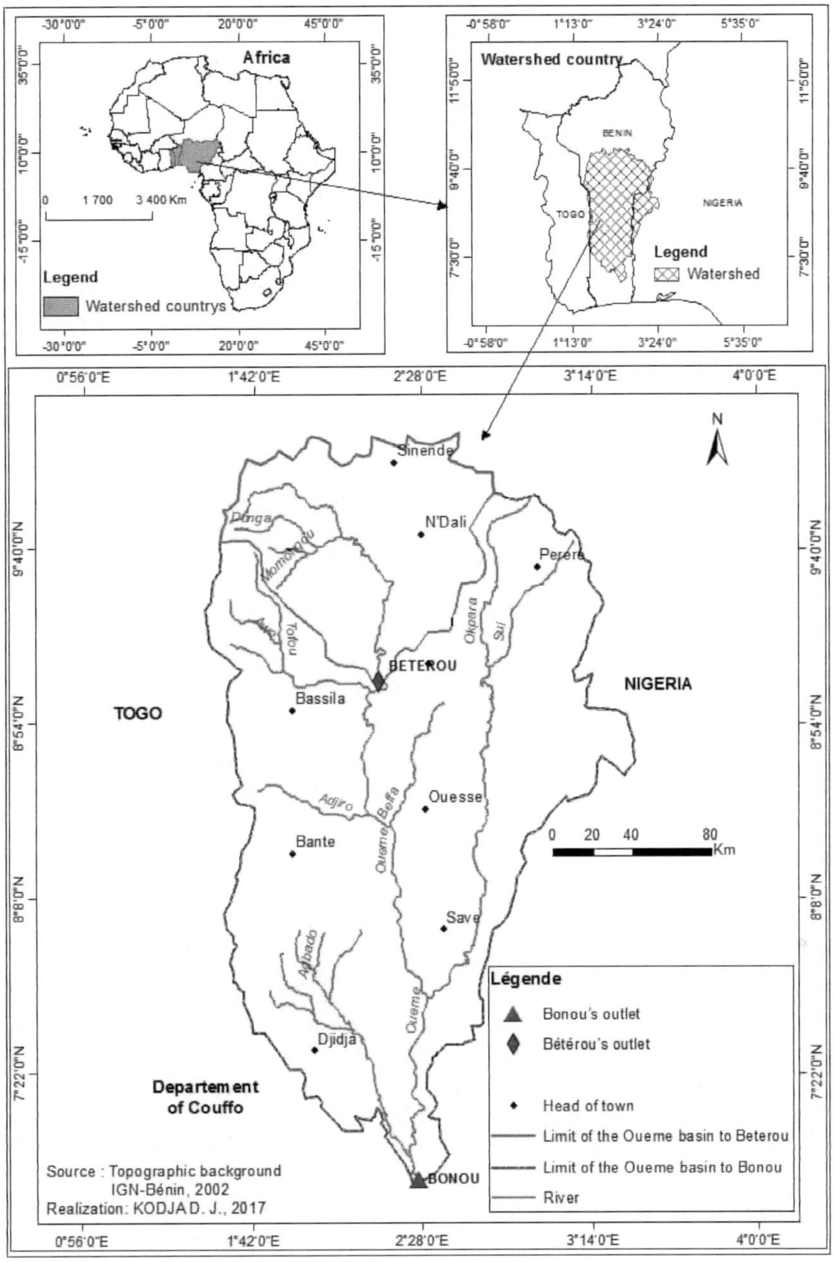

Figure 1. Geographical location.

2.3 GR4J Hydrological Model

The GR4J is a global conceptual rainfall-runoff model applied with only four parameters, at a daily time step and using few variables (Perrin et al., 2003). The GR4J model makes it possible to obtain the simulation of daily flows in a hydrological basin, to carry out the forecast of floods and low water levels, to reproduce the functioning of the basin and to obtain the interpolation and extrapolation of the climatic conditions of the basin (Perrin et al., 2003).

2.3.1 Method of calibration/validation of the model

To calibrate and to validate the GR4J model, it has been retained two sub-periods (calibration: 1971–1987; validation: 1988–2010). The homogeneity of these sub-periods will be

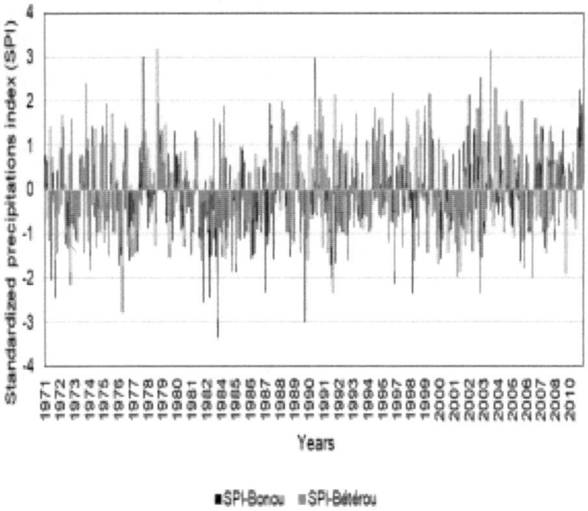

Figure 2. Rainfall index at Bétérou and Bonou (1971–2010).

discussed through the stationarity rupture analysis in the rainfall and flow chronicles.

2.3.2 Quality criteria of calibration and validation performances

The calibration of the GR4J model is based on the criteria of optimization and performance of the model to reproduce the flows with the criteria of Nash-Sutcliffe (1970) designated by NSE and whose formulas are Eq. (3):

$$\mathrm{NSE}(Q) = \frac{\sum(Q_{\mathrm{sim}} - Q_{\mathrm{obs}})}{\sum(Q_{\mathrm{obs}} - Q_{\mathrm{obs}}}$$

calculated on high water flows;

$$\mathrm{NSE}(\sqrt{Q}) = \frac{\sum(\sqrt{Q_{\mathrm{sim}}} - \sqrt{Q_{\mathrm{obs}}})}{\sum(\sqrt{Q_{\mathrm{obs}}} - \sqrt{Q_{\mathrm{obs}}})}$$

calculated on average flows;

$$\mathrm{NSE}(\ln(Q) = \frac{\sum(\ln(Q_{\mathrm{sim}}) - \ln(Q_{\mathrm{obs}}))}{\sum(\ln(Q_{\mathrm{obs}}) - \ln(Q_{\mathrm{obs}}))}$$

calculated on low flow. (3)

These 3 quality criteria are used in this study and all their values are between [$-\infty$ to 1], (Dezetter, 2008).

3 Results

3.1 Variability of interannual rainfall on the watersheds of Ouémé at Béterou and Ouémé at Bonou

The analyse of the evolution of annual rainfall in the areas of this study (Fig. 2) reveals the positive rainfall indices characterize wet periods or sequences, while negative value indices indicate periods of drought in the study environment.

Among the years corresponding to the wet sequences, there are 1963, 1968, 1978, 1990, 2002 and 2010. As for the years marked by drought, there are 1971, 1974 1976, 1983, 1987, 1989. These observations have been highlighted by many studies over the study area (Le Barbé et al., 1993; Amoussou et al., 2015).

In addition, the analysis of the rupture in pluviometric stationarity revealed by the Pettitt's non-parametric method and Hubert's segmentation of Hubert, a rupture in 1987, dividing the series into two homogeneous sub-periods (1971–1987 and 1993–2010) in the two sub-basins. Thus, in the Ouémé basin at Bétérou, the 1971–1987 sub-period recorded an average annual rainfall of 1091 mm against 1202 mm over the 1988–2010 sub-period. This observation leads us to say that the 1971–1987 sub-period is pluviometrically deficient compared to the 1988–2010 sub-period, i.e. an increase of 10 %. Similar situations are observed in Bonou between the two sub-periods, with an increase in heights of more than 15 % during the last sub-period. This situation can have repercussions on hydrological dynamics, activities and uses related to water in the study area.

3.2 Variability of interannual flows in the Ouémé Basin at Bétérou and at Bonou

Both at the Ouémé-Bétérou watershed and at the Ouémé-Bonou watershed (Fig. 3), the annual flow variability is marked by wet periods illustrated by positive anomalies in 1973, 1991, 2010 while the negative anomalies explain the deficit years of water flows during the years 1974, 1977, 1983, 1987, 1989, 2005. Moreover, the rainfall rupture is taken into account to better analyzing at the impact of this rainfall variability on flows in the study area. Indeed, it is found at Ouémé-Bétérou that the average annual flows in the sub-period 1971–1987 are 47.5 $\mathrm{m^3\,s^{-1}}$ while those in the sub-period 1988–2010 are 54.5 $\mathrm{m^3\,s^{-1}}$, an increase of 14.8 %. As in the watershed at Ouémé-Bétérou, it is also noted in the Ouémé-Bonou's watershed that the sub-period 1971–1987 (111 $\mathrm{m^3\,s^{-1}}$) is hydrologically deficit in the 1988–2010 (200 $\mathrm{m^3\,s^{-1}}$) with a difference of 78.3 %.

Based on these results, it should be noted that rainfall variability has impacted the average flow in the study area, as already studied by Totin et al. (2016). This finding is consistent with previous studies by Mahé et al. (1995), who indicated that from the 1990s, there is a slight recovery of rainfall in West Africa, in an increasingly anthropogenic environment that has led to an increase in water levels in watercourses and water bodies compared to the 1970s and 1980s, which can lead to flood events in the watershed (Amoussou et al., 2015).

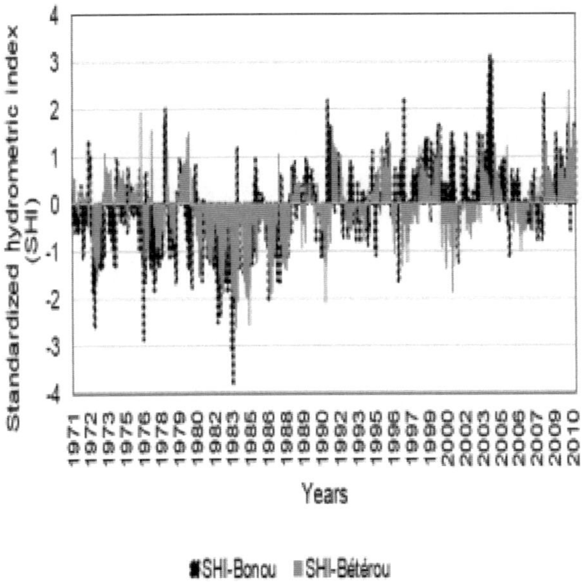

Figure 3. Flow index at Bétérou and Bonou (1971–2010).

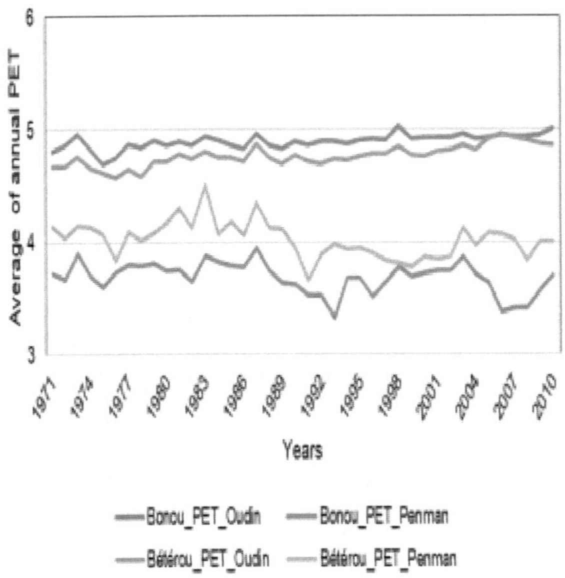

Figure 4. Comparison of annual average evolution of daily PET at Bétérou and at Bonou calculated by two methods, from 1971 to 2010.

3.3 Interannual evolution of the average potential evapotranspiration of Penman Monteith and Oudin

The evolution of the PET estimated by the method of Oudin has almost the same rhythm as the PET calculated with the method of Penman Monteith at Bétérou and Bonou (Fig. 4).

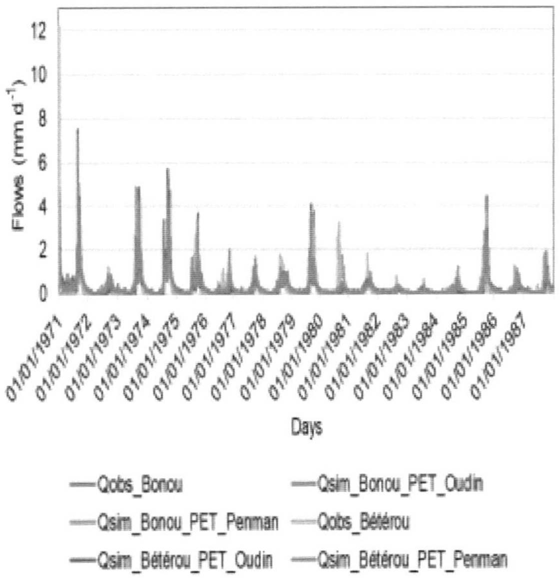

Figure 5. Observed and simulated flow variability for GR4J model calibration with Oudin and Penman PET.

Moreover, on the 1971–2010 series, the annual average of the PET at Bétérou is $4.8\,\mathrm{mm\,d^{-1}}$ for Oudin method while that of Penman Monteith is $4.0\,\mathrm{m\,d^{-1}}$ at Bétérou. As for Bonou, the PET average of Oudin method is $4.9\,\mathrm{mm\,d^{-1}}$ while that of Penman Monteith is $3.7\,\mathrm{mm\,d^{-1}}$. In addition, the discrepancies between the Oudin and Penman-Monteith PET are respectively $-15.6\,\%$ at Bétérou and $-24.7\,\%$ at Bonou. It is therefore appropriate for this analysis to say that for the two sub-basins the Oudin method of PET calculation underestimates the PET value in comparison of the PET of Penman-Monteith method.

3.4 Variability of the flows observed, simulated and calibrated by the GR4J model with the PET estimated by the Oudin and Penman-Monteith methods

The variability of the flows of observed and simulated for calibration with PET using the Oudin and Penman-Monteith methods by the GR4J model (Fig. 5), have almost the same rhythm as both Bonou and Bétérou.

However, the analysis of the biases reveals between the observed and simulated flows a difference of 8.1 % and 15.5 % respectively with the flows simulated with the Oudin and Penman-Monteith PET. It is the same for the simulated flow rates at Bétérou, giving a spread of 3.1 % with Oudin PET and 7.3 % with the Penman-Monteith PET. As already found by Amoussou et al. (2015),we can conclude that Oudin PET allows a flow simulation with a smaller bias than the Penman-Monteith PET

Table 1. Nash values with Oudin and Penman Monteith PET.

PET	Oudin			Penman Monteih		
Subperiods	NSE $\sqrt{Q}$	NSE(ln(Q))	NSE(Q)	NSE $\sqrt{Q}$	NSE(ln(Q))	NSE(Q)
Ouémé at Bonou						
Cal_71–87 (DP)	0.91	0.83	0.83	0.91	0.83	0.83
Val_93–10 (WP)	0.88	0.56	0.76	0.87	0.56	0.76
Cal_93–10 (WP)	0.90	0.90	0.85	0.90	0.90	0.85
Val_71–87 (DP)	0.85	0.60	0.75	0.85	0.60	0.75
Ouémé at Bétérou						
Cal_71–87 (DP)	0.64	0.70	0.52	0.64	0.70	0.52
Val_93–10 (WP)	0.75	0.78	0.59	0.75	0.78	0.59
Cal_93–10 (WP)	0.78	0.79	0.61	0.78	0.79	0.61
Val_71–87 (DP)	0.61	0.70	0.49	0.61	0.70	0.49

DP = Dry Period; WP = Humid Period.

However, optimization criteria of the model is also an indicator to better appreciate the model's performance, the calibration of the GR4J model from potential evapotranspiration estimates, including the Penman-Monteith and Oudin PET in the basin.

3.5 Values of calibration criteria and validation with the GR4J model in the catchment areas of Ouémé at Bétérou and Ouémé at Bonou

The results show that the values of Nash (NSE) calculated on the square roots of the flows, on the logarithms of the flows and on the high water flows in the watersheds of Ouémé at Bétérou and Ouémé at Bonou, give the same values corresponding to each criterion for both the Oudin PET and the Penman Monteith PET (Table 1).

Furthermore, Evapotranspiration (PET) is an important process in the water cycle. However, despite the difference between Oudin's PET and Penman-Monteith's PET reported on the analysis of Figs. 4 and 5, it is observed that the formulations of these PET do not have as much influence on the values of the Nash in calibration (in dry period the values of Nash vary from 0.83 to 0.91 at Bonou and from 0.52 to 0.70 at Bétérou) as in validation (Nash vary from 0.56 to 0.88 at Bonou and from 0.59 to 0.78 at Bétérou) with the GR4J model on the two sub-watersheds, under different rainfall regimes (Le Barbé et al., 1993). This result is similar to the work of Dezetter (1991); Oudin et al. (2005) and recently Ben Khediri and Drugs (2015); Dallaire (2019). They demonstrated that the rainfall-runoff models that use PET as input are not much sensitive to the estimation method of the PET calculation for the model performance.

4 Conclusions

Despite a great variability between the two different formulations of PET Oudin and Penman-Monteith used in this research, it should be noted that the results show the similar trends with the Nash values. Thus, it should be concluded that the formulation of PET has few impacts on the results of the GR4J model for these tropical basins. However, in the context of global changes, and in particular the changes that the area of study is facing, this result must be taken into account in the analysis of current hydrological processes and the revision of hydrological standards in tropical basins.

Data availability. The data are not publicly accessible: The climatological data are the properties of the National Directorate of Meteorology (NDM) at Cotonou in Benin, and an Environmental System of Information on Water Resource and Modeling (SIEREM)/HydroSciences Montpellier (HSM), UMR 5569. The hydrometric data, are the properties of the Directorate General of Water (DGEau)/Benin and an Environmental System of Information on Water Resource and Modeling (SIEREM)/HydroSciences Montpellier (HSM), UMR 5569.

Author contributions. All of the authors contributed effectively to the collection, processing and analysis of the data. They also contributed to the writing and proofreading of the paper.

Competing interests. The authors declare that they have no conflict of interest.

Special issue statement. This article is part of the special issue "Hydrological processes and water security in a changing world". It is a result of the 8th Global FRIEND-Water Conference: Hydrological Processes and Water Security in a Changing World, Beijing, China, 6–9 November 2018.

Acknowledgements. The Authors thank the organizers of 8th Global FRIEND-Water Conference and IHP-UNESCO. The Authors also thank the Benin Weather Agency and the SIEREM/HSM database for providing the data. The authors would also like to thank the reviewers of this paper.

References

Afouda, A., Aguè, A. I., and Lanhoussi, F.: Etude comparative d'un modèle conceptuel global (GR4J) et d'un modèle semi-distribue (GEOSFM) sur le bassin versant de l'Ouémé à Savè (Bénin, Afrique de l'Ouest), Revue LJEE, 24–25, 2015.

Allen, R. G., Pereira, L. S., Raes, D., and Smith, M.: Crop evapotranspiration- Guidelines for computing crop water requirements-FAO Irrigation and drainage paper 56, Rome, Italy 15 pp, 1998.

Amoussou, E., Totin Vodounon, H., Houessou, S., Tramblay, Y., Camberlin, P., Houndenou, C., Boko, M., Mahe, G., and Paturel, J. E.: Application d'un modèle conceptuel à l'analyse de la dynamique hydrométéorologique des crues dans un bassin-versant en milieu tropical humide: cas du Fleuve Mono. XXVIIIe Colloque de l'Association Internationale de Climatologie, Liège 2015, 17–24, 2015.

Andréassian, V., Perrin, C., and Michel, C.: Impact of imperfect potential evapotranspiration knowledge on the efficiency and parameters of watershed models, J. Hydrol., 286, 19–35, 2004.

Ben Khediri, W. and Drogue, G.: Quel est l'impact de l'échantillonnage spatial des précipitations et de l'évapotranspiration potentielle sur le pouvoir prédictif d'un modèle hydrologique empirique?, Climatologie, 12, 24 pp., 2015.

Boyer, J. F., Dieulin, C., Rouché, N., Crès, A., Servat, E., Paturel, J. E., and Mahé, G.: SIEREM: an environmental information system for water resources, in: Water Resource Variability: Hydrological Impacts. Proc. of the 5th FRIEND World Conference, La Havana, Cuba, IAHS Publ., 308, 19–25, 2006.

Crétat, J., Richard, Y., Pohl, B., Rouault, M., Reason, C. J. C., and Fauchereau, N.: Recurrent daily rainfall patterns over South Africa and associated dynamics during the core of the austral summer, Int. J. Climatol., 2, 261–273, 2012.

Dallaire, G.: Étude de l'incertitude liée à la modélisation de l'évapotranspiration potentielle dans les études d'impact des changements climatiques sur les ressources en eau au Canada, aux États-Unis et au Mexique, Doctoral dissertation, École de Technologie Supérieure de Québec, 144 pp., 2019.

Dezetter, A.: Modélisation globale de la relation pluie débit. Application en zone de savanes soudanaises (Nord-Ouest de la Côte d'Ivoire), Thèse de Doctrat, Université de Montpellier II, France, 428 pp., 1991.

Dezetter, A., Girard, S., Paturel, J. E., Mahé, G., Ardoin-Bardin, S., and Servat, E.: Simulation of runoff in West Africa: Is there a single data-model combination that produces the best simulation results?, J. Hydrol., 354, 203–212, 2008.

Kay, A. L. and Davies, H. N.: Calculating potential evaporation from climate model data: A source of uncertainty for hydrological climate change impacts, J. Hydrol., 358, 221–239, 2008.

Le Barbé, L., Alé, G., Millet, B., Texier, H., Borel, Y., and Gualde, R.: Les ressources en eaux superficielles de la République du Bénin, Edition ORSTOM, 540 pp., 1993.

Mahé, G. and Olivry, J. C.: Variations des précipitations et des écoulements en Afrique de l'Ouest et Centrale de 1951 à 1989, Sécheresse, 6, 109–117, 1995.

Nash, J. E. and Sutcliffe, J. V.: River flow forecasting through conceptual models, Part I – A discussion of principles, J. Hydrol., 10, 282–290, 1970.

Oudin, L.: Recherche d'un modèle d'évapotranspiration potentielle pertinent comme entrée d'un modèle pluie-débit global, Thèse de doctorat, ENGREF (AgroParisTech), 496 pp, 2004.

Oudin, L., Hervieu, F., Michel, C., Perrin, C., Andréassian, V., Anctil, F., and Loumagne, C.: Which potential evapotranspiration input for a lumped rainfall–runoff model?: Part 2 – Towards a simple and efficient potential evapotranspiration model for rainfall–runoff modelling, J. Hydrol., 303, 290–306, 2005.

Paturel, J. E., Ouedraogo, M., Mahe, G., Servat, E., Dezetter, A., and Ardoin, S.: The influence of distributed input data on the hydrological modelling of monthly river flow regimes in West Africa, Hydrol. Sci. J., 48, 881–890, 2003.

Perrin, C., Michel, C., and Andreassian, V.: Improvement of a parsimonious model for stream flow simulation, J. Hydrol., 279, 275–289, 2003.

Sighomnou, D.: Analyse et redéfinition des régimes climatiques et hydrologiques du Cameroun : perspectives d'évolution des ressources en eau, Yaoundé: Université de Yaoundé 1, multigr. Th.: Sci. de l'Eau, Université de Yaoundé, 1, 289 pp., 2004.

Totin, V. S. H., Amoussou, E., Odoulami, L., Boko, M., and Blivi, B. A.: Seuils pluviométriques des niveaux de risque d'inondation dans le bassin de l'Ouémé au Bénin (Afrique de l'Ouest), in: Actes de Colloque de l'Association Internationale de Climatologie, Besançon, 369–374, 2016.

Vissin, E. W.: Impact de la variabilité climatique et de la dynamique des états de surface sur les écoulements du bassin béninois du fleuve Niger, Thèse de Doctorat de l'Université de Bourgogne, Dijon, France, 280 pp., 2007.

Proc. IAHS, 383, 171–183, 2020
https://doi.org/10.5194/piahs-383-171-2020

Hydropluviometric variability in non-Sahelian West Africa: case of the Koliba/Corubal River Basin (Guinea and Guinea-Bissau)

Saly Sambou[1], Honore Dacosta[1], Rene Ndimag Diouf[2], Ibrahima Diouf[3], and Alioune Kane[1]

[1]Faculty of Letters and Human Sciences, Department of Geography, Cheikh Anta Diop University, Dakar (UCAD), BP: 5005, Dakar-Fann, Senegal
[2]Faculty of Technologies of Education and Training Sciences, Didactic Department History and Geography, Cheikh Anta Diop University, Dakar, Senegal
[3]NOAA Center for Weather and Climate Prediction, 5830 University Research Court, College Park, Maryland, USA

Correspondence: Saly Sambou (sambousaly@gmail.com)

Published: 16 September 2020

Abstract. The Koliba/Corubal River watershed is poorly documented due to the hydrometric measurements shutdown and gaps in the very short hydropluviometric timeseries. The purpose of this study is to analyze the variability of rainfall in the Basin, by simulating flows using the GR2M rain-flow model and extending the discharge timeseries. From the regional vector method, the rainfall timeseries were homogenized, and the gaps filled by the estimated values. The rank correlation and Pettitt test on annual rainfall amounts (1924–2015) indicate breaks in 1958, 1967 and 1969, leading to rainfall deficits ranging from 9.7 % to 20.2 %. For some stations, the segmentation method shows a recovery of rainfall towards the end of the 1980s (Gaoual, Mali) and the early 2000s (Gabu). The analysis of the temporal distribution of the Monthly Rainfall Coefficients shows an improvement of the contributions of a few months during the period after rupture. From a hydrological point of view, the correlation between the mean annual rainfall and the runoff has allowed to extend the flow timeseries. The mean monthly rainfall calculated using the inverse square of distance method, the Potential Evapotranspiration, and the flow rates were used to calibrate and validate the model to determine the parameters that better transform rainfall in flow. The values of the Nash criteria close to 100 have made it possible to extend the monthly flow data from Koliba/Corubal to Gaoual, Cade and Tche-Tche until 2015.

1 Introduction

In West Africa, many recent studies (Sène and Ozer, 2002; Ali et al., 2008; Niang, 2008; Sarr and Lona, 2009; Ouoba, 2013; Panthou, 2013; Descroix et al., 2013, 2015; Bodian, 2014) showed a recovery in rainfall towards the end of the 1990s, after a long period of severe drought in the 1970s and 1980s. The study of Lebel and Ali (2009) indicated that the eastern and central parts of the Sahel experienced an increased rainfall situation, while the western part remains in deficit for the period 1990–2007. This improvement in rainfall conditions coupled with high interannual variability, can be appreciated in terms of the number of rainfall events with high runoff coefficients leading to floods. The manifestation of these events is heterogeneous and variable in time and space. The Sahelian zone more affected by the rainfall deficit (Dai et al., 1998) seems to be the most documented (Hufty, 1994). Some West African Basins such as the Koliba/Corubal River (a Sudanese area) are still poorly investigated due to lack of hydropluviometric data. Moreover, discontinuous hydrological datasets have many shortcomings related to the lack of adequate monitoring of measurement networks due to limited human and financial resources. According to Claude (1980), the creation of Water and Environment Service at the General Direction of Natural Resources of Guinea-Bissau is very recent (only from 1977). The hydrological regime and the streamflow are poorly known. The hydrometric network of Guinea is very obsolete, the first stations were implanted around 1910, but its development was slow and subjected to many hazards. The current network has been restructured starting in 1967. However, some newly created stations are not calibrated due to difficult access (Claude, 1980). This is an obstacle to any hydrological study, especially the assessment of the impact of rainfall variability on the streamflow.

The Koliba/Corubal Basin is a transboundary Basin, shared between Guinea and Guinea-Bissau. It is part of the intervention area of the Gambia River Basin Organization (in French, Organisation pour la Mise en Valeur du Fleuve Gambie, OMVG hereafter), whose mission is to undertake a rehabilitation of rainfall and hydrometric networks for the realization of its projects.

However, the availability of long-term timeseries of rainfall (annual and monthly), although sometimes incomplete, is an asset to provide a solution to the lack of data, through the modeling of the rain-flow relationship. Among the existing models, the GR2M of Cemagref (Mouelhi, 2003; Mouelhi et al., 2006), Rural Engineerig at the monthly time step and with two parameters (X_1 and X_2), has been selected. It is a simple and robust model much applied in Africa (Paturel et al., 1995; Mahé et al., 2005; Bodian et al., 2012). The objective of this study is to analyze the variability of rainfall in the Koliba/Corubal River Basin. Due to discontinuous and incomplete hydrological data, a rain-flow modeling approach will allow, on the basis of existing data, to extend the discharge timeseries for a better quantification of water resources for rational management and the satisfaction of different uses.

2 Data and methods

2.1 The study area

The Koliba/Corubal River watershed lies between 11° N and 12°30′ N and between 12° W and 14°30′ W. It is shared between Guinea (84.5 %) and Guinea-Bissau (15.5 %) and covers an area of 20 876.4 km^2 at the Tche-Tche hydrometric station (Fig. 1).

The River has its source from the West of Fouta Djalon, in Middle Guinea, in the region of Labe. It is a junction of two rivers: the Tomine which has its source from the Sangale and the Komba which takes its source from Madina Wora. These two rivers confluence at Gaoual to give the Koliba. After a course, going West, more than 200 km with many meanders, it forms the border between the two countries for a few kilometers, before entering Guinea-Bissau where it is called Corubal. It joins the Kayanga/Geba near Xime in a flat and marshy zone where the tide goes very far inside the continent, to form the estuary of Geba (SOFRECO, 1993). Although, its Basin occupies a smaller area, the River concentrates the most important surface water resource in Guinea-Bissau. The very dense hydrographic network is simplified (Fig. 1) for more readability.

The vegetation is composed of dense forests, degraded forest altitude, dry forest covered annually by bush fires, wooded savannah and woodland, crops, fallow field, etc. (Zoumanigui, 2003). The Koliba/Corubal Basin belongs to a tropical climate characterized by a single rainy season that generally lasts five months in the north and six months in the south of the Basin and a dry season which however records a few millimeters of water (April and November). Rainfall decreases from the south to the north following the West African monsoon flux. The monthly maximum temperature average is between 26.0 and 33.4 °C at Labe, 31.1 and 40.20 °C at Koundara, respectively in August and April. The average minimum temperatures range from 10.2 and 18.3 °C at Labe and 15.0 and 24.2 °C at Koundara in December and May.

2.2 Physiographic data

The cartographic documents covering a large area of study are not always easy to access. Moreover, even if they are available, a problem of the heterogeneity of the scales arises. Thus, for some reasons of homogeneity, the Shuttle Radar Topography Mission (SRTM) data of 30 m resolution NASA (National Aeronautics and Space Administration) were used. They are used for the characterization of the relief and the determination of the morphometric characteristics of the Koliba/Corubal Basin (Table 1 and Fig. 2). The Digital Terrain Model (DTM) realized with the Surfer software shows that the mountainous terrain is in the northeast part of the Basin with altitudes exceeding 1400 m. It is at these levels that the tributaries Bantala and Komba take their source. In the downstream part, altitudes decrease to less than 100 m (Fig. 2). The Basin and its various sub-basins have an elongated shape as shown by the compactness coefficients (Table 1), resulting in a long period of concentration of runoff water. This is nuanced because the influence of relief is even more important.

2.3 Rainfall data

Rainfall data come from the National Meteorology Direction of Guinea (DNM), the National Meteorological Service of Guinea-Bissau and the OMVG. The length of the timeseries is very variable from one source to another, because the shortest are found in Guinea-Bissau over a period of 12 years (Beli and Tche-Tche). Table 2 lists the rainfall stations selected by country and available data, while Fig. 1 shows their spatial distribution in the Basin. Given their number, only the results of a few will be presented. The choice will be based on the sample size, the quality of the data (small gaps in the timeseries) and the geographical position in the Basin.

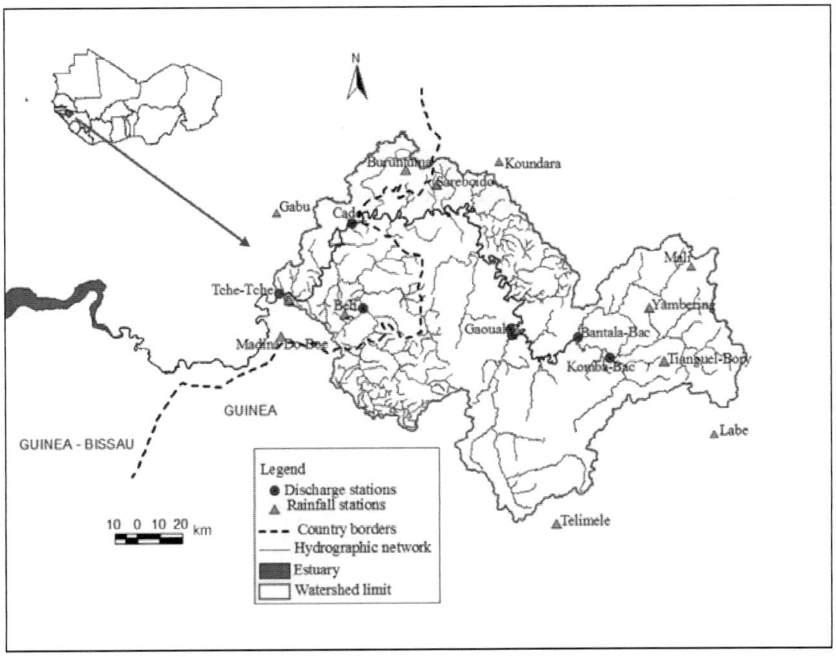

Figure 1. Location of the Koliba/Corubal River watershed in West Africa and spatial distribution of rainfall and hydrometric networks.

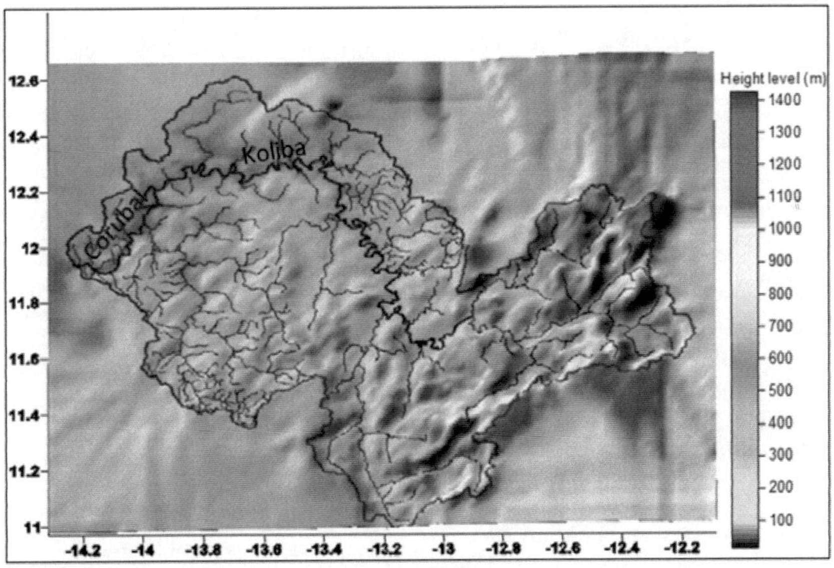

Figure 2. 3D Digital Terrain Model of the Koliba/Corubal Basin at Tche-Tche.

2.4 Potential evapotranspiration data

Potential Evapotranspiration data (PET) from the Labe station are calculated using the Penman (1948) formula. They come from the DNM of Guinea and cover the period 1975–2000. This station being far from Cade and Tche-Tche, Koundara's maximum and minimum temperature data (1975–2015) are used to calculate the PET using the CROPWAT 8.0 software (Smith, 1992) and based on the formula Penman–Monteith (Monteith, 1965).

Table 1. Morphometric characteristics of the Basin and its sub-basins.

River	Station	Area (km^2)	Perimeter (km)	K_C	L (km)	l (km)
Koliba/Corubal	Tche-Tche	20 876.4	1417.7	2.75	678.1	30.8
Koliba/Corubal	Cade	15 658.9	1278.5	2.86	613.7	25.5
Koliba/Corubal	Gaoual	9726.9	868.4	2.47	410.5	23.7
Fefine	Beli	2237.3	359.6	2.13	166.4	13.4
Tomine	Gaoual	3319.4	532.5	2.59	253.1	13.1
Bantala	ferry	1535.7	347.0	2.48	164.1	9.4
Komba	ferry	2895.6	497.0	2.59	236.2	12.3

K_C: Gravelius compactness index; L (km): length of the equivalent rectangle; l (km): width of the equivalent rectangle.

Table 2. List of selected rain gauge for this study.

Country	Station	Altitude (m)	Start year	End year	% gaps	Average annual rainfall (mm)
Guinea	Gaoual	500	1925	2006	18.3	1776.1
	Koundara	76	1928	2015	*	1197.9
	Labe	1025	1921	2015	3.2	1629.3
	Mali	1464	1922	2015	19.1	1659.7
	Sareboïdo	520	1961	1976	*	1364.4
	Telimele	650	1924	2003	11.3	2195.7
	Tianguel-Bory	650	1960	1991	*	1678.0
	Yambering	868	1967	1981	*	1602.1
Guinea-Bissau	Beli	70	1978	1989	8.3	1583.4
	Buruntuma	100	1950	1992	*	1266.0
	Gabu	63	1941	2005	*	1321.6
	Madina Do Boe	75	1950	1967	*	1905.7
	Tche-Tche	59	1978	1989	8.3	1468.0

* complete timeseries; av. rain.: average annual rainfall.

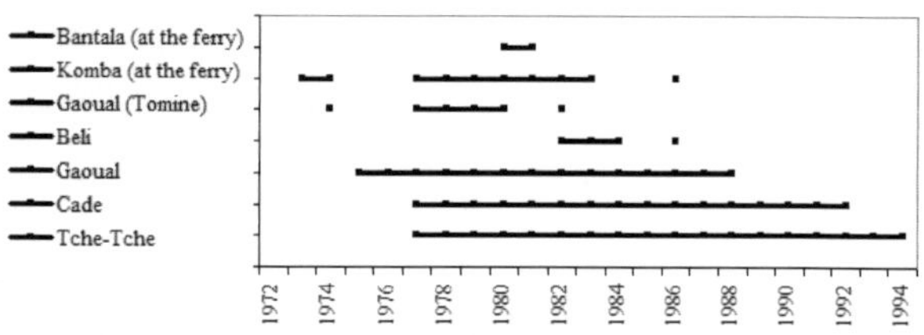

Figure 3. Chronogram of available discharge according to the stations.

2.5 Data and hydrological model

2.5.1 Hydrological data

The Koliba/Corubal hydrometric network at Tche-Tche is composed of seven (7) stations; three (3) of them are on the mainstream and four (4) are on the Bantala, Komba, Tomine and Fefine tributaries (Fig. 1). The hydrological data come from the OMVG database. Figure 3 shows the chronogram of available discharge, the timeseries of which are sometimes interspersed with gaps (Bantala, Komba, Gaoual on the Tomine and Beli). Due to failures, these stations were excluded from the hydrological analysis. Only the stations of Gaoual (14 years), Cade (16 years) and Tche-Tche (18 years) are chosen to simulate the flows and thus extend the discharge timeseries.

2.5.2 Presentation of the hydrological model GR2M

The GR2M model is a global conceptual hydrological model (Fig. 4) that operates on a monthly scale and uses rainfall and potential evapotranspiration as inputs (Mouelhi, 2003; Mouelhi et al., 2006). The discharges are used for setting parameters and evaluating performance. It has two parameters X_1 and X_2 that can be optimized: the first one is evolved in the production function and the second one in the transfer function. The production function is organized around a soil reservoir "S" which represents its maximum capacity. This reservoir regulates the quantitative aspect of modeling (Gogien, 1998). The transfer function is used to translate the way in which the rainfall for the flow reaches the outlet of the basin by specifying the time course (Mouelhi, 2003). This is the fraction of the net rain that has not been used to feed the soil reservoir.

3 Methods

3.1 Critical and homogenization of rainfall data

The critical and the homogenization of the data were done by the Regional Vector Method (Brunet-Moret, 1977), integrated into the Hydraccess software (Vauchel, 2004). The Regional Vector is defined as a chronological series of annual or monthly rainfall indices derived from the extraction of the most "probable" information in the most frequent sense contained in the dataset of observation posts in a homogeneous climatic region (Hiez, 1969, 1977; Brunet-Moret, 1979; L'Hôte, 1986; Hiez et al., 1987, 1992). Among the two usual methods, the one of Brunet-Moret (1979) was chosen because it relies on all the stations and the calculation of the deviations from the regional average, makes it possible to de-

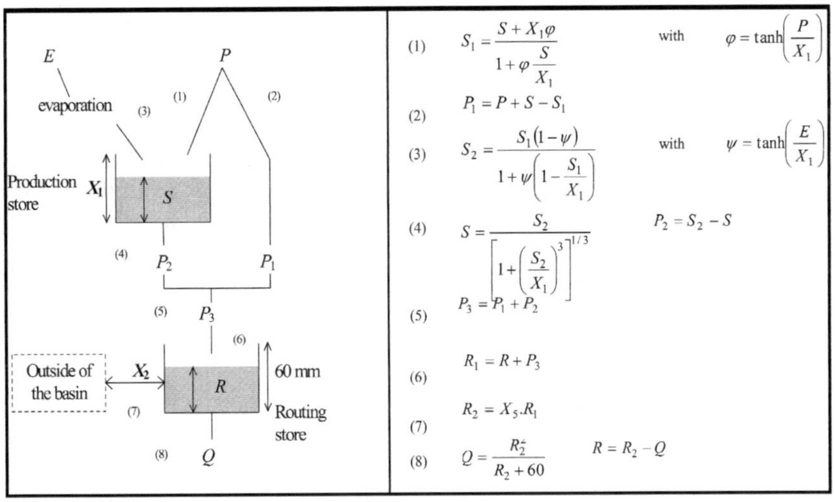

Figure 4. Architecture of the GR2M model (Mouelhi et al., 2006).

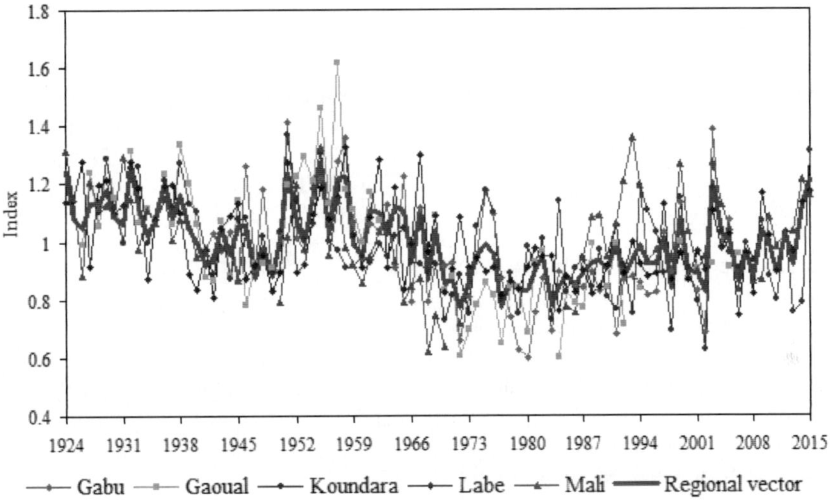

Figure 5. Annual vector of rainfall index over the period 1924–2015 and index of Gabu, Gaoual, Koundara, Labe and Mali stations.

tect the doubtful data. For each rainfall station, a timeseries of annual indices is calculated, the mean of which represents the regional vector (Fig. 5). These indices indicate the rainfall of the year compared to the regional average. The timeseries are homogenized over the period 1924–2015 and the calculated values were used to fill the gaps.

3.2 Detection of breaks within the annual rainfall timeseries

Given the high variability of rainfall, statistical tests integrated into the KhronoStat software (IRD, 1998) were used to detect stationarity breaks in the timeseries. A break can be compared to a change in the probability law of the timeseries at a given moment, most often unknown (Lubès et al., 1994). Among the existing break tests, three (3) are selected for this study because of their robustness and sensitivity to a change in the average. They are widely used in the climate variability analysis in Africa (Servat et al., 1998, 1999; Paturel et al., 1998; Haidu, 2006; Kouassi et al., 2010; Bodian, 2011, 2014; Sambou et al., 2018). It is the rank correlation test (Spearman, 1904) that analyzes the randomness of a timeseries and proposes an alternative hypothesis of trend. The Mann–Whitney test modified by Pettitt (1979) consists in dividing the main timeseries into two sub-series having different statistical distributions, separated by a break at time t. The null hypothesis H_0 corresponds to the absence of break (homogeneous timeseries), whereas if it is rejected, a break date is determined. The segmentation procedure (Hubert and Carbonnel, 1987; Hubert et al., 1989) has the principle of dividing the timeseries into several segments whose averages are significantly different.

3.3 Recent variability of monthly rainfall and calculation of mean rainfall

The analysis of the temporal distribution of rainfall on a monthly scale is relevant in the context of the recent evolution characterized by improved rainfall conditions. The missing values were reconstructed with the regional vector method, considering each month of the year separately, like an annual value. To do this, a calculation of the Monthly Rainfall Coefficients (MRC, hereafter) was performed at the periods pre-break and post-break. They indicate the contribution of each month to the annual total rainfall and they are obtained from the report of the monthly total of the annual cumulation. For the purposes of the discharge simulation, the average rainfall over the Basin upstream from Gaoual, Cade and Tche-Tche stations was calculated according to three interpolation methods (Thiessen method, inversing of the square of the distance method and Kriging method) using the Hydraccess software. It is about going from a punctual rainfall to a surface rainfall. The results of the best method are considered. This average rainfall is one of the input data of the model.

3.4 Calculation of potential evapotranspiration

Oudin (2004) proposed an evapotranspiration formula based on few data (temperature), and having a high spatial robustness optimized to meet the needs of hydrological modeling. For reasons of consistency and to avoid possible differences, PET was calculated with the Penman–Monteith formula in Koundara over the period 1975–2015. The Labe data already available until 2000 have been extended using the same formula to harmonize with the rainfall data. According to Quentin (2015, in Bell et al., 2011), the Penman–Monteith formula is often used in evapotranspiration extrapolation for its "physical basis", which is more sensitive to the multiple effects of climate change. Table 3 presents the average values of PET at Labe and Koundara over the period 1975–2015. It has a periodicity with a maximum in the dry season and a minimum in the rainy season.

3.5 Hydropluviometric correlation

In terms of short timeseries and gaps, a correlation approach between rainfall and runoff is adopted in order to lengthen the annual discharge timeseries. For this, the average annual rainfall over the Basin was determined using three (3) methods mentioned above, and the annual modules transformed into a runoff. Then, the good correlation between the two variables, with a coefficient of determination of 0.6 and 0.8, makes it possible to appreciate the relation. The linearity of the relationship leads to the extension of the timeseries of runoff from the average rainfall before finally being converted into an annual module.

3.6 Calibration of the model

The objective of the calibration phase is to find the parameter settings that bring the behavior of the model as close as possible to that of the modeled basin, the similarity of the behaviors being quantified by the Nash criterion (Nash and Sutcliffe, 1970; Ardoin-Bardin, 2004) defined by the Eq. (1):

$$\text{Nash} = 100\left[1-\frac{\sum_i (Q_\text{o}^i - Q_\text{c}^i)^2}{\sum_i (Q_\text{o}^i - Q_\text{m}^i)^2}\right]. \quad (1)$$

Q_o^i is the observed monthly discharge, Q_c^i is the calculated monthly discharge, Q_m^i is the average monthly discharge observed and i is the month considered.

The Nash criterion allows appreciating the quality of the simulations compared to the observations. The best fit corresponds to a criterion value close to 100 %, while a criterion below 60 % indicates a poor relationship between simulations and observations (Ardoin-Bardin, 2004).

Generally, global conceptual models require a phase of parameter specification which must be done on sufficiently long timeseries for which it must be assumed that there has been no change in the hydrological or physiographic characteristics of the basin (Diello, 2007). Like most West African basins, the Koliba/Corubal experienced a decrease in rainfall between 1960s and 1970s. This decrease has influenced the hydrological functioning. The brevity of discharge data does not facilitate their division into two parts for calibration and validation. In addition, the covered periods covered by the data are included in the phase of post-break. These periods can be considered as relatively homogeneous; knowing that the recovery of rainfall observed in most stations was not yet effective. According to Ardoin-Bardin (2004), the shortest periods may depend on more extreme conditions affecting the considered period. This justifies the choice of all the periods available for calibration and validation. Thus, for each of them, an adjustment of the parameters X_1 and X_2 was carried out and the model validated. These parameters will be applied to the chronicles to simulate the flows of the Koliba/Corubal River from the input data.

4 Results

4.1 Homogenization of annual rainfall

The application of the regional vector method enables to estimate values to fill gaps in the timeseries of stations. Figure 6 shows the observed and estimated rainfall over the period 1924–2015. The correlation coefficients between observed and estimated values are 0.649 in Mali, 0.812 in Gabu and 0.832 in Gaoual.

Table 3. Average monthly potential evapotranspiration at Labe and Koundara.

Station	J	F	M	A	M	J	J	A	S	O	N	D
Labe	144.0	148.0	183.6	172.7	154.1	125.0	108.1	105.6	110.0	121.3	128.4	134.4
Koundara	162.0	166.3	201.0	202.1	195.8	158.6	143.0	135.4	132.9	139.6	144.9	156.3

Table 4. Synthesis of the results of Pettitt test.

Station	Year of	Average (mm)		Standard deviation		Coefficient of variation		Deficit
	break	Before	After	Before	After	Before	After	(%)
Gabu	1967	1501.6	1242.2	180.3	213.2	12.0	17.2	17.3
Gaoual	1967	1983.9	1583.2	294.3	227.1	14.8	14.3	20.2
Koundara	1969	1285.7	1125.9	158.0	189.0	12.3	16.8	12.4
Labe	1969	1729.3	1526.9	241.6	166.8	14.0	10.9	11.7
Mali	1958	1748.4	1579.7	229.3	254.1	13.1	16.1	9.7

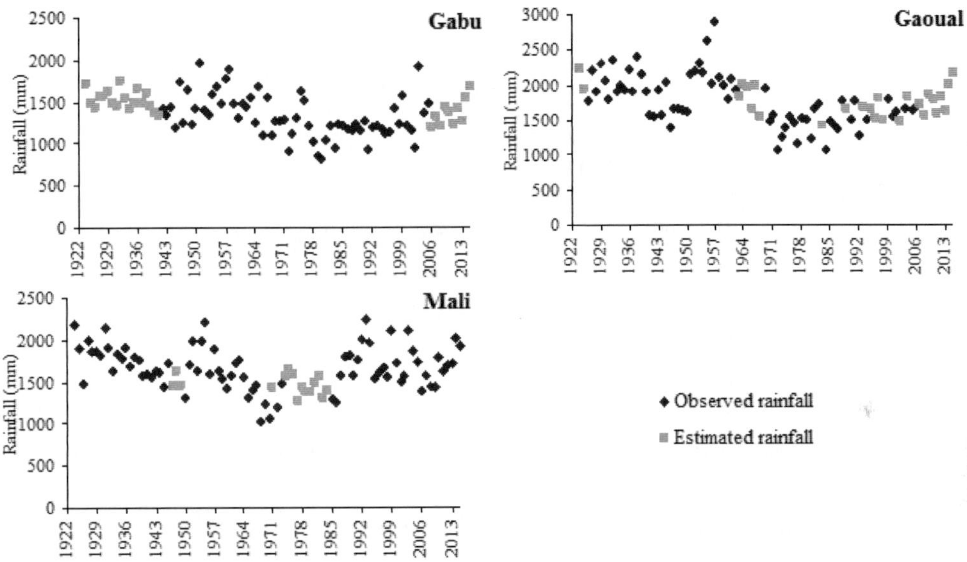

Figure 6. Annual observed and estimated rainfall at Gabu, Gaoual and Mali stations over the period 1924–2015.

4.2 Homogeneity tests on annual rainfall

4.2.1 Rank correlation and Pettitt test

The results of the rank correlation test indicate a decrease in rainfall. The null hypothesis of no change is rejected at the 99 %, 95 % and 90 % confidence thresholds at all stations except Mali where it is accepted at the different thresholds.

The Pettitt tests shown in Table 4 exhibit that all timeseries show early break for Mali (1958) and later for Labe and Koundara (1969). For most stations, the break occurred in 1967 and 1969, which confirms the results of the ICCARE program (Servat et al., 1998, 1999) in West and Central Africa. The rainfall post-break deficit is between 9.7 % (Mali) and 20.2 % (Gaoual). The study of Bodian and al. (2011) (for the period 1923–2005), indicated a deficit of 13 % in Mali and 24 % in Labe. It has decreased to 9.7 % and 11.7 % respectively over the period 1924–2015.

4.2.2 Segmentation method

The segmentation method presented in graphical, show one or more breaks in the timeseries (Fig. 7). Koundara and Labe timeseries show a single break in 1955 and 1969 respectively. In Gabu, the decrease in rainfall recorded since 1965 reduced in 2002, the year that marks the end of the dry period and the beginning of a rainfall recovery. The Mali timeseries had a succession of two periods of low rainfall, before returning to more favorable rainfall conditions from 1987. The Gaoual station presents a weird situation with five break events sep-

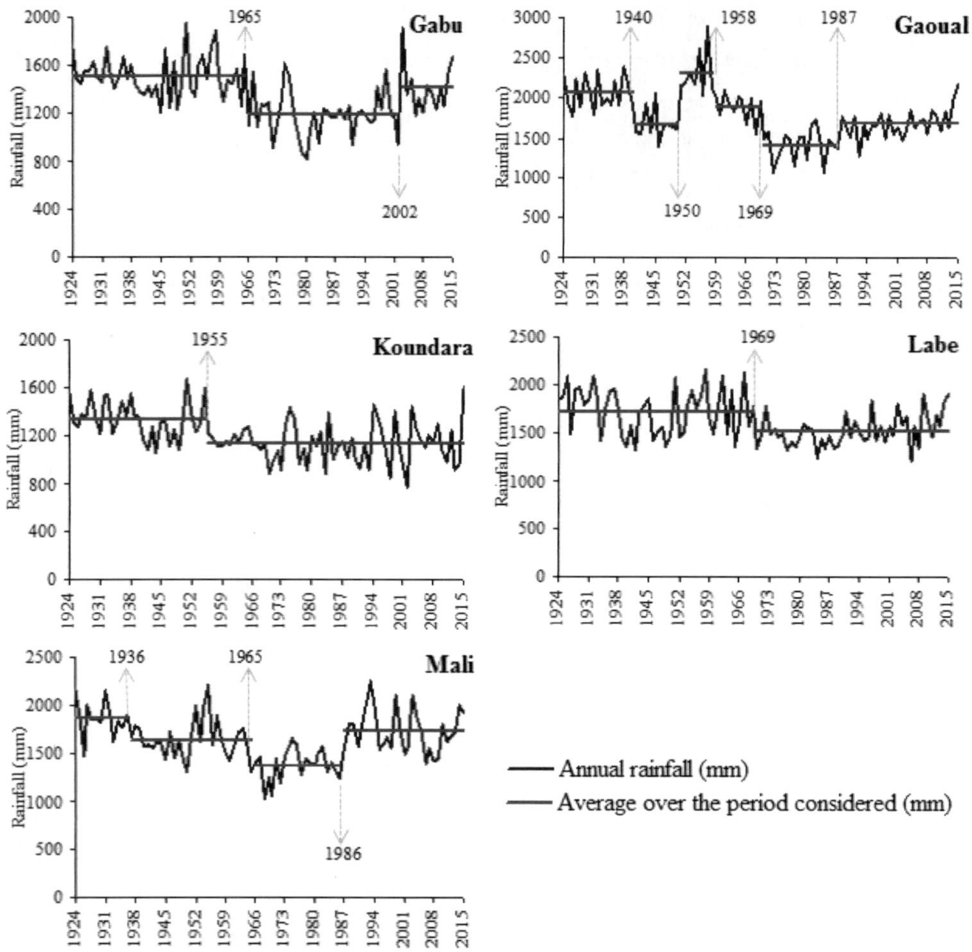

Figure 7. Variability and segmentation of annual rainfall in Gabu, Gaoual, Koundara, Labe and Mali from 1924 to 2015.

arating six unequal periods. Comparison is difficult due to inequality of the sub-periods.

5 Evolution in monthly rainfall according to Pettitt's test

The correlation coefficients between the observed and estimated monthly rainfall have deteriorated. They vary between 0.531 (Mali) and 0.696 (Koundara) in May; 0.579 (Gabu) and 0.731 (Mali) in June, 0.565 (Mali) and 0.795 (Koundara) in July, 0.520 (Koundara) and 0.744 (Gabu) in August, 0.621 (Gabu) and 0.745 (Gaoual) in September, 0.687 (Gaoual) and 0.747 (Koundara) in October.

The temporal distribution of the MRC shows that the contribution of the monthly to annual total has undergone a significant evolution during the post-break period. In Gabu, the contributions from June to September are greater than those of the first period, whereas in Gaoual, Koundara (Fig. 8), Labe and Mali, they concern three months. The increase of MRC is correlated with that of rainfall. Although nuanced, these results support the recovery of rainfall observed in West Africa in the late 1990s and early 2000s (Sène and Ozer, 2002; Descroix et al., 2015; Sambou et al., 2018). The MRCs for October remain in favor of the pre-break period at all stations.

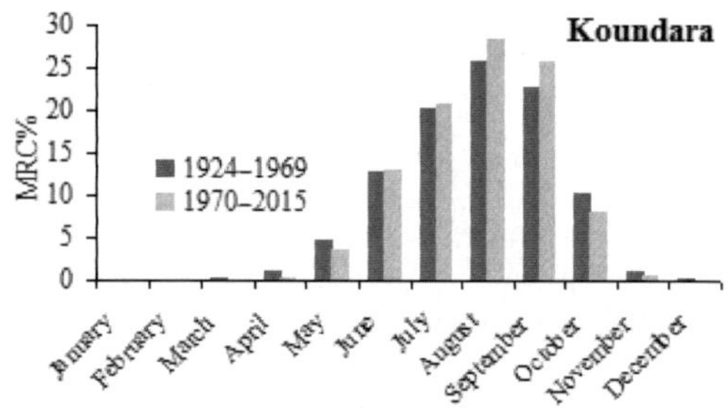

Figure 8. Temporal evolution of MRC during pre-break and post-break at Koundara (1924–2015).

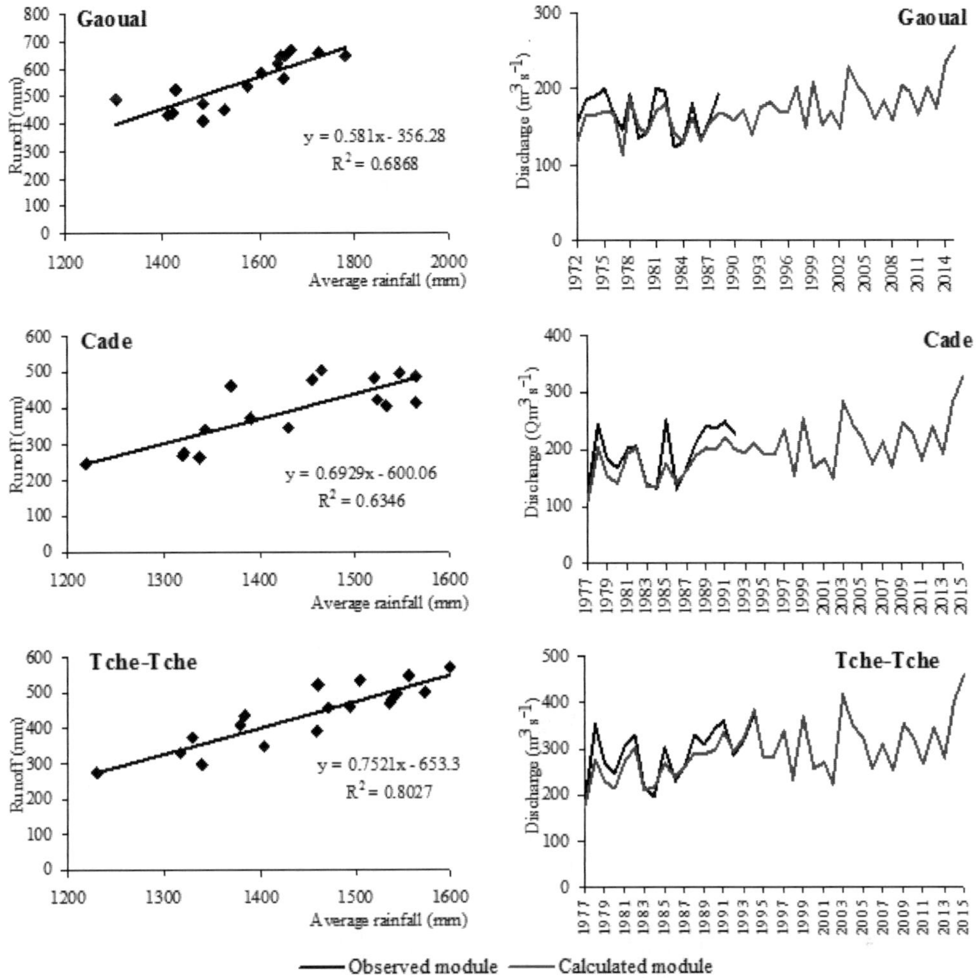

Figure 9. Relation between runoff and annual mean rainfall (left panels) and comparative evolution of observed and calculated annual modules (right panels).

6 Hydropluviometric correlation results

The average annual rainfall calculated using the inverse square of distance method is used because, according to Vauchel (2000–2004), it gives better results, especially in the case where rainfall stations are not well distributed in space. The results of the correlation indicate a good relationship between the average annual rainfall and the runoff. The determination coefficients (r^2) of 0.687 (Gaoual), 0.635 (Cade) and 0.803 (Tche-Tche) show that the runoff is function of rainfall (Fig. 9). But it is noted that some of the average rainfall will be used to reconstitute the water reserves while the other feeds the evapotranspiration. The correlation coefficients (r) are 0.829 at Gaoual, 0.797 at Cade and 0.896 at Tche-Tche respectively. The quality of the relationship allows to extend the timeseries of the runoff, which are then translated into annual modules. The Gaoual modules is extended from 1971–1972 to 2014–2015 and those of Cade and Tche-Tche from 1976–1977 to 2014–2015. Figure 9 shows the variation compared between the annual modules observed and calculated.

7 Modeling results

7.1 Model performance in calibration and validation

The results in Table 5 show the model parameters that have been optimized and the efficiency criteria obtained after the use in the "Solver" function. The values of the X_1 parameters, which are larger, vary from 7.55 to 7.6 while those of X_2 range from 1.06 to 1.13. The values of the Nash criteria are sufficiently high in general for the Gaoual (91.6 %), Cade (96.1 %) and Tche-Tche (95.7 %) basins with correlation coefficients of 0.934, 0.944 and 0.973 respectively.

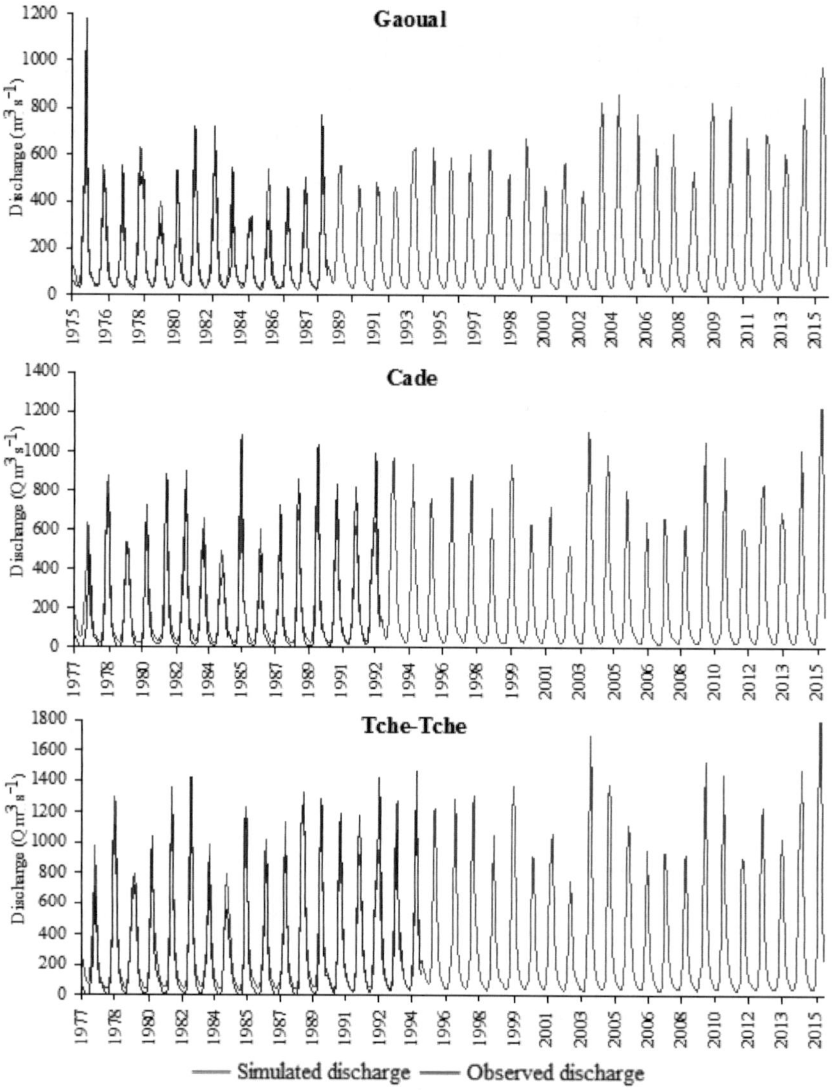

Figure 10. Comparison of observed and simulated monthly discharge.

Table 5. Calibration and validation of results.

Station	Calibration period	Parameter X_1	Parameter X_2	Efficiency criterion
Gaoual	1975–1988	7.6	1.10	91.6
Cade	1977–1992	7.58	1.06	96.1
Tche-Tche	1977–1994	7.55	1.13	96.3

7.2 Simulation of monthly discharge

The model's performance being established, the average monthly rainfall obtained by the inverse square distance method and the PET allowed to simulate the monthly discharges from Koliba/Corubal to Gaoual (1975–2015), Cade and Tche-Tche (1977–2015). The comparison of the observed and simulated discharge allows appreciating the quality of calculated discharge (Ardoin et al., 2002). A first qualitative analysis shows a good reproduction of the observed discharge shape at the three stations (Fig. 10). There is a temporal concordance of the peak discharge with sometimes an underestimation of the simulated discharge for some years with strong runoff. This is observed in 1975, 1981–1982 and 1988 in Gaoual, 1985, 1989 and 1992 in Cade, 1981–1982, 1992 and 1994 in Tche-Tche. Low water levels are more or less well reproduced at Gaoual with the exception of the hydrological years 1983–1984 and 1984–1985, where a slight shift is observed during the dry period. In Cade and Tche-Tche, they are overestimated respectively until 1990 and 1989 before being superimposed with observed discharge.

Considering the underestimation of discharge, a relationship between observed and simulated peak monthly flows is search. The correlation coefficients of 0.796 (Gaoual), 0.781 (Cade) and 0.703 (Tche-Tche) indicate a positive linear relationship between the maximum observed discharge and the maximum simulated discharge. Also, the calculation of the observed and the simulated volumes was carried out. At Gaoual, the correlation coefficient of 0.831 shows that they are close. In Cade and Tche-Tche, however, the correlation coefficient of 0.574 and 0.576 show a deterioration of the relationship. The simulated volumes are overestimated by the model.

In general, the simulated discharge from 1989 to 2015 at Gaoual seem to follow the logic of the recovery of rainfall with a clear improvement in flows in the early 2000s (Fig. 10). At Cade and The-Tche, a strong variability of the flows is observed. The model followed the signal from the dry period 1975–2002.

8 Conclusion and perspectives

This study of the hydropluviometric variability of the Koliba/Corubal River watershed highlights interesting results. The use of the regional vector method made it possible to homogenize the data and to calculate values that were used to reconstruct the gap data during some years. The Pettitt test detected breaks in 1958 (Mali), 1967 (Gabu, Gaoual) and 1969 (Koundara, Labe), over the period 1924–2015. The rainfall deficit of the period after break is between 9.7 % in Mali and 20.2 % in Gaoual. The segmentation method identifies a positive break in Gabu, Gaoual and Mali thus reflecting a return to better rainfall conditions, with however a strong interannual variation. The analysis of the MRC after the reconstitution of the missing values is in agreement with the resumption of the rainfall since the contribution of certain months to the annual total evolved greatly during the period post-break.

The hydropluviometric correlation approach gave coefficients of determination of 0.687 for Gaoual, 0.635 for Cade and 0.803 for Tche-Tche. This means that the runoff is function of the average rainfall. The part contributing to a reconstitution of the water reserves and the evapotranspiration not being quantified and considered, can constitute a source of discrepancies between the two variables. The correlation coefficients ranging from 0.797 (Cade) to 0.896 (Tche-Tche) encourages to extend the timeseries of runoff that are finally expressed in annual modules.

The use of the GR2M rain-flow model allowed us to first determine the model parameters that better reflect the transformation of the rain into flow with Nash criteria close to 100. Then, these parameters applied to the data of average rainfall and PET inputs simulated monthly discharge and finally extended the timeseries from Koliba/Corubal to Gaoual, Cade and Tche-Tche. The temporal evolution of the discharge permits to distinguish a phase with a weak flow, and another supported by the resumption of the rainfall.

These results constitute a first step in the knowledge of the Koliba/Corubal River flows with a global conceptual model. They can be useful to Guinea and Guinea-Bissau and especially to the Gambia River Basin Organization for the realization of common development projects and to ensure integrated and concerted transboundary management of resources. This study does not use Bantala, Komba, Tomine and Fefine tributaries due to discrepancies noted at the installed hydrometric stations in these locations. At the end of the analysis of the monthly discharge observed and calculated, other parameters such as river regime, floods, low flows, flow coefficients, can be studied. As perspectives, it would be interesting to extend the study area to the Saltinho station and to refine the flow knowledge with other data and techniques. Access to data at finer scales is difficult; the availability of satellite estimates can be an asset for the simulation of daily discharge which, however, requires historical knowledge that will serve as a reference for the validation of their quality.

Data availability. Data sets are available at

- National Meteorology Direction of Guinea (contact: meteogui@leland-gn.org),
- National Meteorological Service of Guinea-Bissau (contact: dgmeteobissau@yahoo.fr) and
- OMVG (contact: contact@pe-omvg.org).

Author contributions. SS and HD defined the title, realized the experimental activities, treated the hydrological data, analyzed all the results and wrote the initial manuscript. RND elaborated the mapping from physiographic data and treated the rainfall and evapotranspiration data. ID provided advice and translation of the manuscript into English. AK provided advice and discussions on the context of hydropluviometric variability in West Africa. All authors discussed the results and helped to improve the manuscript.

Competing interests. The authors declare that they have no conflict of interest.

Special issue statement. This article is part of the special issue "Hydrological processes and water security in a changing world". It is a result of the 8th Global FRIEND–Water Conference: Hydrological Processes and Water Security in a Changing World, Beijing, China, 6–9 November 2018.

Acknowledgements. We are grateful to our reviewers for their suggestions and comments. We also thank all the structures which provided us the data (National Meteorology Direction of Guinea, National Meteorological Service of Guinea-Bissau, and OMVG) and without which this work would not have been realized. We thank FRIEND-Water, UNESCO-IHP, IAHS, IWHR.

References

Ali, A. Lebel, T., and Amani, A.: Signification et usage de l'indice pluviométrique au Sahel, Sécheresse, 19, 227–235, 2008.

Ardoin, S. Dezetter, A. Servat, E., and Bocquillon, C.: Redéfinition des fonctions de production des modèles globaux de relation pluie-débit en milieu semi-aride africain. Regional Hydrology: Bridging the Gap between Research and Practice, IAHS Publ. no. 274, in: Proceedings of the Fourth International FRIEND Conference held at Cape Town, South Africa, 197–203, 2002.

Ardoin-Bardin, S.: Variabilité hydro-climatique et impacts sur les ressources en eau de grands bassins hydrographiques en zone soudano-sahélien, Thèse de doctorat, Université de Montpellier II, Montpellier, p. 440, 2004.

Bell, V. A., Gedney, N., Kay, A. L., Smith, R. N. B., Jones, R. G., and Moore, R. J.: Estimating Potential Evaporation from Vegetated Surfaces for Water Management Impact Assessments Using Climate Model Output, J. Hydrometeorol., 12, 1127–1136, 2011.

Bodian, A.: Approche par modélisation pluie-débit de la connaissance régionale de la ressource en eau: Application au haut bassin du fleuve Sénégal, Thèse de doctorat, Université Cheikh Anta Diop de Dakar, Dakar, p. 288, 2011.

Bodian, A.: Caractérisation de la variabilité temporelle récente des précipitations annuelles au Sénégal (Afrique de l'Ouest), Physio-Géo – Géographie Physique et Environnement, 8, 297–312, https://doi.org/10.4000/physio-geo.4243, 2014.

Bodian, A. Dezetter, A., and Dacosta, H.: Apport de la modélisation pluie-débit pour la connaissance de la ressource en eau: Application au haut bassin du fleuve Sénégal, Climatologie, 9, 109–125, 2012.

Brunet-Moret, Y.: Test d'homogénéité, Cahiers ORSTOM, série Hydrologie, XIV, 119–129, 1977.

Brunet-Moret, Y.: Homogénéisation des précipitations, Cahiers ORSTOM, série Hydrologie, XVI, 147–170, 1979.

Claude, J.: Données hydrologiques nécessaires pour une campagne de lutte contre l'onchocercose dans la région du bassin du fleuve Sénégal, Rapport préliminaire, OMS, ORSTOM, Centre ORSTOM de Ouagadougou, p. 57, 1980.

Dai, A., Trenberth, K. E., and Karl, T. R.: Global variations in droughts and wet spells: 1900–1995, Geophys. Res. Lett., 25, 3367–3370, 1998.

Descroix, L., Diongue Niang, A., Dacosta, H., Panthou, G., Quantin, G., and Diédhiou, A.: Evolution des pluies de cumul élevé et recrudescence des crues depuis 1951 dans le bassin du Niger Moyen (Sahel), Climatologie, 10, 37–49, 2013.

Descroix, L., Diongue Niang, A., Panthou, G., Bodian, A., Sané, Y., Dacosta, H., Malam Abdou, M., Vandervaere, J.-P., and Quantin, G.: Évolution récente de la pluviométrie en Afrique de l'Ouest à travers deux régions: La Sénégambie et le bassin du Niger Moyen, Climatologie, 12, 25–43, 2015.

Diello, P.: Interrelations Climat–Homme–Environnement dans le Sahel Burkinabé : impacts sur les états de surface et la modélisation hydrologique, Thèse de doctorat, Université de Montpellier II, Montpellier, p. 395, 2007.

Gogien, F.: Application d'un modèle pluie-débit à un bassin versant de Tunisie Centrale: essai de prise en compte de la variabilité spatio- temporelle de la pluie, Mémoire de DAA, LP-SNGR/ENSA de Rennes, Rennes, p. 119, 1998.

Haidu, I.: Comparaison entre tests statistiques concernant le changement climatique global, in: XIXe Colloque de l'Association Internationale de Climatologie, Actes du colloque d'Épernay, 6–9 September 2006, d'Épernay, 282–287, 2006.

Hiez, G.: Homogénéisation des données pluviométriques, Instructions de calculs pour la programmation, Éit. SUDENE-UNESCO, p. 19, 1969.

Hiez, G.: Homogénéisation des données pluviométriques, in: Cahiers ORSTOM, série Hydrologie, Vol. XIV, 1977.

Hiez, G., Rancan, L., Barros, M. C., and Pedrollo, O.: Le vecteur régional et les fluctuations climatiques, in: Actes des Deuxièmes journées hydrologiques de l'ORSTOM (Montpellier, 1986), Édit. ORSTOM, collection "Colloques et séminaires", Paris, 101–116, 1987.

Hiez, G. Cochonneau, G. Sechet, P., and Fernandez, U. M.: Application de la Méthode du Vecteur Régional à l'analyse de la pluviométrie annuelle du bassin amazonien, Veille climatique satellitaire, 43, 39–52, 1992.

Hubert, P. and Carbonnel, J. P.: Approche statistique de l'aridification de l'Afrique de l'Ouest, J. Hydrologie, 95, 165–183, 1987.

Hubert, P., Carbonnel, J. P., and Chaouche, A.: Segmentation des séries hydrométéorologiques : application à des séries des précipitations et de débits de l'Afrique de l'ouest, J. Hydrologie, 110, 349–367, 1989.

Hufty, A.: Orientations et vocabulaire de la climatologie (1988–1992), Publications de l'Association Internationale de Climatologie, 7, 15–23, 1994.

IRD: KhronoStat version 1.0, Logiciel, IRD, Montpellier, availabl at: http://www.hydrosciences.org/spip.php?article239 (last access: 12 March 2009), 1998.

Kouassi, A. M., Kouamé, K. F., Kofi, Y. B., Dje, K. B., Paturel, J. E., and Oularé, S.: Analyse de la variabilité climatique et de ses influences sur les régimes pluviométriques saisonniers en Afrique de l'Ouest: cas du bassin versant du N'zi (Bandama) en Côte d'Ivoire, Cybergeo: European J. of Geography, Environnement, Nature, Paysage, available at: http://cybergeo.revues.org/23388 (last access: 12 July 2018), 2010.

Lebel, T. and Ali, A.: Recent trends in the Central and Western Sahel rainfall regime (1990–2007), J. Hydrol., 375, 52–64, 2009.

L'Hôte, Y.: Contrôle de la qualité et de l'homogénéisation des données pluviométriques par la méthode du vecteur des indices annuels de précipitations, Zürcher Geographische Schriften, 23, 37–51, 1986.

Lubès, H., Masson, J. M., Servat, E., Paturel, J.-E., Kouamé, B., and Boyer, J. F.: Caractérisation de fluctuations dans une série chronologique par application de tests statistiques, Etude bibliographique, Programme ICCARE, Rapport no. 3, ORSTOM, Hydrologie, Montpellier, France, 1994.

Mahé, G., Paturel, J. E., Servat, E., Conway, D., and Dezetter, A.: Impact of land use change on soil water holding capacity and river modeling of the Nakambe River in Burkina-Faso, J. Hydrol., 300, 33–43, 2005.

Monteith, J.: Evaporation and environment, Symp. Soc. Exp. Biol., 19, 205–234, 1965.

Mouelhi, S.: Vers une chaîne cohérente de modèles pluie-débit conceptuels globaux aux pas de temps pluriannuel, annuel, mensuel et journalier, Thèse de Doctorat, ENGREF, Cemagref Antony, France, p. 323, 2003.

Mouelhi, S., Michel, C., Perrin, C., and Andréassian, V.: Stepwise development of a two-parameter monthly water balance model, J. Hydrol., 318, 200–214, https://doi.org/10.1016/j.jhydrol.2005.06.014, 2006.

Nash, J. E. and Sutcliffe, J. V.: River flow forecasting through conceptual models. Part I – A discussion of principles, J. Hydrol., 10, 282–290, 1970.

Niang, A. J.: Les processus morphodynamiques, indicateurs de l'état de la désertification dans le Sud-Ouest de la Mauritanie. Approche par analyse multisource, Thèse de Doctorat, Université de Liège, Liège, Belgique, p. 286, 2008.

Oudin, L.: Recherche d'un modèle d'évapotranspiration potentielle pertinent comme entrée d'un modèle pluie-débit global, Thèse de doctorat, ENGREF (AgroParisTech), Paris, p. 496, 2004.

Ouoba, A. P.: Changements climatiques, dynamique de la végétation et perception paysanne dans le Sahel burkinabè, Thèse de Doctorat Unique, Université de Ouagadougou, Ouagadougou, Burkina Faso, p. 305, 2013.

Panthou, G.: Analyse des extrêmes pluviométriques en Afrique de l'Ouest et de leur évolution au cours des 60 dernières années, Thèse de doctorat, UJF – Grenoble 1, Grenoble, p. 282, 2013.

Paturel, J. E., Servat, E., and Vassiliadis, A.: Sensitivity of conceptual rainfall-runoff algorithms to errors in input data – case of the GR2M model, J. Hydrol., 168, 11–125, 1995.

Paturel, J. E., Servat, É., Lubès-Niel, H., Kouamé, B., Travaglio, M., Lubès, H., Marieu, B., and Masson, J. M.: Variabilité des régimes pluviométriques et hydrologiques en cette fin de siècle en Afrique de l'Ouest et centrale non sahélienne, Sud-Sciences et Technologies, 2, 44–52, 1998.

Penman, H. L.: Natural Evaporation from open water, bare soil and grass, P. Roy. Soc. Lond. A, 193, 120–145, 1948.

Pettitt, A. N.: A non-parametric approach to the change-point problem, Appl. Stat., 28, 126–135, 1979.

Quentin, A.: Etude de la variabilité interannuelle de l'évaporation potentielle : conséquence sur les prévisions de l'impact hydrologique du changement climatique, Master 2, Sciences de l'Univers, Environnement, Ecologie, Université Pierre et Marie Curie, p. 50, 2015.

Sambou, S. Dacosta, H., and Paturel, J. E.: Variabilité spatiotemporelle des pluies de 1932 à 2014 dans le bassin versant du fleuve Kayanga/Gêba (République de Guinée, Sénégal, Guinée-Bissau), Physio-Géo, 12, 61–78, 2018.

Sarr, B. and Lona, I.: Les fortes pluies enregistrées au Sahel au cours de l'hivernage 2007: variabilité et/ou changement climatique, 14ème Colloque International, SIFEE "Changement climatique et évaluation environnementale" Outils et enjeux pour l'évaluation des impacts et l'élaboration des plans d'adaptation, Centre Régional Agrhymet, Département Formation et Recherche, Niamey, Niger, p. 9, 2009.

Sène, S. and Ozer, P.: Évolution pluviométrique et relation inondations-événements pluvieux au Sénégal, Bulletin de la Société Géographique de Liège, 42, 27–33, 2002.

Servat, É. Paturel, J. E., Kouamé, B., Travaglio, M., Ouedraogo, M., Boyer, J. F., Lubès-Niel, H., Fritsch, J. M., Masson, J. M., and Marieu, B.: Identification, caractérisation et conséquences d'une variabilité hydrologique en Afrique de l'Ouest et Centrale, in: Water Resources Variability in Africa during the XX thème 123 siècle: Century (Abidjan'98 Conference), Publication AISH no. 252, edited by: Servat É., Hugues, D. Fritsch, J. M., and Hulme, M., 323–337, 1998.

Servat, É. Paturel, J. E., Lubès-Niel, H., Kouamé, B., Masson, J. M., Travaglio, M., and Marieu, B.: De différents aspects de la variabilité de la pluviométrie en Afrique de l'Ouest et Centrale, Revue des Sciences de l'Eau, 12, 363–387, 1999.

Smith, M.: CROPWAT a computer program for irrigation planning and management, FAO Irrigation and Drainage, Paper no. 46, FAO, Rome, Italy, 1992.

SOFRECO: Etude du Schéma Directeur d'aménagement et de développement intégré des bassins des fleuves Kayanga/Gêba et Koliba/Corubal, Ressources et besoins en eau et option d'aménagement hydraulique et hydro-agricole, in: Vol. 3, Annexe B, Senegal, 1993.

Spearman, C.: The proof and measurement of association between two things, Am. J. Psychol., 15, 72–101, 1904.

Vauchel, P.: Logiciel Hydraccess. En ligne, availalb eat: http://www.ore-hybam.org/index.php/fre/Software/Hydraccess (last access: 23 June 2008), 2004.

Zoumanigui, K.: Base de données géoréférencées sur l'utilisation des terres en Guinée, Rapport final, FAO, Irag, p. 32, 2003.

Proc. IAHS, 383, 185–192, 2020
https://doi.org/10.5194/piahs-383-185-2020

Urban flood modelling in Qiqihar city based on MIKE flood

Jiahong Liu[1,2], Zejin Li[1], Weiwei Shao[1], Dianyi Yan[3], and Chao Mei[1]

[1]State Key Laboratory of Simulation and Regulation of Water Cycle in River Basin, China Institute of Water Resources and Hydropower Research, Beijing, 100038, China
[2]School of Transportation and Civil Engineering and Architecture, Foshan University, Guangdong, 528000, China
[3]Yangtze River Eco-Environment Engineering Research Cente, China Three Gorges Corporation, Beijing, 100038, China

Correspondence: Jiahong Liu (liujh@iwhr.com)

Published: 16 September 2020

Abstract. Qiqihar is a significant city on the Nen River in China, which is the main stream of the Songhua River basin. The length of the return period of Qiqihar's flood control design standard is fifty years. If a 100-year flood event happened, Qiqihar would face the risk of a burst levee. To quantitatively evaluate flood risk to the city from a burst levee or proactive flood diversion, a model for analysing flood submergence from a burst levee in the City of Qiqihar is established based on MIKE Flood. The model integrates one- and two-dimensional hydrodynamic models to implement coupled simulation. The terrain data are from city elevation data on a scale of 1 : 10 000. Following local modifications made based on survey data, such as on levees, roads, and buildings, a 20 m × 20 m grid of terrain data was formed as the terrain input of the model. The model simulates the water level of Nen River and the flood path, submerged time/depth/area, and duration in floodplain under three scenarios: baseline, proactive downstream flood diversion, and an upstream levee burst under a flood with a one hundred-year return period. Proactive downstream flood diversion can reduce the maximum water level by 0.068 m and correspondingly decrease peak flood flow by 1120 $m^3 s^{-1}$. These results provide basic information to support urban flood risk analysis and flood dispatching and management across the whole river basin.

1 Introduction

Flood protection levees are a vitally important hydraulic structure; if a levee bursts, it will bring immeasurable losses. The evolution of levee-bursting floods is an important factor in evaluating urban flooding risk. There are two main types of models that simulate urban flooding: static and dynamic models. Static models focus on statistical representation of simulation results while dynamic models focus on the changes that take place during the whole flooding period (Siddiqui et al., 2011). Two-dimensional hydrodynamic models are the most widely used type of model, are based on exact physical significance, and can dynamically predict and simulate the flood process (Shirvan et al., 2013). Hydrodynamic models include: the MIKE models developed by the Danish Hydraulic Institute (DHI) (Danish Hydraulic Institute, 2012a, b, c, d; Wang et al., 2017), the SWMM model developed by the USA EPA (Beck, 2016), and the LISFLOOD model developed by the European Commission Joint Research Centre (Van der Knijff et al., 2010), amongst others (Teng et al., 2017). Many researchers have applied numerical simulation techniques and proposed new perspectives and concepts (Soares-Frazão et al., 2008; Chen et al., 2009). Based on two-dimensional Saint-Venant equations, different discrete methods are used to obtain different calculation stabilities, accuracies, and applications. Among them, the finite volume method has been widely used due to its conservativeness and geometric flexibility (Caleffi et al., 2003). In addition, some researchers have drawn on the methods and techniques of other academic fields, such as neural networks (Kia et al., 2012) and cellular automata (Guidolin et al., 2016).

These methods are beneficial for decreasing calculation complexity and reducing calculation time. This paper dynamically couples the MIKE 11 HD and MIKE 21 HD models. The one- and two-dimensional models are linked through setting up a virtual river. Use of both models allows exploitation of their advantages whilst avoiding some problems – such as simulation efficiency and accuracy – that arise when the models are applied separately. Based on a 100 year return period designed flood event, the flooding process is simulated and analysed in Qiqihar City.

2 Methodology

The urban flood simulation system is established based on the MIKE 11 and MIKE 21 hydrodynamic (HD) models. The coupled model is called MIKE FLOOD and is used to simulate the inundation processes, duration, and inundation area within urban areas, which can provide decision support for flood management.

2.1 MIKE 11 hydrodynamic model

MIKE 11 HD is mainly used for flood forecasting, dispatching measures, canal/irrigation system design, and dispatch and storm surge research. The basic aim of MIKE 11 HD is to provide hydrological factors, such as river water level and discharge process of each river section. It is solved by one-dimensional unsteady flow Saint-Venant equations based on vertical integration. Water level and discharge at each grid point are calculated alternately by the Abbott-Ionescu six points implicit scheme (Fig. 1), and the discrete equations are solved using the chasing method.

$$\frac{1}{B}\frac{\partial Q}{\partial x}+\frac{\partial H}{\partial t}=q_{\mathrm{L}} \tag{1}$$

$$\frac{\partial u}{\partial t}+\frac{\partial u}{\partial x}+g\frac{\partial H}{\partial x}+g\frac{u\left|u\right|}{C^2R}=0 \tag{2}$$

where H is the water level of the channel section, Q is discharge, u is average velocity, B is the width of the river cross-section, q_{L} is lateral flow, R is hydraulic radius, C is the Chezy coefficient, and x and t are position and time coordinates.

2.2 MIKE 21 hydrodynamic model

MIKE 21 HD is mainly used to simulate two-dimensional free surface flow. This module can simulate changes in water level and discharge due to the impact of various forces in lakes, river mouths, and coastal areas. When the user provides terrain data, Manning coefficients, wind fields, and hydrodynamic boundary conditions, the model can simulate changes in water level and discharge in each grid. The model uses the finite difference method of alternating-difference implicit (ADI) second-term accuracy to solve the mass equation and momentum equation of dynamic flow. This model

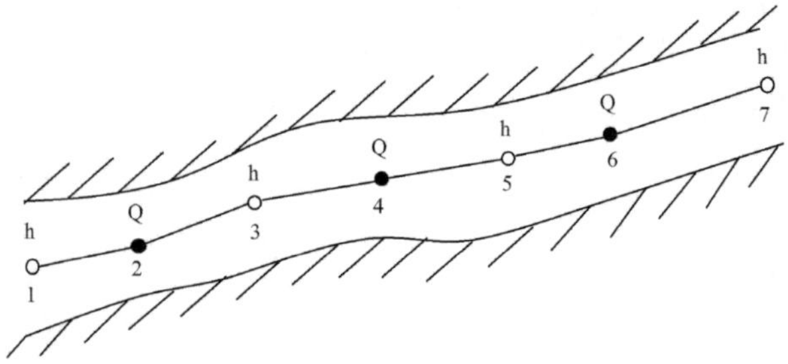

Figure 1. Water points and discharge points of the Abbott scheme.

can describe various hydraulic phenomena, such as tidal exchange, tidal flow, storm surge, whirlpools, surface undulation, dam bursts, and tsunamis. The equations take into account many factors, such as Coriolis force (Ω), wind force (V), and atmospheric pressure (P_{a}). The variables ζ, d, h, p, and q represent water level, water depth, elevation, and discharge in the x- and y-directions respectively.

$$\frac{\partial \zeta}{\partial t}+\frac{\partial p}{\partial x}+\frac{\partial q}{\partial y}=\frac{\partial d}{\partial t} \tag{3}$$

$$\begin{aligned}&\frac{\partial p}{\partial t}+\frac{\partial}{\partial x}\left(\frac{p^2}{h}\right)+\frac{\partial}{\partial y}\left(\frac{pq}{h}\right)+gh\frac{\partial \zeta}{\partial x}\\&+\frac{gp\sqrt{p^2+q^2}}{C^2h^2}-\frac{1}{\rho_w}\left[\frac{\partial}{\partial x}(h\tau_{xx})+\frac{\partial}{\partial y}(h\tau_{yy})\right]\\&-\Omega_q-fVV_x+\frac{h}{\rho_x}\frac{\partial}{\partial y}(p_{\mathrm{a}})=0\end{aligned} \tag{4}$$

$$\begin{aligned}&\frac{\partial q}{\partial t}+\frac{\partial}{\partial y}\left(\frac{q^2}{h}\right)+\frac{\partial}{\partial x}\left(\frac{pq}{h}\right)+gh\frac{\partial \zeta}{\partial y}\\&+\frac{gq\sqrt{p^2+q^2}}{C^2h^2}-\frac{1}{\rho_w}\left[\frac{\partial}{\partial y}(h\tau_{yy})+\frac{\partial}{\partial x}(h\tau_{xx})\right]\\&+\Omega_p-fVV_y+\frac{h}{\rho_x}\frac{\partial}{\partial xy}(p_{\mathrm{a}})=0\end{aligned} \tag{5}$$

2.3 MIKE Flood model

The MIKE Flood model dynamically couples the one-dimensional (MIKE 11 HD) and two-dimensional (MIKE 21 HD) models. The coupled model can take advantage of the benefits of both models, and can solve problems of spatial resolution and computation that often occur when the two models are used separately. In this study, the MIKE Flood model was used to simulate the two-dimensional flood evolution path, submerged area, and duration in urban areas as well as one-dimensional river flood overtopping and levee bursts.

MIKE Flood provides two different approaches to link one-dimensional and two-dimensional models, namely standard and lateral links, which are fitted for different occasions. An important part of MIKE Flood model application is link selection and creation.

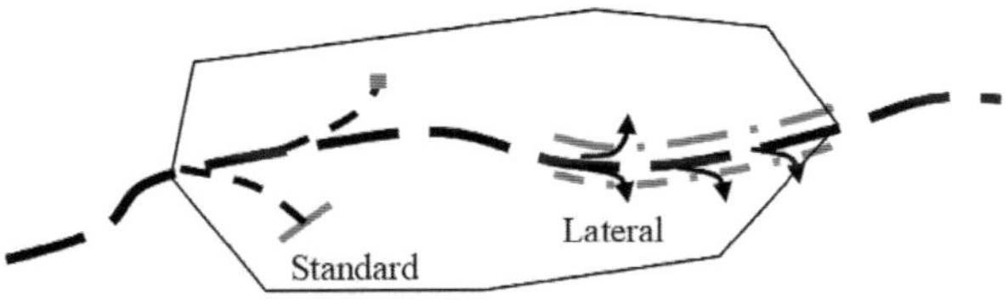

Figure 2. Standard and lateral links.

2.3.1 Standard links

Standard links in MIKE Flood correspond to one or more MIKE 21 grids linked to the beginning or end of MIKE 11 sections. This link can simulate submersion processes in urban inflow areas after levee bursts, by setting up a virtual river canal (Fig. 2). The section of MIKE 11 is connected to grids of MIKE 21 by Weir flow formula.

2.3.2 Lateral links

Lateral links are a series of grids linked with part or all of MIKE 11 side by side (Fig. 2). Flow through the lateral links is calculated using hydraulic structure formulae or water level-discharge relations. This type of link is particularly suitable for flood simulation from river to floodplain.

3 Datasets for urban flood modelling in Qiqihar

3.1 Terrain data

Basic terrain data of Qiqihar are available on a 1 : 10 000 scale, and are supplemented with road, house, and urban residence community datasets (Fig. 3). The datasets were converted into 20 m × 20 m grids using ArcGIS 9.3 software. The grid terrain data was input to the MIKE Flood model to simulate the urban submersion process in Qiqihar City.

3.2 Manning roughness coefficient data

Manning roughness coefficients (MRC) for urban terrain are discussed in DHI's global urban flood simulation projects in the past decades (Wang and Hartnack, 2006). Choice of different values of MRC will affect the flood peak time and to some extent the flow path within the urban area, but has little impact on the inundation area and flood depth (for a given design flood event, within the given flood duration the total breach volume could reduce as MRC increases). Roughness values are evaluated mainly on the basis of land use types. For instance, the roughness coefficient is set to 0.08 for developed areas, such as the leisure square with low buildings, gardens, and dense vegetation area; and in open space comprising grassland, sparse vegetation, streets, and pavements is set to 0.06. In similar projects in Germany, roughness coefficients in urban areas, streets, and gardens are set at 0.040–0.067 (considering the influence of pavements and vehicles parked on roads, the roughness coefficient of paths is larger); for vegetation are set at 0.05–0.2 (0.2 in dense young forest and 0.05 in open agricultural areas); and in other areas are set at 0.033–0.1, according to landscape features. Roughness coefficients in this study are not less than 0.033 (Fig. 4).

Based on the analysis above and field investigations, the principles of determining all-terrain roughness coefficients for Qiqihar are as follows:

1. Urban roads in the simulated city are generally wider, some main roads have central reservations, and some have green verges on both sides of the road. There are generally pavements on both sides and varying amounts of vehicles on the road at any one time. Therefore, the roughness coefficient of the main roads is set at 0.06.

2. Vegetation in parks is generally dense, so the roughness coefficient is set at 0.083.

3. In suburbs and other open spaces, there are no buildings and land cover is mainly grass or crops, so the roughness coefficient is set at 0.06.

4. Buildings usually interrupt water flow, which can store water inside. The roughness coefficient is set at 0.2 for buildings (houses and urban residence communities).

3.3 Hydrological data

Basic hydrological data were used for scenario simulation, including the once-in-a-century design flood process upstream of the Qiqihar hydrological station, the Q-H curve in the downstream section (Jiangqiao hydrological station) and design flood/peak/discharge in Nen River (Fig. 5).

4 Simulation scenarios

There are three simulation scenarios, which are shown in Table 1. In scenario 2 and 3, there are two burst points. The burst point of Daang embankment (Water Gate 2 in Fig. 8) is set for flood diversion, and the burst point at "7 km point of west embankment" (Water Gate 1 in Fig. 8) is set for flood damage simulation. The burst point of Daang embankment is select according to the Qiqihar City Flood Prevention Scheme. Because the outside of Daang embankment is a reserved flood storage area, and the main stream of Nenjiang River turns from the southward to the westward, which is the best point for flood diversion. The burst point at "7 km point of west embankment" is select by "the most unfavorable principle". In other words, if there is a collapse at this point, the city will suffer the greatest flooding. The "7 km point of west embankment" is located at the upstream of Qiqihar. The flood through this point will flood over throughout the whole city, thus it is the most appropriate point by "the most unfavorable principle".

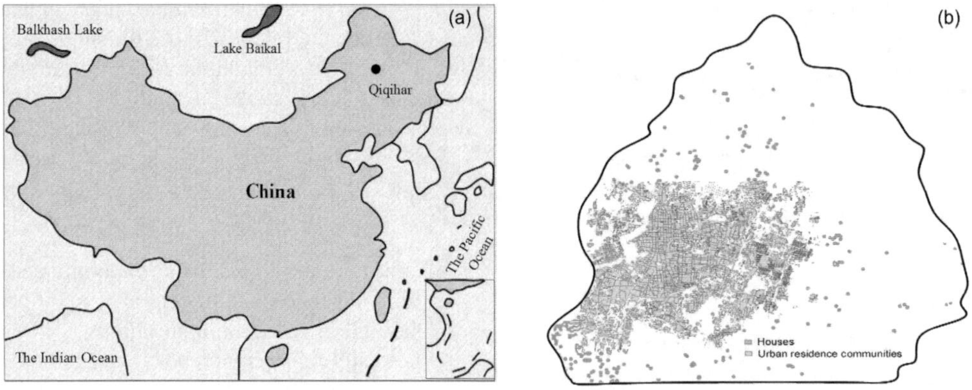

Figure 3. Location and residence data of Qiqihar (the black line around is MIKE Flood simulation grid boundary).

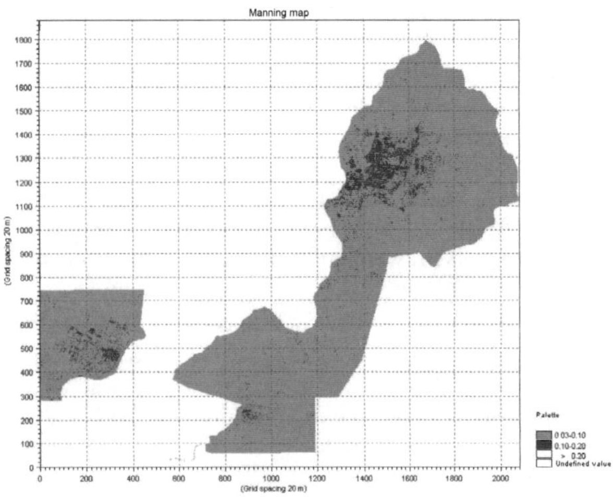

Figure 4. Distribution of Manning roughness coefficients.

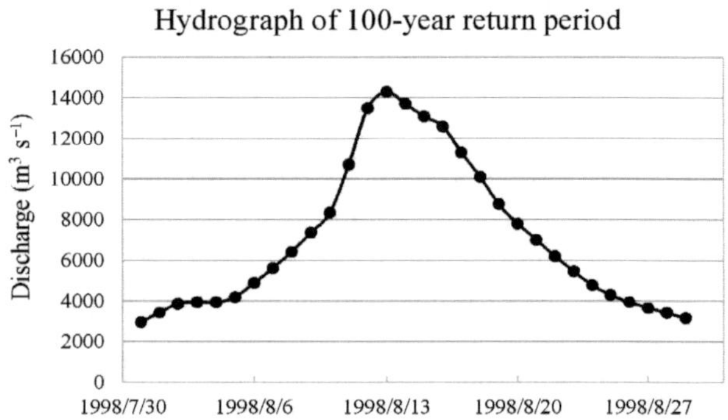

Figure 5. Design flood hydrograph of 100-year return period.

4.1 MIKE 11 HD basic parameters

Qiqihar City lies on the bank of the Nen River, which is upstream of Songhua River. The 1-D hydrodynamic model is established between CS31 and CS64, where the main stream of Nen River flows from 169 437 to 320 987 m (Fig. 6). The section position identification, such as 169 437 and 320 987 m, is a relative distance. The start point is at the dam of Nierji reservoir, which is a large and very important reservoir in the upper reaches of the Nenjiang River.

The upstream boundary condition of the model is the discharge of a 100-year return flood that occurred in August 1998. The downstream boundary condition of the model is the Q-H curve at Jiangqiao hydrological station in section CS64, which is more than 150 km from Qiqihar City and has little effect on the upstream water level.

4.2 MIKE 21 HD basic parameters

After considering the topography and elevation of Qiqihar City, the real land elevation is defined as 165 m, which means regions where elevation is higher than 165 m are not included in the model. In order to satisfy the requirements of lateral links, the river is filled with land (because the river is calculated by a one-dimensional model and is not included in the two-dimensional simulation process).

The Nen River runs through the whole of Qiqihar City and divides the city into the main urban and Hulan Ergi regions. The model uses a 20 m × 20 m grid and the total calculated areas are $2081 \times 1881 \times 400\,\mathrm{m}^2 = 1565.7\,\mathrm{km}^2$. The actual calculation areas cover about $430\,\mathrm{km}^2$, after removing the areas outside the boundary and inside the river, and urban areas that are not involved in simulation, as shown in Fig. 7. The boundary is set along the levee in the research area. The levee is surveyed and mapped by Songliao Water Conservancy Commission and provided in shape file of ArcGIS.

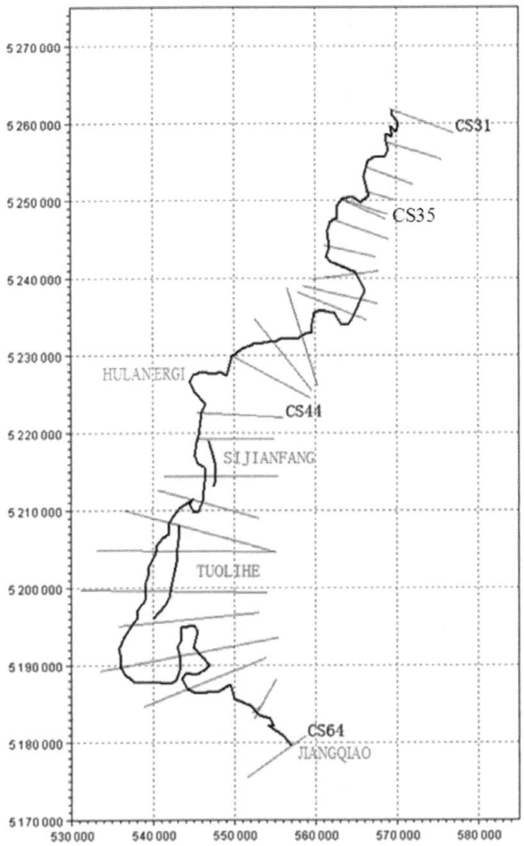

Figure 6. Main sections of one-dimensional model.

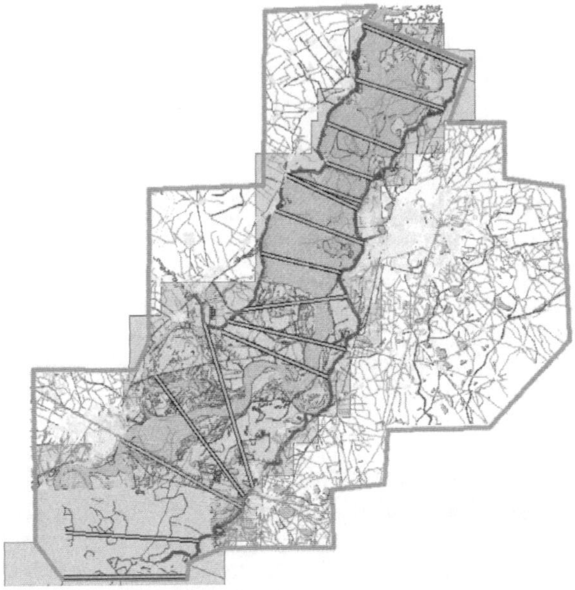

Figure 7. Simulation area of two-dimensional model in Qiqihar.

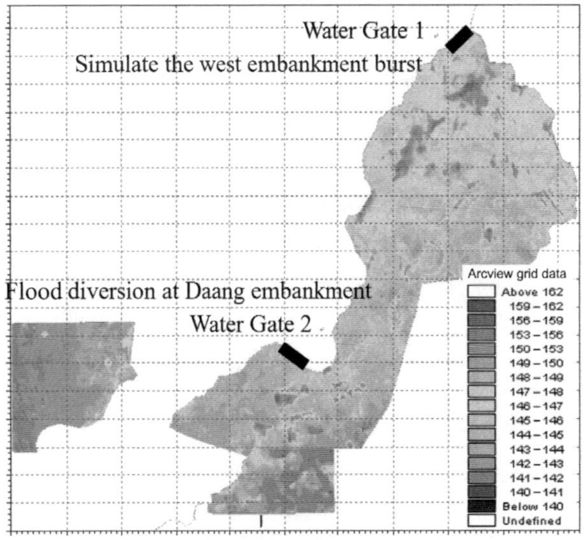

Figure 8. Simulation area of MIKE Flood and setting model links.

4.3 MIKE Flood basic parameters

The complete design flood process over one month is simulated, with a time step of 3 s. In order to simulate the defence level of the Qiqihar City levee in resisting the Nen River flood and to analyse locations and times where dangerous overtopping in levees may occur, all urban levees are linked with the main stream of Nen River by lateral links.

4.4 Levee burst simulation parameters

The simulation of urban levee burst outfall is mainly used to analyse the urban inundation range, duration, and water depth distribution after the most dangerous levee section is broken. As there is no module in the MIKE Flood model designed especially for levee bursts, a water gate is used to simulate the burst process. When the water level in the river rises to the levee burst control threshold, the gate is opened, which corresponds to the levee having burst. The gate will not close until the flood is over. There are two water gates in this simulation, one to simulate the bursting of the west embankment in Qiqihar, and the other to simulate the flood diversion at Daang embankment (Fig. 8). MIKE 11 HD and MIKE 21 HD are linked by adding the virtual river at the two water gates.

5 Results

Three scenarios were simulated (Table 1). The first is a baseline (normal) for comparison. The second simulates the flood control effect of proactive downstream flood diversion. The third simulates flooding processes when the west embankment of Qiqihar bursts.

Table 1. Simulation scenarios for urban flooding in Qiqihar.

No.	Flood Scheme (return period)	Flood Control Measures	Levee Bursts
1	100-year	None	No
2	100-year	Flood diversion at Daang embankment	No
3	100-year	None	Bursts at 7 km point of west embankment

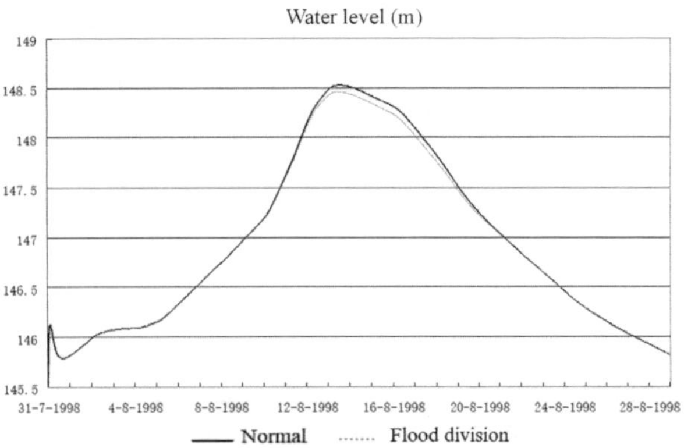

Figure 9. Water level at Qiqihar hydrological station ($P = 1\,\%$).

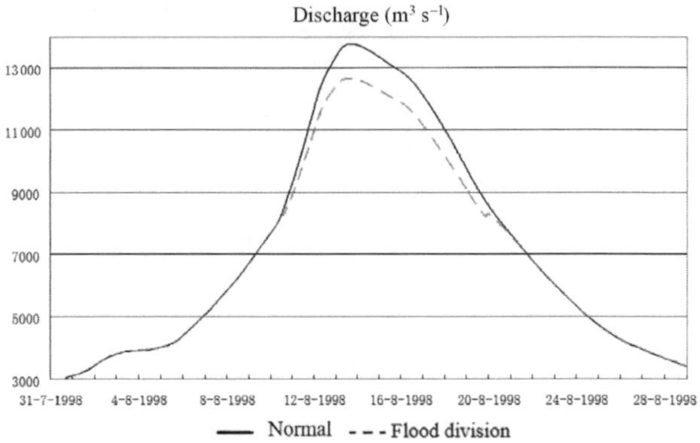

Figure 10. Discharge at Qiqihar hydrologic station ($P = 1\,\%$).

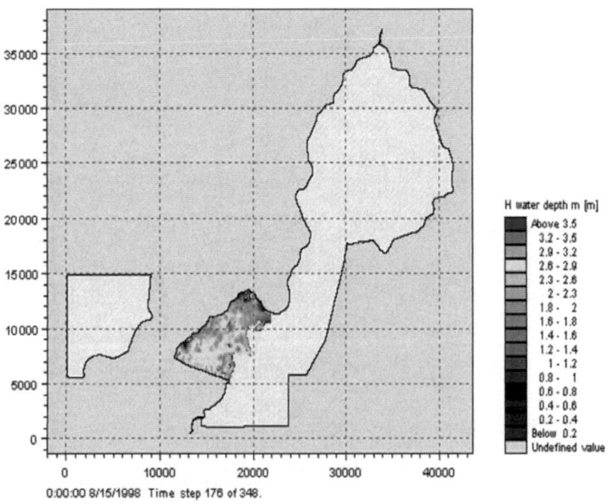

Figure 11. Submerged area in scenario 2.

In scenario 2, the highest water level decreases by 0.068 m and peak flow discharge decreases by $1120\,m^3\,s^{-1}$ after flood diversion. The water level and discharge at Qiqihar hydrological station (CS35) are shown in Figs. 9 and 10. The submerged area reaches a maximum 5 d after implementation of flood diversion measures. The maximum submerged area is about $50\,km^2$, and the maximum submerged depth is 3.5 m, as shown in Fig. 11.

In scenario 3, it is assumed that the west embankment of Qiqihar bursts when the water level reaches the threshold of a 30-year return flood. The simulation results show that the discharge when the levee is broken is $328\,m^3\,s^{-1}$ and the water level decreases by 0.05 m after the levee bursts. About 7 d after the burst, the submerged area reaches a maximum (nearly $100\,km^2$), or about 1/4 of the calculation area. The maximum depth of submersion is 3.2 m. Figure 12 shows the submerged area and depth 7 d after the burst.

6 Discussion and conclusion

This paper has established an urban flood model based on MIKE Flood, which integrates MIKE 11 HD and MIKE 21 HD models. MIKE Flood software is suitable for urban flood simulation, but still presents some problems, such as computational stability and the simulation of water inflow in buildings. Since the urban flood model simulates the process of levee bursts, flow discharge changes suddenly and rapidly and the time step must be short in order to ensure stability and accuracy of the simulation. For the two-dimensional model, a time step of 3 s significantly decreases the calculation speed of the urban inflow process. With an increase in grid points, simulation time increases obviously. However, if the variable time step method is used, results of the two-dimensional model must be further processed following simulation.

The model simulates the flood inflow of buildings by increasing roughness in building areas, which reflects the blocking and backing effect of buildings to water. However, this method is subjective and cannot reflect actual building inflow processes. Furthermore, the roughness coefficients are sensitive to discharge and velocity calculation, which may introduce random errors.

This study has simulated three scenarios of urban flooding in Qiqihar, despite the above problems. The result is reason-

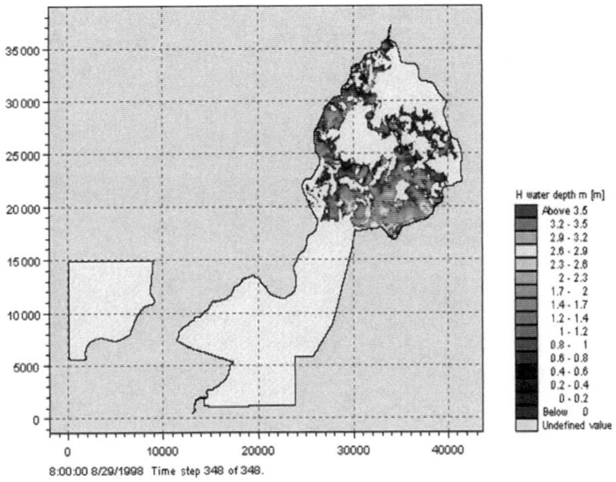

Figure 12. Submerged area in scenario 3.

able and provides key technical support for urban flood control in Qiqihar. Comparing the proactive flood diversion scenario and upstream levee burst scenario, the former has less submerged area and a better effect on flood control. In addition, the submerged areas in the former scenario are mainly in rural regions, reducing losses caused by submersion. Therefore, proactive flood diversion is an effective measure to reduce losses resulting from urban floods.

Data availability. Cross-sections geometry from Songhua River Project (2006), DEM data available on http://gdex.cr.usgs.gov/gdex/, last access: 13 June 2018 and Digital orthophotography purchased from China Centre for Resources Satellite and Application (http://www.cresda.com/CN/, last access: 20 April 2019).

Author contributions. Methodology, JL and WS; datasets preparation, DY and CM; software and simulation, ZL and CM; writing, JL and ZL.

Competing interests. The authors declare that they have no conflict of interest.

Special issue statement. This article is part of the special issue "Hydrological processes and water security in a changing world". It is a result of the 8th Global FRIEND–Water Conference: Hydrological Processes and Water Security in a Changing World, Beijing, China, 6–9 November 2018.

Acknowledgements. This study was supported by the Chinese National Natural Science Foundation (no. 51739011), the National Key Research and Development Program of China (2016YFC0401401 and 2018YFC1508203), and the Research Fund of the State Key Laboratory of Simulation and Regulation of Water Cycle in River Basin, China Institute of Water Resources and Hydropower Research (SKL2020ZY03).

Financial support. This research has been supported by the National Natural Science Foundation of China and Research Fund of the State Key Laboratory (grant nos. 51739011 and SKL2020ZY03). It is also supported by the National Key Research and Development Program of China (grant nos. 2016YFC0401401 and 2018YFC1508203).

References

Beck, J.: Comparison of three methodologies for quasi-2D river flood modeling with SWMM5, J. Water Manag. Model., C402, https://doi.org/10.14796/JWMM.C402, 2016.

Caleffi, V., Valiani, A., and Zanni, A.: Finite volume method for simulating extreme flood events in natural channels, J. Hydraul. Res., 41, 167–177, https://doi.org/10.1080/00221680309499959, 2003.

Chen, J., Hill, A. A., and Urbano, L. D.: A GIS-based model for urban flood inundation, J. Hydrol., 373, 184–192, https://doi.org/10.1016/j.jhydrol.2009.04.021, 2009.

Danish Hydraulic Institute: MIKE 11 User Manual, Danish Hydraulic Institute, Copenhagen, 2012a.

Danish Hydraulic Institute: MIKE 21 User Manual, Danish Hydraulic Institute, Copenhagen, 2012b.

Danish Hydraulic Institute: MIKE Flood User Manual, Danish Hydraulic Institute, Copenhagen, 2012c.

Danish Hydraulic Institute: MIKE Zero User Manual, Danish Hydraulic Institute, Copenhagen, 2012d.

Guidolin, M., Chen, A. S., Ghimire, B., Keedwell, E. C., Djordjević, S., and Savić, D. A.: A weighted cellular automata 2D inundation model for rapid flood analysis, Environ. Model. Softw., 84, 378–394, https://doi.org/10.1016/j.envsoft.2016.07.008, 2016.

Kia, M. B., Pirasteh, S., Pradhan, B., Mahmud, A. R., and Sulaiman, W. N. A.: An artificial neural network model for flood simulation using GIS: Johor River Basin, Malaysia, Environ. Earth Sci., 67, 251–264, https://doi.org/10.1007/s12665-011-1504-z, 2012.

Shirvan, M. K. M., Alamatian, E., Bafti, F. G., and Kheradmand, Y.: Flow simulation of dam break and determining flooding zones using MIKE 21, Ecol. Environ. Conserv., 19, 679–684, 2013.

Siddiqui, Q. T. M., Hashmi, H. N., Ghumman, A. R., and Mughal, H. R.: Flood inundation modeling for a watershed in the Pothowar region of Pakistan, Arab. J. Sci. Eng., 36, 1203–1220, 2011.

Soares-Frazão, S., Lhomme, J., Guinot, V., and Zech, Y., Two-dimensional shallow-water model with porosity for urban flood modelling, J. Hydraul. Res., 46, 45–64, https://doi.org/10.1080/00221686.2008.9521842, 2008.

Teng, J., Jakeman, A. J., Vaze, J., Croke, B. F. W., Dutta, D., and Kim, S.: Flood inundation modelling: A review of methods, recent advances and uncertainty analysis, Environ. Model. Softw., 90, 201–216, https://doi.org/10.1016/j.envsoft.2017.01.006, 2017.

Van der Knijff, J. M., Younis, J., and de Roo, A. P. J.: LISFLOOD: a GIS-based distributed model for river basin scale water balance and flood simulation, Int. J. Geograph. Inform. Sci., 24, 189–212, https://doi.org/10.1080/13658810802549154, 2010.

Wang, F. and Hartnack, J. N.: Simulation of flood inundation in Jilin City, Songhua River Project, City, 10, 18–31, 2006.

Wang, X., Wang, W., and Huang, G.: Simulation research of urban dam break flood based on MIKE Flood model, Hydrosci. Eng., 5, 67–73, 2017 (in Chinese).

Proc. IAHS, 383, 193–199, 2020
https://doi.org/10.5194/piahs-383-193-2020

Evolution of Urban Flooding in China

Dianyi Yan[1,2], Jiahong Liu[2,3,4], Weiwei Shao[2,4], and Chao Mei[2]

[1]Yangtze River Eco-Environment Engineering Research Center, China Three Gorges Corporation, Beijing 100038, China
[2]State Key Laboratory of Simulation and Regulation of Water Cycle in River Basin, China Institute of Water Resources and Hydropower Research, Beijing 100038, China
[3]School of Transportation and Civil Engineering & Architecture, Foshan University, Guangdong, 528000, China
[4]Engineering and Technology Research Center for Water resources and Hydroecology of the Ministry of Water Resources, Beijing 100038, China

Correspondence: Jiahong Liu (liujh@iwhr.com)

Published: 16 September 2020

Abstract. With the rapid development of cities in China, urban flooding becomes a growing problem, which not only affect the living quality of the public, but also threatens urban safety. Recently, urban flooding is attached more attention. By analyzing and summarizing urban flooding data from recent decades, which includes frequency, scope of influence, and losses, the causes and status quo of urban flooding problems in China is generalized from varied perspectives of urban climate change, urban planning, urban construction, urban management and policy. The data shows that the number of cities suffered from urban flooding increased by approximately 30 % since 1980s, while the average frequency of urban flood increased from nearly 1 time in every 2 years to more than 3 times per year in some major cities that face serious urban flooding risks. Even for those cities located in west-northern and semi-arid regions, such as Taiyuan in Shanxi province, they also suffered from urban flooding. Prevention and control measures like the constriction of sponge city, the upgrade of drainage system, and the promotion of relative policies are suggested.

1 Introduction

Throughout history, flooding took place frequently in China. As recorded, 1092 times of flooding has happened in China from 206 BC to 1949 (Zhang et al., 1992). Among all the flooding events, urban flooding, which is a phenomenon occurs when the amount of rain precipitated is more than the drainage capacity of cities (Konrad, 2016), becomes a kind of disasters that affect cities in China the most. With the rapid development of these cities, urban flooding becomes a growing problem. Recently, urban flooding has occurred in more than 60 % of cities in China; approximately 140 cities were flooded for more than 3 times between 2008 and 2010, which reflects the problems of outdated infrastructure construction and incomplete city functions (Huang, 2013; Zhang et al., 2016). Begin from 2000, 200 times of urban flooding happens in China every year in average (Zhang et al., 1992), which severely influences the living quality of the public and causes loss of life and property; in addition, it also threatens urban safety. The status quo of urban flooding in China is that heavy rainstorm causes severe stagnant water on the road, and the major problems it generates include washing away buildings and public utilities, leading to traffic snarls, flooding places like parking lots and subway stations, causing cable breakdown for water supply, electricity and telecommunication, and creating social panic. As a result, more attention is attached to urban flooding in China now.

2 Facts of flooding events in China

Before 2006, there were no clear record for urban flooding in China. On the contrary, some data of flood was collected carefully. Based on the "Bulletin of Flood and Drought Disaster in China (2007–2016)" (MWR, 2007–2016), the scope

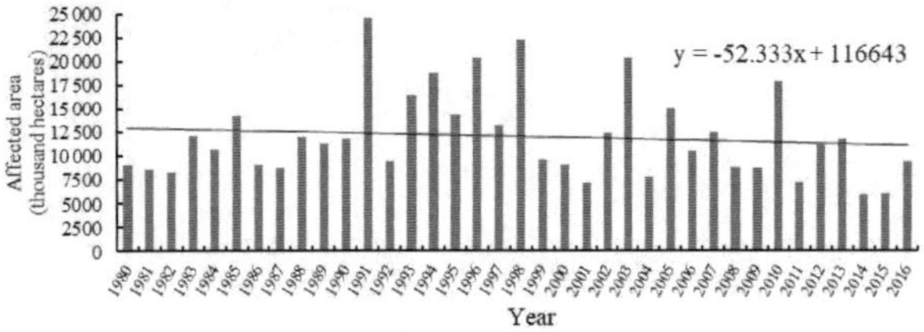

Figure 1. Data of flood-affected area from 1980–2016 (MWR, 2007–2016).

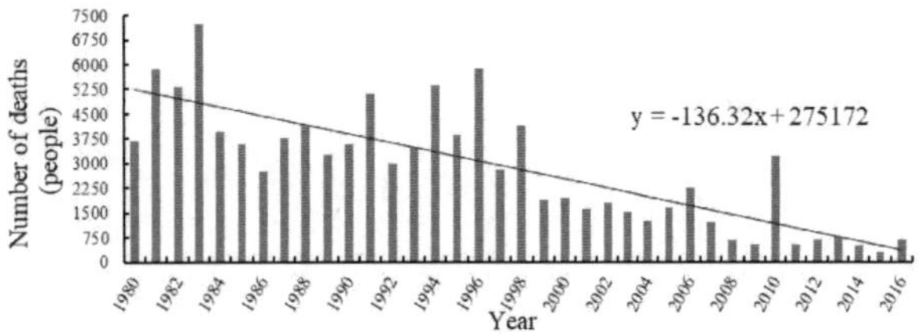

Figure 2. Number of deaths caused by flood from 1980–2016 (MWR, 2007–2016).

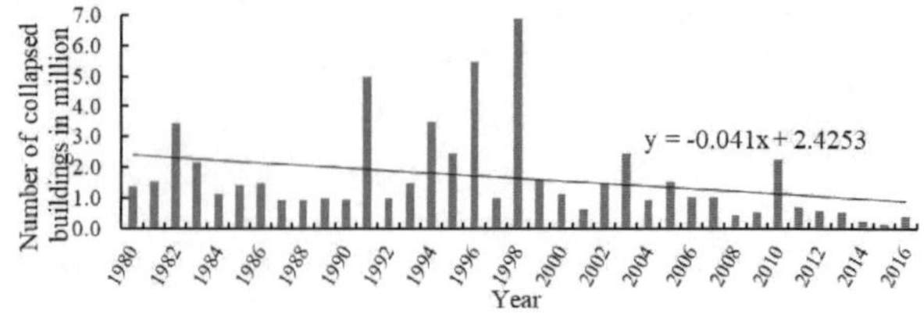

Figure 3. Number of collapsed buildings after flood from 1980–2016 (MWR, 2007–2016).

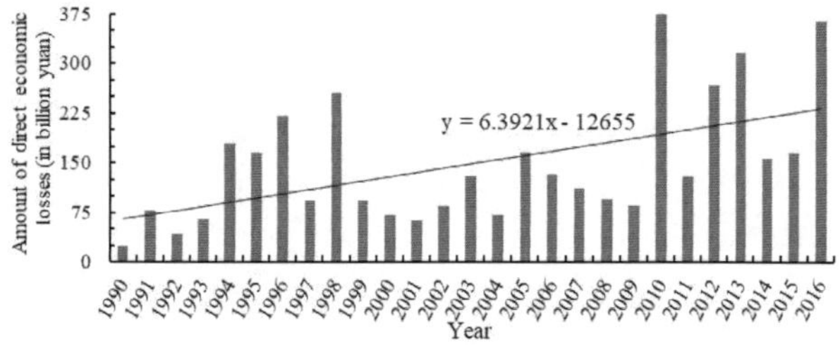

Figure 4. Direct economic losses caused by flood from 1990–2016 (MWR, 2007–2016).

of flood, which includes areas affected by flood and the number of deaths, and losses caused by flood, which includes the number of collapsed buildings and direct economic losses in China from 1980 to 2016 which can be summarized as follows.

The average area that affected by flood in 1980s was approximately 10.4 million hectares, and the average area that affected by flood from 2010 to 2016 was around 9.9 million hectares. By adding a trendline on the graph (Fig. 1), the area affected by flood in China turns to slowly decrease from 1980s to 2016.

As shown in Fig. 2, it is obvious that the number of deaths decreased relatively rapidly from 1980s to 2016. In average, there are about 4349 people died per year because of flood in 1980s, and 954 people died per year from 2010 to 2016, which can be a strong evidence to show the successful development of economic society in China because the government has strove to create a much safer living environment for the public.

The change of the number of collapsed buildings (Fig. 3) is similar as the change of the number of deaths (Fig. 2). The number of collapsed buildings caused by flooding was reduced from 1980s to 2016. The average numbers of collapsed buildings in 1980s and from 2010 to 2016 are 154 and 70 buildings, which strongly proves that the construction quality in China has soared.

Due to the information gap before 1990, only the data of direct economic losses caused by flood from 1990 to 2016 are in record. The trend of the change of direct economic losses caused by flood in these 27 years keeps raising. Nowadays, economy is highly concentrated in cities. Almost half of these economic losses were contributed by cities. It was proved that the frequency of urban flooding increased from nearly 1 times in every 2 years to more than 3 times per year in some major cities (Z. Zhang et al., 1992; J. Zhang et al., 2016).

To explore the potential relationship, the data of flood-affected area is adjusted by a factor of 45, the data of deaths and direct economic losses caused by flood is adjusted by a factor of 10. As shown in Fig. 5, begin from 1990, the trend of the change of flood-affected area, deaths, collapsed buildings, and direct economic losses caused by flood were similar, which indicates that population and economics in China are highly concentrate in places vulnerable to floods, and there was a strong chain reaction between factors displayed in Fig. 5.

According to the collection of flood and drought distribution data from 1980 to 2000 (Zhang, 2003), the number of cities that are potential to be flooded is summarized as shown in Fig. 5, which covered the whole range of China and included data collected from 120 sites, but it was not updated after 2000.

Due to the lack of data collection of exact number of cities where urban flooding has happened in 1980s, the trend of increasing number of flooded cities can be roughly estimate by data displayed in Fig. 6. By applying a trendline, the average number of potential flooded cities in 1980s is calculated as 37, and the average number of potential flooded cities from 2010 to 2017 is calculated as 48, which indicates that the number of potential flooded cities increased by approximately 30 % from 1980s to now.

Begin from 2006, the Ministry of Water Resources of the People's Republic of China (MWR) began to collect the

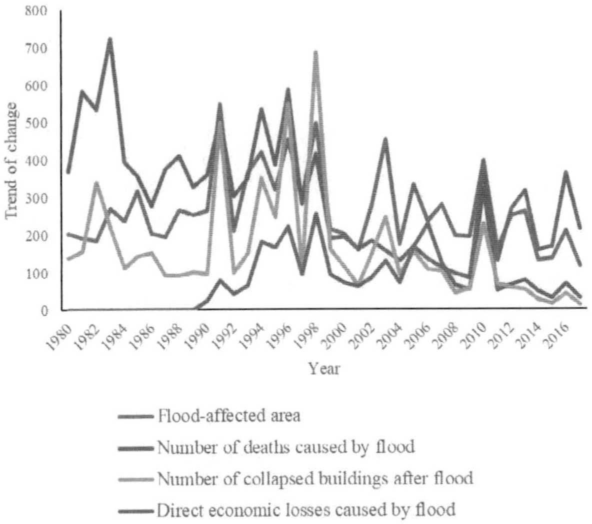

Figure 5. Trend of change (MWR, 2007–2016).

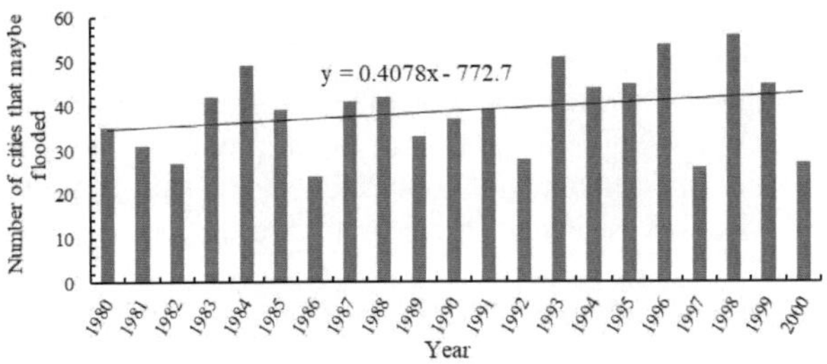

Figure 6. Number of cities that have potential to be flooded (Zhang, 2003).

number of cities that were flooded in each year. Data from 2006 to 2016 is shown in Table 1. There are over 650 cities in China. As recorded by MWR, approximately 25 % of cities in China were flooded in each year from 2006 to 2016. Even for those cities located in west-northern and semi-arid regions, such as Taiyuan in Shanxi province, they also suffered from urban flooding. Thus, urban flooding is become as one of the most urgent problems in China right now.

3 Causes of urban flooding in China

With further development of China's reform and opening up, the urbanization rate in China changed from 10.64 % in 1949 to 57.4 % in 2016, and it was predicted to be 75 % in 2050 (National Bureau of Statistics of the People's Republic of China, 2017). Although the rapid development of cities brings people a better living environment, it also created some problems that affect the living quality of people. One of the problems that people in China have paid more attention on is urban flooding, and this problem is caused by

Table 1. Data of Flooded Cities from 2006 to 2016 in China (MWR, 2007–2016).

Year	Number of Flooded City*
2006	118
2007	109
2008	115
2009	136
2010	258
2011	136
2012	184
2013	243
2014	125
2015	168
2016	192

* It includes county-level cities, prefecture-level cities, and municipalities.

factors in five major aspects, which are geographic position and urban climate, urban planning, urban construction, urban management, and policy.

3.1 Geographic Position and Urban Climate

3.1.1 Geographic Position (Zhang et al., 1992)

China is located in the east coast of the Eurasia, so monsoon circulation has a great influence on the climate in the country. Most of the major cities in China are seated close to river bank, estuarine, or coastal, and some of the other cities are distributed in places like intermountain basin. Thus, once there were floods, these cities are easy to be drowned, and the urban order would be easy to be disrupted.

3.1.2 Urban Climate (Zhang et al., 2016)

Most of buildings in cities are made from reinforced concrete. Those buildings have a specific heat that is much lower than greenbelts, which means that greenbelts can absorb more heat than buildings. In the other words, buildings will increase temperature in cities more than greenbelts. In addition, buildings always block wind, which causes the temperature inside cities is always 1 to 3 °C higher than the temperature in rural area, which makes cities like an island that is full of heat. This phenomenon is named as urban heat island effect (EPA). Coupling with those air pollution particles, the amount of rainfall condensation nucleus is increased. As a result, the frequency of heavy rainstorm in cities significantly soars as cities develop faster, which creates another phenomenon called urban rain island effect. Take Shanghai as an example, it was proved that the amount of annual precipitation has increased by 5 % in Shanghai than before, and the total precipitation volume and frequency has raised more than 10 %. Combined with factors from urban planning, construction, management and policy aspects, urban flooding becomes a common urban disaster in China.

3.2 Urban Planning

One of the salient problems that China's urban planning has is that people focus more on what to do on the ground instead of caring about what happens under the ground. Fancy buildings and advanced infrastructure construction are visible, and it is easy for people to notice the benefits these facilities can bring to the society. However, on the contrary, the significance of the function of drainage system is always ignored because it is indiscernible. The design of drainage system in China's cities is lack of foresight, the capacity was designed mainly based on the average precipitation at the time when the system was built. In average, the flood control standard for drainage system in most of cities in China is aim to reduce runoff caused by urban flooding that may happen in every one to two years. On the contrary, cities in developed country like the United States, Japan, and France have a much higher flood control standard (Tencent News Encyclopaedia, 2016). The flood control standard for drainage system in New York is for preventing flood that may happen in every 10 to 15 years. As for Tokyo, the drainage system was designed for flood that may occur in every 5 to 10 years. As for Pairs, it was designed for flood that may take place in every 5 years. What is more, when roads in cities were upgraded, the increasing demand of drainage capacity was not fully considered, most of drainage facilities were not synchronously upgraded. Meanwhile, some original facilities even got damaged. Besides, aged pipe in the drainage system is another urgent problem (Yang, 2011). Due to the high frequency of heavy rainstorm, the amount of rain water goes into the drainage system is always exceed its capacity, which directly leads to severe stagnant water on the ground. When the situation gets worse, urban flooding occurs.

Except for problems raised by drainage system, the damage of natural environment is another factor that leads to urban flooding. In the past, people did not have the awareness of the importance of green environment. Ditches and ponds were used to play a significant role on water division and absorption; however, they were filled into roads or buildings. Along with the disappearance of ditches and ponds, cities become fragile in terms of floods. Combined with underdeveloped drainage system, negative effects brought by urban flooding becomes increasingly worse in China.

3.3 Urban Construction (Li, 2013)

With the acceleration of urbanization, the surface hardening rate raises, the area of impervious surface grows in places like rooftop and outdoor parking lots. These impervious surfaces, which were made from materials like cement or asphalt, prevent rainwater and snowmelt infiltrating into the ground. It directly increases the surface runoff coefficient (U.S. Geological Survey). The higher the surface runoff coefficient is, the less water can be absorbed by the ground. Furthermore, compared with greenbelts, the construction of roads and buildings significantly reduces the roughness coefficient of the ground, which shortens the concentration time of surface runoff (Zhang et al., 2016). In addition, some construction processes block brooks. Plenty of roads, railways, buildings even occupied lands that were used to be water bodies like rivers or lakes without considering sustainable development of cities. All these factors severely affect the drainage capacity in cities in China.

3.4 Urban Management (Li, 2013)

Urban management, especially for drainage management, is weak in most of cities in China. There is no sound urban flooding management department, and the existing method for management is relatively backward. What is more, there is no real-time monitoring system for urban flooding. Thus, people cannot receive precisely information when urban flooding occurs. Coupled with the lack of administrative staff for drainage facilities and weak management system, urban flooding cannot be timely controlled. Besides, the maintenance of drainage system is always not in time. Funding is not enough. Additionally, the drainage equipment is backward when compared with cities in some developed countries.

3.5 Policy (Liu, 2016)

In China, there is a regulation called "Land Finance", which allows local governments to sell land-use right for fiscal expenditure. This regulation accelerates the reduction of natural surface area with great flood discharging function. Those surface areas that is low-lying and easy to be waterlogged are especially cheap, the demolition of these areas is less difficult, and the cost recovery will be faster. When compared with the low price of these lands, the potential risk of urban flooding is ignored. Besides, the financial investment from the central government for drainage pipelines is insufficient. Approximately 90 % of the investment is from local government. As the local government's debts grows, it turns arduous for local government to invest more for urban flooding preventing work.

4 Measures to prevent future urban flooding in china

4.1 Reinforcement from Government

Drainage system is one of the critical indispensable infrastructures in cities. Nowadays, water surface ratio and seepage rate of grounds keep decreasing. Along with an increasing frequency of heavy rainstorm, the capacity of most build-up drainage systems in China cannot afford the actual amount of water input due to less governmental investment and aged pipelines (Jiang, 2011). Thus, both central government and local governments must reinforce the construction of drainage systems in cities and increase governmental input to reduce the frequency of urban flooding. Besides, existing urban flooding related standards and regulations are just on a technical level. When relative departments plan to supervise the planning and construction of infrastructures, it is lack of legal ground. Therefore, laws related to urban drainage and urban flooding prevention must be established based on China's national conditions. In the United States, some states announced an ordinance which regulates that the flow of rainwater in newly developed area must not exceed its original flow before this area was developed. It means that local governments must avoid the excessive flow of rainwater by developing and improving drainage facilities, and it also enforces governments to increase the financial input in this field (Li, 2013).

4.2 Improvement of Urban Planning for Newly Developed Area

The major reason why urban flooding recently turns much easier to occur in China is mainly because that nowadays most of cities are lack of resilience, which means that cities cannot automatically mitigate problems by themselves (Wang et al., 2015). During construction and development, cities lost their ecological functions because roads were covered by impervious material like cement or asphalt, and lakes and rivers were filled for farmland or buildings. Thus, as for future urban planning for newly develop area, ecological rehabilitation should be the priority. With outstanding ability to store floodwater and drain flooded fields, existing watercourse, ditches, and ponds should be protected instead of being filled for more ground area for urban construction. Large low-lying land could be developed as wetland park. In proper area, impounding reservoir could be established to optimal utilize underground space. When designing roads, hardening area should be reduced. Instead of applying cement or asphalt to cover the ground, pervious material must be applied to increase water and air permeability of the road and reduce stagnant water on the ground. Moreover, it is a must to expand green areas because they not only can beautify the environment but also can absorb and store stormwater to assist to prevent urban flooding (Yang, 2011; Liu, 2016). In addition, when preparing flood protection measures, cities' affordability for floods should be considered and studied for a more precise planning. Besides, although the drainage capacity depends on drainage system, but it is also affected by coordination work between planning department, department of water, department of municipal administration, and department of urban construction. Each involved department should establish a sound operating network. Instead of designing drainage system in cities in a district level, it should be taken into full account as a whole to avoid the unbalance between drainage capacity and actual amount of stormwater needed to be discharged into the system (Li, 2013). Most importantly, design return period of urban flooding must be expanded because most of drainage systems were design for flood that may happen only in every 20 years (Yang, 2011), which is extremely not enough for recent precipitation condition. As for cities where people are living now, problems left over by history, such as substandard drainage system, should be solved and improved properly.

Another cause of frequent urban flooding is due to the lack of financial support and attention for the construction of underground facilities. Therefore, when reforming or developing new areas for human activity, underground construction must be put in the first place.

4.3 The Do's and Don'ts for Urban Construction

Most of urban embankments in China have a long history with low quality, and their foundations were not properly treated when they were built (Zhang et al., 1992). Besides, the average flood control standard is relatively low in China. As a result, when floods take place, it is easy for river to burst its banks and cause severe urban flooding, so it is vital to strengthen the construction of floodwall and embankments. Additionally, the expansion of cities increases the risk of urban flooding. During urban construction, water permeability of underlying surface must be ensured, and the upgrade of existing drainage system in old neighbourhoods should be concurrently upgraded. Besides, underground space should be fully utilized. Reservoir could be a proper option to build under the ground. During construction process, permeable material should be fully applied. When choosing construction materials and plants for green area like gardens and wetlands, local precipitation condition need to be fully considered.

4.4 Improvement of Management and Early Warning System

To effectively remit the negative influences caused by urban flooding, drainage management must be taken seriously. As time goes by, those pipelines for drainage gradually become aged, and their capacity cannot meet the actual amount of rainwater or snowmelt that needs go into them. Thus, it is significant to renovate old pipelines, especially in those old districts. As for those intricate pipes, they must be located systematically and scientifically so that when problems, such as leakage, occur in these pipes, it is easy to timely find out where it happens and come up with proper solutions. Besides, when heavy rainstorm occurs, all the drainage system must be operated instantly to drain as much water as it can. What is more, regular clean-up must be done for drainage system to avoid unnecessary blocking, especially for the sewers, watercourse, and river channels. As for natural water bodies, it must be prohibited to fill them for more land to use or dump trash into them to en smooth drainage when heavy rainstorm occurs. Another important measure here is the setup of advanced early warning system. The government must attach importance to response mechanism (Jiang, 2011). Meteorological department should forecast and forewarn the happening of heavy rainstorm and instantly send feedback information to the other involved departments. Water department must activate necessary equipment, such as water suction pumps, for drainage system in time. Transportation department must be ready to disperse excess traffic with more traffic management coordinators. Public security department should take the responsibility to keep order in the streets to avoid over-crowdedness. As for those locations that is easy to be flooded, such as underground parking lots, drainage equipment like water suction pumps and sandbag should be prepared in advance. Moreover, before heavy rainstorm comes, water level in controllable rivers and canals should be lowered to assist to drain and store water (Yang, 2011). In addition, the corporation between government departments and social media should be strengthened. Real-time update of urban flooding condition in cities through text message, radio, and television should be promoted for the public.

After heavy rainstorm, all relevant government departments should check each drainage system and collect statistic data to summarize if there were anything need to be improved for a more effectively and quickly response to future urban flooding. As for places that suffered from severe flood, it should be insisted on the accountability within relative departments.

4.5 Construction of Sponge City (Citiscope, 2016; Go Chengdu, 2015)

In China's the 12th and 13th five-year plans, they both indicated that the construction of sponge city should play a significant role in China's future development. There are 30 pilot sponge cities in China right now. "Sponge city" means to reform cities as sponges to absorb and store water. The most important components in sponge cities are green buildings, permeable road surface, bio-retention facilities, wetlands, lakes, rivers, rain gardens, and forests. The construction of sponge city can promote the water permeability of roads, expand green areas, increase the utilization rate of underground space, reduce urban heat island effect to lower urban air temperature, raise the utilization rate and promote purification of rainwater, decrease the occurrence rate of urban flooding, and raise city's resilience (Liu, 2016). Due to the rainwater circulation system created by the construction of sponge city, when heavy rainstorm occurs, each component in a sponge city can help to reduce the amount of surface runoff, which can effectively prevent urban flooding.

4.6 Rising Awareness of Urban Flooding (Jiang, 2011)

Urban heat island effect is one of the major causes of heavy rainstorm. Artificial heat emission is the major factor that aggravates this effect. Thus, Citizens should raise the awareness of reducing artificial heat emission. Instead of driving personal cars, public transportation should be advocated to reach low-carbon emission. Besides, citizens' social sense should be raised as well. Everyone should contribute to the prevention of urban flooding.

4.7 Learning from Successful Foreign Experience

Developed countries, such as the United States, France, and Japan, are ahead of the curve in urban flooding prevention area. The United States have announced an ordinance which regulates that the flow of rainwater in newly developed area must not exceed its original flow before this area was developed. France has developed sophisticated large-scale drainage system and effective early warning system for flooding risks, which distinctly decreased the threat of urban flooding. Japan used 14 years, from 1992 to 2006, established the Metropolitan Area Outer Underground Discharge Channel, which is also known as "The Underground Temple". In this huge drainage system, 59 18-meter-long massive pillars hold a huge reservoir with a width of 78 m and a length of 177 m, and this system can pump up to 200 t of water into the Edo River per second for preventing urban flooding in Tokyo (Risa, 2017). China should learn from their experience to establish proper systems or laws to mitigate the negative influences brought by urban flooding to cities and the public.

5 Conclusions

Urban flooding has strongly affected people's livelihood and social development in China. Although the areas influenced by urban flooding, the number of deaths, and the number of collapsed buildings keeps decreasing from 1980s to now, the direct economic losses and the number of flooded cities grows. The number of cities where urban flooding has taken place is approximately 25 % of the total number of cities in China, so urban flooding is still an urgent problem needed to be solved. Close-to-water geographic position, warming urban climate, unbalanced urban planning and construction, unsubstantial urban management, and deficient policies are the major factors that cause frequently urban flooding in China. The most significant tasks are to upgrade old drainage systems and promote the construction of sponge city to increase cities' resilience to mitigate urban flooding in China. Besides, proper investment and policies are also required.

Data availability. Data analyzed in this article is from "Bulletin of Flood and Drought Disaster in China (2007–2016)", the data of the scope of flood, which includes areas affected by flood and the number of deaths, and losses caused by flood, which includes the number of collapsed buildings and direct economic losses in China from 1980 to 2016 was applied in the analysis. All the data are open accesses, and the URL (in Chinese) is http://www.mwr.gov.cn/sj/tjgb/zgshzhgb/, last access: 20 July 2017.

Author contributions. DY is the major author of this paper, this author contributed to conceive the whole article, collect and analyze data, and write the article. JL contributed significantly to analysis and manuscript preparation. WS contributes to perform the analysis with constructive discussions. Chao Mei contributes to embellish the article.

Competing interests. The authors declare that they have no conflict of interest.

Special issue statement. This article is part of the special issue "Hydrological processes and water security in a changing world". It is a result of the 8th Global FRIEND–Water Conference: Hydrological Processes and Water Security in a Changing World, Beijing, China, 6–9 November 2018.

Acknowledgements. This study was supported by the Chinese National Natural Science Foundation (nos. 51522907 and 51739011), the National Key Research and Development Program of China (2016YFC0401401), and the Research Fund of the State Key Laboratory of Simulation and Regulation of Water Cycle in River Basin, China Institute of Water Resources and Hydropower Research (no. 2017ZY02).

Financial support. This research has been supported by the Chinese National Natural Science Foundation (grant nos. 51522907 and 51739011).

References

Can "sponge cities" solve China's urban flooding problem?, Citiscope, available at: http://citiscope.org/story/2016/can-sponge-cities-solve-chinas-urban-flooding-problem, last access: 28 July 2016.

Chengdu plans to become "sponge city", Go Chengdu, available at: http://www.gochengdu.cn/news/Highlights/chengdu-plans-to-become-sponge-city--a1287.html, last access: 2 March 2015.

Effects of impervious surfaces on streamflow, U.S. Geological Survey, available at: https://water.usgs.gov/edu/impervious.html, last access: 2017.

Heat Is land Effect, U.S. Environmental Protection Agency, available at: https://www.epa.gov/heat-islands, last access: 2017.

Huang, J.: The influence of climate change like flooding on urban roads drainage and the measures, Urban Construction Theory Research, 28, 1–4, 2013.

Jiang, D.: Major causes of urban flooding and countermeasures in China, Commercial Culture, Academic Edition, 3, p. 351, 2011.

Konrad, C. P.: Effects of urban development on floods, Department of the Interior, U.S. Geological Survey, available at: https://pubs.usgs.gov/fs/fs07603/, last access: 29 November 2016.

Li, J.: Causes of urban flooding in larger and medium-sized cities in China and the analysis of countermeasures, Urban Construction Theory Research, 2, 123–126, 2013.

Liu, J.: The urban expansion problems caused by frequent urban flooding in China, available at: http://finance.sina.com.cn/roll/2016-08-04/doc-ifxutfyw0531051.shtml, last access: 4 August 2016.

National Bureau of Statistics of the People's Republic of China, The People's Republic of China national economic and social development statistical bulletin in 2017, available at: http://www.stats.gov.cn/tjsj/zxfb/201802/t20180228_1585631.html, last access: 28 February 2018.

Risa, Y.: "Beauty" of Japan's huge concrete utilities pulls in tourists, The Asahi Shimbun, available at: http://www.asahi.com/ajw/articles/AJ201703100008.html, last access: 10 March 2017.

MWR: The Ministry of Water Resources of the People's Republic of China, Bulletin of Flood and Drought Disaster in China, 2007–2016.

Wang, W., Wang, Q., Lin, H., Gong, D., and Zhang, S.: The review of China's urban flooding research and the outlook, Urban Problems, 5, 24–28, 2015.

Why is Wuhan always flooded?, Tencent News Encyclopaedia, available at: http://news.qq.com/cross/20160706/48K6xRR1.html{#}0, last access: 6 July 2016.

Yang, L.: Reasons of urban waterlogging and its countermeasures, Anhui Architecture, 1, 127–158, 2011.

Zhang, D., Xiaoquan, L., and Youye, L.: The renewal of "the distribution data collection in recent 500 years in China" (1993–2000), J. Appl. Meteorol. Sci., 14, 379–388, 2003.

Zhang, J., Wang, Y., He, R., Hu, Q., and Song, X.: Discussion on the urban flood and waterlogging and causes analysis in China, Adv. Water Sci., 27, 485–491, 2016.

Zhang, Z., Wang, C., and Zhao, Z.: On city flood cause in China and reduction measures, China Population Resources and Environment, 2, 21–24, 1992.

Proc. IAHS, 383, 201–211, 2020
https://doi.org/10.5194/piahs-383-201-2020

Analysis of the specific water level of flood control and the threshold value of rainfall warning for small and medium rivers

Yuzhong Hu[1], Zhaoxia Wang[2], Bin Zhou[2], and Shiqi Jiang[3]

[1]Hydrology Bureau of Anhui Province, Hefei, 230022, China
[2]Hydrology Bureau of Anqing, Anqing, 230022, China
[3]Office of Shitai County Flood Control and Drought Relief Headquarters, Chizhou, 247100, China

Correspondence: Yuzhong Hu (jxxbc@163.com)

Published: 16 September 2020

Abstract. In order to fully use the newly-built hydrologic stations in small and medium rivers for flood warning and prevention, based on the technical line of the investigation and evaluation of flash flood disasters, different historical flood information of the warning river section was investigated, while the elevations of riverain houses were measured. The relationship between the stations and affected population has been established for the study region. According to integrated flood control capability of riverside residents, the characteristic water levers and rainfall threshold values were determined in this case. The results indicate that the warning and safety levels are 54.0 and 55.3 m, respectively. The analysis results show that the warning should be issued when the net rainfall intensity is 70 mm h^{-1} or 110 mm for 6 h condition. And considering the effectiveness of disaster avoidance, variable characteristic water levers of the hydrologic station for flood warning in different village were determined on the basis of the flood control capability of each village. The relationship between hydrologic numerical value and inundating influence was established in a one-to-*n* way.

1 Background

Since the Construction of small and medium-sized rivers hydrology monitoring system, density of hydrological monitoring sites was improved evidently. However, there are two shortcomings that make real-time hydrological elements playing a limited role in flood hazard mitigation. Firstly, there is no historical data in the new hydrological stations, and the historical ranking comparison cannot be realized (Yao et al., 2014). No matter the rainfall or water level, the severity cannot be known according to the value. Secondly, there is no characteristic water level. The correlation between site elements and early warning objects is lacking. There is no connection between water level and flood impact, and it cannot play an effective early warning role. If these two problems cannot be solved, the new hydrological stations are useless for flood prevention. From 2016, Chinese Office of State Flood Control and Drought Relief Headquarters and Waterway Bureau of Water Resources began to promote the analysis of the characteristic water level of small and medium-sized rivers. The paper takes the new hydrological station in the central district of Shitai county in the southern Anhui province as an example with technical route which called "The research and application of investigation and evaluation on mountain torrent disaster as well as early warning key technology" (Hu et al., 2016), through the survey of historical flood and elevation of the houses along the banks of the river (MWR, 2015). Then it evaluates the flood protection capacity of the early warning object and establishes the link between hydrologic and social economic factors, to determine the rainfall warning threshold and characteristics of water level.

2 Regional overview

Shitai county is located in the northwest of the southern Anhui mountains which belongs to qiupu watershed of a tributary of Changjiang River. The qiupu watershed has the history of frequent floods and waterlogging disasters. Typical floods include the flood of 14 July 1991, the flood of 28 May 1995, the flood of 19 June 1996 and the flood of 26 June 1998. Shitai county was flooded six times in 1996, and the water depth of urban area was as deep as 6 m, and the water depth of urban area was as deep as 7 m in 1998. After real-time flood control engineering, during the flood of 21 June 2002, the flood of 10 July 2007, the flood of 24 July 2015, the flood of 3 July 2016, the flood of 24 June 2017, Shitai county in addition to the scope of protection of the flood control works from submerged, while other areas were still affected by flooding for many times, and often repeated many times within a year. It is a typical area where mountain torrent disasters take place frequently. So it is necessary to constructe the non-engineering measures of flood protection. Since the Construction of small and medium-sized rivers hydrology monitoring system, hydrological bureau of anhui provincial has set up a gauging station here with an area of 818 km^2 above the hydrological station since 2016.

3 Technical route

3.1 General idea

There are three conditions needing to be met for the construction of the early-warning index system for flood hazard mitigation of small and medium-sized rivers: firstly, setting the warning range (river section); secondly, mastering the influence of historical flood and the existing defense capability of early warning objects (population concentration areas); finally, establishing the relationship between the early warning source (hydrological station) and the early warning object (person), and determining the early warning index system according to the defensive ability. Generally speaking, it means to know yesterday (historical flood), evaluate today (present defense capability), and lock in tomorrow (earlywarning indicator system), the core is human security. Therefore, the specific implementation path is "taking villages (market towns) as the object, household registration as the unit, and people as the center", then establish the relationship between the early warning object which includes changes in the height of houses and the hydrological elements of the river. According to business logic, there are three stages:

- The first stage is establishing the evaluation system of hydrological factors and flood affected population through investigation and analysis. This stage is carried out in accordance with "two main lines and integration". Main line 1 is obtaining the surface line of the historical typical flood in the river section associated with the early warning object through investigation. It also integrates the housing foundation elevation and housing registered population of the residential group along the river, and estimates the impact of typical flood to identify the affected houses by the flood depth and relates the registered population with the houses. Then according to the assumption which means different scale of the flood water surface slope is same, it is got that flood effects of different grade flood with translating the water line, and gathers the flood affected population and water level which on the representative cross section of the matching flood water line, to establish the relationship line between water level and flooded affected population. Main line 2 is establishing the relationship between water level and flow. This part of the content is directly referred to the hydrological data compilation specifications. It needs to be emphasized that in addition to the actual measurement, it also needs to make up for the insufficient variation of the measured relationship between water level and flow through the hydrological survey. The relationship between water level and flow was established by obtaining the factors of gradient, vertical and cross section, bed roughness and other related factors. "Regression and integration" refers to the water level of the representative section as the cohesive factor, that is, water level takes the same section value among the relationship flood affects population, water level and flow. Build the core foundation called "Flow-Water Level- Flood Affects Population" "Two Coaxial Line", which determine the disaster water level and disaster flow, as the basis of flood impact assessment and early warning index system analysis.
- The second stage is assessing the risk level (recurrence period) of the disaster flow corresponding to the disaster water level of the early-warning object by calculating the design flood, and according to the disaster water level, the flow, as well as the regularity of flood rising of the dependent land, the alarm and the guarantee water level are recommended.
- The third stage is determining the distribution of rainstorm schedule and calculating the various characteristics time of critical rainfall (net rainfall threshold) by disaster flow and the calculation method of design flood, that is, how long and how much net rain intensity will be disaster. At the same time, according to the recurrence period of the disaster flow, the same frequency of rainstorm and flood is assumed, the rainfall threshold values of different periods in the same period as the recurrence period of the disaster flow are obtained, so as to establish the rainfall warning index system.

3.2 Core technology

3.2.1 Determination of relationship between water level and population

In the technical link of investigation, evaluation and early warning, the most important thing is how to establish the connection between flood and population, which is not only the technical difficulty, but also the core of disaster prevention and early warning. Especially in mountainous area, channel slope and fall head of water surface of warning channel is big, not like a lake, reservoir, whose relationship elevation- population and water level- population are the same, as shown in Fig. 1a, but needs to provide an algorithm that can change the waterline along the river into the surface of water slope, in order to determine both longitudinal and lateral variation of the building elevation layout and water slope height relations, as shown in Fig. 1b. The author put forward the idea "locking rivers trajectory, calculating origin, appointing representative section and flood mark, building the shortest distance double object plane projection" through the method "Relationship and assessment between non-static water level of mountains and population affected by flooded" (Hu, 2016a), to set up the business logic of the recognition system named "The ternary homologous" of the flood waterline of warning river, vertical slope line of river, housing base elevation, as shown in Fig. 2, the specific operation is divided into four steps:

- Step 1: according to the measurement results of the longitudinal section of the river segment associated with the early-warning objects (including elevation and coordinate), the river trajectory line can be drawn and the upstream starting point is marked as the origin.

- Step 2: the investigation of the flood marks is projected onto the river track according to their coordinates, the starting distance of the flood mark projection point relative to the origin was extracted from the river track line, match the water level of the flood mark point and establish the coordinate system of water level change along the course.

- Step 3: the house is also projected onto the river track according to its coordinates, the starting distance of the projection point of the house on the track line of the river was obtained, match the building base elevation, it is incorporated into the coordinate system of water level change along the path. The relationship between the water line and the house flood is realized, property of housing registered population is extracted, population affected by subsurface inundation is calculated.

- Step 4: different flood populations can be obtained by assuming different water surface lines (flood of different orders of magnitude). The relationship between water level and flood affected population can be established by determining the water level on the representative section of a flood water surface line of different orders of magnitude.

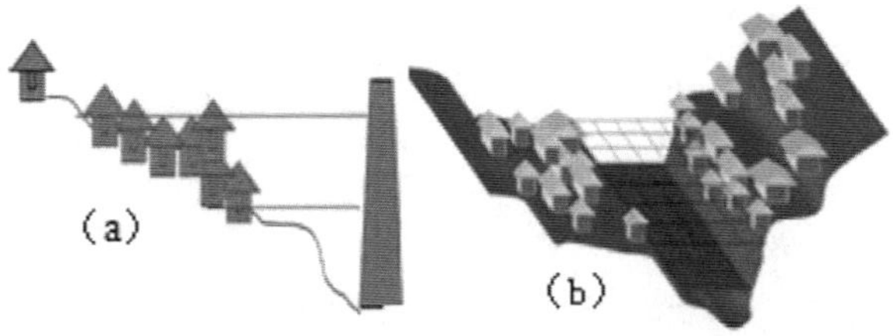

Figure 1. Comparison of effects of static and non-static water flooding.

In the figure, the water surface line is the line when the first household is affected by disaster, and the water level of the disaster water surface line on the representative section is the disaster water level, and its corresponding flow is the disaster water flow.

3.2.2 Critical rainfall threshold analysis method

The analysis methods of rainfall threshold have three types: back extrapolation, same frequency and event statistics. This paper mainly introduces the back extrapolation and the same frequency.

The calculation of rainfall threshold by back extrapolation is based on the inverse calculation of the disaster flow to the catchment flow, under the condition of known confluence parameters, three undetermined elements are also involved. The first is the amount of precipitation, the second is the time distribution of precipitation, and the third is the soil moisture in the early stage (Yao et al., 2019). Under the combination of three uncertain factors, in order to form the same peak value, the solutions will be infinite, which is equal to "no solution". To solve the problem in the absence of solution, conditions must be set. The solution is that the early effects can be solved by the flow generation, that is, the formation of net rain. The remaining is net rainfall and its distribution of the schedule problem. It still has various combination. To have a solution, the schedule distribution needs to be set, and then according to the distribution of rain type characteristics of a given period of time to set the value of rainfall, calculate the concentration flow. If the peak value of the flood process is less than the disaster flow, the set rainfall value should be increased. Otherwise, decrease it until the calculated peak value is equal to the disaster flow rate, which is the threshold value of the characteristic time period. In this topic, the distribution of rainfall schedule chooses the results of designed rainstorm schedule distribution or the formula of rainstorm attenuation index, as shown in Eq. (1):

$$H_{t,p} = 24^{n-1} H_{24,p} t^{1-n} \quad (1)$$

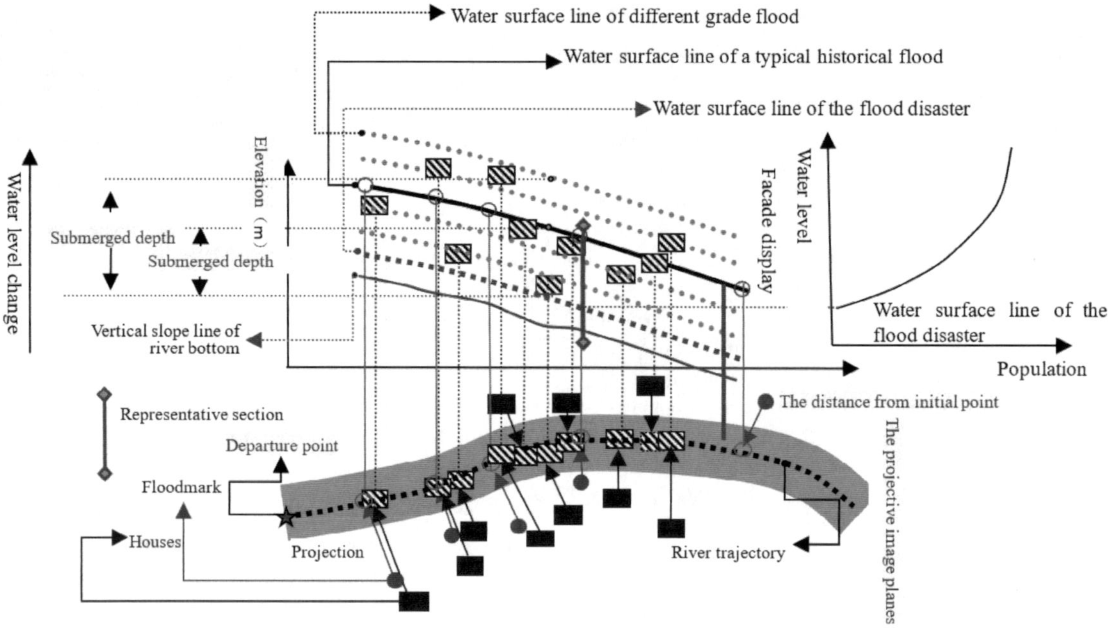

Figure 2. Schematic diagram of double projection water level and flooding affect population.

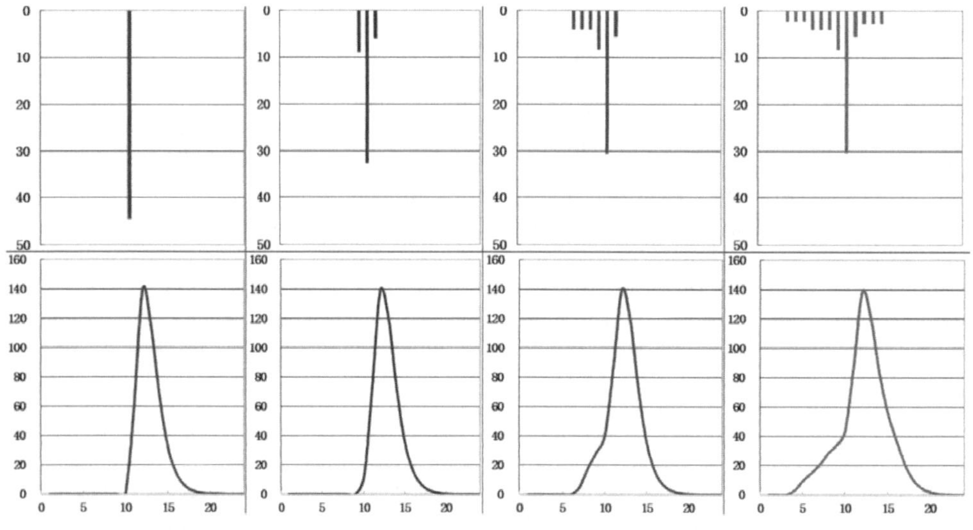

Figure 3. Schematic diagram of the back extrapolation of net rain threshold value.

where n is Rainstorm attenuation index, $0 < n < 1$; $H_{t,p}$ is the amount of rain that lasts for t and frequency p.

Special attention, it is not feasible to directly calculate the rainfall in the 24 h period and then extract the rainfall in different periods as the threshold value. It is a trial run at different times. In conclusion, if the calculation of preproduction flow and the limitation of precipitation process distribution are given, the solution can be obtained by deducting the net rain threshold, as shown in Fig. 3.

The same frequency is based on the assumption of the same frequency of rainstorm and flood, only fits in the small watershed. It determined the disaster flow by investigation stage, according to design way to determine return period of disaster flow, then analysis the different periods and frequency of precipitation based on design storm statistical results. It is important to note that this method is in accordance with the basin theory. If time session is greater than or equal to basin confluence, the same frequency achievement is valid, otherwise rainfall results in the same frequency characteristic

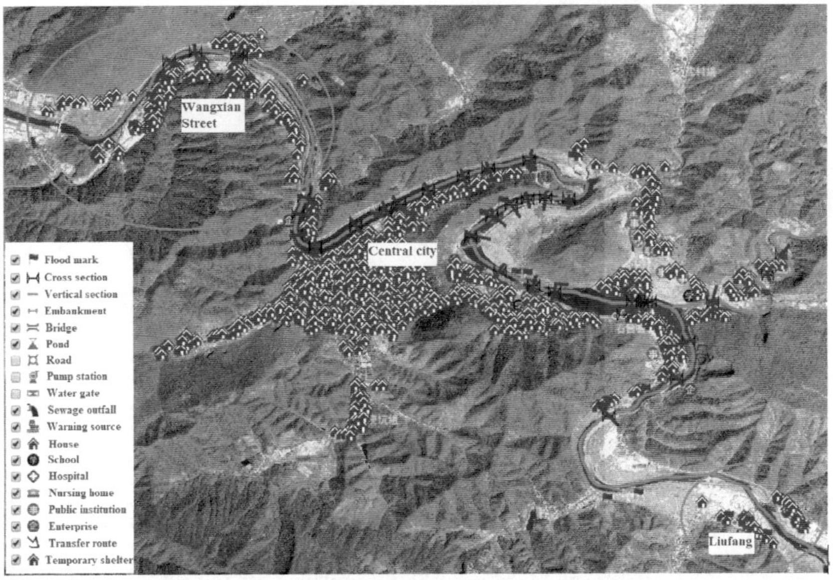

Figure 4. Distribution diagram of Shitai mountain flood impact survey and measurement (map data © MapWorld Anhui, http://anhui.tianditu.gov.cn, last access: August 2020).

period will be smaller. The advantage of this method is that the calculated threshold is gross rainfall omitting the calculation of production flow, which can be directly used for early warning.

4 Investigation and evaluation

4.1 Measurement area layout

In order to accurately and effectively reflect the correlation between hydrological factors and urban flood control ability, hydrology bureau of Anhui province organized and carried out an urban mountain flood impact survey in shitai county according to the above technical route (Hu, 2015). Hydrology bureau collected the flood mark of 1998, 2016 and 2017 along the river, surveyed the residence building foundation elevation along the both side of river, formed the survey measurement area including three parts area, as shown in Fig. 4. A total of 36 large sections and 43 historical flood marks were measured in the survey area. In particular, the flood marks of 2016 and 2017 were marked immediately after the flood, and the results are reliable.

4.2 Relationship between water level and flow

Shitai is a new hydrological station, automatic measuring and reporting of water level are put into use from 2016 flood season. In order to meet the needs of analysis of flood warning characteristic water level of shitai, flow test is carried out in 2017.There is no cable facilities at this station, the bridge is used as a river crossing equipment. The station measures the flow with ADCP and radio current meter. In 2017, the measured maximum flow is $1660\,m^3\,s^{-1}$, and the corresponding water level is 53.96 m (yellow sea). The flood peak water level of "2016.7.3" is 56.03. The measured maximum water level is 58.10 m. When the measured flow rate does not meet the requirements for determining the relationship between water level and water level, the combination of measured flow rate and hydrological investigation results is adopted. According to the conditions of manning's formula, under the condition of known section and stable section, the main factors influencing the determination of high flood flow rate are specific drop and roughness rate. According to the historical flood survey, the water surface lines along the shitai river are basically parallel under different flood conditions of different orders of magnitude. The water surface drop is stable at about 0.9 ‰, as shown in Fig. 5. Large section of shitai station, as shown in Fig. 6a, water level area relation of shitai station, as shown in Fig. 6b. In addition, the roughness rate can be deduced by measured flow, as shown in Fig. 6c. The corresponding roughness rate at the high water level of shitai station is between 0.035 and 0.045, so as to determine the flood peak flow rate of the flood investigated, as shown in Table 1, and the relationship line between water level and flow is determined comprehensively, as shown in Fig. 6d.

4.3 Determination of disaster water level and flow

Shitai investigation area divides into three parts: upstream liufang village test area, central urban area and downstream wangxian street area. In view of the differences of watershed area, water surface slope and elevation layout of population

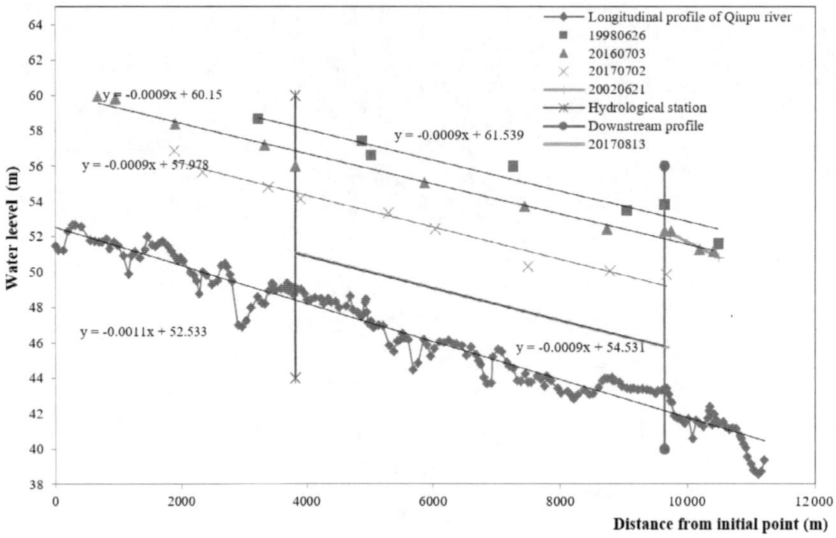

Figure 5. Comparison diagram of flood water surface ratio drop of different magnitude (field) in the river section of Shitai.

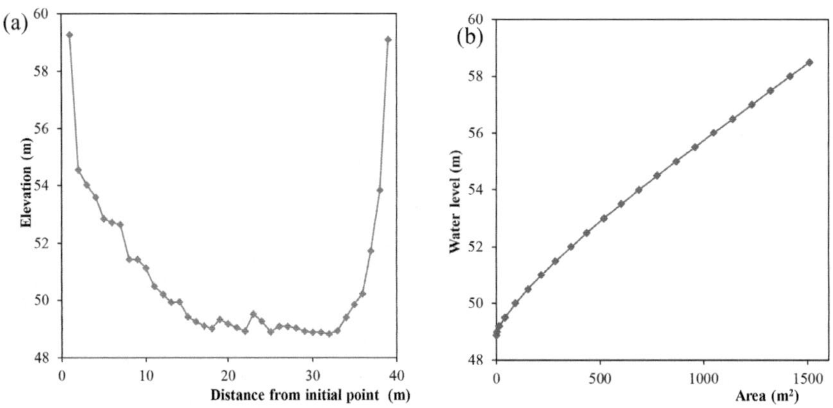

Figure 6. Diagram of analysis results of relationship between water level and flow at Shitai station.

Table 1. Results of flood peak flow to investigation flood.

The number of flood	Water level	Gradient	Roughness	Flow
"1998.06.26"	58.10	0.9 ‰	0.042	3930
"2016.07.03"	56.03	0.9 ‰	0.037	2720

Note: river bottom constraint is 48.85, corresponding flow is 0.

on both sides of the river between each area, determine the disaster water level in the aggregate measurement area according to subarea and short board effect principle.

4.3.1 Determination of disaster water level in the zone

With the flood level distribution of flood "2016.07.03", mix the foundation elevation of houses on both sides of river, according to the projection method each division measurement area has analyzing submerged. Flood submerged effects on both sides of the residents including liufang, central urban are *a* and wangxian street direction are shown in Fig. 7. According to this, the disaster water level of the liufang survey area is determined to be 60.90 m, corresponding to the disaster flow is 1540 $m^3 s^{-1}$; the disaster water level and flow in wangxian street area are 45.80 m and 1480 $m^3 s^{-1}$ respectively; although central urban area is protected by flood walls,

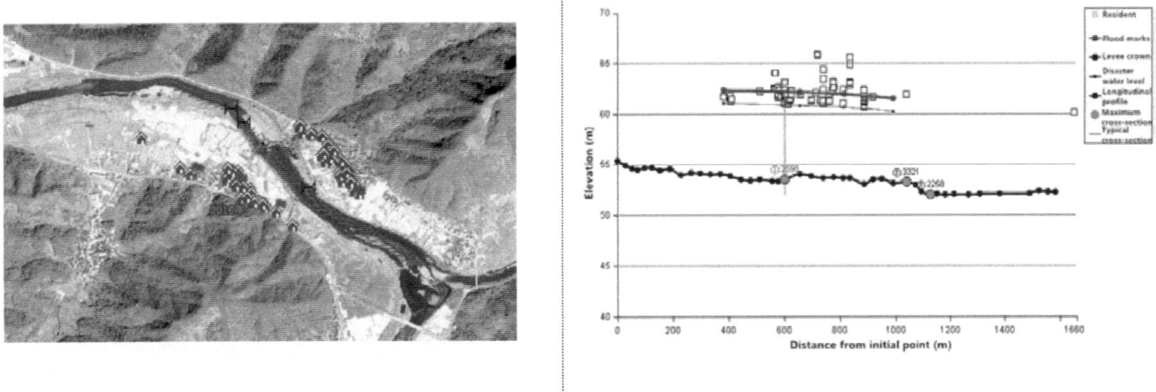

(a) Distribution and projection of measured area in Liu Fang Village

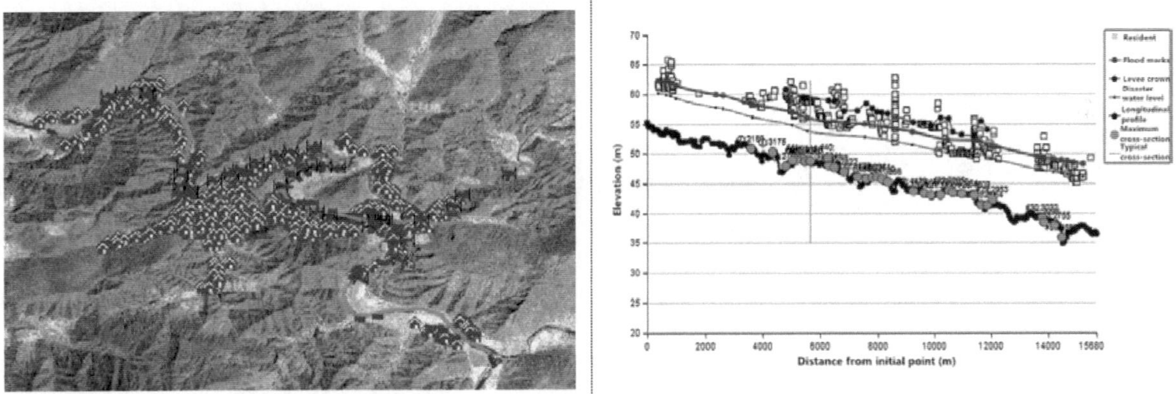

(b) Distribution and projection of measured area in central urban area

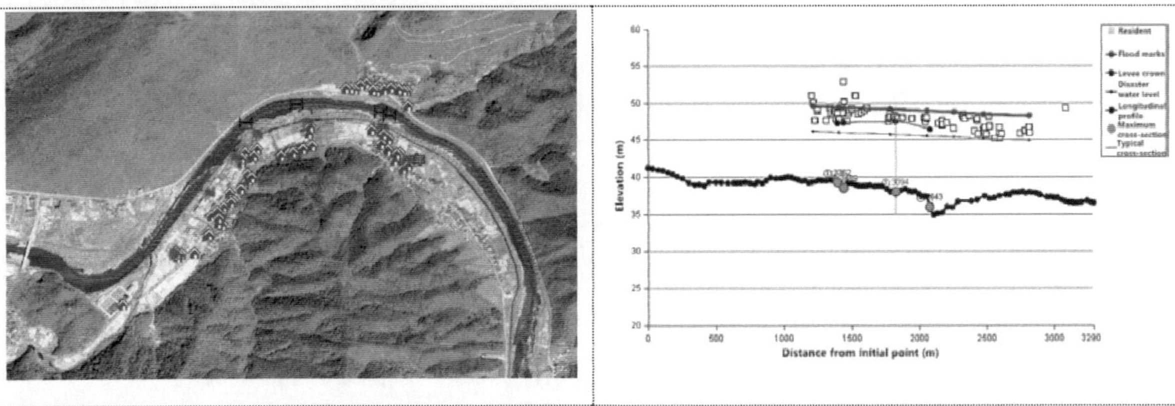

(c) Distribution and projection of measured area in Wang Xian Street

Figure 7. Diagram of water level and height projection of building foundation in each measuring area (map data © MapWorld Anhui, http://anhui.tianditu.gov.cn, last access: August 2020).

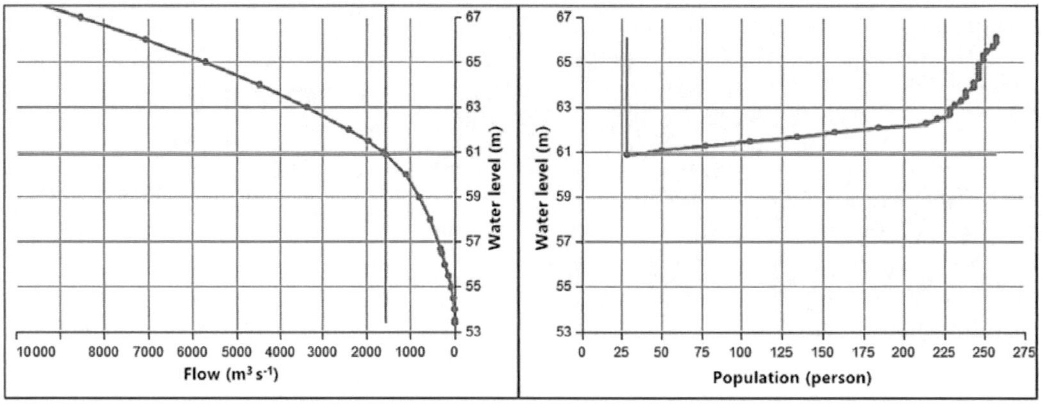

(a) Liu Fang Village

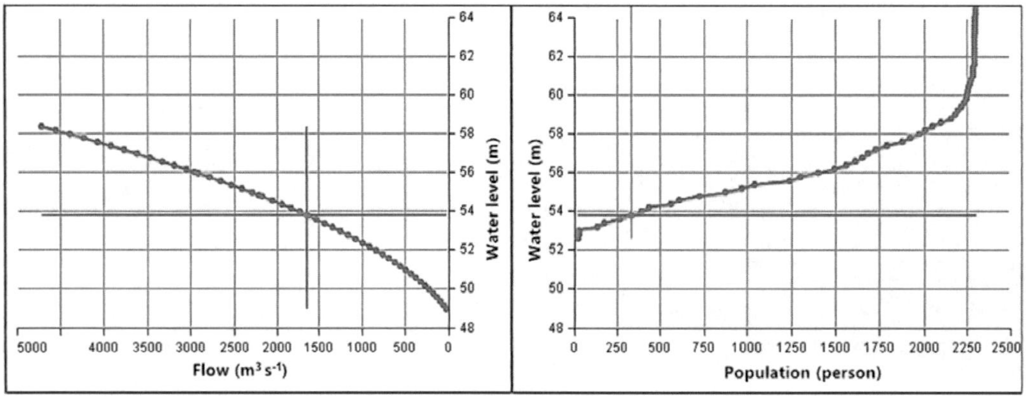

(b) Central urban area

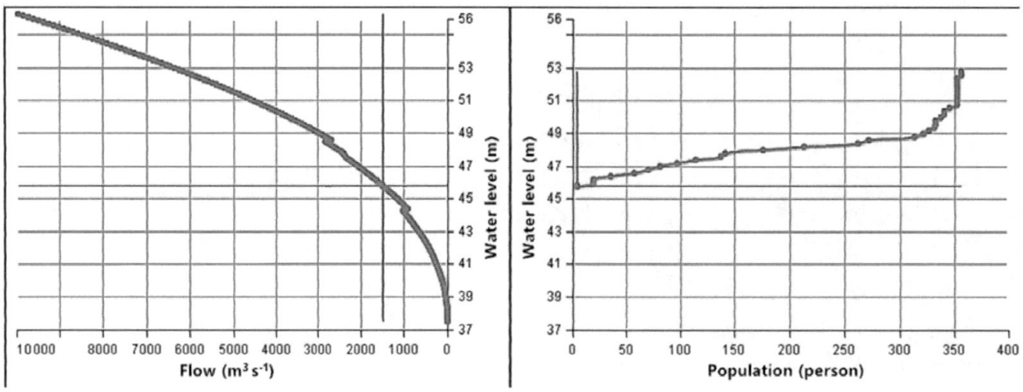

(c) Wang Xian Street

Figure 8. Results of the investigation points of upstream and downstream associated river sections in Shitai area on the same axis and two lines.

Table 2. Analysis results of disaster water level in downtown area of Shitai and villages related to upstream and downstream.

Location	Residential area	Representative section	Z_{disaster} (m)	Q_{disaster} ($m^3\,s^{-1}$)	i (‰)	$\Delta L_{\text{section}}$ (m)	$Z_{\text{disaster,section}}$ (m)	$\Delta L_{\text{station}}$ (m)	$Z_{\text{disaster,station}}$ (m)
Upstream of main area	Liufang village	Sect. 1	60.9	1521	0.9	4795	54.7	5740	49.53
Main urban area	Main area	Sect. 8	53.8	1730	0.9	0	53.8	5740	48.63
Downstream of main area	Wangxian Street	Sect. 3	45.8	1480	0.9	8514	53.5	5740	48.33

Note: The disaster water level of the main urban area is in the downstream of the urban flood wall which is not protected.

in its protected areas, the water level higher than the height of the housing foundation on both sides of the river is not necessarily a disaster. As shown in Fig. 8b, the corresponding place of residence in the part below the disaster water level in the figure on the right. However, the left bank of the fuhu river and the confluence of the fuhu river and the qiupu river outside the protection area of the flood wall are vulnerable to waterlogging. Therefore, the disaster water level in the central urban area is subject to the population distribution in this section, the disaster water level and flow rate are respectively 53.8 m and 1730 $m^3\,s^{-1}$. The relationship between flow, water level and population in each test area is shown in Fig. 8.

4.3.2 Determination of comprehensive disaster water level

Based on the disaster water level analysis results of each division, water level difference between the representative section and the early-warning section (shitai hydrological station) in each section is calculated, and the water level of each disaster water level corresponding to the early-warning section (shitai hydrological station) is determined (Hu et al., 2015). According to Fig. 3 analysis results, water surface slope of central urban area is 9 ‰, the distance between behalf section and early warning section of each part is shown in Table 2. The calculated results of the disaster water level in each measured area corresponding to the disaster water level in the early-warning section are shown in Table 3. According to each division of the relationship between water level and the corresponding water flow, it is estimated that the disaster flow of each residential concentration group is between 1500 and 1700 $m^3\,s^{-1}$. Finally, according to the short board effect principle, the comprehensive disaster water level is determined as follows: 53.50 m, and the disaster flow is 1500 $m^3\,s^{-1}$.

4.4 Impact assessment

Shitai county is non-data area, so the design flood uses the existing methods of territorial and peripheral areas (ASDIWCH, 1984; Zhang et al., 2015). The area comparison method takes the gaotan hydrological station downstream of shitai of qiupu river as the reference station. The empirical formula is based on the huangshan hydrology manual. As we know from the results, there is a big difference between the methods. Therefore, the multimethod mean is taken as the comprehensive result, as shown in Table 3. Compared with 1500 $m^3\,s^{-1}$ of the disaster flow and the results of Table 3. The current defense capability is less than once in 5 years.

Table 3. Design flood results of Shitai water level station.

Return period	Area of analogy	84 method	Manual of anhui	Empirical formula	Synthesis
100	3750	5650	4712	3213	4330
50	3260	4629	4070	2874	3710
20	2714	3202	3290	2423	2910
10	2282	2307	2660	2061	2330
5	1821	1483	2055	1671	1760

5 Early warning indicators

5.1 Rainfall warning threshold

Based on the 1500 $m^3\,s^{-1}$ of the disaster flow, using Xin'anjiang Model and the method of designing flood in mountainous area of anhui province, the rainfall warning threshold (net rain) of different time periods is deduced by trial calculation and back extrapolation method. At the same time, according to the recurrence period of the disaster flow in each part, the rainfall warning threshold (gross rainfall) in the characteristic period is determined by assuming the same frequency of rainstorm flood (Hu, 2016b), as shown in Table 4. It needs to be explained that the results of the backward extrapolation method need to be calculated in advance, the warning time of the same frequency method is greater than the watershed convergence time, the convergence time is about 6 h in this watershed.

Table 4. Rainfall warning threshold of downtown area in Shitai and upstream and downstream related villages.

Location		Cluster	1 h	2 h	3 h	6 h	12 h	24 h
Back extrapolation	"84" method	Liufang village	73	74	74	77	86	107
		Main area	81	81	82	85	94	116
		Wangxian street	68	68	69	71	78	97
	Xin'anjiang model	Main area	73	73	74	108	116	153
Same frequency method	"84" method rainstorm statistics	Liufang village				111	155	181
		Main area				121	167	197
		Wangxian street				110	151	151

5.2 Characteristic water stage

Artificial observation scale is set up in the lower reaches of 5740 m of the hydrological station in the central district of shitai. The warning water level is 47.00 m, which corresponds to the water level of the hydrological station is 52.20 and the corresponding flow is 990 $m^3 s^{-1}$. Directly from the section, the existing warning water level has not been on the beach, and the warning water level is low. However, the basin above shitai is a typical river, and the flood rises up and down sharply. In terms of the flood "2016.6.24", the maximum hourly increase reaches 1.35 m, with only 1 h and 15 min from the alert to the disaster water level. The warning water level of shitai water level station is 52.20 m and the guaranteed water level is 53.50 m. At the same time, in order to take account of the consistency of the water level elevation system in the Changjiang river basin, wusong base surface is adopted for the water level of shitai hydrological station. The elevation difference between wusong and huanghai is 1.83 m. The warning water level of shitai hydrological station is 54.0 m and the guaranteed water level is 55.3 m by taking an integer.

6 Conclusion

Using the technical route and analysis method of survey and assessment about mountain torrent disaster for reference, this paper takes the new site stone stage of small and medium-sized rivers for example, to carry out the survey of flood and elevation layout of live group along the river, and sets up warning source (hydrologic station) associated with early warning object's layout, carry out the early warning defense capability evaluation object, provided by flow, water level determines the hydrological site characteristics, and on the basis of the submitted by flow characteristic time critical rainfall warning threshold, effectively solves the small and medium-sized rivers new site without historical reference, without warning associated "stand naked" problem, achieved by providing hydrological values to hydrology factors effect. The technical route provided in this paper can provide reference for the construction of similar site early-warning index system.

In this paper, the actual flood control capacity of shitai county and related villages is comprehensively considered, and the conclusion is drawn that the overall defense capacity of the early-warning objects is less than once in 5 years. The warning water level is 54.0 m and the guaranteed water level is 55.3 m.The average net rainfall intensity of the 1 h watershed is 70 mm h^{-1}, or the average rainfall of the 6 h surface is 110 mm, which is the threshold value of the critical rainfall causing disasters. This result can be used as a reference basis for flood prevention of shitai.

Data availability. Data generated or analyzed during the study are available from the corresponding author by request.

Author contributions. YH designed and directed the study and wrote the paper. BZ and SJ processed the data. YH and ZW carried out software and investigation for this study. YH, ZW and BZ analyzed the data.

Competing interests. The authors declare that they have no conflict of interest.

Special issue statement. This article is part of the special issue "Hydrological processes and water security in a changing world". It is a result of the 8th Global FRIEND–Water Conference: Hydrological Processes and Water Security in a Changing World, Beijing, China, 6–9 November 2018.

Financial support. This work was supported by the National Key Research and Development Program of China (grant no. 2018YFC1508106).

References

ASDIWCH – Anhui Survey and Design Institute of Water Conservancy and Hydropower: Calculation method of design floods in mountainous area of Anhui Province (The "84" method), ASDIWCH Research Report, 35 pp., 1984.

MWR – Ministry of Water Resources of the People's Republic of China: Code for hydrologic investigation (SL 196-2015), China Water & Power Press, 52 pp., 2015.

Hu, Y. Z.: Questions and answers on core technology of field investigation of torrential flood impact assessment, China Flood Drought Manage., 25, 26–33, 2015.

Hu, Y. Z.: Evaluation method of relationship between non static water level and inundated population in mountain area, Chinese Patent No. ZL201410803764.4, 2016a.

Hu, Y. Z.: Questions and answers on key technologies of torrential flood impact assessment and early warning threshold analysis, China Flood Drought Manage., 26, 23–28, 2016b.

Hu, Y. Z., Yao, X. B., Zhang, C. X., Fang, H., and Fang, G. J.: Investigation and Assessment of Flash Flood Effect and Research on Construction of Early Warning System: Take Luxi Reach of Chang River as A Case, J. China Hydrol., 35, 20–25, 2015.

Hu, Y. Z., Xu, Y. P., and Ye, C. L.: Research and application of key technologies of investigation, evaluation and early warning of flash flood disasters, Hydrology Bureau of Anhui Province Research Report, 235 pp., 2016.

Yao, C., Zhang, K., Yu, Z., Li, Z., and Li, Q.: Improving the flood prediction capability of the Xinanjiang model in ungauged nested catchments by coupling it with the geomorphologic instantaneous unit hydrograph, J. Hydrol., 517, 1035–1048, 2014.

Yao, C., Ye, J., He, Z., Bastola, S., Zhang, K., and Li, Z.: Evaluation of flood prediction capability of the distributed Grid–Xinanjiang model driven by weather research and forecasting precipitation, J. Flood Risk Manage., 12, e12544, https://doi.org/10.1111/jfr3.12544, 2019.

Zhang, C., Zhang, S., Xia, S., and Cheng, Q.: Huangshan Hydrology Manual, Hydrology Bureau of Huangshan Report, 155 pp., 2015.

Proc. IAHS, 383, 213–222, 2020
https://doi.org/10.5194/piahs-383-213-2020

Open Access
PIAHS

Ensemble flood simulation for the typical catchment in humid climatic zone by using multiple hydrological models

Jie Wang[1,2,3], Jianyun Zhang[1,2], Guoqing Wang[1,2], Xiaomeng Song[1,2,3], Xiaoying Yang[1,4], and Yueyang Wang[2,5]

[1]State Key Laboratory of Hydrology–Water Resources and Hydraulic Engineering, Nanjing Hydraulic Research Institute, Nanjing 210029, China
[2]Yangtze Institute for Conservation and Development, Nanjing 210098, China
[3]School of Resource and Earth Science, China University of Mining and Technology, Xuzhou 221116, China
[4]Department of Environmental Sciences and Engineering, Fudan University, Shanghai 200433, China
[5]Geographical Science School, Nanjing University of Information Science and Technology, 210029, China

Correspondence: Guoqing Wang (gqwang@nhri.cn)

Published: 16 September 2020

Abstract. A good performance of hydrological model for flood simulation is of critical importance for flood forecasting. Taking Yandu River catchment, as the study area, three hydrological models (i.e. Xin'anjiang model, TOPMODEL, artificial neural network model) and a multi-model ensemble simulation method (i.e. entropy-based method) were applied to simulate the hydrological processes of 30 flood events occurring in 1981–1987. The performance of the ensemble members and multi-model ensemble simulation method was evaluated by comparing indicators of Nash-Efficiency coefficient, errors in root mean square, peak occurrence time, and relative errors of flood peak discharge, event runoff depth. Results show that the three hydrological models perform well for hydrological simulation of all 30 storm floods with Nash and Sutcliffe Efficiency coefficient of above 0.75 and relative error of less than 10 %. However, different model exhibits a difference in simulation errors of peak discharge and peak occurrence time. For example, BP model has the smallest error of 3.78 % for peak discharge simulation while that of Xin'anjiang model and TOPMODEL are 20.9 % and 24.7 % respectively. The entropy-based ensemble simulation method improved flood simulation accuracy to some extent for all evaluation criteria comparing to the three hydrological models. It is feasible to apply entropy-based ensemble approach for improving accuracy of flood forecasting in humid regions of China.

1 Introduction

In the context of global warming, the probability and intensity of extreme precipitation are increasing (IPCC, 2014), which further aggravates the risk of flood disasters. It is therefore critical to improve flood forecasting technologies (Pitt, 2007). Hydrological modeling has been one of the commonly used tools of flood forecasting (Werner, 2005; Song and Kong, 2010). Tens of hydrological models with different structures have been successfully developed and applied globally during the past decades. Bao (2009) considered the 1950s as an important time node to divide hydrological model development into two stages: experience-based stage and model study stage. At the former stage, statistical methods are used to analyze long-term observation records to reveal the relationships between hydrological elements and the change regular, such as unit hydrograph method (Lin, 2003), corresponding stage/discharge method (Rui, 2004) and so on. The latter one produces with the development of theoretical technologies including computer technology, 3S technology, and geographic information systems, such as Xin'anjiang (XAJ), Shanbei model, Mixed Runoff yield model in China (Zhao, 1984), SAC (Burnash, 1995) and SSARR models

(Rockwood, 1968) in the US, Tank model (Sugawara, 1972) in Japan, CLS model in Italy (Craft et al., 1996), etc.

At the same time, models are divided into different types according to various criteria. Yang divides flood forecasting technology into five genres: black box genre, concept genre, residual genre, filtering genre and statistical genre (Yang, 1996). However, difference exists more or less among these models, such as model structure, model type, parameters, generalization methods and so on, causing simulation results to perform variously (Vrugt et al., 2006). While it is possible for one model to poorly simulate individual flood event, it is rare for multiple models to all yield poor performance since the deficiency of certain model may be made up by other models (Roulin, 2007). This idea fits well with the concept of multi-model ensemble. This concept was first proposed in the economic and meteorological fields in the 1960s. It refers to the use of multiple means or methods to obtain the forecast value of certain factor, and then the various forecast values are used to calculate the optimal forecasting scheme. Bates and Granger (1969), Epstein (1969) and Leith (1974), respectively, are considered to be the first to propose ensemble ideas in the economic and meteorological fields. Since then, some scholars have specialized in ensemble forecasting of meteorology. For example, Andrew et al. proposed a Bayesian ensemble method in several common circulation models (GCM) for seasonal precipitation ensemble forecasting in 2003 (Andrew, 2003). Takemasa and Masaru (2011) developed the WRF-LETKF system to forecast precipitation by combining a mesoscale numerical weather prediction model (WRF) and a filtering algorithm (LETKF). Mudasser et al. (2014) used 12 different weather models to simulate precipitation in the rainy season in New Zealand and found that the choice of ensemble method was more important than the number of ensemble members to affect simulation accuracy.

In view of the successful application of ensemble methods in meteorology, hydrologists have tried to introduce this method into flood forecasting (Balint et al., 2006). Cloke and Pappenberger (2009) proposed that Ensemble Forecasting System (EPS) in flood forecasting, which is based on the Monte Carlo structure, consisted of control prediction and disturbance prediction. Up till now, scholars have applied different ensemble methods to further improve the flood forecasting scheme. For example, Jasper et al. (2002) forecasted the inflow of Lake Maggiore using five models ensemble. Davolio (2008) forecasted floods in northern Italy by using six different rainfall-runoff models. Diks and Vrugt (2010) and Arsenault et al. (2005) used a variety of ensemble methods to simulate runoff processes in different catchments in the United States, and both found that the Granger-Ramanathan ensemble method had the highest accuracy. Arsenault et al. (2017) improved the traditional ensemble method and proposed a new idea combining multi-inputs and multi-model ensemble. They used 12 ensemble members combined with three hydrological models and four climate data, to simulate the runoff process in 424 catchments in the United States. The results show that 70 % of the catchments have greatly improved the forecasting accuracy through the ensemble method.

Flood forecasting scheme is different for different climate zone due to different hydrological characteristics (Hamill et al., 2004; Guan et al., 2018). Yangtze River is the first largest river in China in terms of its drainage area and river length. Effective flood forecasting is of critical importance for flood control of this river basin. However, there are rare studies of flood forecasting by using ensemble method of multiple models, particularly for tributaries of the Yangtze River. In this paper, taking Yandu River catchment, a tributary of upper Yangtze River, as a study case, three hydrological models (e.g. Xin'anjiang model, TOPMODEL, artificial neural network model) were used to simulate flood events in 1981–1987. Then the entropy-based method is used to ensemble multiple models so as to improve the forecasting scheme and the accuracy of flood forecasting, which can provide preliminary data support for the further promotion of hydrological models ensemble application research. The remainder of this paper is organized as follows: Sect. 2 contains a brief description of the study area, three hydrological models, the ensemble method and the evaluation criteria. The results of individual model and multi-model ensemble are described in Sects. 3 and 4 gives conclusions of the study.

2 Data and methods

2.1 Study area

The Yandu River catchment is located in Badong County, Hubei Province, China, in the upper tributary of the Yangtze River, with a catchment area of 601 km^2. The terrain in the catchment is mostly mountainous and covered with dense vegetation of forest and grass. The catchment situates in typical monsoon climate zone with mean annual temperature of 11.5 °C, and mean annual precipitation of about 1650 mm. The flood season mostly starts from May and ends in September. There are five rain stations in the catchment. The outlet hydrometric station of Yandu station was established in 1958. The river system and locations of rain gauges and hydrometric station are shown in Fig. 1. Thirty flood events in 1981–1987 were used to evaluate performance of hydrological model for flood simulation. Rainfall and discharge data with temporal resolution of 1 h were collected from Hydrological Yearbook published by Hydrology Bureau (Ministry of Water Resources of China, 1981–1987).

2.2 Hydrological models

2.2.1 XAJ model

Xin'anjiang model (XAJ) is a conceptual rainfall-runoff model developed by HoHai University (Zhao, 1984). The XAJ model is based on mechanism of saturation excess.

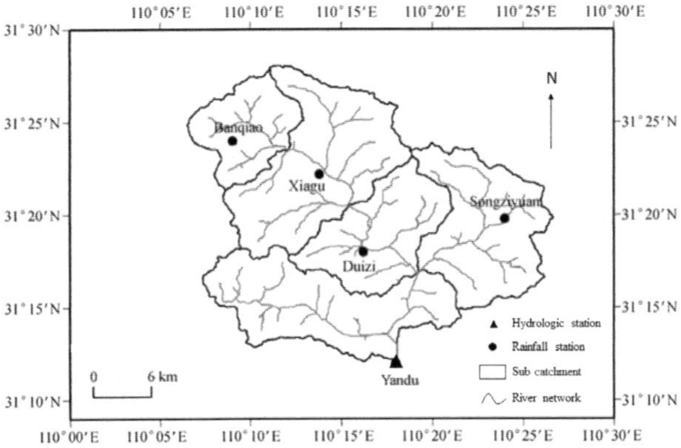

Figure 1. River networks and observational station distribution.

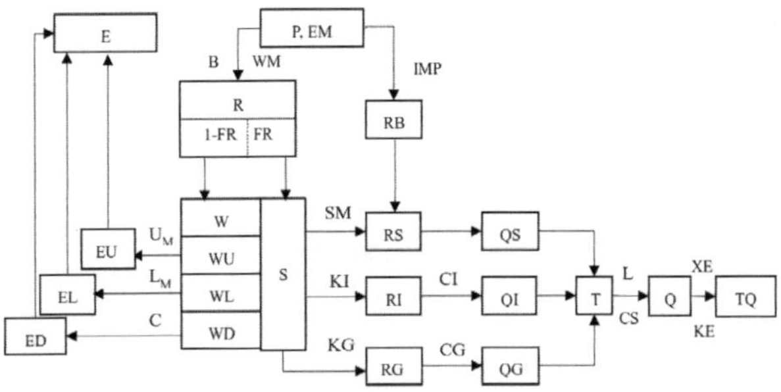

Figure 2. Framework of the Xin'anjiang Model. Note: K: Reduction coefficient of evaporation; IMP: Impervious area proportion; B: Exponent of storage capacity curve; WM: Average water storage capacity in the catchment; WUM: Storage capacity of the upper layer; WLM: Storage capacity of the lower layer; C: Evapotranspiration coefficient of deeper layer; SM: Storage capacity of free water; EX: Exponent of free water storage capacity curve; KG: Outflow coefficient of free water storage to groundwater; KI: Outflow coefficient of free water storage to interflow; CG: Regression coefficient of groundwater storage; CS: Regression coefficient of surface runoff; CI: Regression coefficient of interflow storage; L: Coefficient of hysteresis calculation; XE: Muskingum parameter; KE: Muskingum parameter.

The model is mainly composed of four parts, namely runoff yield, evapotranspiration, water source partition and confluence. For flood simulation, hourly rainfall series are needed to drive XAJ model. The model structure and parameters are shown in Fig. 2.

2.2.2 TOPMODEL

TOPMODEL is a semi-distributed watershed hydrological model proposed by Beven and Kirkby (Xu, 2009). The model is based on the concept of variable flow generation with consideration of catchment topographical features, soil texture, etc. The model divides soil layer into three aquifer zones, vegetation root zone, unsaturated soil zone, and saturated soil zone. The inputs of TOPMODEL not only include rainfall series, but also include catchment topographic index $\ln(\alpha/\tan\beta)$. Digital Elevation Model (DEM) data is therefore needed for the model application. Total runoff is the sum of interflow and saturated slope flow. The conceptual framework of the TOPMODEL is shown in Fig. 3. There are five parameters in TOPMODEL need to calibrate.

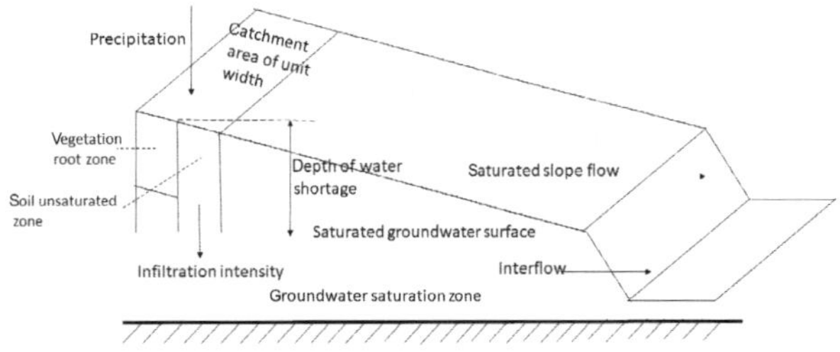

Figure 3. Conceptual framework of TOPMODEL.

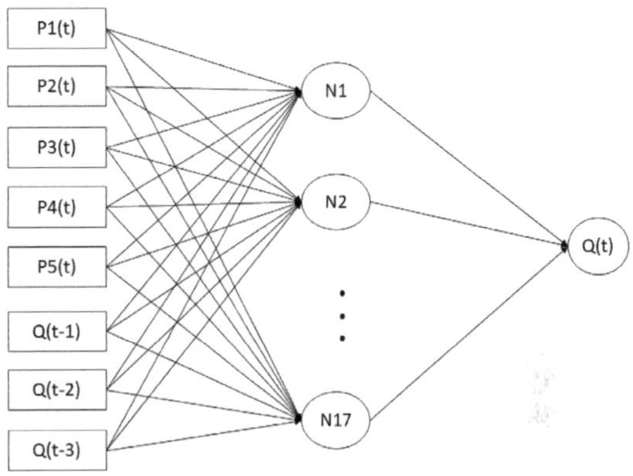

Figure 4. BP model flowchart. Note: $P1(t)$, $P2(t)$, $P3(t)$, $P4(t)$, P5(t): five precipitation station data; $Q(t-1)$, $Q(t-2)$, $Q(t-3)$: discharge data for the first three moments; $N1$–$N17$: the intermediate layer nodes; $Q(t)$: discharge at the corresponding time of the basin outlet.

2.2.3 BP model

The BP model is a multi-layered feedforward neural network with a strong self-learning ability. It can transmit corrected errors in reverse order (Zhao, 1996). Its hierarchy includes the input layer, the hidden layer, and the output layer. This paper uses a three-layer BP model with only one hidden layer. The structure of the constructed BP model is 8-17-1, shown in Fig. 4. It means BP model has 8 inputs, including data from 5 precipitation station data and discharge data for the first three moments. The only output of the model is the discharge at the corresponding time of the basin outlet. The intermediate layer is connected to the input and output and is calculated as 17 layers by the empirical formula, representing the complexity of the established model.

2.3 Entropy-based ensemble method

In ensemble methods, it is important to identify the weight coefficients. The basic idea of the entropy method is that the variation of the error between the simulated and observed results is inversely proportional to the weight coefficient.

Firstly, calculate the errors between the simulated and observed results and normalize the errors:

$$E_{it} = \begin{cases} 1, \frac{|a_t - a_{it}|}{a_t} \geq 1 \\ \frac{|a_t - a_{it}|}{a_t}, 0 \leq \frac{|a_t - a_{it}|}{a_t} \leq 1 \end{cases} \quad (1)$$

$$B_{it} = \frac{E_{it}}{\sum\limits_{t=1}^{m} E_{it}}, t = 1, 2, \ldots m \quad (2)$$

Secondly, calculate the entropy value of the relative error in the model simulation:

$$H_i = -\sum_{t=1}^{m} \frac{B_{it} \ln B_{it}}{\ln m}, i = 1, 2, \ldots n \quad (3)$$

Thirdly, calculate the variation index and weight coefficient:

$$D_i = 1 - H_i, t = 1, 2, \ldots n \quad (4)$$

$$Q_i = \frac{1 - \frac{D_i}{\sum_{i=1}^{n} D_i}}{m - 1}, i = 1, 2, \ldots n \quad (5)$$

Assume a_t, $t = 1, 2, 3, \ldots, m$ is the sequence of simulation objects and there are n kinds of single models to simulate, then the simulation value of the method i at time t is $a_{it}, i = 1, 2, \ldots, n$. Where E_{it} and B_{it} represent the relative errors and normalized errors of the method i at the time t, a_t represent the observed value at time t, while H_i, D_i, Q_i mean the entropy value, variation index and weight coefficient of model i.

2.4 Evaluation criteria

Five evaluation criteria of Nash and Sutcliffe efficiency coefficient (so called Nash-Efficiency Coefficient, NEC), errors in root mean square (RSME), error in time to flood peak (ETFP), relative error in flood peak discharge (REFPD) and relative error in event runoff depth (REERD), were selected to evaluate performance of hydrological model for flood simulation. Details about the five evaluation criteria could be found in manual guideline of flood forecasting issued by the Ministry of Water Resources of the People's Republic of China (2008). Good performance of hydrological model for flood simulation will have NEC approaching to 1 and RSME, ETFP, REFPD, and REERD being close to 0.

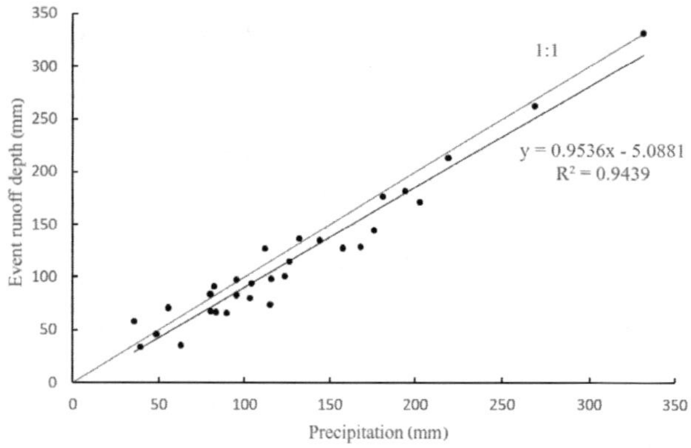

Figure 5. Event runoff depth against precipitation of 30 floods.

3 Results and discussion

3.1 Flood characteristics

Being influenced by Asian monsoon climate, floods in the Yandu River mostly occur in period from May to September with 2–19 d duration. Statistical results of the selected 30 floods indicated that flood peak discharge ranges from 300 to $1200\,m^3\,s^{-1}$ while the corresponding event rainfall varies in range of 35.7–331.7 mm. The hydrograph of flood is highly influenced by the spatiotemporal distribution pattern of rainfall. 56.7 % of flood events have multiple peaks. Flood runoff depth against rainfall in 30 flood events were plotted in Fig. 5.

According to the trend line of the precipitation-runoff (P-R) point group of 30 floods, the slope is 0.95, which is very close to the 1 : 1 line and is below the 1 : 1 line. The slope represents the runoff coefficient here and its value is close to 1, indicating that the loss of the event flood is relatively small overall. This indicates that the study area is humid and antecedent soil moisture is abundant. The P-R relationship of about 7 floods falls above the 1 : 1 line. The runoff coefficient of seven floods is more than 1 due to the influence by previous rainfall. Previous runoff had not completely regressed before the next flood occurred, causing higher runoff than rainfall.

3.2 Model calibration and flood simulation

Three individual models were used to simulate hourly flow. Experience method and manual-trail-error method were used for parameter calibration in the XAJ model, and the results were shown in Table 1. As for the TOPMODEL, topographic index of Yandu River catchment was calculated with ArcGIS (Fig. 6) and then used as model inputs. Its parameters were calibrated by manual-trial-error method (Table 2). In BP model, automatic calibration method (Levenberg-Marquardt method, Levenberg, 1944; Marquardt, 1963) was used for model parameter calibration, where 153 weights and

Table 1. The results of parameters calibration in XAJ hourly model.

Parameters	K	WM	WUM	WLM	B	C	SM	EX
Results	0.45	120.0	20.0	60.0	0.3	0.15	43	1.5
Parameters	KI	KG	CS	CI	CG	K_e	X_e	L
Results	0.3	0.4	0.58	0.89	0.982	1.0	0.28	0

Note: K: Reduction coefficient of evaporation; IMP: Impervious area proportion; B: Exponent of storage capacity curve; WM: Average water storage capacity in the catchment; WUM: Storage capacity of the upper layer; WLM: Storage capacity of the lower layer; C: Evapotranspiration coefficient of deeper layer; SM: Storage capacity of free water; EX: Exponent of free water storage capacity curve; KG: Outflow coefficient of free water storage to groundwater; KI: Outflow coefficient of free water storage to interflow; CG: Regression coefficient of groundwater storage; CS: Regression coefficient of surface runoff; CI: Regression coefficient of interflow storage; L: Coefficient of hysteresis calculation; XE: Muskingum parameter; KE: Muskingum parameter.

Table 2. The calibrated parameters of TOPMODEL.

Parameters	M	$\ln(T_0)$	SR_{max}	SR_{init}	ChVel
Results	0.013	5.7	0.06	0	408

Note: m: Rate parameter of soil infiltration intensity decreasing exponentially; $\ln(T_0)$: Natural logarithm of soil effective conductivity reaching saturation; SR_{max}: Maximum water storage capacity in root zone; SR_{init}: Volume of initial water shortage; ChVel: Effective surface convergence speed.

18 thresholds need to be determined. Due to the length of the article, BP model parameters are not shown in this paper.

Evaluation statistics of the three models are shown in Table 3. There are differences in the simulation results among the three models. Combined with the multi-objective evaluation results, the BP model based on the self-memory pattern has the highest total qualified rate (90.0 %), but each model shows a large difference under different evaluation conditions. XAJ model has the highest peak occurrence time qualified rate (93.3 %); BP model has the highest qualified rate at the flood peak discharge (100 %); TOPMODEL has the highest runoff depth qualified rate (90 %), the highest Nash-Efficiency coefficient (0.840) and the lowest average *RSME* (39.8).

3.3 Ensemble flood simulation with multiple models

The entropy method was used to calculate the weight coefficients of the three models. In simulating different flood, each model is assigned with different weight coefficient as shown in Table 4. For the 30 floods, the average weight coefficients of the XAJ model, TOPMODEL and BP model are 0.347, 0.299, and 0.354, respectively. This indicates that the three models contribute differently to the ensemble results. The order of the three models is BP model, XAJ model, and TOPMODEL based on its weight from high to low. This indicates that, to some extent, simulation results of the BP model are better.

Figure 7 compared the hydrograph of the 820716 flood yielded by the three individual models and the ensemble. As seen from the figure, the ensemble flood hydrograph is closer to the observed flood hydrograph than all those yielded by single models.

Figure 8 gives the scatterplots between observed and simulated flood discharge by three single models and the multi-model ensemble. With R^2 greater than 0.8, simulated discharge by all three models has a good linear relationship with observations. The trend lines of the XAJ model and BP model are close to the 1 : 1 line, indicating their better discharge simulation performance. The trend line of the TOPMODEL is above the 1 : 1 line, indicating its simulated values tend to be higher than observed values. Higher than all three single models, the R^2 value of the multi-model ensemble reaches 0.935.

The average results of six indicators in the single model and the multi-model ensemble were compared, as shown in Table 5. It can be seen that compared with the single model, the multi-model ensemble method effectively reduces various errors and improves the Nash-Efficiency coefficient and the qualified rate. Among them, the ensemble results are better than the single model except the flood peak discharge and peak occurrence time. Although the relative errors of flood peak discharge is larger than that of BP model, relative to XAJ model and TOPMODEL is significantly reduced. The ensemble peak error is higher than TOPMODEL, but it is lower than the XAJ model and the BP model. Only two items are lower than the ensemble results, and the overall improvement is more significant.

In order to more intuitively understand the distribution of the improvement degree of the multi-model ensemble results relative to the single model, the box line diagrams of the improvement degree in each evaluation objective function were drawn, as shown in Fig. 9. As seen from Fig. 9a, the me-

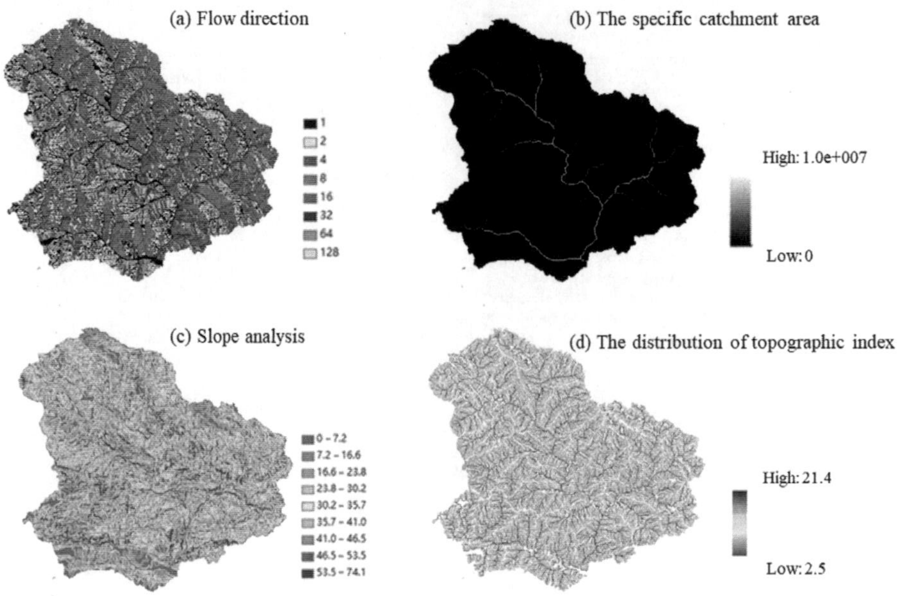

Figure 6. Topographic index calculation process.

Table 3. Comparison of results derived by single hydrological model for flood simulation.

Qualified rate	XAJ	TOPMODEL	BP model
Runoff depth qualified rate (%)	90.0	90.0	90.0
Flood peak discharge qualified rate (%)	60.0	56.7	100.0
Peak occurrencetime qualified rate (%)	93.3	93.3	80.0
Total qualified rate (%)	81.1	80.0	90.0
Average Nash-Efficiency coefficient	0.785	0.840	0.753
Average RSME	42.1	39.8	48.9

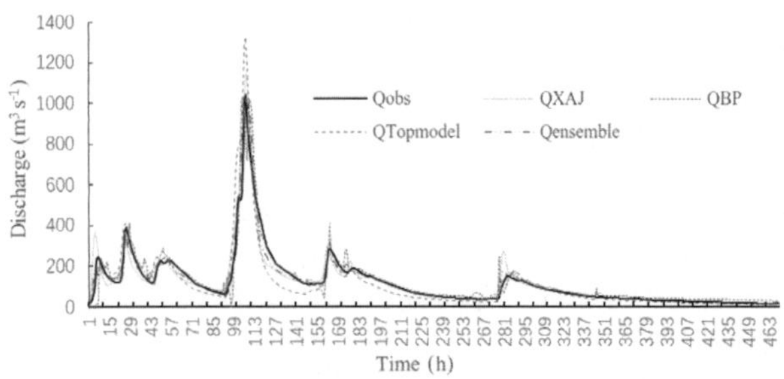

Figure 7. Observed and simulated hydrographs of 820 716 flood in the Yandu River Catchment.

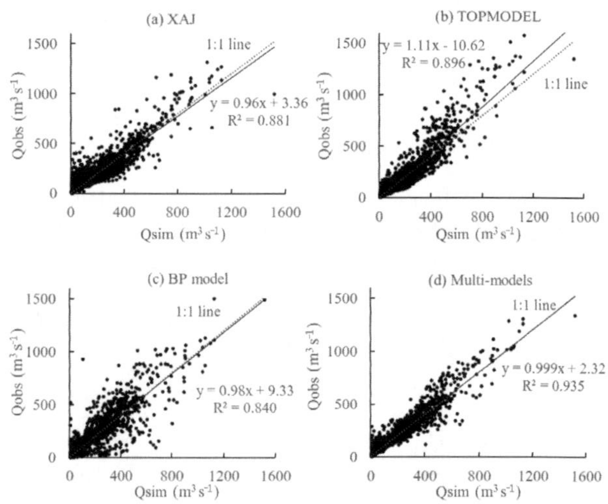

Figure 8. Comparison of the observed and simulated runoff of models.

dian, mean and interquartile range of the boxplots are similar, indicating that the ensemble plays a similar role in reducing the relative error of the event runoff depth for the three models. For the peak discharge, the XAJ model and the TOPMODEL box are basically in the positive range, while the BP model box is in the negative range, indicating that the ensemble has significant improvements for the first two

Table 4. Weight coefficients assigned to the three models based on the entropy method.

Flood number	810 623	810 714	810 810	810 824	820 716	820 820	820 908	830 623	830 721	830 822
XAJ	0.265	0.507	0.249	0.353	0.477	0.100	0.542	0.100	0.482	0.401
TOPMODEL	0.580	0.100	0.346	0.181	0.100	0.657	0.100	0.742	0.100	0.252
BP	0.155	0.393	0.405	0.466	0.423	0.243	0.358	0.158	0.418	0.347
Flood number	830 906	830 922	831 004	831 017	840 612	840 703	840 723	840 909	850 424	850 603
XAJ	0.554	0.467	0.486	0.100	0.190	0.528	0.419	0.141	0.533	0.252
TOPMODEL	0.100	0.100	0.100	0.878	0.549	0.100	0.100	0.489	0.100	0.390
BP	0.346	0.433	0.414	0.022	0.261	0.372	0.481	0.370	0.367	0.358
Flood number	850 621	860 615	860 714	860 909	870 511	870 622	870 627	870 719	870 821	870 827
XAJ	0.267	0.340	0.311	0.367	0.440	0.570	0.499	0.232	0.241	0.009
TOPMODEL	0.482	0.233	0.457	0.100	0.137	0.120	0.002	0.354	0.373	0.638
BP	0.251	0.427	0.232	0.533	0.422	0.310	0.499	0.414	0.386	0.353

Table 5. Statistics of evaluation indexes.

Average	Relative errors of runoff depth $\%^{-1}$	Relative errors of flood peak discharge $\%^{-1}$	Errors in time to flood peak h^{-1}	Nash-Efficiency coefficient	RSME	Qualified rate $\%^{-1}$
XAJ	9.31	20.94	2.1	0.785	42.09	85.9
TOPMDOEL	8.25	24.72	1.1	0.840	39.78	85.0
BP model	9.07	3.78	3.0	0.753	48.92	91.7
Ensemble	6.31	12.67	1.5	0.900	30.50	92.5

Note: (1) The bold font in the table indicates the item whose ensemble result reduces the accuracy of the single model result. (2) The error average value in the table is the average value of the absolute value of the relative error of each field.

models (Fig. 9b). The ensemble has the greatest improvement on TOPMODEL, with the box basically above the x-axis. However, the peak discharge accuracy of the BP model is decreased after multi-model ensemble. Moreover, for the peak time (Fig. 9c), the error in time to flood peak of BP is reduced the most, with an average decrease of 1 h. The XAJ model has a single flood error reduced by 24 h, which is the model with the largest error reduction. However, the error in time to flood peak of the other floods generally increased after multi-ensemble. As for Nash-Efficiency coefficient (Fig. 9d), the BP model has the greatest improvement, and the Nash-Efficiency coefficient increases by an average of 0.14. In conclusion, after the multi-model ensemble, the models were improved to different degrees in the accuracy except the individual models and several floods.

According to the improvement of the evaluation criteria above results, the number of ensemble results was count in this paper, as shown in Table 6. On the average, the multi-model ensemble performed better than all three models in simulating 43.4 % of the 30 flood events. It shows the most improvement in terms of the Nash-Efficient coefficient and *RSME* by yielding better results for more than 15 floods than all three single models. In addition, the multi-model ensemble givers better overall performance than two models for 34.68 % of the floods, and one model for 18.67 % of the floods. The ensemble accuracy of the event runoff depth for only one flood, the flood peak discharge for two floods, and the peak occurrence time for one flood is lower than the single model, accounting for a lower proportion. So it can be considered that the accuracy results obtained by multi-model ensemble are significantly improved compared to the single model.

3.4 Discussion

XAJ model and TOPMODEL both perform well, the manual-trial-error method was used to calibrate the parameters, which made the results of some floods relatively unsatisfactory. Because it takes into account the influencing factors such as basin topography and soil properties, TOPMODEL performs slightly better than the XAJ model. Based on a self-learning algorithm, the BP model is the best among the three models in discharge simulation. However, it is essentially a statistical model with little representation of physical mechanisms.

The purpose of ensemble method is to determine the weight and methods are various. Entropy-based ensemble method in this paper has many advantages comparing to the simple weighted average. The closer the simulated value and measured value, the larger the weight. Each flood has a set

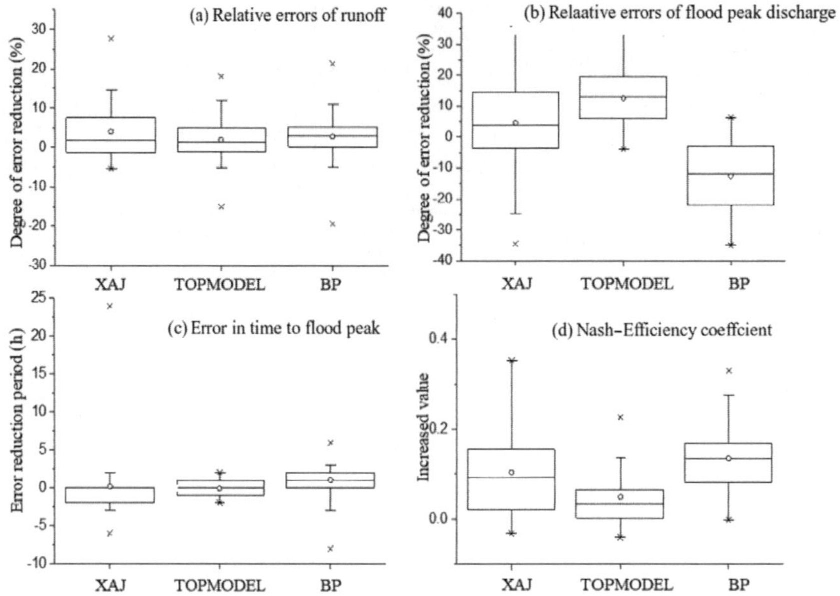

Figure 9. Multi-model ensemble changes relative to single model.

Table 6. Multi-model ensemble improvement.

Percentage of flood event $\%^{-1}$	Better than 3	Better than 2	Better than 1	Worse than 3
Relative errors of runoff depth	36.67	36.67	23.33	3.33
Relative errors of flood peak discharge	20.00	36.70	33.30	10.00
Errors in time to flood peak	46.70	20.00	30.00	3.30
Nash-Efficiency coefficient	56.67	43.33	0.00	0.00
RSME	56.70	36.70	6.70	0.00
Average	43.35	34.68	18.67	3.33

of weights, which can get the best ensemble scheme in each flood. In the future, more methods can be chosen for comparative discussion.

The central idea of this paper is to carry out flood simulation. This is the preliminary work of flood forecasting. For forecasting, it may be considered to classify the floods according to the characteristics of the floods, determine the set of weight parameters for each type, and then carry out the forecasting work, which needs further improvement and verification in the future.

4 Summary and conclusions

The XAJ model, TOPMODEL and BP model all perform well in the simulation of floods in the Yandu River catchment. Overall, the BP model has the highest forecast qualified rate (90 %). The Nash-Efficiency coefficient, runoff depth and peak time accuracy of the XAJ model and TOPMODEL are relatively high, while the flood peak discharge accuracy of the BP model is relatively high. Taking into account the overall simulation performance, the parameter failed to take care of all flooding floods. So the simulation accuracy of few floods is not ideal.

The entropy method was used to calculate the weight coefficients of the three models in the ensemble. Comparison of five model evaluation statistics had shown that the multi-model ensemble had improved flood simulation to various degrees. On the average, the multi-model ensemble reduces the relative errors of the runoff depth by 3.9 %, the relative errors of the flood peak discharge by 1.5 %, the error in time to flood peak by 0.4 h and the RSME by 10.9. At the same time, it increases the average Nash-efficiency coefficient by 0.1. Eighty percent of the flood ensemble results are better than at least two single model results.

Data availability. Data is available based on request to the corresponding authors.

Author contributions. JW contributed results analysis and drafting the manuscript, JZ and Guoqing Wang structured the manuscript and contributed results discussion, XS contributed methodology of the work, XY contributed analysis on results reasonability, YW contributed data collection.

Competing interests. The authors declare that they have no conflict of interest.

Special issue statement. This article is part of the special issue "Hydrological processes and water security in a changing world". It is a result of the 8th Global FRIEND–Water Conference: Hydrological Processes and Water Security in a Changing World, Beijing, China, 6–9 November 2018.

Acknowledgements. We are thankful to anonymous reviewers and editors for their helpful comments and suggestions.

Financial support. This research has been supported by the National Key Research and Development Program of China (no. 2016YFA0601501), the National Natural Science Foundation of China (grant nos. 41830863, 41330854, 51879162, 51609242, 51779146, 41601025), and the State Key Laboratory of Hydrology – Water Resources and Hydraulic Engineering (grant nos. 2017490211, 2015490411)

References

Andrew, W. R., Upmanu, L., Stephen, E. Z., and Balaji, R.: Categorical climate forecasts through optimal combination of multiple GCM ensembles, World Water and Environmental Resources Congress, 130, 1792–1811, https://doi.org/10.1061/40685(2003)184, 2003.

Arsenault, R., Gatien, P., Renaud, B., Brissette, F., and Martel, J. L.: A comparative analysis of 9 multimodel averaging approaches in hydrological continuous streamflow simulation, J. Hydrol., 529, 754–767, https://doi.org/10.1016/j.jhydrol.2015.09.001, 2005.

Arsenault, R., Gilles, R. C., and Francois, P. B.: Improving hydrological model simulations with combined multi-Input and multimodel averaging frameworks, J. Hydrol. Eng., 22, 66–76, https://doi.org/10.1061/(ASCE)HE.1943-5584.0001489, 2017.

Balint, G., Csik, A., Bartha, P., Gauzer, B., and Bonta, I.: Application of meterological ensembles for Danube flood forecasting and warning, in: Transboundary floods: Reducing risks through flood management, edited by: Marsalek, J., Stancalie, G., and Balint, G., Springer, NATO Science Series, Dordecht, the Netherlands, 57–68, 2006.

Bao, W.: Hydrological forecasting (4th Edn.), China Water Conservancy and Hydropower Press, China, 141–142, 2009.

Bates, J. M. and Granger, C. W. J.: The combination of forecasts, Oper. Res. Aquart., 20, 451–468, 1969.

Burnash, R. J. C.: The NWS river forecast system-catchment modeling, in: Computer models of watershed hydrology, edited by: Singh, V. P., Water Resour. Publ. Am., 4, 311–366, 1995.

Cloke, H. L. and Pappenberger, F.: Ensemble flood forecasting: A review, J. Hydrol., 375, 613–626, 2009.

Craft, T. J., Launder, B. E., and Suga, K.: Development and application of a cubic eddy-viscosity model of turbulence, J. Heat Fluid Flow, 17, 108–115, https://doi.org/10.1016/0142-727X(95)00079-6, 1996.

Davolio, S., Miglietta, M. M., Diomede, T., Marsigli, C., Morgillo, A., and Moscatello, A.: A meteo-hydrological prediction system based on a multi-model approach for precipitation forecasting, Nat. Hazards Earth Syst. Sci., 8, 143–159, https://doi.org/10.5194/nhess-8-143-2008, 2008.

Diks, C. and Vrugt, J.: Comparison of point forecast accuracy of model averaging methods in hydrologic applications, Stoch. Env. Res. Risk A., 24, 809–820, https://doi.org/10.1007/s00477-010-0378-z, 2010.

Epstein, E. S.: Stochastic dynamic prediction, Tellus, 21, 739–759, https://doi.org/10.1111/j.2153-3490.1969.tb00483.x, 1969.

Guan, X., Liu, Y., Jin, J., Liu, C., Liu, Y., and Wang, G.: Hydrological variation characteristics of typical watersheds in different climate regions of China, Journal of North China University of Water Resources and Electric Power (Natural Science Edition), 39, 13–17, 2018.

Hamill, T., Whitaker, J., and Wei, X.: Ensemble reforecasting: improving medium-range forecast skill using retrospective forecasts, Mon. Weather Rev., 132, 1434–1447, 2004.

IPCC (Intergovernmental Panel on Climate Change): Climate change 2014: Synthesis report. Geneva Switzerland: Contribution of working groups I, II and III to the fifth assessment report of the intergovernmental panel on climate change, 85–88, 2014.

Jasper, K., Gurtz, J., and Lang, H.: Advanced flood forecasting in Alpine watersheds by coupling meteorological observations and forecasts with a distributed hydrological model, J. Hydrol., 267, 40–52, https://doi.org/10.1016/s0022-1694(02)00138-5,2002.

Leith, C. E.: Theoretical skill of Montre Carlo forecast, Mon. Weather Rev., 102, 409–418, 1974.

Levenberg, K.: A method for the solution of certain nonlinear problems in least squares, Q. Appl. Math.s, 2, 164–168, https://doi.org/10.1090/qam/10666, 1944.

Lin, S.: Hydrological forecasting, 2nd Edn., China Water Conservancy and Hydropower Press, Beijing, China, 58–62, 2003.

Marquardt, D. W.: An algorithm for least squares estimation of nonlinear parameters, J. Soc. Ind. Appl. Math., 11, 431–441, 1963.

Ministry of Water Resources of the People's Republic of China: Hydrological yearbook published by Hydrology Bureau, China Water Conservancy Press, China, 1981–1987.

Ministry of Water Resources of the People's Republic of China: Hydrological information forecasting specification, China Water Conservancy Press, China, 5–6, 2008.

Mudasser, M. K, Asaad, Y. S., and Brucem, W. M.: Impact of ensemble size on forecasting occurrence of rainfall using TIGGE precipitation forecasts, J. Hydrol. Eng., 19, 732–738, https://doi.org/10.1061/(ASCE)HE.1943-5584.0000864, 2014.

Pitt, M.: Learning lessons from the 2007 floods: an independent review by Sir Michael Pitt: interim report, UK, 32–33, 2007.

Rockwood, D. M.: Application of streamflow synthesis and reservoir regulation – SSARR-program to the lower Mekong River. The use of Analog and Digital Computers in Hydrolog, 377–378, https://doi.org/10.1016/0002-9378(93)90010-G, 1968.

Roulin, E.: Skill and relative economic value of medium-range hydrological ensemble predictions, Hydrol. Earth Syst. Sci., 11, 725–737, https://doi.org/10.5194/hess-11-725-2007, 2007.

Rui, X.: Hydrology principle, China Water Conservancy and Hydropower Press, China, 3–4, 2004.

Song, X. and Kong, F.: Application of Xin'anjiang model coupling with artificial neural networks, Bulletin of Soil and Water Conservation, 30, 135–138, 2010.

Sugawara, M.: Analysis method of valley discharge, Koritsu publisher, Janpan, 110–116, 1972.

Takemasa, M. and Masaru, K.: The local ensemble transform Kalman Filter with the weather research and forecasting model: experiments with real observations, Pure Appl. Geophys., 169, 321–333, https://doi.org/10.1007/s00024-011-0373-4, 2012.

Vrugt, J. A., Gupta, H. V., Nuallain, B., and Bouten, W.: Real-time data assimilation for operational ensemble streamflow forecasting, J. Hydrometeorol., 7, 548–565, https://doi.org/10.1175/JHM504.1, 2006.

Werner, M.: FEWS NL Version 1.0 – Report Q3933, Delft Hydraulics, Delft, 2005.

Xu, Z.: Hydrological models, Science Press, China, 318–340, 2009.

Yang, X.: A review of real-time flood forecasting methods, Chinese Hydrology, 65, 9–16, 1996.

Zhao, R.: Hydrological simulation – Xin'anjiang model and Shanbei model, Water Power Press, Beijing, China, 106–118, 1984.

Zhao, Z.: The Foundation and Application of Fuzzy Theory and Neural Network, Tsinghua University Press, China, 125–129, 1996.

Proc. IAHS, 383, 223–228, 2020
https://doi.org/10.5194/piahs-383-223-2020

Operational flood forecasting and warning under the changing environment in China

Zhiyu Liu

Information Center, Ministry of Water Resources of China, Beijing, 100053, China

Correspondence: Zhiyu Liu (liuzymwr@163.com)

Published: 16 September 2020

Abstract. Flooding has been the most severe hazard in China from time immemorial, due to its special geographical, climatic and socio-economic conditions. In recent years, against the background of global climate change and rapid urbanization, extreme hydro-meteorological events have obviously increased in China, in turn affecting the sustainable social and economic development. This paper analyses how modern flood risk management has evolved from the early "build and protect" flood control approach to a broader flood risk management approach by analyzing what has happened in China and the role of major floods in this evolution. The development of flood forecasting models and systems in terms of how they have informed decision-making as flood risk management has evolved over the years. The challenging of recognizing and dealing with forecast uncertainty and flood risk in decision-making is also analyzed in the paper.

1 Introduction

Floods are the most frequent and serious natural disasters in China. Although a few floods in west China are subject to the mixed-type floods from snow and glacier melting as well as local storms, the majority of floods in China are caused by rainstorms in combination with the coastal storm surges. According to historical records, big flood disaster events averaged once every two years before 1949. Beginning from the twentieth century, many flood fatalities have been recorded in China. In 1931, disastrous floods hit the Huaihe River basin and the Yangtze River basin, affected over 51 million people and caused 400 000 deaths (SFCDRHO, 2010). The average annual death toll in recent floods indicates a decreasing trend from 8571 (1950s) to 875 (2010s) as flood control systems have been improved significantly since 1950 (MWR, 2018).

It is the exponentially increasing impact of flooding that has raised the profile of the practice of flood forecasting and warning. Since the late 1980s there has been a move away from the primacy of major structural measures for flood control towards a more integrated approach, of which flood forecasting and warning is a component. The Integrated Flood Management Concept encourages a shift from the traditional, fragmented and localized approach, towards the use of the resources of a river basin as a whole, while at the same time provides protective measures against losses due to flooding (Arduino et al., 2005; Linnerooth-Bayer and Amendola, 2003).

Due to global warming, social and environment changes, and climate systems, flooding is becoming more extreme, more widespread and more frequent in China, increasing the risks and the costs of flood disasters (Wu et al., 2013). This has given rise to increased emphasis on the improvement of operational flood forecasting and warning, and the enhancement and refinement of flood-risk (understood as a product of probability and consequences) management systems.

This paper presents an overview of Chinese experiences in coping with floods and gives lessons from disastrous flood events. The paper also describes the development of hydrological forecasting and prediction technology and methods that support flood management in China, focusing on recent advances in flood forecasting and warning. Furthermore, challenges and expectations in China's flood forecasting and warning are also presented in the paper.

Published by Copernicus Publications on behalf of the International Association of Hydrological Sciences.

2 Evolution of flood control and management under the changing environment in China

For thousands of years, China has suffered greatly from flooding on large and small rivers. Before the founding of the P. R. China in 1949, big flood occurred in China once in every two years with serious loss. In 1933, the Yellow River levee was breached at over 50 locations flooding and more than 3.6 million people suffered, of which18 thousand people died. After 1949 big floods occurred in large rivers (e.g. the Yangzte-Huaihe Rivers Flood in 1954, the Songhua River Flood in 1957, the Yellow River Flood in 1958, the Haihe River Flood in 1963, the Upper Huaihe River Flood in 1975, the Great Flood in 1998, and so on), yet the losses were far less serious because of the role of the completed flood control projects and that great efforts were made in flood fighting. For example, in the summer of 1954, big floods occurred over the Yangtze River Basin. The water level in the main stream of the Yangtze reached the highest record. The Jingjiang Flood Diversion Works and Dongting lake were to divert and store flood water. The city of Wuhan was protected.

Since 1949, the government has attached high significance to the prevention and control of floods. Continuous efforts have been exerted to summarize experiences and lessons from major flood events and the role played by innovative work mechanisms, enhancement of structural development, improvement of non-structural measures and this has led to remarkable progress in capacity building for flood prevention and management. The accomplishments of China in this regard during the past 7 decades encompass the following three phases of flood control and management.

2.1 Start-up phase: before the mid-1970s

The 1950s and 1960s saw increasing efforts to harness major rivers, control and prevent floods. In 1950, the State Flood Control Headquarters and the Yellow River Flood control Headquarters were formally established, followed by the setting up of the Yangtze River Flood Control Headquarters and some key provincial flood control headquarters in succession. During this period, based on the principles of attaching equal importance to enhancing flood-water storage and discharge capacity, with the emphasis on the latter, the comprehensive management planning for the Huaihe River, the Yellow River, the Yangtze river, and the Haihe River basin have been successively carried out, which laid a solid foundation for harnessing rivers afterwards. Moreover, an engineering system for flood control has been preliminarily constructed, including the construction of trunk embankments for protecting major rivers, building numbers of large and medium-sized reservoirs, completing flood storage and detention areas on major rivers and progressively training major rivers.

2.2 Development phase: from the mid-1970s to 1998

In August of 1975, the Great Flood of the Huaihe River caused by the heavy rainstorm and dam-failures, devastated Henan province, killing 26 000 people (SFCDRHO, 2010). A subsequent profound reflection on the "75.8" flood led to the transition towards a broader, comprehensive flood management approach where the emphasis shifted from engineering measures to a broad portfolio of structural and non-structural measures for managing flood risk. From this time onward, the national specification on design flood estimation was revised, and the standard of flood control projects has been significantly enhanced. Besides, investment on the water conservancy has been progressively increased, with the emphasis on dyke reinforcement and renovation on major rivers, danger control and reinforcement of dangerous reservoirs by stages and in groups, construction of flood storage and detention areas with a view to guaranteeing the safety of people living there. By 1998, continuous efforts have been made to improve the non-structural measures including development of early warning and forecasting systems, formulating flood disaster preparation plans, improvement of the legal framework, establishing the administrative leader responsibility system, strengthening the organizational and institutional system, and setting up flood relief rescue teams.

2.3 Improvement phase: from 1998 to the present day under the changing environment

In 1998, the most devastating floods occurred in some major rivers such as the Yangtze, Songhuajiang, Pearl and Min Rivers, which left over 4000 people dead and caused material damage of USD 30 billion (SFCDRHO, 2010). In the aftermath of the Great Flood of 1998, the central government launched the strategic plan encompassing post-disaster reconstruction, renovation of rivers and lakes, start construction of water conservancy, and developed the water governance policy, namely overall planning and comprehensive treatment, promoting what is beneficial and abolish what is harmful, attaching equal importance to increasing income and reducing expenditure, and putting equal emphasis on flood-control and anti-drought. Thus, flood control work entered a new period of development.

The flood risk concept was introduced and gradually applied in flood management in China. In 2003, the China Ministry of Water Resources proposed to re-direct flood prevention from flood control to flood management, which aims to enhance understanding about interconnecting systemic issues and risk awareness. The flood prevention efforts thus shifted from attempts to eliminate floods forward to building capacity to endure floods with certain degrees of risk. In flooding areas, community management was introduced to regulate human activities. Meanwhile, efforts were also taken to utilize storm water as a complementary resource for water supply.

In recent years, against the background of global climate change and rapid urbanization, extreme hydrometeorological events have obviously increased in China. Local storms, sudden flash floods, urban water-logging, extreme high temperature drought and super typhoon events occurred in a sudden, more frequent and concurrent manner. For example, from 9 to 11 August 2007, affected by two typhoons, an extreme storm event hit the Leizhou Peninsula on the western part of Guangdong Province; the 24-hourly rainfall reached 1188.2 mm, a new heavy rainstorm record in China. Floods happened in 2010 were the worst since 1998, which big investment in flash-flood prevention and harnessing of medium and small rivers. On 12 July 2012, a rarely-seen rainstorm hit Beijing. The daily rainfall reached one quarter (1/4) to one third (1/3) of the yearly total. The mean areal rainfall is 170 mm, a historical record since the year 1949. In the rainstorm center, Hebei Town of Fangshan District, the rainfall reached 541 mm, and the recurrence interval is 500 years. Moreover, from 7 to 12 July 2013, an extreme rainfall process appeared in northern Sichuan Province of China, and the accumulated rainfall is 1151 mm in Dujiangyan City, with a recurrence interval of 100 years. These extreme rainstorms resulted in disastrous floods of riverine floods, flash flooding and urban water-logging.

Research shows that the increased frequency and intensity of flooding in typical southern river basins with high flood risk, and small and medium-sized rivers became prominent, and urban flooding and water-logging caused by storms in metropolises showed an increased trend. Heavy rain, floods and other extreme events may increase in occurrence and in intensity in future in parts of China, so as to increase the risk of flood disasters and accelerate the difficulty of flood-prevention and dispatch command, and thus causing unfavorable influences on the economic and social sustainable development.

3 Development of operational flood forecasting and warning in China

In response to the changing environment, the hydrological services in China have taken efforts in enhancing the monitoring network, strengthening monitor over emergency water events, and improving the monitoring, forecasting and warning in support of flood risk management in the country.

3.1 Hydrological monitoring

Hydrological monitoring stations in China are divided into two types: state level basic stations and special stations. By the end of 2018, there are 113 245 hydrometric stations maintained by the water sector in China, including 7102 hydrological stations (3148 state-level basic hydrological stations), 13 579 water level stations, 54 477 rainfall stations, 2571 soil moisture stations, 19 147 groundwater monitoring stations, among which there are 59 104 flood-reporting stations and 2100 flood forecasting sites.

At present, hydrological observation items include precipitation, evaporation, stage, discharge, sediment, water quality, ground water, water temperature, ice slush etc. The types of hydrological stations in China include perennial stationary gauging, stationary gauging during flood season, full-year touring gauging and entrust gauging etc. The discharge measurement is done mainly by using cableway and hydrometric boat. Current-meter method is commonly used for discharge measurement while the float method, discharge calculation by hydraulic method, flow measurement by hydraulic structure are used in the case that ordinary hydrometry facilities are destroyed due to major floods. Presently, some new techniques and methods such as the moving-boat method by using Acoustic Doppler Current Profiler (ADCP) are being applied to flow measurement in China.

3.2 Hydrological data transmission

Real-time hydrological information are transmitted by means of GSM, GPRS and telephone, satellite, transmitter-receiver to local information centers, and then transferred to regional information centers and the Information Center of MWR, through Wide Area Network (WAN). The information are automatically received, processed and stored by computer system.

Hydrological information transmission technology in China has been improved. By 1995, the traditional telegraph transmission way has been replaced by public packet switched data network (CHINAPAC X.25) to achieve a countrywide real-time hydrological information transmission by WAN. By 2005, based a wide band network of 2 Mbps, using hydrological information code way, the real-time hydrological information has been transmitted rapidly. By 2011, the Hydrological Information Transmission System replaces the hydrological information code way to realize a comprehensive and efficient transmission for hydrological information. At present, over 80 % of the hydrological stations measure and report information in an automatic way, about 98 % of the hydrological information from the hydrological stations can be transmitted to Information Center, MWR, within 15 min.

3.3 Hydrological forecasting method and model

The constraints of hydrological data availability and computer science in the past limited China's hydrological forecasting methods to those based on experience. In the 1960s, hydrologic models started to be developed and applied following the introduction of mathematical simulation into the field of hydrology. China focused on bringing in suitable hydrologic models and developing those to be applicable to local hydrological conditions. Currently there are basically two types of hydrological forecasting methods: empirical hydrological forecasting method and that using conceptual hydrologic models.

The empirical hydrological forecasting method comes from the experiences that hydrological practitioners gathered over a long time, and this proves to be very effective in operational hydrological forecasting. At present, the 7 major river basin authorities in China have compiled complete empirical hydrological forecasting schemes, and the 600 plus hydrological forecasting stations in China have nearly 1000 schemes. In general, the empirical hydrological forecasting schemes apply such methods as antecedent precipitation index (API), corresponding stage (discharge) method, resultant discharge method, stage (discharge) fluctuating rate method, multi-factors combined axes correlation method, rainfall runoff correlation method, and so on.

China has extensive territories, with varied underlying surface conditions and different climates such as humid, dry, semi-humid and semi-dry climates. For decades, Chinese hydrologists have made efforts to learn from international hydrologic models and succeeded in developing a series of river basin hydrologic models applicable to local conditions. The Xin'anjiang model proposed by Zhao Renjun in 1973 is the representative model that was developed. The main feature of the model is the concept of runoff formation on repletion of storage, which means that runoff is not produced until the soil moisture content of the aeration zone reaches the field capacity, and thereafter runoff equals the rain fall excess without further loss. The model can be used for real time flood forecasting, investigations for gaining understanding of hydrological processes and study of the impacts of land use change on runoff, etc. Nowadays, hydrologic models used in operational hydrological forecasting in China comprise three types: those developed by Chinese professionals, international models and adapted international models. Locally developed models include examples of the Xin'anjiang model (Zhao et al., 1980; Zhao, 1992; Zhao and Liu, 1995), the double excess runoff yield model, the Hebei storm flood model, the Jiangwan Bay runoff model and the double attenuation curve model. Tank, Sacramento, NAM and SMAR are major international models that are used, while the continuous API model and SCLS model were also introduced and modified to adapt to local conditions (WMO, 2011).

In recent years, environmental changes brought challenges to hydrological forecasting and warning. Distributed hydrological models based on underlying terrain geographic information were developed and started to be used in flood forecasting and warning for flash floods as well as medium-small river flooding. Besides, the concept of dynamic critical rainfall index based on soil moisture was developed, and has been applied to medium-small rivers for flash flood early warning.

3.4 Hydrological forecasting systems

The computer science and network technology are indispensable for the development and application of Chinese hydrological forecasting systems. In the early 1980's, China developed the real-time flood forecasting system run on the VAX machines and on a single computer. As the consequence of the limited database management and software development technology at that time, data were mostly managed in a file system and on-line operational forecasting was at a very low level with very simple functions, mainly programming existing forecasting schemes and models. From 1990 onward, hydrologists in China learned from peers in developed countries with more mature and advanced technologies, and achieved progress in developing forecasting systems with functions of database management, model parameter calibration and optimized calculation, operational forecasting, real-time forecast updating, and visualized data and forecasting results. After the 1998 Great Flood, the Chinese hydrological sector made efforts to enhance the hydrological monitoring capacity and develop the universal flood forecasting system software platform: China National Flood Forecasting System (CNFFS) (Zhang and Liu, 2007). The CNFFS consists of three major components in a calibration system, operational forecast system and extended stream flood prediction system, which are shown in Fig. 1. The CNFFS contains common forecasting models and a method library allowing prompt building of forecasting schemes. Many different flood forecasting models are employed in the system, including the Xinanjiang model, and other models such as API, Sacramento, Tank, SMAR, and the synthetic constrained linear system (SCLS), as listed in Table 1.

Table 1. Flood forecasting models employed in the CNFFS.

No.	Name
1	Xin'anjiang Model
2	API Model
3	Jiangwan Runoff Model
4	Hebei Storm Flood Model
5	Shanbei Model
6	Xin'anjiang Model for Semiarid Area
7	Liaoning Model
8	Double Attenuation Curve Model
9	Double Excess Runoff Yield Model
10	SMAR Model
11	NAM Model
12	Tank Model
13	Sacramento Model
14	SCLS Model
15	Index Recession Method
16	Recession Curve Method
17	Unit Hydrograph Method

Most large reservoirs and key medium-sized reservoirs built for flood prevention purpose have established telemetry systems, flood forecasting and regulation systems. The CNFFS has been successfully used in 33 flood forecast cen-

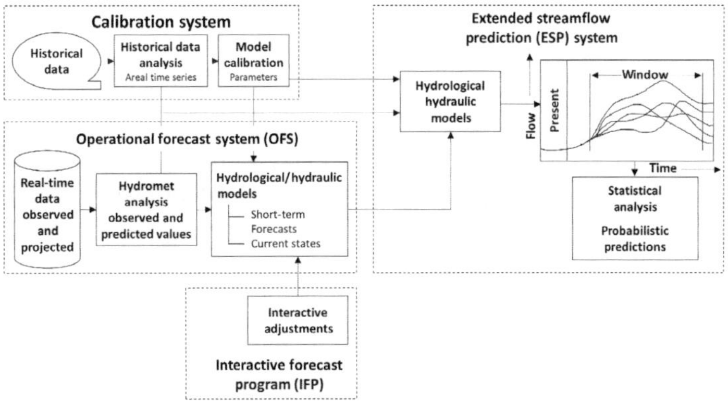

Figure 1. Overall structure chart of China National Flood Forecasting System.

ters at national, river-basin and provincial levels in support of flood management in the country.

3.5 Application of new technology in flood forecasting and warning

The scientific and technological development in recent years enables China to apply new and advanced technology to the hydrological forecasting, which improves the forecasting software and operational forecasting accuracy. The following are three representative examples.

3.5.1 Application of NWP-Based QPF in flood forecasting and reservoir operation

In real-time operational forecasting, the rainfall and flood joint forecast technology applies satellite and radar images for quantitative precipitation estimation (QPE), NWPs for QPF. In the first step, the meteorological department provides data needed by the hydrological department for rainfall forecasts and prediction in terms of location, quantity and time. The next step is the flood forecast. The results of the two steps are then combined to improve the lead time and prediction accuracy of hydrological forecasting and prediction. Recently, the availability of multiple NWP models from various weather services such as China Meteorological Administration, Japan Meteorological Agency, European Centre for Medium-Range Weather Forecasts (ECMWF), offers an opportunity for the development of hydro-meteorological coupling and flood forecasting and operation coupling systems in China, and this approach has been proved successful in operational application. The Three Gorges reservoir experienced three major floods in the summer of 2010. In July 2010, it reached a flood peak of $70\,000\,\mathrm{m^3\,s^{-1}}$ exceeding the peak in 1998 Yangtze River Floods. During that time, national hydrological forecasting service issued a timely, accurate flood forecasting 2 d in advance, which was performed by combining NWPs with national flood forecasting system.

3.5.2 Application of distributed hydrological models for flash flood forecasting

More recently, deterministic spatially distributed hydrologic models have been gradually used in China. Based upon equations representing the storage and the movement of water in the soil and on the surface, distributed models are potentially easier to calibrate by relating the values for their physically meaningful parameters to additional information provided in the form of Digital Elevation Maps, Soil Maps and Land Use Maps. These models account for the spatial distribution of rainfall (now available at pixels of between 1×1 and $10 \times 10\,\mathrm{km}$ from RADAR images or in terms of QPF from NWP models) and are better equipped to address the problem of extrapolating the parameter values to the un-gauged sub-catchments, a problem that has not yet found a proper solution. All these potential properties, combined with the increased efficiency of computers, have given rise to a number of fully distributed rainfall-runoff models, some of which, for instance, the TOPKAPI model (Liu et al., 2008), have been incorporated into real time flood forecasting systems, and applied to flash flood forecasting and warning in a number of medium and small sized rivers in Jiangxi, Hunan, Hubei, Anhui, Zhejiang, Henan, Shanxi provinces (Liu et al., 2015).

3.5.3 Application of RS technology in flood monitoring and forecasting

Recently, RS technology with medium and high-resolution data has been widely used in China in water resources monitoring and evaluation and management, flood and drought monitoring and prediction, and soil and water conservation monitoring and evaluation. Recently, RS data of MODIS, HJ, HS-1, NOAA/AVHRR, NPP/VIIR, SPOT, Landsat-2 have been of operational use in China, e.g. in the Great Flood of Heilongjiang River in 2013. After the levee breach happened, water levels at the downstream of the River stopped rising. The question raised for forecasters is "how big is the discharge from the breach, and will the water level on downstream will rise further? If it goes up again, how much?" In this case, the accurate water forecast analysis is directly related to the flood control scheme and the scientific decision-making. Through RS technology and DEM data analysis, the hydrological department continuously monitors the submerged area, water depth and water volume, estimates the discharge from the breath and inundated water-volume, and then predicts the flood peak water level of the downstream, providing important information support for the decision-making and deployment for flood control and emergency response.

4 Conclusions and outlook

Flood disasters in the past decades have become more frequent and devastating in China. The Chinese Government gives high priority to flood management through both structural measures and non-structural measures with the aim of improving capacity and level of flood disaster prevention and mitigation. As the most significant non-engineering measure for flood control and disaster reduction, flood forecasting and warning plays an important role in support of decision-making for masses evacuation, rescue and relief, reservoir optimum operation, emergency response etc., bringing eminent economic and social benefits.

Entering 21st century, flood forecasting and warning shall be further applied to agriculture and rural areas, city construction, traffic transportation, tourism and other industries, for flood disaster prevention and mitigation, water resources management, economic and social development and other water-related activities to provide more comprehensive and accurate information services.

It is anticipated that impact-based flood forecasting and warning will be developed in the future in China. Flood forecasting must be done to predict the possible impacts of flood and water-logging on rivers, lakes and reservoirs, cities, transportation and agriculture, etc., realizing the transition from hydrological forecasting to disaster impact forecasting, from making hydrological forecasts at sites, to simulating hydrological cycle process at sites, flood propagation along rivers, and flood inundation mapping. The probability of occurrence of flood and water-logging disasters will be forecasted, and the flood security risk of reservoirs, industry and agriculture production, city transportation and other objectives can be assessed, linking together hydrological forecasts of discharge with high resolution flood mapping, and realizing the transition from the threshold-based flood warning to the risk-based flood warning. Meteo-hydrological ensemble prediction and multi-model based probability forecast technology will be employed for this purpose.

Data availability. No data sets were used in this article.

Competing interests. The authors declare that they have no conflict of interest.

Special issue statement. This article is part of the special issue "Hydrological processes and water security in a changing world". It is a result of the 8th Global FRIEND–Water Conference: Hydrological Processes and Water Security in a Changing World, Beijing, China, 6–9 November 2018.

Financial support. This study was supported by the National Key Research and Development Program of China – study and application of key technologies for flood control and emergency management of small and medium-sized rivers (grant no. 2018YFC1508100).

References

Arduino, G., Reggiani, P., and Todini, E.: Recent advances in flood forecasting and flood risk assessment, Hydrol. Earth Syst. Sci., 9, 280–284, https://doi.org/10.5194/hess-9-280-2005, 2005.

Linnerooth-Bayer, J. and Amendola, A.: Introduction to special issue on flood risks in Europe, Risk Anal., 23, 537–543, 2003.

Liu, Z., Tan,B., Tao, X., and Xie, Z.: Application of a distributed hydrologic model to flood forecasting in catchments of different conditions, J. Hydrol. Eng., 13, 378–384, https://doi.org/10.1061/(ASCE)1084-0699(2008)13:5(378), 2008.

Liu, Z., Hou, A., and Wang, X.: Flood forecasting technology for medium and small rivers based on distributed hydrological model, J. China Hydrol., 35, 1–6, 2015.

MWR – Ministry of Water Resources: Bulletin of flood and drought disasters in China 2017, China Water Power Press, Beijing, 82–84, 2018.

SFCDRHO – The State Flood Control and Drought Relief Headquarters Office: Training materials for specialized cadres in flood control and drought relief, China Water Power Press, Beijing, 16–17, 2010.

WMO: Manual on Flood Forecasting and Warning, WMO Publication No. 1072, World Meteorological Organization, 142, 48–49, https://doi.org/10.1016/j.transproceed.2010.08.027, 2011.

Wu, Z. Lu, G., Liu, Z., Wang, J., and Xiao, H.: Trends of extreme flood events in the pearl river basin during 1951–2010, Adv. Clim. Change Res., 4, 110–116, 2013.

Zhang, J. and Liu, Z.: Hydro-meteorological monitoring and operational hydro-systems for flood management in China, in: IAHS Publ. 311, Proceedings of the International Symposium on Methodology in Hydrology, Nanjing, China, 3–9, 2007.

Zhao, R. J.: The Xinanjiang model applied in China, J. Hydrol., 135, 371–381, https://doi.org/10.1016/0022-1694(92)90096-E, 1992.

Zhao, R. J. and Liu, X. R.: The Xinanjiang model, in: Ccomputer models of watershed hydrology, edited by: Singh, V. P., Water Resources Publications, 215–232, 1995.

Zhao, R. J., Zhuang, Y. L., Fang, L. R., Liu, X. R., and Zhang, Q. S.: The Xinanjiang model, in: Hydrologica Forecasting Proceedings Oxford Symposium IAHS, 129, 351–356,1980.

Proc. IAHS, 383, 229–235, 2020
https://doi.org/10.5194/piahs-383-229-2020

Open Access PIAHS

Extreme hydrological phenomena in the forest steppe and steppe zones of Ukraine under the climate change

Valeriya Ovcharuk[1], Eugene Gopchenko[1], Nataliya Kichuk[1], Zhannetta Shakirzanova[1], Liliia Kushchenko[1], and Mariia Myroschnichenko[2]

[1]Hydrometeorological Institute, Odessa State Environmental University, Odessa, 65016, Ukraine
[2]Department of Ukrainian studies and social sciences, Odessa State Environmental University, Odessa, 65016, Ukraine

Correspondence: Valeriya Ovcharuk (valeriya.ovcharuk@gmail.com)

Published: 16 September 2020

Abstract. Nowadays, during the period of the global climate change, scientists around the world have noticed an increased incidence of extreme natural disasters. The authors of the study suggested methods of using climate change as a part of a genetic model of maximal floods runoff. This model makes possible the introducing of "climate changes" directly through the maximal stocks of snow and precipitation during the spring flood and runoff coefficients. The object of study is the basin of the Southern Bug – one of the largest rivers in Ukraine, which flows within two geographical zones – forest-steppe and steppe. Overall results using scenario RCP 4.5 showed a decrease of runoff by the end of 2050 from 20 % in the north part of the basin (the forest-steppe zone) to 50 % – in the south (the steppe part of basin). On the other hand, the characteristics of minimal river flow in the Southern Bug basin – winter and summer low waters had been analyzed. One option for low water flow simulation is to study the connection between the drought index (e.g. SPEI) and the minimal water. Studies show that such correlation exists, so knowing the forecast of drought index it is possible to project the value of minimal water discharge, which is the object of the research.

1 Introduction

Global climate change has been studied in recent decades, practically all over the world. Thus, according to the leading experts (CRED, 2019), natural disasters causing the negative impact on the development of the world economy, have recently become more common on our planet. For the period from 2007 to 2016 the number of natural disasters has been only 354, whereas in 2017 it rose to 335! In particular, in Europe, according to the European Environment Agency (EEA, 2019), in the EEA member states, the total economic losses from climate-related extremes over the period 1980–2017 reached approximately EUR 453 billion. The average annual economic losses in the EEA member states varied from EUR 7.4 billion in 1980–1989 to EUR 13.0 billion in 2010–2017. The most expensive climate-related extremes in the EU member states included floods in Central Europe in 2002 (over EUR 21 billion), drought and heat waves in 2003 (almost EUR 15 billion), the 1999 Lothar winter storm and floods in Italy and France in October 2000 (both were about EUR 13 billion). Thus, the considerable economic losses and potential danger from climate-related extremes are of great interest to scientists all over the world and they need to be studied, systematized and modeled for probabilistic and operational forecasting. Among the results of the last 5 years the Chinese scientists research on the effects of solar activity on floods and droughts (Li et al., 2015), an assessment of the effects of climate change on hydrological extremes – floods and droughts within the territory of Poland (Romanowicz, 2017); a thorough corporate study by European scientists on the effects of climate change on the water resources of Europe's largest water artery – the Danube Basin (Bisselink et al., 2018) should be noted. Similar researches are done about the rivers of Algeria (Kouidri et al., 2019) and the Canadian Prairies (Akhter and Azam, 2019). In Ukraine, the research of the water regime of rivers and its changes nowadays is actively done by the scientists of Taras Shevchenko National University and their colleagues from the Potsdam Institute for Climate Impact Study (Lobanova et al., 2018), representatives of the Ukrainian Hydrometeorological Institute (Zabolotnia et al., 2019) and Odessa State Environmental University (Loboda et al., 2019).

2 Methodology and data

Mathematical models are very important in planning water resources management and, consequently, different types of models with varying degrees of complexity have since been developed and still improved by scientists from different countries (Zhao et al., 1980; Singh, 1988; Bergstorm, 1992; Kan et al., 2017). Nowadays, scientific literature presents more advanced conceptual models with distributed parameters and even models using artificial neuronal networks (Sajikumar and Thandaveswara, 1999; Tokar and Markus, 2000), which makes it possible to determine the rivers runoff at separate catchment's areas with sufficient accuracy, using a large amount of source information, but for ungauged catchments, such models are not applicable. The basis of the calculation of the maximal flow of floods in the considering territory is the theory of channel isochrones, implemented in the form of a two-operator model of a river runoff formation, developed at the Odessa State Environmental University. Its advantage is that it is universal both in terms of the genetic type of floods, and the size of catchment's areas.

The operator model (of Gopchenko and Ovcharuk, 2006) makes it possible to determine separately the characteristics of the slope inflow (the depth of flood runoff, the duration of the slope inflow and the temporal non uniformity of slope inflow to the channel network) and to take into account the natural process of precipitation transformation into the river bed in the structure of "slope inflow – channel runoff".

Ovcharuk and Gopchenko (2015), modified that operator model next which allowed to take into account "climatic correlations" directly at maximal snowfalls, precipitation and runoff coefficients during the period of floods. The block diagram of this model is presented in Fig. 1.

The modified version of the operator model is proposed as a calculation method determining the maximal runoff of ungauged rivers during the spring floods period, and the calculated equation is:

$$q_p = q'_{1\,\%}\psi(t_p/T_0)\varepsilon_A r \lambda_p k_{ch}, \quad m^3\,(s\,km^2)^{-1}, \tag{1}$$

Where r – is the coefficient which takes into account the influence on q_m circulating water bodies (lakes, water reservoirs, ponds); ε_A – is the coefficient of channel and flood plain regulation that depends on catchments' size; $\psi(t_p/T_0)$ – is the transformation function influenced by channel lag; λ_p – is the transition coefficient from the 1 % probability of excess to any other; $q'_{1\,\%}$ – is the maximal slope modulus that

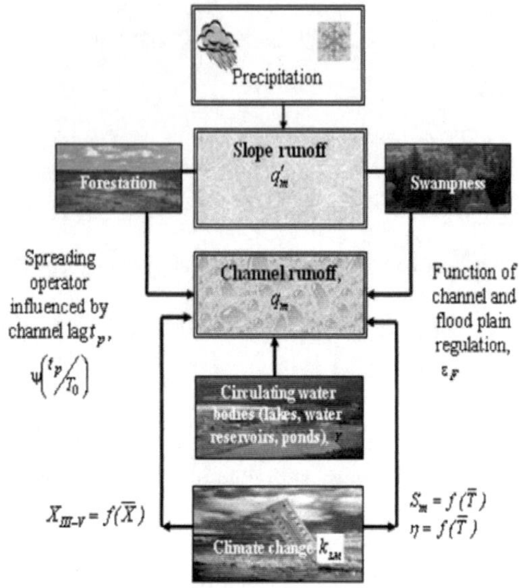

Figure 1. Block diagram of maximal runoff formation.

is equal to

$$q'_{1\,\%} = 0.28\frac{n+1}{n}\frac{1}{T_0}(S_m + \Sigma P)_{1\,\%}\eta, \quad m^3\,(s\,km^2)^{-1}, \tag{2}$$

where $(n+1)/n$ – is the coefficient of time nonuniformity of slope inflow; T_0 – is the duration of slope influx, hours; η – is the runoff coefficient; S_m – is the maximal stock of snow, mm; ΣP – is the precipitation during the spring flood, mm; k_{ch} – is the coefficient of climate change that is equal to

$$k_{ch} = \frac{\left(\left(\overline{S}_m + \Sigma\overline{P}\right)\cdot\eta\right)\text{forecast.}}{\left(\left(\overline{S}_m + \Sigma\overline{P}\right)\cdot\eta\right)\text{modern.}}. \tag{3}$$

In Eq. (3) the average forecast values of the average maximal stocks of snow, precipitations and spring floods coefficient over a longtime period, are determined using their dependencies, from the forecasted temperature and precipitation with using any scenario and model for the selected calculating period. The same actual values are obtained from the results of calculations for available time series of meteorological observations.

Standard methods of statistical processing and their generalization in space in the form of isoline maps have been used to determine the calculating characteristics of the dry weather flow. In order to study the relation between the minimum river runoff and hydrological droughts, the standardized precipitation-evapotranspiration index (SPEI), proposed in the works of Vicente-Serrano has been used (Vicente-Serrano et al., 2010). The winter and summer–autumn dry weather flow has been considered, and the statistical relation between the SPEI index and the runoff parameters has been estimated using the linear Pearson correlation coefficient (Semenova et al., 2014).

 https://doi.org/10.5194/piahs-383-229-2020

Figure 2. Location of the Southern Bug basin.

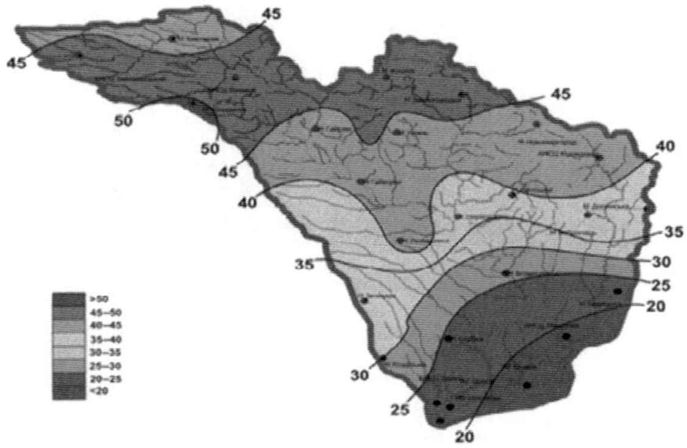

Figure 3. Distribution of average long-term maximal stocks of snow in the basin of the Southern Bug River, mm.

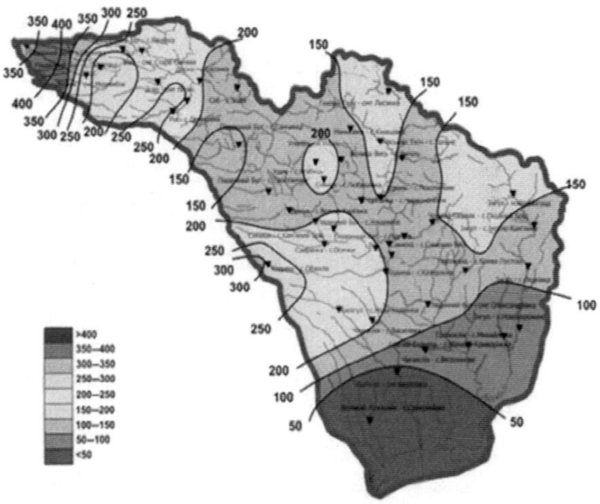

Figure 4. Distribution of the duration of sloping influx during the spring flood in the basin of the Southern Bug, hours.

The object of the study is the territory of the Southern Bug river basin, which lies within two natural zones (the upper part of the basin is the forest-steppe zone, and the lower part is the steppe) and includes the main watercourse – the Southern Bug river, its main tributaries – the rivers Riv, Sob, Savranka, Kodima, Sinukha, Ingul, and others (Fig. 2).

Output data on the characteristics of maximal runoff have been used as long-term data series for 39 water gauging station (WGS), and for the minimal runoff characteristics the data of 35 WGS have been used, which have long time series of runoff observations from their beginning to 2010. The range of catchment's areas covers basins in the size from 36.5 km^2 (river Southern Bug – Chorniava village) to 46 200 km^2 (river Southern Bug – Oleksandrivka village).

3 Results

3.1 The maximal runoff

During the spatial-temporal generalizations of the maximal runoff characteristics, the cycle of the fluctuations of the maximal runoff of spring flood has been analyzed and homogeneity of the initial information has been estimated. Statistical processing of the initial time-series of maximal snow supplies, maximal discharges and runoff depths has been carried out. For the determination of precipitation in the spring period, the regional calculation formula is proposed, and the maximal stocks of snow (Fig. 3) and the coefficients of their variation are generalized in a form of the map.

The runoff coefficients are determined through the coefficients of runoff formation, which are generalized in a form of the map and taking into account the coefficients of the influence of the size of the catchments on the losses of the runoff in the spring flood period. The characteristics of the slope influx, which are an important component of the calculation scheme, are represented by the maximal slope modulus, which in turn is determined by the coefficient of unevenness of the sloping influx, the duration of the flow into the channel network and the total water supplies to the catchments. All listed parameters are valid for the studied territory, in particular, the duration of the sloping influx represented on Fig. 4.

The transformation of the maximal slope modulus is represented by functions that take into account the channel time, flood-plain regulation and the impact of lakes and reservoirs. For the determination of the transformation function and the coefficients of flood-plain regulation, the equations of exponential form are derived, with separate parameters in natural zone Eqs. (4)–(5)

$$\psi(t_p/T_0) = e^{-\gamma(\lg(A+1))}, \quad (4)$$

where γ – is the exponent, the value of which for the forest-steppe zone is 0.43, and for the steppe is 0.48.

$$\varepsilon_A = e^{-0.28\lg(A+1)}, \quad (5)$$

where A – the catchments areas, km^2.

The calculation of the climate change coefficients for the rivers of the Southern Bug basin by the model RACMO2,

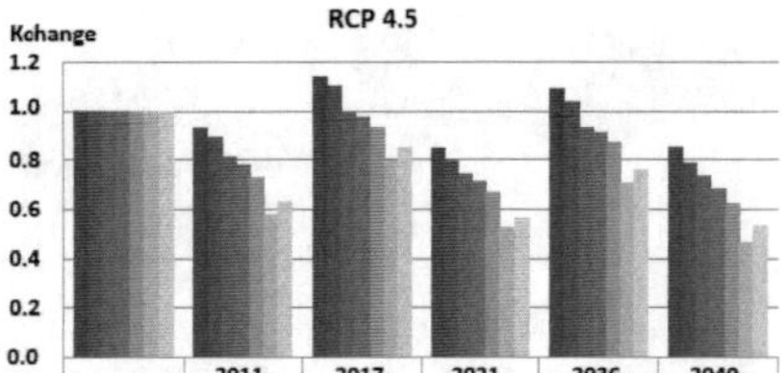

	2010	2011–2016	2017–2020	2021–2035	2036–2039	2040–2050
Southern Bug - v.Pyrohivtsi	1	0.93	1.14	0.84	1.09	0.85
Sob -v.Zoziv	1	0.89	1.10	0.79	1.04	0.79
Southern Bug - v. Oleksandrivka	1	0.81	0.99	0.74	0.93	0.73
Hnylyy Tikych-v. Lysyanka	1	0.78	0.97	0.71	0.91	0.68
Ingul- city Kirovograd	1	0.73	0.93	0.66	0.87	0.62
Hromokliya-v.Mykhaylivka	1	0.58	0.79	0.52	0.70	0.46
Tylihul-v.Berezivka	1	0.63	0.85	0.56	0.75	0.53

Figure 5. Expected average annual values k_{ch} for taking into account the influence of climate change on the maximal runoff of spring flood of the rivers of the Southern Bug basin using the model RACMO2, scenario RCP4.5

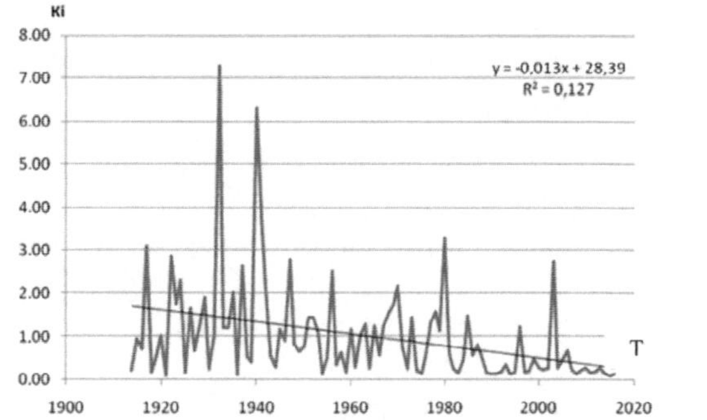

Figure 7. The time series of maximal water discharges of spring floods on the river Southern Bug – village Oleksandrivka, $A = 46\,200\,\mathrm{km}^2$.

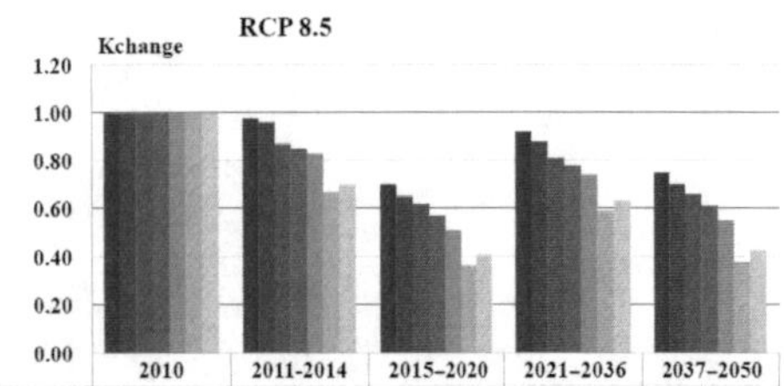

	2010	2011-2014	2015-2020	2021-2036	2037-2050
Southern Bug - v.Pyrohivtsi	1	0.98	0.70	0.92	0.75
Sob -v.Zoziv	1	0.96	0.65	0.88	0.70
Southern Bug - v. Oleksandrivka	1	0.87	0.62	0.81	0.66
Hnylyy Tikych-v. Lysyanka	1	0.85	0.57	0.78	0.61
Ingul- city Kirovograd	1	0.83	0.51	0.74	0.55
Hromokliya-v.Mykhaylivka	1	0.67	0.36	0.59	0.38
Tylihul-v.Berezivka	1	0.70	0.41	0.63	0.43

Figure 6. Expected average annual values k_{ch} for taking into account the influence of climate change on the maximal runoff of spring flood of the rivers of the Southern Bug basin using the model RACMO2, scenario RCP8.5

RCP 4.5 and RCP 8.5 scenarios is presented in the form of diagrams (Figs. 5 and 6).

Consequently, analyzing the obtained results, we can conclude that in the basin of the river Southern Bug the value of runoff during spring floods can significantly change in the future. The analysis of the results obtained in the scenario RCP 4.5 shows that there is a certain cycle in the climate change dynamics. The $k_{ch} = 0.93$ indicates a spring runoff decrease for the period 2011–2016 in the upper part of the basin (the Southern Bug river – Pirogovtsy village), and $k_{ch} = 0.58$ in the lower part of the basin (the river Gromokliya – Mikhailivka village), that means that spring runoff is expected to decrease almost twice.

For the period (2017–2020), the runoff decrease is forecasted, compared with the previous period, so $k_{ch} = 1.14$ and $k_{ch} = 0.79$, respectively. For the period 2021–2035 the climate change coefficients equal $k_{ch} = 0.75$ (river Southern Bug – village Pirogovtsy) and $k_{ch} = 0.38$ (river Hromokliya – Mikhailivka village), which suggests a spring runoff decrease. For the period 2036–2039 k_{ch} is 1.09 and 0.70, respectively, that indicates a runoff increase compared to the previous period. For the period 2040–2050 the climate change coefficients are $k_{ch} = 0.85$ (the river Southern Bug – Pirogovtsy village) and $k_{ch} = 0.46$ (the river Hromokliya – Mikhailivka village). In general, according the scenario RCP 4.5, there is a runoff decrease by the end of 2050 from 20 % in the north and up to 50 % in the south of the studied area. Similar results are obtained using calculations of climate change in the scenario RCP 8.5 (Fig. 6): the climate change coefficients indicate a runoff decrease for the periods 2011–2014 and 2015–2020, with average $k_{ch} = 0.84$ in the basin, i.e. the 16 % spring runoff decrease is forecasted.

For the next period (2021–2036), the climate change coefficients indicate the runoff increase compared to the previous periods. It equals $k_{ch} = 0.92$ in the upper part of the basin (the river Southern Bug – Pirogovtsy village) and $k_{ch} = 0.59$ in the lower part of the basin (the river Hromokliya – Mikhailivka village).

For the period 2037–2050 the coefficients of climate change equal $k_{ch} = 0.75$ (the river Southern Bug – Pirogovtsy village) and $k_{ch} = 0.38$ (the river Hromokliya – Mikhailivka village), that suggests a runoff decrease. In general, there is a spring runoff decrease of 60 % in the scenario RCP 8.5.

For the purpose of preliminary estimation of the performed forecasts, the time series of the maximal spring flood discharges for the station with the longest number of observations of the river Southern Bug – village Oleksandrivka (Fig. 7) have been investigated.

The chronological diagram of the maximal water discharges (in modulus coefficients) according to the long-term observations (until 2016), indicating the continuation of the trend towards a runoff decrease shows the likeliness of tendency so the forecast about decrease, approximately 20 %, of the maximal water discharges during the period of spring floods for the period 2011–2016, in both scenarios is confirmed.

These results are also in good agreement with recent studies of European scientists. In the last decades of the retrospective period, a gradual warming of climate has been in progress as a main factor in changes in water resources in European river basins (Best, 2019; Stagl and Hattermann, 2015; Bisselink et al., 2018). Those changes include an increase in the frequency of occurrence of extreme rainfalls (in short periods of time), the formation of catastrophic floods from them, and a decrease in the water resources of large territories with an increasing evaporation (Best, 2019; Loboda et al., 2019).

As the average annual air temperature rises, especially in the winter months, there is a decrease in the amount of snow in the annual rainfall. It leads to general trends in river runoff in the spring (from snowmelting) and its increase in autumn and winter (from rainfalls) (Bormann, 2010; Blöschl et al., 2017, 2019).

However, as it was noted by the authors (Bolbot and Grebin, 2019), in the present period of the water regime (after 2008), the characteristics of the minimal runoff (both summer and winter) also tend to decrease and continuation of monitoring is needed.

3.2 The minimal runoff

The analysis of minimal runoff long-term fluctuations and the most dry water periods, which are characterized by the lowest water losses per year, has been performed according to the average monthly summer and winter minimal water discharges data, as the most stable characteristics, as well as the minimal daily water discharges data during these water periods. In general, the lowest dry weather flow is typical for the summer–autumn period in the basin of the Southern Bug, when there is a depletion of groundwater forming the runoff. Sometimes the summer low flow is broken with sudden water level rise caused by rainfalls and the formation of floods on the rivers. The winter low flow is higher, as a result of the underground water flow increase, the reserves of which are formed from the autumn period, as well as due to the supply of groundwater with melt water and rainwater during the winter thaw.

In general, as it is known, the minimal river runoff is caused by the geographical zonality, but the fact that the hydrogeological conditions of the catchment can have a significant impact on its values must be considered. The influence of karst can be manifested by the values of the small rivers runoff (on the areas less than 1000 km^2). For such water catchments, the dry weather runoff can be different from the average one, typical for that climate zone, as a result of mismatches between surface and underground catchments areas.

As it was illustrated in the Fig. 8, during the winter low water period, the 30 d minimal modulus of 80 % probability

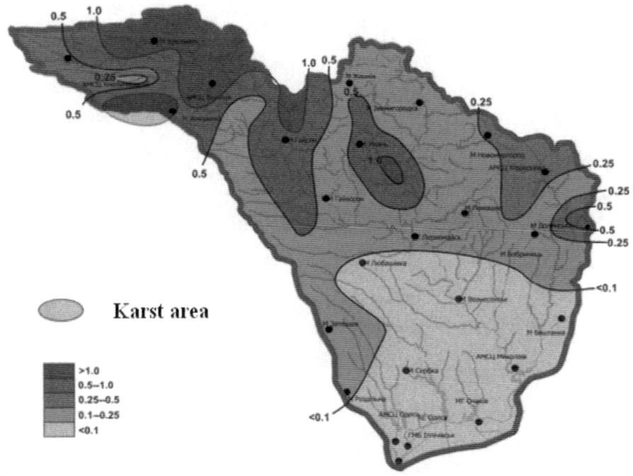

Figure 8. Distribution of 30 d minimal runoff modulus with 80 % probability during the winter period with stable freezing in the basins of the Southern Bug River L $(s\,km^2)^{-1}$.

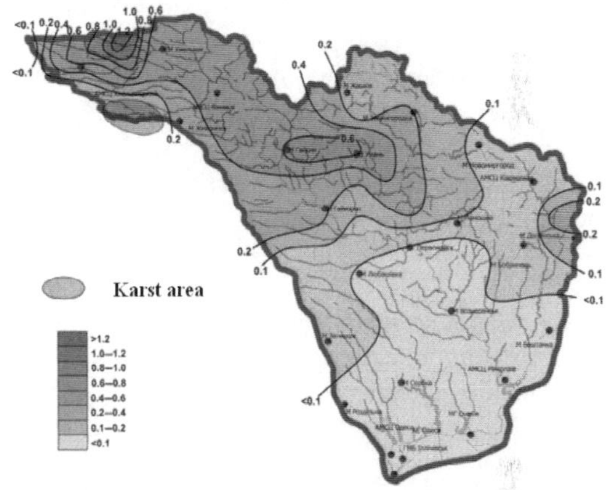

Figure 9. Distribution of 30 d minimal runoff modules of 80 % supply for the period of open channel with stable freezing in the basin of the Southern Bug, L $(s\,km^2)^{-1}$.

(in conditions of a stable freezing) reduces in the direction from the northwest (from 0.5–1.0 L $(s\,km^2)^{-1}$) to almost zero values – in the south-east.

The increased values (up to 1.71 L $(s\,km^2)^{-1}$ belong to the basin of the river Riv – Demidovka village, with karst rock formation. For this area, we propose the adjustment coefficient (k_{kars}). The adjustment coefficient for the rivers of the studied territory is 3.5. The spatial distribution of $(q_{30})_{80\,\%}$ for the period of the open channel (in the presence of stable freezing on the rivers) is characterized by the same regularity, but the modules decreases in the direction from the northwest to the northeast and the south – from 0.8 L $(s\,km^2)^{-1}$ to almost zero values (Fig. 9).

As it has been mentioned before, the largest modulus $(q_{30})_{80\,\%}$ are observed in the basin of the river Riv – Demidivka village, but the adjustment coefficient k_{karst} reaches 5.85,

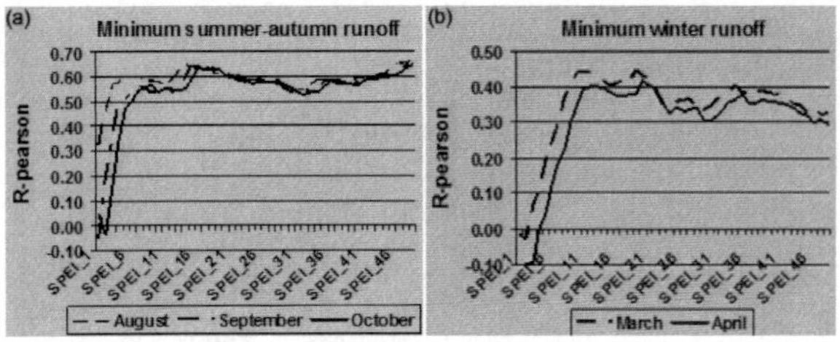

Figure 10. Correlation between the minimal river runoff and SPEI on scales of 1–48 months for the river Southern Bug – village Oleksandrivka.

which is almost 2 times higher than the same coefficient for the winter low water.

As it was noted above, one of the objectives of this study was to find the relation between runoff characteristics during the mean water period and the SPEI drought index. Figure 10 shows the correlation dependencies between the SPEI index, for different time scales and the value of the minimal runoff using the example of the river Southern Bug – village Oleksandrivka (catchment's area $A = 46\,200\,\text{km}^2$). Analyzing the results of the minimal runoff (Fig. 10a, b), you can note that for the summer–autumn low water runoff for SPEI on the scale of 1–2 months, the values of the correlation coefficient are in most cases insignificant, but from 3 and up to 48 months R is significant and changes slightly in the range from 0.55 to 0.65.

These relations are evidently explained by the fact that the summer–autumn mean water runoff is mainly determined by the water content of the year, but it can be broken by separate summer rain floods, therefore SPEI_3 for August characterizes the minimal runoff during 3 summer months and has $R = 0.57$. The index values for September and October become significant on scales beginning from 4–5 months and characterize the whole autumn-summer period.

Concerning the winter low water runoff, the distribution is characterized with a shift towards an increase in the scale at which the interrelation becomes significant up to 9–12 months. It is worth noting that during the winter mean water period, the runoff regime is determined mainly by seasonal underground supply, the reserves of which have been formed since autumn, but during the warm years which have been prevalent over the few last decades, the melt and rain waters of winter thaws also became a part of it. Apparently, this combination of factors leads to lower interrelation coefficients ($R_{\text{max}} = 0.44$–0.45); they are significant for scales from 9 to 46 months.

4 Conclusions

The review of the normative framework in the field of calculations of maximal runoff showed that despite the vast experience gained by scientists in this issue, the problem is still far from its solution due to the multifactority of the investigated phenomenon and regional features of the forming of maximal runoff on the rivers.

The authors have proposed a new modified version of the operator model for determining the maximal runoff of spring flood, which allows taking into consideration the possible impact of climate change on the estimated values of the maximal modules 1 % probability of exceeding.

Climate change is taken into account by introducing a separate coefficient, based on a comparison of the main parameters of the method obtained on the basis of current data (maximal snow supply at the beginning of the spring flood, precipitation during the spring flood and runoff coefficients), and similar values obtained from climatic modeling data.

For the territory of forest-steppe and steppe zone of Ukraine on the example of the Southern Bug basin the author's modified version of the calculating method for determining the characteristics of spring flood in climate change conditions has implement. The implementation of the proposed calculation option using different models and scenarios has shown that the results differ significantly, but in practically all cases up to 2050 a significant decrease in the runoff of spring flood (from 10 %–20 % in the forest-steppe and 40 %–50 % in the steppe zone) is forecasted.

The estimation of the statistical connection between SPEI in different time scales and minimal river runoff of the Southern Bug River during the period 1950–2010 showed that the obtained dependencies are significant, that gives an opportunity to use the drought index in the hydrological processes modeling in the forest-steppe and steppe zones of Ukraine.

Data availability. Data are available upon request to the contact author.

Author contributions. EG and VO developed mathematical model, NK, ZS and LK performed the simulations, VO analyzed the data and prepared the manuscript with contributions from all co-authors, MM is responsible for translation and redaction of manuscript.

Competing interests. The authors declare that they have no conflict of interest.

Special issue statement. This article is part of the special issue "Hydrological processes and water security in a changing world". It is a result of the 8th Global FRIEND–Water Conference: Hydrological Processes and Water Security in a Changing World, Beijing, China, 6–9 November 2018.

References

Akhter, A. and Azam, S.: Flood-Drought Hazard Assessment for a Flat Clayey Deposit in the Canadian Prairies, Journal of Environmental Informatics Letters, 1, 8–19, 2019.

Bergstorm, S.: The HBV model – its structure and applications, SMHI Reports RH, 4, Norrkping, Sweden, 35 pp., 1992.

Best, J.: Anthropogenic stresses on the world's big rivers, Nat. Geosci., 12, 7–21, 2019.

Bisselink, B., Bernhard, J., Gelati, E., Adamovic, M., Jacobs, C., Mentaschi, L., Lavalle, C., and De Roo A.: Impact of a changing climate, land use, and water usage on water resources in the Danube river basin, JRC Technical Reports, European Commission, 70 pp., https://doi.org/10.2760/89828, 2018.

Blöschl, G., Hall, J., Parajka, J., Perdigao, R. A. P., Merz, B., Arheimer, B., Aronica, G.-T., Bilibashi, A., Bonacci, O., Borga, M., Čanjevac, I., Castellarin, A., Chirico, G.-B., Claps P., Fiala K., Frolova, N., Gorbachova, L., Gül, A., Hannaford, J., Harrigan, S., Kireeva, M., Kiss, A., Kjeldsen, T., Kohnová, S., Koskela, J.-J., Ledvinka, O., Macdonald, N., Mavrova-Guirguinova, M., Mediero, L., Merz, R., Molnar, P., Montanari, A., Murphy, C., Osuch, M., Ovcharuk, V., Radevski, I., Rogger, M., Salinas, J.-L., Sauquet, E., Šraj, M., Szolgay, J., Viglione, A., Volpi, E., Wilson, D., Zaimi, K., and Živković, N.: Changing climate shifs timing of European foods, Science, 357, 588–590, https://doi.org/10.1126/science.aan2506, 2017

Blöschl, G., Hall, J., Viglione, A., Perdigão, R.-A.-P., Parajka, P., Merz, B., Lun, D., Arheimer, B., Aronica, G.-T., Bilibashi, A., Boháč, M., Bonacci, O., Borga, M., Čanjevac, I., Castellarin, A., Chirico, G.-B., Claps, P., Frolova, N., Gorbachova, L., Gül, A., Hannaford, J., Harrigan, S., Kireeva, M., Kiss, A., Kjeldsen, T., Kohnová, S., Koskela, J.-J., Ledvinka, O., Macdonald, N., Mavrova-Guirguinova, M., Mediero, L., Merz, R., Molnar, P., Montanari, A., Murphy, C., Osuch, M., Ovcharuk, V., Radevski, I., Rogger, M., Salinas, J.-L., Sauquet, E., Šraj, M., Szolgay, J., Volpi, E., Wilson, D., Zaimi, K., and Živković, N.: Changing climate both increases and decreases European river floods, Nature, 573, 108–111, https://doi.org/10.1038/s41586-019-1495-6, 2019.

Bormann, H.: Runoff regime changes in German rivers due to climate change, Erdkunde, 64–63, 257–279, 2010.

Bolbot, H. and Grebin, V.: Long-term dynamics and current trends in fluctuations of the flow characteristics of the Siverskyi Donets river basin, in: Proceeding of Conference "Monitoring of Geological Processes and Ecological Condition of the Environment", 12–15 November 2019, Kyiv, Ukraine, 5–6, 2019.

CRED (Centre for Research on the Epidemiology of Disasters): Natural Disasters 2018, available at: https://www.emdat.be/publications (last access: 25 April 2020), 2019.

EEA: Economic losses from climate-related extremes in Europe, available at: https://www.eea.europa.eu/downloads/efb8ac2f19e7464db01757ef0627e6a2/1554216645/assessment-2.pdf (last access: 25 April 2020), 2019.

Gopchenko, E. D. and Ovcharuk, V. A.: Theoretical ground of normative base for calculation of the characteristics of the maximal runoff and its practical realization. Transboundary Floods: Reducing Risks through Flood Management, NATO Science Series, IV Earth and Environmental Sciences, Vol. 72, Lightning Source UK Ltd., United Kingdom, 91–99, 2006.

Kan, G., He, X., Ding, L., Li, J., Liang, K., and Hong, Y. Study on Applicability of Conceptual Hydrological Models for Flood Forecasting in Humid, Semi-Humid Semi-Arid and Arid Basins in China, Water, 9, 719, https://doi.org/10.3390/w9100719, 2017.

Kouidri, S., Megnounif, A., and Ghenim, A. N.: Long-term seasonal characterization and evolution of extreme drought and flooding variability in northwest Algeria, Meteorology, Hydrology and Water Management, 7-2, 63–71, 2019.

Li, H.-J., Xue, L.-J., and Wang, X.-J.: Relationship between solar activity and flood/drought disasters of the Second Songhua river basin, J. Water Clim. Change, 6, 578, https://doi.org/10.2166/wcc.2014.053, 2015

Lobanova, A., Liersch S., Nunes, J. P., Didovets, I., Stagl, J., Huang, S., Koch, H., Rivas López, M. del R., Maule, C. F., Hattermann, F., and Krysanova, V.: Hydrological impacts of moderate and high-end climate change across European river basins, Journal of Hydrology: Regional Studies, 18, 15–30, https://doi.org/10.1016/j.ejrh.2018.05.003, 2018.

Loboda, N. S., Kuza, A. M., and Kozlov, O. M.: Assessment of possible changes of water resources of the rivers belonging to the Kuyalnytskyi Liman catchment at the beginning of the 21st century (2021–2050) according to the models of the climatic scenario RCP4.5, Ukrainian Hydrometeorological Journal, 23, 42–53, https://doi.org/10.31481/uhmj.23.2019.05, 2019.

Ovcharuk, V. and Gopchenko, E.: Method of determining characteristics maximal spring flood runoff in Ukraine in view of global and regional climate change, in: Proceedings of 26th IUGG General Assembly, Prague, 22 June–2 July 2015, 38 pp., 2015.

Romanowicz, R.: The Influence Of Climate Change On Hydrological Extremes: Floods & Droughts, Science Trends, https://doi.org/10.31988/SciTrends.3899, 2017.

Sajikumar, N. and Thandaveswara, B.: A nonlinear rainfall runoff model using an artificial neural network, J. Hydrol., 216, 32–35, 1999.

Semenova, I., Ovcharuk, V., and Shakhirzanova, J.: On use of drought indexes in modeling hydrological processes, in: Conference Proceedings SGEM 2014. Geoconference on Water Resources: Hydrology and Water Resources, 17–26 June 2014, Albena, 1, 503–510, https://doi.org/10.5593/SGEM2014/B31/S12.065, 2014.

Singh, V. P.: Hydrologic Systems. Rainfall-Runoff Modeling, Vol. I, Englewood Cliffs, Prentice Hall, 480 pp., 1988.

Stagl, J. C. and Hattermann, F. F.: Impacts of climate change on the hydrological regime of the Danube River and its tributaries using an ensemble of climate scenarios, Water, 7, 6139–6172, 2015.

Tokar, A. S. and Markus, M.: Precipitation runoff modelling using artificial neural networks and conceptual models, J. Hydrol. Eng., ASCE, 5, 156–161, 2000.

Vicente-Serrano, S. M., Beguería, S., and López-Moreno, J. I.: A multiscalar drought index sensitive to global warming: the standardized precipitation evapotranspiration index, J. Climate, 23, 1696–1718, 2010.

Zabolotnia, T., Gorbachova, L., and Khrystiuk, B.: Estimation of the long-term cyclical fluctuations of snow-rain floods in the Danube basin within Ukraine, Meteorology Hydrology and Water Management, 7, 1–12, 2019.

Zhao, R., Yilin, Z., and Lerun, F.: The Xinanjiang model, J. Proc. Oxf. Symp. Hydrol. Forecast, IAHS Publ., 135, 371–381, 1980.

Proc. IAHS, 383, 237–240, 2020
https://doi.org/10.5194/piahs-383-237-2020

Rainfall measurement from mobile telecommunication network and potential benefit for urban hydrology in Africa: a simulation framework for uncertainty propagation analysis

Maxime Turko[1,2,3], **Marielle Gosset**[1], **Christophe Bouvier**[4], **Nanee Chahinian**[4], **Matias Alcoba**[1], **Modeste Kacou**[5], and **Apoline Yappi**[5]

[1]Géosciences Environnement Toulouse, GET (UMR 5563 CNRS, IRD, UTIII), Observatoire Midi-Pyrénées, Toulouse, 31400, France
[2]Centre National d'Etudes Spatiales, CNES, Toulouse, 31400, France
[3]Centre National de Recherches Météorologiques, CNRM, Toulouse, 31057, France
[4]HydroSciences Montpellier, HSM (UMR 5569 CNRS-IRD-UM), Montpellier, 34000, France
[5]LAPA, Université Félix Houphouët-Boigny (UFHB), Abidjan, Cote d'Ivoire

Correspondence: Marielle Gosset (marielle.gosset@ird.fr)

Published: 16 September 2020

Abstract. Urban floods due to intense precipitation is a major problem in many developing countries, especially in Africa. There are few available solutions for effective and yet affordable flood early warning systems for these regions. Weather radar is common in industrialized countries but too costly for most tropical cities. Satellite provides interesting information in real time but not yet quantitative enough at the space and time scales needed for urban flood monitoring. Rainfall measurement using commercial microwave links (CML) from cellular communication networks is a cost effective alternative to conventional methods. The method is based on measuring rain induced fluctuations between telecommunication antennas; if the operator provides this information rain maps can be produced and used for hydrological prediction. Many CML studies have been carried out in Europe and Israel. Recently IRD implemented pilot sites in Africa in order to test this rainfall estimation technique and quantify the uncertainties. After reviewing the method principles and providing an overview of the current research on CML, we present a simulation framework to analyse the propagation of CML rainfall uncertainties in an urban hydrological model.

1 Rainfall measurement from commercial microwave links (CML) principle and applications

Urban floods in West Africa engender every year much human and material damage. To monitor these risks, flood alert systems based on reliable rainfall estimation are needed. In industrialized countries, alert systems are usually based on weather radar networks that are not yet available in developing countries. Satellite provides interesting information in real time but not yet quantitative enough at the space and time scales needed for urban flood monitoring. In the last decade, studies have demonstrated the possibility to use mobile telecommunication networks to estimate rainfall.

The method exploits the fluctuation of the microwave signal due to attenuation by rainfall. This method is cost effective because an existing network is used rather then costly investment in new instruments. The high density of the mobile network in cities is advantageous for urban applications. The IRD (Institut de Recherche pour le Développement) initiated the Raincell (for Rain measurement from Cellular phone network) project to promote and demonstrate the use of commercial microwave links (CML) for rainfall measurement in West Africa (Gosset et al., 2016; film "Raincell Africa:

Alert signal" available at: https://www.youtube.com/watch?v=QM9GHchJIf4&feature=youtu.be, last access: 12 August 2020). Pilot experiments have been set up in several African cities: Ouagadougou (Burkina-Faso; Doumounia et al., 2014), Bamako (Mali), Niamey (Niger) and Yaounde and Douala in Cameroun.

1.1 Rainfall measurement with commercial microwave links (CML)

Rainfall measurement by CML is based on the robust relationship between microwave signal attenuation and rainfall intensity over the link (between two telecommunication antennas). This principle is summarized in Eq. (1):

$$\mathrm{PIA} = aR^b \times L, \tag{1}$$

the Path Integrated Attenuation (PIA in dB) measured by the link of length L, can be related to the Rainfall intensity over the link (R), through a simple power law relation – the coefficients a and b depend on the operation frequency and can be estimated easily. a increases with frequency. b is close to 1 (0.5 to 2) for common telecom frequencies and the relation is quasi-linear near 30 GHz. If the PIA is provided by the mobile operator, Eq. (1) can be inverted to estimate rainfall all over the network.

The accuracy in rainfall measurement using the CML network (from Eq. 1) depends on many factors: the variability of the coefficients a and b in Eq. (1); the density and length of the links; the accuracy of the PIA itself. The latter depends on the method used by the operator to collect and record the raw power levels (typically accuracy of 1 dB or 0.1 dB is common).

1.2 Current research

Research on rainfall measurement from CML started in 2003 in the Netherlands and Israel, later followed by several international research groups. The first papers focused on the method feasibility at link scale (Messer et al., 2006) and on the uncertainties (Leijnse et al., 2007; Berne and Uijlenhoet, 2007). These research groups have shown the possibility to use this method in various environment and region in the world, and at different spatial scales, from a single link to an agglomeration or a country. The state of the art about the CML technique and its applications is well described in Uijlenhoet et al. (2018).

In the next section we introduce a simulation framework develop to analyse the potential of this technique for urban hydrology, accounting for rainfall uncertainty propagation.

2 A simulation framework to analyse the propagation of CML rainfall uncertainties into a hydrological modelling

Several studies have analysed and quantified the uncertainties in CMLs' rainfall estimation, for different network configuration and rainfall regime (Berne and Uijlenhoet, 2007; Leijnse et al., 2007, 2008a, b, 2010; Zinevich et al., 2010; Schleiss et al., 2013; van Leth et al., 2017). Here a simulation framework has been set up in order to analyse how typical errors in CMLs rainfall estimation may propagate within a hydrological model. Building on previous studies, the considered uncertainties relate to: the accuracy of attenuation and rainfall estimation for each link ; the network configuration (density; length; frequency).

The city of Ouagadougou where the first Raincell pilot started (Doumounia et al., 2014) is taken as an example. A simulator of a network of links was created based on radar rainfall fields used as a reality proxy. The radar rain fields were obtained from a campaign that took place in Ouagadougou in 2012 and 2013 (http://meghatropiques.ipsl.polytechnique.fr/the-ouagadougou-super-site/, last access: 12 August 2020).

2.1 A link simulator to generate pseudo estimation and their uncertainties

Each link is emulated as the weighted mean of the radar map pixels crossed by the link:

$$R_{\mathrm{link}} = \frac{\sum_{i=1}^{n} l_i R_{\mathrm{pix}_i}}{L_{\mathrm{link}}}, \tag{2}$$

where R_{link} is the mean rainfall intensity over the link of length L_{link}. R_{pixi} is the rainfall intensity of a pixel i weighted by the link size l_i in this pixel. Using Equation 1 the PIA over the link is then computed. To take into account measurement accuracy the PIA is rounded at in 1 dB or at 0.1 dB, which are typical values for the data provided by the operators. Several network configurations can be simulated by varying the a, b and L parameters in Eq (1).

2.2 Analysis of the uncertainties in the simulated discharge

An urban distributed hydrological model (Athys, Bouvier et al., 2018) is used to analyse the propagation of the CML rainfall uncertainties in the simulated discharge. The discharge uncertainty analysis is performed at several discharge points within the watershed in order to study its variability. The sensitivity of the discharge to various sources of uncertainty in the rain field is analysed with numerical experiments: using the original rain map as a forcing field, or using the simulated CMLs derived rain maps, with different network configurations and PIA accuracy. As an example, Fig. 1 compares the

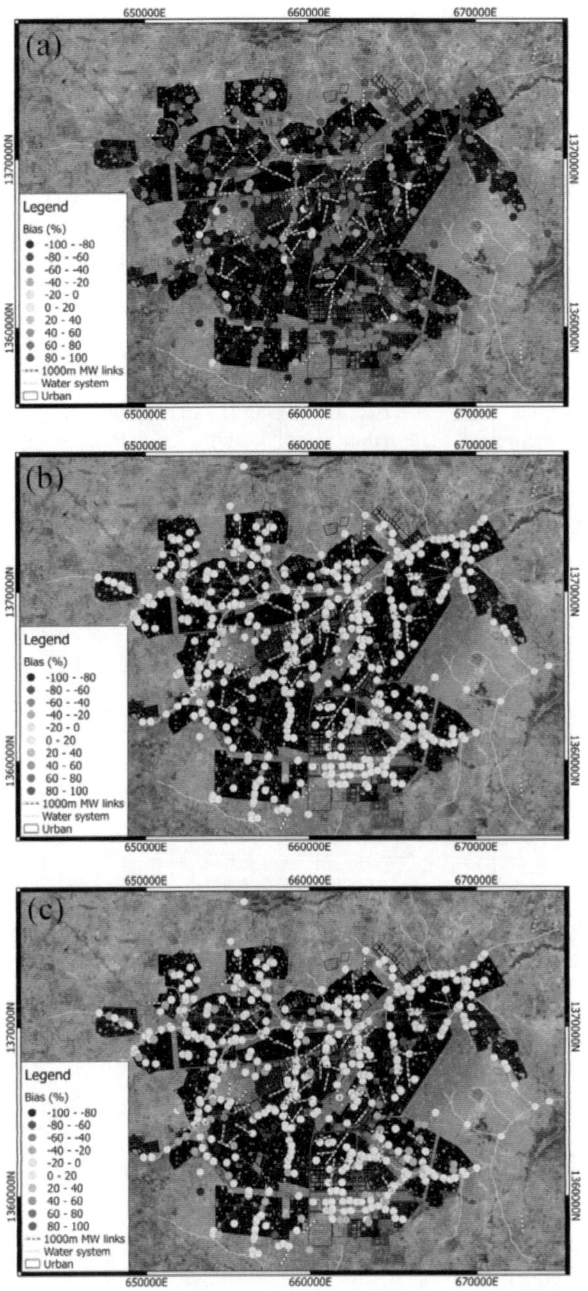

Figure 1. Map of the bias (%) between the simulated discharge when the CML network provides perfect accuracy on the attenuation (reference) and the simulated discharge with less accurate measurement of CML attenuation (1 dB or 0.1 dB). **(a)** CML configuration with 6 GHz frequency and 0.1 dB precision. **(b)** CML configuration with 30 GHz and 0.1 dB. **(c)** CML configuration with 30 GHz and 1 dB. Map background information provided by © Google Maps and © OpenStreetMap contributors 2016. Distributed under a Creative Commons BY-SA License.

bias (in %) in the discharge for a CMLs network with coarse accuracy (1 dB) with the same network but with full accuracy (no error on the attenuation PIA). Two frequencies (6 and 30 GHz) and two signal accuracy hypothesis (0.1 and 1 dB) are displayed. In this example all links are 1000 m long. The 6 GHz and 1 dB configuration is not represented because the bias was homogenous (−100 %). The figure highlights the importance of the raw signal accuracy, especially when the network frequency is low. At, 6 GHz frequency, 1 dB accuracy on a 1 km link rainfall below 100 mm h^{-1} would not be detected; with 0.1 dB accuracy th threshold would be 28.4 mm h^{-1}. At higher frequency, the attenuation for a given rain rate is higher; at 30 GHz frequency, 0.1 dB over 1 km is equivalent to 0.4 mm h^{-1} and 1 dB to 4.4 mm h^{-1}; in both cases the amount of rainfall missed by the network is low, and the bias in the discharge is therefore small (Fig. 1b and c) whatever the accuracy.

The presented CML simulation framework coupled with a hydrological model can be used to test various sensitivities (network configuration and density within the basin). It is a useful tool to demonstrate to operators the potential use of the information collected on their network and the importance of raw data accuracy.

Data availability. Data available from the corresponding author upon request.

Author contributions. MT run the simulations and provided all plots and analysis; MG supervised the study and wrote the English text; CB and NC provided and configured the hydrological model; MK, MA and AY participated in the technical discussions on the CML processing.

Competing interests. The authors declare that they have no conflict of interest.

Special issue statement. This article is part of the special issue "Hydrological processes and water security in a changing world". It is a result of the 8th Global FRIEND–Water Conference: Hydrological Processes and Water Security in a Changing World, Beijing, China, 6–9 November 2018.

References

Berne, A. and Uijlenhoet, R.: Path-averaged rainfall estimation using microwave links: Uncertainty due to spatial rainfall variability, Geophys. Res. Lett., 34, 1–5, https://doi.org/10.1029/2007GL029409, 2007.

Bouvier, C., Chahinian, N., Adamovic, M., Cassé, C., Crespy, A., Crès, A., and Alcoba, M.: Large-Scale GIS-Based Urban Flood Modelling: A Case Study on the City of Ouagadougou,

Adv. Hydroinformatics, 703–717, https://doi.org/10.1007/978-981-10-7218-5_50, 2018.

Doumounia, A., Gosset, M., Cazenave, F., Kacou, M., and Zougmore, F.: Rainfall monitoring based on microwave links from cellular telecommunication networks: First results from a West African test bed: wireless networks for rainfall in Africa, Geophys. Res. Lett., 41, 6016–6022, https://doi.org/10.1002/2014GL060724, 2014.

Gosset, M., Kunstmann, H., Zougmore, F., Cazenave, F., Leijnse, H., Uijlenhoet, R., Chwala, C., Keis, F., Doumounia, A., Boubacar, B., Kacou, M., Alpert, P., Messer, H., Rieckermann, J., and Hoedjes, J.: Improving Rainfall Measurement in Gauge Poor Regions Thanks to Mobile Telecommunication Networks, B. Am. Meteorol. Soc., 97, ES49–ES51, https://doi.org/10.1175/BAMS-D-15-00164.1, 2016.

Leijnse, H., Uijlenhoet, R., and Stricker, J. N. M.: Hydrometeorological application of a microwave link: 2. Precipitation: Microwave Link, 2, Water Resour. Res., 43, 1–9, https://doi.org/10.1029/2006WR004989, 2007.

Leijnse, H., Uijlenhoet, R., and Stricker, J. N. M.: Microwave link rainfall estimation: Effects of link length and frequency, temporal sampling, power resolution, and wet antenna attenuation, Adv. Water Resour., 31, 1481–1493, https://doi.org/10.1016/j.advwatres.2008.03.004, 2008a.

Leijnse, H., Uijlenhoet, R., and Holleman, I.: Sources of error in microwave link rainfall estimation, Proceedings of the International Symposium on Weather Radar and Hydrology, Grenoble, France, 10–15 March 2008, p. 4, 2008b.

Leijnse, H., Uijlenhoet, R., and Berne, A.: Errors and Uncertainties in Microwave Link Rainfall Estimation Explored Using Drop Size Measurements and High-Resolution Radar Data, J. Hydrometeor., 11, 1330–1344, https://doi.org/10.1175/2010JHM1243.1, 2010.

Messer, H., Zinevich, A., and Alpert, P.: Environmental Monitoring by Wireless Communication Networks, Science, 312, 713–713, https://doi.org/10.1126/science.1120034, 2006.

Schleiss, M., Rieckermann, J., and Berne, A.: Quantification and Modeling of Wet-Antenna Attenuation for Commercial Microwave Links, IEEE Geosci. Remote Sens., 10, 1195–1199, https://doi.org/10.1109/LGRS.2012.2236074, 2013.

Uijlenhoet, R., Overeem, A., and Leijnse, H.: Opportunistic remote sensing of rainfall using microwave links from cellular communication networks, WIREs Water 5, 1–15, https://doi.org/10.1002/wat2.1289, 2018.

van Leth, T. C., Overeem, A., Leijnse, H., and Uijlenhoet, R.: A measurement campaign to assess sources of error in microwave link rainfall estimation, Atmos. Meas. Tech., 11, 4645–4669, https://doi.org/10.5194/amt-11-4645-2018, 2018.

Zinevich, A., Messer, H., and Alpert, P.: Prediction of rainfall intensity measurement errors using commercial microwave communication links, Atmos. Meas. Tech., 3, 1385–1402, https://doi.org/10.5194/amt-3-1385-2010, 2010.

Proc. IAHS, 383, 241–248, 2020
https://doi.org/10.5194/piahs-383-241-2020

Estimating extreme dry spell risk in Ichkeul Lake Basin (Northern Tunisia): a comparative analysis of annual maxima series with a Gumbel distribution

Majid Mathlouthi[1] **and Fethi Lebdi**[2]

[1]Research Laboratory in Sciences and Technology of Water at National Agronomic Institute of Tunisia, Tunis, 1082, Tunisia
[2]National Agronomic Institute of Tunisia,Tunis, 1082, Tunisia

Correspondence: Majid Mathlouthi (majid_mathlouthi@yahoo.fr)

Published: 16 September 2020

Abstract. This paper analyses a 42 year time series of daily precipitation in Ichkeul Lake Basin (northern Tunisia) in order to predict extreme dry-spell risk. Dry events are considered as a sequence of dry days separated by rainfall events from each other. Thus the rainy season is defined as a series of rainfall and subsequent dry events. Rainfall events are defined as the uninterrupted sequence of rainy days, when at last on one day more than a threshold amount of rainfall has been observed. A comparison of observed and estimated maximum dry events (42 year return period) showed that Gumbel distribution fitted to annual maximum series gives better results than the exponential (E) distribution combined with partial duration series (PDS). Indeed, the classical Gumbel approach slightly underestimated the empirical duration of dry events. The AMS–G approach was successfully applied in the study of extreme hydro-climatic variable values. The results reported here could be applied in estimating climatic drought risks in other geographical areas.

1 Introduction

A proper simulation of precipitation is important. Precipitation is a very important element of climate that affects both the natural environment and human society. Events ranging from prolonged droughts to short-term, high intensity floods are often associated with devastating impacts both to society and the environment (Hui et al., 2005). An alternative to the Markov chain process which is typically used to simulate the occurrence of precipitation is to use a wet-dry spell model or alternating renewal model, that is, to simulate wet and dry spells separately by fitting their durations to an appropriate probability distribution. Among the study using the wet-dry spell approach one can cite, for example, Bogardi and Duckstein (1993); Wilks (1999); Mathlouthi (2009); Mathlouthi and Lebdi (2008, 2009, 2017); Dunxian et al. (2015); Konjit et al., 2016.

It is well known that dry spells cause major economic and human losses, and numerous studies have highlighted the need for drought prevention and mitigation plans (Vicente-Serrano and Beguería, 2003). The spatial and temporal assessment of dry spells is necessary in order to protect agriculture, water resources and other socio economic concerns, and areas at risk from droughts of long duration and great intensity need to be determined. Sivakumar (1992) point up the importance of admitting partial patterns of extreme drought, which can then be used in the management of cultivated areas (crop selection, irrigation planning, etc.) and water resources management.

The analysis of extremes in dry-spell series has been examined classically using annual maximum series (AMS) adjusted to a Gumbel distribution (Gupta and Duckstein, 1975; Lana and Burgueño, 1998). The AMS are constructed by determining the maximum dry spell for each year, so the series length equals the number of years for which records are available. However, the main drawback is the loss of the second, third, etc. largest annual dry spells, which might exceed the maximum dry spells of other years. An option approach is the partial duration series (PDS), which is constructed using the values above a selected threshold regardless of the year in which they occurred (Hershfield, 1973; Vicente-Serrano and Beguería, 2003). Typically, the generalized Pareto (GP) distribution has been used to model PDS (Bobée and Rasmussen, 1995). Although the PDS approach has obvious advantages over the AMS approach (Cunnane, 1973), it has been used only infrequently in precipitation dry-spell analysis (Vicente-Serrano and Beguería, 2003).

In virtue, this paper is focused on the modelling of rainfall occurrences under Mediterranean climate by wet-dry spell approach. The intended objective is to determine whether the use of AMS with the Gumbel distribution (AMS–G approach) is suitable for modelling extreme dry-spell risk; to analyse if PDS with a probability distribution, that best fits the data set, is adequate for modelling extreme dry-spell risk and finally to compare both approaches with the observed maximum dry spells to determine the most suitable estimation of drought risk.

The study area is the Ichkeul basin (Northern Tunisia) with several dams for irrigation, drinking water and water transfer to other regions of the country. The precipitation irregularity and the frequent dry spells are major restrictive factors in crop growth and water demand satisfaction imposed on dams. For this reason, this area is particularly suitable for examining this approach.

2 Data

Daily precipitation records at rain gauges in the basin of Ichkeul Lake located in Northern Tunisia were used in this analysis (Fig. 1). The total area of this region calculated by GIS is 2120 km^2. The rainy season starting at September and lasting until the beginning of May. The mean of annual rainfall is 600 mm; the coefficient of variation is 0.25. The climate of this area is classified as sub-humid; the average annual rainfall is below 40 % of the total annual potential evaporation. Except in occasional wet years, most precipitation is confined to the winter months in this basin. The dry season lasts from May to August. Daily values of precipitation are quite variable. There is also considerable variation from year to year. Ten time series of daily precipitation exist for the period from 1968 to 2010.

3 Method

3.1 Extract series of dry and rainfall events

In the *wet-dry spell* approach, the time-axis is split up into intervals called *wet periods* and *dry periods*. A rainfall *event* is an uninterrupted sequence of wets periods. The definition of event is associated with a rainfall threshold value which defines *wet* (Fig. 2). As this limit 3.6 mm d^{-1} has been selected. This amount of water corresponds to the expected daily evapotranspiration rate, marking the lowest physical limit for considering rainfall that may produce utilizable surface water resources. In this approach, the process of rainfall occurrences is specified by the probability laws of the length of the wet periods, and the length of the dry periods (time between storms or inter-event time).

The rainfall event r in a given rainy season n will be characterized by its duration $D_{n,r}$, the temporal position within the rainy season, the dry event or inter-event time $Z_{n,r}$ and by the cumulative rainfall amounts of $H_{n,r}$ of $D_{n,r}$ rainy days (Fig. 2).

$$r = f(D_{n,r}, H_{n,r}, Z_{n,r}) \tag{1}$$

Where f is the function defined on R_+^*, which to each event r associates a value D, H and Z themselves, real discrete random variables.

$$Z_{n,r} = x_{n,r} - x_{n,r-1} \tag{2}$$

$$H_{n,r} = \sum_{k=1}^{D_{n,r}} h_k \tag{3}$$

Where h_k represents the total daily rainfall in mm. Let $h_k > 0$ and at least a value of $h_k > 3.6$ mm.

The varying duration of the events requires that the cumulative rainfall amounts corresponding to each event should be conditioned by the duration of the event. The identification and fitting of conditional probability distributions to rainfall amounts may be problem especially in the case of short records and for events with extreme (long) durations (Foufoula-Georgiou and Georgakakos, 1991). The number of rainfall events per rainy season n is N_n and the length L_n of this last, of random duration, is defined as the time span between the start of the first and the end of the last rainfall event.

$$L_n = \sum_{r=1}^{N_n} D_{n,r} + \sum_{r=1}^{N_n-1} Z_{n,r} \tag{4}$$

The length of the climatic cycle C_n is determined as the time lapsed between the onsets of two subsequent rainy seasons.

$$C_n = x_{n+1,r_1} - x_{n,r_1} \tag{5}$$

3.2 Extreme dry event modelling with annual maximum series and Gumbel distribution

The distribution introduced by Gumbel is very useful for extreme dry event frequency modelling using the AMS–G approach (Gumbel, 1958; Vicente-Serrano and Beguería-Portugués, 2003). The Gumbel distribution is a two-parameter distribution with constant skewness. It is a particular case of the three-parameter generalized extreme value (GEV) distribution, i.e. the limit distribution for maxima series. The Gumbel is usually preferred to the GEV because of its ease of calculation. Its probability density function is

$$f(x) = \frac{1}{\alpha} e^{\{-[(x-\beta)/\alpha]\}^{-e^{\{-[(x-\beta)/\alpha]\}}}} \tag{6}$$

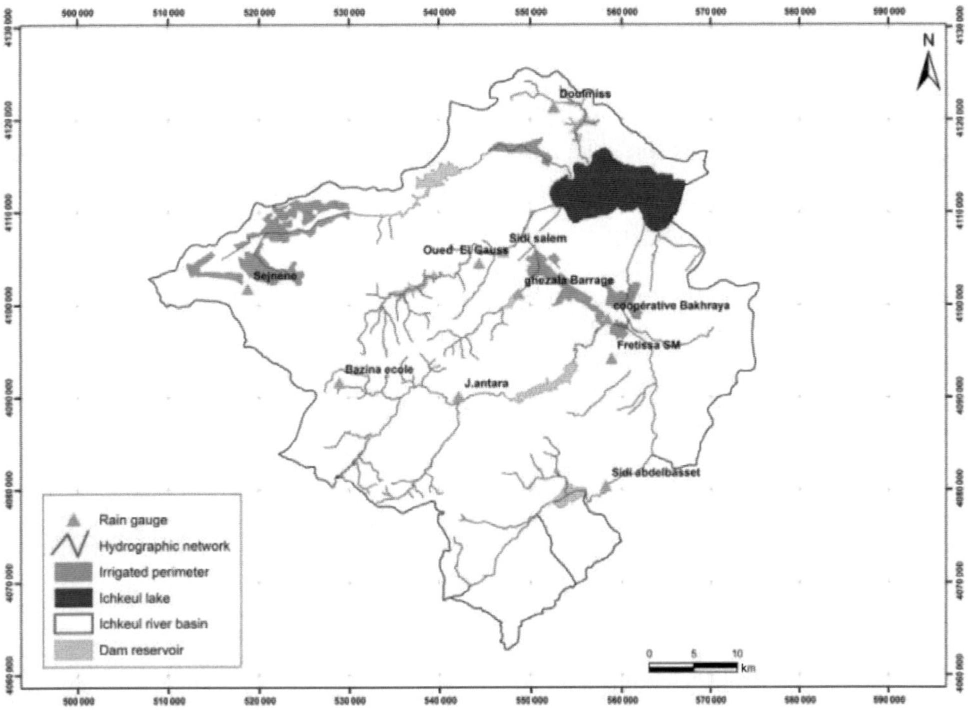

Figure 1. Study area.

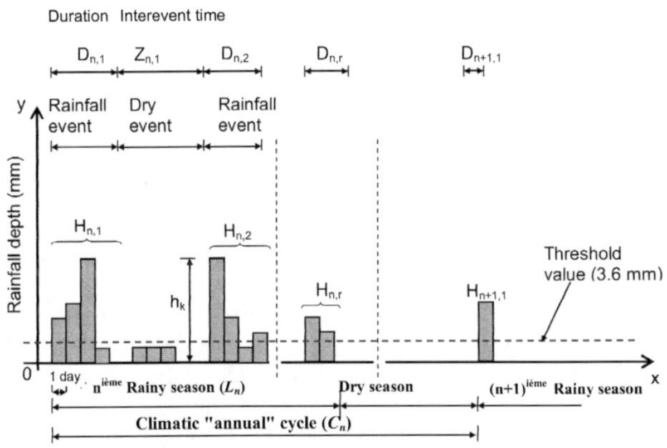

Figure 2. Event representation of the climatic cycle.

and its cumulative distribution function is expressed by

$$F(x) = e^{-e^{-\left[\frac{x-\beta}{\alpha}\right]}} \tag{7}$$

where x is the value of the variable, and α and β are scale and location parameters of the distribution, respectively. The mean and the variance are $\mu = \beta + 0.5772\alpha$ and $\sigma^2 = \frac{\pi^2\alpha^2}{6}$, accordingly.

The prospective maximum dry event for a T year period X_T can be calculated using

$$X_T = \beta - \alpha \ln\left\{-\ln\left[1 - (1/T)\right]\right\} \tag{8}$$

3.3 Estimation of extreme dry event using PDS

3.3.1 Characteristics of PDS

Although the preceding method has been widely used in the study of extreme dry spells, in the analysis of other hydrological and climatic variables (e.g. extreme rainfall, floods) many studies prefer to use PDS or series of peaks over an upper limit. Given the dry spell series $a = \{a_1, a_2, \ldots, a_n\}$, for the station a, where a_i is the duration of a given dry spell, the PDS $b = \{b_1, b_2, \ldots, b_j\}$ consists of all the values of the original series that exceed a predetermined upper limit a_0:

$$b_j = a_i - a_0 \ \forall \ a_i > a_0 \tag{9}$$

The size of the series obtained depends, therefore, on the upper limit a_0. For this reason, PDS use the information contained in the original sample more efficiently, and permit the inclusion of more than one event per year, if they satisfy the conditions established in defining an extreme event (Chow et al., 1988; Vicente-Serrano and Beguería-Portugués, 2003).

3.3.2 Probability distributions used to adjust PDS

Many probability distributions have been adjusted to PDS hydrological series, including lognormal, Pearson III, Gamma, GEV, Weibull, etc. (Bobée et al., 1993; Vicente-Serrano and Beguería-Portugués, 2003). In this study, we evaluated the continuous probability distributions given by Hyfran software, and we found that the Exponential law is the best fitting probability distribution to PDS. The parameter estimates is performed by the method of moments. A Chi-Squared goodness-of-fit Test is used to determine how well the theoretical distribution fits the empirical distribution obtained from the sample data. The exponential (E) distribution function is

$$f(t) = b \cdot e^{-b(t-m)} \quad m < t < +\infty \tag{10}$$

where b parameter of the exponential distribution, can be estimated as the reciprocal mean $\bar{t}$ of the sample of times observed:

$$b = \frac{1}{\bar{t}} \tag{11}$$

and its cumulative distribution function is expressed by

$$G(t) = 1 - e^{-b(t-m)} \quad m < t < +\infty \tag{12}$$

The event X_T in a period of T years is obtained using

$$X_T = m - \frac{1}{b}\ln\left(1 - \frac{1}{T}\right) \tag{13}$$

A major problem in using PDS is the selection of the lower bound a_0. This value should be low enough to ensure the inclusion of as much relevant information as possible, without violating the assumption of independence of the peaks. Various methods have been proposed to determine the most appropriate lower bound (Ashkar and Rouselle, 1987; Madsen et al., 1997). However, according to Vicente-Serrano and Beguería-Portugués (2003) Beguería (2003) has shown that the parameters and quantile estimations vary randomly with the threshold value, and no single value is entirely adequate. For this reason, in this paper, the maximum dry event in the 42 year period was calculated using different lower bounds in the PDS–E approach. These bounds were defined using the percentiles of the dry event series every 0.5 from percentiles 90 to 99.5. Dry event were considered extreme above the 90th percentile.

3.4 Comparison of the AMS–G and PDS–E approaches

The maximum dry event observed in each series in the period 1968–2010 was extracted. These were compared with the 42 year estimates using the AMS–G and PDS–E approaches. It is clear that the maximum dry event observed in a 42 year period does not necessarily correspond to a return period of 42 years. This limitation was partially overcome by using several rain gauges in the same region. The goodness of fit was tested by means of the root-mean-square error (RMSE) (Willmott, 1982), the lowest value indicating the best estimation:

$$\text{RMSE} = \sqrt{\frac{1}{n}\sum_{i=1}^{n}\left(z_i - \hat{z}_i\right)^2} \tag{14}$$

Where z_i is the observed value and $\hat{z}_i$ the estimated value using annual maximum or partial duration series; n is the number of rain gauges.

4 Results

4.1 Selection of the lower bound in using PDS

The main problem in using PDS involves the selection of the lower bound. In theory, the method is invariant to the variation in the lower bound. In practice, however, the results may vary greatly, especially with the sample sizes that are common in hydro climatic studies. This is exemplified in Fig. 3, in which the maximum dry spells expected in 42 years are shown for five rain gauges, in relation to the lower bound used. Whereas this value was expected to be similar independently of the lower bound chosen, it showed great random variation, being as 21 % compared to the average in some cases. Here, we assumed that the average of the different values would provide a good estimate of the unknown true value, this being less uncertain than using a unique, arbitrary, threshold.

4.2 Comparison of maximum dry event estimations using the AMS–G and PDS–E approaches with the observed maximum dry event

Figures 4 and 5 compare AMS–G and PDS–E estimates with the observed maximum dry events. The AMS–G method estimated adequately in the majority of cases the duration of the observed maximum dry events. The underestimation did not exceed 9 d, which prudent use of this method. The PDS–E clearly overestimated the maximum dry events duration. The difference between predicted and observed values varies from −5.4 % to 25.7 %.

The RMSE between the observed and estimated values is also highly indicative of the better performance of the AMS–G distribution. There was a better adjustment for the dry event series (RMSE = 4.7 versus 9.2).

Figure 6 shows the spatial distribution of the maximum dry events observed in the study area between 1968 and 2010, along with the estimations using the PDS–E and AMS–G approaches. The longest dry events are located in the southern areas, with values over 81 consecutive days of precipitation below 3.6 mm. A negative southwestern gradient of the maximum dry event duration is established. The same

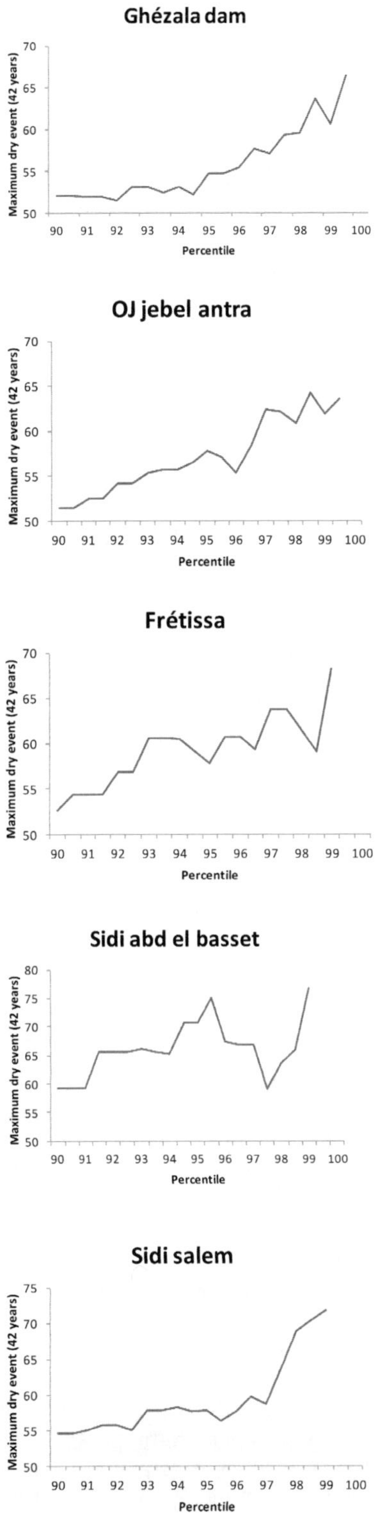

Figure 3. Oscillation of the maximum dry event (days) estimations as a function of the selected percentile in the creation of the dry event PDS. Five representative rain gauges are shown.

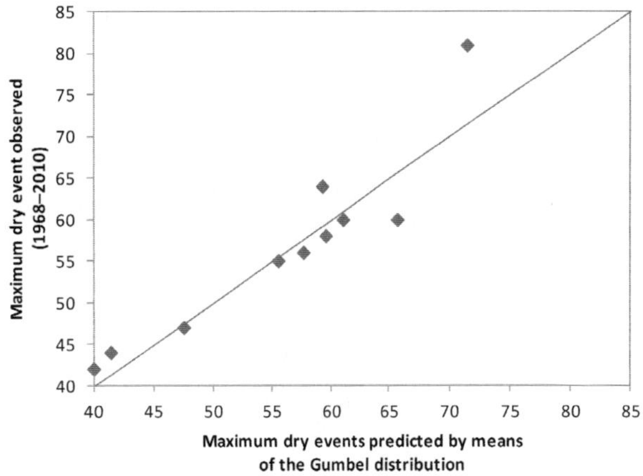

Figure 4. Differences between maximum dry events observed and estimated using AMS modelled using Gumbel distribution.

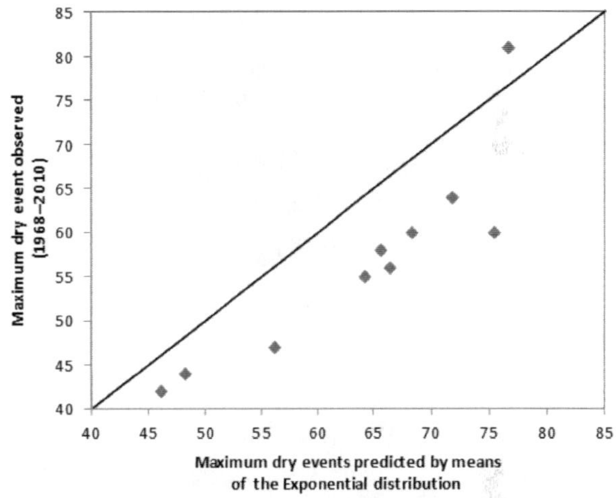

Figure 5. Differences between maximum dry events observed and estimated using PDS modelled using the Exponential distribution.

pattern is revealed by both estimations. There were significant contrasts between the south and west, with differences about 40 d. The AMS–G map shows a much closer match to the observed data. The Exponential estimation is clearly little higher than the observed figure.

The absolute errors of the estimations are shown in Fig. 7. The high magnitude of the errors resulting from the PDS–E approach is evident. Here, the positive errors indicate the underestimation provided by this approach. By contrast, the errors of the AMS–G approach include low negative values and the estimation is, in general, better.

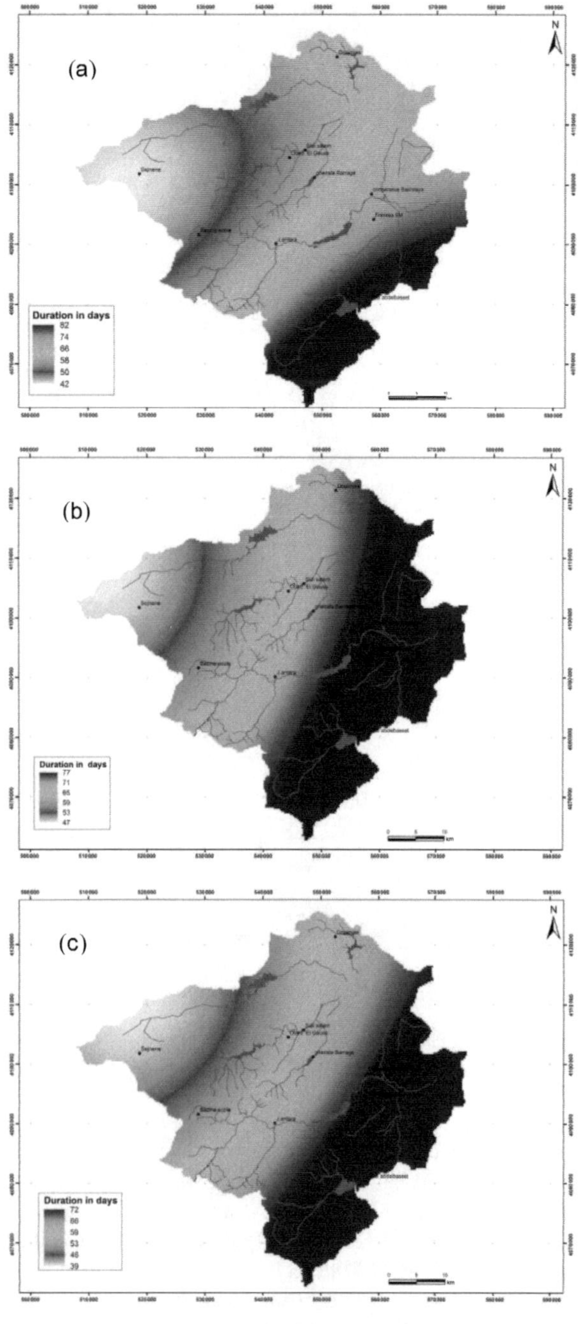

Figure 6. Maximum dry events: **(a)** observed in 42 years (1968–2010); **(b)** predicted using the E distribution; **(c)** predicted using Gumbel distribution.

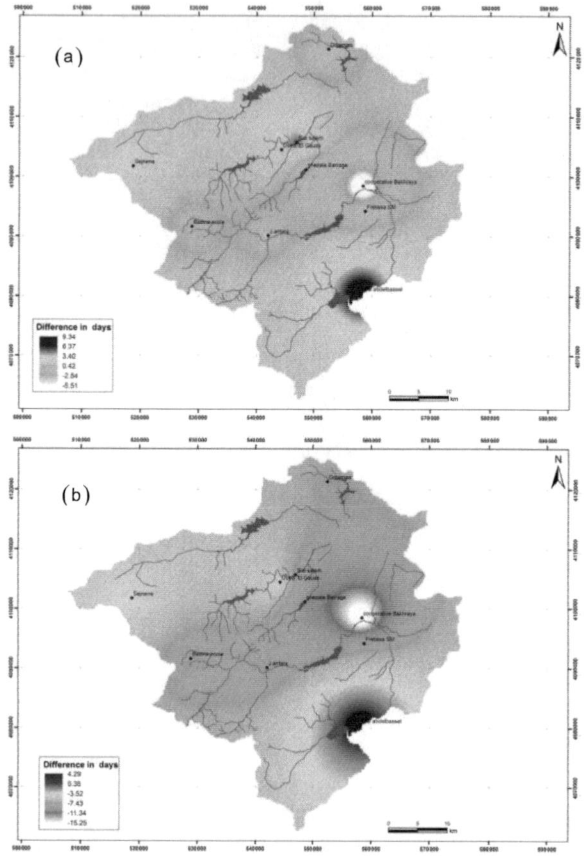

Figure 7. Differences between maximum dry spells observed and estimated: **(a)** observed-estimated using Gumbel distribution; **(b)** observed-estimated using E distribution.

5 Discussion and conclusions

In this paper, we have used a PDS sampling in conjunction with an Exponential distribution. The results obtained have been compared with those obtained when adopting the AMS–G approach for the maximum dry event series observed in the study area.

Different probability distributions can be used to fit both AMS and PDS. The Gumbel distribution is a two parameter extreme values distribution widely used in modelling AMS. It has been compared with the one parameter Exponential distribution fitted to PDS. It is a particular case of the gamma distribution. It is the continuous analogue of the geometric distribution. It is obvious that a two parameter distribution would fit the observed data better than a one parameter one. Nevertheless, the need to estimate a greater number of parameters introduces an extra source of uncertainty that can affect the final estimates. Here, we find that the use of AMS-G is more efficient than PDS-E, contrary to what has been reported in several other studies. In this sense Moreno and Roldán (1999), Mkhandi et al. (2000) and Vicente-Serrano and Beguería (2003) indicated that the use of PDS for the stochastic modelling of extremes has yielded good results in the analysis of hydrological variables, whereas numerous studies have pointed out that AMS produces a significant loss of data for extreme modelling (Cunnane, 1973; Madsen et al., 1997; Vicente-Serrano and Beguería, 2003). Accordingly, the RMSE obtained by the AMS–G is lower than that obtained by the PDS–E when analysing the empirical maximum dry events for a 42 year time series.

One shortcoming of the PDS method is the selection of the upper limit used to define the PDS. We found that the final quantile estimates vary significantly when only small changes are made in the upper limit used. This result has been reported previously by the study of Vicente-Serrano and Beguería, 2003. To cope with this problem, as proposed by Vicente-Serrano and Beguería, (2003), the use of different upper limits when constructing a set of PDS, and then taking the average quantile estimates obtained with them. A set of PDS with limits ranging from percentiles 90 to 99.5, rising by 0.5 steps, was used in this paper. This proved to stabilize the variability of the quantile estimates. However, if this methodology is used on a more general scale, the upper limit range should be defined more precisely because it may differ for each set of data.

This paper has revealed that the widely used AMS–G approach estimates adequately the observed extreme dry-spell risk in the study area, by contrast with the PDS-E.

The results obtained here are of potential importance for agrarian planning. The method used is of potential importance for agrarian planning and of benefit in crop management. It facilitates the drawing of risk maps and the drafting of preventive and palliative plans for the mitigation of the effects of drought.

Data availability. The data are printed in paper documents stored (archived) in office of the General Directorate of Water Resources and the Division of Dam Operation of Extreme North and Ichkeul of the Ministry of Agriculture of Tunisia, 2019a, b (http://www.agriculture.tn/). These data are the property of this organization, and are available in situ.

Author contributions. Discussed the results and contributed to the final version of the manuscript: FL; Developed and performed the design and implementation of the research, the computations, the analysis of the data and the results and the writing of the manuscript: MM.

Competing interests. The authors declare that they have no conflict of interest.

Special issue statement. This article is part of the special issue "Hydrological processes and water security in a changing world". It is a result of the 8th Global FRIEND-Water Conference: Hydrological Processes and Water Security in a Changing World, Beijing, China, 6–9 November 2018.

Acknowledgements. The authors acknowledge the help of General Directorate of Water Resources and Division of Dam Operation of Extreme North and Ichkeul of the Ministry of Agriculture of Tunisia. The authors thank the reviewers for their relevant remarks that contributed to the improvement of this article.

References

Ashkar, F. and Rouselle, J.: Partial duration series modeling under the assumption of a Poissonian flood count, J. Hydrol., 90, 135–144, 1987.

Beguería, S.: Identificación y características de las fuentes de sedimento en áreas de montaña: erosión y transferencia de sedimento en la cuenca alta del río Aragón, PhD thesis (unpublished), University of Zaragoza, 2003.

Bogardi, J. J. and Duckstein, L.: Evénements de période sèche en pays semi-aride, Revue des Sciences de l'Eau, 6, 23–44, 1993.

Bobée, B., Cavadias, G., Ashkar, F., Bernier, J., and Rasmussen, P.: Towards asystematic approach to comparing distributions used in flood frequency analysis, J. Hydrol., 142, 121–136, 1993.

Chow, V. T., Maidment, D. R., and Mays, L. W.: Applied Hydrology, McGraw-Hill, New York, NY, 572 pp., 1988.

Cunnane, C.: A particular comparison of annual maxima and partial duration series methods of flood frequency prediction, J. Hydrol., 18, 257–271, 1973.

Dunxian, S., Ashok, K. M., Jun, X., Liping, Z., and Xiang, Z.: Wet and dry spell analysis using copulas, Int. J. Clim., 36, 476–491, https://doi.org/10.1002/joc.4369, 2015.

Foufoula-Georgiou, E. and Georgakakos, K. P.: Hydrologic advances in space-time precipitation modeling and forecasting, in: Recent advances in the modeling of hydrologic systems, edited by: Bowles, D. S. and O'Connell, P. E., NATO ASI Series, Serie C: mathematical and physical sciences, Kluwer Academic Publishers, Dordrecht, The Netherlands, 345, 47–65, 1991.

Gumbel, E. J.: Statistics of Extremes, Columbia University Press, 1958.

Gupta, V. K. and Duckstein, L.: A stochastic analysis of extreme droughts, Water Resour. Res., 11, 221–228, 1975.

Hershfield, D. M.: On the probability of extreme rainfall events, B. Am. Meteorol. Soc., 54, 1013–1018, 1973.

Hui, W., Xuebin, Z., and Elaine, M. B.: Stochastic modelling of daily precipitation for Canada, Atmos. Ocean, 43, 23–32, https://doi.org/10.3137/ao.430102, 2005.

Konjit, S., Fitsume, Y., Asfaw, K., and Shoeb, Q.: Wet and dry spell analysis for decision making in agricultural water management in the eastern part of Ethiopia, West Haraghe, Int. J. Water Res. Environ., 8, 92–96, https://doi.org/10.5897/IJWREE2016.0650, 2016.

Lana, X. and Burgueño, A.: Spatial and temporal characterisation of annual extreme droughts in Catalonia (NE Spain), Int. J. Clim., 18, 93–110, 1998.

Madsen, H., Pearson, C. P., and Rosbjerg, D.: Comparison of annual maximum series and partial duration methods for modelling extreme hydrologic events, 1. At-site modelling, Water Resour. Res., 33, 759–769, 1997.

Mathlouthi, M.: Optimisation des règles de gestion des barrages réservoirs pour des évènements extrêmes de sècheresse, Thèse de Doctorat, Institut National Agronomique de Tunisie, Tunis, Tunisie, 162 pp., 2009.

Mathlouthi, M. and Lebdi, F.: Event in the case of a single reservoir: the Ghèzala dam in Northern Tunisia, Stochastic Environ. Res. Risk Assess., 22, 513–528, https://doi.org/10.1007/s00477-007-0169-3, 2008.

Mathlouthi, M. and Lebdi, F.: Analyse statistique des séquences sèches dans un bassin du nord de la Tunisie, Hydrol. Sci. J., 54, 442–455, https://doi.org/10.1623/hysj.54.3.442, 2009.

Mathlouthi, M. and Lebdi, F.: Frequency and severity of dry spell phenomenon in Ghezala Dam reservoir (Tunisia), European Water, 60, 255–261, 2017.

Ministry of Agriculture of Tunisia: General Directorate of Water Resources, available at: http://www.agriculture.tn/, Daily rain gauge observations, 2019a.

Ministry of Agriculture of Tunisia: General Directorate of Dams and Major Hydraulic Works, available at: http://www.agriculture.tn/, Division of Dam Operation of Extreme North and Ichkeul, Hydraulic database of the Ghézala dam, 2019b.

Mkhandi, S. H., Kachroo, R. K., and Gunasekara, T. A. G.: Flood frequency analysis of southern Africa: II. Identification of regional distributions, Hydrol. Sci. J., 45, 449–464, 2000.

Moreno, F. and Roldán, J.: Regional daily precipitation stochastic model parameters. Application to the Guadalquivir valley in southern Spain, Phys. Chem. Earth Pt. B, 24, 35–47, 1999.

Sivakumar, M. V. K.: Empirical analysis of dry spells for agricultural applications in West Africa, J. Climate, 5, 532–539, 1992.

Vicente-Serrano, S. M. and Beguería Portugues, S.: Estimating extreme dry-spell risk in the middle Ebro Valley (Northeastern Spain): a comparative analysis of partial duration series with a General Pareto distribution and annual maxima series with a Gumbel distribution, Int. J. Clim., 23, 1103–1118, https://doi.org/10.1002/joc.934, 2003.

Wilks, D. S.: Interannual variability and extreme value characteristics of several stochastic daily precipitation modes, Agric. Meteorol., 93, 153–169, 1999.

Willmott, C. J.: Some comments on the evaluation of model performance, B. Am. Meteorol. Soc., 63, 1309–1313, 1982.

Proc. IAHS, 383, 249–254, 2020
https://doi.org/10.5194/piahs-383-249-2020

Urban flood vulnerability mapping of part of the Lagos metropolis

Olusegun Adeaga, Olufemi Taiwo Oyeneye, and Olutosin Akinbaloye
Department of Geography, University of Lagos, Lagos, 101017, Nigeria
Correspondence: Olusegun Adeaga (oadeaga@unilag.edu.ng)

Published: 16 September 2020

Abstract. Flooding is the most common of all environmental hazards with vast devastating effect due to enormous losses to lives and properties world-wide, annually. Thus, the effect of extreme flooding is dramatic, not only at the individual household level, but in the country as a whole. This study focuses on flood risk assessment based on urban and coastal flood estimation due to the physical peculiarity of the Lagos region and her urbanization characteristics. It involves creating a flood change detection using GIS approach with the aid of a Cellular Automation Framework to simulate flooding in Lagos Metropolitan area, mapping out the flood prone areas and generating a flood vulnerability map for flood management and planning purposes. At twenty minutes intervals the discharge distribution for the different flow regimes (Low, Medium and Extreme) display flow duration between 60–120 min while the time of concentration is attained at 1 h 20 min (80 min).

1 Introduction

Recurrence of extreme precipitation anomalies resulting in floods is a normal component of natural climate variability. Flood is an overflow of an expanse of water that submerges land and a temporary covering by water of land that is not normally covered by water. In the sense of "flowing water", the word may also be applied to the inflow of the tides that comes from the overflow of sea, lakes, rivers, canals, sewers or from rainwater (EU, 2007). Floods cause about one third of all deaths, one third of all injuries and one third of all damage from natural disasters (Askew, 1997; Jeb and Aggarwal, 2008). It is the most common of all environmental hazards which regularly claims over 20 000 lives per year and adversely affects around 75 million people world-wide (Smith, 1996). Death and destruction due to flooding therefore continue to be all too common phenomena throughout the world, affecting millions of people annually. Flooding therefore poses as one of the greatest natural risks to sustainable development; reduce the asset base of households, communities and societies through the destruction of standing crops, dwellings, infrastructure, machinery and buildings, quite apart from the tragic loss of life.

Flooding in urban areas is seriously becoming an ecological menace in Nigeria as several coastal areas along the Atlantic ocean, surrounding cities and river valleys are affected by flooding on a yearly basis resulting in land degradation in some other parts of the country (Abbas, 2009; Jeb and Aggarwal, 2008).

The obvious reason for flooding especially in municipalities and coastal areas in Nigeria lies in the wide distribution of river floodplains and low-lying coastal areas and river floodplains, and because these areas have fast become a long standing attractions for human settlement (Ologunorisa and Abawua, 2005). The effect of extreme flooding is dramatic; not only at the individual household level, but in the country as a whole. At least 20 % of the population is at risk from one form of flooding or another (Etuonovbe, 2011). Flooding in various parts of Nigeria have forced millions of people out from their homes, destroyed businesses, polluted water resources and increased the risk of diseases (Adeaga, 2008; Adelekan, 1998).

Flood occurrence in the Lagos metropolis is a threat to lives and properties and its frequency is increasing, due to her low-lying physical terrain and increasingly urbanization pressure on available marginal lands along the floodplains and congested drainage channels. Severe flooding in Lagos destroys among others large scale, infrastructures with recorded deaths and displacement of thousands. Such displacement usually rendered hundreds homeless and loss of

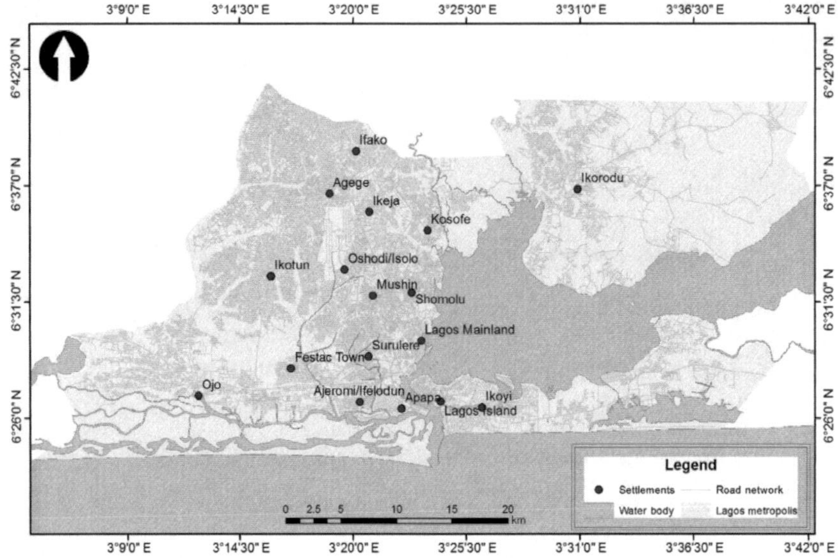

Figure 1. Lagos metropolis area. (@ LABCARS, Dept. of Geography, University of Lagos, Nigeria).

surrounding farmland (IFRC, 2001; Adeaga, 2008; Adeoye et al., 2009). Such losses reduce the asset base of households, communities and societies through the destruction of crops, dwellings, infrastructure, machinery and buildings, quite apart from the tragic loss of life.

Thus, there is therefore the need to provide detailed flood vulnerability map for Lagos mega city, if annual losses attributed to flooding can be effectively addressed.

In recent years, remote sensing and Geographic information systems have been embedded in the evaluation of the geo-environmental hazards. According to Van Westen et al. (2006), the purposes of using remote sensing include: the need "to investigate the susceptibility of the land and the vulnerability of the society, to construct hazard zoning maps and potential damage maps, to monitor potential hazards, and to deal with emergency situations after a disaster. Studies on extensive usage of flood vulnerability mapping using remote sensing data and GIS tools have been extensively used in flood monitoring across globe (Di Martino et al., 2005).

Thus, this study is aimed at creating a flood change detection using GIS approach with the aid of a Cellular Automation Framework. The cellular Automation was used to simulate flooding in the Lagos Metropolitan area towards identification of flood prone areas and flood vulnerability map toward an appropriate Pre-flood preparation and planning processes.

2 Regional setting

Lagos is located on the South-western part of Nigeria on the narrow Coastal flood plain of the Bight of Benin. The State lies approximately between longitudes 2°42′ to 3°42′ E and latitudes 6°22′ to 6°52′ N (Fig. 1).

The Southern Boundary of the State is 180 km along Atlantic Coastline, it is bounded in the north and east by Ogun State; the drainage system of the state is characterized by a maze of lagoons and waterways, which constitute about 22 %, or 787 km^2 of the State total landmass with the Lagos and Lekki Lagoons, Yewa and Ogun Rivers as major water bodies.The Lagos lagoon is one of several lagoon systems in the West African sub region and most extensive. The lagoon is part of the barrier lagoon coasts of Nigeria. The water is shallow and covers an area of about 208 km^2. The lagoon is fed mainly by the rivers of Ogun, Shasha, Oshun, Agboyi and Majidun; the Ogudu creeks and waters of Epe and Lekki lagoons. The Lagoon empties into the Atlantic Ocean via Lagos harbour. The southern margin of the Lagos Lagoon is bounded by the Five Cowrie Creek, the eastern margin by the Palavar Islands and its northern border by Ikorodu. The lagoon is 40–64 km long and has two arms; one connects the Lekki Lagoon while the other leads northward into the hinterland (Allen, 1965). The lagoon is shallow with depths of 1.5–3 m (Ibe, 1988), and made up of muddy and sandy bottom. Its bottom relief is negligible.

The environment of Lagos region is characterized by a tropical climate, with the dominant vegetation of tropical swamp forest, comprising fresh waters and mangrove swamp forests. It has a double rainfall pattern, with two climatic seasons – the dry (November–March) and wet (April–October).

The Lagos metropolis is a densely populated low-lying coastal area, with a dense network of roads and buildings, and a population of over 20 million people (Adeloye and Rustum, 2011). The Lagos metropolis is the biggest city in

Table 1. Data source and characteristics.

S/no	Data	Scale	Year	Source
1	Lagos State Administrative Map	1 : 25 000	2009	GCLME Lab, Unilag
2	SRTM DEM	90 m	2000	http://srtm.csi.cgiar.org (last access: 18 July 2019)
3	Discharge/stage height	Daily/monthly	1998–2008	Ogun-Oshun River Basin Authority (OORBA)

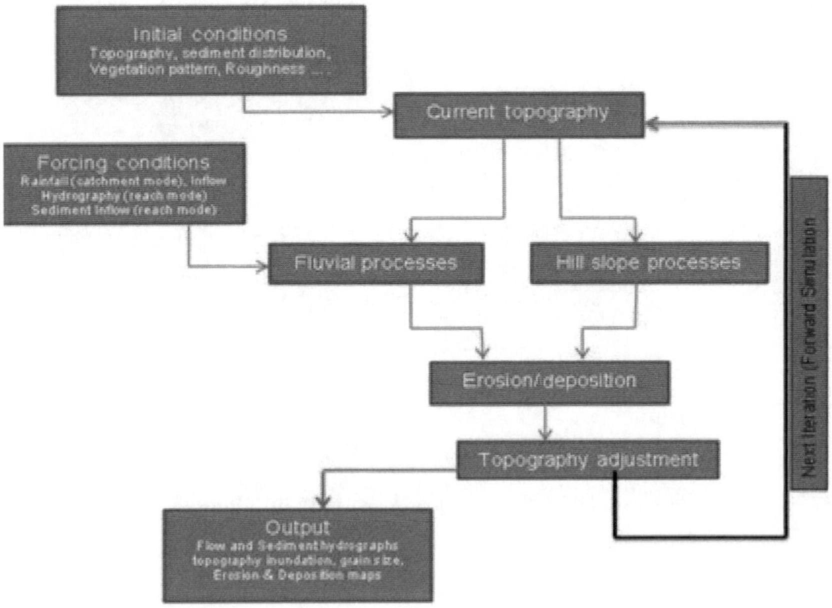

Figure 2. Framework of CEASAR model.

Nigeria, (although the smallest land area), the second largest city in Africa, and the seventh largest city in the world. The population growth rate in the Lagos metropolis is estimated at 3.2 % (World Bank, 2013). Against this background, the United Nations predicts that Nigeria will be one of the eight countries expected to account collectively for half of the total population increase in the world from 2005 to 2050, and will by 2100, have a population of between 505 million and 1.03 billion (UN, 2004). High population density is a major impasse in the Lagos region, subjecting the area to lack of space for the myriad of human activities, which often manifests itself in muddled human settlements, overcrowding, slum envelopments, pollution, illegal structures, and other social and environmental disorder. These factors, either singularly or in combination, have the potential to increase exposure to hazard, especially the risks associated with urban flooding events.

3 Methodology

The study methodology is basically based on integration of cellular automation modelling within GIS framework. In the study the basic data used and their characteristic are depicted in Table 1, in addition to the basic data input of the daily discharge, the Shuttle radar Topographical Mission (SRTM) data was converted to Digital elevation model on the GIS platform using the Arc Hydro tool and Raster Edit. This was incorporated with flow in order to simulate flooding and morphological change within the Lagos metropolis.

Cellular Automation and Evolutionary Slope and River (CAESAR) model which is a two dimensional flow and sediment transport model was used for the flow simulation. CEASER occupies a unique space in fluvial modelling with capability to simulate timescales that are useful to engineers, researchers of fluvial systems (1–100 years) and to simulate flooding and morphological change of pertinent spatial scales (from 2 km reaches to 400 + km catchments). The cellular framework uses a regular mesh of grid cells to represent the river catchment studied. It is based upon the cellular automaton concept, whereby the repeated iteration of a series of rules on each of these cells determines the behaviour of the whole system (Fig. 2). Some factors or rules exist in CAESAR for a hydrological model, hydraulic model (flow routing), fluvial erosion and deposition, and slope characteristics (Van et al., 2007).

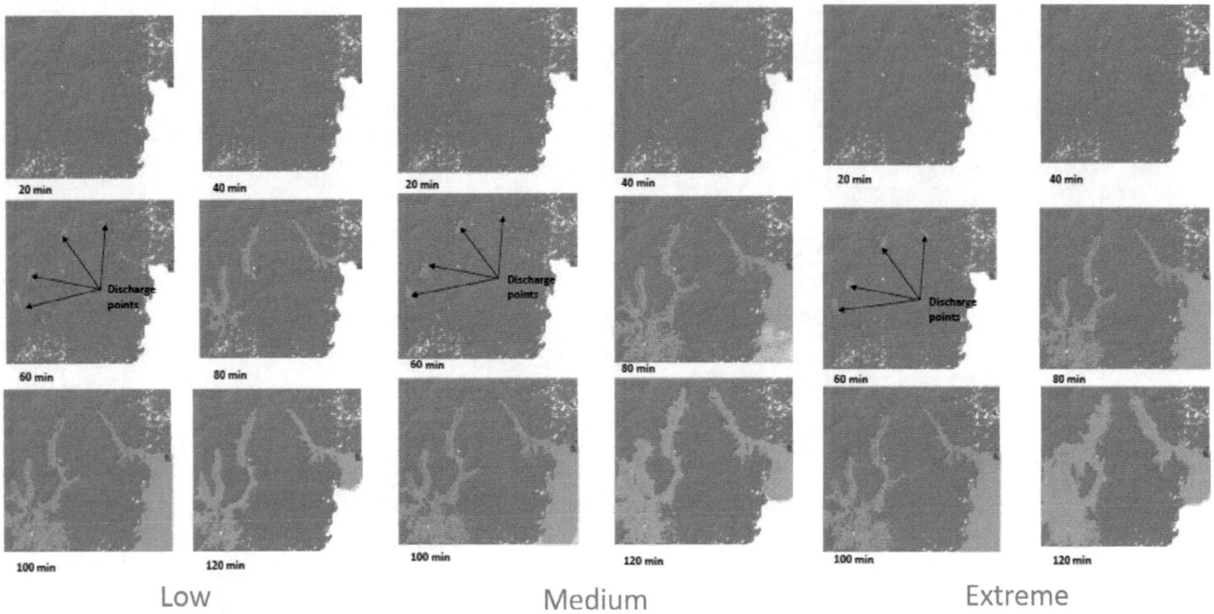

Figure 3. Simulated flow for Lagos metropolis.

In simulation mode, each grid cell is assigned initial values for elevation, water discharge, water depth, and drainage area and grain size fractions. An initial state (time $t = 0$) is selected by assigning a state for each cell and for each cell, a set of cells called its *neighborhood* is defined relative to the specified cell. For each time-step or iteration, these values are updated in relation to the immediate neighbours according to laws applied to every cell. A new *generation* is created (advancing t by 1), according to fixed *rule* (generally, a mathematical function) that determines the new state of each cell in terms of the current state of the cell and the states of the cells in its neighborhood. Typically, the rule for updating the state of cells is the same for each cell and does not change over time, and is applied to the whole grid simultaneously, though exceptions are known, such as the stochastic cellular automaton and asynchronous cellular automaton.

For the simulation, four identified head water points along the channels are Point 1 – (Column 208, Row 44); Point 2 – (Column 129, Ro 60); Point 3 – (Column 56, Row 142) and Point 4 – (Column 21, Row 190) were used (Fig. 3). The input data time step is set to 1440 min at a 42 years, hourly water discharge in daily time steps. The quantity of sediment re-circulated been controlled by proportion re-circulated at a set parameter of 0.5, to re-circulate 50 % of the sediment at the 5000 iterations for simulated low, medium and extreme channel water discharge (flow).

4 Result and analysis

Simulated flow within the channel from the selected head water points shows the discharge distribution at twenty minutes intervals for the different flow regimes (Fig. 3).

For the simulation discharge duration runs between 60–120 min and flow time of initial concentration is attained at 1 h 20 min (80 min) and at peak at 120 min before receeding. Discharge at varied flow regime depicts 27.78, 138.89 and 277.78 $m^3 s^{-1}$ for low (normal), medium and extreme flow events, respectively.

Simulation for the four different discharges at discharge rate of 100 000, 500 000 and 1 000 000 $m^3 h^{-1}$, show that the flows is contained within the channel banks at low discharge while inundation of the immediate adjacent floodplain area occurs during the medium-discharge flows and inundation of the entire floodplain is attained at extreme-discharge (Fig. 4).

Areas liable to flood as depicted by the vulnerability map for the Lagos metropolis for the different discharge/flow regimes shows that Apapa, parts of Kosofe are flooded at the normal flood extent, as a result of the low elevation level and proximity to identified water body identified (Atlantic Ocean and Lagos Lagoonal System) while the extended flows at the extreme extent regime towards the upper part to places like Ikeja, Shomolu, Surulere, Mushin and Lagos Mainland, which were not flooded at normal flood extent, are due to the increased discharges within the channel. Overbank flooding is more extensive on the southern side of the channel in this reach, due to the raised terrain of the alluvial fan to the south.

Floodplain topography also controls the pattern and depth of inundation; low lying areas are more likely to be flooded

 https://doi.org/10.5194/piahs-383-249-2020

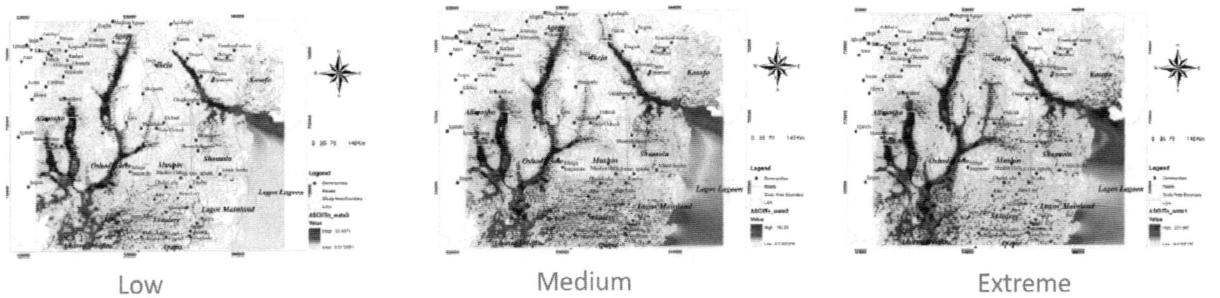

Figure 4. Flood vulnerability of Lagos metropolis.

Table 2. Flood risk in Lagos metropolis.

Flow	Discharge ($m^3\,h^{-1}$)	Area at risk (km^2)	Risk area (%)
Extreme	1 000 000	237.83	42
Medium	500 000	197.32	34
Low	100 000	134.91	24

due to their nearness to the river channel and low topography. It is clear that an increased discharge has more areas flooded and the reverse is the case for the effect of the discharge when the water is less.

Estimation of total area at risk shows the flooded area of the study area. It shows the greatest percentage (42 %) is attained during the extreme flood extent due to a high discharge while the lowest percentage (24 %) occurs at low discharge (Table 2). Although these results may appear trivial, they represent notable improvements for cellular automaton flow algorithms – particularly, the ability to route flow through a high-resolution meandering channel.

5 Conclusion and recommendation

GIS technology provides a wide range of vantage in the generation of the flood vulnerability maps because the output from various platforms were embedded into the GIS environment from which all the various flood regimes and vulnerability scenarios were produced. The CAESAR model approach is relatively simple and has the ability to address some of the shortcomings of other types of numerical model due to its proper flow routing scheme, channel dynamics that easily be accounted for, with the relevant prediction of inundation patterns, and water depths at different scenarios.

It is recommended that effective flood planning and management in Lagos metropolis should entail realistic early flood warnings services, constant monitoring of waterways and drainage pathway and provision of appropriate flood vulnerability map at varied scale (local – state level) for every planning and decision-making purpose while an appropriate adaptable plan should be adhered to.

Data availability. Please note that the sole proprietors of dataset are their respective sources. They are third parties (secondary data) and can be requested or accessed through the various organizations website and address.

Author contributions. OAd is involved in the study conception and design as well as data analysis, interpretation, drafting of manuscript and critical review. OTO partook in the acquisition of data, data analysis and interpretation. OAk engaged in data Analysis, interpretation of data as well as manuscript writing.

Competing interests. The authors declare that they have no conflict of interest.

Special issue statement. This article is part of the special issue "Hydrological processes and water security in a changing world". It is a result of the 8th Global FRIEND–Water Conference: Hydrological Processes and Water Security in a Changing World, Beijing, China, 6–9 November 2018.

Acknowledgements. We would like to thank the Hydrological team and Laboratory for Cartography and Remote Sensing; Department of Geography University of Lagos, Nigeria.

References

Abbas, I. I.: Use of Satellite Remote Sensing and Geographic Information Systems to Monitor Land Degradation along Ondo Coastal Zone, Nigeria, J. Geogr. Environ. Plan., 5, 26–35, 2009.

Adeaga, O.: Flood hazard mapping and risk management in part of Lagos N. E., in: GSDI 10 Conference Proceedings, St. Augustine, Trinidad, available at: http://www.gsdi.org/gsdiconf/gsdi10/papers/TS13.3paper.pdf (last access: 20 July 2019), 2008.

Adelekan, I. O.: Spatio-temporal variations in thunderstorm rainfall over Nigeria, Int. J. Climatol., 18, 1273–1284, 1998.

Adeloye, A. J. and Rustum, R.: Lagos (Nigeria) flooding and influence of urban planning, Proc. Inst. Civ. Eng., 164, 175–187, 2011.

Adeoye, N. O., Ayanlade, A., and Babatimehin, O.: Climate change and menace of floods in Nigerian cities: Socio-economic implications, Adv. Nat. Appl. Sci., 3, 369–377, 2009.

Allen, J. R.: Late Quaternary of the Niger Delta, and Adjacent Areas: Sedimentary Environments and Lithofacies, Bull. Am. Petrol. Geol., 49, 547–600, 1965.

Askew, A. J.: Water in the International Decade for Natural Disaster Reduction, in: Destructive Water: Water-caused Natural Disasters, their Abatement and Control, edited by: Leavesley, G. H., Lins, H. F., Nobilis, F., Parker, R. S., and Schneider, V. R., IAHS, Publ. No. 239, International Association of Hydrological Sciences, IAHS Press, Institute of Hydrology, Wallingford, Oxfordshire, 1997.

Di Martino G., Iodice, A., Pasera, M., Riccio, D., and Ruello, G.: Sensing for Developing countries: Landsat data and GIS, Naples: Ingegneria senza frontier, Napoli, 2005.

Etuonovbe, A. K.: Devastating Effect of Flooding in Nigeria, in: FIG Working Week 2011 Bridging the Gap between Cultures Marrakech, 18–22 May 2011, Morocco, 2011.

EU – European Union: Floods directive, available at: http://ec.europa.eu/environment/water/flood_risk/index.htm (last access: 18 July 2019), 2007.

Ibe, A. C.: Coastline Erosion in Nigeria, Ibadan University Press, Ibadan, Nigeria, 1988.

IFRC – International federation of Red Cross and Red Crescent society: World Disaster Report: Focus on recovery, available at: http://www.ifrc.org/Global/Publications/disasters/WDR/21400_WDR2001.pdf (last access: 20 July 2019), 2001.

Jeb, D. N. and Aggarwal, S. P.: Flood Inundation Hazard Modeling of the River Kaduna Using Remote Sensing and Geographic Information Systems, J. Appl. Sci. Res., 4, 1822–1833, 2008.

Ologunorisa, T. E. and Abawua, M. J.: Flood risk assessment: a review, J. Appl. Sci. Environ. Manage., 9, 57–63, 2005.

Smith, K.: Environmental Hazards, Routledge, London, 1996.

UN – United Nations: World Population to 2300, United Nations, Department of Economic and Social Affairs, Population Division, New York, 2004.

Van, T. P. D., Carling, P. A., Coulthard, T. J., and Atkinson, P. M.: Cellular Automata Approach for Flood Forecasting in a Bifurcation River System, Inst. Geophys., Polish Academy of Science, Warszawa, Poland, 2007.

Van Westen, C. J., Van Asch, T. W. J., and Soeters, R.: Landslide hazard and risk zonation – why is it still so difficult?, Bull. Eng. Geol. Environ., 65, 167–184, https://doi.org/10.1007/s10064-005-0023-0, 2006.

World Bank: World Bank Population Growth Rate 2009–2013, available at: http://data.worldbank.org/indicator/SP.POP.GROW (last access: 10 March 2019), 2013.

Proc. IAHS, 383, 255–260, 2020
https://doi.org/10.5194/piahs-383-255-2020

Improved understanding of the Senegal floodplain socio-hydrosystems with multi-scalar earth observations

Andrew Ogilvie[1,2], **Jean-Christophe Poussin**[1], **Jean-Claude Bader**[1], **Didier Martin**[1], **Gora Ndiaye**[3], and **Djiby Dia**[2]

[1]G-EAU, AgroParisTech, CIRAD, IRD, IRSTEA, Montpellier SupAgro, Univ Montpellier, Montpellier, France
[2]ISRA BAME, Dakar, Senegal
[3]Commission Permanente des Eaux, OMVS, Dakar, BP 3152, Senegal

Correspondence: Andrew Ogilvie (andrew.ogilvie@ird.fr)

Published: 16 September 2020

Abstract. In semi-arid regions, the annual floods of alluvial plains, wetlands and lakes provide essential services to millions of farmers, livestock holders and fishers. The scale, number and dispersion of these water bodies limit however their hydrological understanding, whose complexity is exacerbated by the mutual interactions between water and society. Rising availability of high spatial and temporal resolution remote sensing imagery provide novel opportunities to improve the understanding of the hydrology of multiple water bodies and of their ability to support local livelihoods. Multi-scalar interdisciplinary research combining remote sensing, in situ observations, hydrological modelling and farmer interviews was developed to investigate multiple socio-hydrosystems in the Senegal floodplain. Surface water monitoring with MNDWI on multi-spectral imagery (Landsat 5, 7, 8 and Sentinel-2A) is used to identify intra-annual and inter-annual variations in water availability. Validation against field monitoring ($R^2 = 0.88$) and photointerpretation of drone imagery highlights the benefits of Sentinel's increased resolution to monitor floods of minor surface area and duration and improve the representation of hysteresis in floodplains. Integrated into forthcoming river basin modelling, these results can help simulate the influence of upstream changes on the hydrology and associated services of these complex socio-hydrosystems and improve watershed management.

1 Introduction

The annual floods of large rivers in Africa (Nile, Zambezi, Niger, Senegal, Congo) and Asia (Mekong, Ganges, Indus) support numerous ecosystem services along their alluvial floodplains. In Africa, these seasonal resources support the livelihoods of small-scale farmers, fishers and herders over 30 million ha. In semi-arid areas, these systems are subject to strong hydroclimatic variability and vulnerability. In West Africa, the droughts in the 1970–1980s favoured the development of full control irrigation, however millions of smallholder farmers continue to rely on flood recession farming for subsistence. Besides, the construction of dams to support the development of irrigation and hydropower has modified river regimes, reducing the peak floods while allowing potential support during low flows.

The scale, diversity and flat topography of these floodplains generate difficulties and uncertainties in the hydrological monitoring and modelling (Neal et al., 2012). These are exacerbated by the decline in observation networks, and the difficulties to apprehend and represent human influences on water resources (withdrawals, dam management, land use changes) and in turn the relationships between water resources and associated uses and services. These considerations are at the heart of the reflections developed currently in socio-hydrology (Sivapalan et al., 2012) and the 2013–2022 IAHS Panta Rhei decade (Montanari et al., 2013) devoted to increasing the understanding and representation of anthropic modifications to hydrological regimes. In parallel, the opportunities in quantitative hydrology provided by the recent improvements in low cost, high spatial and temporal resolution imagery have increased significantly. New sensors aboard

Sentinel-2 satellites notably provide free high resolution (up to 10 m) imagery every 10 d since 2015. Their ability to reproduce hydrological dynamics and extrapolate across multiple sites and larger scales must be explored and validated against extensive ground truth field data. Previous research highlighted the difficulties (resulting from size, flooded vegetation, shallow depths, cloud cover during the flood rise), and the need to assess and reduce remote sensing uncertainties in different hydrological settings (Ogilvie et al., 2018).

This paper presents the first results from research focussing on the alluvial floodplain of the Senegal River where field instrumentation, remote sensing and agronomic surveys are being used to improve the hydrological understanding of the flood and its influence on recession farming. Upscaled and integrated into hydrological modelling, these results will help predict climatic and human influences on flooding patterns and associated services in the Senegal floodplain socio-hydrosystems.

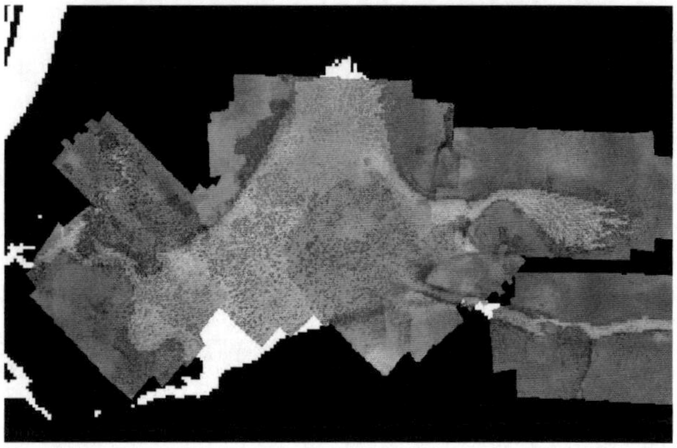

Figure 1. UAV imagery of the flooded areas superposed to the classified Sentinel-2 image showing flooded area in white (5 October 2017).

2 Data and methods

2.1 Floodplains of the Senegal River

The middle valley of the Senegal River is an alluvial floodplain extending over more than 225 000 ha on the border of Senegal and Mauritania. Local rainfall in this Sahelian floodplain remains limited (300 mm on average concentrated in July–September) and the flood is produced by heavy rainfall of the West African monsoon on the Guinean headwaters. The flood peaks occur between mid-September and mid-October depending on the location along the Senegal River. The annual floods are a vital resource for local fishers, herders and farmers, who grow floating rice and traditional crops (sorghum, corn, cowpea) as the flood recedes.

River flows in the Senegal River were modified by the commissioning in 1988 of the Manantali dam in Mali on the Bafing river, built to generate hydroelectricity, improve downstream irrigation and navigation during low flows and support limited flood-recession farming. The Diama dam built in the lower part of the river in 1986 aims to prevent saltwater intrusion (flowing back upstream as far as Richard Toll) and to improve the potential for irrigation and drinking water supply. Annual runoff of the Senegal River also noticeably declined during the 1968–1993 West African drought (over 40 % decline compared to 1950–1968), and has only partly recovered despite the return of high rainfall years (Descroix et al., 2018).

2.2 Satellite optical imagery

443 images over 1984–2017 from sensors aboard Landsat 5 to Landsat 8 satellites were used in this study. Images have a 30 m spatial resolution and a 16 d frequency though the combination of two satellites can reduce observations to every 8 d at certain periods. Images obtained through the U.S. Geological Survey (USGS) ESPA are processed to surface reflectance (Level 2A) using the Landsat Ecosystem Disturbance Adaptive Processing System (LEDAPS) for sensors TM and ETM+ and the Landsat Surface Reflectance Code (LaSRC) for the OLI sensor. A cloud & shadow mask is provided, generated by Cmask, a C version of the Fmask algorithm (Zhu and Woodcock, 2015). 48 Level 2A Sentinel-2 images over 2015–2017 were downloaded from the French Space Agency CNES THEAI data centre. These products include atmospheric corrections based on the MAJA (MACSS-ATCOR Joint Algorithm) treatment chain developed by CNES-CESBIO (Hagolle et al., 2015) and German Aerospace Centre DLR, as well as a cloud and shadow mask. The percentage of clouds, shadows and lost pixels resulting from scan line corrector (SLC) failure after 2003 on ETM+ sensor aboard Landsat 7 were calculated for each image over the region of interest.

Flooded areas over the Podor floodplain (Fig. 1) were determined using the Modified Normalised Difference Water Index (MNDWI) (Xu, 2006) based on the Green and Shortwave Infrared bands, previously used across similar wetlands and shown to provide the greatest accuracy in detecting water in mixed environments (water, flooded vegetation, shallow waters) (Ogilvie et al., 2015, 2018). Calibration of the MNDWI threshold was performed using very high resolution (20 cm) imagery of the 2017 flood acquired with a DJI Phantom 4 UAV (unmanned aerial vehicle). UAV RGB images were mosaicked, orthorectified, resampled to 1 m (to reduce calculation times) and flooded areas were delineated using Maximum likelihood classification and GPS locations of flooded areas.

2.3 Hydrological monitoring at Podor

Monitoring of stage levels (m. NG) of the Senegal River at Podor has been carried out since 1903 and is currently managed by the OMVS (Organisation pour la Mise en Valeur du Fleuve Sénégal). Daily ladder readings over 1999–2017 were used here and compared with the flood amplitude and variations obtained by remote sensing of surface waters in the Podor floodplain. Previous research in the Podor floodplain in 1999 established a correlation $S_{\max} = f(Z_{\text{peak}})$ between the stage values in the river at Podor (Z_{peak}) and the maximum flooded surface area ($S_{\max}$) based on 8 classified SPOT images of the flood peak between 1987 and 1999. The 18-year period of field and remote sensing data are notably used here to assess the ability and accuracy of remotely sensed flooded areas to refine the $S_{\max} = f(Z_{\text{peak}})$ relationship.

3 Results and discussion

3.1 Monitoring surface water variations

Figure 1 illustrates the remarkable resolution in imagery of the flooded area obtained from a series of high altitude flights (300 m) with an UAV. Here, these enable the calibration of the MNDWI to classify water pixels on Sentinel-2 20 m imagery and provide a much greater sample size compared to traditional GPS-based ground truthing. MNDWI allows accurate detection of the open water but also the narrow channels in the presence of significant flooded vegetation and turbidity.

The full series of 443 Landsat images processed provide initial insights into the flood variations over 1984–2017. The peak flooded area reached 1900 ha during high flood years (1999) and only 500 ha during low flood years (2006). Image availability prior to 1999 remains scarce (on average 4 images per year over 1984–1998 compared to 20 images per year over 1999–2015) and insufficient to provide information on flood dynamics. Nevertheless, single images as in October 1988 occasionally provide valuable information on the flood peak.

Comparisons with daily stage data from the nearby river gauge at Podor confirm the aptitude of Landsat long term series to reproduce the annual flood dynamic as well as the interannual variability in the flood amplitudes (Figs. 2 and 3). Over the full period, interannual variations of the peak flooded area are well correlated ($R^2 = 0.88$) with the variations of river levels at the Podor gauging station (compared to $R^2 = 0.81$ with previous SPOT imagery). The variations in amplitude are well reproduced most years, except in 2000 and 2003 where cloud cover prevented detection of the flood peak. After 2015, the combination of Sentinel-2 imagery with Landsat 8 increases the resolution of the flood dynamics reproduced (Fig. 3). Over a single flood, R^2 increases from 0.92 when using Landsat, to 0.98 with Sentinel-2A (and to 0.94 when combining both satellites due to the residual errors from Landsat 7 SLC failure).

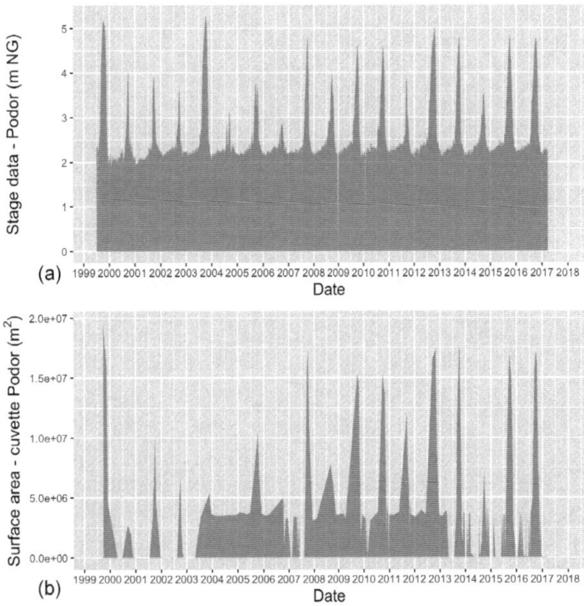

Figure 2. Monitoring surface water variations in the Podor floodplain using Landsat 5, Landsat 7, Landsat 8 and Sentinel-2A satellites **(a)** compared to daily stage data at Podor **(b)**.

The analysis of the percentage of pixel loss from clouds, shadows and SLC failure after 2003 over the study area showed a low sensitivity to clouds and shadows, possibly due to many thin, cirrus clouds, which allowed setting a high threshold (40 %) to exclude cloud and shadow interferences. The repetitivity of images from multiple sources and the gradual flood rise also lead to the flood rise being relatively well represented (Fig. 3), unlike in other areas (Ogilvie et al., 2015).

The benefit of combining radar (e.g. Sentinel-1) and optical imagery during the rainy season to overcome dense cloud cover may then be lower in this setting. Interference from SLC failure (bands with no data between 60 to 420 m wide and reducing at the centre of the image) are however important here, considering the location of Podor on the edge of the Landsat tiles. These affect on average 18 % of the Podor floodplain and are visible in the noise (systematic overestimation) during dry seasons between 2003 and 2013 notably (Fig. 2). Temporal and spatial interpolations approaches must therefore be used (Zeng et al., 2013) to reduce uncertainties from SLC pixel loss. Benefits of pan-sharpening 20 m Sentinel-2 bands to 10 m to detect the numerous narrow inlets and increase accuracy of water bodies of this size may also be investigated.

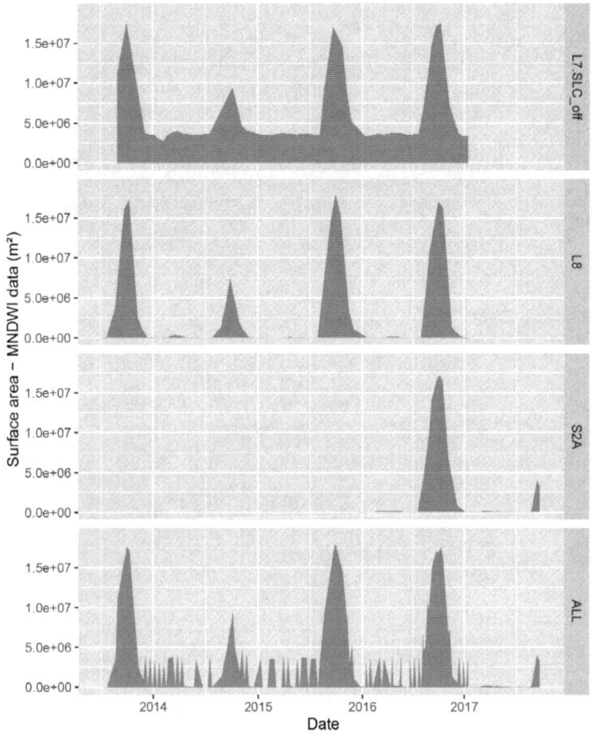

Figure 3. Monitoring of the flood dynamic in the Podor floodplain over 2013–2016 with multispectral imagery from Landsat 7, Landsat 8, Sentinel 2A satellites and their combination.

3.2 Relationship between flooded areas and stage data

Figure 4 illustrates the good correlation between stage levels at Podor gauging station in the Senegal River and the peak flooded area in the Podor floodplain. Values obtained with Landsat and Sentinel-2 are coherent with results from previous research which used 8 SPOT 2 and 4 20 m resolution images of the flood peak during the 1990s. Surface area for the 1999 maximum flooded area are notably close, 2.9 % lower with SPOT than Landsat 7. The additional data allows the relationship to be extended and refined notably for lower floods. Similar relations using discharge rather than stage levels can also be developed.

Importantly, Fig. 5 highlights the opportunity to refine this relationship and distinguish between the relationship during the flood rise and during the flood decline. The high temporal resolution allows us to discern the importance of the hysteresis process, resulting in a slower decline of flooded areas as water remains captive by naturals weirs in the floodplain (cf. Fig. 5), which must be accounted for in hydrological modelling. Additional field data over years of low (2017) and high floods (2018) will help further refine these relationships.

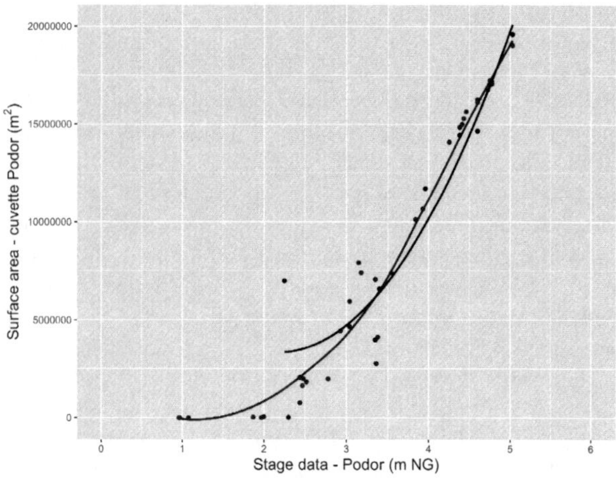

Figure 4. Relationship between stage data at Podor and remotely sensed flooded areas in the Podor floodplain during flood peak, (i) with 8 SPOT images (black line), (ii) with Landsat and Sentinel-2A imagery (blue line).

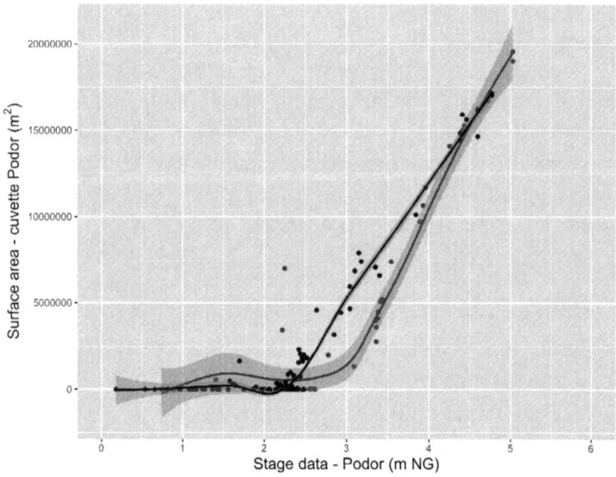

Figure 5. Relationship between stage data at Podor and remotely sensed flooded areas in the Podor floodplain as the flood rises (red line) and recedes (blue line), revealing the strong hysteresis.

4 Conclusions and applications to agricultural water management

The analysis of 443 Landsat 5–8 images between 1999–2017 and 48 Sentinel-2A images between 2015–2017 confirms the possibility of monitoring interannual and intra-annual floods within the depressions of the Senegal River floodplain over 1999–2017. The MNDWI threshold was successfully calibrated against photointerpretation of very high-resolution UAV imagery (20 cm, 600 ha) of surface water. Further research to assess the accuracy and efficiency of automatic photointerpretation attempts are required, as well as to investigate the benefits in terms of accuracy and treatment time (automation) of multispectral UAV imagery (e.g. Sequoia sensors). Remote sensing analysis reproduces the amplitude and dynamic of surface water variations in the floodplain and values are coherent with hydrological data acquired on site. The availability of a higher number of remote sensing observations. It also identifies the hysteresis phenomena and accordingly distinguishes a different $S = f(Z)$ relationship during the rise and decline phase of the flood (Fig. 5).

Ongoing observations will provide additional field data to refine this research and quantify the benefits of pan-sharpening 20 m bands to 10 m, and from raised image availability (every 5 d with the S2A and S2B constellation of satellites). The benefits of Radar imagery to increase image availability when cloud cover is dense will also be explored. Long term series will then be used to provide spatialised information of the timing and duration of the flood across the floodplain. Combined with observations of soil humidity and agricultural practices in situ and by remote sensing, this work will seek to establish relationships between the flooded areas and the agricultural benefits as well as identify strategies to support farmers (sowing dates, suitable crops, water management ...). High-resolution topography of the floodplain will also allow water depths to be estimated from remotely sensed surface areas, providing valuable information to stakeholders to understand how water management in these heavily cultivated areas may be optimised based on weirs and sluice gates to reduce water intake or increase its retention.

Replicated to multiple lakes in the 250 000 ha Senegal River floodplains, this approach can then provide essential insights into the agricultural benefits provided by the annual flood in these depressions. Ultimately, these findings will be integrated into a hydrological model of the Senegal River to understand how upstream changes in land use, water management (dams, releases) and climate are modifying the flood regime and the services provided by these socio-hydrosystems.

Data availability. Landsat surface reflectance imagery is available from USGS EarthExplorer. Level 2A Sentinel-2 imagery can be accessed via CNES THEIA (http://theia.cnes.fr, last access: 24 July 2020).

Author contributions. AO performed the analyses and interpretation and wrote the manuscript. DD, DM, GN, JCB, JCP, contributed to data acquisition, project management and improving the manuscript.

Competing interests. The authors declare that they have no conflict of interest.

Special issue statement. This article is part of the special issue "Hydrological processes and water security in a changing world". It is a result of the 8th Global FRIEND–Water Conference: Hydrological Processes and Water Security in a Changing World, Beijing, China, 6–9 November 2018.

Acknowledgements. The authors are grateful for the hydrological data supplied by the OMVS.

Financial support. This research has been supported by the ANR LEAP-AGRI (grant no. ANR-18-LEAP-0002) and the FSP Agricora Generia project.

References

Descroix, L., Guichard, F., Grippa, M., Lambert, L. A., Panthou, G., Mahé, G., Gal, L., Dardel, C., Quantin, G., Kergoat, L., Bouaïta, Y., Hiernaux, P., Vischel, T., Pellarin, T., Faty, B., Wilcox, C., Malam Abdou, M., Mamadou, I., Vandervaere, J.-P., Diongue-Niang, A., Ndiaye, O., Sané, Y., Dacosta, H., Gosset, M., Cassé, C., Sultan, B., Barry, A., Amogu, O., Nka Nnomo, B., Barry, A., and Paturel, J.-E. : Evolution of surface hydrology in the Sahelo-sudanian strip: An updated review, Water, 10, 748, https://doi.org/10.3390/w10060748, 2018.

Hagolle, O., Huc, M., Pascual, D., and Dedieu, G.: A Multi-Temporal and Multi-Spectral Method to Estimate Aerosol Optical Thickness over Land, for the Atmospheric Correction of FormoSat-2, LandSat, VENUS and Sentinel-2 Images, Remote Sens., 7, 2668–2691, https://doi.org/10.3390/rs70302668, 2015.

Montanari, A., Young, G., Savenije, H. H. G., Hughes, D., Wagener, T., Ren, L. L., Koutsoyiannis, D., Cudennec, C., Toth, E., Grimaldi, S., Blöschl, G., Sivapalan, M., Beven, K., Gupta, H., Hipsey, M., Schaefli, B., Arheimer, B., Boegh, E., Schymanski, S. J., Di Baldassarre, G., Yu, B., Hubert, P., Huang, Y., Schumann, A., Post, D., Srinivasan, V., Harman, C., Thompson, S., Rogger, M., Viglione, A., McMillan, H., Characklis, G., Pang, Z., and Belyaev, V.: Panta Rhei-Everything Flows: Change in hydrology and society-The IAHS Scientific Decade 2013–2022, Hydrolog. Sci. J., 58, 1256–1275, https://doi.org/10.1080/02626667.2013.809088, 2013.

Neal, J., Schumann, G., and Bates, P.: A subgrid channel model for simulating river hydraulics and floodplain inundation over large and data sparse areas, Water Resour. Res., 48, W11506, https://doi.org/10.1029/2012WR012514, 2012.

Ogilvie, A., Belaud, G., Delenne, C., Bailly, J. S., Bader, J.-C., Oleksiak, A., Ferry, L., and Martin, D.: Decadal monitoring of the Niger Inner Delta flood dynamics using MODIS optical data, J. Hydrol., 523, 368–383, https://doi.org/10.1016/j.jhydrol.2015.01.036, 2015.

Ogilvie, A., Belaud, G., Massuel, S., Mulligan, M., Le Goulven, P., and Calvez, R.: Surface water monitoring in small water bodies: potential and limits of multi-sensor Landsat time series, Hydrol. Earth Syst. Sci., 22, 4349–4380, https://doi.org/10.5194/hess-22-4349-2018, 2018.

Sivapalan, M., Savenije, H., and Bloschl, G.: Socio-hydrology: A new science of people and water, Hydrol. Process., 26, 12700–1276, https://doi.org/10.1002/hyp.8426, 2012.

Xu, H.: Modification of normalised difference water index (NDWI) to enhance open water features in remotely sensed imagery, Int. J. Remote Sens., 27, 3025–3033, https://doi.org/10.1080/01431160600589179, 2006.

Zeng, C., Shen, H., and Zhang, L.: Recovering missing pixels for Landsat ETMC SLC-off imagery using multi-temporal regression analysis and a regularization method, Remote Sens. Environ., 131, 182–194, https://doi.org/10.1016/j.rse.2012.12.012, 2013.

Zhu, Z. and Woodcock, C. E.: Automated cloud, cloud shadow, and snow detection in multitemporal Landsat data: An algorithm designed specifically for monitoring land cover change, Remote Sens. Environ., 152, 217–234, https://doi.org/10.1016/j.rse.2014.06.012, 2015.

Proc. IAHS, 383, 261–266, 2020
https://doi.org/10.5194/piahs-383-261-2020

Open Access
PIAHS

A review of the adaptability of hydrological models for drought forecasting

Zikang Xing[1,2,3], **Miaomiao Ma**[2], **Zhicheng Su**[2], **Juan Lv**[2], **Peng Yi**[1,3], **and Wenlong Song**[2]

[1]State Key Laboratory of Hydrology-Water Resources and Hydraulic Engineering, Hohai University, Nanjing, China

[2]Research center on Flood & Drought Disaster Reduction of the Ministry of Water Resources, China Institute of Water Resources and Hydropower Research, Beijing, China

[3]College of Hydrology and Water Resources, Hohai University, Nanjing, China

Correspondence: Zikang Xing (xzk@hhu.edu.cn)

Published: 16 September 2020

Abstract. Drought intensity and frequency are increasing in recent years in multiple regions across the world due to global climate change and consequently drought forecasting research has received more and more attention. Previous studies on drought forecasting mostly focus on meteorological drought based on precipitation and temperature. However, the trend of predicting agriculture and hydrological drought, which consider soil moisture and runoff, have developed rapidly in recent years. Hydrological drought forecasting is based on the hydrological models and the model structure plays a role to improve predictions. This study scrutinized more than 50 hydrological models, including lumped models, semi-distributed models, distributed models, surface water and groundwater coupled models, to explore the adaptability of hydrological models in drought simulation and forecasting. The advantages and disadvantages of typical models, such as DTVGM, GWAVA, and HEC-HMS models were analyzed to provide valuable reference for drought forecasting model development. Future work aims at improving the hydrological models to simulate the drought processes and make better prediction.

1 Introduction

Drought intensity and frequency are increasing in past and recent years in multiple across the globe due to global climate change (Wilhite and Glantz, 1985). In this context, methods to predict drought has received more and more attention (Edwards and McKee, 1997). Previous study of drought forecasting mostly focused on meteorological drought based on precipitation and temperature, while the trend prediction of agriculture and hydrological drought, which consider soil moisture and runoff, have developed rapidly in recent years (Wilhite, 2000; Wanders et al., 2019). There are two main types of methods involved in the hydrological drought forecasting. One is the statistical method that tries to develop construct the relationship between hydrological characteristics and drought events, such as the gray forecasting method (Vishnu and Syamala, 2012), Markov chain method (Paulo and Pereira, 2007), Error back Propagation neural network (Raju et al., 2011), and correlation analysis. The second method is the methods based on hydrological models and coupled atmospheric-hydrological models (Mishra and Singh, 2011). For the latter method, the model structure plays a significant role to improve predictions.

Most hydrological models have been developed for flood simulation and prediction (Huggins and Monke, 1970). Generally, flood generation tends to focus on fast processes and peaks can form in a few days or even hours, while the process of drought development is much slower covering several months or years. Therefore, flood forecasting is a short-term high flow prediction, but the drought forecasting focuses on medium and long-term low flow prediction (Mishra and Singh, 2010). Most hydrological models could improve on drought prediction in humid watersheds as they put emphasis on peak flow simulation rather than low flow (Yu et al., 1999). If these models are used for the drought forecasting, they might suffer from the limitations of structure. Two scientific questions are addressed: what are the limitations of existing hydrological models when used for the drought forecasting, and how to improve the structure in order to make the drought prediction better? In this study, we scrutinized more than 50 hydrological models, including lumped models, semi-distributed models, distributed models, integrated surface water and groundwater models, to explore the adaptability of hydrological models in drought simulation and forecasting. Afterwards, improvements of the model structure are proposed.

2 Methods

2.1 Model selection and classification

More than 50 hydrological models were involved in this study. The main source was from the model review paper by Singh and Woolhiser (2002), some seldom used models were abandoned, and other recent representative hydrological models were also included (20 models). The baseline of model selection is that the model has been widely used and is easy to get the manual. According to the evolution of hydrological models, these selected models were classified into four types: lumped model, semi-distributed model, distributed model and integrated surface water and groundwater model. The selected models include: 13 lumped models (Fig. 1), 6 semi-distributed models (Fig. 2), 29 distributed models (Fig. 3) and 10 integrated surface water and groundwater models (Fig. 4).

2.2 Evaluation method for assessment of the adaptability of hydrological models for drought forecasting

2.2.1 Evaluation criteria

Considering the differences between flood and drought forecasting, their corresponding model structure need to be distinct. For drought forecasting, emphasis on low flows, therefore, the simulation of evapotranspiration, soil water moisture and the recession process need to be well developed. Snow melting, vegetation interception, groundwater and surface water exchange cannot be neglect any more. The influences of crops and soil water dynamics should be considered in the low flow simulation and prediction. In addition, human impacts, including water storage and drainage from reservoirs, water transfer, agricultural irrigation, groundwater pumping or artificial recharging, water consumption, also need to be included in drought forecasting, because they have significant effect on drought in some areas (Van Loon et al., 2016). Given that the drought forecasting needs longer lead times for prediction, hydrological models should be linked directly to the climate forecasting results in order to extend the lead times. Since drought forecasting deals with low flow, the energy balance should be considered to a certain extent to supplement the water balance calculation. To sum up, 15 evaluation criteria (Table 1) were used to explore the adaptability of hydrological models for drought forecasting in this study.

Table 1. Summary of evaluation indicators.

No.	Evaluation indicators
1	Special human impact module
2	Water storage and drainage from reservoirs
3	Water transfer
4	Irrigation simulation
5	Groundwater pumping and artificial recharging
6	Water consumption
7	Evapotranspiration simulation
8	Canopy interception
9	Snowmelt simulation
10	Soil water moisture simulation
11	Low flow recession process
12	Surface water and groundwater exchange
13	Connection to climate forecasts
14	Energy balance
15	Crop growth interaction with soil moisture

2.2.2 Evaluation method

A rank evaluation method was used in this study. Three grades were distinguished: 0, 1 and 2. "0" stands for no consideration of the function or module; "1" stands for simple calculation of the function or module; "2" stands for mature or powerful calculation of the function or module.

3 Results

3.1 Lumped model evaluation

The adaptability of lumped hydrological models for drought forecasting is shown in Fig. 1. Most of selected lumped models simulate evaporation calculation except CLS model, whereas the XAJ and VM show the powerful function with a three layers evaporation calculation method. 70 % of the selected models simulate soil water moisture with simple equations, while the Sacramento model divides the soil into different layers and calculate soil moisture in each layer. Five models include simulation of canopy interception and snowmelt. Three models (i.e. EPIC, ARM and SVAT) consider the energy balance together with the water balance. Only the EPIC model has irrigation module, while the EPIC and ARM consider the influences of crop growth.

3.2 Semi-distributed model evaluation

The evaluation results of six selected semi-distributed models are shown in Fig. 2. Results exhibit that all the selected semi-distributed models simulate evaporation and soil water

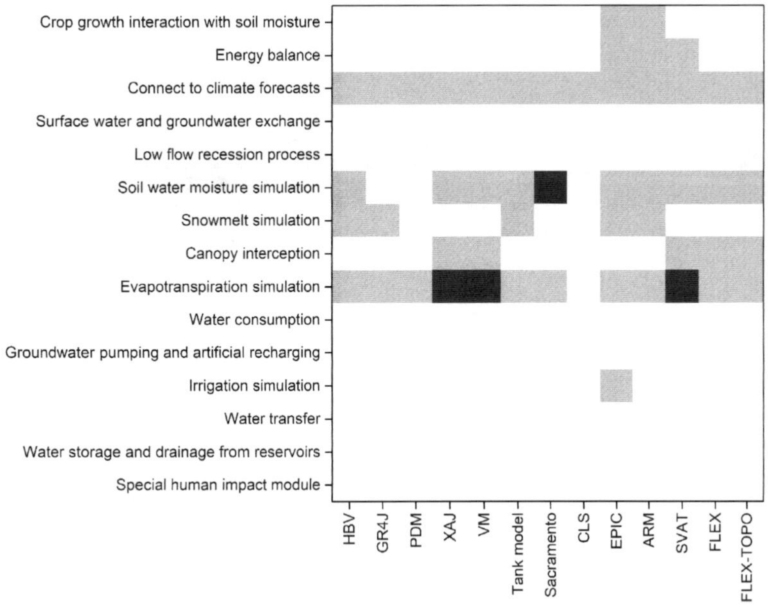

Figure 1. Evaluation of the adaptability of lumped hydrological models for drought forecasting. Where white color stands for grade 0 (no consideration of the function or module); gray color stands for grade 1 (simple simulation of the function or module); black color stands for grade 2 (powerful simulation of the function or module).

moisture, whereas the LASCAM and ARNO appear powerful soil water moisture simulation. The ARNO model involves the spatial probability distribution of soil moisture to dynamically change the saturation contribution area. The SWRRB and SLURP consider the effects of human activities, such as water storage and drainage from reservoirs. In particular, the SLURP model includes simulation of several human activities, including water transfers, irrigation and groundwater pumping and artificial recharging. The ARNO model simulates water consumption, while the Top model takes the exchange of groundwater and surface water into consideration. All semi-distributed models only simulate water balance without taking the energy balance into account. None of the selected models consider low flow recession and crop growth influences on soil water moisture and evapotranspiration.

3.3 Distributed model evaluation

The evaluation results of the 29 selected distributed models are shown in Fig. 3. All the models simulate evapotranspiration and soil water moisture, where SWAT, WISTOO, CASC2D, PARCHED-THIRST, VIC and PRMS approach the soil water moisture simulation in different ways. SWAT, GWAVA, MIKE-SHE and DTVGM consider most of the human activities that impact the hydrological system. The four human activity modules we put forward, are ranked as follows: water storage and drainage from reservoirs, water transfer, irrigation, groundwater pumping and artificial recharging, water consumption. The impact of the reservoirs is most frequently addressed by model developers. Nearly half of the selected models consider canopy interception and snowmelt modules, which often can be explained for which catchment the model originally has been developed. Half of selected models consider groundwater and surface water exchange, among which SWIM, SHETRAN and HEC-HMS containing specialized groundwater simulation modules. Five models consider the simulation of low flow conditions, while night models (i.e. HYDROTEL, CREST, HSPF, GWAVA, VIC, Wasim-ETH, MIKE-SHE, DHSVM and GBHM) can be connected to weather forecasts well, because of their grid-based structure. Four models (i.e. SWAT, PARCHED-THIRST, SWIM, and Mike-SHE) investigate crop growth influences on the hydrological system. Compared with lumped and semi-distributed models, many distributed models start to consider the energy balance to supplement the water balance simulation.

3.4 Integrated surface water and groundwater model evaluation

The evaluation results of 10 selected integrated surface water and groundwater models are shown in Fig. 4. All the model studied consider evaporation, canopy interception, soil water moisture and the exchange between surface water and groundwater. The ability to simulate groundwater flow and groundwater pumping and artificial recharging makes this type of models more powerful than the other types of models.

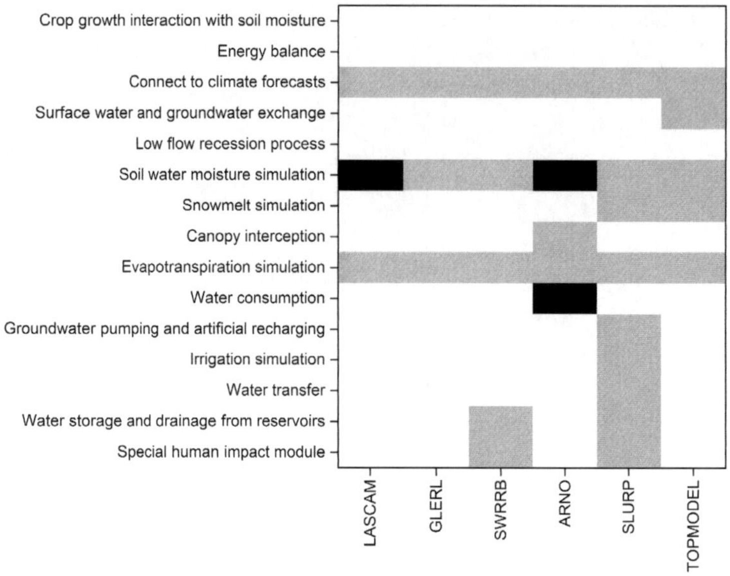

Figure 2. Evaluation of the adaptability of semi-distributed hydrological models for drought forecasting. Where white color stands for grade 0 (no consideration of the function or module); gray color stands for grade 1 (simple simulation of the function or module); black color stands for grade 2 (powerful simulation of the function or module).

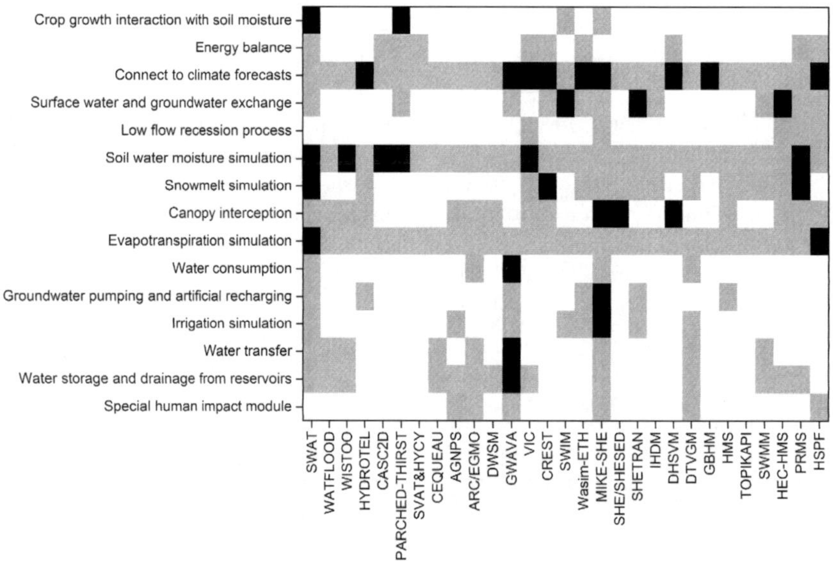

Figure 3. Evaluation of the adaptability of distributed hydrological models for drought forecasting. Where white color stands for grade 0 (no consideration of the function or module); gray color stands for grade 1 (simple simulation of the function or module); black color stands for grade 2 (powerful simulation of the function or module).

MODHMS, HydroGeoSphere and Parflow are fully coupled models of groundwater and surface water. They use partial differential equations to describe surface water and groundwater flow processes. SWATMOD, and MODHMS perform well in simulating human activities. The Parflow model is a hydrological model for parallel computing that takes the energy balance into account and can be efficiently linked to climate models. The SWATMOD as the coupled model between SWAT and Modflow consider the influence of crop growth on the hydrological system.

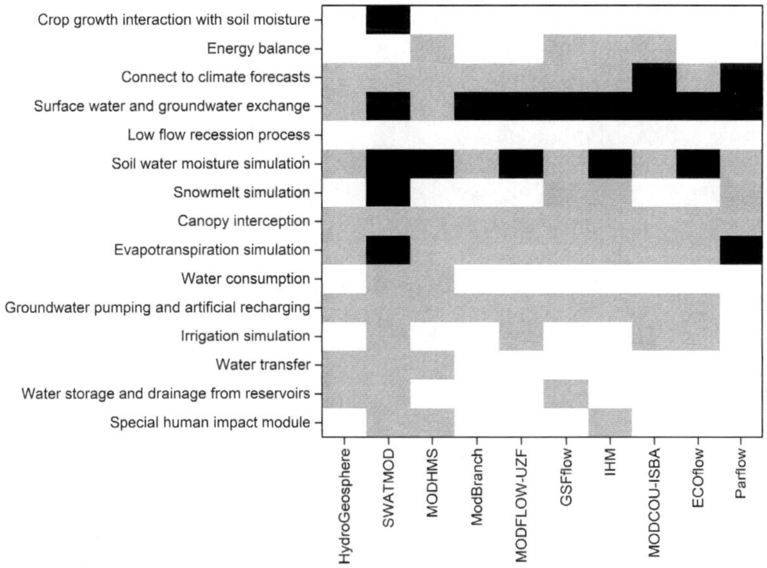

Figure 4. Evaluation of the adaptability of integrated surface water and groundwater models for drought forecasting. Where white color stands for grade 0 (no consideration of the function or module); gray color stands for grade 1 (simple simulation of the function or module); black color stands for grade 2 (powerful simulation of the function or module).

4 Discussion

The model structure of lumped hydrological model seems a bit simpler compared with the other types of models, and they seldom consider human impact (Bergstrom, 1995). While distributed models investigate more human impact in the hydrological cycle, such as the water storage and drainage from reservoirs, water transferring, water consumption, irrigation. The integrated surface water and groundwater model couple the advantages of surface water and groundwater simulations. They consider the exchange between surface water and groundwater, and enable to simulate the groundwater pumping and artificial recharging. However, they are too sophisticated to be used for drought forecasting, because they need too many data to run (Beven and Binley, 1992). In summary, distributed models are more suitable for drought forecasting compared with lumped and semi-distributed models. Results indicate that SWAT, GWAVA, MIKE-SHE and DTVGM are the most powerful in the category of the distributed models because they simulate the human impact on the hydrological system, and they solve the energy balance together with water balance.

The MIKE-SHE model covers ranges of physical processes with high requirements on parameters and data, and hence its operation seems more sophisticated than the other three above-mentioned distributed models (Refsgaard and Storm, 1995). The SWAT model is an open source model and it is updated by world researchers, which have developed several modules for human impacts and crop growth. The SWAT model has been widely used in non-point source pollution simulation (Jayakrishnan et al., 2005), but seldom has been used for drought forecasting. When applied to drought forecasting, it is better to improve the low flow simulations and the connection to climate forecasting. The advantage of GWAVA model is the human impact module that can simulate water demand, water transfer, agricultural irrigation and population distribution as a driver for human activities. However, the simulation modules for human impacts seems a bit simple. In contrast, the DTVGM model was developed based on the characteristics of Chinese Yellow River basin, with emphasis on water conservation, water consumption, reservoirs and low flow simulations (Ning et al., 2016). The highlight of HEC-HMS is reflected in the flexibility of its model structure, which can be adapted to the catchments for different natural conditions (Hydrologic Engineering Center, 2000). To sum up, combination of the model structures of SWAT, GWAVA and DTVGM might be a good solution for drought forecasting, using the flexible structure of the HEC-HMS model.

5 Conclusions

This study scrutinized more than 50 hydrological models, including lumped models, semi-distributed models, distributed models, surface water and groundwater coupled models, to explore the adaptability of hydrological models for drought simulation and forecasting. In this paper, we discuss the limitations of the wide range of selected hydrological models to be used for the drought forecasting and to propose improvement of model structures to make better fit for drought prediction. Results indicate that distributed models are more suitable for drought forecasting compared with the other types of models. A combination of SWAT, GWAVA and DTVGM might be a good solution for drought forecasting using the flexible structure of the HEC-HMS model.

Data availability. All relevant data are within the paper.

Author contributions. PY and WS designed the research; ZS and JL collected the data; ZX and MM analyzed the data and wrote the paper.

Competing interests. The authors declare that they have no conflict of interest.

Special issue statement. This article is part of the special issue "Hydrological processes and water security in a changing world". It is a result of the 8th Global FRIEND–Water Conference: Hydrological Processes and Water Security in a Changing World, Beijing, China, 6–9 November 2018.

Financial support. This research has been supported by the IWHR Research & Development Support Program (grant no. JZ0145B582017), the the National Natural Science Foundation of China (grant no. 51609257), the the Hydraulic science and technology in Hunan Province (grant no. [2017]230-36), and the the National Key R&D Program of China (grant no. 2017YFC1502406).

References

Bergstrom, S.: Chapter 13: The HBV model, Computer models of watershed hydrology, edited by: Singh, V. P., Water Resources Publications, Littleton, Colorado, USA, 1995.

Beven, K. J. and Binley, A. M.: The future of distributed models: model calibration and predictive uncertainty, Hydrol. Process., 6, 279–298, 1992.

Edwards, D. C. and McKee, T. B.: Characteristics of 20th century drought in the United States at multiple time scales, in: Climatology Report 97-2; Department of Atmospheric Science, Colorado State University, Fort Collins, CO, USA, 1997.

Huggins, L. F. and Monke, E. J.: Mathematical simulation of hydrologic events of ungaged watersheds, Water Resources Research Center, Tech. Rep. No. 14, Purdue Univ., West Lafayette, Ind., USA, 1970.

Hydrologic Engineering Center (HEC): Hydrologic modeling system HEC-HMS user's manual, version 2, Engineering, U.S. Army Corps of Engineers, Davis, Calif., USA, 2000.

Jayakrishnan, R., Srinivasan, R., Santhi, C., and Arnold, J. G.: Advances in the application of the SWAT model for water resources management, Hydrol. Process., 19, 749–762, 2005.

Mishra, A. K. and Singh, V. P.: A review of drought concepts, J. Hydrol., 391, 202–216, 2010.

Mishra, A. K. and Singh, V. P.: Drought modeling – A review, J. Hydrol., 403, 157–175, 2011.

Ning, L., Xia, J., Zhan, C., and Zhang, Y.: Runoff of arid and semi-arid regions simulated and projected by CLM-DTVGM and its multi-scale fluctuations as revealed by EEMD analysis, J. Arid Land, 8, 506–520, 2016.

Paulo, A. A. and Pereira, L. S.: Prediction of SPI Drought Class Transitions Using Markov Chains, Water Resour. Manag., 21, 1813–1827, 2007.

Raju, M. M., Srivastava, R. K., Bisht, D. C. S., Sharma, H. C., and Kumar, A.: Development of artificial neural-network-based models for the simulation of spring discharge, Adv. Art. Int., 2011, 1–11, https://doi.org/10.1155/2011/686258, 2011.

Refsgaard, J. C. and Storm, B.: Chapter 23: MIKE SHE, Computer models of watershed hydrology, edited by: Singh, V. P., Water Resources Publications, Littleton, Colorado, USA, 1995.

Singh, V. P. and Woolhiser, D. A.: Mathematical Modeling of Watershed Hydrology, J. Hydrol. Eng., 7, 270–292, 2002.

Van Loon, A. F., Gleeson, T., Clark, J., Van Dijk, A. I. J. M., Stahl, K., Hannaford, J., Di Baldassarre, G., Teuling, A. J., Tallaksen, L. M., Uijlenhoet, R., Hannah, D. M., Sheffield, J., Svoboda, M., Verbeiren, B., Wagener, T., Rangecroft, S., Wanders, N., and Van Lanen, H. A. J.: Drought in the Anthropocene, Nat. Geosci., 9, 89–91, https://doi.org/10.1038/ngeo2646, 2016.

Vishnu, B. and Syamala, P.: Grey Model for Stream Flow Prediction, Aceh International Journal of Science and Technology, 1, 14–19, available at: https://doaj.org/article/806ab2d8b3134e98a512ddf3409ca26f (last access: 5 August 2020), 2012.

Wanders, N., Thober, S., Kumar, R., Pan, M., Sheffield, J., Samaniego, L., and Wood, E. F.: Development and evaluation of a Pan-European multimodel seasonal hydrological forecasting system, J. Hydrometeorol., 20, 99–114, https://doi.org/10.1175/JHM-D-18-0040.1, 2019.

Wilhite, D. A. (Ed.): Droughts as a natural hazard: concepts and definitions, in: DROUGHT, A Global Assessment, vol. I, Routledge Hazards and Disasters Series, Routledge, London, UK, 3–18, 2000.

Wilhite, D. A. and Glantz, M. H.: Understanding the drought phenomenon: The role of definitions, Water Int., 10, 111–120, 1985.

Yu, Z., Lakhtakia, M. N., Yarnal, B., White, R. A., Miller, D. A., Frakes, B., Barron, E. J., Duffy, C., and Schwartz, F. W.: Simulating the river-basin response to atmospheric forcing by linking a mesoscale meteorological model and a hydrologic model system, J. Hydrol., 218, 72–91, 1999.

Proc. IAHS, 383, 267–272, 2020
https://doi.org/10.5194/piahs-383-267-2020

Open Access
PIAHS

Linking drought indices to impacts in the Liaoning Province of China

Miaomiao Ma[1], Juan Lv[1], Zhicheng Su[1], Jamie Hannaford[2], Hongquan Sun[1], Yanping Qu[1], Zikang Xing[1,3], Lucy Barker[2], and Yaxu Wang[1]

[1]Research center on Flood & Drought Disaster Reduction of the Ministry of Water Resources, China Institute of Water Resources and Hydropower Research (IWHR), Beijing, China
[2]Center of Ecology and Hydrology, Wallingford, UK
[3]State Key Laboratory of Hydrology-Water Resources and Hydraulic Engineering & College of Hydrology and Water Resources, HOHAI University, Nanjing, China

Correspondence: Miaomiao Ma (mamm@iwhr.com)

Published: 16 September 2020

Abstract. Drought is an inherent meteorological characteristic of any given region, but is particularly important in China due to its monsoon climate and the "three ladder" landform system. The Chinese government has constructed large-scale water conservation projects since 1949, and developed drought and water scarcity relief frameworks. However, drought still causes huge impacts on water supply, environment and agriculture. China has, therefore, created specialized agencies for drought management called Flood Control and Drought Relief Headquarters, which include four different levels: state, provincial, municipal and county. The impact datasets they collect provide an effective resource for drought vulnerability assessment, and provide validation options for hydro-meteorological indices used in risk assessment and drought monitoring. In this study, we use the statistical drought impact data collected by the Liaoning province Drought Relief Headquarter and meteorological drought indices (Standardized Precipitation Index, SPI and Standard Precipitation Evaporation Index, SPEI) to explore a potential relationship between drought impacts and these indices. The results show that SPI-24 and SPEI-24 are highly correlated to the populations that have difficulties in obtaining drinking water in four out of the six cities studied. Three impacts related to reservoirs and the availability of drinking water for humans and livestock exhibit strong correlations with SPI and SPEI of different accumulated periods. Results reveal that meteorological indices used for drought monitoring and early warning in China can be effectively linked to drought impacts. Further work is exploring how this information can be used to optimize drought monitoring and risk assessment in the whole Liaoning province and elsewhere in China.

1 Introduction

Drought is an inherent meteorological characteristic of any given region, but is particularly important in China due to monsoon climate and "three ladder" landform system. In China, the uneven distribution of water resources and large population load results in issues of water scarcity and recurring unpredictable drought events in many regions. Water scarcity issues are exacerbated by physio-geographic conditions, including the fact that the South has 69 % of available water supplies, while the North contains 64 % of arable land for agriculture. Rainfall distribution follows the same North-South distribution, ranging from 320 $mm\,yr^{-1}$ in the northern river systems (i.e., China north of the Yangtze River) to 1200 $mm\,yr^{-1}$ in the southern river systems (Quinn et al., 2014). Although there have been massive efforts to transfer water from the South to the North in recent years, the Northern provinces have experienced drought problems due to low rainfall and huge water consumption.

The Chinese government has constructed large-scale water conservancy projects since the People's Republic of China was established in 1949, and has developed drought and water scarcity relief frameworks. In general, people no longer

Published by Copernicus Publications on behalf of the International Association of Hydrological Sciences.

die as a result of drought in China and those affected will receive relief from the government, such as free drinking water delivery service from drought relief service teams. Although many efforts have been made by the government, drought still causes huge impacts on water supply, agriculture and ecological environment (e.g. Huang et al., 2013; Li, 1995). According to the official statistics (State Flood Control and Drought Relief Headquarters of Ministry of Water Resources, 2017), there have been 33 extreme severe drought disasters in China since 1949. The most critical drought disaster occurred in 2000 when two thirds of China was affected and crop yield loss reached to 60 billion kilograms – about 12 % of national total crop yield in that year, costing around CNY 51 billion (EUR 6.8 billion). About 28 million of the rural population and 17 million large cattle had temporary drinking water difficulties, while 620 towns had water supply shortages and 26.4 million of the urban population were affected by this disaster (State Flood Control and Drought Relief Headquarters of Ministry of Water Resources, 2017). On average, every year drought occurs in different areas in China and the drought disaster has caused millions of dollars of economic losses. The Chinese government therefore pays close attention to drought disaster reduction and has created specialized agencies for drought management called Flood Control and Drought Relief Headquarters (FCDRHs), which include four different levels: state, provincial, municipal and county.

The FCDRHs at various levels are responsible for drought relief planning work, taking early action risk reduction and mitigation measures based on predictions, delivering drought early warning, making decisions for drought emergency plan, and drought impact assessment etc. The state FCDRH shall be responsible for organizing and leading the work of nationwide drought management. Local FCDRH at provincial/municipal/county level shall be responsible for organizing and directing the drought management work within their jurisdiction. This "top to bottom" drought management system enables complete drought impact statistics to be collected. The impact datasets they collect provide an effective resource for drought vulnerability assessment, and provide opportunities to validate hydro-meteorological indices used in risk assessment and drought monitoring.

Hydrological and meteorological characterization of drought is well-known but there is a lack of knowledge on impacts (Bachmair et al., 2016a). Observed impacts reflect the real situation of a drought disaster on the ground and are key to understand baseline vulnerability. Bachmair et al. (2016a) point out that drought indices are proliferating, but they don't consider whether the indices represent the local drought conditions. Bachmair et al. (2015, 2016b) provide a quantitative analysis to objectively appraise drought indices and model European drought impacts. Compared with European drought impact data, which are collected from range of sources (Stahl et al., 2016), Chinese drought impact data are collected more systematically and are more comprehensive, which can be attributed to the "top to bottom" drought management system. In some areas, these data have been collected since 1972, and have been collected nationally from 1990. Eleven indices (Table 1) used to describe the drought impacts are counted every ten days and then gathered in the drought reporting system. These impact datasets offer huge potential for studying vulnerability and for evaluating indices in a way that is rarely possible elsewhere.

Table 1. Drought impact indices and their abbreviations.

Drought impact indices	Abbreviation
Total area of crops that have been affected by drought	Impact-1
Area of crops that have been affected by light drought	Impact-2
Area of crops that have been affected by severe drought	Impact-3
Area of crops that have been affected by extreme drought	Impact-4
Agriculture water deficit Total area	Impact-5
Paddy fields water deficit area	Impact-6
Dry land water deficit area	Impact-7
Populations that have difficulties in drinking water	Impact-8
The number of Livestock that have difficulties in drinking water	Impact-9
The number of reservoirs that have dried up	Impact-10
The number of well in water shortage	Impact-11

This paper aims to understand the correlations between drought indices and impacts based on the Liaoning systematic statistical drought impact data. We use the statistical drought impact data collected by the Liaoning province Drought Relief Headquarters and meteorological drought indices SPI/SPEI, to explore the potential relationship between drought impacts and these indices.

2 Methods and Data

2.1 Study Area

The study area is located at the northwest part of the Liaoning province, including six cities, Chaoyang, Fuxin, Huludao, Jinzhou, Shenyang and Tieling, with a total area of 76 8000 km^2 and a population of 22 million people. Figure 1 shows the six cities in the study area and the locations of the selected meteorological stations. Northwest part of the Liaoning province is the typical severe drought area of China. Annual average precipitation is about 400 mm, much lower than the provincial average of 680 mm, and it has the least rainfall in the Liaoning province. The water surface evaporation capacity reaches to 1000–1300 mm yr^{-1}. Therefore, there are high water deficits in this area and it is the most frequent drought affected area in the Liaoning Province. Spring drought is the main drought type, occupying 30 %–65 % of drought events, followed by the spring–summer continuous drought.

2.2 Data

There were two main types of data used in this research: meteorological data and drought impact data. Meteorological data were used to calculate the drought indices SPI and SPEI, while the drought impact data were used to calculate the drought impact indices. Six cities in the Liaoning province were included in this study, and one meteorological station was selected in each city to calculate the SPI and SPEI (shown on Fig. 1). These six stations had more than 60 years observed daily data from 1950 to 2016. We used these daily data to calculate the SPI with accumulation periods of 1–12, 15, 18, 21, 24 and 36 months, and 1–12 and 24 months for SPEI. The SPI and SPEI are valid indices of drought monitoring and have been widely recognized internationally. They are easier to be calculated than Palmer Drought Severity Index as they have few data requirements. In this study, we calculated SPI and SPEI of different accumulation periods. The SPI and SPEI were computed using the standardized Drought Analysis Toolbox Package developed by IWHR Drought Mitigation Research Team.

Drought impact data from 1999 to 2016 were from the Liaoning Province drought statistics reporting system. They were counted every ten days by the county FCDRH, and then reported to higher level FCDRH before being gathered in the State FCDRH. In this study, we employ the arithmetic mean method to transfer them into average monthly data constituent with the time scale of SPI and SPEI. Eleven drought impact indices are chosen and listed at Table 1. SPI and SPEI data from 1999 to 2016 were selected corresponding to the available impact data.

2.3 Correlation Analysis

Spearman correlations were undertaken to explore the relationship between SPI and SPEI of different accumulation periods and the eleven drought impact indices (Table 1). The correlation strength is defined as follows: 0–0.1 (no correlation), 0.1–0.3 (weak), 0.3–0.6 (moderate), 0.6–0.9 (strong) and 0.9–1 (very strong) (Bachmair et al., 2016b).

3 Results

For the six cities, most accumulations periods of SPI and SPEI showed negative correlations with drought impact indices, and the stronger correlations occurring for longer time scales of SPI and SPEI, such as SPI-21, SPI-24, SPI-36, and SPEI-24 (Figs. 2–3). In contrast, Tieling exhibited more negative correlations around SPEI of 7 or 8 months (Fig. 3).

Table 2. Spearman correlation coefficients between drought impact indices in Table 1 to SPI and SPEI with different accumulation periods.

City Name	Indices	Impact-8	Impact-9
Fuxin	SPI-21	−0.623	−0.667
	SPI-24	−0.698	−0.794
	SPEI-24	−0.63	−0.761
Huludao	Indices	Impact-8	
	SPI-24	−0.612	
	SPEI-24	−0.624	
Jinzhou	Indices	Impact-8	Impact-10
	SPI-24		−0.631
	SPEI-24	−0.67	−0.679
Shenyang	Indices	Impact-4	Impact-8
	SPI-4	−0.615	
	SPEI-24		−0.633
Tieling	Indices	Impact-8	
	SPEI-8	−0.625	

Table 2 showed the Spearman correlation coefficients between drought impact indices to SPI and SPEI of different accumulation periods. Five of the cities studied showed strong negative correlations between impact indices and SPI/SPEI except Chaoyang (Table 2). Four of the six cities indicated that SPI-24 and SPEI-24 were highly correlated to the populations that had difficulties in drinking water (Impact-8). The other two impact indices exhibited strong correlations with SPI and SPEI of different accumulated periods, such as the number of livestock that had difficulties in drinking water (Impact-9) and the number of reservoirs that had dried up (Impact-10). The area of crops that have been affected by extreme drought (Impact-4) had a strong correlation with SPI-4 in Shenyang, while SPEI-8 highly correlated to Impact-8 in Tieling. In Fuxin, SPI-21 exhibited strong correlations with Impact-8 and Impact-9 and their coefficients were −0.623 and −0.667, respectively. The strongest correlation coefficient of all impact indices at accumulation periods of SPI or SPEI was 0.794 in Fuxin between SPI-24 and Impact-9. In summary, the three impact indices related to reservoirs and the availability of drinking water for humans and livestock (i.e. Impact-8, Impact-9 and Impact-10) exhibited strong correlations with SPI and SPEI of different accumulated periods.

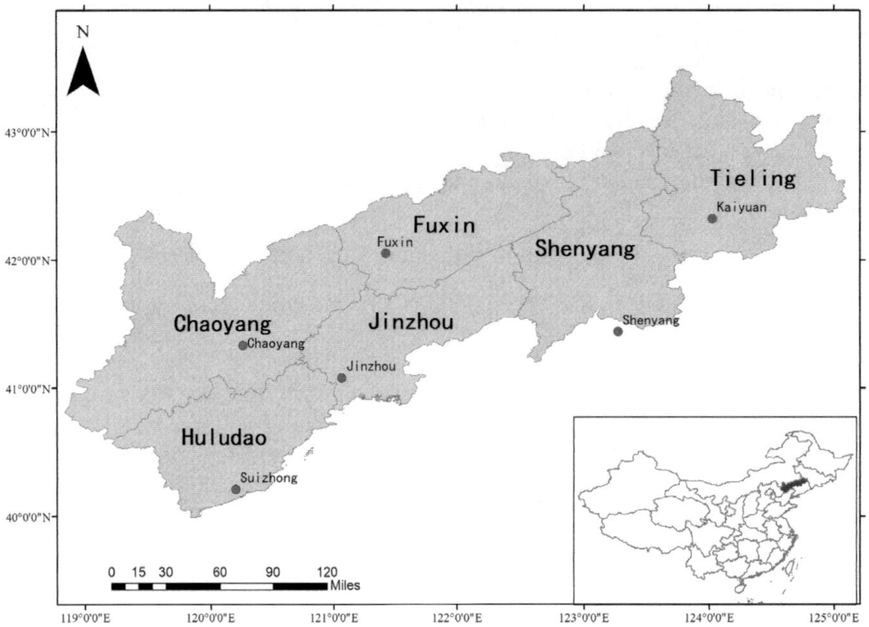

Figure 1. Study area and the location of the selected meteorological stations.

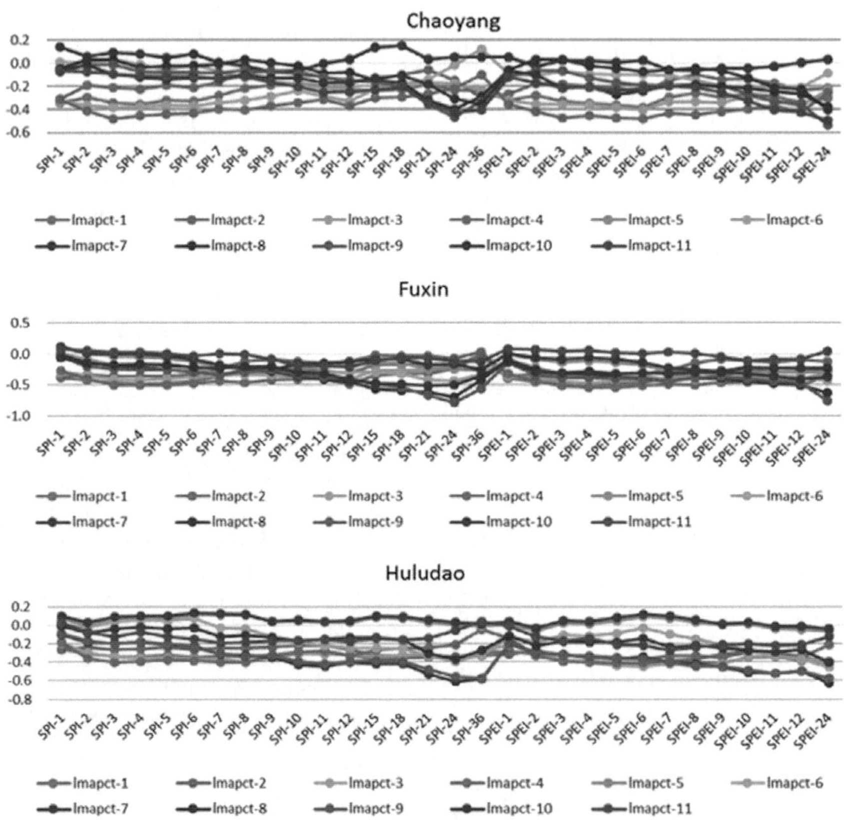

Figure 2. Correlations between SPI/SPEI with different accumulation periods and the 11 drought impact indices in Chaoyang, Fuxin and Huludao.

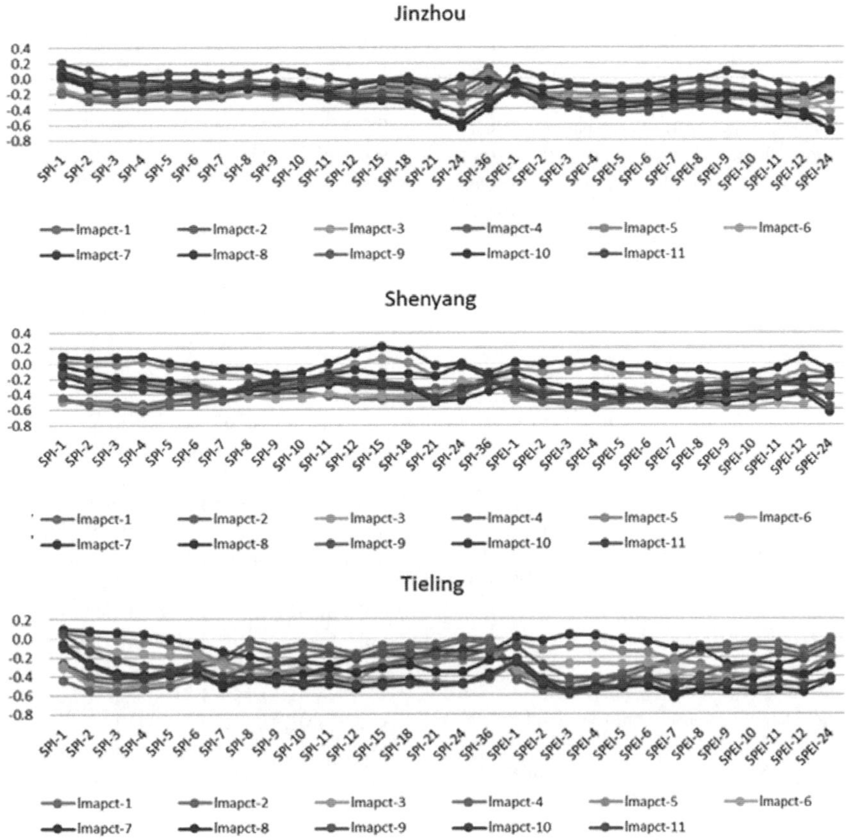

Figure 3. Correlations between SPI/SPEI with different accumulation periods and the 11 drought impact indices in Jinzhou, Shenyang and Tieling.

4 Discussions

In this study, the SPI and SPEI calculations were based on the meteorological station monitoring data in six cities. Due to the small number of meteorological stations used here, it was not possible to reflect the average precipitation over the whole region. In China, precipitation is also monitored at hydrological stations, especially for the flood season; If the denser and more comprehensive hydro-meteorological datasets were used, it is likely that stronger relationships between drought indices and impacts would be found.

Similarly, correlations between drought indices and impacts depend on the quality and accuracy of impact data. Therefore, the drought impact data need to be properly verified. Although the impact data was reported by various levels of FCDRHs, and notably are systematic and complete by international standards, there are still always some subjective factors involving in estimating impact statistics. For example, the mild drought is difficult to be identified at the early stages; moderate drought conditions have often already started when people realize the drought is occurring. In such situation, the statistical values may underestimate the actual drought condition.

Results show that SPI-24 and SPEI-24 are highly correlated to the populations that have difficulties in drinking water in four of the six cities studied. This means two years periods of SPI and SPEI better represent these drought impacts in the northwest of the Liaoning province.

5 Conclusions

This paper uses the statistical drought impact data and meteorological drought indices (i.e. SPI and SPEI) to explore the relationship between drought indices and impacts. The results indicate that SPI and SPEI with accumulation periods of two years best represent the drought impacts in the northwest of the Liaoning province. Meteorological indices used for drought monitoring and early warning in China can be effectively linked to drought impacts. Further work is exploring how this information can be used to optimize drought monitoring and risk assessment in the whole of the Liaoning province and elsewhere in China.

Data availability. The drought impact data of study area cannot be open according to the data licensing agreement.

Author contributions. MM design the whole research and carried them out. She prepared the manuscript with contributions from all co-authors. JL and ZS contacted with the data agency and collected the drought impact data. JH and LB revised the manuscript and proposed some suggestions. HS and YQ calculated the SPI and SPEI of study area. ZX and YW handled with the data preprocessing.

Competing interests. The authors declare that they have no conflict of interest.

Special issue statement. This article is part of the special issue "Hydrological processes and water security in a changing world". It is a result of the 8th Global FRIEND–Water Conference: Hydrological Processes and Water Security in a Changing World, Beijing, China, 6–9 November 2018.

Financial support. This research has been supported by the National Natural Science Foundation of China (grant no. 51609257), the IWHR Research & Development Support Program (grant no. JZ0145B582017), NC-ODA Sustainable Use of Natural Resources to Improve Human Health and Support Economic Development (SUNRISE) project NE/R000131/1, and the Hydraulic science and technology in Hunan Province (grant no. [2017]230-36).

References

Bachmair, S., Kohn, I., and Stahl, K.: Exploring the link between drought indicators and impacts, Nat. Hazards Earth Syst. Sci., 15, 1381–1397, https://doi.org/10.5194/nhess-15-1381-2015, 2015.

Bachmair, S., Stahl, K., Collins, K., Hannaford J., Acreman, M., Svoboda, M., Knutson, C., Simth, K. H., Wall, N., Fuchs, B., Crossman, N. D., and Overton, I. C.: Drought indicators revisited: the need for a wider consideration of environment and society, Wiley Interdisciplin. Rev.: Water, 3, 516–536, https://doi.org/10.1002/wat2.1154, 2016a.

Bachmair, S., Svensson, C., Hannaford, J., Barker, L. J., and Stahl, K.: A quantitative analysis to objectively appraise drought indicators and model drought impacts, Hydrol. Earth Syst. Sci., 20, 2589–2609, https://doi.org/10.5194/hess-20-2589-2016, 2016b.

Huang, X., Li, X., Mu, X., Wang, K., Tian, K., Duan, C., and Yu, F.: The severe drought occurred in south China in 2010: An examination from perspective of ecology, Research of Soil and Water Conservation, 20, 282–292, 2013 (in Chinese with English abstract).

Li, C.: The Effects of Flood Waterlogging and Drought on Environment, Journal of Shanxi Water Power, 11, 57–61, 1995 (in Chinese with English abstract).

Quinn, L., Nieuwaal, M., Howarth, S., Lv, J., Yu, M., Su, Z., Wu, Y., Qu, Y., Wang, Y., and Sun H.: TA8185: Pilot Implementation of the Drought Management Strategy, People's Republic of China: Pilot Implementation of the Drought Management Strategy, Asian Development Bank Technical Assistance Consultant's Final Report, Mott Macdonald, Beijing, China, 2014.

Stahl, K., Kohn, I., Blauhut, V., Urquijo, J., De Stefano, L., Acácio, V., Dias, S., Stagge, J. H., Tallaksen, L. M., Kampragou, E., Van Loon, A. F., Barker, L. J., Melsen, L. A., Bifulco, C., Musolino, D., de Carli, A., Massarutto, A., Assimacopoulos, D., and Van Lanen, H. A. J.: Impacts of European drought events: insights from an international database of text-based reports, Nat. Hazards Earth Syst. Sci., 16, 801–819, https://doi.org/10.5194/nhess-16-801-2016, 2016.

State Flood Control and Drought Relief Headquarters of Ministry of Water Resources, China: The statistics of Flood and drought disaster bulletin in China (2000–2017), China Water & Power Press, available at: http://www.mwr.gov.cn/sj/tjgb/zgshzhgb/201808/t20180806_1044770.html (last access: 6 August 2018), 2017.

Proc. IAHS, 383, 273–279, 2020
https://doi.org/10.5194/piahs-383-273-2020

Drought monitoring and early warning in China: a review of research to pave the way for operational systems

Lucy J. Barker[1], Jamie Hannaford[1,2], and Miaomiao Ma[3]

[1]UK Centre for Ecology & Hydrology, Wallingford, UK
[2]Irish Climate and Research UnitS (ICARUS), Maynooth University, Maynooth, Ireland
[3]Research Centre on Flood & Drought Disaster Reduction of the Ministry of Water Resources, China Institute of Water Resources and Hydropower Research, Beijing, China

Correspondence: Lucy J. Barker (lucybar@ceh.ac.uk)

Published: 16 September 2020

Abstract. Geographic and socio-economic factors combined make China particularly vulnerable to droughts. Here we review academic literature to assess publication trends on the topics of drought monitoring and early warning in China, exploring the common themes and recent advances presented. Literature searches for the 1970–2017 period were made for a range of search terms relating to drought monitoring. It was found that publications on these topics started to appear sporadically in 1989 and have increased rapidly since the early 2000s, with the rate of publication increasing over the last eight years. The scope of studies varied, often with the spatial scale considered: at the national scale, studies focused on evaluating existing indices, while new indices were developed and tested at the regional scale – relating, in general, to monitoring agricultural and meteorological droughts. At the catchment scale, novel monitoring techniques were developed, often incorporating other data types such as modelled data or soil moisture measurements. National scale operational drought monitoring in China is currently being improved and updated. However, although operational systems are discussed in the literature, there is a still a disconnect between theory and practice with the most recent advances not yet operationalised. Here, we identify the methods and approaches which can be translated from the experimental case study scale to the national operational scale.

1 Introduction

Unlike most other natural hazards, droughts can occur in any climatic region, are (often) slow to develop and can affect a large area. The wide range of spatial and temporal scales, in addition to the different parts of the hydrological cycle that can be affected, means impacts are wide ranging and can be costly (e.g. Mishra and Singh, 2010; Stahl et al., 2016; Tallaksen and Van Lanen, 2004; Wilhite, 2000). The interplay between socio-economic and physical factors (including the spatial distribution of water resources, population and agriculture) means China is particularly vulnerable to droughts. Over the past 60 years, drought affected on average 29 million people and 206 000 km^2 of farmland per year in China (State Flood Control and Drought Relief Headquarters of Ministry of Water Resources China, 2016). Agriculture is the sector worst affected by drought in China (He et al., 2011), with annual average crop yield losses of 16.3 billion kilograms. But there are also many severe socio-economic and environmental impacts listed on the Flood and Drought Disaster Bulletin for China each year (e.g. State Flood Control and Drought Relief Headquarters of Ministry of Water Resources China, 2016).

Effective monitoring and early warning (MEW) systems are integral to mitigating impacts and improving resilience to droughts. Drought management in China (including monitoring activities and the coordination of management and relief efforts) is led by the State Flood Control and Drought Relief Headquarters (FCDRHs) with support from additional ministries, local government and the FCDRH at various levels (state, provincial, municipal and county). Current drought monitoring tools in China are fed by a range of data sources and deliver drought monitoring products, including derived drought indices (Sayers et al., 2016). The China Drought and Flood Bulletin is published by the Ministry of Water Resources, and gives a retrospective review of the year.

Here, we review the proliferation of academic studies concerned with drought MEW in China in terms of publication trends, the common themes and recent advances. We end by outlining the steps required in order to make recommendations of methods and approaches which may be translated from experimental case studies to improve operational drought MEW in China.

2 Data and methods

In order to investigate publication trends on the topic of drought MEW in China (Sect. 3), several literature searches on publication topic and title were made using the following phrases using Web of Science (English) and the China Knowledge Resource Integrated Database (Chinese):

- drought monitoring China;
- drought indicators/index/indices China;
- drought early warning China.

The English literature was searched using Web of Science (http://webofknowledge.com, last access: 28 July 2020) and publications in the core collection between 1970 and 2017. The results of each search were then filtered to return only article publications (thereby excluding proceedings papers). The Chinese literature was also searched using the China Knowledge Resource Integrated Database (https://www.cnki.net/, last access: 28 July 2020) using the same search terms with data up to the 25 May 2018 when the search was conducted. Results for the search "drought indicators/index/indices China" is hereafter referred to as "drought indices etc. China". Having reviewed some of the most highly cited papers on drought MEW in China in addition to some more recent papers, the common themes and recent advances in drought monitoring techniques and methods are discussed in Sect. 4 Common themes and Sect. 5 Recent advances. For brevity, we do not include all details of the literature surveyed here. Further details of the papers reviewed in this study can be found in Table 2 of Barker (2018) and Ma and Wang (2018), for the English and Chinese searches, respectively.

3 Publication trends

Table 1 shows the results in terms of numbers of publications of these search terms whilst Fig. 1 shows the cumulative number of publications per year returned for each search term.

A clear increase in the number of publications over time can be seen in Fig. 1, with the rate of publication on the topic of drought indices increasing dramatically in 2014. Figure 1 shows the total numbers of publications for each search term peaking in 2015 ("Drought early warning China") and 2017 ("Drought monitoring China" and "Drought Indices China") for all search results (all Publications), whilst when only articles were searched, for all three, the number of publications peaked in 2017.

There is a clear change in the proportion of results returned between all results and articles only results. The drought monitoring and drought indices searches begin with more proceedings paper and book chapters, with the number of articles increasing over time to a greater proportion of the total number of publications.

However, an overview of the titles of the returned publications do not always specifically mention these phrases. When publication title is searched, Fig. 1 shows that the number of publications returned reduces substantially. A much higher proportion of publications on the topic of drought indices also included the phrase in the title than those on the topic of drought monitoring or drought early warning.

A review of the Chinese literature revealed similar publication trends as those in the English academic literature. The majority of publications (when either topic or title was searched) related to drought monitoring, followed by drought indices. As with the search of the English literature, drought early warning returned the least results (Table 2). The first entries in the CNKI database were published in the late 1980s with a large increase in number after 2005, with numbers peaking in 2014 for drought monitoring and drought indices, and 2009 for drought early warning (Ma and Wang, 2018).

4 Common themes

Common themes, namely the purpose and spatial scale, and types of drought, identified from the papers are discussed in more detail below.

4.1 Purpose and spatial scale

Although the papers reviewed in Barker (2018) aimed to improve drought MEW through a better understanding of the drought hazard (such as a better understanding of the timing of drought propagation), the use of a new indicator or method, few/if any of the methods presented are used operationally in China (with the exception of Yan et al. (2016), although the Drought Watch reports are not yet publically available). This suggests that operational Chinese drought monitoring activities may be written up in the Chinese literature, and/or that there may remain a step of translating drought monitoring research into operational practices and that this research area has been driven by scientific endeavour rather than operational users.

Table 1. Results from Web of Science for drought monitoring in China and associated search terms between 1970 and 2017. The Max. column indicates the year with the highest number of publications for each search term, the number of publications is given with the year in brackets.

Search term	Topic: all publications		Topic: article only		Title: all publications	Title: article only
	Total	Max.	Total	Max.	Total	Total
Drought monitoring China	485	90 (2017)	334	82 (2017)	60	30
Drought early warning China	57	15 (2015)	40	13 (2017)	4	2
Drought indices etc. China	1602	304 (2017)	1371	289 (2017)	99	76

Table 2. Results from CNKI researches for drought monitoring in China and associated search terms (after Ma and Wang, 2018). The Max. column indicates the year with the highest number of publications for each search term, the number of publications is given with the year in brackets.

Search term	Topic		Title	
	Total	Max.	Total	Max.
Drought monitoring China	1417	147 (2014)	712	69 (2014)
Drought early warning China	296	51 (2009)	175	32 (2009)
Drought indices etc. China	1440	137 (2014)	666	63 (2014)

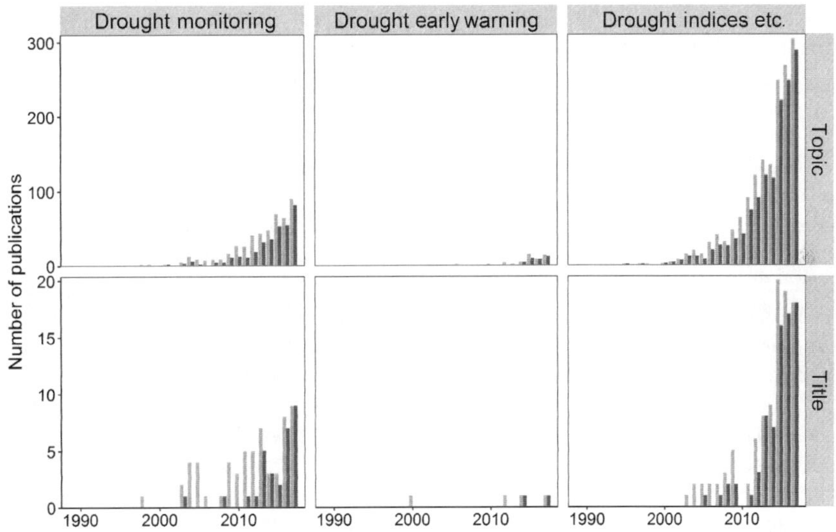

Figure 1. Number of publications each year in the period 1989–2017 for Web of Science search terms by both topic and title and for all publications from the search results and article only search results. Note the y axis scale is different for each row to best show the range of data.

Six papers reviewed here used data for the whole of China, the remaining studies focussed on smaller regions (of varying sizes). Eight papers considered large-scale regions of China (such as northern China and south-west China), four papers the province level, whilst Wu et al. (2001) researched the best indicator for monitoring moisture conditions in four cities (Urumqi, Fuzhou, Beijing and Wuhan). Ten papers were concerned with drought monitoring at the catchment (river or lake) scale, and others geographical area such as the Guanzhong Plain or Loess Plateau (e.g. Bai et al., 2017; Zhao et al., 2015, respectively).

Those considering the national scale tended to investigate the effectiveness of indices or remote sensing data for real-time drought monitoring (e.g. Guo et al., 2016; Wu et al., 2015), whilst Yu et al. (2014) assessed the spatio-temporal variation in drought characteristics across China and determined triggers and thresholds for monitoring and early warning. Regional studies mostly aimed to develop and test a new indicator (Hao et al., 2015; Wu et al., 2013; Zhang and Jia, 2013; Liu et al., 2015; Meng et al., 2016), with some regional and area studies aiming to analyse the performance of indices for monitoring agricultural drought (Bai et al., 2017; Wang et al., 2016). Catchment scale studies focussed on a range of issues, including: propagation timescales from meteorological to hydrological droughts (Wu et al., 2017; Ye et al., 2016), monitoring lake levels (Ding and Li, 2011), developing new indices for identifying regional/areal droughts (Fluixá-Sanmartín et al., 2018), developing thresholds for management (Zhao et al., 2016) and testing the operational Drought Watch system (Yan et al., 2016).

4.2 Types of drought

The complexity of defining drought has led to the definition of "types" of drought which relate to the compartment of the hydrological cycle in which deficits occur (Wilhite and Glantz, 1985). Deficits in precipitation are characterised as meteorological drought, soil moisture as agricultural drought and river flow or groundwater as hydrological drought. It recommended that meteorological drought indices should not be used to monitor other types of drought due to the non-linear responses of hydrological processes to climate inputs (Van Loon and Van Lanen, 2012; Van Lanen et al., 2013). As such, drought-type specific indices should be used for operational monitoring (e.g. meteorological drought and precipitation; agricultural drought and soil moisture/vegetation condition; hydrological drought and streamflow/groundwater levels) for a full understanding of the drought hazard.

Given the importance of the agricultural sector in China, many papers in this brief review focussed on monitoring agricultural drought. Meteorological drought was the next most commonly monitored drought type, with most papers using observed rainfall data and one study (Zhang and Jia, 2013) using microwave data to monitor precipitation.

Hydrological drought (river flow) was not considered by many of the papers reviewed here, however, there were studies which compared the lag between meteorological and hydrological droughts (Ye et al., 2016; Wu et al., 2017) but did not discuss the use of the hydrological drought data directly for monitoring. However, these studies did provide an indication of timescales over which precipitation should be monitored in these catchments to reflect any deficits in streamflow where there may be a lack of easily available streamflow data for monitoring purposes. Wu et al. (2017) also illustrated how the operation of a reservoir may impact the lag between meteorological and hydrological drought. Despite the reliance upon groundwater sources for water supply in northern China, only Long et al. (2014) included an assessment of groundwater storage (in a total water storage assessment using GRACE data), and so groundwater (hydrological) drought.

5 Recent advances

5.1 Data sources

Although many of the papers summarised in Barker (2018) used observed measurements of precipitation (in some cases gridded products) and temperature (with some using soil moisture or river flow observations), most did so in order to validate the results/indices produced using remotely sensed data. The most commonly used remote sensing data were MODIS (Normalised Difference Vegetation Index – NDVI – and Land Surface Temperature – LST) and AVHRR (NDVI and LST), with other datasets including AMSR-E and GRACE. This perhaps reflects the wider availability and ease of accessing these data in comparison to point observations, particularly for large areas, or where observation points are scarce or data are not easily accessible. Zhang and Jia (2013) used microwave sensors to improve meteorological drought monitoring although remote sensing techniques have more commonly been used to monitor agricultural drought.

Modelled data were used by Zhang and Zhou (2015) in order to calculate the Standardised Moisture Anomaly Index (SZI) enabling water budget simulations to be incorporated within the indicator. This meant that the SZI agreed with observed streamflow better than meteorological drought indices like the Standardised Precipitation Index (SPI) and Standardised Precipitation Evapotranspiration Index (SPEI). Elsewhere, modelled data have been used to monitor regions with sparse observations, for example modelled hydrological data are used in the High Resolution South Asia Drought Monitor (although does not include China; https://sites.google.com/a/iitgn.ac.in/high_resolution_south_asia_drought_monitor/streamflow-monitoring, last access: 28 July 2020). Li et al. (2015) used the Xin'anjiang model to derive variables to calculate the Multivariate Drought Index (MDI), however, they note that many of the required variables could be obtained from observations if available (e.g. precipitation, soil moisture and runoff).

The input and calibration data needed to calculate the SZI and other modelled monitoring outputs, in addition to the computational effort of processing the data (an issue also applicable to the use of remotely sensing data for monitoring) should be considered for an operational monitoring system in need of reliable, regular updates. The simple calculation of SPI, requiring only one data input, was a contributing factor to the World Meteorological Organisation recommending the use of the SPI for monitoring meteorological droughts (Hayes et al., 2011). Table 2 includes Ding and Li (2011) where satellite imagery was used to calculate lake water area, which was used with observed lake level data to calculate the water storage change. Long et al. (2014) developed a method whereby GRACE data could be used to monitor (and forecast) total water storage availability, but few studies used observed hydrological (river flow/groundwater) data directly for monitoring. In other cases, soil moisture observations were used (e.g. Bai et al., 2017; Su et al., 2003), Fluixá-Sanmartín et al. (2018) used a catalogue of drought impacts to verify droughts identified from indicator time series, whilst Wang et al. (2016) used winter wheat yields to identify the best indicator for agricultural drought monitoring.

5.2 Combined indices

Of the papers reviewed here which proposed new indices (nine papers and 11 new indices – marked with an asterisk in Table 2 of Barker, 2018), all proposed a combined indicator designed to monitor several aspects of the water cycle, or combine data and information to better characterise a particular type of drought. Most aimed to improve agricultural drought monitoring, but some (e.g. Li et al., 2015) aimed to improve drought monitoring of meteorological, agricultural and hydrological droughts using the MDI. The optimised combined indices developed by Hao et al. (2015) performed better and were more highly correlated with SPI-1 and SPEI-3 (derived from point observations) than individual remotely sensed indices. The SZI developed by Zhang and Zhou (2015) was more comparable to streamflow and the NDVI as it included evapotranspiration, runoff and soil moisture changes in its combination of the Palmer Drought Severity Index and the SPEI. The integrative aim of these combined indices is laudable, however, the added complexity to interpret and computational effort of calculating such combined indices should be assessed for an operational drought MEW system which requires fast, regular updates.

6 Conclusion

It is clear there is a large volume of material in the English and Chinese literature aiming to improve drought monitoring in China, with dramatic increases in the numbers of publications since the mid-2000s – particularly on the topic of drought indices. The importance of agricultural drought was reflected in the number of papers addressing this type of drought, specifically through the development of new indices attempting to improve agricultural drought monitoring. The increasing use of remote sensing data is evident, as is its use in the calculation of combined drought indices. The searches for "drought early warning China" returned few results in both the English and Chinese literature (Tables 1 and 2), and the papers reviewed here did not discuss the application of indices in early warning activities in order to forecast drought hazard or impacts.

In order to provide recommendations for operational drought MEW services, it will be necessary to undertake a broader literature search and assimilate results from searches of the English and Chinese literature in addition to a review of the tools and indices currently used to monitor droughts in China. Following these review activities, it will also be necessary to understand stakeholder needs (including local government, different levels of the FCDRHs etc.) for monitoring in terms of spatial and temporal resolution of data, time scales, the types of drought and indices and finally data visualisation. Crucial to this is understanding what data are available in real-time to support drought monitoring in China. However, an effective drought MEW system should provide information at a range of spatial scales (preferably from national to local scales, as well as integrating targeted local monitoring information into national scale drought alert and management systems) and a range of time scales (short term, for monitoring agricultural drought which can start quickly, to long term for monitoring hydrological droughts which may take longer to emerge). Understanding how drought indices link to drought impacts (e.g. Bachmair et al., 2016) is particularly important to ensure MEW tools can be used and interpreted in a way that enables appropriate actions to be taken in order to mitigate impacts. Through ongoing collaborations between Chinese and UK partners, these steps will be addressed going forwards to provide recommendations for operational drought MEW in China.

Data availability. No data sets were used in this article.

Author contributions. LJB, JH and MM discussed and developed the aims of the paper. LJB was responsible for the data analysis and visualisation. LJB prepared the original paper with contributions from JH and MM.

Competing interests. The authors declare that they have no conflict of interest.

Special issue statement. This article is part of the special issue "Hydrological processes and water security in a changing world". It is a result of the 8th Global FRIEND–Water Conference: Hydrological Processes and Water Security in a Changing World, Beijing, China, 6–9 November 2018.

Acknowledgements. This study is an outcome of the NC-ODA Sustainable Use of Natural Resources to Improve Human Health and Support Economic Development (SUNRISE) project (NE/R000131/1) and the National Key R & D Program of China (2017YFC1502406).

Financial support. This research has been supported by the Natural Environment Research Council (grant no. NE/R000131/1) and the National Key R & D Program of China (grant no. 2017YFC1502406).

References

Bachmair, S., Svensson, C., Hannaford, J., Barker, L. J., and Stahl, K.: A quantitative analysis to objectively appraise drought indicators and model drought impacts, Hydrol. Earth Syst. Sci., 20, 2589–2609, https://doi.org/10.5194/hess-20-2589-2016, 2016.

Bai, J.-J., Yu, Y., and Di, L.: Comparison between TVDI and CWSI for drought monitoring in the Guanzhong Plain, China, J. Integr. Agricult., 16, 389–397, https://doi.org/10.1016/S2095-3119(15)61302-8, 2017.

Barker, L. J.: SUNRISE: Drought Monitoring in China A Brief Review, Centre for Ecology & Hydrology, Wallingford, UK, 2018.

Ding, X. and Li, X.: Monitoring of the water-area variations of Lake Dongting in China with ENVISAT ASAR images, Int. J. Appl. Earth Obs. Geoinf., 13, 894–901, https://doi.org/10.1016/j.jag.2011.06.009, 2011.

Fluixá-Sanmartín, J., Pan, D., Fischer, L., Orlowsky, B., García-Hernández, J., Jordan, F., Haemmig, C., Zhang, F., and Xu, J.: Searching for the optimal drought index and timescale combination to detect drought: a case study from the lower Jinsha River basin, China, Hydrol. Earth Syst. Sci., 22, 889–910, https://doi.org/10.5194/hess-22-889-2018, 2018.

Guo, H., Bao, A. M., Liu, T., Chen, S., and Ndayisaba, F.: Evaluation of PERSIANN-CDR for Meteorological Drought Monitoring over China, Remote Sens., 8, 379, https://doi.org/10.3390/rs8050379, 2016.

Hao, C., Zhang, J., and Yao, F.: Combination of multi-sensor remote sensing data for drought monitoring over Southwest China, Int. J. Appl. Earth Obs. Geoinf., 35, 270–283, https://doi.org/10.1016/j.jag.2014.09.011, 2015.

Hayes, M., Svoboda, M., Wall, N., and Widhalm, M.: The Lincoln declaration on drought indices: universal meteorological drought index recommended, B. Am. Meteorol. Soc., 92, 485–488, 2011.

He, B., Lü, A., Wu, J., Zhao, L., and Liu, M.: Drought hazard assessment and spatial characteristics analysis in China, J. Geogr. Sci., 21, 235–249, https://doi.org/10.1007/s11442-011-0841-x, 2011.

Li, Q., Zeng, M., Wang, H., Li, P., Wang, K., and Yu, M.: Drought assessment using a multivariate drought index in the Huaihe River basin of Eastern China, in: Extreme Hydrological Events, edited by: Cudennec, C., Eicker, A., Pilon, P., Stoffel, M., Viglione, A., and Xu, Z., Proc. Int. Assoc. Hydrolog. Sci., 369, 61–67, 2015.

Liu, M., Yang, S. Q., Huang, H., He, H. X., Li, S. J., and Cui, Y.: Joint use of soil moisture and vegetation growth condition by remote sensing on the agricultural drought monitoring, in: MIPPR 2015: Remote Sensing Image Processing, Geographic Information Systems, and Other Applications, edited by: Liu, J. and Sun, H., International Society for Optics and Photonics, 9815, 98151T, 2015.

Long, D., Shen, Y., Sun, A., Hong, Y., Longuevergne, L., Yang, Y., Li, B., and Chen, L.: Drought and flood monitoring for a large karst plateau in Southwest China using extended GRACE data, Remote Sens. Environ., 155, 145–160, https://doi.org/10.1016/j.rse.2014.08.006, 2014.

Ma, M. and Wang, Y.: Drought monitoring research in China: a brief review based on the CNKI database analysis, China Institute of Water Resources and Hydropower Research, Beijing, China, 2018.

Meng, L. K., Dong, T., and Zhang, W.: Drought monitoring using an Integrated Drought Condition Index (IDCI) derived from multi-sensor remote sensing data, Nat. Hazards, 80, 1135–1152, https://doi.org/10.1007/s11069-015-2014-1, 2016.

Mishra, A. K. and Singh, V. P.: A review of drought concepts, J. Hydrol., 391, 202–216, 2010.

Sayers, P., Yuanyuan, L., Moncrieff, C., Jianqiang, L., Tickner, D., Xiangyu, X., Speed, R., Aihua, L., Gang, L., and Bing, Q.: Drought risk management: a strategic approach, UNESCO on behalf of WWF, Paris, 2016.

Stahl, K., Kohn, I., Blauhut, V., Urquijo, J., De Stefano, L., Acácio, V., Dias, S., Stagge, J. H., Tallaksen, L. M., Kampragou, E., Van Loon, A. F., Barker, L. J., Melsen, L. A., Bifulco, C., Musolino, D., de Carli, A., Massarutto, A., Assimacopoulos, D., and Van Lanen, H. A. J.: Impacts of European drought events: insights from an international database of text-based reports, Nat. Hazards Earth Syst. Sci., 16, 801–819, https://doi.org/10.5194/nhess-16-801-2016, 2016.

State Flood Control and Drought Relief Headquarters of Ministry of Water Resources China: The statistics of Flood and Drought Disaster Bulletin in China, Beijing, 2016.

Su, Z., Yacob, A., Wen, J., Roerink, G., He, Y., Gao, B., Boogaard, H., and van Diepen, C.: Assessing relative soil moisture with remote sensing data: theory, experimental validation, and application to drought monitoring over the North China Plain, Phys. Chem. Earth Pt. A/B/C, 28, 89–101, https://doi.org/10.1016/S1474-7065(03)00010-X, 2003.

Tallaksen, L. M. and Van Lanen, H. A. (Eds.): Hydrological drought: processes and estimation methods for streamflow and groundwater, in: Vol. 48, Elsevier, 2004.

Van Lanen, H., Wanders, N., Tallaksen, L., and Van Loon, A.: Hydrological drought across the world: impact of climate and physical catchment structure, Hydrol. Earth Syst. Sci., 17, 1715–1732, https://doi.org/10.5194/hess-17-1715-2013, 2013.

Van Loon, A. F. and Van Lanen, H. A. J.: A process-based typology of hydrological drought, Hydrol. Earth Syst. Sci., 16, 1915–1946, https://doi.org/10.5194/hess-16-1915-2012, 2012.

Wang, H. S., Vicente-Serrano, S. M., Tao, F. L., Zhang, X. D., Wang, P. X., Zhang, C., Chen, Y. Y., Zhu, D. H., and El Kenawy, A.: Monitoring winter wheat drought threat in Northern China using multiple climate-based drought indices and soil moisture during 2000–2013, Agr. Forest Meteorol., 228, 1–12, https://doi.org/10.1016/j.agrformet.2016.06.004, 2016.

Wilhite, D. A.: Drought as a natural hazard: concepts and definitions, in: A Global Assessment, Vol. I, chap. 1, edited by: Wilhite, D. A., Routledge, London, 3–18, 2000.

Wilhite, D. A. and Glantz, M. H.: Understanding: the drought phenomenon: the role of definitions, Water Int., 10, 111–120, 1985.

Wu, H., Hayes, M. J., Weiss, A., and Hu, Q.: An evaluation of the Standardized Precipitation Index, the China-*Z* Index and the statistical *Z*-Score, Int. J. Climatol., 21, 745–758, https://doi.org/10.1002/joc.658, 2001.

Wu, J., Zhou, L., Liu, M., Zhang, J., Leng, S., and Diao, C.: Establishing and assessing the Integrated Surface Drought Index (ISDI) for agricultural drought monitoring in mid-eastern China, Int. J. Appl. Earth Obs. Geoinf., 23, 397–410, https://doi.org/10.1016/j.jag.2012.11.003, 2013.

Wu, J., Zhou, L., Mo, X., Zhou, H., Zhang, J., and Jia, R.: Drought monitoring and analysis in China based on the Integrated Surface Drought Index (ISDI), Int. J. Appl. Earth Obs. Geoinf., 41, 23–33, https://doi.org/10.1016/j.jag.2015.04.006, 2015.

Wu, J., Chen, X., Yao, H., Gao, L., Chen, Y., and Liu, M.: Non-linear relationship of hydrological drought responding to meteorological drought and impact of a large reservoir, J. Hydrol., 551, 495–507, https://doi.org/10.1016/j.jhydrol.2017.06.029, 2017.

Yan, N. N., Wu, B. F., Boken, V. K., Chang, S., and Yang, L. D.: A drought monitoring operational system for China using satellite data: design and evaluation, Geomat. Nat. Hazards Risk, 7, 264–277, https://doi.org/10.1080/19475705.2014.895964, 2016.

Ye, X. C., Li, X. H., Xu, C. Y., and Zhang, Q.: Similarity, difference and correlation of meteorological and hydrological drought indices in a humid climate region – the Poyang Lake catchment in China, Hydrol. Res., 47, 1211–1223, https://doi.org/10.2166/nh.2016.214, 2016.

Yu, M., Li, Q., Hayes, M. J., Svoboda, M. D., and Heim, R. R.: Are droughts becoming more frequent or severe in China based on the Standardized Precipitation Evapotranspiration Index: 1951–2010?, Int. J. Climatol., 34, 545–558, https://doi.org/10.1002/joc.3701, 2014.

Zhang, A. and Jia, G.: Monitoring meteorological drought in semiarid regions using multi-sensor microwave remote sensing data, Remote Sens. Environ., 134, 12–23, https://doi.org/10.1016/j.rse.2013.02.023, 2013.

Zhang, L. and Zhou, T.: Drought over East Asia: A Review, J. Climate, 28, 3375–3399, https://doi.org/10.1175/jcli-d-14-00259.1, 2015.

Zhao, H., Gao, G., An, W., Zou, X., Li, H., and Hou, M.: Timescale differences between SC-PDSI and SPEI for drought monitoring in China, Phys. Chem. Earth Pt. A/B/C, 102, 48–58, https://doi.org/10.1016/j.pce.2015.10.022, 2015.

Zhao, J., Xu, J., Xie, X., and Lu, H.: Drought monitoring based on TIGGE and distributed hydrological model in Huaihe River Basin, China, Sci. Total Environ., 553, 358–365, https://doi.org/10.1016/j.scitotenv.2016.02.115, 2016.

Proc. IAHS, 383, 281–290, 2020
https://doi.org/10.5194/piahs-383-281-2020

Hydrological Drought Characteristics Based on Groundwater and Runoff Across Europe

Samuel Jonson Sutanto and Henny A. J. Van Lanen

Hydrology and Quantitative Water Management, Environmental Sciences Group, Wageningen University and Research, Droevendaalsesteeg 3a, 6708 PB, Wageningen, the Netherlands

Correspondence: Samuel Jonson Sutanto (samuel.sutanto@wur.nl)

Published: 16 September 2020

Abstract. Hydrological drought often gets less attention compared to meteorological drought. For water resources managers, information on hydrological drought characteristics is prerequisite for adequate drought planning and management. Therefore, the aim of this study is to analyse hydrological drought characteristics in the pan-European region based on past drought events from 1990 to 2017. The annual average drought duration, deficit volume, onset, termination, and intensity during drought years were calculated using daily runoff and groundwater data. All data were simulated with the LISFLOOD hydrological model (resolution 5×5 km) fed with gridded time series of observed weather data. Results based on runoff and groundwater data show that regions in Northeast to Southeast Europe, which stretched out from Poland to Bulgaria, were identified as profound regions to severe hydrological drought hazards. The most severe droughts during our study period were observed in 1992 to 1997, where on average Europe experienced drought events, which lasted up to 4 months. Long average drought durations up to 4 and 8 months in runoff and groundwater occurred in a few parts of the European regions (around 10 % area). Longer drought durations and a lower number of drought events were found in groundwater drought than in runoff, which proved that slow responding variables (groundwater) are better in showing extreme drought compared to fast responding variables (runoff). Based on our results, the water managers can better prepare for upcoming drought and foster drought adaptation actions.

1 Introduction

Drought by definition is a sustained and regionally-extensive period of below-average natural water availability. The spatial and temporal characteristics of drought vary from one region to another (Tallaksen and Van Lanen, 2004). Compared to other flashy natural hazards, such as floods, flash floods, and landslides, drought events develop slower and sometimes unnoticed, but have comparable devastating impacts in terms of economy, society, and environment (Mishra and Singh, 2010). Based on its types, droughts are generally classified into four categories: meteorological drought, soil moisture drought or agriculture drought, hydrological drought, and socio-economic drought (Tallaksen and Van Lanen, 2004; Mishra and Singh, 2010; Van Loon, 2015). Meteorological drought is defined as a lack of precipitation, possibly combined with increased evapotranspiration, over a region for a period of time. Soil moisture drought occurs when the supply of moisture to vegetation decreases leading to yield reduction or even crop failure. Hydrological drought is related to negative anomalies in surface and subsurface water (including groundwater). Socio-economic drought refers to the failure of water resources systems associated with the impacts of the three aforementioned drought types.

All these four drought types in general are triggered by atmospheric conditions. A prolonged lack of precipitation (meteorological drought) can propagate through the soil moisture, which in turn, affects the whole hydrological systems, resulting in soil moisture drought and hydrological drought, respectively (Peters et al., 2003; Tallaksen and Van Lanen, 2004; Van Loon and Van Lanen, 2012). This might be a reason that droughts often are viewed from a climatic perspective, by only looking at anomalies in climatic variables, such as precipitation and temperature (Herring et al., 2015; Heim, 2015). Another reason is data availability and access.

Published by Copernicus Publications on behalf of the International Association of Hydrological Sciences.

Climatic data are largely available compared to hydrological data, such as discharge, runoff, and groundwater. Based on these reasons, hydrological droughts have received lesser attention than meteorological drought.

Although meteorological drought has ben intensively studied in the past (Llyod-Hughes and Saunders, 2002; Stagge et al., 2017; Bachmair et al., 2018), many recent studies showed that droughts could not be simply characterized by a lack of precipitation alone (Teuling et al., 2013; Haslinger et al., 2014). Minor meteorological drought may not develop into a hydrological drought, but a series of meteorological droughts can evolve into one long hydrological drought. Specific hydrological drought studies also indicate that hydrological drought diverges from meteorological drought due to its complex phenomenon and different catchment characteristics (Peters et al., 2003; Vidal et al., 2010; Van Dijk et al., 2013; Van Loon and Van Lanen, 2012; Van Lanen et al., 2013; Wanders et al., 2017). Moreover, drought impacts, such as water supply, water-borne transport, and hydropower energy production are more closely linked with hydrological drought rather than with meteorological ones. Understanding hydrological drought characteristic, therefore, is a key for the management and assessments of water resources under drought (Van Lanen et al., 2016). Hence, information on hydrological drought characteristics is prerequisite for adequate drought planning and management.

In this study, hydrological drought characteristics, such as drought duration, deficit volume, onset, termination, and intensity, in the pan-European region were analysed based on past drought events from 1990 to 2017. While many studies have mainly focused on precipitation or soil moisture drought (Spinoni et al., 2016; Haslinger et al., 2014), this study analysed drought characteristics based on runoff and groundwater data. Groundwater is slower responding than, e.g., precipitation and streamflow. It is one of the most important components for the development of hydrological drought (Van Lanen et al., 2013). Thus, the information derived from the hydrological drought characteristics will give additional value to better understanding of droughts in Europe and support water resources managers for managing and planning drought in their regions.

2 Data and Method

2.1 Data

Groundwater and runoff data for the pan-European region were simulated using the LISFLOOD hydrological model fed with gridded time series of observed weather data (thereafter called the LISFLOOD simulation forced with observations, SFO) (De Roo et al., 2000; Van der Knijff, 2008). The model calculates a complete water balance every 6 h or daily for every grid-cell (5 by 5 km in total 259 023 land cells) within the model domain in Europe. Spatially-distributed thematic data on soil, vegetation, and land use obtained from European datasets were used to set up the LISFLOOD model. A calibration was carried out using river discharge data in 258 catchments by tuning parameters that control snowmelt, overland flow, river flow, infiltration, residence times in the soil and subsurface (Feyen and Dankers, 2009; Forzieri et al., 2014). This spatially-distributed rainfall runoff model has initially been used within the European Flood Awareness System (EFAS) to provide data for flood monitoring and warnings (Thielen et al., 2009; Van der Knijff et al., 2010). Later it has been extended to a drought early warning system within the EU ANYWHERE project (Sutanto et al., 2020). The model was run from 1990 to present. In the study, however, we only used the model data from 1990 to 2017 for runoff with missing data in 2013 and 2014 for runoff, and in 2013 and 2017 for groundwater. For groundwater data, we summed groundwater data from upper layer and deep layer.

2.2 Method

Drought events were determined using the variable threshold method (Yevjevich, 1967; Hisdal et al., 2004). Drought starts when the groundwater or runoff data fall below a certain threshold value and ends when it is equal or rises above the threshold (Eq. 1). In this study, the threshold values were derived from the 80th percentile of the duration curve (Q80, Hisdal et al. (2004); Van Loon and Van Lanen (2012)).

$$D_k(x,y,t) = X_k(x,y,t) - \tau_k(x,y,t) \quad (1)$$

where $D_k(x,y,t)$ is the drought occurrence for the hydrological variable k, grid cell x and y, and day t (time step). $X_k(x,y,t)$ is the targeted hydrological variable, and $\tau_k(x,y,t)$ is the 30 d moving average of the variable threshold. $D_k(x,y,t)$ is a binary variable; one if a drought event is found (negative, red color in Fig. 1) and zero if no drought event is found (positive, blue color in Fig. 1).

Using the threshold method, the following drought characteristics can be quantified for each drought event: drought duration (day), drought deficit volume for runoff (mm), onset (day), termination (day), and intensity (mm d^{-1}) (see also Fig. 1). The duration of a drought event is defined as the number of uninterrupted time steps (day) with a hydrological variable below the threshold value. In our study, the duration of drought is calculated as total duration of all drought events for each year.

$$\mathrm{Du}_k(x,y,z) = \sum_{i=0}^{i} D_k(x,y,t) \quad (2)$$

where $\mathrm{Du}_k(x,y,z)$ is the total drought duration (day) for hydrological variable k, grid cell x and y, and year z. i is the day number in the particular year z, which is 365 for none-leap year and 366 for a leap year.

The deficit volume of a drought event is defined as the sum of the deficit volumes over a number of uninterrupted time steps (day) with a hydrological variable below the threshold value. We also calculated the drought deficit volume as total deficit volume of all drought events for each year (in mm).

$$\mathrm{Dv}_k(x,y,z) = \mathrm{ABS}\left(\sum_{i=0}^{i} X_k(x,y,t) - \tau_k(x,y,t)\right),$$
$$\text{for } X_k(x,y,t) < \tau_k(x,y,t) \quad (3)$$

where $\mathrm{Dv}_k(x,y,z)$ is the total deficit volume in runoff drought in mm for hydrological variable k, grid cell x and y, and year z.

Drought onset and termination are defined as the starting date and ending date of drought event over a number of uninterrupted time steps with a hydrological variable below the threshold value. In this study, we selected the onset and termination date from the longest drought event (duration) in each year.

Drought onset and termination are defined as the starting date and ending date of drought event over a number of uninterrupted time steps with a hydrological variable below the threshold value. In this study, we selected the onset and termination date from the longest drought event (duration) in each year.

3 Results

During the period from 1990 to 2017, many parts of Europe suffered from noticeable droughts, such as in 1991–1995, 1996–1997, 2000, 2003, 2004–2007, 2010, 2011, 2015 (Stagge et al., 2013; Spinoni et al., 2015; Van Lanen et al., 2016). Those events are captured well in the hydrological drought analyses using groundwater droughts in South-West Germany obtained from the LISFLOOD SFO model output from 1991–1997, as an example (Fig. 1). In that period, at least two major drought events clearly showed up. The first drought event occurred from the end of 1991 to the beginning of 1994 and then the next major drought event occurred from the beginning of 1996 to the end of 1998. The capability of the LISFLOOD model to simulate hydrological droughts (here runoff and groundwater) opens an opportunity to determine drought characteristics across Europe that will be presented in the next sub chapters.

3.1 Drought duration

Annual drought duration was calculated from the total number of days when the hydrological variable was below the predefined threshold in each year. These durations were then averaged in order to obtain the spatially-distributed average drought duration in Europe over the period 1990–2017 (see Fig. 2a and b for runoff and groundwater, respectively). This was only done for the drought years, that is, years with at least one day in drought. Thus, the analysis excludes non-drought years. The number of drought years may differ per cell and per variable (groundwater, runoff). Our results show that the groundwater drought has a longer duration than the runoff drought. In many parts of Europe, the duration of drought in groundwater can be twice longer than in runoff. The longest drought duration for both runoff and groundwater are detected in the east of Europe (starting from Poland stretching out to Bulgaria), as well as in Northwest Russia, with duration from 200 to 300 d for runoff (7–10 months, Fig. 2a) and from 300 to 330 d for groundwater (10–11 months, Fig. 2b). North, West, and South Europe in general have similar drought durations, with durations around 90 d for runoff (3 months) and 150 d for groundwater (5 months).

The longest drought duration show up in groundwater compared to runoff, which indicates drought propagation (Peters et al., 2003; Van Lanen and Tallaksen, 2007; Van Loon and Van Lanen, 2012; Van Loon et al., 2012). Drought events become longer when moving from precipitation, soil moisture to subsurface or groundwater storage, associated with fast and slow responding variables (Tallaksen et al., 2009; Van Lanen et al., 2013). Runoff usually takes an intermediate position. The fast responding variables, such as precipitation, and to a lesser extent such as, runoff and discharge, in principle, have shorter drought durations, but a higher number of drought events. On the other hand, slowly responding variables, such as groundwater, have longer drought duration but lesser drought occurrences (Fig. 3). This can be seen, for instance, in Eastern Europe, such as Latvia, Lithuania, Belarus, and Ukraine, where the drought duration is high ($\sim$ 200 d, Fig. 2b), the number of drought events in groundwater is low ($\sim$ 10 events from 1990–2017, Fig. 3b). Runoff shows lesser drought duration ($\sim$ 60 d, Fig. 2a) but higher number of drought years ($\sim$ 20 events, Fig. 3a).

3.2 Drought deficit volume and intensity

Similar like drought duration, the highest drought deficit volumes based on runoff are situated in the east of Europe (starting from Poland stretching out to Bulgaria), as well as in Northwest Russia (Fig. 4a). This was expected, since the longer the drought, the more water would be lost. In these regions on average, availability of runoff water is around 20 to 40 mm lower than normal during the drought years (Fig. 4a). In addition, the eastern European regions also suffer from high groundwater drought intensity (Fig. 4b). On average, more than 100 mm d^{-1} of groundwater is lost during the drought years in these regions.

One interesting phenomenon was found in Iceland. The island, which has similar drought durations as West Europe (3 months for drought in runoff, Fig. 2a), has higher drought deficit volumes. Drought events in Iceland are frequent, but have short drought durations. Hydrological variables fluctuate more in West Europe, resulting in higher threshold values and a larger deviation from the threshold. Early and/or late snow accumulation and melting may also play a significant role for droughts in this region e.g., cold and warm snow

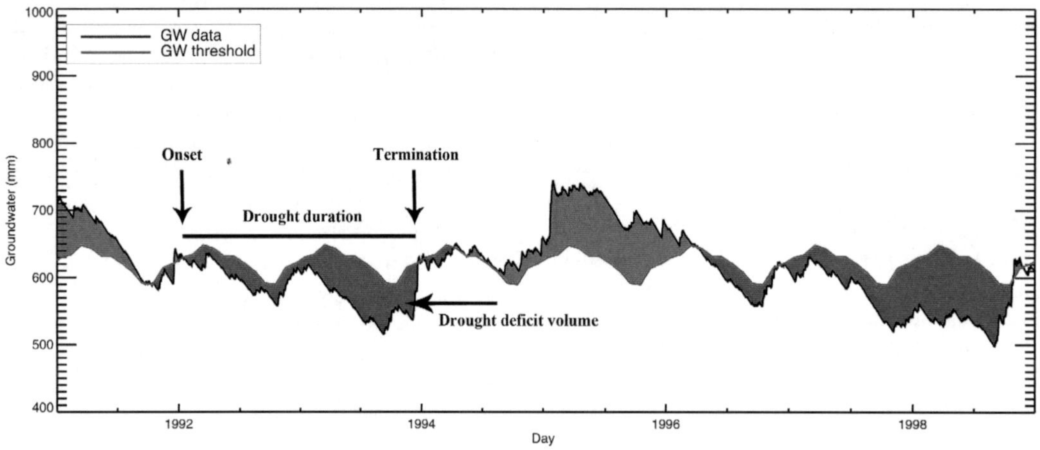

Figure 1. Groundwater storage modelled by LISFLOOD SFO in South-West Germany from 1990–1997.

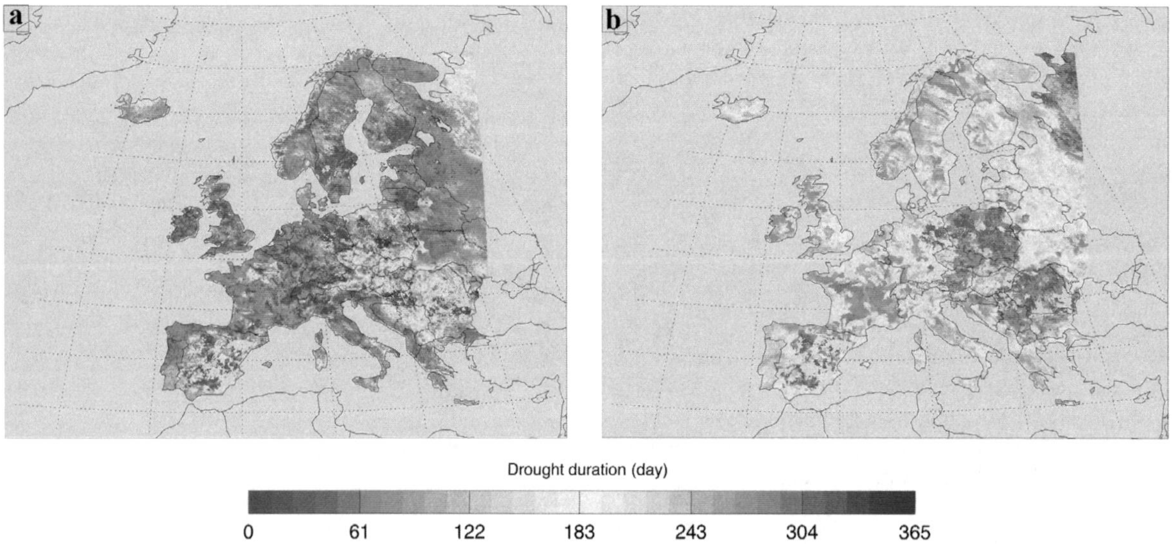

Figure 2. Spatial distribution of the average annual drought duration (days per year in drought years) derived from drought in runoff **(a)** and groundwater **(b)**, over the period 1990–2017. Non-drought years are excluded.

season drought (Van Loon and Van Lanen, 2012; Van Loon et al., 2012).

3.3 Drought onset and termination

The calculation of drought onset and termination is different to that of the calculation of drought duration and deficit volume. For drought onset and termination, we determine the onset and termination day only for the longest drought event in each year, meaning that we neglected droughts in a certain year with shorter duration. This means that the difference between onset day and termination day does not reflect the aforementioned drought duration (Sect. 3.1). One should keep in mind that drought duration and deficit volume were calculated from all drought events occurring in a year.

Figure 5a and b show the average onset and termination day, respectively, obtained from the longest drought event in each year for runoff over the years 1990–2017. In general, Europe can be divided into two regions based on the drought onset. Eastern Europe has earlier drought onsets (spring) than West Europe (summer). However, most of Europe has a similar drought termination day, which is in autumn. This implies that Eastern Europe has more extended drought durations for the longest drought than the West. If we deduct the termination day from the onset day, the most extended drought du-

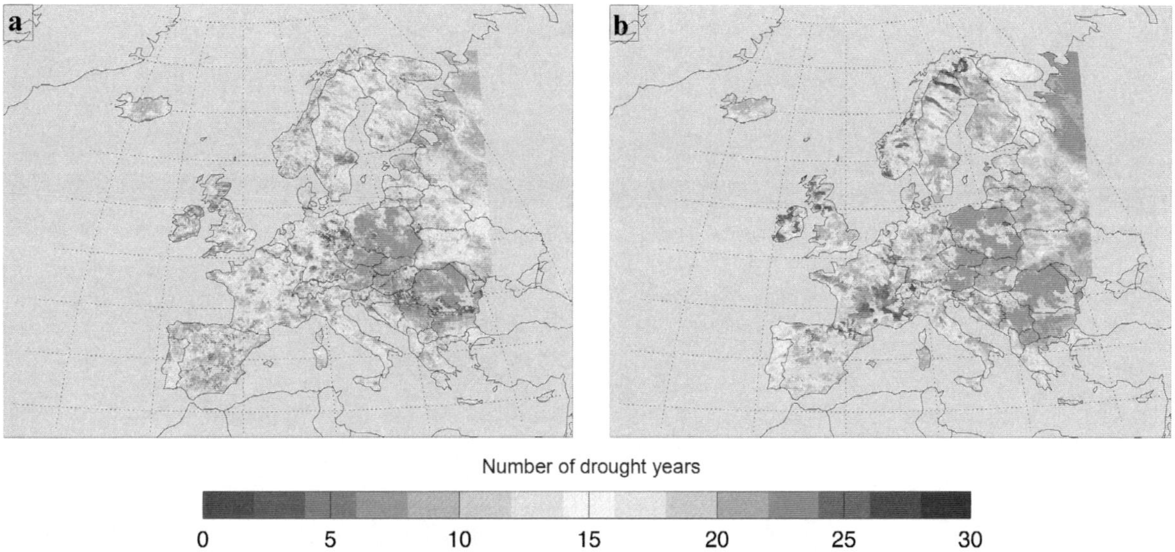

Figure 3. Spatial distribution of the number of drought years derived from drought in runoff **(a)** and groundwater **(b)** over the period 1990–2017.

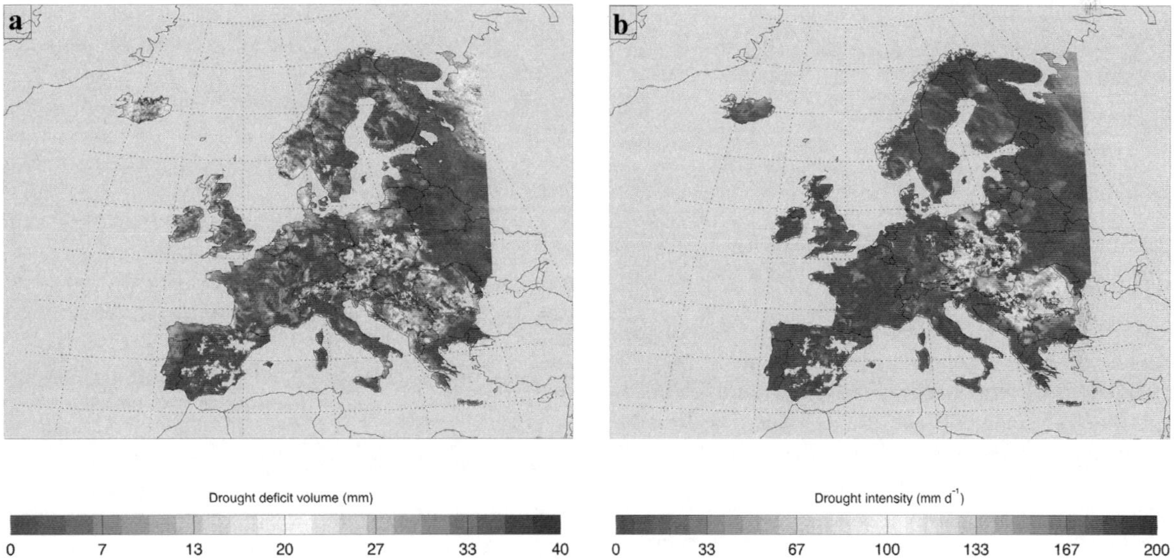

Figure 4. Spatial distribution of the average annual drought deficit volume (mm per year in drought years) derived from drought in runoff **(a)** and the average annual drought intensity (mm per day in drought years) derived from drought in groundwater **(b)**, over the period 1990–2017. Non-drought years are excluded.

ration is situated in the same regions as those regions having the lengthiest total drought duration (Fig. 2a).

Compared to runoff, the average onset day of the longest drought in each year for groundwater is earlier with a onset day of 150 (May) for most of European regions and of 60 (February) for Northeast to Southeast European regions (Fig. 6a). However, the average termination day of the longest drought in each year for groundwater is later than for runoff, with day numbers of 210 (July) to 365 (December) (Fig. 6b). Again, if we deduct the termination day from the onset day, the most extended drought duration of the longest drought per year is situated in the same regions as those having the maximum total drought duration (Fig. 2b). However, average drought duration of the longest drought is close to the average total annual drought duration. This indicates that

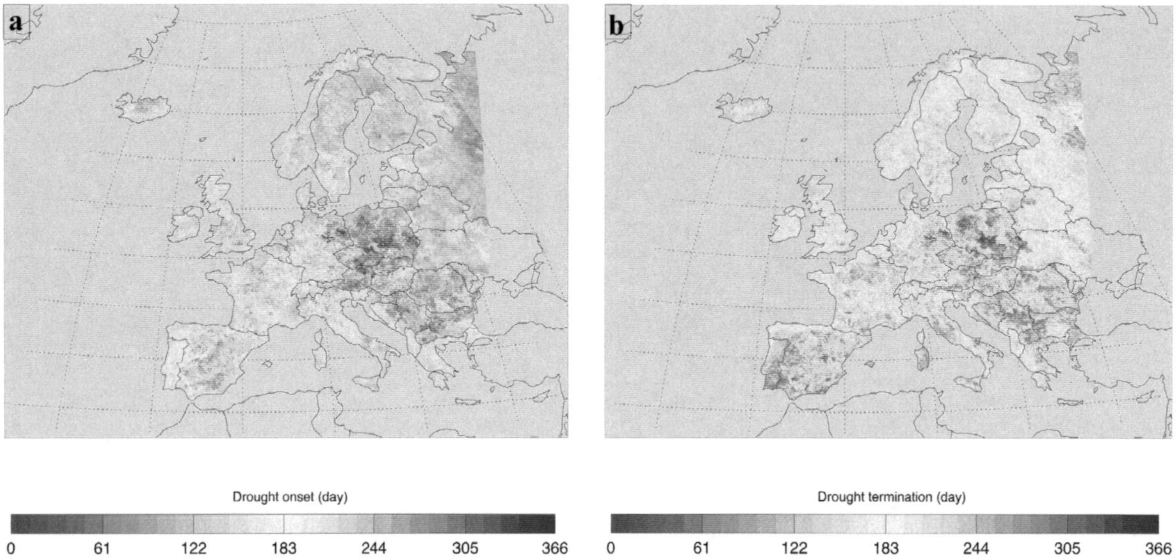

Figure 5. Spatial distribution of the average drought onset **(a)** and termination **(b)** of the longest drought per year (in drought years) derived from drought in runoff over the period 1990–2017. Non-drought years are excluded.

groundwater drought tends to have longer drought durations and a lower number of drought events than drought in runoff.

3.4 Summary of European hydrological drought characteristics

In the previous sections, we presented spatially-distributed average annual drought characteristics for Europe. In this section, drought characteristics for each year and averaged for the whole Europe are presented. Table 1 shows the spatially-averaged drought duration, deficit volume (only in runoff), onset and termination of the longest drought, and drought intensity in runoff and groundwater. The longest total drought durations occurred in 1992–1993 and 1996 in runoff, and in 1992–1993 and 1995–1997 in groundwater. During these drought years, drought occurred up to 2.5 months in runoff and up to 4 months in groundwater. Slightly lower drought durations were found for the longest single drought event in a year, which were derived from the difference between termination and onset values. Based on this drought analysis using runoff and groundwater data, the 2003 European drought was not categorized as the longest drought event (41 and 92 d in runoff and groundwater, respectively) although it has longer drought durations than the average (37 and 68 d in runoff and groundwater, respectively).

The highest drought deficit volumes for drought in runoff were found in 1992 and 1993, with deficit volumes of 5.5 and 6.5 mm, respectively. In average, the volume loss during drought events is 3 mm.

The earliest drought event was in the winter period on day 45 for runoff in 2012 and on day 38 for groundwater in 2008. However in those years, early drought occurrences were also followed by early drought termination. The drought events starting in the winter and terminating in the beginning of spring is likely related to a cold snow season drought (Van Loon and Van Lanen, 2012; Van Loon et al., 2012). These drought events may occur due to cold temperatures below normal during the winter period and thus the groundwater levels drop just below the threshold level due to no recharge from the snowmelt, which normally would happen. This type of drought ends when the snow melts during spring. The longest drought duration in runoff (56 d) started in spring (day 73) and ended in the end of spring (day 130), while the longest drought duration in groundwater (107 d) also started in spring (day 75), but ended in summer (day 182). One should note that the yearly drought onset and termination in this study were averaged for the whole Europe (Table 1). On average, the onset of drought events occurred on day 74 and 60 and terminated on day 101 and 110 for runoff and groundwater, respectively.

The longest drought duration obtained for Europe does not reflect the highest drought intensity. The highest drought intensity in groundwater (16.4 $\mathrm{mm\,d^{-1}}$) occurred in 1995, while the drought duration was only 93 d. The longest drought duration in groundwater (1996, 136 d) only resulted in a drought intensity of 14.7 $\mathrm{mm\,d^{-1}}$. Beside averaging all grid cells over the whole Europe, drought characteristics were also calculated by averaging drought characteristic values from 1990 to 2017 (Table 2) for the 10th driest area in Europe (called 10th percentile area). This has been repeated for 25th, median, 75th and 90th percentile areas. At least 10 % of European regions experienced drought with minimum du-

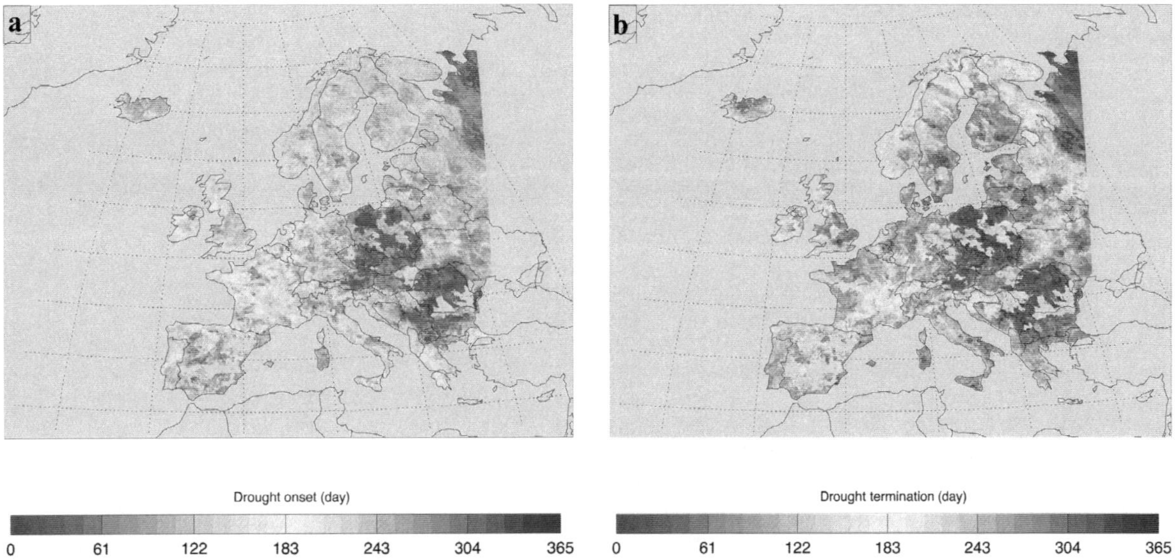

Figure 6. Spatial distribution of the average drought onset **(a)** and termination **(b)** of the longest drought per year (in drought years) derived from drought in groundwater over the period 1990–2017. Non-drought years are excluded.

ration of 121 d (4 months) and 251 d (8 months), deficit volume of 8.5 mm, and intensity of 0.1 mm and 16.6 mm d^{-1} for drought in runoff and groundwater, respectively. However, half of European regions (50th percentile area) only experienced short drought durations. Based on drought onset and termination analyses, 10 % of the European regions (10th percentile area) had drought in the end of summer, which lasted up to the autumn. Early droughts during spring to summer struck a quarter of European area. Table 2 also shows that drought never occurred in almost all European regions (above 75th percentile area) at the same time.

4 Discussion and Conclusion

In this paper, we analysed hydrological drought characteristics across Europe, represented by drought in runoff and groundwater. The results are presented as: (1) spatially-distributed average annual hydrological drought characteristics maps for the period 1990–2017, and (2) summary statistics of hydrological drought characteristics for each year, for 10th, 25th, 50th, 75th and 90th driest areas in Europe. The results show that regions in the eastern part of Europe, starting in Poland and stretching out to Bulgaria, are profound regions to severe drought hazards. These regions, as well as Northwest Russia experienced the longest drought duration and highest deficit volume compared to other regions. Early drought onset and late drought termination were also found in these regions.

Based on summary statistics of drought characteristics, the most severe droughts occurred from 1992 to 1997. The European 2003 drought was considered severe, but it appeared not the most extreme, in term of duration and water deficits. During this period, the groundwater drought lasted up to 4 months. Please keep in mind that this analysis was carried out by averaging over the whole of Europe. In some regions, the drought may last longer than 4 months (see Fig. 2b). Moreover, our driest area analysis also indicated that around 10 % of the European regions experienced groundwater drought for more than 8 months. We need to remark that all analyses in this study were carried out by considering only drought events occurring in the same year. Drought events passing the turn of the year were not considered. This drawbacks our analysis, since drought event may last up to more than one year, i.e. multiple year droughts (Tallaksen and Van Lanen, 2004; Mishra and Singh, 2010; Van Loon, 2015).

Longer drought duration and a lower number of droughts were found in groundwater than in runoff, which proves that slowly responding variables (e.g., groundwater) are better in showing extreme droughts compared to fast responding variables (e.g., precipitation). Small amount of precipitation can alleviate the drought by putting the hydrological variable above the threshold and end the drought, in a fast responding variable. The slowly responding variables are suited to be used for drought early warning system since they have a longer memory due to the influence of catchments storage.

Our study gives an insight in hydrological drought characteristics and is anticipated to provide useful information for water management practices in Europe. It is already used as benchmark for drought forecasting in the EU ANYWHERE project (http://anywhere-h2020.eu/, last access: 1 June 2020). Based on our results, water managers, and also emergency centres, can better prepare for an upcoming

Table 1. Spatially-averaged drought characteristics for each year over the period 1990–2017, which were obtained from drought in runoff (R) and groundwater (GW).

Year	Duration (day)		Deficit volume (mm)	Onset (day)		Termination (day)		Longest Duration (day)		Intensity (mm)	Intensity ($mm\,d^{-1}$)
	R	GW	R	R	GW	R	GW	R	GW	R	GW
1990	38	56	2.9	88	77	117	119	29	42	0.1	2.3
1991	52	84	4.4	92	74	127	135	35	61	0.1	6.2
1992	70	110	5.5	91	62	143	148	52	86	0.1	10.5
1993	71	111	6.5	73	45	130	138	56	93	0.1	14.9
1994	53	91	4.1	68	60	110	133	42	73	0.1	16.2
1995	49	93	3.2	116	72	151	146	35	74	0.1	16.4
1996	64	136	6.0	71	75	116	182	45	107	0.1	14.7
1997	48	118	4.4	82	77	115	161	33	84	0.1	9.8
1998	31	78	2.3	52	45	73	102	22	57	0.1	6.3
1999	30	52	2.9	57	54	79	95	22	41	0.1	4.7
2000	28	50	2.4	52	43	72	83	20	40	0.1	4.7
2001	22	40	1.6	85	51	102	80	16	29	0.1	2.7
2002	44	51	3.4	96	69	126	106	30	37	0.1	3.3
2003	41	92	2.7	64	68	94	134	30	66	0.1	5.4
2004	15	50	0.6	58	54	69	85	12	31	0.0	2.8
2005	27	43	1.6	82	73	101	101	19	28	0.1	2.1
2006	33	78	2.4	65	67	88	118	24	51	0.1	4.8
2007	25	44	1.9	80	60	97	89	17	29	0.1	2.7
2008	15	38	0.9	55	38	66	64	11	26	0.1	2.5
2009	19	37	1.2	61	58	76	84	14	26	0.1	2.2
2010	18	32	2.1	47	40	60	62	13	22	0.1	2.3
2011	29	49	2.4	81	62	102	95	21	33	0.1	3.2
2012	27	56	2.5	45	48	66	88	20	40	0.1	3.3
2013	–	–	–	–	–	–	–	–	–	–	–
2014	–	43	–	-	58	–	88	–	30	–	3.1
2015	37	61	2.7	96	71	124	118	28	47	0.1	4.5
2016	33	71	3.2	115	54	138	109	23	55	0.1	6.4
2017	41	–	4.9	63	–	92	–	29	–	0.1	–
Average	37	68	3.0	74	60	101	110	27	50	0.1	6.1

Table 2. Drought characteristics based on percentiles areas in Europe (e.g. 10th percentile means the 10 % driest area in Europe).

No.	Percentile	Duration (day)		Deficit volume (mm)	Onset (day)		Termination (day)		Intensity (mm)	Intensity ($mm\,d^{-1}$)
		R	GW	R	R	GW	R	GW	R	GW
1	10th	121	251	8.5	250	235	317	351	0.1	16.6
2	25th	84	4.4	92	74	127	135	35	61	0.1
3	50th	110	5.5	91	62	143	148	52	86	0.1
4	75th	111	6.5	73	45	130	138	56	93	0.1
5	90th	91	4.1	68	60	110	133	42	73	0.1

drought since they know where the drought prone regions are located, when the drought event may emerge and recover, and what is the average duration of drought. All these information will help water managers in the drought prone regions to foster drought resilience actions.

Data availability. The runoff and groundwater data were obtained from the ECMWF. All data used in the study can be obtained from the authors (samuel.sutanto@wur.nl).

Author contributions. All authors conceived and implemented the research. Data analyses, model output analyses, and all figures have been performed by SJS.

Competing interests. The authors declare that they have no conflict of interest.

Special issue statement. This article is part of the special issue "Hydrological processes and water security in a changing world". It is a result of the 8th Global FRIEND–Water Conference: Hydrological Processes and Water Security in a Changing World, Beijing, China, 6–9 November 2018.

Acknowledgements. The research is supported by the ANYWHERE project, which is funded within EU's horizon 2020 research and innovation program (http://anywhere-h2020.eu/, last access: 1 June 2020). This research is part of the Wageningen Institute for Environment and Climate Research (WIMEK-SENSE) and it supports the work of the EURO-FRIEND and Panta Rhei program.

Financial support. This research has been supported by the European Commission (Horizon 2020 (grant no. 700099)).

References

Bachmair, S., Tanguy, M., Hannaford, J., and Stahl, K.: How well do meteorological indicators represent agricultural and forest drought across Europe, Environ. Res. Lett., 13, 03042, https://doi.org/10.1088/1748-9326/aaafda, 2018.

De Roo, A. P. J., Wesseling, C. G., and Van Deursen, W. P. A.: Physically based river basin modelling within a GIS: the LISFLOOD model, Hydrol. Process., 14, 1981–1992, 2000.

Feyen, L. and Dankers, R.: Impact of global warming on streamflow drought in Europe, J. Geophys. Res., 114, D17116, https://doi.org/10.1029/2008JD011438, 2009.

Forzieri, G., Feyen, L., Rojas, R., Flörke, M., Wimmer, F., and Bianchi, A.: Ensemble projections of future streamflow droughts in Europe, Hydrol. Earth Syst. Sci., 18, 85–108, https://doi.org/10.5194/hess-18-85-2014, 2014.

Haslinger, K., Koffler, D., Schöner, W., and Laaha, G.: Exploring the link between meteorological drought and streamflow: effects of climate-catchment interaction, Water Resour. Res., 50, 2468–2487, https://doi.org/10.1002/2013WR015051, 2014.

Heim Jr., R. R.: An overview of weather and climate extremes-products and trends, Weather and Climate Extremes, 10, 1–9, https://doi.org/10.1016/j.wace.2015.11.001, 2015.

Herring, S. C., Hoerling, M. P., Kossin, J. P., Peterson, T. C., and Stott, P. A.: Explaining extreme events of 2014 from a climate perspective, B. Am. Meteorol. Soc., 96, S1–S172, 2015.

Hisdal, H., Tallaksen, L. M., Clausen, B., Peters, E., and Gustard, A.: Hydrological Drought Characteristics, in: Tallaksen, L. M. and Van Lanen, H. A. J., Hydrological Drought, Processes and Estimation Methods for Streamflow and Groundwater, Development in Water Science 48, Elsevier Science B.V., 139–198, 2004.

Llyod-Hughes, B. and Saunders, M. A.: A drought climatology for Europe, Int. J. Climatol., 22, 1571–1592, https://doi.org/10.1002/joc.846, 2002.

Mishra, K. K. and Singh, V. P.: A review of drought concepts, J. Hydrol., 391, 202–216, 2010.

Peters, E., Torfs, P. J. J. F., Van Lanen, H. A. J., and Bier, G.: Propagation of drought through groundwater – a new approach using linear reservoir theory, Hydrol. Process., 17, 3023–3040, 2003.

Peters, E.: Propagation of drought through groundwater systems. Illustrated in the Pang (UK) and Upper-Guadiana (ES) catchments, PhD thesis, Wageningen University, The Netherlands, 2003.

Spinoni, J., Naumann, G., Vogt, J. V., and Barbosa, P.: Meteorological Droughts in Europe. Events and Impacts Past Trends and Future Projections, Publications Office of the European Union, Luxembourg, EUR 27748 EN, https://doi.org/10.2788/450449, 2016.

Spinoni, J., Naumann, G., Vogt, J. V., and Barbosa, P.: The biggest drought events in Europe from 1950 to 2012, J. Hydrol. Reg. Stud., 3, 509–524, https://doi.org/10.1016/j.ejrh.2015.01.001, 2015.

Stagge, J. H., Kingston, D. G., Tallaksen, L. M., and Hannah, D. M.: Observed drought indices show increasing divergence across Europe, Scientific Reports, 7, 14045, https://doi.org/10.1038/s41598-017-14283-2, 2017.

Stagge, J. H., Tallaksen, L. M., Kohn, I., Stahl, K., and Van Loon, A. F.: European drought reference (EDR) database: design and online implementation, Technical report no.12, Drought R&SPI, FP7 program, 2013.

Sutanto, S. J., Van Lanen, H. A. J., Wetterhall, F., and Llort, X.: Potential of pan-European seasonal hydro-meteorological drought forecasts obtained from a Multi-Hazard Early Warning System, B. Am. Meteorol. Soc., 101, 368–393, https://doi.org/10.1175/BAMS-D-18-0196.1, 2020.

Tallaksen, L. M., Hisdal, H., and Van Lanen, H. A. J.: Space-time modelling of catchment scale drought characteristics, J. Hydrol., 375, 363–372, 2009.

Tallaksen, L. M. and Van Lanen, H. A. J. (Eds.): Hydrological Drought. Processes and Estimation Methods for Streamflow and Groundwater, Developments in Water Science, 48, Elsevier Science B.V., Amsterdam, p. 579, 2004.

Teuling, A. J., Van Loon, A. F., Seneviratne, S. I., Lehner, I., Aubinet, M., Heinesch, B., Bernhofer, C., Grünwald, T., Prasse, H., and Spank, U.: Evapotranspiration amplifies European summer drought, Geophys. Res. Lett., 40, 2071–2075, https://doi.org/10.1002/grl.50495, 2013.

Thielen, J., Bartholmes, J., Ramos, M.-H., and de Roo, A.: The European Flood Alert System – Part 1: Concept and development, Hydrol. Earth Syst. Sci., 13, 125–140, https://doi.org/10.5194/hess-13-125-2009, 2009.

Van Dijk, A. I. J. M., Beck, H. E., Crosbie, R. S., De Jeu, R. A. M., Liu, Y. Y., Podger, G. M., Timbal, M., and Viney, N. R.: The Millennium Drought in southeast Australia (2001–2009): natural and human causes and implications for water resources, ecosystems, economy, and society, Water Resour. Res., 49, 1040–1057, https://doi.org/10.1002/wrcr.20123, 2013.

Van der Knijff, J. M., Younis, J., and De Roo, A. P. J.: LISFLOOD: a GIS-based distributed model for river basin scale water balance and flood simulation, International

Journal of Geographical Information Science, 24, 189–212, https://doi.org/10.1080/13658810802549154, 2010.

Van der Knijff, J.: LISVAP: Evaporation pre-processor for the LISFLOOD water balance and flood simulation model, EUR 22639 EN/2, 2008.

Van Lanen, H. A. J., Laaha, G., Kingston, D. G., Gauster, T., Ionita, M., Vidal, J-P., Vlnas, R., Tallaksen, L. M., Stahl, K., Hannaford, J., Delus, C., Fendekova, M., Mediero, L., Prudhomme, C., Rets, E., Romanowicz, R. J., Gailliez, S., Wong, W. K., Adler, M-J., Blauhut, V., Caillouet, L., Chelcea, S., Frolova, N., Gudmundsson, L., Hanel, M., Haslinger, K., Kireeva, M., Osuch, M., Sauquet, E., Stagge, J. H., and Van Loon, A. F.: Hydrology needed to manage droughts: the 2015 European case, Hydrol. Process., 30, 3097–3104, https://doi.org/10.1002/hyp.10838, 2016.

Van Lanen, H. A. J., Wanders, N., Tallaksen, L. M., and Van Loon, A. F.: Hydrological drought across the world: impact of climate and physical catchment structure, Hydrol. Earth Syst. Sci., 17, 1715–1732, https://doi.org/10.5194/hess-17-1715-2013, 2013.

Van Lanen, H. A. J. and Tallaksen, L. M.: Hydrological drought, climate variability and change, in: Climate and Water, edited by: Heinonen, M., Proc. of the Third Int. Conf. on Climate and Water, 3–6 September 2007, Helsinki, Finland, 488–493, 2007.

Van Loon, A. F.: Hydrological drought explained, WIREs Water, 2, 359–392, https://doi.org/10.1002/wat2.1085, 2015.

Van Loon, A. F. and Van Lanen, H. A. J.: A process-based typology of hydrological drought, Hydrol. Earth Syst. Sci., 16, 1915–1946, https://doi.org/10.5194/hess-16-1915-2012, 2012.

Van Loon, A. F., Van Huijgevoort, M. H. J., and Van Lanen, H. A. J.: Evaluation of drought propagation in an ensemble mean of large-scale hydrological models, Hydrol. Earth Syst. Sci., 16, 4057–4078, https://doi.org/10.5194/hess-16-4057-2012, 2012.

Vidal, J.-P., Martin, E., Franchistéguy, L., Habets, F., Soubeyroux, J.-M., Blanchard, M., and Baillon, M.: Multilevel and multiscale drought reanalysis over France with the Safran-Isba-Modcou hydrometeorological suite, Hydrol. Earth Syst. Sci., 14, 459–478, https://doi.org/10.5194/hess-14-459-2010, 2010.

Wanders, N., Van Loon, A. F., and Van Lanen, H. A. J.: Frequently used drought indices reflect different drought conditions on global scale, Hydrol. Earth Syst. Sci. Discuss., https://doi.org/10.5194/hess-2017-512, in review, 2017.

Yevjevich, V.: An objective approach to definition and investigations of continental hydrologic droughts, Hydrology Papers 23, Colorado State University, Fort Collins, USA, 1967.

Proc. IAHS, 383, 291–295, 2020
https://doi.org/10.5194/piahs-383-291-2020

Open Access
PIAHS

The challenges of hydrological drought definition, quantification and communication: an interdisciplinary perspective

Kerstin Stahl[1], Jean-Philippe Vidal[2], Jamie Hannaford[3], Erik Tijdeman[4], Gregor Laaha[5], Tobias Gauster[5], and Lena M. Tallaksen[6]

[1]Faculty of Environment and Natural Resources, University of Freiburg, Freiburg, Germany
[2]INRAE, UR RiverLy, Centre de Lyon-Grenoble, Villeurbanne, France
[3]Centre for Ecology and Hydrology, Wallingford, UK
[4]Institute of Geography, Heidelberg University, Heidelberg, Germany
[5]University of Natural Resources and Life Sciences, Vienna, Austria
[6]Department of Geoscience, University of Oslo, Norway

Correspondence: Kerstin Stahl (kerstin.stahl@hydrology.uni-freiburg.de)

Published: 16 September 2020

Abstract. Numerous indices exist for the description of hydrological drought. The EURO FRIEND-Water Low flow and Drought Group has repeatedly discussed changing paradigms in the perception and use of existing and emerging new indices for hydrological drought identification and characterization. Group members have also tested the communication of different indices to stakeholders in several national and international transdisciplinary research projects. This contribution presents the experience gained with regard to the purpose and applicability of different classes of drought indices. A recent paradigm shift is the use of anomalies, traditionally from climatology, in hydrology. For instance, anomaly-based indices, such as the Standardized Streamflow Index (SSI) and the variable threshold level method to define streamflow deficiencies, are used increasingly for real-time monitoring. How these indices relate to low flows and their impacts may have become less clear as a result. Assessments of the severity of a particular drought may also differ depending on whether return periods based on traditional low flow or drought frequency analyses or whether SSI time series index values are used. These experiences call for a systematic comparison, classification and evaluation of different low flow and drought indices and their usages.

1 Introduction

Numerous indices exist for the definition and description of hydrological drought, which here refers primarily to streamflow drought. Hydrological drought indices have been developed and used for a number of purposes. These include, but are not limited to, planning of water resources/flow allocations, analysis of temporal trends and the influence of climate change, analysis of spatial patterns and regionalization of drought characteristics (estimation at the ungauged site), large-scale comparative monitoring (and forecasting) of river flows and drought conditions, and drought vulnerability and risk analysis. The purposes these indices serve therefore vary and may also have changed over time.

The EURO FRIEND-Water Low flow and Drought Group has repeatedly discussed changing paradigms and advantages and disadvantages of existing and emerging new indices for hydrological drought identification and characterization. Group members have also tested the communication of different types of drought indices to stakeholders in various national and international transdisciplinary research projects on drought risk management. The objective of this contribution is to review some of the experience gained with regard to the origin of well-known drought indices, their statistical basis and changing purposes and applications over time. The review is a collection of thoughts and not a complete overview. It aims to spawn further discussion and to provide a starting point for a more comprehensive classification of the multitude of indices and their recommended usage.

2 Drought indices based on streamflow – in brief

A comprehensive description of low flow and streamflow drought indices can be found e.g., in Tallaksen and van Lanen (2004) and WMO (2005). Hydrological drought indices are mostly based on time series of gauged or simulated streamflow (catchment scale) and occasionally groundwater levels; but also modelled runoff (grid cell scale) has been used. A wide range of approaches exists that allow to determine the occurrence, magnitude and frequency of a drought or extreme low flow event at a given site. There are many potential ways of classifying these approaches.

Tallaksen and van Lanen (2004) define a low flow or drought index as a "single-value index", noting that alternative terms used include "statistic", "measure", "parameter" and others. Such an index can be derived in different ways, (i) sampled at regular intervals, e.g. one value per year, and then averaged, which is commonly done to derive low flow indices (Fig. 1a) or (ii) sampled as event characteristics with a particular set of criteria that define the event, which is often done to obtain drought event characteristics (Fig. 1b, c). Besides averaging, also probability distributions may be fitted to a sample of low flow values or drought event characteristics to derive an index value corresponding to a particular return period T.

Low flow indices include statistics such as a particularly rare (hence extreme) quantile of the flow duration curve, e.g. the flow exceeded 95 % of the time (Q95), the mean annual n-day minimum flow, MAM(n-day), or the value associated with a given return period (AM-n-day$_T$). Drought event characteristics are identified using the threshold level method or the sequent peak algorithm. Resultant indices include mean values or return periods of drought duration, deficit volume (severity), and intensity (severity/duration) defined from events below the threshold (τ). Often a low flow index is used as τ to define drought event characteristics. The threshold may be constant (Fig. 1b) or variable (Fig. 1c).

Not only the interdisciplinary tradition of using the term "index" for a single value, in drought research and management, also index-time series are common. Several such drought index time series can be created, some of which stem from single value indices. Streamflow series may be transformed and classified according to a given drought status based on the distribution of the entire time series (Fig. 1g, h) or with respect to the time of the year, i.e. a particular calendar month (week or day) (Fig. 1d). Similar to the single-value indices, this distinction is important for the interpretation of the index (Sect. 3). Indices derived from streamflow (or runoff) drought time series include the use of empirical percentiles. Furthermore, standardized streamflow or runoff indices, SRI, SSI and their variations, such as SFI for modelled streamflow by Vidal et al. (2010) or SMRI based on rain and snowmelt by Staudinger et al. (2014), have been developed. Theoretically, these indices could be calculated for different accumulation periods of n months as "SSI-n", similar to the standardized precipitation indices SPI-n. As streamflow already integrates the hydrometeorology of the past months, we only consider monthly streamflow (sometimes called SSI-1, but for simplicity here called SSI). Figure 1g and h show that the shapes of the resulting transformed hydrographs may differ. Including a drought threshold into the definition allows to derive time series expressed, for example, as a binary drought/no-drought variable (Fig. 1e), which have been used e.g. to study changes over time due to regulation (Tijdeman et al., 2018), and map the area affected by drought (Tallaksen and Stahl, 2014). Alternatively, a cumulative duration or deficit can be derived (Fig. 1f) as for example for the Low Flow Index (LFI) of the European Drought Observatory (Cammalleri et al., 2017). For spatially simulated runoff, a timeseries of the "area in drought" within a particular region may be derived, and many others.

In summary, low flow and drought indices may be single value indices or time series indices that are derived from streamflow characteristics sampled at regular or irregular (event-based) intervals. The indices also differ regarding the conditions they are being related to (next section).

3 Drought or anomaly?

Drought is often broadly defined as a temporary deviation of normal water availability. As described in the previous section, "normal" can be defined with respect to different thresholds, either derived from the entire year's streamflow record or derived from a seasonal window (day of year, week of year, month of year, etc.). The early tradition in hydrology is the former. It involves defining drought by a distinct constant threshold of streamflow and is therefore directly linked to low flow situations.

For hydrological regimes with distinct seasonal variation and/or seasonally different water uses, it may be useful to vary the threshold to define drought throughout the year. Tallaksen and van Lanen (2004) describe both daily and monthly varying thresholds, stressing that the derived characteristics describe relative "streamflow deficiencies or anomalies", which is distinctly different from the absolute "drought deficit" in the original threshold level method. The varying threshold level method has later been modified and deficiencies are now often also termed "drought events" (e.g. van Loon and van Lanen 2012, among many others), even though the event may occur during the wet season. However, impacts on the water availability for a particular use may not be felt until later in the year during the low flow season, or not at all if the deficit is recovered or of limited importance.

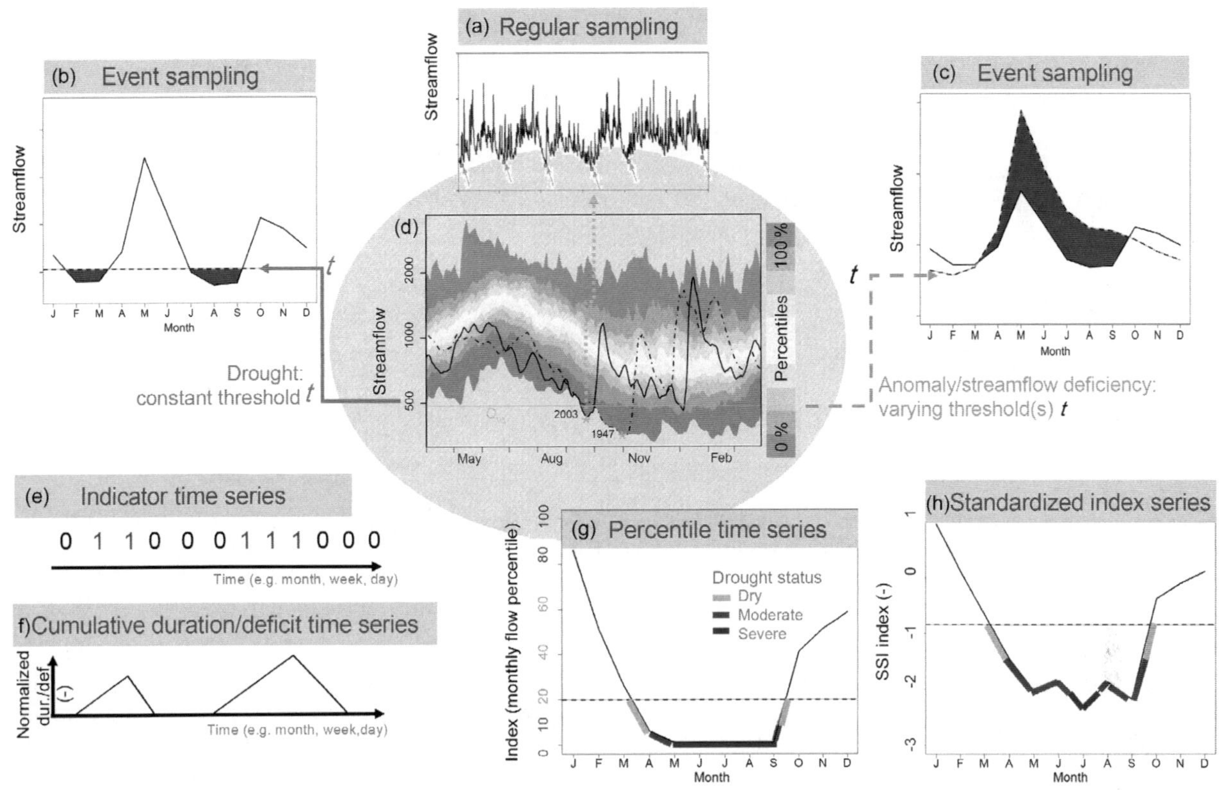

Figure 1. Schematic illustration of low flow and drought indexing: the upper part shows indices derived with **(a)** regular sampling (e.g. annual minima) and **(b)** event sampling with constant thresholds (e.g. Q95 of the period-of-record flow duration curve) or **(c)** with a varying threshold; the lower part shows time series of a drought or low flow status indexed with **(e)** binary indicators for flow below a threshold, **(f)** cumulative durations/deficits derived from those, **(g)** timeseries transformed into percentiles and **(h)** standardized indices. Graphs are schematic for illustrative purpose and based on different data and sources, with an example of underlying streamflow data given in **(d)**.

The use of daily, weekly or monthly anomalies that are categorized to create index time series for drought, is a relatively recent development within hydrology. The paradigm of drought as an anomaly with respect to the time of the year, has been adopted from the discipline of climatology and climate modelling. In addition, the use of a Standardized Streamflow Index (SSI) may have been spawned by WMO's recommendation of the Standardized Precipitation Index (SPI) as a uniformly applicable index. Van Loon and van Lanen (2012) describe typical processes leading to such anomaly-defined below-threshold events in different regimes. Anomalies have the advantage of often being more consistent regionally as they filter local catchment-specific modifications to meteorological input. Caillouet et al. (2017) capitalized on the advantages of both approaches by using a combination of fixed and variable thresholds for the reconstruction of historic extreme low flows in France. So far, however, the comparison between the two approaches and the implications of a paradigm shift for hydrology have not been given much systematic thought. In summary, a more integrated view on drought across the disciplines and their conventions appears useful. There may be certain demands on the definition of hydrological drought given its purpose, which suggest a careful evaluation of the drought and anomaly paradigms. Most practitioners highlight that they use particular flow levels rather than anomalies as triggers for management actions. Clarifying the various interpretations of drought events as defined using different approaches and their link to major environmental, economic and societal impacts and drought management aspects, are therefore important.

4 Discussion of changing paradigms

Members of the EURO FRIEND-Water Low Flow and Drought Group have over time experienced changing paradigms in low flow and drought indices, including and beyond the use of anomalies. Particularly cross-disciplinary and transdisciplinary projects have added new demands and viewpoints. Collaborations with climatologists and large-scale hydrology modelers on climate change projects have, for example, promoted the use of monthly anomalies simi-

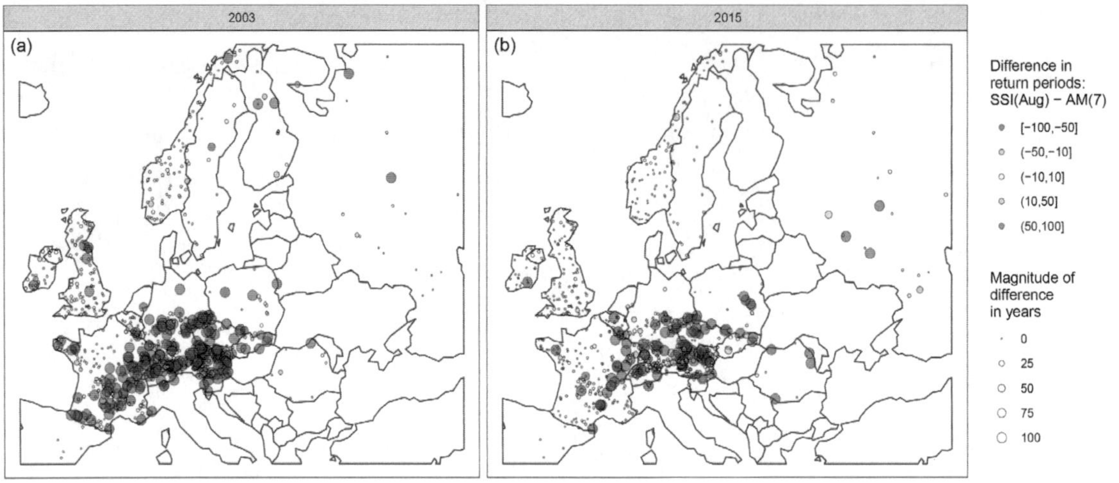

Figure 2. Difference in the return periods T of the droughts in 2003 **(a)** and 2015 **(b)** estimated by SSI(August) and AM(7) with differences given in Years as: $T_{\mathrm{SSI(August)}} - T_{\mathrm{AM(7)}}$, resulting in positive differences (blue colors), where T_{SSI} estimates a higher T (more severe event) and negative differences (red colors), where $T_{\mathrm{AM(7)}}$ estimates a more severe event. Methods and data are based or an extension of the streamflow drought dataset from Laaha et al. (2017).

lar to precipitation and the derivation of spatial and temporal drought characteristics below a threshold as described above. The approach is robust as large-scale models tend not to represent absolute streamflow values very accurately (e.g. Gudmundsson et al., 2014). It also allows to communicate future relative changes. Nevertheless, it presents a change from the traditional viewpoint of hydrological drought as extreme low flow with immediate impacts, e.g. on river habitats and water use restrictions, to a relative lower seasonal flow, which may not have simultaneously impacts. Another change in paradigm has been the use of less extreme thresholds for drought definition, i.e. from a commonly used threshold of 10 % or lower daily streamflow exceedance probabilities, to monthly quantiles in the 20 % range. Whether this may result in an altered perception of the extremeness of drought overall needs to be investigated.

Another recent development is that many regions or countries have – or are in the process of – establishing drought monitoring and early warning online systems as an important component of drought risk management. Such continuous monitoring, based on observation or simulation data, requires time series indices and the result is often displayed spatially on maps. In their review of existing systems, Bachmair et al. (2016) found that only few use hydrological drought indices. Those that do tend to use anomaly-based time series indices such as monthly percentiles (e.g. US Drought Monitor), SSI (e.g. UK Drought Portal), or the LFI (European Drought Observatory). Still, national river flow monitoring is often separate and has traditionally focused on floods. There appears to be no consensus how to monitor low flows continuously as part of drought monitoring. Many different approaches exist. Some of the national mapping-based river flow monitoring in Europe simply mark stations with daily flow below the mean or below the 25th percentile or use a combination of percentiles and mean annual minimum flows (e.g. Bavaria's Low flow information service website https://www.nid.bayern.de/, last access: 2 August 2020).

The estimation of return periods is another paradigm that has a long tradition in hydrological planning, mainly for design flood estimation. Return periods for low flows and drought have been studied extensively and e.g., allowed a comparison of the severity of particular drought events (e.g. Laaha et al., 2017). So far little work has been done to compare the probability of a low flow or drought index derived from regular or event-based sampling with for instance the SSI in a particular month of a drought. Figure 2 shows such a comparison, i.e. the spatial distribution of the differences between return periods estimated from a traditional low flow frequency analysis of the AM(7) and the SSI of the month of August, for the droughts of 2003 and 2015 based on a large sample of streamflow records in Europe. For simplicity, August was assumed here to have been the most severe month everywhere, which may cause some of the differences. Nevertheless, August was one of the peak drought months in both 2003 and 2015 and notable differences between return periods can be seen. Further details on this study, including data and methods, are given in Laaha et al. (2017). The study suggests that similar comparisons may provide useful insights into how to interpret the severity of droughts using different indices.

Another inter- and transdisciplinary task supporting drought management, is the mapping of drought vulnerability and risk. Such maps may help to set regional priorities and are often constructed by indexing vulnerability factors that relate to drought sensitivity, adaptive capacity and exposure. FRIEND members have experienced that in particular the popular SPI-based time series indices, perhaps because they are widely known from drought information systems, are often falsely considered suitable for such a purpose. The question of distinguishing regions that are more or less prone to drought (exposed to drought) require single-value indices that are able to reveal differences in real water availability.

These examples only provide a small selection of paradigm shifts and emerging issues noticed by the group. Many further experiences exist and should be thoroughly discussed and evaluated in terms being 'fit for purpose' and easy to communicate.

5 Conclusions

The high diversity of drought definitions and related indices appears to have further increased over the past years. Some shifts in paradigms have taken place as a consequence of changing use and emerging applications: Indices used for streamflow monitoring purposes, for example, and indices used as reference values for climate change adaptation studies, need to be different and fit for purpose. This may require either comparative anomalies or quantitative targets. Members of the EURO FRIEND-Water Low flow and Drought group have experienced different preferences in the use of drought indices among stakeholders in different disciplines and countries. Overall there are some indications of shifting preferences towards more use of anomaly-based definitions and less extreme thresholds, though robust conclusions will require a better survey of practices of index usage. A guide to the interpretation, comparability and recommended usage of the wealth of indices is therefore needed.

Data availability. Details on streamflow data collection are given in the cited source: Laaha et al. (2017). Most data is available from the GRDC portal https://portal.grdc.bafg.de (last access: 2 August 2020), which contains the former FRIEND European Water Archive.

Author contributions. All co-authors contributed to the manuscript design, content and editing. KS, ET, GL and TG made the figures.

Competing interests. The authors declare that they have no conflict of interest.

Special issue statement. This article is part of the special issue "Hydrological processes and water security in a changing world". It is a result of the 8th Global FRIEND-Water Conference: Hydrological Processes and Water Security in a Changing World, Beijing, China, 6–9 November 2018.

Financial support. This research has been supported by the Deutsche Forschungsgemeinschaft (grant no. STA632/3-1) and the Natural Environment Research Council (grant no. NE/R016429/1 (UK-SCAPE)) and Austrian Climate Research Program ACRP, project DALF-Pro (GZ B464822).

References

Bachmair, S., Stahl, K., Collins, K., Hannaford, J., Acreman, M., Svoboda, M., Knutson, C., Smith, K., Wall, N., Fuchs, B., Crossman, N., and Overton, I.: Drought indicators revisited: the need for a wider consideration of environment and society, Wiley Interdisciplinary Reviews Water, https://doi.org/10.1002/wat2.1154, 2016.

Caillouet, L., Vidal, J.-P., Sauquet, E., Devers, A., and Graff, B.: Ensemble reconstruction of spatio-temporal extreme low-flow events in France since 1871, Hydrol. Earth Syst. Sci., 21, 2923–2951, https://doi.org/10.5194/hess-21-2923-2017, 2017.

Cammalleri, C., Vogt, J., and Salamon, P.: Development of an operational low-flow index for hydrological drought monitoring over Europe, Hydrol. Sci. J., 62, 346–358, https://doi.org/10.1080/02626667.2016.1240869, 2017.

Gudmundsson, L., Wagener, T., Tallaksen, L. M., and Engeland, K.: Evaluation of nine large-scale hydrological models with respect to the seasonal runoff climatology in Europe, Water Resour. Res., 48, W11504, https://doi.org/10.1029/2011WR010911, 2012.

Laaha, G., Gauster, T., Tallaksen, L. M., Vidal, J.-P., Stahl, K., Prudhomme, C., Heudorfer, B., Vlnas, R., Ionita, M., Van Lanen, H. A. J., Adler, M.-J., Caillouet, L., Delus, C., Fendekova, M., Gailliez, S., Hannaford, J., Kingston, D., Van Loon, A. F., Mediero, L., Osuch, M., Romanowicz, R., Sauquet, E., Stagge, J. H., and Wong, W. K.: The European 2015 drought from a hydrological perspective, Hydrol. Earth Syst. Sci., 21, 3001–3024, https://doi.org/10.5194/hess-21-3001-2017, 2017.

Staudinger, M., Stahl, K., and Seibert, J.: A drought index accounting for snow, Water Resour. Res., 50, 7861–7872, https://doi.org/10.1002/2013WR015143, 2014.

Tallaksen, L. M. and van Lanen, H. A. J. (Eds): Hydrological drought – Processes and Estimation Methods for Streamflow and Groundwater Development in Water Sciences no. 48, Elsevier Publ., The Netherlands, 579 pp., 2004.

Tallaksen, L. M. and Stahl, K.: Spatial and temporal patterns of large-scale droughts in Europe: model dispersion and performance, Geophys. Res. Lett., 41, 429–434, https://doi.org/10.1002/2013GL058573, 2014.

Tijdeman, E., Hannaford, J., and Stahl, K.: Human influences on streamflow drought characteristics in England and Wales, Hydrol. Earth Syst. Sci., 22, 1051–1064, https://doi.org/10.5194/hess-22-1051-2018, 2018.

Van Loon, A. F. and Van Lanen, H. A. J.: A process-based typology of hydrological drought, Hydrol. Earth Syst. Sci., 16, 1915–1946, https://doi.org/10.5194/hess-16-1915-2012, 2012.

Vidal, J.-P., Martin, E., Franchistéguy, L., Habets, F., Soubeyroux, J.-M., Blanchard, M., and Baillon, M.: Multilevel and multiscale drought reanalysis over France with the Safran-Isba-Modcou hydrometeorological suite, Hydrol. Earth Syst. Sci., 14, 459–478, https://doi.org/10.5194/hess-14-459-2010, 2010.

Proc. IAHS, 383, 297–305, 2020
https://doi.org/10.5194/piahs-383-297-2020

Open Access
PIAHS

The Groundwater Drought Initiative (GDI): Analysing and understanding groundwater drought across Europe

Bentje Brauns[1], Daniela Cuba[1], John P. Bloomfield[2], David M. Hannah[3], Christopher Jackson[1], Ben P. Marchant[1], Benedikt Heudorfer[3], Anne F. Van Loon[3,a], Hélène Bessière[4], Bo Thunholm[5], and Gerhard Schubert[6]

[1]British Geological Survey, Keyworth, NG12 5GG, UK
[2]British Geological Survey, Wallingford, OX10 8BB, UK
[3]School of Geography, Earth and Environmental Sciences, University of Birmingham, Birmingham, B15 2TT, UK
[4]BRGM – Bureau de Recherches Géologiques et Minières, 45060 Orléans, France
[5]Geological Survey of Sweden, 751 28 Uppsala, Sweden
[6]Geological Survey of Austria, 1030 Vienna, Austria
[a]now at: Institute for Environmental Studies, Vrije Universiteit Amsterdam, the Netherlands

Correspondence: Bentje Brauns (benaun@bgs.ac.uk)

Published: 16 September 2020

Abstract. In Europe, it is estimated that around 65 % of drinking water is extracted from groundwater. Worryingly, groundwater drought events (defined as below normal groundwater levels) pose a threat to water security. Groundwater droughts are caused by seasonal to multi-seasonal or even multi-annual episodes of meteorological drought during which the drought propagates through the river catchment into the groundwater system by mechanisms of pooling, lagging, and lengthening of the drought signals. Recent European drought events in 2010–2012, 2015 and 2017–2018 exhibited spatial coherence across large areas, thus demonstrating the need for transboundary monitoring and analysis of groundwater level fluctuations. However, such monitoring and analysis of groundwater drought at a pan-European scale is currently lacking, and so represents a gap in drought research as well as in water management capability. To address this gap, the European Groundwater Drought Initiative (GDI), a pan-European collaboration, is undertaking a large-scale data synthesis of European groundwater level data. This is being facilitated by the establishment of a new network to co-ordinate groundwater drought research across Europe. This research will deliver the first assessment of spatio-temporal changes in groundwater drought status from $\sim$ 1960 to present, and a series of case studies on groundwater drought impacts in selected temperate and semi-arid environments across Europe. Here, we describe the methods used to undertake the continental-scale status assessment, which are more widely applicable to transboundary or large-scale groundwater level analyses also in regions beyond Europe, thereby enhancing groundwater management decisions and securing water supply.

1 Introduction

Groundwater is the primary source for public water supply, agricultural irrigation and industry in many countries. It also supports flow and water levels in rivers, lakes, and wetlands, maintaining healthy ecosystems, and providing amenity value and supporting livelihoods associated with these areas. In the European Union (EU), about three quarters of public water supplies come from groundwater, but some areas, such as Denmark, are entirely dependent on groundwater as drinking water source (EEA, 1999). When managed appropriately, groundwater generally is a reliable resource that can sustain water demand throughout the year, and that is relatively independent from short-term climatology. However, recharge of this valuable resource depends on natural year-on-year variations in rainfall (particularly meteorological conditions during the wet season), and drought signals from major meteorological droughts may propagate into the subsurface, causing groundwater heads to fall below normal (Van Lanen and Peters, 2000). These periods of below normal groundwater levels, termed groundwater droughts, have a number of unwelcome effects, such as reduced production of groundwater from boreholes, and the drying up of groundwater-dependent ecosystems with resulting implications for wildlife and livelihoods.

Major episodes of drought often affect large areas across national boundaries. For example, the long groundwater drought in the UK from 2010–2012 showed the same drought effects over large parts of continental Europe, particularly France. Increasing awareness of transboundary drought events in Europe has resulted in a series of international research initiatives. For example, Phase 8 of the UNESCO IHP Programme has a focus on water security in a changing environment. In Europe, projects such as the ongoing EURO FRIEND-Water Low Flow and Drought (http://ne-friend.bafg.de/servlet/is/7402/, last access: 28 July 2020) and the related WATCH (2007–2011, http://www.eu-watch.org/, last access: 28 July 2020), DEWFORA (2011–2014), DROUGHT-R&SPI (2011–2014, https://www.wur.nl/en/show/drought-r-spi.htm, last access: 28 July 2020), and DrIVER (2014–2017, https://www.drought.uni-freiburg.de/, last access: 28 July 2020) have all focussed on water resources and drought. In the UK, the Natural Environment Research Council (NERC) commissioned the UK Drought and Water Scarcity Programme (2013–2019, https://nerc.ukri.org/research/funded/programmes/droughts/, last access: 28 July 2020), consisting of several projects that will support decision-making regarding droughts and water scarcity. The EURO FRIEND-Water Low Flow and Drought project has subsequently established an enduring Low Flow and Drought network (LFD network) with a focus on surface water droughts in Europe. However, there is currently no coordination of relevant groundwater information or groundwater drought research across Europe, and no groundwater drought status is monitored, or its impacts assessed at the continental scale.

In this context, a new pan-European project, the Groundwater Drought Initiative (GDI) has been established to bring together data holders of existing groundwater level data and researchers with an interest in groundwater drought across Europe. The main objectives of the GDI are to (i) assess groundwater drought status across Europe from 1960 to the present, (ii) assess and analyse the impacts of major groundwater droughts at the European scale, and to (iii) establish enduring international collaborations and partnerships to enable the development of impactful research into groundwater drought. This paper describes the background to the GDI and introduces the methodological approach that is being developed to assess groundwater status at the European scale (i.e. the first objective of the GDI), using selected groundwater level data from across Europe.

2 Background

2.1 Groundwater drought and impacts of groundwater droughts

Hydrological droughts propagate from a meteorological signal (precipitation deficit), through soils (soil moisture deficit), and, via reduced recharge, cause lowered groundwater levels. The spatio-temporal characteristics of the drought signal change as it passes through the terrestrial water cycle (Changnon, 1987; Peters et al., 2003; Tallaksen and Van Lanen, 2004; Van Loon, 2015). Due to the potentially large storage changes in aquifers, groundwater systems act to attenuate and add lags to meteorological drought signals, thus mitigating the effects of relatively short monthly to subseasonal droughts (Tallaksen and Van Lanen, 2004). However, groundwater resources are highly susceptible to prolonged (multi-seasonal to multi-annual) episodes of drought, where pooling, lagging and lengthening of multiple consecutive sub-seasonal or sub-annual drought periods can lead to extended drought events in groundwater systems (Van Loon and Van Lanen, 2012; Van Loon, 2015). The efficacy of these pooling, lagging and lengthening effects depends on catchment and aquifer characteristics (Van Lanen et al., 2013) and may be impacted by anthropogenic influences such as groundwater abstraction (Bloomfield and Marchant, 2013; Bloomfield et al., 2015; Tallaksen et al., 1997; Tallaksen and Van Lanen, 2004; Van Loon and Laaha, 2015; Van Loon et al., 2016).

The impacts of groundwater drought are wide-ranging and have important consequences for society and ecosystems. For example, public water supply, energy production and industry, agriculture and livestock farming, freshwater and terrestrial ecosystems, tourism and recreation may all be adversely impacted by major droughts (Lange et al., 2017; Stahl et al., 2016). Specifically, a reduction in deployable output from boreholes due to lower than normal groundwater levels during major droughts may also lead to costly restrictions to public supply, or to reduced abstractions for agriculture and industry, thus causing economic strain to society (Van Loon, 2015). Good examples of this are the last major groundwater drought in the UK in 2010–2012 when seven water companies in south and east England had to impose temporary use bans (colloquially known as "hosepipe bans") on about 20 million people, including the farming sector in spring of 2012, and the 2015 drought in Europe with dried up wells and boreholes leading to issues providing drinking water for cattle (e.g. in Germany; Van Lanen et al., 2016).

2.2 The challenge of developing a Europe-wide groundwater drought assessment

Improved understanding of drought propagation in similar aquifers (such as the Chalk of the Anglo-Paris Basin) and in a wider range of hydrogeological settings across Europe would provide insights into generic groundwater drought processes, opening the possibility for a systematic analysis of drought propagation from meteorological, soil moisture and surface water to groundwater droughts at the continental scale (Van Lanen et al., 2016), and thus contribute to improved preparedness and more informed decisions associated with the use of drought triggers.

During the 2015 event, the EURO FRIEND-Water LFD-network collated and analysed data to characterise meteorological and streamflow drought conditions, and impacts across Europe (Ionita et al., 2017; Laaha et al., 2017; Van Lanen et al., 2016). However, an equivalent data synthesis and analysis of the 2015 groundwater drought situation was not possible due to the lack of co-ordination of groundwater level data and research across Europe (Laaha et al., 2017; Van Lanen et al., 2016; Van Loon et al., 2017). This highlights a major gap in European drought research capability. In addition, there have recently been initiatives to better inform policy makers, water managers, and the general public of drought status, for example in Europe through on-line monitoring services such as the European Drought Observatory (EDO, https://edo.jrc.ec.europa.eu/edov2/php/index.php?id=1000, last access: 28 July 2020) and the European Drought Centre (EDC, http://europeandroughtcentre.com/, last access: 28 July 2020). The EDO is mapping droughts and drought indicators but does not include groundwater droughts. EDC hosts the European Drought Reference (EDR) database, a collection of data on historic European meteorological and streamflow droughts, and a web-based viewer of meteorological drought status. EDC also hosts the European Drought Impact Report Inventory (EDII), an account of documented impacts of droughts. However, the EDII does not distinguish impacts of groundwater droughts and up to date, there is no attempt to establish a stand-alone database for the groundwater domain.

In summary, despite some early attempts to highlight the importance of research into groundwater droughts across Europe (Hisdal et al., 2004; Peters et al., 2001; Van Lanen and Peters, 2000) recent projects and initiatives mentioned above lack a systematic approach of clearly integrating groundwater droughts into their drought analysis. With few exceptions (Van Loon et al., 2013), they focus instead on droughts in the driving meteorology and consequences of droughts only in the soil and surface water compartments of the terrestrial water cycle. Therefore, the following research gaps can been identified: (i) although the skills are present across Europe, there is a major gap in European drought research capability related to groundwater droughts, primarily due to a lack of co-ordination between hydrogeologists and hydrologists with appropriate expertise (Staudinger et al., 2019); (ii) there is a gap in research capability to provide a timely, integrated and consistent overview of groundwater level data at the continental scale under conditions of drought, and hence to assess the status of groundwater drought in a manner comparable with other hydrometeorological drought signals; and, (iii) there has to date been no systematic attempt to undertake an assessment of the impacts of groundwater droughts at the continental scale.

3 The European Groundwater Drought Initiative

The GDI aims to deliver the first pan-European assessment of groundwater drought status ($\sim$ 1960 to present), and will analyse the most recent European groundwater droughts, thereby addressing the research gaps identified above. The GDI is collaborating on a state-by-state basis with dataholders and researchers to collect available groundwater level data and site characteristics (including human influence) from unconfined aquifers across Europe. In this section, we introduce the GDI workflow for groundwater drought assessment with examples from four different European countries.

3.1 Study sites

Sixteen sites from four countries (Austria, France, UK, and Sweden) have been chosen for analysis and to illustrate the GDI workflow. The main selection criteria for these sites were the availability of relatively long, mostly continuous series of groundwater level data from unconfined aquifers and the representation of different hydrogeological settings. For a brief comparison between aquifer settings, we additionally chose two sites from confined aquifers (SW3 and SW4). The spatial distribution of the chosen sites, Fig. 1, covers different parts of France and Austria (though with one pair of neighbouring boreholes respectively, namely FR1 and FR2 in France, and AU3 and AU4 in Austria), and the southern parts of the UK and Sweden. The aquifer types differ across the sites (Table 1), and include fissured igneous and metamorphic rocks, fractured limestone aquifers such as the Chalk of the Anglo-Paris basin, consolidated sandstones and superficial glacio-fluvial and alluvium deposits. All groundwater level data were obtained from publicly available databases of groundwater levels for each country respectively (Austria: https://ehyd.gv.at/, last access: 28 July 2020, France: https://ades.eaufrance.fr/Recherche/Index/Piezometre?g=d5acb7 (last access: 28 July 2020), Sweden: https://www.sgu.se/produkter/geologiska-data/oppna-data/grundvatten-oppna-data/grundvattennivaer-tidsserier/ (last access: 28 July 2020), UK: https://www.bgs.ac.uk/products/hydrogeology/WellMaster.html, last access: 28 July 2020).

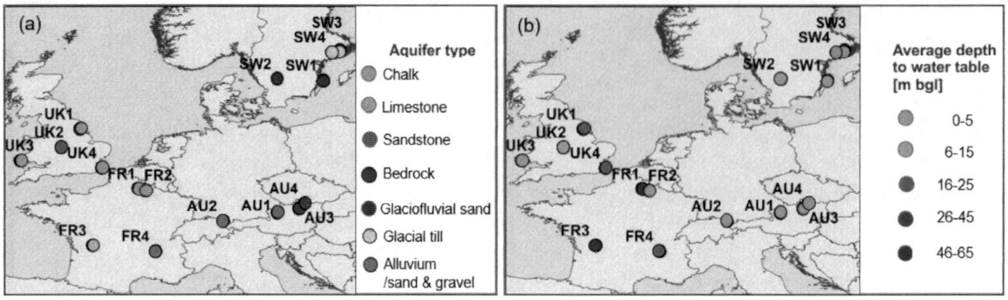

Figure 1. Site map of groundwater gauges categorized by **(a)** aquifer type, and **(b)** average depth to water table in metres below ground level (m b.g.l.). The average depth to water table was calculated based on simulated values of the site's groundwater levels from 1955–2018. Map image is the intellectual property of Esri and is used herein under license. Copyright © 2019 Esri and its licensors. All rights reserved.

Table 1. Table showing aquifer types per site, average water level in metres below ground level (m bgl; average depth was calculated based on the interpolated groundwater levels 1955–2018), maximum cross-correlation (CC) between SPI and SGI, autocorrelation range (m_{max}) of the SGIs, and the accumulation period (q_{max}) with maximum cross-correlation between SPI and SGI.

ID	Aquifer type	Average WL (m b.g.l.)	Max CC	m_{max} (months)	q_{max} (months)
FR1	Chalk	61	0.78	20	14
FR2	Chalk	14	0.56	28	8
FR3	Limestone	48	0.76	13	4
FR4	Alluvium (glaciofluvial)	25	0.74	21	20
AU1	Gravel & sand	3	0.83	5	3
AU2	Gravel & sand	2	0.6	28	2
AU3	Fissured igneous & metamorphic rock	2	0.75	18	11
AU4	Fissured igneous & metamorphic rock	5	0.64	20	11
SW1	Glaciofluvial sand	6	0.27	28	14
SW2	Glaciofluvial sand	8	0.65	28	15
SW3	Glacial till	2	0.73	8	7
SW4	Glacial till	1	0.63	13	11
UK1	Chalk	20	0.73	24	33
UK2	Sandstone	11	0.59	28	6
UK3	Limestone	14	0.79	5	3
UK4	Chalk	19	0.81	14	12

3.2 Methodology

Figure 2 shows the overall workflow of the GDI. It is based on the analysis method developed by Marchant and Bloomfield (2018) that enables irregularly observed groundwater level data from a range of locations and variety of periods of observations to first be regularised into a common analysis frame, then standardised, and finally then to be used for characterisation of spatio-temporal variation in extracted episodes of groundwater drought. The raw time series of groundwater level data for each site is first pre-processed to provide a regularised monthly groundwater level time series using newly developed R-scripts run on a SHINY-interface. The regularisation step entails harmonisation of the observed groundwater levels to regular monthly time steps and stochastic modelling of groundwater levels (Marchant et al., 2016) to infill data gaps using the temporal correlation amongst the data and a modelled relationship with monthly precipitation data, obtained from the E-OBS gridded dataset (Haylock et al., 2008). This is achieved by estimating and predicting from a linear mixed model of the groundwater levels where the fixed effects are an impulse response function of rainfall plus a seasonal term. Random effects are temporally correlated according to a Matérn function – see Marchant and Bloomfield (2018) for a fuller description.

A variant of the Standardised Precipitation Index (SPI), the Standardised Groundwater level Index (SGI) approach (Bloomfield and Marchant, 2013; McKee et al., 1993; Svoboda and Fuchs, 2016) is then used to standardise the groundwater level data. The SGI is a non-parametric approach, which assigns a standardised value to the reconstructed groundwater levels for each month, thereby resulting in normalised monthly indices that allow for comparison across the sites (see Bloomfield and Marchant, 2013). Additionally,

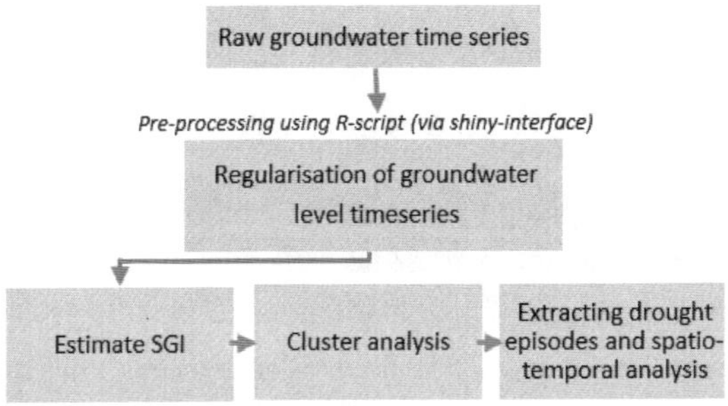

Figure 2. Methodology of data analysis.

cluster analysis using the k-means approach is undertaken to obtain information about similarities between hydrographs across the study area. Due to the limited number of sites that have been modelled and analysed in the present study, only two clusters have been identified to illustrate the approach. As a more extensive set of hydrographs becomes available, the number of clusters used may be increased to help identify spatially coherent groundwater hydrographs and hence coherent groundwater drought behaviour. Simple statistics are applied to each site for comparison of drought frequency and duration across the sixteen sites. Drought is thereby defined as intervals with SGI < 0, in which there is at least one occurrence of SGI ≤ -1, following the convention of the World Meteorological Organization (Svoboda et al., 2012).

3.3 Preliminary results

Data availability for the selected sites varied and ranged between near-complete time series from 1955–2018 (UK1) to datasets with observations starting only in the 1990s (AU3, AU4, UK3). An example of the driving precipitation data, the observed groundwater levels, and the regularised groundwater level time series for three of the sites is shown in Fig. 3, demonstrating different degrees of flashiness (response of the aquifer to precipitation events) across the sites.

Table 1 shows cross-correlation (CC) between the estimated SGI for each of the sixteen sites and the corresponding SPI for the period 1955 to 2018. There was a reasonably strong correlation (> 0.5) between the SGI time series and the corresponding SPI series accumulated over an appropriate number of months for all sites except SW1. These accumulation periods ($q_{\max}$), which maximise the correlation, were determined for each site and ranged 2–33 months. The CC can potentially be improved by selecting a shorter time period with a later onset, and by choosing sites with more homogenously distributed observations across time, thereby avoiding decadal gaps in the data. Table 1 also shows the maximum autocorrelation range ($m_{\max}$) in the estimated SGI, a measure of the "memory" of groundwater levels at each site, ranging 5–28 months.

Table 2. Summary table of number of droughts and drought duration (in months) for each site for the period 1955–2018.

ID	Number of droughts	Shortest drought (months)	Longest drought (months)	Median duration (months)
FR1	13	2	48	19
FR2	14	3	43	14
FR3	26	2	48	7
FR4	13	9	44	16
AU1	35	2	23	6
AU2	34	1	48	6
AU3	15	3	50	24
AU4	22	5	52	9
SW1	4	34	81	64
SW2	10	4	49	24
SW3	28	1	24	7
SW4	22	2	24	8
UK1	13	5	64	11
UK2	5	24	95	38
UK3	40	1	22	6
UK4	10	4	59	22

The SGI time series for the sixteen sites (Fig. 4a) demonstrate the different response patterns to driving meteorology. Some of the sites are characterised by fewer, but longer droughts (e.g. SW1, SW2, UK2) and others exhibit more frequent, but shorter drought periods (e.g. AU1, UK3). This is also reflected in differences in $m_{\max}$ and $q_{\max}$, which are lower for the sites with frequent, short droughts, and higher for the sites with less frequent, longer droughts (Table 1).

The k-means clustering in two groups (Fig. 5) highlights the divide between slow- and fast-responding systems (Fig. 4b), Cluster 1 and Cluster 2 respectively. Even though only a limited number of sites are included in this analysis, there appears to be some spatial coherence of the hydrographs based on the two clusters. For example, there appears to be spatial coherence among the Austrian sites. However, the identification of spatial coherence in the clusters needs to be tested with the inclusion of more hydrographs in future analyses. In addition to the clustering capturing systematic differences in the responsiveness or flashiness of the hydrographs, the clusters also appear to reflect a systematic difference in the degree to which sites in each cluster were impacted by the drought in the early 1990s, with Cluster 1 being highly impacted and Cluster 2 showing no significant impact of drought. This can be seen both in the individual SGI hydrographs (Fig. 4a) and in the differences in the mean hydrographs of each cluster (Fig. 4b). In contrast, a major drought episode in the summer of 1976 is present in both clusters and most of the sites.

Drought statistics related to drought frequency, duration and intensity can be extracted from the prepared SGI series. For illustration, Table 2 summarises drought frequency and

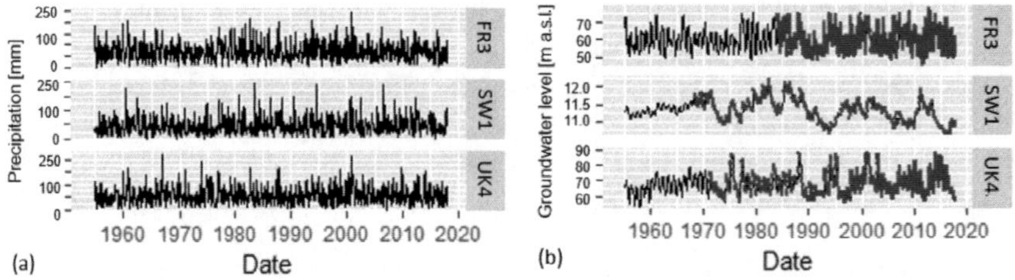

Figure 3. **(a)** Time series of precipitation (mm/month), and **(b)** observed (red) and simulated (black) time series of groundwater levels for selected sites (FR3, SW1, and UK4) in metres above sea level (m a.s.l.). Daily precipitation data was extracted from the E-OBS dataset and used to calculate the monthly accumulated rainfall for each site.

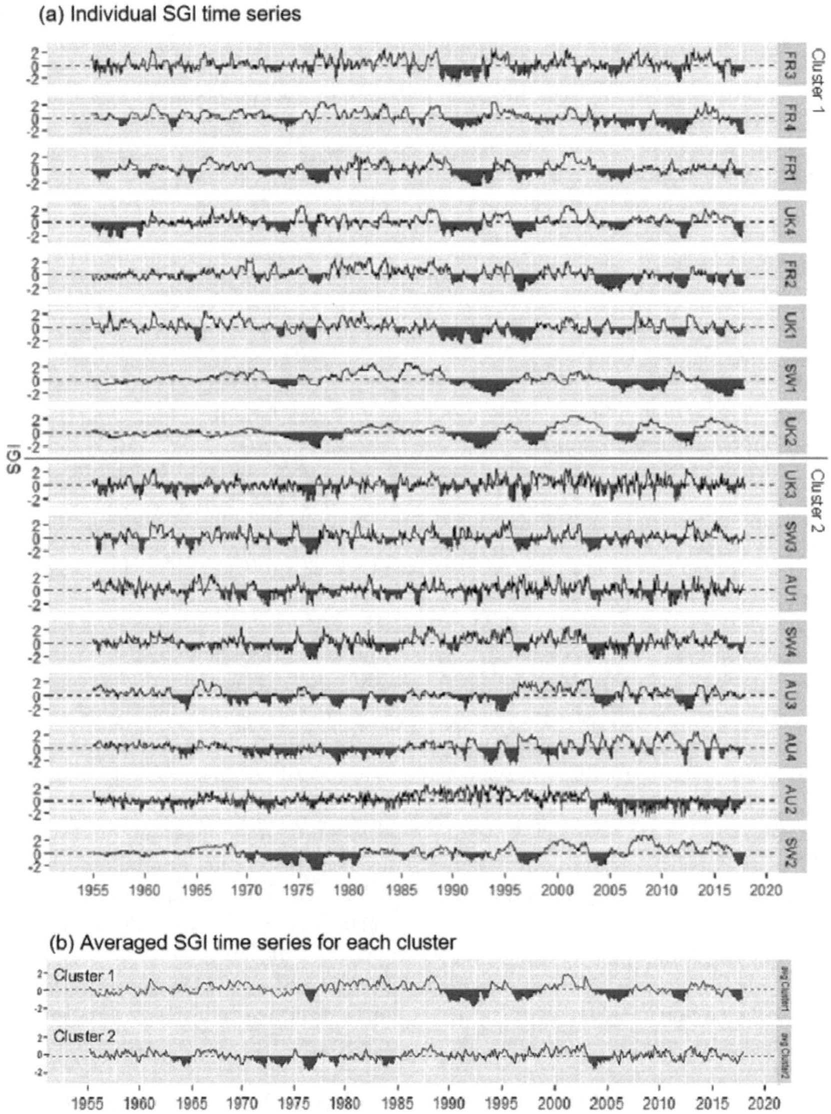

Figure 4. Combined graphic showing **(a)** the Standardized Groundwater level Index (SGI) for each site, and **(b)** the time series of calculated mean SGI values for each cluster, deriving from k-means clustering.

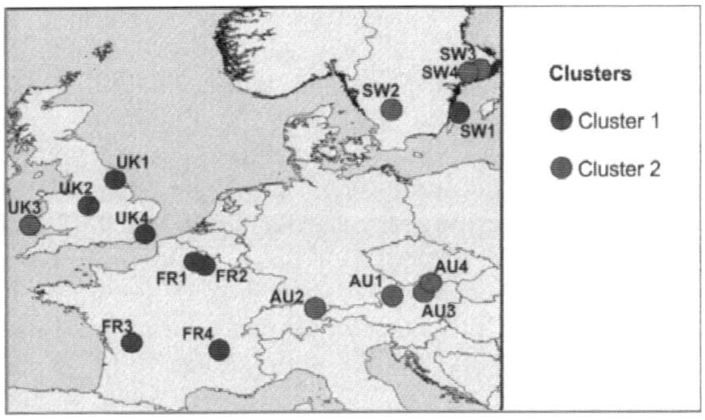

Figure 5. Geographical distribution of determined clusters. Map image is the intellectual property of Esri and is used herein under license. Copyright © 2019 Esri and its licensors. All rights reserved.

duration for each site. It shows a total number of droughts per site between as few as four 4 (SW1) to as many as 40 (UK3). It is noteworthy that the median drought duration differs substantially between aquifer types, ranging for example from only 6–7 months at limestone (FR3 and UK3) and gravel/sand sites (AU1 and AU2), to 11–22 at Chalk sites (FR1, FR2, UK1, and UK4), and 24–64 months in glaciofluvial sand aquifers (SW1, SW2); thereby confirming the importance of aquifer type in drought propagation as previously observed in other studies (Bloomfield and Marchant, 2013; Bloomfield et al., 2015; Tallaksen et al., 1997; Tallaksen and Van Lanen, 2004; Van Lanen et al. 2013; Van Loon and Laaha, 2015; Van Loon et al., 2016).

4 Summary and future outlook

In this paper, the methodological approach for a pan-European groundwater drought assessment was presented and applied to selected sites across four countries and different hydrogeological settings across Europe. The analysis shows that groundwater level time series can be successfully regularised using a stochastic model, and that the standardisation procedure using the Standardised Groundwater level Index (SGI) allows for comparison between sites. Varying strength of observed relationships between drought characteristics of the sites were found in terms of geographical proximity and between common types of aquifers; thereby supporting the theory that hydrogeological settings play an important role in groundwater drought propagation.

Further improvement to the analysis procedure will be undertaken in the future workflow of the GDI, such as using more advanced clustering techniques and advances in the methodology for groundwater drought event extraction from the dataset. This can be done for example by using sequent peak algorithms to collate separate drought events during which the aquifer has not fully recovered yet from the first drought episode. An improved analysis of drought events, for example by using mid-points of droughts and calculating the percentage of sites in drought at a given time, will allow for a better description of their spatio-temporal distribution and evolution.

Although the focus of the analysis is on sites relatively unaffected by abstraction or irrigation returns, identification of sites with poor correlation between precipitation and groundwater will present an opportunity for research into anthropogenic effects on groundwater droughts to allow distinction between climate and human affected droughts. In addition, it is hoped that comparative studies of the impacts of groundwater drought between semi-arid regions, e.g. southern Europe, and more temperate north-western Europe could provide additional, complementary insights into to the impacts of extremely low groundwater levels during droughts, thus informing future drought management planning across Europe.

The methodology used in the GDI is generic and could also be employed beyond Europe to other geographic regions. The approach is equally applicable to high groundwater level stands. Consequently, generic insights from the GDI will be of interest to the international groundwater research community, particularly to those interested in better understanding of extreme events in groundwater.

Data availability. The data presented here are from an ongoing and yet incomplete project. The datasets generated during the current study are available on request from the corresponding authors.

Author contributions. JPB, DMH and AFVL conceptualised overarching research aims and secured funding, BPM, JPB and DC developed methods and models, DC performed numeric modelling and statistical analysis, HB, BT, and GH provided data and expertise, BB collated and visualised data and wrote the original draft, all co-authors reviewed and provided feedback.

Competing interests. The authors declare that they have no conflict of interest.

Special issue statement. This article is part of the special issue "Hydrological processes and water security in a changing world". It is a result of the 8th Global FRIEND-Water Conference: Hydrological Processes and Water Security in a Changing World, Beijing, China, 6–9 November 2018.

Acknowledgements. Bentje Brauns, John P. Bloomfield, Christopher Jackson, Ben P. Marchant, and Daniela Cuba publish this paper with the permission of the Executive Director of the British Geological Survey (UK Research and Innovation, UKRI). We acknowledge the E-OBS dataset from the EU-FP6 project ENSEMBLES (http://ensembles-eu.metoffice.com/, last access: 28 July 2020) and the data providers in the ECA&D project (http://www.ecad.eu, last access: 28 July 2020).

Financial support. This research has been supported by the Natural Environment Research Council (grant no. NE/R004994/1).

References

Bloomfield, J. P. and Marchant, B. P.: Analysis of groundwater drought building on the standardised precipitation index approach, Hydrol. Earth Syst. Sci., 17, 4769–4787, https://doi.org/10.5194/hess-17-4769-2013, 2013.

Bloomfield, J. P., Marchant, B. P., Bricker, S. H., and Morgan, R. B.: Regional analysis of groundwater droughts using hydrograph classification, Hydrol. Earth Syst. Sci., 19, 4327–4344, https://doi.org/10.5194/hess-19-4327-2015, 2015.

Changnon, S. A.: Detecting drought conditions in Illinois, Illinois State Water Survey, Champaign, USA, available at: https://www.isws.illinois.edu/pubdoc/C/ISWSC-169.pdf (last access: 24 July 2020), Circular, 169, 1987.

European Environment Agency (EEA): Groundwater quality and quantity in Europe, European Environment Agency, Copenhagen, Denmark, 123 pp., 1999.

Haylock, M. R., Hofstra, N., Klein Tank, A. M. G., Klok, E. J., Jones, P. D., and New, M.: A European daily high-resolution gridded dataset of surface temperature and precipitation, J. Geophys. Res., 113, D20119, https://doi.org/10.1029/2008JD010201, 2008.

Hisdal, H., Clausen, B., Gustard, A., Peters, E., and Tallaksen, L. M.: Event definitions and indices, in: Hydrological droughtprocesses and estimation methods for streamflow and groundwater, Developments in water sciences 48, edited by: Tallaksen, L. M. and Van Lanen, H. A. J., Elsevier Sciences B.V., Amsterdam, Netherlands, 139–198, 2004.

Ionita, M., Tallaksen, L. M., Kingston, D. G., Stagge, J. H., Laaha, G., Van Lanen, H. A. J., Scholz, P., Chelcea, S. M., and Haslinger, K.: The European 2015 drought from a climatological perspective, Hydrol. Earth Syst. Sci., 21, 1397–1419, https://doi.org/10.5194/hess-21-1397-2017, 2017.

Laaha, G., Gauster, T., Tallaksen, L. M., Vidal, J.-P., Stahl, K., Prudhomme, C., Heudorfer, B., Vlnas, R., Ionita, M., Van Lanen, H. A. J., Adler, M.-J., Caillouet, L., Delus, C., Fendekova, M., Gailliez, S., Hannaford, J., Kingston, D., Van Loon, A. F., Mediero, L., Osuch, M., Romanowicz, R., Sauquet, E., Stagge, J. H., and Wong, W. K.: The European 2015 drought from a hydrological perspective, Hydrol. Earth Syst. Sci., 21, 3001–3024, https://doi.org/10.5194/hess-21-3001-2017, 2017.

Lange, B., Holman, I., and Bloomfield, J. P.: A framework for a joint hydro-meteorological-social analysis of drought, Sci. Total Environ., 578, 297–306, https://doi.org/10.1016/j.scitotenv.2016.10.145, 2017.

Marchant, B. P., Mackay, J., and Bloomfield, J. P.: Quantifying uncertainty in predictions of groundwater levels using formal likelihood methods, J. Hydrol., 540, 699–711, https://doi.org/10.1016/j.jhydrol.2016.06.014, 2016.

Marchant, B. P. and Bloomfield, J. P.: Spatio-temporal modelling of the status of groundwater droughts, J. Hydrol., 564, 397–413, https://doi.org/10.1016/j.jhydrol.2018.07.009, 2018.

McKee, T. B., Doesken, N. J., and Kleist, J.: The relationship of drought frequency and duration to time scales, in: Proceedings of the 8th Conference on Applied Climatology, Anaheim, USA, 17–22 January 1993, 179–184, 1993.

Peters, E., Torfs, P. J. J. F., Van Lanen, H. A. J., and Bier, G.: Propagation of drought through groundwater – a new approach using linear reservoir theory, Hydrol. Process., 17, 3023–3040, https://doi.org/10.1002/hyp.1274, 2003.

Stahl, K., Kohn, I., Blauhut, V., Urquijo, J., De Stefano, L., Acácio, V., Dias, S., Stagge, J. H., Tallaksen, L. M., Kampragou, E., Van Loon, A. F., Barker, L. J., Melsen, L. A., Bifulco, C., Musolino, D., de Carli, A., Massarutto, A., Assimacopoulos, D., and Van Lanen, H. A. J.: Impacts of European drought events: insights from an international database of text-based reports, Nat. Hazards Earth Syst. Sci., 16, 801–819, https://doi.org/10.5194/nhess-16-801-2016, 2016.

Staudinger, M., Stoelzle, M., Cochand, F., Seibert, J., Weiler, M., and Hunkeler, D.: Your work is my boundary condition!: Challenges and approaches for a closer collaboration between hydrologists and hydrogeologists, J. Hydrol., 571, 235–243, https://doi.org/10.1016/j.jhydrol.2019.01.058, 2019.

Svoboda, M., Hayes, M., and Wood, D.: Standardized precipitation index user guide, World Meteorological Organization, Geneva, Switzerland, 24 pp., 2012.

Svoboda, M. and Fuchs, B. A.: Handbook of drought indicators and indices, in: Integrated Drought Management Programme (IDMP), integrated Drought Management Tools and Guidelines Series 2, World Meteorological Organization (WMO) and Global Water Partnership (GWP), Geneva, Switzerland, 52 pp., 2016.

Tallaksen, L. M., Madsen, H., and Clausen, B.: On the definition and modelling of streamflow drought duration and deficit volume, Hydrol. Sci. J., 42, 15–33, https://doi.org/10.1080/02626669709492003, 1997.

Tallaksen, L. M. and Van Lanen, H. A. J. (Eds.): Hydrological drought: processes and estimation methods for streamflow and groundwater, Elsevier, Oxford, UK, 579 pp., 2004.

Van Lanen, H. A. J. and Peters, E.: Definition, effects and assessment of groundwater droughts, in: Drought and Drought Mitigation in Europe, Advances in Natural and Technological Hazards Research, edited by: Vogt, J. V. and Somma, F., Springer, Dordrecht, Netherlands, 14, 49–61, https://doi.org/10.1007/978-94-015-9472-1_4, 2000.

Van Lanen, H. A. J., Wanders, N., Tallaksen, L. M., and Van Loon, A. F.: Hydrological drought across the world: impact of climate and physical catchment structure, Hydrol. Earth Syst. Sci., 17, 1715–1732, https://doi.org/10.5194/hess-17-1715-2013, 2013.

Van Lanen, H. A. J., Laaha, G., Kingston, D. G., Gauster, T., Ionita, M., Vidal, J. P., Vlnas, R., Tallaksen, L. M., Stahl, K., Hannaford, J., Delus, C., Fendekova, M., Mediero, L., Prudhomme, C., Rets, E., Romanowicz, R. J., Gailliez, S., Wong, W. K., Adler, M. J., Blauhut, V., Caillouet, L., Chelcea, S., Frolova, N., Gudmundsson, L., Hanel, M., Haslinger, K., Kireeva, M., Osuch, M., Sauquet, E., Stagge, J. H., and Van Loon, A. F.: Hydrology needed to manage droughts: the 2015 European case, Hydrol. Process., 30, 3097–3104, https://doi.org/10.1002/hyp.10838, 2016.

Van Loon, A. F. and Van Lanen, H. A. J.: A process-based typology of hydrological drought, Hydrol. Earth Syst. Sci., 16, 1915–1946, https://doi.org/10.5194/hess-16-1915-2012, 2012.

Van Loon, A. F. and Van Lanen, H. A. J.: Making the distinction between water scarcity and drought using an observation-modeling framework, Water Resour. Res., 49, 1483–1502, https://doi.org/10.1002/wrcr.20147, 2013.

Van Loon, A. F.: Hydrological drought explained, Wiley Interdiscip. Rev. Water, 2, 359–392, https://doi.org/10.1002/wat2.1085, 2015.

Van Loon, A. F. and Laaha, G.: Hydrological drought severity explained by climate and catchment characteristics, J. Hydrol., 526, 3–14, https://doi.org/10.1016/j.jhydrol.2014.10.059, 2015.

Van Loon, A. F., Gleeson, T., Clark, J., Van Dijk, A. I., Stahl, K., Hannaford, J., Di Baldassarre, G., Teuling, A. J., Tallaksen, L. M., and Uijlenhoet, R.: Drought in the Anthropocene, Nat. Geosci., 9, 89, https://doi.org/10.1038/ngeo2646, 2016.

Van Loon, A. F., Kumar, R., and Mishra, V.: Testing the use of standardised indices and GRACE satellite data to estimate the European 2015 groundwater drought in near-real time, Hydrol. Earth Syst. Sci., 21, 1947–1971, https://doi.org/10.5194/hess-21-1947-2017, 2017.

Proc. IAHS, 383, 307–314, 2020
https://doi.org/10.5194/piahs-383-307-2020

Open Access
PIAHS

Trends in national and regional scale drought in New Zealand

Daniel G. Kingston and Eleanor J. Treadwell

Department of Geography, University of Otago, Dunedin, P.O. Box 56, New Zealand

Correspondence: Daniel G. Kingston (daniel.kingston@otago.ac.nz)

Published: 16 September 2020

Abstract. Drought is a critical natural hazard in New Zealand, affecting both agricultural production and hydroelectric generation. However, the detection and characterisation of drought events are complicated by the range of different drought metrics available. Here, time-series variation in drought is investigated for New Zealand using two commonly used indices: the Standardized Precipitation Index (SPI) and the Standardized Precipitation Evapotranspiration Index (SPEI). Both the SPI and SPEI are calculated from WATCH Forcing ERA-Interim data (a bias-corrected version of ERA-Interim). Analyses are based on the proportion of New Zealand experiencing at least "moderate" drought (i.e. index values at the threshold of -1 or lower). There is no statistically significant difference in area in drought between the two indices averaged over the time series, and no national-scale trends in drought occurrence are identified. However, a statistically significant trend of divergence between the SPI and SPEI is present when examining drought at moderate spatial coverage (10 % and 20 % coverage based on SPEI), with the SPEI showing increasingly greater drought coverage. These typically become more apparent either as a major drought event is building or terminating. As such, the SPEI generally indicates an extended duration (rather than higher peak extent) of drought events. The spatial expression of the largest SPI-SPEI differences reveals "hotspots" in terms of both average and divergent trends in drought extent, results which are reflected in a regional drought analysis using a cluster analysis of the SPI and SPEI area in drought. In particular, SPEI drought coverage is greater in the South Island east coast and central/southwestern North Island regions. Conversely, SPI drought extent is greater in the eastern North Island. Overall, the differences and similarities in drought extent between the SPI and SPEI can be linked to the highly varied nature of New Zealand climate, with the instances of greater SPEI drought located in regions of driest and most continental climate.

1 Introduction

The identification and characterisation of drought events is a challenging research topic. There are multiple ways to characterise drought occurrence, which leads to uncertainty in the exact magnitude of individual events and makes comparison across different studies, places and times both uncertain and complex. In turn, these difficulties defining the nature of a drought event lead to further challenges understanding the causes of drought initiation, severity, duration and cessation. Characterisation of drought events and subsequent analysis of the causes of drought is a particularly important research challenge given the multi-sectoral and economically costly impacts of this natural hazard (e.g. Van Lanen et al., 2016). In New Zealand, past drought events have had substantial impacts on generation of electricity from hydropower (the 1992 "electricity crisis"; Fitzharris, 1992), on the water supply of the largest city (Auckland in 1994), and on agricultural production – the 2012/2013 drought depressed GDP by 0.7 %–0.9 % (MPI, 2013).

Perhaps the first aspect of drought characterisation is its category: meteorological (a lack of precipitation), hydrological (a lack of river water or other surface water resource), or agricultural (a lack of soil moisture) (Wilhite, 2000; Tallaksen and Van Lanen, 2004; Mishra and Singh, 2010). Henceforth only meteorological drought will be the focus here – although (initially at least) the least severe in terms of absolute water shortage, meteorological drought often has the most widespread and complete data availability for drought characterisation. However even when only considering meteorological drought, multiple indices have been used to characterise drought – each with their own particular focus in terms of aspect of drought that is captured. For example, the Standardised Precipitation Index (SPI) has been used widely. By using standardised precipitation the index is easily comparable between areas of different precipitation climate, and the index is also multiscalar – i.e. it can be calculated at different timescales. The SPI has been subject to rigorous investigation and has previously been recommended by the World Meteorological Organization due to its simplicity and multiscalar nature (WMO, 2012). However, whilst the standardised and multiscalar nature of the SPI may provide a widely applicable and representative drought index, the SPI does not include representation of evaporation, thus omitting a potentially important contributor to meteorological drought. A more recent development in terms of drought characterisation is the Standardised Precipitation Evaporation Index (SPEI, Vicente-Serrano et al., 2010). As with the SPI, the SPEI is multi-scalar and readily comparable across climate zones, and by including standardised precipitation minus potential evaporation this index includes a representation of the meteorological water balance more equivalent to the PDSI. Notwithstanding uncertainties associated with the method used to calculate PET (e.g. Sheffield et al., 2012), the SPEI has become a valuable addition to the range of available meteorological drought indices.

The implications of including PET in any measure of drought is thought to be particularly important within the context of climate change, where under a more energetic climate system evaporation may play an increasing role in the water balance – and so drought occurrence – for many regions. As such, there is potential for increasing levels of divergence between drought indices that do and do not include PET. There is evidence that indicates such divergence is already occurring for European-scale drought (Stagge et al., 2017), but this possibility has not been widely explored (particularly outside Europe). Here, this topic is advanced through an investigation of national- and regional-scale time series variation in drought occurrence as measured by the SPI and SPEI in New Zealand. New Zealand represents a particularly interesting case study of the relative performance of these two indices, given that it encompasses a wide range of climatic zones, from subtropical to polar, alpine to lowland, maritime to continental, and on a national scale has experienced changes in mean annual temperature approximately similar to the global average (Reisinger et al., 2014).

2 Data and methods

2.1 Climate data

The gridded SPI and SPEI time series data were generated following the procedure described by Stagge et al. (2015). Both indices were generated using the Watch Forcing ERA-Interim data set (WFDEI) which was created by applying a bias-correction to the ERA-Interim reanalysis data at a spatial resolution of $0.5 \times 0.5°$ latitude/longitude (Weedon et al., 2014). At this resolution, a total of 164 grid cells cover the three main islands of New Zealand (North, South and Stewart). The SPI and SPEI data were calculated for an accumulation period of six monthly time steps, that is SPI-6 and SPEI-6, over the 1979–2010 period. This follows previous (Europe-based) analysis, which indicated that this time step generally approximates best hydrological drought caused by both summer rainfall deficits and winter snow-based drought (Van Loon and Van Lanen, 2012), drought types which are important across New Zealand.

2.2 Drought identification

The lack of quantifiable characteristics and non-structural nature of drought (Spinoni et al., 2017; Mishra and Singh, 2010) makes the agreement of a universally accepted definition impractical (Wilhite, 2000). In order to establish a water balance deficit, pre-defined standard thresholds from McKee et al. (1993) were used to identify drought events, beginning when the index falls below 0. Different threshold levels of SPI and SPEI values were analysed, with -1 (a "moderate drought", according to McKee et al., 1993) determined as the most appropriate threshold in terms of maximizing the sample of drought events available for analysis whilst restricting the sample to covering only the more substantial events.

In order to identify national-scale drought events for the SPI and SPEI, binary occurrence was used for the presence/absence of drought within each cell using the -1 threshold level. The total number of cells (max. 164) with the SPI(t) or SPEI(t) ≤ -1 ($t = 1$, 384 months) was used as the basis for a national time series of area in drought at time t. This time series was the focus of all subsequent analysis.

2.3 Spatial variation in drought occurrence

In addition to the national-scale analysis, regions of similar SPI and SPEI time series variation were identified, and drought occurrence was subsequently analysed on a region-by-region basis. Regional analysis can be particularly insightful in New Zealand due to the latitudinal range of the country combined with diverse orography, resulting in strong regional variation in both temperature and precipitation (e.g. Salinger and Mullan, 1999). Hierarchical cluster analysis using the Ward's clustering algorithm was used to establish regions of similar time series variation. The final clustering solution was determined following analysis of the clustering dendrogram, scree plots of the clustering agglomeration coefficient, as well as the physical interpretability of the spatial distribution of the regions. Once identified, the -1 SPI and SPEI threshold was applied to each grid cell in each cluster to generate an area in drought time series for each cluster which then formed the basis for subsequent analysis.

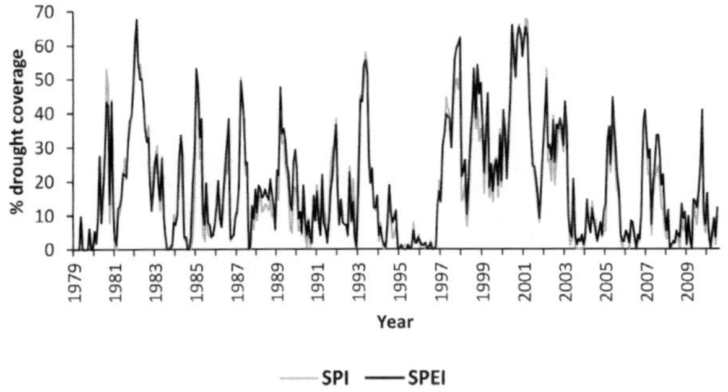

Figure 1. New Zealand percent area covered by SPI and SPEI values ≤ -1.

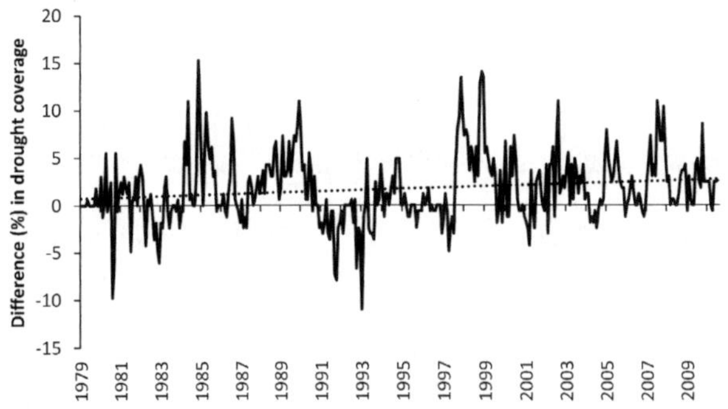

Figure 2. SPEI-SPI difference in percentage area in drought. The dashed line indicates the linear trend fitted to the time series.

3 Results

3.1 National area in drought

There is strong year-to-year variation in the spatial extent of drought conditions (ranging 0 % to 68 % coverage), with a relatively drought-free period between 1994–1996 followed by a longer period of enhanced drought extent (Fig. 1). There is no statistically significant (i.e. $p \leq 0.05$) trend in percentage area in drought for either the SPI or SPEI based on the Mann-Kendall test. Neither is there a statistically significant difference in the average area in drought between the indices over the whole 1979–2010 time series (Mann-Whitney test). However, when considering droughts of differing minimum spatial coverage, more substantial differences occur: for times with at least 10 % and 20 % of New Zealand in drought according to the SPEI, the median SPEI area in drought is significantly higher than the equivalent SPI coverage (by 3.7 % and 2.4 % respectively). There are no significant differences between indices for area in drought with a minimum coverage of 30 %, 40 % or 50 % of New Zealand.

In addition to trends in droughts of different minimum coverage, there is a trend of increasing divergence between the entire SPEI and SPI drought area time series (Fig. 2), although the average over entire time series is not significantly different. The SPEI – SPI difference in percentage drought area increases over time, a trend that is statistically significant at $p < 0.01$. Equivalent trends in divergence are also evident for droughts with at least 10 %, 20 % and 30 % coverage (but not > 40 % or 50 % coverage). The average difference of area in drought between indices also approximately doubles between the first and second halves of the data set (a difference of 1.10 % for the first 15 full years, 2.34 % for the second 15 years). Furthermore, there is a reduction in the magnitude of events where SPI drought area is greater than SPEI area. For example, there are seven months where SPI drought covers > 5 % more of New Zealand than the SPEI in the first half of the time series, and none in the second half (Fig. 2).

In addition to the increasing divergence over the study period, there are a series of relatively large event-specific differences. The greatest difference between indices occurred during the 1985–1986 drought event, and specifically during the development rather than at the peak of this event. The SPEI event began one month earlier than for the SPI, leading to a 15 % difference in national area in drought by the second (first) month of the SPEI (SPI) drought event. There are ten further instances of > 10 % difference in drought coverage: these are most commonly associated with either SPEI events beginning earlier or retreating more slowly than SPI events. In addition, there are two instances where the peak coverage of the drought event is substantially greater for the SPEI.

There is a relatively consistent spatial pattern of differences between occurrence and magnitude of drought across all eleven drought events where SPEI coverage is more than 10 % greater than for the SPI. Specifically, the east coast of the South Island is generally drier according to the SPEI, as is the central-southern North Island, whereas drought in the far south and the central West Coast are typically less severe for the SPEI compared to SPI (Fig. 3). This pattern is highly consistent across all eleven of these high-difference events.

3.2 Regional variation

Analysis of clustering dendrograms, agglomeration coefficients and spatial expression led to the identification of six clusters for both the SPI and SPEI datasets (Fig. 4). Although both indices result in the same number of clusters, there are some differences in spatial expression. There are three particular differences of note. Firstly, the East Cape area of the North Island comprises a separate cluster (Cluster 6, C6) for the SPI, whereas for the SPEI C6 extends from this eastern region to the far north of the island. Secondly, the South Island West Coast cluster extends further inland for the SPI compared to the SPEI (C2), indicating a split at the Main Divide of the Southern Alps for the SPEI, but also including the downwind precipitation "spillover" zone under the SPI. Finally, the South Island East Coast is split between a northern and southern half for the SPEI (C3 and C4), but is a single cluster in the SPI classification (C3).

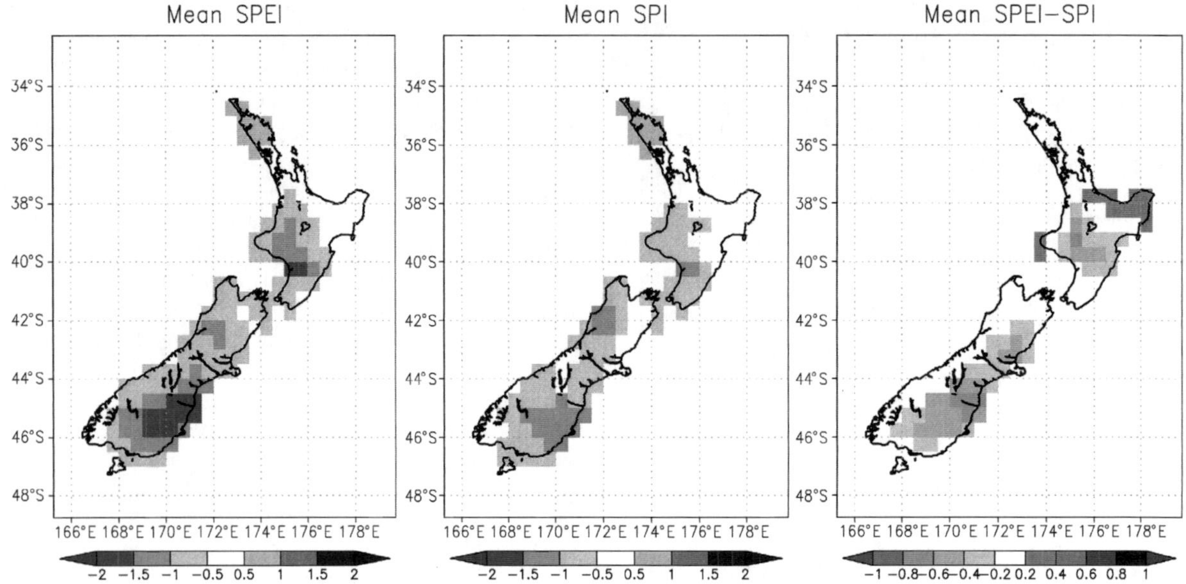

Figure 3. Mean SPEI, SPI and SPEI-SPI difference over drought events where the SPEI area in drought is more than 10 % greater than for the SPI.

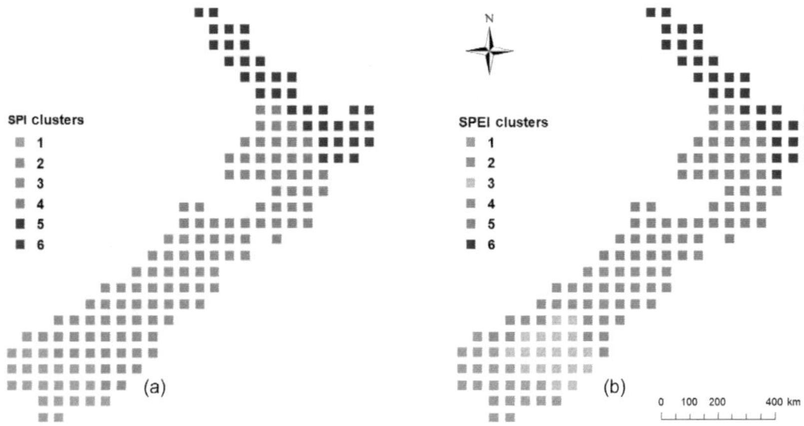

Figure 4. Spatial expression of clusters for the SPI **(a)** and SPEI **(b)**.

Each cluster shows some evidence of drought rich and drought poor periods (Fig. 5), ranging from 43 consecutive drought free months (SPI C6: August 2003 to February 2007) to 85 consecutive months with at least part of SPEI C3 in drought (May 1997 to May 2004), with an average of 66 % of this region in drought over this period, including 16 months at 100 % coverage. This drought-rich period for SPEI C3 corresponds to the drought-rich period identified at the national scale, and to a lesser extent is also present in all clusters except for the far north of the North Island (SPEI C6, SPI C5 and C6). Notwithstanding these differences in time series variation, there is very limited evidence of time series trends, with only SPEI C3 displaying a statistically significant rising trend, and SPEI C6, SPI C5 and 6 displaying weak but significant downward trends. Some clusters show a greater propensity for regionally extensive drought than others, with SPEI C5 and SPI C4 (both lower North/upper South Islands) each only showing one single month 100 % drought coverage event, compared to SPEI C3, which has a total of 22 months at 100 % drought coverage. Average area in drought across the study period also varies substantially between clusters, ranging from 12.5 %–30.4 % for the SPEI C1 and C4, and 12.4 %–19.1 % for SPI C1 and C4.

Overall, clusters with similar spatial expression between SPI and SPEI have very similar time series variation (i.e. the South Island southern and west coast clusters (C1 and C2 for both SPEI and SPI), and the upper South/lower North Cluster (SPEI C5 and SPI C4). Given the different spatial expres-

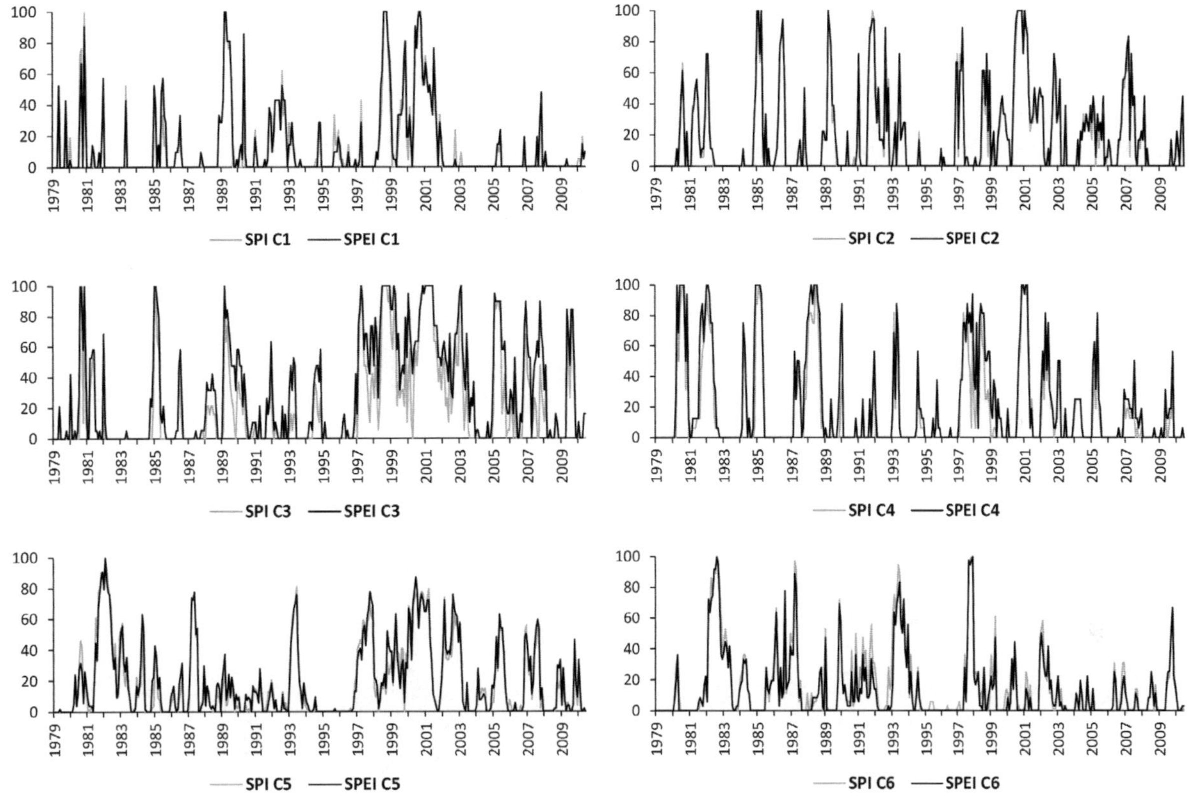

Figure 5. Drought extent for SPEI cluster spatial groupings according to both SPEI and SPI data.

sion of drought regions in the east of the South Island and north of the North Island, it is instructive to calculate area in drought within the spatial extent of each of SPEI C3, C4 and C6 according to both the SPI and SPEI, and vice versa. Accordingly, it is possible to directly compare the frequency and extent of regional drought occurrence between the SPEI and SPI indices (Fig. 6). Use of the SPEI results in greater drought extent in the area of SPEI C1–5. Differences are relatively small for C1 (southern South Island), C2 (South Island west coast) and C5 (upper South Island to lower and mid North Island). Much larger differences are present in the two South Island east coast clusters, and particularly for the more southerly C3, where on average an additional 10 % of the region is in drought for the SPEI vs. SPI, and a peak event-based difference of 63 %. This is followed by the more northern South Island east coast cluster, where the mean drought coverage is 5.5 % higher for the SPI, with a peak event-based difference of 62.5 %. The difference is still large when the east coast is combined into a single cluster (as under the SPI-based classification). In this case of the SPI C3 extent, there is a mean of 7.5 % additional drought coverage according to the SPEI vs. SPI. In contrast to C1–C5, for SPEI C6 the SPEI indicates a slightly lower area in drought compared to the SPI. Reversing the calculation to consider SPEI drought coverage within the two northern SPI clusters, the SPEI again results in a lower spatial extent of drought across both clusters, with this difference slightly greater for the (eastern) C6.

Just as there are weak or absent trends for area in drought at a national scale, there are mostly similarly weak trends for divergence between the two indices at the regional scale. The only exception to this is for the SPEI C3 region, where both indices show increasing drought, but the SPEI increase is greater than corresponding increase for SPI, resulting in a statistically significant divergence between the indices.

4 Discussion

The proportion of New Zealand in drought (SPEI and SPI ≤ -1, "moderate drought") is generally similar for both indices (Fig. 1). The two times series are highly correlated, and the mean difference across the time series is small (+1.72 %) and not statistically significant. Combined with the additional uncertainty that PET adds to a drought index (both from PET method and application in water limited situations), it is initially questionable whether the SPEI offers much net benefit as a drought index compared to the SPI. However, differences between the indices do become statistically significant once the times with very low and high drought coverage have

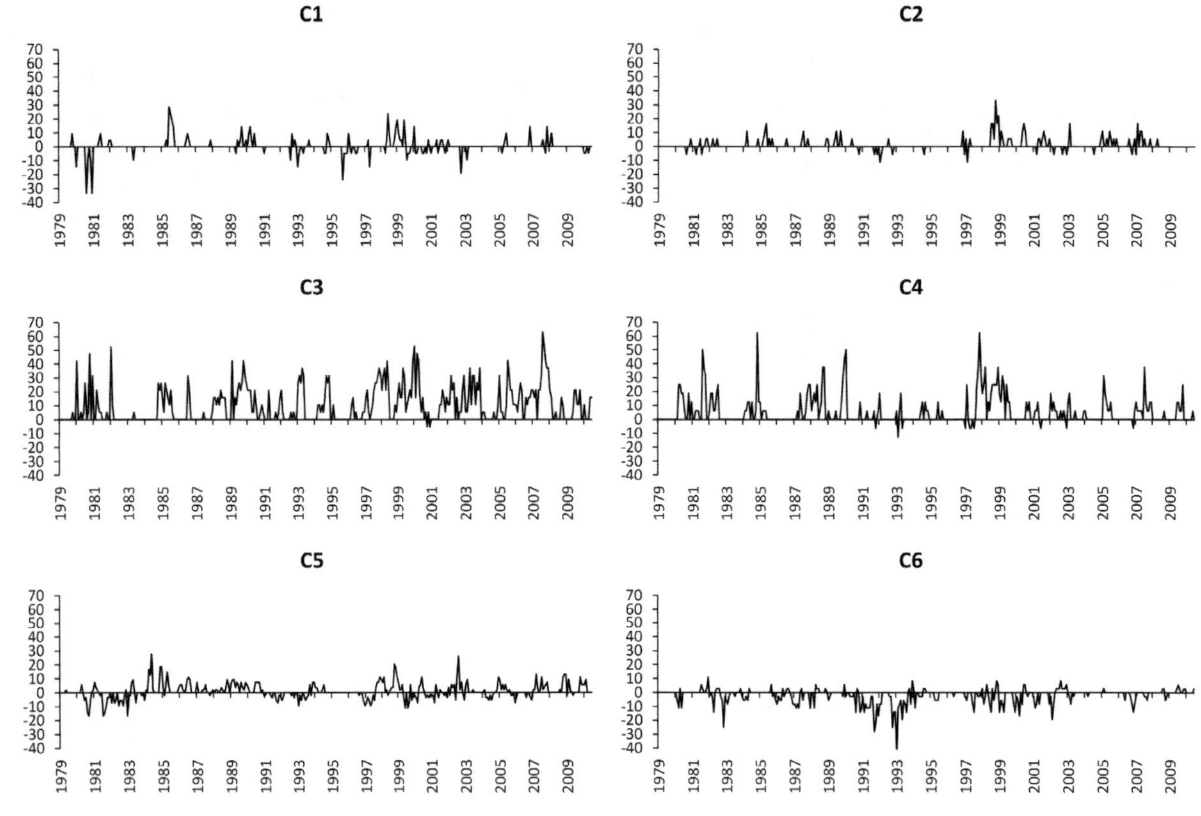

Figure 6. Difference in drought extent for SPEI cluster spatial groupings (Fig. 4b) between SPEI and SPI data.

been removed (i.e. coverage less than 10 % and greater than 30 %) indicates that use of the full meteorological water balance (i.e. precipitation and PET) is most important for identifying drought occurrence of intermediate national coverage. These findings are further emphasised by analysis of the eleven instances where the SPEI coverage is > 10 % greater than for the SPI. With just two instances of higher peaks for the SPEI (where SPI drought coverage instead reaches a plateau: e.g. May 1998), the majority of instances occur at times of intermediate drought coverage, either where development of a SPEI event has started earlier compared to the SPI (e.g. May 1985), or where a SPI event has declined more rapidly than the SPEI (e.g. November 1984). As such, inclusion of the full meteorological water balance appears to have most impact for identifying both initial drought development and drought termination. In consequence, when the influence of PET is included, drought onset is typically earlier and drought termination more gradual. In contrast, inclusion of PET (on average) has a non-significant effect on the peak spatial extent of large drought events, indicating that PET is ultimately a second-order driver of national-scale drought event coverage in New Zealand – peak spatial extent of large droughts is driven primarily by rainfall deficit. This may reflect the precipitation-dominated climate in many areas (e.g. South Island west coast, where *P* is approximately four times higher than PET), together with the absence of marked seasonality in rainfall totals in some locations and importance of seasonal snow storage and melt in others. A focus on water deficits at shorter timescales than six months (i.e. more seasonally-confined, and closer to the typical timescale of meteorological drought) may also result in an increased influence of PET on the identification of water deficits.

In addition to the differences in spatial extent of moderate droughts averaged over the time series, these differences show statistically significant increases over the course of the time series. In this instance, it is only the most severe droughts (> 40 % coverage) that do not show a statistically significant trend of divergence between SPI and SPEI coverage. So, whilst the spatial extent of the largest droughts may be driven primarily by rainfall, PET appears to be increasingly important as a cause of drought at moderate spatial extents. This corresponds to findings elsewhere, for example in southern Europe where increasing PET combined with decreasing precipitation has caused increasing meteorological water balance drought (Stagge et al., 2017; Spinoni et al., 2015, 2017).

The largest differences in drought coverage between the SPI and SPEI are focussed on central Otago, the central and southwestern North Island, and eastern parts of the North Island (Figs. 3 and 6). For the latter region use of the SPEI typically results in reduced drought compared to the SPI, with the reverse true for the former two regions. These differences are consistent with the results of the cluster analysis. Firstly, by different regional groupings: under the SPEI, separate regions are identified along the east coast of the South Island, but not for the SPI. Conversely, a single region is present from the north of the North Island across to the far east of the North Island for the SPEI, whereas this eastern area is part of a separate region with the SPI. Secondly, by differences between SPEI and SPI time series of drought for each region. Here, locations where precipitation or precipitation and PET dominate as drivers of drought become clear, and largely follow regional differences in spatial coverage of drought as identified in Fig. 3. Accordingly, the two South Island east coast SPEI regions (C4 and especially C3) record both greater drought than the SPI for this area and increasing divergence between the indices (Figs. 3 and 6). These relatively strong (and increasing) differences are likely linked to this being the region with the most continental climate in New Zealand – i.e. hot summers, cold winters, and low annual precipitation (< 350 mm in places). Accordingly, this is an area where high PET rates occur, together with a substantial climatological water balance deficit, and also where precipitation deficits are most likely to be accompanied by high temperatures (and so PET). Again, these results bear comparison to Europe, where it is the hotter and drier regions (i.e. southern Europe) where drought has increased the most, and where divergence between the SPI-SPEI indices is strongest (Stagge et al., 2017).

The central and southwestern North Island is another prominent region where the SPEI often indicates more widespread drought compared to the SPI, but here there are negligible differences in cluster spatial expression. Furthermore, time series variation of the area in drought for this cluster (SPEI C5, SPI C4) is relatively similar between the two indices, especially compared to the South Island east coast clusters (Fig. 6). Closer analysis of differences between the C5/C4 time series reveals that, unlike for the South Island east coast clusters, the number of instances where SPEI drought is more widespread is relatively well balanced against events SPI drought is greater. So in some instances, inclusion of PET leads to greater extent of drought being identified, and in other instances in the inclusion of PET ameliorates drought extent for this region.

The different spatial expression of clusters in the eastern North Island reflects the apparent amelioration of drought coverage by the inclusion of PET, with relatively few and weak instances where SPEI coverage is greater than for the SPI. Instead, more frequent and wider coverage (up to 60 %) events occur for the SPI. The increased severity of SPI droughts indicates that precipitation deficits in this location occur at times of low PET.

5 Conclusions

At the national scale for New Zealand there are few overall differences in the area in drought whether or not standardised PET is included alongside precipitation in the drought metric. Closer examination reveals some differences in drought at moderate spatial extents (> 10 % and < 40 % of New Zealand land surface area), which typically occurs during drought development or termination, rather than peak drought extent. As such it appears that inclusion of the full meteorological water balance in the drought index is more important for the duration of drought events, rather than their peak spatial extent. The finding of regional hotspots of SPI – SPEI differences, and trends in such differences, reveals that New Zealand is subject to similar changes in the role of PET for drought occurrence as have been found elsewhere in the world – except that these are confined to particular parts of the highly spatially variable New Zealand climate. Overall, this variation in climate, the dominance of precipitation over PET in many parts, and the overall temperate maritime climate regime mean that PET remains more of a secondary driver of drought characteristics than has been found for many other locations.

Data availability. James H. Stagge (University of Oslo – now University of Ohio) provided the SPI and SPEI data, using the data and methods described in Stagge et al. (2015).

Author contributions. DGK designed the analyses and EJT carried them out. DGK prepared the manuscript, with assistance from EJT.

Competing interests. The authors declare that they have no conflict of interest.

Special issue statement. This article is part of the special issue "Hydrological processes and water security in a changing world". It is a result of the 8th Global FRIEND–Water Conference: Hydrological Processes and Water Security in a Changing World, Beijing, China, 6–9 November 2018.

References

Fitzharris, B. B.: The 1992 electricity crisis and the role of climate and hydrology, New Zeal. Geogr., 48, 79–83, 1992.

McKee, T. B., Doesken, N. J., and Kleist, J.: The relationship of drought frequency and duration to time scales, Proceedings of the 8th Conference on Applied Climatology vol. 17, no. 22, American Meteorological Society, Boston, MA, USA, pp. 179–183, 1993.

Mishra, A. K. and Singh, V. P.: A review of drought concepts, J. Hydrol., 391, 202–216, 2010.

MPI: Four Degrees of Global Warming: Effects on the New Zealand Primary Sector, MPI Technical Information Paper No: 2013/49, prepared for the Ministry for Primary Industries, 180 pp., 2013.

Reisinger, A., Kitching, R. L., Chiew, F., Hughes, L., Newton, P. C. D., Schuster, S. S., Tait, A., and Whetton, P.: Australasia, in: Climate Change 2014: Impacts, Adaptation, and Vulnerability. Part B: Regional Aspects. Contribution of Working Group II to the Fifth Assessment Report of the Intergovernmental Panel on Climate Change, edited by: Barros, V. R., Field, C. B., Dokken, D. J., Mastrandrea, M. D., Mach, K. J., Bilir, T. E., Chatterjee, M., Ebi, K. L., Estrada, Y. O., Genova, R. C., Girma, B., Kissel, E. S., Levy, A. N., MacCracken, S., Mastrandrea, P. R., and White, L. L., Cambridge University Press, Cambridge, United Kingdom and New York, NY, USA, pp. 1371–1438, 2014.

Salinger, J. M. and Mullan, A. B.: New Zealand climate: temperature and precipitation variations and their links with atmospheric circulation, Int. J. Climatol., 19, 1049–1071, 1999.

Sheffield, J., Wood, E. F., and Roderick, M. L.:Little change in global drought over the past 60 years, Nature, 491, 435–438, 2012.

Spinoni, J., Naumann, G., Vogt, J., and Barbosa, P.: European drought climatologies and trends based on a multi-indicator approach, Glob. Planet. Chang., 127, 50–57, 2015.

Spinoni, J., Naumann, G., and Vogt, J.: Pan-European seasonal trends and recent changes of drought frequency and severity, Glob. Planet. Chang., 148, 113–130, 2017.

Stagge, J. H., Tallaksen, L. M., Gudmundsson, L., Van Loon, A. F., and Stahl, K.: Candidate distributions for climatological drought indices (SPI and SPEI), Int. J. Climatol. 35, 4027–4040, 2015.

Stagge, J. H., Kingston, D. G., Tallaksen, L. M., and Hannah, D. M.: Observed drought indices show increasing divergence across Europe, Sci. Rep.-UK, 7, 14045, https://doi.org/10.1038/s41598-017-14283-2, 2017.

Tallaksen, L. M. and Van Lanen, H. A. J. (Eds.): Hydrological Drought. Processes and Estimation Methods for Streamflow and Groundwater, Developments in Water Science, 48, Elsevier Science B.V., Amsterdam, 579 pp., 2004.

Van Lanen, H. A. J., Laaha, G., Kingston, D. G., Gauster, T., Ionita, M., Vidal, J.-P., Vlnas, R., Tallaksen, L. M., Stahl, K., Hannaford, J., Delus, C., Fendekova, M., Mediero, L., Prudhomme, C., Rets, E., Romanowicz, R. J., Gailliez, S., Wong, W. K., Adler, M.-J., Blauhut, V., Caillouet, L., Chelcea, S., Frolova, N., Gudmundsson, L., Hanel, M., Haslinger, K., Kireeva, M., Osuch, M., Sauquet, E., Stagge, J. H., and Van Loon, A. F.: Hydrology needed to manage droughts: the 2015 European case, Hydrol. Process., 30, 3097–3104, https://doi.org/10.1002/hyp.10838, 2016.

Van Loon, A. F. and Van Lanen, H. A. J.: A process-based typology of hydrological drought, Hydrol. Earth Syst. Sci., 16, 1915–1946, https://doi.org/10.5194/hess-16-1915-2012, 2012.

Vicente-Serrano, S. M., Beguería, S., and López-Moreno, J. I.: A multi-scalar drought index sensitive to global warming: the standardized precipitation evapotranspiration index – SPEI, J. Climate, 23, 1696–1718, 2010.

Weedon, G. P., Balsamo, G., Bellouin, N., Gomes, S., Best, M. J., and Viterbo, P.: The WFDEI meteorological forcing data set: WATCH Forcing Data methodology applied to ERA-Interim reanalysis data, Water Resour. Res., 50, 7505–7514, https://doi.org/10.1002/2014WR015638, 2014.

Wilhite, D.: Drought: A Global Assessment, Vol I & II, Routledge Hazards and Disasters Series, Routledge, London, UK, 2000.

WMO (World Meteorological Organisation): Standardized Precipitation Index – User Guide, WMO-No. 1090, Geneva, 2012.

Proc. IAHS, 383, 315–318, 2020
https://doi.org/10.5194/piahs-383-315-2020

Past and future drought in Northwestern Algeria: the Beni Bahdel Dam catchment

Senna Bouabdelli[1], Ayoub Zeroual[1], Mohamed Meddi[1], Fateh Djelloul[2], and Ramdane Alkama[3]

[1]Ecole Nationale Supérieure d'Hydraulique de Blida, L-GEE, 09000, Blida, Algeria
[2]Ecole Nationale Supérieure d'Hydraulique de Blida, L-MVRE, 09000, Blida, Algeria
[3]European Commission, JRC, Directorate D – Sustainable Resources, Bio-Economy Unit, TP124 Via E. Fermi, 2749, 21027 Ispra (VA), Italy

Correspondence: Senna Bouabdelli (s.bouabdelli@ensh.dz)

Published: 16 September 2020

Abstract. In last decades, the impact of climate change started to appear in the semi-arid regions of the Mediterranean Basin. The severity and frequency of drought events in Northwestern Algeria have affected water resources availability and agriculture. This study aims to evaluate the temporal evolution of drought events characteristics, such as drought duration, frequency and severity, of the Beni Bahdel Dam catchment, Northwestern Algeria. Drought characteristics have been derived from the Standardized precipitation index (SPI) computed for the period from 1941 to 2100 using precipitation data from observations and simulations of the regional climate model RCA4 (Rossby Centre Atmosphere model, version 4). The RCA4 model was forced by the global circulation model MPI-ESM-LR under two Representative Concentration Pathways (RCPs) scenarios. The ability of the model simulations was firstly assessed to reproduce the drought characteristics from observed data (1951–2005). Then, future changes in drought characteristics over the twenty-first century were investigated under the two scenarios (RCP4.5 and RCP8.5). Results show an amplification of drought frequencies and durations in the future under the RCP8.5 scenario.

1 Introduction

In last decades, the impact of climate change started to appear in the Mediterranean basin. This change have caused enormous socio-economic damages, including water scarcity and drawdown of groundwater levels that has put rain-fed crops at risk. Marcos-Garcia et al. (2017) reported an increase in drought severity and duration in the southern part of the Mediterranean basin in the second half of the 20th century. Several studies also published drying trends in the Western Mediterranean region starting at different times, i.e. from the mid-1920's to 1950 and since the mid-1970's (Brunet et al., 2007; Zeroual et al., 2019). The vast plains of the western region of Algeria are an important agricultural production and farming zone. However, the Northwestern Algeria is particularly vulnerable area to climate change (Achour et al., 2020; Bouabdelli et al., 2020; Meddi et al., 2010; Zeroual et al., 2017). The severity and frequency of drought events in this region have affected water resources availability and agriculture and the trends to drier conditions seem to intensify. The western plains have suffered from a drastic reduction of rainfall by about 30 % since 1975 (Meddi et al., 2010). This reduction caused negative impacts on water resources availability in several periods. The Beni Bahdel Dam, which is principally designed for the irrigation of the western plains and drinking supply, has attained its minimum level in several years failing to satisfy the water demand. This situation likely will intensify in the future, where there is consensus in projections by the International Panel on Climate Change (IPCC) about future decreases in rainfall (Christensen et al., 2007). Zeroual et al. (2019) have also confirmed this for the Northwestern part of the country using the output of 15 regional climate models (RCMs), as part of the CORDEX-Africa program (Coordinated Regional climate Downscaling Experiment). The climate in Northwestern Algeria was humid in 1951, semi-arid since 1973 and it is projected to be arid in 2100. Bouabdelli et al. (2020) also found that hydrological drought risk determined with frequency analysis likely will rise especially under RCP8.5 with higher severities and durations.

The above issues motivate this study, which is aimed to assess drought occurrence during the last decades and investigate their projected changes under future climate. The results can be helpful to establish an accurate warning tool for drought monitoring over the Beni Bahdel Dam catchment.

2 Material and methods

In this study, the Standardized Precipitation Index (SPI), Mckee et al. (1993) accumulated over 12 months was used to characterize drought events. The SPI-12 is considered to be an approximation of the state of the surface water volume (dams) and the groundwater level (aquifers) (Khan et al., 2008). The observed precipitation data used for this are coming from the Beni Bahdel basin located in the Northwest Algeria (Fig. 1). These data cover the period 1941–2010 and have a monthly time step. We used the computation procedure for SPI as defined by Mckee et al. (1993). For drought events characteristics, the period, during which the SPI value is continuously negative and reaches at least value of (-1) or less is defined by Dracup et al. (1980) as drought duration (D). The absolute sum of SPI values during this period (D) was called drought severity (S) or drought magnitude (Shiau, 2006).

We have also used in this study, simulations of the regional climate model RCA4 of SMHI at the horizontal resolution of roughly 0.44°, forced by the MPI-ESM-LR global circulation model and turned on the African region (Jones et al., 2011). The Simulated monthly precipitation data are provided by the CORDEX-Africa (Coordinated Regional climate Downscaling Experiment) initiative of the CMIP5 project for the World Climate Research Program (Giorgi et al., 2009).

Once the time series of the monthly precipitation were extracted from RCA4-MPI-ESM-LR model from 1951 to 2100, we proceeded to the future changes in drought characteristics according to the two following stages: the first step is to evaluate the ability of the RCA4-MPI-ESM-LR model simulations to capture drought events characteristics derived from the monthly observed precipitation, including, severity (S) and duration (D) during the period from 1951 to 2005. We should note here that for this comparison, the monthly precipitation was not bias corrected. As a second step, we calculate the SPI-12 series and their characteristics for the far future (2045–2100) based on the projected monthly precipitation, under the two RCP scenarios (RCP4.5 and RCP8.5). The monthly-modeled precipitation used in this step has been corrected using the quantile–quantile (QQ) bias correction algorithm (Teutschbein and Seibert, 2012). The mathematical equations of the quantile–quantile (QQ) bias correction algorithm are detailed in Teutschbein and Seibert (2012).

Table 1. Number and mean severity and duration of drought events, and maximum severity and duration.

	Number of events	D mean (month)	S mean	D max (month)	S max
Observed 1951–2005	25	14.04	11.75	59	68
Historical 1951–2005	30	10.69	8.95	59	75
Observed 1941–2010	36	11.36	9.11	59	68
RCP4.5 2045–2100	38	10.34	8.92	47	63
RCP8.5 2045–2100	49	7.45	8.4	58	70

3 Results and discussion

The drought duration and severity have been derived from SPI-12 time series. Results are shown for the observed period 1941–2010 (Fig. 2a). As can be seen, there is trends towards dry conditions in the meteorological records since 1970, i.e. more negative SPI-12 values. The Beni Bahdel basin has experienced more severe drought episodes with longer durations since mid-1970s. The drought events attained its maximum severities during the periods 1982–1987 and 1995–2003.

Table 1 presents the number of drought events, the mean and maximum values of the duration and the severity of past and future drought events. The Beni Bahdel basin experienced 36 meteorological drought episodes; the maximum duration was 59 months for the drought that started in June 1996. The drought with the maximum severity (68) started in December 1981.

For the historical period, the SPI-12 data derived from the RCA4-MPI-ESM-LR climate model show similar drying trend in the period 1951–2005. Compared to the observed data, the model overestimates the number of drought events (30 versus 25 observed events) but with less severity (8.95 versus 11.75) and duration events (10.69 versus 14.04 months) (Table 1). In addition, the model underestimated the median of the annual precipitation by 15 %. These results confirm the importance of noise investigation and bias correction in climate simulations (Alderlieste et al., 2014).

We should note that, for the future period, the SPI-12 obtained from the model simulation are calculated after bias correction for the two scenarios.

In the future, the number of drought events is predicted to remain about equal (38 months, RCP4.5). The mean number events under the most pessimistic scenario (RCP8.5) is projected to rise to 49 months, which means an increase by

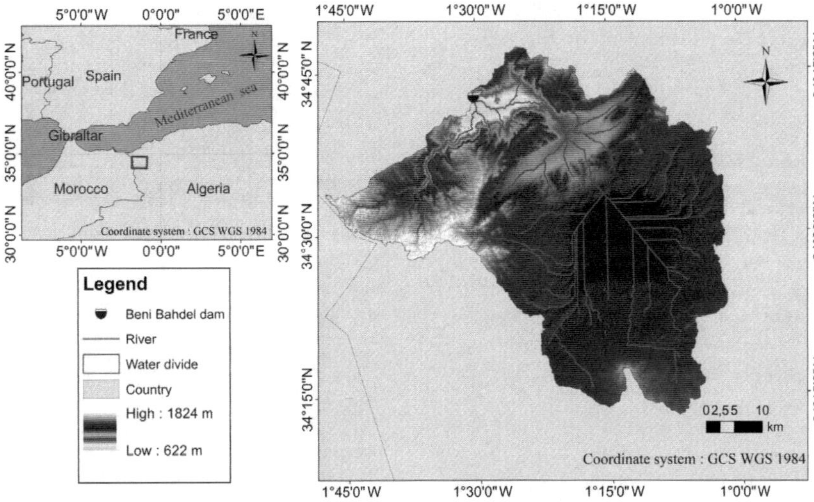

Figure 1. Location of the Beni Bahdel Dam catchment.

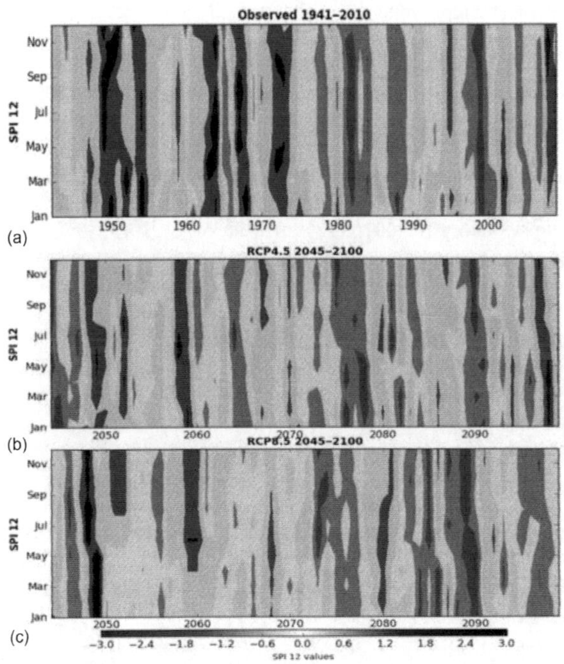

Figure 2. Temporal variation of meteorological drought (SPI 12) in the Beni Bahdel Dam catchment. **(a)** SPI-12 derived from observed precipitation (1941–2010), and **(b, c)** SPI-12 derived from the projected precipitation by the RCA4-MPI-ESM-LR climate model for the RCP4.5 and RCP8.5 scenario, respectively, period 2045–2100.

36 %. The maximum duration is expected to hardly change under the RCP8.5 scenario (58 months) or even to decline (RCP4.5).

The future SPI-12 are presented in Fig. 2b and c for the two RCP scenarios. Results show that the drought severity and duration likely will intensify in the future period, especially under the RCP8.5 scenario.

4 Conclusions

The temporal evolution of drought events using the SPI-12 was analyzed for the period from 1941 to 2100 in the Beni Bahdel Dam catchement situated in the Northwest of Algeria. The SPI12 was calculated from the oberved rainfall (1941–2005) and from the RCA4-MPI-ESM-LR climate model (1951–2100). The SPI-12 derived from the observations shows that there has been a clear drying trend in the past, particularly since the mid 1970s. The climate model predicts an increase of the mean number, severity and the duration of the drought events in the future under the most pessimistic scenario of greenhouse gas emissions (RCP8.5). Under the RCP4.5 scenario the drought is projected to remain about the same, or even will become less extreme. The current drought episodes experienced in the Beni Bahdel basin and the anticipated increase of these dry trends in the future under the extreme emissions sceanrio will cause numerous challenges, particularly to esnure food safety in Algeria. This finding must be taken into account in the elaboration of future socio-economic plans of the developement.

Data availability. The in-situ data are provided from the Algerian National Agency of Water Resources (ANRH). The projected data from RCA4-MPI-ESM-LR are provided from Coordinated Regional climate Downscaling Experiment modeling groups over Africa https://esg-dn1.nsc.liu.se/search/cordex/ (last access: 23 June 2018) (SMHI, 2018).

Author contributions. BS, ZA and MM made the conceptualization of the work; BS and ZA done the methodology. BS done the data analysis; wrote the codes and the original draft; have made the review and editing of the paper. ZA supervised the work and participated in writing-review and editing. MM supervised and presented the work in Beijing during the 8th Global FRIEND–Water Conference: Hydrological Processes and Water Security in a Changing World. DF helped in the data treatment and analysis. AR provided climate model projections and bias correction; and supervised the work.

Competing interests. The authors declare that they have no conflict of interest.

Special issue statement. This article is part of the special issue "Hydrological processes and water security in a changing world". It is a result of the 8th Global FRIEND–Water Conference: Hydrological Processes and Water Security in a Changing World, Beijing, China, 6–9 November 2018.

Acknowledgements. The authors wish to thank the National Agency of Water Resources (ANRH) of Algeria for providing the data on which reported analyses are based. We acknowledge the World Climate Research Program's Working Group on Regional Climate the CORDEX-Africa (Coordinated Regional climate Downscaling Experiment), initiative of the CMIP5 project for the World Climate Research Program. We also thank the Rossby Center for producing and making available their regional atmospheric model, RCA4 output. Lastly, we thank the reviewers for their constructive comments.

References

Achour, K., Meddi, M., Zeroual, A., Bouabdelli, S., Maccioni, P., and Moramarco, T.: Spatio-temporal analysis and forecasting of drought in the plains of northwestern Algeria using the standardized precipitation index, J. Earth Syst. Sci., 129, 42, https://doi.org/10.1007/s12040-019-1306-3, 2020.

Alderlieste, M. A. A., Van Lanen, H. A. J., and Wanders, N.: Future low flows and hydrological drought: How certain are these for Europe?, IAHS-AISH Proc. Rep., 363, 60–65, 2014.

Bouabdelli, S., Meddi, M., Zeroual, A., and Alkama, R.: Hydrological drought risk recurrence under climate change in the karst area of Northwestern Algeria, J. Water Clim. Change, https://doi.org/10.2166/wcc.2020.207, in press, 2020.

Brunet, M., Jones, P. D., Sigró, J., Saladié, O., Aguilar, E., Moberg, A., Della-Marta, P. M., Lister, D., Walther, A., and López, D.: Temporal and spatial temperature variability and change over Spain during 1850–2005, J. Geophys. Res.-Atmos., 112, D12117, https://doi.org/10.1029/2006JD008249, 2007.

Christensen, J. H., Hewitson, B., Busuioc, A., Chen, A., Gao, X., Held, R., Jones, R., Kolli, R. K., Kwon, W. K., Laprise, R., Magana Rueda, V., Mearns, L., Menendez, C. G., Räisänen, J., Rinke, A., Sarr, A., Whetton, P., Arritt, R., Benestad, R., Beniston, M., Bromwich, D., Caya, D., Comiso, J., de Elia, R., and Dethloff, K.: Regional climate projections. Climate Change, 2007: The Physical Science Basis, in: chap. 11, Contribution of Working group I to the Fourth Assessment Report of the Intergovernmental Panel on Climate Change, Cambridge University Press, Cambridge, 847–940, 2007.

Dracup, J. A., Lee, K. S., and Paulson, E. G.: On the definition of droughts, Water Resour. Res., 16, 297–302, https://doi.org/10.1029/WR016i002p00297, 1980.

Giorgi, F., Jones, C., and Asrar, G. R.: Addressing climate information needs at the regional level: the CORDEX framework, Bull.-World Meteorol. Organ., 58, 175–183, 2009.

Jones, C., Giorgi, F., and Asrar, G.: The Coordinated Regional Downscaling Experiment: CORDEX – an international downscaling link to CMIP5, Clivar Exch., 56, 34–40, 2011.

Khan, S., Gabriel, H. F., and Rana, T.: Standard precipitation index to track drought and assess impact of rainfall on watertables in irrigation areas, Irrig. Drain. Syst., 22, 159–177, https://doi.org/10.1007/s10795-008-9049-3, 2008.

Marcos-Garcia, P., Lopez-Nicolas, A., and Pulido-Velazquez, M.: Combined use of relative drought indices to analyze climate change impact on meteorological and hydrological droughts in a Mediterranean basin, J. Hydrol., 27, R713–R715, https://doi.org/10.1016/j.jhydrol.2017.09.028, 2017.

Mckee, T. B., Doesken, N. J., and Kleist, J.: The relationship of drought frequency and duration to time scales, Water, 179, 17–22, 1993.

Meddi, M., Assani, A. A., and Meddi, H.: Temporal Variability of Annual Rainfall in the Macta and Tafna Catchments, Northwestern Algeria, Water Resour. Manage., 24, 3817–3833, https://doi.org/10.1007/s11269-010-9635-7, 2010.

Shiau, J. T.: Fitting drought duration and severity with two-dimensional copulas, Water Resour. Manage., 20, 795–815, https://doi.org/10.1007/s11269-005-9008-9, 2006.

SMHI – Swedish Meteorological and Hydrological Institute: ESGF@LiU/CORDEX, available at: https://esg-dn1.nsc.liu.se/search/cordex/, last access: 23 June 2018.

Teutschbein, C. and Seibert, J.: Bias correction of regional climate model simulations for hydrological climate-change impact studies: Review and evaluation of different methods, J. Hydrol., 456–457, 12–29, https://doi.org/10.1016/j.jhydrol.2012.05.052, 2012.

Zeroual, A., Assani, A. A., and Meddi, M.: Combined analysis of temperature and rainfall variability as they relate to climate indices in Northern Algeria over the 1972–2013 period, Hydrol. Res., 48, 584–595, https://doi.org/10.2166/nh.2016.244, 2017.

Zeroual, A., Assani, A. A., Meddi, M., and Alkama, R.: Assessment of climate change in Algeria from 1951 to 2098 using the Köppen–Geiger climate classification scheme, Clim. Dynam., 52, 227–243, https://doi.org/10.1007/s00382-018-4128-0, 2019.

Proc. IAHS, 383, 319–325, 2020
https://doi.org/10.5194/piahs-383-319-2020

Extreme flow variability analysis at the Bianouan hydrometric station on the Bia River watershed in Côte d'Ivoire

Kan Martin Kouassi[1,2], Koffi Blaise Yao[1], Kouakou Lazare Kouassi[2,3], Jean Biemi[1], and Nagnin Soro[1]

[1]Hydrogeology Laboratory of Earth Sciences and Mining Resource department , Félix Houphouet Boigny University, Abidjan Cocody, 01 PB V4 Côte d'Ivoire
[2]Ecology Research Center, Abidjan, 08 PB 109, Côte d'Ivoire
[3]Lorougnon Guede University, Daloa, 12 PB V25, Côte d'Ivoire

Correspondence: Kan Martin Kouassi (kanstephane@yahoo.fr)

Published: 16 September 2020

Abstract. Extreme hydrological phenomena recurrence in West Africa in recent decades is a problem in the anthropized watershed management. Link analysis between extreme hydrological phenomena and climate variability is necessary to guard against their consequences. However, qualitative data accessing difficulties and explanatory variable definition of extremes hydrological phenomena limit extreme flows rate studies. This study proposes the analysis the variability of the floods and the low flows at the Bianouan hydrometric station. Thus, five (5) floods characteristics variables (QCX5, QXJA, F90p, F95p and F99p) and three (3) for, the low flows (VCN10, QJNA and F10p) were extracted daily flows from the Bianouan hydrometric station from 1 January 1962 to 31 December 2005. The variability of these extremes has been studied by the trend analysis (linear regression) and of the stationarity (Pettitt and Hubert tests). The results show that for flood variables the negative trend is very significant and ruptures are observed in 1980. For low flow variables the negative trend is significant and the ruptures are observed in 1981. From this study, we can deduce that the ruptures observed in the rains and flows average in the end 1960s and early 1970s have affected later the extreme flows around the 1980s.

1 Introduction

Floods and low flows provoke natural phenomena which qualify extreme that affect the functioning of antropized watersheds. In the face of climate change, these extreme phenomena have caused disasters in many parts of the world. They occur with increasingly higher frequency and their socio-economic impacts on the populations are more considerable. According to an OMM (2014) report, from 1970 to 2012 in Africa, 1319 disasters resulted in 698 380 casualties and losses of USD 26.6 billion. In Côte d'Ivoire, the low-flows of the years 1983–1984, caused selective power cuts which had negative consequences on the national and sub-regional West African economy. The low-flows of in 2010 caused a second period of selective power cuts with considerable financial losses for national and sub-regional companies and a financial deficit of 57 billion CFA francs to the Ivorian Company of Electricity (CIE, 2010). The transboundary basin of the Bia River with which the present study is concerned does not escape of the phenomena. Recurrent low-flows have affected Ayame dams functioning, causing downstream drying up to such an extent that it was impossible for the managers to run any turbine. Faced with these situations, link analysis between extreme hydrological phenomena and climate variability is necessary to protect against these events consequences However, qualitative data accessing difficulties and explanatory variable definition of extremes hydrological events limit extreme flows rate studies. This study proposes to analyze floods and low flows variability at the Bianouan hydrometric station. It emphasizes the intensity and duration notions in extreme flows variables choice to explain extreme events (Goubanova, 2007).

2 Study area presentation

Bia is a transboundary river between Côte d'Ivoire and Ghana; it flows between latitude 5°30′ and 5°50′ N and between longitude 3 and 3° 15′ W (Fig. 1). Its source is in Kérimasso in northern Ghana and flows into the Aby lagoon in southern Côte d'Ivoire (Reizer, 1967). The catchment area of the Bia is elongated and covers an area of 9650 km^2 of which $2/3$ are located in the Ghanaian territory. The rainfall analysis done by Meledje (2016), makes it possible to distinguish two (2) climatic zones on the basin of the Bia. A humid tropical zone (1460 mm) corresponding to the northern part of the basin with a single rainy season and a subequatorial zone (1870 mm) corresponding to the south of the basin with two distinct rainy seasons.

3 Data and materiel

For this study we have daily discharge data provided by the Human Hydraulics sub-directorate at the Bianouan Hydrometric Station from 1 January 1962 to 31 December 2005.

4 Methodology

The methodology adopted in this study is divided into two steps: choice and study of the variability of the variables' characteristics of the extreme flows.

4.1 Extremes flows characteristics variables choice

In this methodology, we define the multi-duration variables, the MINAN and MAXAN variables and extreme flow indices.

4.1.1 Multi-duration variables choice

These variables make it possible to take into account the duration or threshold notion in the variables choice which characterize the flows extremes. They are chosen from two value of the graded flow curve; the $Q_{15\,\%}$ and $Q_{85\,\%}$ which are the flows respectively exceeding 15 % and 85 % of the values of the graded flow curve. Thus on a moving average d days series (with d = 1, 3, 5, 10, 15, 20, 60 and 90 d), the VCNd and VCXd represent respectively the minimum and maximum values of each year (Galéa and Javelle, 2000). QCXd is the annual maximum flow greater than $Q_{85\,\%}$ and QCNd is the annual minimum flow lower than $Q_{15\,\%}$. The Kendall rate was used to select a low flow characteristic variable among the VCNd and QCNd variables and then the high flow rate variable among the VCXd and QCXd variables.

Table 1. Extremes flow index.

Identification	Index Name	Definition
F10p (day)	Low flow	Number of days with flow < 10th percentile
F90p (day)	High flow	Number of days with flow > 90th percentile
F95p (day)	Extremely high flow	Number of days with flow > 95th percentile
F99p (day)	Exceptional flow	Number of days with flow > 99th percentile

4.1.2 MAXAN and MINAN flow variables

It is a matter of selecting for each year, the maximum flow (MAXAN) as well as the minimal flow (MINAN) which characterize respectively the high flows and the low flows.

4.1.3 Extreme flow index

Flow rate indices are variables introduced recently by the IPCC to characterize extreme hydrological events. In this study, thresholds are defined in relation to the 10th, 90th, 95th and 99th percentiles (Table 1). The percentiles are calculated over 30 years (1956–1985) according to the recommendations of the World Meteorological Organization.

4.2 Hydrometric extremes variability study

The extreme flow and rain variability may be more damaging than a simple mean variation (Lemaître, 2002 and Karimou et al., 2015). This section objective is to evaluate the extremes flow rate variability through extreme flows characteristic variables trend and stationarity study.

4.2.1 Hydrometric extremes trend analysis

Hydrometric extremes trend analysis was carried out by linear regression between the flow rate extremes and time (in years). Slopes thus estimated were grouped into positive or negative significant trends' classes. The limit of the classes considered is defined from the Student's t-statistic used to test the hypothesis of a slope equal to 0. The trend is qualified according to the probability p of the test t applied to the regression slope.

- If $p < 0.01$: the trend is qualified as very significant.
- If $0.01 \leq p < 0.05$: the trend is qualified as significant.
- If $p \geq 0.05$: the trend is considered insignificant.

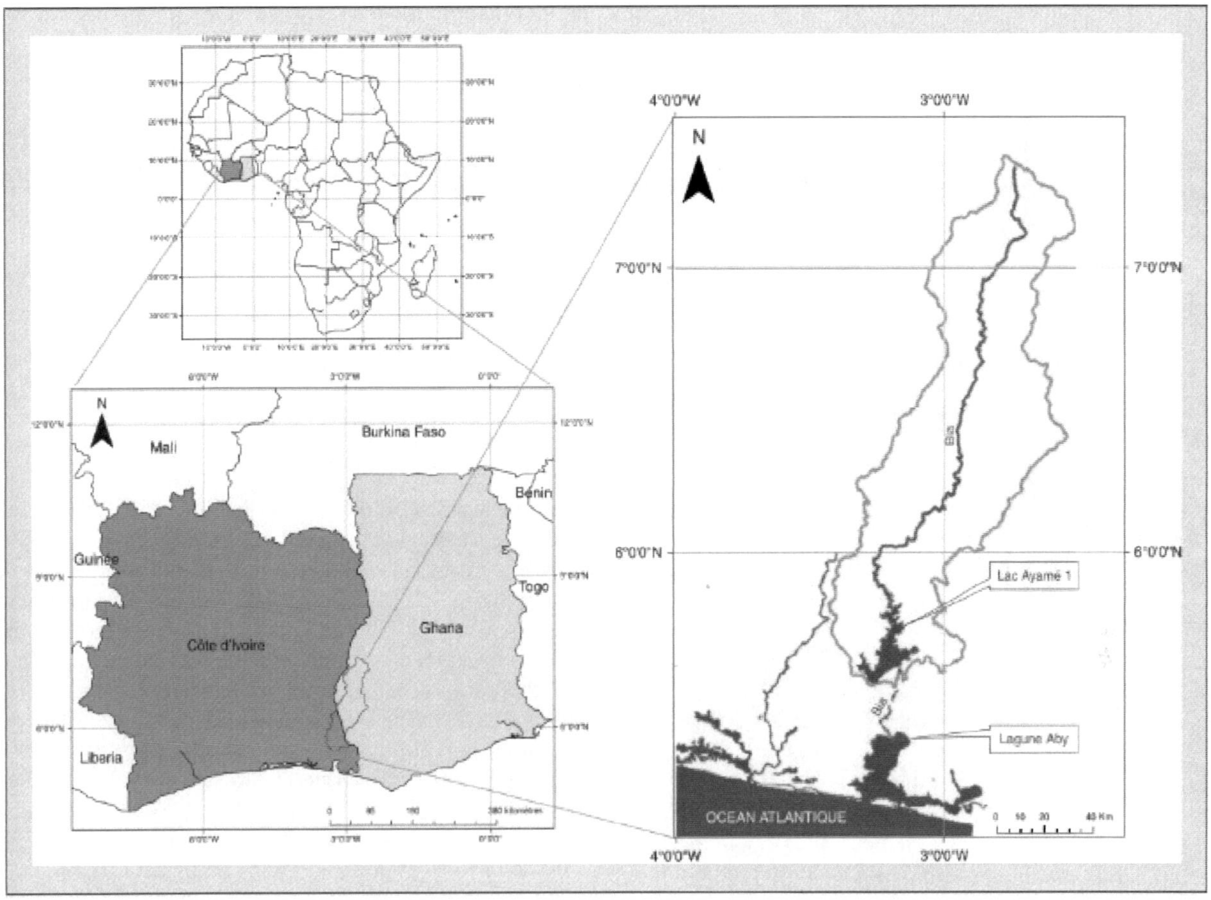

Figure 1. Location of the study area (Meledje et al., 2015).

4.2.2 Hydrometric extremes stationarity analysis

For extremes flow stationarity study, Pettitt test (Pettitt, 1979) and Hubert segmentation procedure are used to detect respectively unique and multiples ruptures in the extremes flow variables. The applicability of these two methods has been successfully verified in many studies in the West African climate context (Lubes et al., 1994).

5 Results

5.1 Variables characteristics of extremes hydrometrics

The extremes of the selected flow rates are multi-duration variables, flow indices and MAXAN as well as MINAN.

5.1.1 Multi-duration variables choice

The Fig. 2 shows the curve of rated flows. From this curve, the flow rates $Q_{85\,\%}$ and $Q_{15\,\%}$ necessary for the choice of QCXd and VCXd variables are respectively 68.90 and $1.25\,m^3\,s^{-1}$.

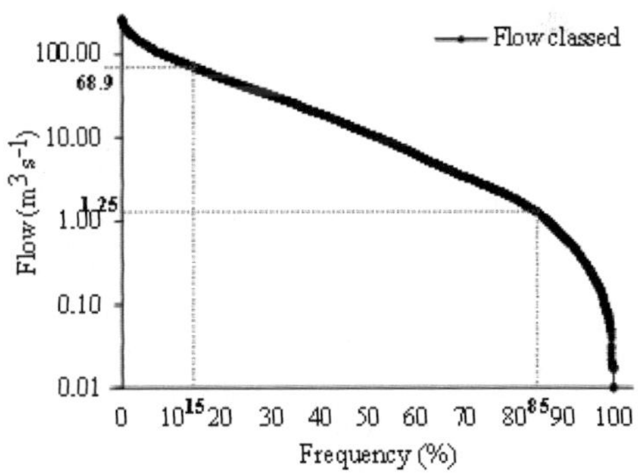

Figure 2. Flows duration curve on the Bianouan hydrometric station (1962–2005).

The Fig. 3 summarizes the average Kendall rates calculated on the different multi-duration variables. The analysis of this figure made it possible to choose QCX5 and VCN10 flow rates to characterize the flow extremes at the Bianouan hydrometric station from 1962 to 2005, insofar as

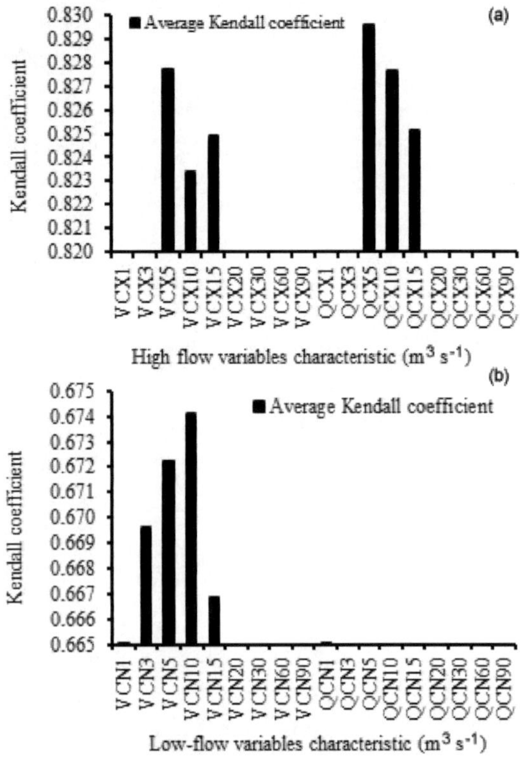

Figure 3. Kendall coefficient of high flow **(a)** and Low Flow **(b)** Variables.

Table 2. Percentile calculated from 1956 to 1985.

Percentile	Values ($m^3 s^{-1}$)
10	0.47
90	76.84
95	106
99	160.44

these two variables record the highest Kendall rates. Indeed, the QCX5 flow rate is the maximum annual flow rate above $68.29\,m^3\,s^{-1}$ calculated on a moving average of 5 consecutive days, while the VCN10 flow represents annual minimum flow over a moving average of 10 consecutive days.

5.1.2 Extremes flow index choice

The Table 2 summarizes percentiles calculated over 30 years. The exceptional (F99p) and extremely high (F95p) flow rates are defined from the respective values of 160.4 and $106\,m^3\,s^{-1}$, while the high (F90) and low (F10) flow rates are chosen from 76.84 and $0.47\,m^3\,s^{-1}$.

In addition, the maximum (MAXAN) and minimum (MINAN) flow rate for each is chosen from 1 January 1962 to 31 December, 2005, are respectively noted QJXA and QJNA.

Table 3. Extreme flow characteristic variables average (1962–2005).

Extremes flow variables	Identification	Average	Variation Coefficient
Flow index	F10p (day)	28.46	1.22
	F90p (day)	317.46	0.12
	F95p (day)	26.91	1.24
	F99p (day)	9.71	2.24
MAXAN	QJXA ($m^3 s^{-1}$)	155.27	0.31
MINAN	QJNA ($m^3 s^{-1}$)	0.56	1.26
Multi-duration	QCX5 ($m^3 s^{-1}$)	144.761	0.37
	VCN10 ($m^3 s^{-1}$)	0.78	1.29

Table 3 summarizes the flow extremes used in this study. This table shows that, on average, the QJNA and VCN10 variables remain below $1\,m^3\,s^{-1}$ with coefficients of variation exceeding 1. QCX5 and QJXA, on the other hand, they exceed $100\,m^3\,s^{-1}$ but are associated with coefficients of variation lower than 1. For the extreme flow indices, the high flow (F90p) averages 317.46 d while the exceptional flow (F99p) which remains lower than 10 d (9.7 d) is recorded. The low flow (F10p) and very high flow (F95p) are on average 28.46 and 26.91 d respectively. Exceptional flow has noted a greatest variability (2.24) whereas a contrary observation is made with the strong flow (0.12).

In the rest of our analysis, the hydrological extremes retained at the Bianouan hydrometric station from 1962 to 2005 are summarized in the variables characterizing the low flow or low flows (VCN10, QNJA and F10p) and those characterizing the strong flows or floods. (QCX5, QXJA, F90p, F95p and F99p). This variability was assessed by analyzing the trend (linear regression) and stationarity (Pettitt's test and the Hubert segmentation procedure) of these variables.

5.2 Hydrological extremes variability study

5.2.1 Hydrological extremes trend analysis

The results in the Table 4 show that an average trend of 0.5 % per decade is observed. This trend which has been very significant for high flow variables proves to be negative with variables QJXA, QCX5, F95p and F99p and positive with variable F90p. The significant trend observed at low flow have been negative for the variables VCN10 and QJNA and then positive with the variable F10p.

5.2.2 Hydrological extremes stationarity

Low-flow variables stationary analysis (Table 5) shows that the hypothesis of a rupture is accepted by the Pettitt test in 1981 for the variables (QJNA and VCN10) and in 1982 for the variable F10p.

Table 4. Trend (% by decade) of high flow and low flow variables.

Characteristic variables	Identification	Trend (% per decade)
Low flow	VCN10 ($m^3 s^{-1}$)	−0.5*
	QJNA ($m^3 s^{-1}$)	−0.5*
	F10p (jour)	0.51
High flow	QCX5 ($m^3 s^{-1}$)	−0.49**
	QJXA ($m^3 s^{-1}$)	−0.48**
	F90p (day)	−0.56**
	F95p (day)	−0.5**
	F99p (day)	−0.49**

* significant trend, ** very significant trend.

Table 5. Stationarity of low flow and high flow variables.

Characteristic	Identification	Method	
variables		Pettitt	Hubert
Low Flow	VCN10 ($m^3 s^{-1}$)	1981	1969
	QJNA ($m^3 s^{-1}$)	1981	1970 and 2003
	F10p (day)	1982	1982 and 1990
High flow	QCX5 ($m^3 s^{-1}$)	1980	1980 and 1995
	QJXA ($m^3 s^{-1}$)	1980	1980 and 1995
	F90p (day)	1980	1971
	F95p day)	1983	1968
	F99p (day)	1980	1968

In this last variable and the QJNA variable record each a double rupture respectively in 1982 and 1990 then in 1970 and 2003 according to the Hubert test. In the high flow variables, the hypothesis of no rupture is rejected by the Pettitt test in 1980 with four (4) variables (QCX5, QJXA, F90p and F99p) and then in 1983 in the variable F95p. Similarly, Hubert's test accepts the hypothesis of a double rupture (1980 and 1995) in the variables QCX5 and QJXA.

6 Discussion

The maximum (QJXA) and minimum (QJNA) annual flows are very often used to describe, respectively, floods and low flows. The integration of the duration notion in the choice of the extreme flow characteristic quantities shows that at the Bianouan hydrometric station on the Bia watershed, the flood can also be described by the QCX5 flow. That is to say, on a moving average 5 consecutive days, the maximum flow for each year higher than 68.9 $m^3 s^{-1}$. This value represents the value exceeding 85 % or not exceeded by 15 % of the values of the classed flows curve. These results corroborate the findings already reached on the choice of variables to characterize floods. Indeed, previous studies (Lemaître, 2002; Renard, 2006) have shown that the notion of flood is linked to a threshold but not to a maximum value of the each year. Even Better, the CEMAGREF for the QdF model implementation (flow-duration – frequency) recommends the use of characteristic variable of flood related to the notion of flow higher than a threshold. Therefore, the one-year maximum (VCXd) values used in this study do not appear to be representative of floods because the maximum value of a year is not necessarily a flood (Lemaître, 2002). It is rather the variable above a threshold over a given period that better characterizes the floods. Thus, the period of 5 consecutive days also called "pentad" or amplitude of the rainy season, seems the interesting period to characterize the floods. This period is also the minimum period to describe the succession of wet periods. It is therefore a good indicator of the agrometeorological value of the rainy season (Ozer et al., 2009).

In addition, low flows can be described with variable VCN10, which is defined as the annual minimum of flows calculated on a moving average of 10 d. This variable is coherent with that determined by Renard (2006) in northern France. Similarly, VCN10 is the regulatory value in many countries, including the United States (Kroll and Vogel, 2002). Also, VN10 is recommended for low-flow characterization (WMO, 2008). However, the low-flow variables relating to a threshold on a moving average of d days seem not be representative of low water levels. This means that the notion of low flow on the Bia watershed from 1962 to 2005 is linked to a minimum value but not to a threshold. This notion of a minimum value appears even in the definition of low flow. Laborde (1984) defines low water as VCN10 variables. Moreover, according to Lang (2011), the definition of low flow must take into account the criterion of duration that this factor is determining in the characterization of the impacts of the phenomenon. On the Bia watershed, a complete description of the low-water phenomenon must therefore be apprehended in terms of minimum flow over a moving average of 10 consecutive days. This 10 d duration is used in the prediction of low-flow flows as part of the PRESAGES study (Lang and Gilles, 2006). Also, the consecutive period of 10 d, also called decade, is involved in the calculation of most climatic indices used in agronomy (Sarr et al., 2012).

The stationarity analysis records late ruptures both in low-flow variables (1981–1982) and floods (1980–1983). These observations are different from those observed in most studies conducted in the West African context (Ouédraogo et al., 2002; Mahé, 2006). This difference could be justified by the fact that the studies that led to these previous results focused on flow modules while the present studies were conducted on the Bia watershed using the variables characterizing floods and low flows. Thus, the use of these variables would provide additional information regarding the variability of flows in the Bia basin. Indeed, the ruptures observed between the end of the 1960s and the beginning of the 1970s affected the extremes flow rate (low water and floods) after 1980. These rupture dates are conform with those observed on the extreme rain by Kouassi et al. (2018) thus confirming that there could exist a link between extreme rains and extreme flows.

7 Conclusions

Various variables are used to characterize floods and low flows. The integration of the duration and threshold notion in the choice of flood and low flow variables in this study has shown that in the Bia watershed, the notion of flood is linked to a threshold (the flow rate exceeding 85 % of the values of the classed flows curve) over an observation period of 5 consecutive days. However, the low flow notion is linked to a minimum value observed over a period of 10 d. The variables thus chosen were grouped into variables of low flow (VCN10, QNJA and F10p) and flood (QCX5, QXJA, F90p, F95p and F99p). These variables' variability analysis shows that the trend of 0.5 % per decade which is mostly negative, has been very significant for the flood variables and significant for the low flow variables. In addition, the stationarity analysis reveals late ruptures in the flood and low flow variables. These results show that the ruptures observed around the 1970s in the flow and rainfall modules affected the hydrometric extreme variables around the 1980s.

Data availability. The flow data used in this study were provided by Ivory Coast Human Hydraulics Sub-directorate. These data are not accessible to the general public as they are subject to a fee.

Author contributions. The KKM conceived the idea for this article and wrote the first draft. YKB, KKL, BJ and SN provided comments and suggestions for quality improvements.

Competing interests. The authors declare that they have no conflict of interest.

Special issue statement. This article is part of the special issue "Hydrological processes and water security in a changing world". It is a result of the 8th Global FRIEND-Water Conference: Hydrological Processes and Water Security in a Changing World, Beijing, China, 6–9 November 2018.

References

CIE: Compagnie Ivoirienne d'Electricité, Rapport annuel d'activités, résultats financiers développement durable, 1–34, 2010.

Galéa, G. and Javelle, P.: Modèles débit durée fréquence de crue en Guadeloupe, Rapport d'étude, protocole cemagref-Lyon DIREN Guadeloupe et météo – France, cemagref, Lyon, 1–25, 2000.

Goubanova, K.: Une étude des événements climatiques extrêmes sur l'Europe et le bassin Méditerranéen et de leur évolution future, Thèse de doctorat ès science, Université Paris VI (France), 1–121, 2007.

Karimou, B. M., Ambouta, K., Sarr, B., and Tychon, B.: Analyse des phénomènes climatiques extrêmes dans le sud-est du Niger. XXVIIIe Colloque de l'Association Internationale de Climatologie, 1–6, 2015.

Kouassi, K. M., Kouassi, K. L., Yao, K. B., Meledje, N. H., Biemi, J., Lasm T., and Nathalie, R.: Variabilité des extrêmes pluviométriques sur le bassin versant de la rivière Bia (Sud-Est, Cote d'Ivoire), Eur. Sci. J., 14, 1857–7881, 2018.

Kroll, C. N. and Vogel, R. M.: Probability Distribution of Low Series in the United States, J. Hydrol. Eng., 7, 137–146, 2002.

Laborde, J. P.: Analyse des données et cartographie automatique en hydrologie, Éléments d'hydrologie Lorraine, Thèse d'État, I.N.P.L, 1–484, 1984.

Lang, C. D.: Sévérité des étiages et indigence des précipitations: une relation évidente mais rarement simple – Exemple en régime pluvial océanique, Revue Géographique de l'Est, 54, 3–4, 2011.

Lang, C. and Gilles, E.: Une méthode d'analyse du tarissement des cours d'eau pour la prévision des débits d'étiage, Norois, 201, 31–43, 2006.

Lemaître, F.: Recensement des tests de détection de tendances ou de ruptures l'analyse de stationnarité des régimes de crues en France, Rapport de fin d'études, Entpe, Cemagref-Lyon (France), 1–94, 2002.

Lubes, H., Masson, J. M., Servat, E., Paturel, J. E., Kouamé, B., and Boyer, J. F.: Caractérisation de fluctuations dans une série chronologique par applications de tests statistiques, Etude bibliographique, Programme ICCARE, Rapport 3, ORSTOM, Montpellier, France, 1–21, 1994.

Mahé, G.: Variabilité pluie-débit en Afrique de l'Ouest et Centrale au 20ème siècle: changements hydroclimatiques, occupation du sol et modélisation hydrologique, Habilitation à Diriger des Recherches, Université des Sciences et Techniques Montpellier, 2, 1–160, 2006.

Meledje, N. H.: Modélisation de la dynamique hydrologique et du flux des sédiments dans le lac du barrage hydroélectrique d'Ayamé 1. Thèse unique de doctorat, Université Nangui Abrogoua, Côte d'Ivoire, 1–197, 2016.

Meledje, N. H., Kouassi, K. L., N'go, Y. N., and Savane, I.: Caractérisation des occurrences de sécheresse dans le bassin hydrologique de la Bia transfrontalier entre la Côte d'Ivoire et le Ghana: contribution des chaînes de Markov, CahAgric, 24, 186–197, https://doi.org/10.1684/agr.2015.0755, 2015.

OMM: Organisation météorologique Mondiale: Déclaration sur l'état du climat mondial, 1152, 1–24, 2014.

Ouédraogo, M., Paturel, J. E., Mahé, G., and Servat, E.: Conséquences des déficits pluviométriques observés depuis le début des années 1970 en Afrique de l'Ouest et Centrale : normes hydrologiques, gestion et planification des ressources en eau, IAHS Pub., 274, 149–156, 2002.

Ozer, P., Hountondji, Y. C., and Laminou, M. O.: Évolution des caractéristiques pluviométriques dans l'Est du Niger de 1940 à 2007, Geo-Eco-Trop, Revue de Géographie et d'Écologie Tropicale, 33, 11–30, 2009.

Pettitt, A. N.: A non-parametric approach to the change-point problem, Appl. Stat., 28, 126–135, 1979.

Reizer, C.: Aménagement piscicole du lac artificiel d'Ayamé, CTFT, 30, 1–108, 1967.

Renard, B.: Détection et prise en compte d'éventuels impacts du changement climatique sur les extrêmes hydrologiques en France, INP Grenoble, 1–364, 2006.

Sarr, B., Atta, S., and Kafando, L.: Revue des indices climatiques utilisés dans les systèmes d'assurances agricoles indicielles en Afrique, Sécheresse, 23, 255–260, 2012.

WMO: World Meteorological Organization (2008), Guidelines on analysis of extremes in a changing climate in support of informed decisions for adaptation, WCDMP-No. 72, Geneva, Switzerland, 1–52, 2008.

Proc. IAHS, 383, 327–339, 2020
https://doi.org/10.5194/piahs-383-327-2020

Water replenishment for ecological flow with E-WAS framework: a case study of the Longgang River Basin, Shenzhen, China

Ziqi Yan, Zuhao Zhou, Xuefeng Sang, Hao Wang, and Yangwen Jia

State Key Laboratory of Simulation and Regulation of Water Cycle in River Basin,
China Institute of Water Resources and Hydropower Research, Beijing 100038, China

Correspondence: Ziqi Yan (yanzq@iwhr.com)

Published: 16 September 2020

Abstract. With rapid urbanization, there will be more conflict between human systems and the riverine ecological system, and therefore, ecological operations, practices and research must involve the ecological water replenishment of entire river basins with new modeling tools. In this study, we establish an ecological flow-oriented water resource allocation and simulation framework (E-WAS). Virtual reservoirs and ecological units are added to the water resources network. With new water balance equations for virtual reservoirs and ecological units, the E-WAS can simulate the ecological replenishment process in a river basin and can provide a recommended water replenishment scheme that considers optimization principles. The E-WAS was applied in the Longgang River Basin, Shenzhen, China. 17 ecological units and 45 water supply nodes are considered in the model. A water replenishment scheme that used water from 31 reservoirs and reclaimed water from 7 water sewage plants was selected. This scheme significantly increased the satisfactory degree of ecological water demand and efficiently supported the formulation of a control scheme for the water environment of a basin. The E-WAS framework is similar to model plug-ins but helps to avoid the large workload that is required for model redevelopment and can expand the functions of models quickly.

1 Introduction

Replenishing ecological water through reservoir operation has become a major metric for the restoration of river ecosystems worldwide. In general, an ecological replenishment scheme is implemented to restore the natural ecosystem of a river basin consists by adjusting reservoir release and restoring the natural rhythm of the river flow as much as possible, at the same time ensuring flood control and no significant decrease in benefit (e.g., power generation and irrigation) (Higgins and Brock, 1999; Symphorian et al., 2003; Gippel et al., 2002; Dong, 2007). Existing research regarding ecological reservoir operation mainly focuses on the theories of and calculation methods for the ecological water requirements of river channels, optimization of hydraulic and hydropower engineering systems, water and sediment control in reservoirs, ecological flood control, water quality protection, habitat improvement for organisms in reservoirs and downstream river channels, the evaluation of ecological operation schemes and legislation for ecological operation (Junk, 1982; Petts, 1996; Hughes and Hannart, 2003). Many researchers have studied methods of improving river ecology (Wang et al., 2014; Galat et al., 1998; Day et al., 2012), and noted that the water replenishment strategy is effective for river protection. Although reservoir construction affects the original hydrological form of a river, reservoirs are also used for flood regulation, which can mitigate the imbalance of water in rivers. The provision of supplemental water to rivers by reservoirs during emergencies has been discussed by Yang et al. (2008). Several scenarios have been initiated in China to release reservoir water into adjacent rivers. For example, water was released from the Nenjiang reservoir into the Zhalong wetlands (Zhou et al., 2007), from the Xiaolangdi hydropower reservoir into the Yellow River (Cui et al., 2009), and from upstream reservoirs into the Tarim River (Huang and Pang, 2010).

In this study, based on a basin-scale water resource allocation and simulation model (WAS), we establish an ecological flow-oriented water resource allocation and simulation framework (E-WAS) to expand the functions of the WAS model. E-WAS can account for ecological flows, in which multiple water resources are allocated to meet the water requirements for various purposes. This study will help decision-makers formulate ecological water replenishment schemes for river basins at the planning level. The remainder of this paper is structured as follows. Section 2 describes the WAS model and derives the E-WAS framework, Sect. 3 describes the application of the E-WAS framework in the Longgang River Basin, and Sect. 4 discusses the results. Section 5 concludes the paper.

2 The E-WAS framework

The E-WAS (ecological flow-oriented WAS model) framework (Yan et al., 2018) is based on the traditional water resource allocation model (WAS), which can dually simulate natural-artificial water cycles that are influenced by both nature and humans (Wang et al., 2014; Zhai et al., 2017; Sang et al., 2010). In order to control multiple types of water sources and dynamically allocate water resources to replenish the ecological water of the river in the model, a virtual reservoir is used in the E-WAS system network and is used in conjunction with the ecological units. The storage capacity of the virtual reservoir is set to "zero" so it does not regulate and store water. The function of the virtual reservoir is to receive the water that is released from the reservoir upstream of the ecological units, the inter-zonally generated water and the water that is exported from external sources. All the water that is imported into the virtual reservoir will be directly supplied to the ecological units to meet their ecological water requirements. Because the water consumption for the ecological unit is also set to "zero", the ecological water that is supplied to the present unit can all be released to the downstream river channel. In addition, because its storage capacity is "zero", the surplus water in the virtual reservoir will also be directly released to the downstream river channel.

Here, an example is presented to illustrate the new water resource allocation network in E-WAS (Fig. 1). The water system consists of a reservoir A, a social water consumption unit B and a control section of ecological flow C at the location where a tributary flows into the main stream (Fig. 1a). First, an ecological unit C′ is set by generalizing the ecological flow section C (Fig. 1b). Then, a virtual reservoir D is set upstream of the ecological unit C′ (Fig. 1c) and is used in conjunction with ecological unit C′. The storage capacity of virtual reservoir D is set to "zero". The function of virtual reservoir D is to receive the water that is released from the upstream reservoir A and the inter-zonally generated water from the area between A and C′, where it does not regulate and store water. All of the water that is imported into virtual

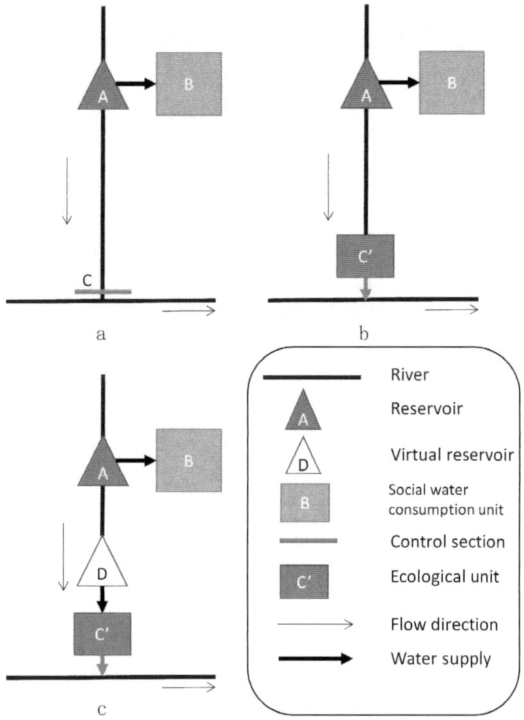

Figure 1. Water resource allocation network in E-WAS (Yan et al., 2018).

reservoir D is directly supplied to ecological unit C′ to meet its ecological water requirements. Because there is no water consumption in ecological unit C′, all of the ecological water that is supplied to ecological unit C′ is released to the downstream river channel. Because its storage capacity is "zero", the surplus water in virtual reservoir D will also be directly released to the downstream river channel.

The computation within the E-WAS framework should follow the following rules: The water (flood water during the flood season, water released from the reservoir, etc.) that is released from the reservoir upstream of the virtual reservoir and the inter-zonal confluent water all flows into the virtual reservoir. The inflow water is first supplied to the ecological unit, and the surplus water is released. If the inflow water from the upstream reservoir and the inter-zonal confluent water cannot meet the water requirements of the ecological unit, a water transfer command will be sent to the upstream reservoir. Upon receiving this command, the upstream reservoir will transfer water to the virtual reservoir to replenish the downstream ecological unit with water.

The E-WAS framework is capable of coupling reservoir operations for ecological flow and water resource allocation optimization. It can provide an optimal water allocation scheme with ecological flow requirements by improving the computational procedure in WAS and supplementing it with a virtual reservoir and an ecological unit, without imparting significant changes in the internal core algorithm of the WAS model. The equations under the E-WAS framework undergo the following changes.

1. Water balance equation for a virtual reservoir:

$$\mathrm{QR} = 0 \tag{1}$$

$$Q_{\mathrm{rin}} + Q_{\mathrm{rin0}} + Q_{\mathrm{rin1}} = Q_{\mathrm{rko}} + Q_{\mathrm{rfo}} \tag{2}$$

where QR is the amount of water stored in the virtual reservoir ($10\,000\,\mathrm{m}^3$); Q_{rin} is the inter-zonal water inflow; Q_{rin0} is the amount of water released from the upstream physical reservoir ($10\,000\,\mathrm{m}^3$); Q_{rin1} is the amount of water imported from the upstream physical reservoir ($10\,000\,\mathrm{m}^3$); Q_{rko} is the amount of water supplied by the virtual reservoir ($10\,000\,\mathrm{m}^3$); and Q_{rfo} is the amount of water released from the virtual reservoir ($10\,000\,\mathrm{m}^3$). Q_{rko} refers to the amount of water supplied to meet ecological water requirements. When the inter-zonal water inflow and the water that is released from the upstream reservoir and received by the virtual reservoir cannot meet the corresponding water requirements of the ecological unit, the virtual reservoir will send a water transfer command to the upstream physical reservoir. If there is surplus water in the upstream physical reservoir after it completes the water supply task, it will transfer water to the virtual reservoir, and the amount of water transferred is Q_{rin1}. The amount of water imported from the physical reservoir can then be treated as the water transferred for ecological replenishment, i.e., the water is specifically released by the physical reservoir to meet the ecological flow requirements of the river channel. While the water released from the upstream reservoir, or Q_{rin0}, also forms an ecological flow, it is not actively released by the upstream reservoir to meet the ecological target.

2. Water balance equation for an ecological unit:

$$\mathrm{QU} = \mathrm{QU}_{\mathrm{resu}} + \mathrm{QU}_{\mathrm{wru}} + \mathrm{QU}_{\mathrm{oresu}} \tag{3}$$

where QU is the water used by the ecological environment within the time period ($10\,000\,\mathrm{m}^3$); $\mathrm{QU}_{\mathrm{resu}}$ is the water supplied to the virtual reservoir ($10\,000\,\mathrm{m}^3$); $\mathrm{QU}_{\mathrm{wru}}$ is the water supplied by reclaimed water sources (if there is a water purification plant in this river segment, the released tail water can be used as the water supplied by reclaimed water sources, and the quality of reclaimed water must meet the high standard for ecological flow) ($10\,000\,\mathrm{m}^3$); and $\mathrm{QU}_{\mathrm{oresu}}$ is the water imported into the reservoir from external sources ($10\,000\,\mathrm{m}^3$).

3. Objective functions and their solutions:

Regional water supply security and equity are the core objectives of water resource allocation. The WAS model uses optimum equity and a minimum water supply deficiency ratio as the objective functions for water resource allocation optimization (Yang et al., 2016).

a. Equity objective:

$$\mathrm{Min}\,F(x) = \sum_{y=1}^{\mathrm{myr}} \sum_{n=1}^{12} \sum_{h=1}^{\mathrm{mh}} q_h \cdot \mathrm{GP}(X_h) \tag{4}$$

$$\mathrm{GP}(X_h) = \sqrt{\frac{1}{\mathrm{mu}-1} \sum_{u=1}^{\mathrm{mu}} \left(x_h^u - \overline{x_h}\right)^2} \tag{5}$$

where $F(x)$ is the equity objective; $\mathrm{GP}(x_h)$ is the equity function; q_h is the sectorial user penalty function; x_h^u is the water deficiency ratio for sectorial user h in unit u; $\overline{x_h}$ is the average water deficiency ratio for sectorial user h in unit u; myr is the number of years within the calculation period; n is the monthly id within the year; mh is the number of sectorial water-use types in the region; and mu is the number of units in the region.

b. Minimum water deficiency ratio objective:

$$\mathrm{Min}\,Y(x) = \sum_{y=1}^{\mathrm{myr}} \sum_{n=1}^{12} \sum_{h=1}^{\mathrm{mh}} q_h \cdot \mathrm{SW}(X_h) \tag{6}$$

$$\mathrm{SW}(X_h) = \frac{1}{\mathrm{mu}} \sum_{u=1}^{\mathrm{mu}} \left| \left(x_h^u - \mathrm{Sob}_h^n\right) \right| \tag{7}$$

where $Y(x)$ is the water supply stress objective; $\mathrm{SW}(x_h)$ is the water supply stress function; q_h is the sectorial user penalty function; x_h^u is the water deficiency ratio for sectorial user h in unit u; Sob_h^n is the ideal value of the water supply stress objective of each month for sectorial user h in the region; myr is the number of years within the calculation period; n is the identification of the month within the year; mh is the number of sectorial water-use types in the region; and mu is the number of units in the region.

3 Application

3.1 Study area and data

The Longgang River Basin is located in northeastern Shenzhen (Fig. 2). The Longgang River Basin is in the South Asian tropical monsoon climate zone. The annual mean temperature is 22.3 °C. The average annual precipitation is 2073.5 mm, precipitation during the flooding season (April–September) accounts for proximately 85 % of the annual precipitation, and precipitation during the dry season (October–March) only proximately 15 %. With rapid economic development, water shortages in the Longgang River basin have become increasingly serious. Historically abundant local water resources are becoming increasingly precious. Thus, no ecological flow is set during the operation scheme of the local reservoirs in this river basin, and all water resources are stored in the reservoirs for urban uses, even if it is unnecessary in wet seasons. Thus, most river channels in the basin are dry. In contrast, some usable reclaimed water resources remain in the Longgang River Basin. Although the water from sewage plants is sufficiently clean, further research is required to determine how it can be used for ecological flow. Thus, it is necessary to perform a systematic analysis of water resource allocation to formulate a scheme to ensure ecological flows in river channels with multiple types of water sources in the Longgang River Basin.

According to the *Statistical Handbook of Water Affairs of Shenzhen* and the *Shenzhen Municipality's Plan for Optimization and Adjustment of Reservoirs*, there are 18 reservoirs in the Longgang River Basin. Long-term hydrological and meteorological data (from 1985 to 2015) were obtained from in the *Statistical Yearbooks of Hydrological Data of Shenzhen*. Information for the water system and river network was extracted based on a 30 × 30 m DEM.

3.2 Control sections and ecological flow requirements

To achieve refined water replenishment in the Longgang River Basin, control sections for ecological flow requirements are determined based on the following principles. (1) Particular consideration is given to the river sections where there are sudden changes in parameters (e.g., the flow in the river segment). (2) The catchment areas that are controlled by each section can be treated as a relatively independent sub-ecological areas that are capable of performing specific ecological functions. According to these principles, 17 sections of the Longgang River are selected as control sections for ecological water requirements, of which 3 are sections of the main stream (in the upper, middle and lower reaches) and 14 are sections of tributaries (Fig. 3).

Ecological flow requirements

There are several methods for calculating the ecological flow in a river, such as the 7Q10 method (Singh and Stall, 1974), the wetted perimeter method (Gippel and Stewardson, 1998), the R2CROSS method (Gregoiy, 1996), IFIM (Gore and Nestler, 1988), and the Tennant method (Tennant, 1976). Based on previous studies that considered the ecological water requirements of Shenzhen, the Tennant method is used to determine the minimum ecological flow requirements. Based on this method, 30 % (non-flood season) or 40 % (flood season) of the average annual runoff is used as the suitable ecological water requirements of each river channel (Mao et al., 2009; Armbruster, 1976; Binns and Eiserman, 1979; King and Louw, 1998). Based on the specific characteristics of the Longgang River Basin, April through October is set as the flood season and November through March of the following year is set as the non-flood season (Table 1).

3.3 Water resource allocation network based on the E-WAS

To achieve ecological water replenishment in the Longgang River Basin, it is necessary to first establish a water resource allocation network under E-WAS, in which social and economic water consumption must be taken into consideration. Particular consideration should be given to ecological flow requirements at the control sections of the rivers. Thus, the Longgang River Basin is divided into 5 computational units, of which one is a social and economic water consumption unit, and 17 are ecological units for ecological water requirements (Table 2).

By analyzing the reservoirs and sewage plants in the Longgang River Basin, 45 water supply nodes are established, of which 31 are reservoir nodes, 17 are virtual reservoir nodes and 7 are reclaimed water supply nodes. All the reclaimed water in this network is derived from treated wastewater used for a certain purpose after strict treatment. Due to the high discharge standards, the reclaimed water can be used for water replenishment in the river, and is treated as clean water in the model. In addition, an external water source must be considered, namely, the Eastern water diversion system. Figure 4 shows the water resource network based on the E-WAS framework. Each tributary is treated as an ecological unit with ecological water requirements. In addition, the upper, middle and lower reaches of the main stream are also treated as an ecological unit with ecological water requirements.

3.4 Model calibration and validation

Model calibration and validation is a key process in verifying model simulation accuracy and calibrating model parameters. The model used in this study involves the water cycle module and the water resource allocation module. First, the water cycle module is validated mainly based on runoff at the key sections. Then, the water resource allocation scheme is examined based on the published amount of water supply for social and economic purposes.

3.4.1 Runoff simulation validation

The model simulation results were evaluated based on the Nash–Sutcliffe efficiency coefficient (NSE) and the regression coefficient (R^2) between the simulated and observed runoff data from 1985 to 2015 from the stations in the lower reaches of the Longgang River. Model simulations were performed for the period from 1961 to 2015. The first 30 years were used for model calibration, and the last 25 years were used for model validation. The runoff simulation results show that the model has an NSE of 0.86 and an R^2 of 0.87 for the calibration period and an NSE of 0.84 and an R^2 of

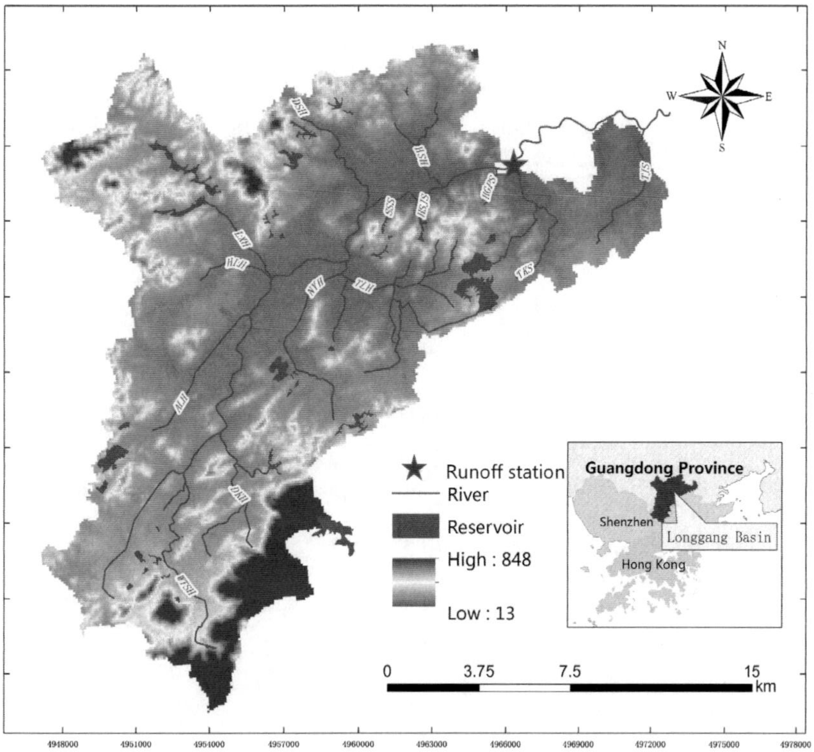

Figure 2. The Longgang River Basin.

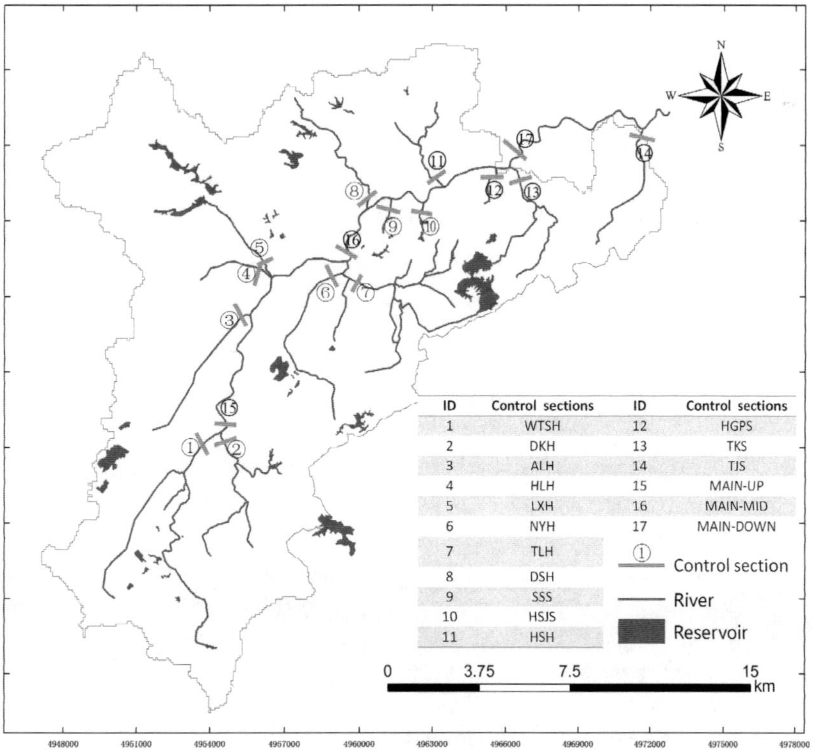

Figure 3. Control sections for ecological flow in the Longgang River Basin.

Table 1. Suitable ecological water requirements (10 000 m^3).

ID	Control sections	Jan	Feb	Mar	Apr	May	Jun	Jul	Aug	Sep	Oct	Nov	Dec
1	WTSH	13.8	22.9	37.2	105.8	170.9	232.1	195.3	206.3	144.2	47.2	16.6	15.0
2	DKH	11.0	18.3	29.6	84.4	136.3	185.1	155.8	164.5	115.0	37.6	13.3	11.9
3	ALH	9.0	14.9	24.2	68.9	111.3	151.1	127.2	134.3	93.9	30.7	10.8	9.7
4	HLH	6.4	10.6	17.2	49.1	79.3	107.8	90.7	95.8	66.9	21.9	7.7	6.9
5	LXH	19.6	32.5	52.7	150.0	242.3	329.1	276.9	292.4	204.4	66.9	23.6	21.2
6	NYH	21.5	35.6	57.8	164.5	265.7	360.9	303.6	320.7	224.1	73.4	25.8	23.3
7	TLH	13.0	21.6	35.1	99.9	161.3	219.1	184.4	194.7	136.1	44.5	15.7	14.1
8	DSH	34.9	57.8	93.8	267.1	431.3	585.8	492.9	520.6	363.8	119.1	42.0	37.8
9	SSS	0.9	1.4	2.3	6.6	10.6	14.5	12.2	12.9	9.0	2.9	1.0	0.9
10	HSJS	0.9	1.4	2.3	6.6	10.6	14.5	12.2	12.9	9.0	2.9	1.0	0.9
11	HSH	18.1	30.0	48.6	138.5	223.6	303.7	255.6	269.9	188.7	61.7	21.8	19.6
12	HGPS	0.9	1.4	2.3	6.6	10.6	14.5	12.2	12.9	9.0	2.9	1.0	0.9
13	TKS	9.0	14.9	24.2	68.9	111.3	151.1	127.2	134.3	93.9	30.7	10.8	9.7
14	TJS	5.1	8.5	13.8	39.2	63.4	86.1	72.4	76.5	53.5	17.5	6.2	5.5
15	MAIN-UP	24.8	41.2	66.8	190.3	307.2	417.3	351.1	370.8	259.2	84.8	29.9	26.9
16	MAIN-MID	94.3	156.5	253.8	722.8	1167.1	1585.2	1333.8	1408.7	984.6	322.2	113.5	102.2
17	MAIN-DOWN	145.6	241.6	391.7	1115.7	1801.6	2447.0	2058.8	2174.5	1519.9	497.4	175.2	157.7

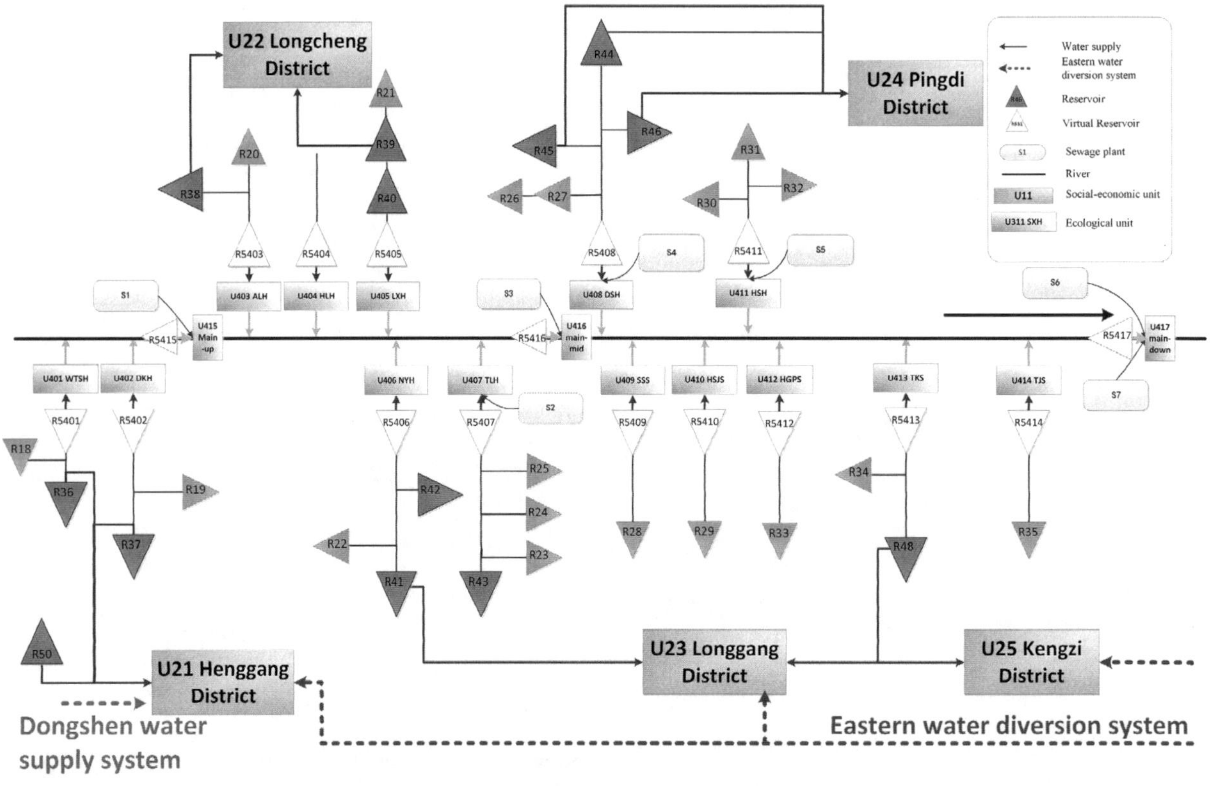

Figure 4. Water resource allocation network of the Longgang River Basin.

Table 2. Relationship between control sections and ecological units.

Control section ID	Control section name	Ecological unit ID
1	WTSH	U401
2	DKH	U402
3	ALH	U403
4	HLH	U404
5	LXH	U405
6	NYH	U406
7	TLH	U407
8	DSH	U408
9	SSS	U409
10	HSJS	U410
11	HSH	U411
12	HGPS	U412
13	TKS	U413
14	TJS	U414
15	MAIN-UP	U415
16	MAIN-MID	U416
17	MAIN-DOWN	U417

0.91 for the validation period. Figure 5 shows the simulated runoff of the Longgang River. From the perspective of water cycle simulation, the model has nearly ideal accuracy.

3.4.2 Validation of water supply for social and economic purposes

The water supply for social and economic purposes in the region is the total amount of water supplied for industrial, domestic and agricultural purposes within the Longgang District, the social and economic unit in the Longgang River Basin. Figure 7 shows a comparison of the simulated water supply for social and economic purposes and statistical data for the water supply that were published in the Shenzhen Water Resource Bulletins (from 1995 to 2015). The error between the simulated water supply and the corresponding statistical data is within 10 % for 80 % of the years between 1995 and 2015 (Fig. 6). The 20-year average water supply that was obtained from the E-WAS was 195 million m^3, which differs from the published average annual water supply (198 million m^3) by 1.5 million m^3; i.e., the relative error is 1.5 %. This indicates that the accuracy of the model in the generalized water resource network in the Longgang River Basin and the simulated water supply are acceptable.

4 Results and discussion

4.1 Water allocation schemes

To analyze the ecological water replenishment scheme for the Longgang River Basin, different operational schemes must be considered to configure the E-WAS framework. Considering multiple combinations of reclaimed water and surface water (reservoirs), the following three schemes are set to identify the ability of water replenishment for river channels with different water resources (Table 3).

Table 3. Scheme for E-WAS calculation.

Scheme	Description
1	Reference scheme without replenishment of ecological flow
2	Replenish ecological flow with reclaimed water
3	Replenish ecological flow with reservoir release

The allocation of water resources under the three schemes is optimized using the E-WAS framework, with suitable ecological flows as the main operation constraints and targets. In this study, the satisfactory degree of ecological water demand (SDE, Eqs. 10–11) is used as an index to judge flow conditions in the river.

$$\mathrm{SDE}(x)_{\mathrm{month}} = 1 - \frac{\max\left(0, \mathrm{de}_i(x) - \mathrm{rep}_i(x)\right)}{\mathrm{de}_i(x)} \tag{8}$$

$$\mathrm{SDE}(x)_{\mathrm{year}} = 1 - \frac{\sum_{i=1}^{12} \max\left(0, \mathrm{de}_i(x) - \mathrm{rep}_i(x)\right)}{\sum_{i=1}^{12} \mathrm{de}_i(x)} \tag{9}$$

where $\mathrm{SDE}(x)_{\mathrm{month}}$ is the monthly satisfactory degree of ecological water demand of river section x; $\mathrm{SDE}(x)_{\mathrm{year}}$ is the annual satisfactory degree of ecological water demand of river section x; $\mathrm{rep}_i(x)$ is the monthly/annual water replenishment of river section x; and $\mathrm{de}_i(x)$ is the monthly/annual ecological water demand of river section x. Table 4 shows the annual SDE (Eq. 9) associated with each scheme as well as the water replenishment capacity (WRC, 10 000 m^3) of multiple water sources in a typical dry year ($P = 75$ %). The WRC is given by the E-WAS framework after long term water allocation calculation.

For the ecological water replenishment of a river basin, it is necessary to account for the extent to which the water requirements are satisfied, as well as various other factors such as engineering and economic factors. Thus, the following principles are proposed for the optimization of the ecological water allocation scheme:

- The reservoirs are prioritized to ensure that the water requirements for social and economic purposes in the Longgang River Basin are satisfied.
- Regarding water replenishment measures for tributaries, the priority is given to (1) reclaimed water from the upper reaches of the tributaries, followed by (2) water released from the reservoirs and (3) reclaimed water from

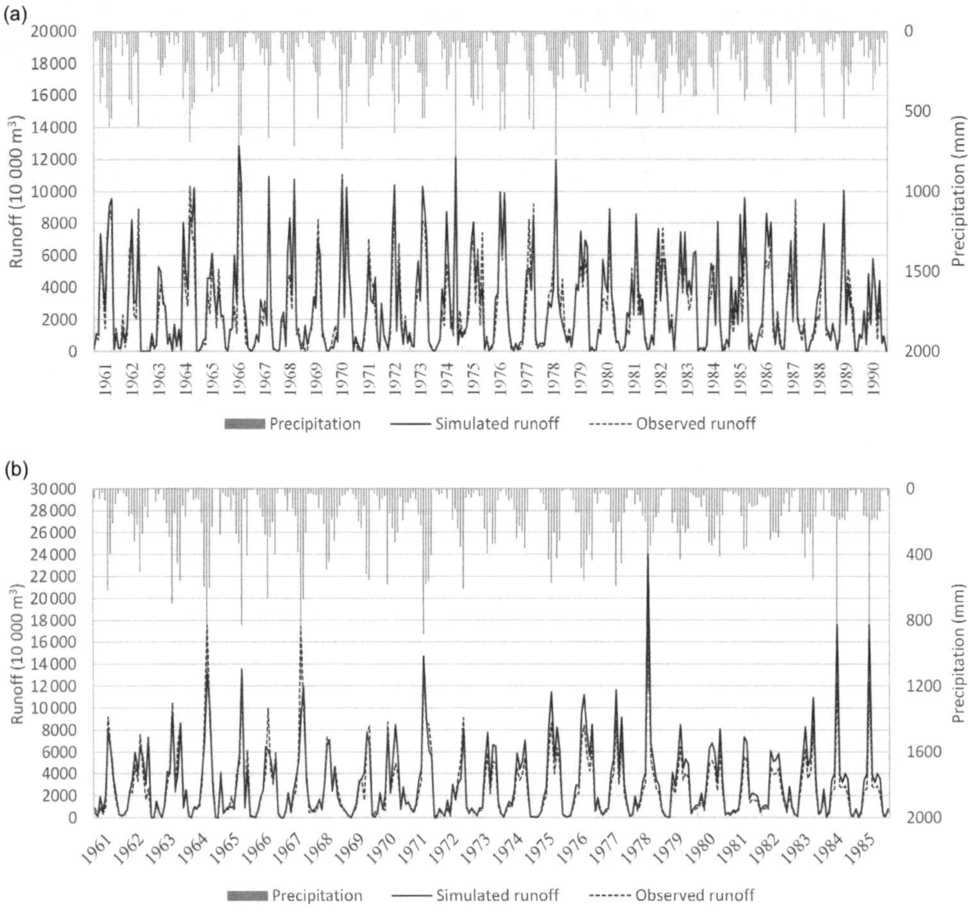

Figure 5. Runoff calibration (a) and validation (b) of the E-WAS framework.

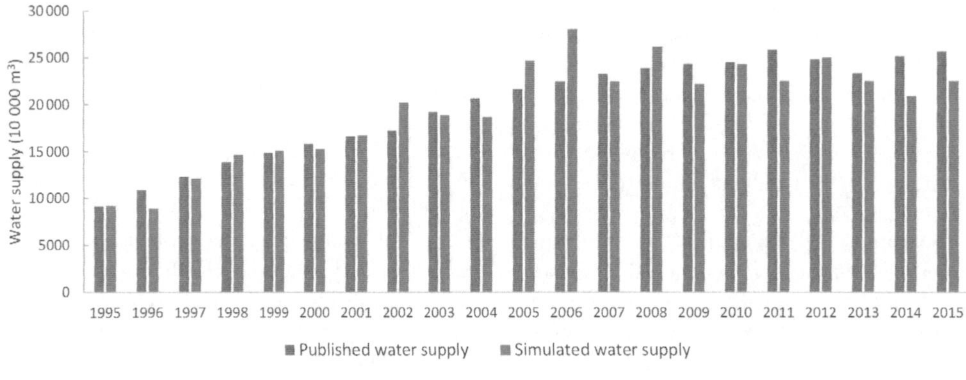

Figure 6. Comparison of water use between simulated and published data.

the lower reaches of the tributaries (water extraction is required).

- After the ecological flow in the tributaries has been increased, the main stream is replenished with water from the tributaries. It is first ensured that the ecological water requirements of each tributary are satisfied, which will indirectly ensure that the ecological water requirements of the main stream are also satisfied.

- For a river with no reclaimed water and reservoir water resources, measures that extract water from the main

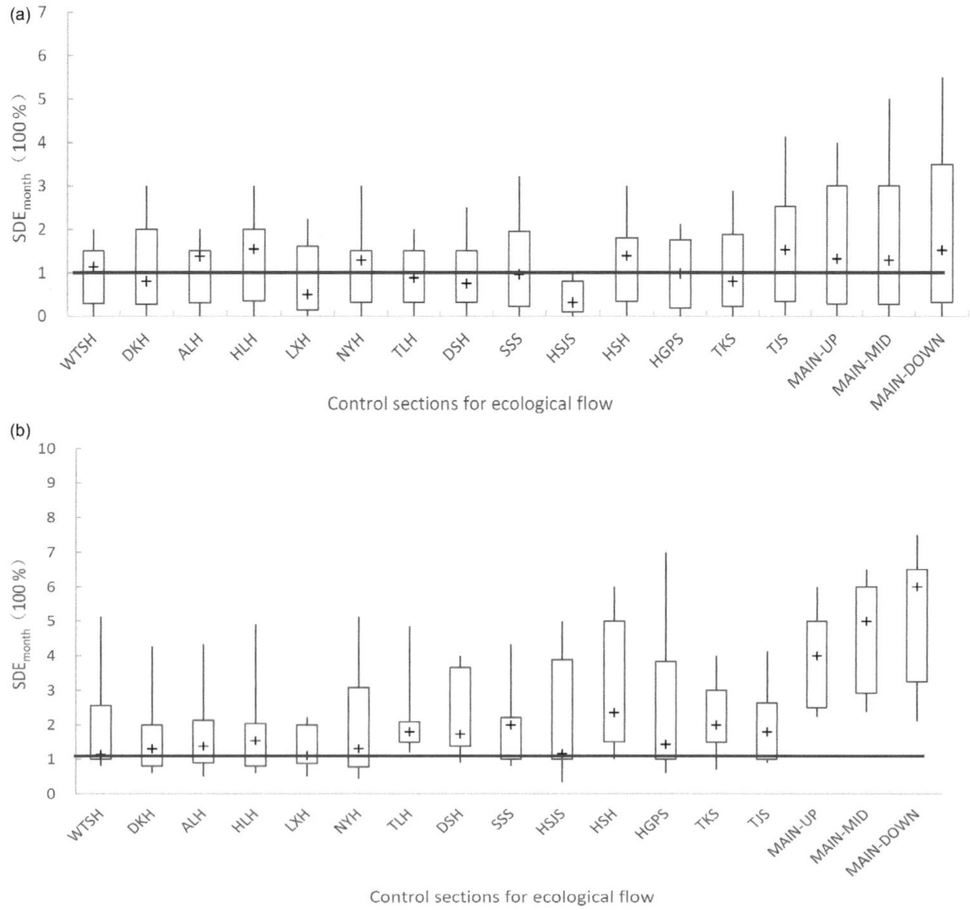

Figure 7. Different SDE_{month} of each section before **(a)** and after **(b)** implementing the recommended water replenishment scheme.

stream or connect water systems to achieve ecological water replenishment will be considered.

Based on the above principles, as well as the water deficiencies and the replenishment capacity of each scheme that is shown in Table 3, an optimized ecological water replenishment scheme for the Longgang River Basin is formulated (Table 5). Reclaimed water generated from the sewage plants is used to replenish tributaries such as the WTSH, the DKH, the DSH, the HSH, the TKS and the TJS. Ten reservoirs (R18, R19, R20, R40, R42, R22, R43, R25, R23 and R33) are used to improve the ecological flows in tributaries such as the WTSH, the DKH, the ALH, the LXH, the NYH, the TLH and the HGPS. The flow in the main stream is mainly indirectly replenished through the cumulative effects achieved by replenishing tributaries; i.e., the main stream is replenished with water from the tributaries. The SDEs of the HLH, SSS and HSJS in Scheme 1 are relatively high (89 %), so there is no need for water replenishment in these two tributaries.

4.2 Effects of the recommended water replenishment schemes

To further analyze the effectiveness of the E-WAS framework, the SDE is analyzed from temporal and spatial perspectives. Boxplots were produced to analyze the monthly SDE (Eq. 8) in each unit in a typical dry year ($P = 75\,\%$) before and after the ecological water replenishment scheme (Fig. 7). Without a replenishment scheme, there are fewer than six months in which the ecological flows at the 17 control sections are ensured and the SDEs of tributaries WTSH, TLH, LXH and HSJS are relatively low. With the recommended water replenishment scheme, the SDE_{year} increases significantly, and there are more than six months in which the ecological flow at each section is ensured. In addition, the SDE_{month} of the aforementioned sections also increases considerably.

The recommended ecological water replenishment scheme mainly focuses on tributaries characterized by ecological water deficiencies. Through the operation of the reservoirs and the use of reclaimed water, the ecological flows in the tribu-

Table 4. SDE and WRC under three E-WAS schemes in a typical dry year ($P = 75\,\%$).

ID	Control sections	Scheme 1		WRC in Scheme 2 ($10\,000\,m^3$)	WRC in Scheme 3 ($10\,000\,m^3$)
		SDE_{year} (%)	Water deficiency ($10\,000\,m^3$)		
1	WTSH	83	205	1080	348
2	DKH	84	154	438	17
3	ALH	84	126	0	89
4	HLH	89	73	0	0
5	LXH	62	650	1800	528
6	NYH	83	319	0	225
7	TLH	83	194	0	268
8	DSH	83	518	2737	86
9	SSS	89	8	0	46
10	HSJS	89	10	0	127
11	HSH	83	269	350	178
12	HGPS	80	13	0	66
13	TKS	77	181	2880	188
14	TJS	84	72	1080	75
15	MAIN-UP	83	359	7200	364
16	MAIN-MID	82	1721	10 800	1470
17	MAIN-DOWN	82	2841	36 360	2261

Table 5. Recommended ecological water replenishment scheme for a typical dry year ($P = 75\,\%$).

ID	Control section	Water resource for replenishment in recommended scheme	Water replenishment ($10\,000\,m^3\,yr^{-1}$)	SDE_{year}
1	WTSH	R18, S1	205	100 %
2	DKH	R19, S1	154	100 %
3	ALH	R20	89	96 %
4	HLH	–	–	89 %
5	LXH	R40	528	93 %
6	NYH	R42,R22	225	100 %
7	TLH	R43, R25, R23	194	100 %
8	DSH	S4	518	100 %
9	SSS	–	–	89 %
10	HSJS	–	–	89 %
11	HSH	S5	269	100 %
12	HGPS	R33	18	100 %
13	TKS	S6	181	100 %
14	TJS	S7	72	100 %
15	MAIN-UP	From tributaries	359	100 %
16	MAIN-MID	From tributaries	1395	96 %
17	MAIN-DOWN	From tributaries	2841	100 %

taries are ensured, which in turn impacts the main stream of the Longgang River and significantly increases the SDE.

4.3 Improvement of the allocation model

The E-WAS framework is an improvement to the WAS model and includes both constraint-type and target-type features in the ecological operation model. The effectiveness of the E-WAS framework in ensuring that the ecological water requirements of the river and allocating water for social and economic purposes is evaluated by comparing the statistical indices of the results obtained from the WAS model and the E-WAS framework. Two schemes are used (Table 6). For scheme 1, the WAS model is employed to simulate the allocation of water resources in the study area. The results that were obtained using the E-WAS framework (Sect. 4.1) are used as the results for scheme 2. The value of the global ob-

Table 6. Different objective function values in WAS and E-WAS.

Objective functions	Model	
	WAS	E-WAS
GOB (global objective function)	90 %	93 %
SDSE (satisfactory degree of social and economic water demand)	93 %	92 %
SDE_{year} (satisfactory degree of ecological water demand)	74 %	95 %

jective function is the mean of the equity objective function (Eq. 4) and the water deficiency objective function (Eq. 6) that were presented in Sect. 2.

In scheme 1, the WAS model, which does not consider the ecological water requirements of the river channels, is employed, and only water consumption for social and economic purposes is considered during the allocation process. By allocating various water resources, the SDSE reaches 93 percent, and the GOB also reaches 90 %; however, the SDE_{year} is relatively low (only 74 %). In scheme 2, the E-WAS computational framework, which is oriented to ecological water replenishment, is used. This framework considers the water requirements for social and economic purposes and considers the ecological flows in the main stream and tributaries. Under scheme 2, the GOB reaches 93 %; in addition, the SDSE decreases slightly to 92 %, and the SDE_{year} reaches 95 %.

5 Conclusions

1. By generalizing the river sections with ecological water requirements, which are treated as ecological units, ecological control targets are included in the E-WAS framework. By establishing links between various water sources and economic and ecologic units, a multi-objective long-term simulation is realized to obtain an allocation scheme that satisfies ecological flows at the control sections of the river channels. Under the E-WAS framework, simulations of ecological replenishment in a river basin considers SDEs and various other factors (e.g., engineering and economic factors). The ecological water replenishment scheme is further optimized based on all the usable water sources and the principle that the main stream and tributaries are coordinated, and the upper and lower reaches are used to replenish one another. Compared with other ecological operation models and ecological water replenishment method, the framework proposed in this paper does not alter the internal computational structure of the model with the aid of virtual reservoirs and units. Thus, it is similar to a model plug-in, which helps researcher to avoid the large workload that is required for model redevelopment and can expand the function of model relatively quickly.

2. A case study of the Longgang River Basin in Shenzhen is performed to investigate the ecological operation with E-WAS. Fourteen ecological units with water requirements and one social-economic unit are set. The suitable ecological water requirements of the 17 ecological units are determined. A total of 45 water supply nodes, including reservoirs and sewage plants are considered in a coordinated manner. Based on the E-WAS, the ecological water requirements at each section are calculated, and the SDE under current conditions can be estimated. By comparing different water replenishment schemes, a water replenishment scheme that uses water from seven reservoirs and reclaimed water from six water sewage plants is selected to replenish the ecological flow at the 17 control sections. This scheme significantly increases the SDE (from 62 % to 89 % under the reference scheme to 89 % to 100 %), and thus efficiently supports the formulation of a control scheme for the water environment of Shenzhen.

3. Computational frameworks such as E-WAS can be flexibly used for various regions. Particularly, under these frameworks, significantly different results with respect to the adjustment of priority levels for the water supply for various sectors will be obtained (in this study, the priority levels for water supply for ecological purposes and for social-economic purposes are the same). This is determined by the levels of importance that are attached to water consumption for social and economic sectors and the ensuring of ecological flows in the river channels in the region. When ecological improvement is set to a high priority level in the E-WAS, the SDE will increase further. Under this condition, due to the constraint of the global objective function, there will not be relatively large deficiencies in water use for social and economic purposes.

4. At present, the E-WAS framework does not include a prediction function, and it is distinguished from a real time reservoir operation model. Thus, the framework is more useful in the planning stage. It can be used as a tool for controlling the use of water resources for water replenishment planning and to provide scheme for regional water control planning to promote rational management of water resources, alleviate contradictions to water demand over supply and improve the flow into the river on the regional and even national levels.

Data availability. All data can be accessed via the references.

Author contributions. The research was designed, the necessary data collected and conclusions drawn by all five authors in collaboration. ZY wrote the main part of the article, ZZ made comments and suggested improvements. YJ and HW served as the primary writer and expert on water replenishment. XS served as expert on model.

Competing interests. The authors declare that they have no conflict of interest.

Special issue statement. This article is part of the special issue "Hydrological processes and water security in a changing world". It is a result of the 8th Global FRIEND–Water Conference: Hydrological Processes and Water Security in a Changing World, Beijing, China, 6–9 November 2018.

Acknowledgements. This study was supported by the National Key Research and Development Program of China (no. 2016YFC0402405), the National Natural Science Foundation of China (nos. 51779270, 51309248), Shenzhen Project (SZCG2016121595B), the Foundation of CWEPC (CSCEC-PSH-2017-03), and the Foundation of SKL-WAC (SKL2018TS04).

Financial support. This research has been supported by the National Key Research and Development Program of China (grant no. 2016YFC0402405), the National Natural Science Foundation of China (grant nos. 51779270 and 51309248), Shenzhen Project (grant no. SZCG2016121595B), the Foundation of CWEPC (grant no. CSCEC-PSH-2017-03), and the the Foundation of SKL-WAC (grant no. SKL2018TS04).

References

Armbruster, J. T.: An infiltration index useful in estimating low-flow characteristics of drainage basins, J. Res. USDS, 4, 533–538, https://doi.org/10.1255/jnirs.866, 1976.

Binns, N. A. and Eiserman, F. M.: Quantification of fluvial trout habitat in Wyoming, T. Am. Fish. Soc., 108, 215–228, https://doi.org/10.1577/1548-8659(1979)108<215:QOFTHI>2.0.CO;2, 1979.

Cui, B. S., Yang, Q. C., Yang, Z. F., and Zhang, K. J.: Evaluating the ecological performance of wetland restoration in the Yellow River Delta, China Ecol. Eng., 35, 1090–1093, https://doi.org/10.1016/j.ecoleng.2009.03.022, 2009.

Day, J., Hunte, R., Keim, R. F., DeLaune, R., Shaffer, G., Evers, E., Reed, D., Brantley, C., Kemp, P., Day, J., and Hunter, M.: Ecological response of forested wetlands withand without Large-Scale Mississippi River input: implications for management, Ecol. Eng., 46, 57–67, https://doi.org/10.1016/j.ecoleng.2012.04.037, 2012.

Dong, Z. R.: Multi-objective ecological operation of reservoirs, Technology of Water Conservancy and Hydropower, 38, 28–32, 2007 (in Chinese).

Galat, D. L., Fredrickson, L. H., Humburg, D. D., Bataille, K. J., Bodie, J. R, Dohrenwend, J., Gelwicks, G. T., Havel, J. E., Helmers, D. L., Hooker, J. B., Jones, J. R., Knowlton, M. F., Kubisiak, J., Mazourek, J., McColpin, A. C., Renken, R. B., and Semlitsch, R. D.: Flooding to Restore Connectivity of Regulated, Large-River Wetlands. BioScience, 48, 721–733, https://doi.org/10.2307/1313335, 1998.

Gippel, C. J. and Stewardson, M. J.: Use of wetted perimeter in defining minimum environmental flows, Regul. River., 14, 53–67, https://doi.org/10.1002/(SICI)1099-1646(199801/02)14:1<53::AID-RRR476>3.0.CO;2-Z, 1998.

Gippel, C. J., Jacobs, T., and McLeod T.: Determining environmental flow needs and scenarios for the River Murray System, Australia, Australian Journal of Water Resources, 5, 61–74, https://doi.org/10.1080/13241583.2002.11465193, 2002.

Gore, J. A. and Nestler, J. M.: Instream flow studies in perspective, Regul. River., 2, 93–101, https://doi.org/10.1002/rrr.3450020204, 1988.

Gregoiy, D. E.: Development of instream flow recommendations in Colorado using R2cross, Dept. Colorado Water Conservation Board, Denver, Colorado, USA, 1996.

Higgins, J. M. and Brock, W. G.: Overview of reservoir release improvement at 20TVA dams, J. Energ. Eng., 125, 1–17, https://doi.org/10.1061/(ASCE)0733-9402(1999)125:1(1), 1999.

Huang, T. M. and Pang, Z. H.: Changes in groundwater induced by water diversion in the Lower Tarim River, Xinjiang Uygur, NW China: evidence from environmental isotopes and water chemistry, J. Hydrol., 387, 188–191, https://doi.org/10.1016/j.jhydrol.2010.04.007, 2010.

Hughes, D. A. and Hannart, P.: A desktop model used to provide an initial estimate of the ecological instream flow requirements of rivers in South Africa, J. Hydrol., 270, 167–181, https://doi.org/10.1016/S0022-1694(02)00290-1, 2003.

Junk, W. J.: Amazonian floodplains: Their ecology, present and potential use, Revue d'Hydrobiologie Tropicale, 15, 285–301, 1982.

King, J. and Louw, D.: In stream flow assessments for regulated rivers in South Africa using the Building Block Methodology, Aquat. Ecosyst. Health, 1, 109–124, https://doi.org/10.1080/14634989808656909, 1998.

Mao, X. L., Tian, K., Li, T. H., and Xu, Y.: Characteristics of Urban Ecological Water Requirement in Baoan, Shenzhen, Acta Scientiarum Naturalium Universitatis Pekinensis, 45, 721–727, https://doi.org/10.13209/j.0479-8023.2009.107, 2009 (in Chinese).

Petts, G. E.: Water allocation to protect river ecosystems, Regul. River., 12, 353–365, https://doi.org/10.1002/(sici)1099-1646(199607)12:4/5<353::aid-rrr425>3.0.co;2-6, 1996.

Sang, X. F, Zhou, Z. H., Wang, H., Qin, D. Y., Zhai, Z. L., and Chen, Q.: Development of soil and water assessment tool model on human water use and application in the area of high human activities, Tianjin, China, J. Irrig. Drain. E., 136, 23–30, https://doi.org/10.1061/(ASCE)IR.1943-4774.0000115, 2010.

Singh, K. P. and Stall, J. B.: Hydrology of 7-day 10-yr low flows, J. Hydraul. Eng., 100, 1753–1771, 1974.

Symphorian, G. R., Madamombe, E., and Van der Zaag, P.: Dam operation for environmental water releases; the case of Osborne dam, Save catchment, Zimbabwe, Phys. Chem. Earth, 28, 985–993, https://doi.org/10.1016/j.pce.2003.08.012, 2003.

Tennant, D. L.: Instream flow regimens for fish, wildlife, recreation and related environmental resources, Fisheries, 1, 6–10, https://doi.org/10.1577/1548-8446(1976)001<0006:IFRFFW>2.0.CO;2, 1976.

Wang, J. H., Sang, X. F., Zhai, Z. L., Liu, Y., and Zhou, Z. H.: An Integrated Model for Simulating Regional Water Resources Based on Total Evapotranspiration Control Approach, Adv. Meteorol., 2014, 345671, https://doi.org/10.1155/2014/345671, 2014.

Yan, Z. Q., Zhou, Z. H., Sang, X. F., and Wang, H.: Water replenishment for ecological flow with an improved water resources allocation model, Sci. Total Environ., 643, 1152–1165, https://doi.org/10.1016/j.scitotenv.2018.06.085, 2018.

Yang, W., Yang, Z. F., and Sun, T.: A review of requirement quantity and allocation of ecological water for wetland, Wetland Sci., 6, 532–535, 2008.

Yang, X., Chen, G., Sang, X. F., Gu, S. X., and Zhou, Z. H.: Water Resources Allocation at Plateau Lakes Based on Interconnected River System Network, China Rural Water and Hydropower, 9, 205–211, 2016.

Zhai, Z. L., Sang, X. F., Chen, J., and Yang, M.: The Total Control of Water Supply and Water Consumption in Tianjin City Based on WAS Model. 2017 3rd International Conference on Green Materials and Environmental Engineering, 22–23 October 2017, Beijing, China, 2017.

Zhou, L. F., Xu, S. G., Li, Q. S., and Liu, D. Q.: Safety threshold of eco-environmental water requirement in wetland, J. Hydraulic Eng., 38, 845–851, https://doi.org/10.3321/j.issn:0559-9350.2007.07.013, 2007 (in Chinese).

Proc. IAHS, 383, 341–346, 2020
https://doi.org/10.5194/piahs-383-341-2020

Impacts of dam construction on river channel evolution: a case of Minjiang River in Southeastern China

Ying Yao[1], Wei Cui[1], Wen Wang[1], Fu-Min Ma[2], and Ben-Yue Chen[2]

[1]State Key Laboratory of Hydrology-Water Resources and Hydraulic Engineering Sciences, Hohai University, Nanjing, China

[2]Bureau of Hydrology and Water Resources of Fujian Province, Fuzhou, China

Correspondence: Wen Wang (w.wang@126.com)

Published: 16 September 2020

Abstract. The Minjiang River is the largest river in Fujian Province. In 1993, the Shuikou Reservoir, which has an effective storage capacity of 700 million m^3, was built at about 161 km above the estuary. The completion of the Shuikou Dam trapped most of the upstream sediment in the reservoir area, resulting in a drastic decrease in sediment in the lower reaches of the Minjiang River. The average annual sand load at the Zhuqi Station (about 45 km below the dam) was reduced about 2/3 after the construction of the reservoir, from 7.42 to 2.55 million t by average, resulting in severe river bed downward cutting. At the same time, the demand for the sand in Minjiang River channel is increasing year by year. The amount of mined river sand is greater than the incoming sediment deposited in the river, which intensified the downcutting of the river bed. The downcutting leads to a continuous upward movement of the tide limit in the river channel especially in the low-flow season. Meanwhile, river embankments and river-related structures are damaged, and the navigation capacity of the Minjiang River is reduced. At present, the river bed of the lower Minjiang River is not yet stable, and the river regime is in a state of constant adjustment.

1 Introduction

The Minjiang River is the largest river in Fujian Province, and 8 large reservoirs with capacity over 100 million m^3 were built in the basin over the last 60 years. Among them, Shuikou Reservoir is the largest one, with a total capacity of 2.6 billion m^3, which controlled 86 % of the drainage area. It is known that the construction of reservoir dams will cause changes in the river runoff process, reducing the peak flows and increasing the low flows (Wang et al., 2011). At the same time, it will result in changes of downstream river flow regimes, riverbed scouring and river bank erosion, causing problems such as salt tide intrusion, river bank and bridge damage (Magilligan and Nislow, 2005). For example, after the closure of the Three Gorges Dam in China, various significant downstream changes are observed, such as sediment reduction (Yang et al., 2007), channel adjustment (Chen et al., 2007; Xia et al., 2016), riverbed erosion (Zheng et al., 2018), delta recession (Luo et al., 2017), flood pulse reduction (Yang et al., 2015), and decline of riverine wetland and lake areas (Zhou et al., 2016). After the completion of Shuikou Dam in 1993, the river channel in the lower reaches of the Minjiang River was continuously eroded, and the salt tide kept moving upward (Huang, 2010). In the present study, we will investigate the downstream hydrological alteration and river channel evolution of the Minjiang River after the construction of Shuikou Reservoir, and the related impacts on the lower reaches.

2 Hydrological characteristics of the Minjiang River

2.1 Introduction about the Minjiang River basin

The Minjiang River flows into the east China sea, with a total length of 541 km and a total drainage area of 60 992 km^2. It originates from the Wuyi mountain range. The topography of the basin is high in the northwest and low in the southeast. The average slope of the river is 5 ‰. The basin is located in the subtropical monsoon climate zone with a warm and

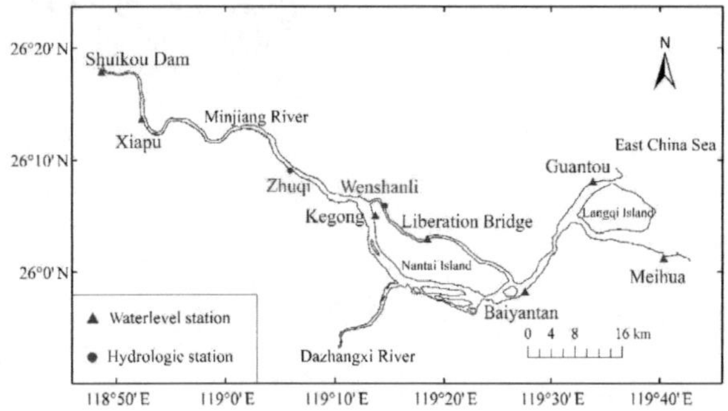

Figure 1. The lower reaches of the Minjiang River.

humid climate. The average annual temperature is about 17–19 °C. The average annual precipitation in the basin is about 1600–1700 mm, declining gradually from upper parts in the mountains to the lower parts along the coast.

The Shuikou Dam is located about 117 km above the Minjiang River estuary. The construction of Shuikou Dam started in March 1987, and its impoundment began in March 1993. The Shuikou Reservoir has a maximum storage capacity of 2.6 billion m^3 and an effective storage capacity of 700 million m^3. The drainage area above the reservoir dam site is 52 438 km^2, accounting for 86 % of the total basin area. The reservoir is mainly used for hydroelectricity, also used for flood control and navigation.

2.2 Hydrological characteristics of the river basin

The Minjiang River Basin has an average annual runoff of about 57.34 billion m^3 and an average runoff depth of 980 mm (China Ministry of Water Resources, 2017). The period during April to September is the flood season, and the runoff takes up about 75 % of the total annual runoff. The maximum monthly runoff occurs in June, when the main flood season in the river basin comes. From October to March of the following year, it is the dry season, and the average annual runoff reaches the minimum in January. The lower reaches of the Minjiang River are often struck by heavy typhoon rains during July to September, causing severe floods. The Minjiang River estuary is a tidal estuary with strong tidal effects, and the tide type is normal semi-diurnal tide. According to the tidal data statistics in 2001, the average high tide level of Baiyantan station is 4.67 m, the average low tide level is 0.82 m, and the average tidal range is 2.55 m. The maximum tidal range of the estuary is 7.04 m, and the average tidal range is 4.1 m. The tidal level increases from the estuary to the upstream, while the tidal amplitude decreases, and the tidal effects decreases.

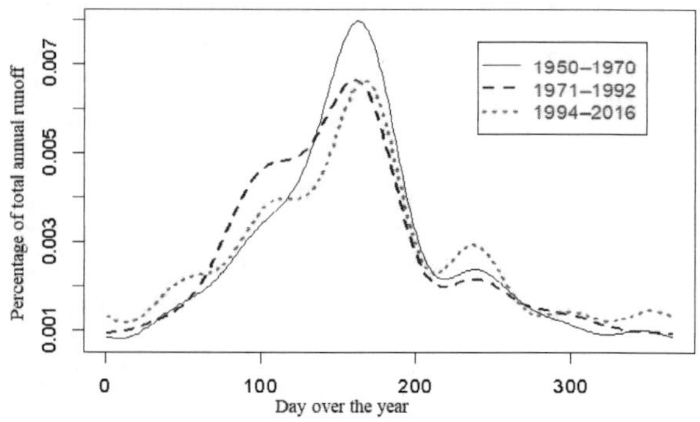

Figure 2. Variation of average daily discharge during three different periods since 1950.

3 Impacts of Shuikou Reservoir on riverflow and sediment process in the lower reaches of Minjiang River

3.1 Flow process change

Because the capacity of Shuikou reservoir takes about only 4.5 % of the average annual runoff of the Minjiang River, Shuikou reservoir has very little impact on the variation of annual runoff of Minjiang River, but the operation of Shuikou reservoir modified the seasonal variation of the runoff. The variation of average daily discharge at Zhuqi Station during three different periods since 1950 is shown in Fig. 2. As shown in Fig. 2, after the construction of Shuikou Dam, during the river flow rising period during March and July, the reservoir stores water, consequently reduces the outflow from the reservoir. During the period when river flow doing down and in low-flow period, the runoff below the Shuikou Dam increased due to increased outflow from the reservoir.

According to the naturalized monthly average flow data at Zhuqi from 1954 to 2004, the minimum monthly average flow before the construction of the Shuikou reservoir was 259 $m^3\,s^{-1}$ in January 1968; and the minimum monthly average flow after the construction of the reservoir was 309 $m^3\,s^{-1}$ in December 2003. In a word, the regulation of Shuikou reservoir makes the runoff tend to be slightly more evenly distributed over the year, but because the reservoir capacity is far less than the total annual runoff, so the effect on the downstream runoff is not obvious.

3.2 Water level change

The lowest water level at several locations along the lower reach of the Minjiang River (see Fig. 3) shows that, the lowest water level of the lower reach has decreased significantly since late 1980s', except at the estuary (Baiyantan). The decline of the water level accelerated in the late 1990s.

A further Comparison of the average water level in 1998, 2006 and 2010 and the corresponding water level changes at

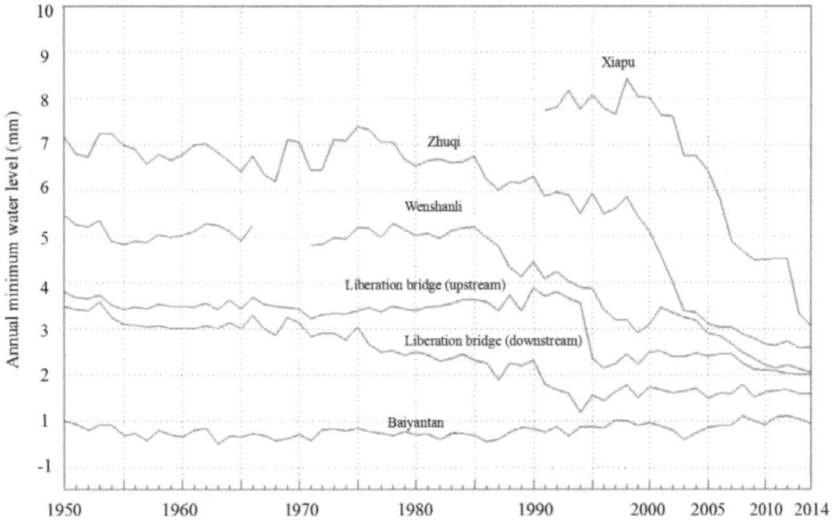

Figure 3. The variation of the lowest water level along the lower reach of the Minjiang River.

Table 1. Comparison of water levels at the same discharge at Zhuqi in different years.

Discharge at Zhuqi ($m^3 s^{-1}$)	Water level in 1998 (m)	Water level in 2006 (m)	Water level in 2010 (m)	Water level difference between 2006 and 1998	Water level difference between 2010 and 1998
5000	7.18	3.91		3.27	
10 000	9.32	6.55	5.00	2.77	4.32
15 000	10.84	8.35	6.70	2.49	4.14
20 000	12.16	9.63	8.20	2.53	3.96
25 000	13.30	10.76	9.67	2.54	3.63

the same discharge at Zhuqi are shown in Table 1. It can be seen that at Liberation Bridge (downstream) and Zhuji there has been a significant and sustained decline in water levels since 1970s'; decline has also occurred at Wenshanli since 1985 and the Liberation Bridge since 1995. Such changes are mostly caused by the gradual change of the riverbed due to the joint effects of reducing sediment by upstream reservoirs, whereas the sudden change in 1995 is caused by the removal of the pier of the old bridge.

3.3 Variation in sediment volume

According to the observation at Zhuqi, before the construction of Shuikou reservoir, the average suspended sediment concentration was 0.129 kg m^{-3} during the period from 1951 to 1992 with a maximum of 0.262 kg m^{-3} in 1962 and the minimum of 0.064 kg m^{-3} in 1991, and the average suspended sediment transport capacity during 1951 to 1975 was 7.42 million t every year. After the completion of Shuikou reservoir, the average suspended sediment concentration at Zhuqi was 0.035 kg m^{-3}, with a maximum of 0.135 kg m^{-3} in 2010 and a minimum of 0.007 kg m^{-3} in 2008, and the average annual suspended sediment transport capacity decreased to 2.55 million t during 1993 to 2016. That is, by average 4.87 million t suspended sediment are deposited in the reservoir every year.

It can be seen from the Table 2 that the proportion of suspended sediment with a size > 0.1 mm significantly reduced after the impoundment of Shuikou Dam. The decrease in the median grain size of sediments in the downstream reaches after dam construction is a common phenomenon observed in many rivers, e.g. the middle reaches of the Yangtze River below the Three Gorges Dam (Chu et al., 2007) and the lower reaches of the Yellow River below the Xiaolangdi Reservoir (Qu et al., 2010), because there are more coarse particles than fine particles deposited in the reservoir.

The ratio of the bed load to the suspended load in the mountainous rivers in Fujian Province is about 0.3, and the average bed load at Zhuqi before the construction of the Shuikou Dam is about 2.2 million t. After the dam construction, due to the reservoir's impounding effect, these bed load sediments are almost all deposited in the upper reaches of the Shuikou Reservoir. Therefore, after the completion of the reservoir, an average of 7.07 million t of sediment per year (including 4.87 million t of suspended sediment and all of the bed load) are deposited in the reservoir.

Table 2. Particle size composition of suspended sediment of Minjiang River at Zhuqi before and after Shuikou reservoir impoundment.

Year	Percentage of particles less than a certain size							
	0.005 (mm)	0.010 (mm)	0.025 (mm)	0.050 (mm)	0.10 (mm)	0.25 (mm)	0.50 (mm)	1.0 (mm)
1992 (Before impoundment)	1.7	6.2	10.8	22.3	94.4	99.0	99.9	100
1994 (After impoundment)	1.3	6.3	11.8	26.9	95.8	99.5	100	

The reduction of sediment downstream is a common phenomenon after dam construction. For example, after the closure of the Three Gorges Dam (TGD) in 2003 the sedimentation in the lower reaches of the Yangtze River decreased by 55 % compared with 1993–2002, and 65 % of the pre- to post-TGD decrease in sediment flux can be attributed to the TGD (Yang et al., 2015).

While the sedimentation below the Shuikou Dam, there is a huge demand for river sand, and large-scale sand mining occurred in the lower reaches of the Minjiang River. The annual river sand consumption in the lower Minjiang is more than 10 million t, which is more than sedimentation of the lower Minjiang. That seriously deepened the riverbed, endangered the stability of the river embankment and the safety of navigation.

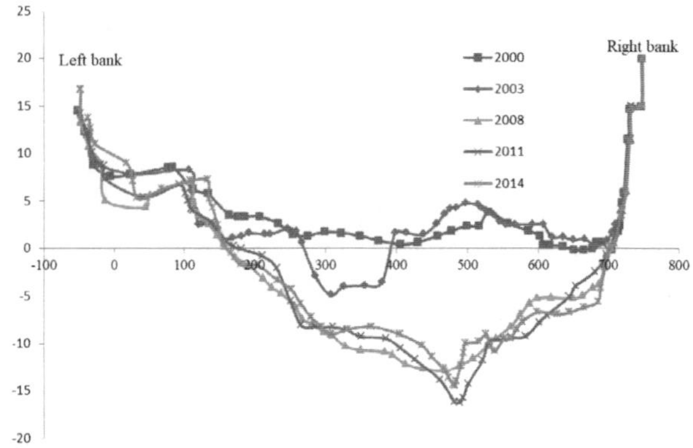

Figure 4. River bed elevation at the cross-section (from the left bank to the right bank) of the Minjiang River at Zhuqi in different years.

4 River channel evolution in the lower reaches of Minjiang River under the influence of Shuikou Reservoir

Due to the changes in discharge and sedimentation caused by the construction of Shuikou Dam, jointly with the influence of human sand mining activities, the shape of the riverbed in the lower reaches of the Minjiang River is constantly changing. Such kind of changes is especially exhibited in the profound changes in the thalweg of Minjiang River.

After the Shuikou Reservoir was completed, the plane position of the thalweg at many river cross-sections in the lower reach of the Minjiang River got deeper generally and oscillated to varying degrees. For example, the riverbed the reach from Shuikou dam site to the south and north distributaries of the lower Minjiang River decreased by 4.64 m from 1989 to 2015. As shown in Fig. 4, the maximum incision depth in the riverbed during 2000 and 2015 exceeds 20 meters at Zhuqi. Figure 5 shows the changes of the thalweg of the lower reaches of the Minjiang River from the Shuikou Dam to the estuary in 2011 and 2015. It can be seen that the riverbed is still in a process of scouring deep 20+ years after the dam was built.

River channel erosion below dams is a common phenomenon all over the world (Basson, 2012). Similar to the impact of the construction of the Shuikou Reservoir in the Minjiang River, many examples of serious channel incision below dams in relation to regional geographic conditions: e.g. 6 m, Davis Dam, Colorado River, USA (Kondolf, 1997); 7.5 m, Hoover Dam, Colorado River, USA (Basson, 2012); 6 m, Saulspoort Dam, Ash River, South Africa (Basson, 2012); 10 m, Three Gorges Dam, China (Zheng et al., 2018). However, bed erosion is not the only possibility and local conditions may determine a variety of outcomes (Lai et al., 2017) in terms of the incision depth or the time to reach a new balance between erosion and deposition. A recent study showed that the new long-term hydro-morphological equilibrium may have been established in the middle and lower Yangtze about 14 years after the closure of TGD (Lai et al., 2017). In the case of the Shuikou Dam, such a new equilibrium is still yet to be achieved. Although after 2008 the overall elevation of the riverbed got more or less stable at Zhuqi (as shown in Fig. 4), the thalweg is still getting deeper generally along the lower Minjiang (Fig. 5).

5 Impacts of river evolution

5.1 Exacerbation of salt tide intrusion

The Minjiang Estuary is a tidal estuary with strong regular semi-diurnal tides. In the estuary, the salt water is mixed up with fresh water, and the salinity of the water varies with runoff and tidal currents. In recent years, as the riverbed has been cut downward year by year, the amount of tides in the riverbed has been increasing, and the tidal current limit and the boundary of the tide have been continuously moving up-

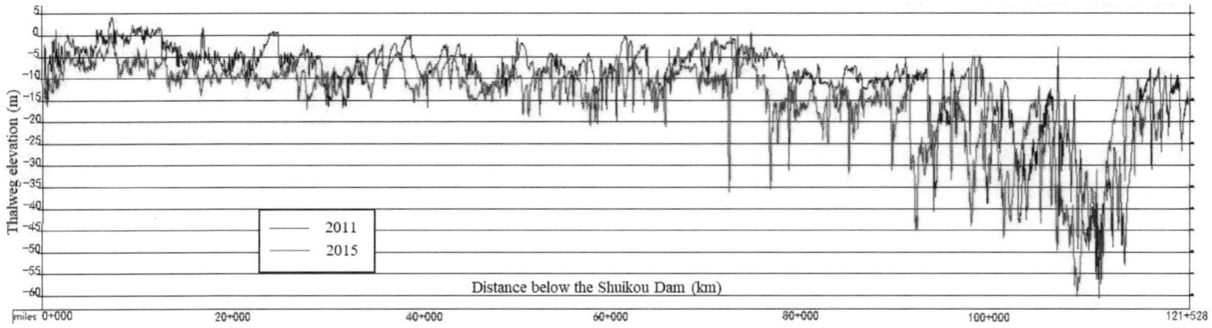

Figure 5. Comparison of thalwegs of the lower Minjiang River in 2011 and 2015.

stream. In the early 1990s, the tidal limit of Minjiang estuary in dry season was near Houguan (68 km from the estuary) and the tidal current limit to Hongshan Bridge (58 km from the estuary). After the completion of the Shuikou Dam in August 1993, according to the measurement in the low-flow period in December 1996 and September 2002, the tidal limit was extended upward to Baitou (72.5 km from the estuary), and the tidal current limit raised to Huai'an (62.5 km from the estuary). In 2009, the tidal limit reached the Shuikou Dam (117 km above the estuary), and the tidal current limit reached Meixikou (about 102 km above estuary). As the tidal current limit kept moving upstream, 11 of the 12 water intakes of the downstream water plants along the lower Minjiang river are affected by the salty tide, and the closer the water plant is located to the river mouth, the more serious is the salt tide impact. At the same time, under the influence of the tidal current, the exchange time of the water body in the estuary is lengthened, the pollutants in the water body stay for a longer time than before. With the discharge of industrial waste and sewage from Fuzhou, which covers both sides of the lower reaches of the Minjiang River, the water environment problem in the lower reaches of the Minjiang River tends to be worsened.

5.2 Damage of river-related structures

After the operation of Shuikou Reservoir, the hydro-dynamic conditions of the rivers in the lower reaches of the Minjiang River have undergone great changes. In addition, the substantial reduction of river bed sediment leads to serious downcutting and riverbank erosion in the downstream river channel, resulting in damages to many bridges and ports, and putting some river embankments in dangers. For example, large scale landslides occurred on the bank near Dukou Village in Minqing County in 2005, near Huangguan Village in Minhou County in 2006, near Beixi Village in Minqing County in 2007. The occurrences of these dangers are great related to the river channel evolution and riverbed incision in the lower reaches of the Minjiang River. The further river bed incision in the lower reaches of the Minjiang River will further damage the riverbank stability and flood control safety.

5.3 Reduction of river navigation capacity

Due to dramatic decrease of the sediment in the lower Minjiang River after the completion of the Shuikou Reservoir and the excessive river sand mining in recent years, the river flow pattern changed, the riverbed geomorphology kept modifying, and the water level in the lower reaches decreased significantly, consequently the navigation was greatly affected. Before 1993, the navigation capacity of the Minjiang River was Class 3 (i.e., navigable for 1000 t ship). The Shuikou Reservoir was originally designed as a waterway with Class 4 standard (i.e., navigable for 500 t ships) but with some room for extra navigation capacity. When the discharge flow rate is 308 $m^3 s^{-1}$, the bottom of the navigation lock is 3 m deep beneath the water level, which allows all ship passing. In recent years, due to the sharp incision of the riverbed of the Minjiang River channel, the navigation flow in the lower reaches of the Minjiang River has been increased year by year from the original design of 308 to 800 $m^3 s^{-1}$ in 2000, 1500 $m^3 s^{-1}$ in 2004, 2000 $m^3 s^{-1}$ in 2005, 2350 $m^3 s^{-1}$ in 2006, and 3800 $m^3 s^{-1}$ in 2017. It is expected that the navigation capacity may be reduced in the future.

6 Conclusion

The construction of the cascade hydropower stations in the upper reaches of the Minjiang River, especially the Shuikou Reservoir began operation in 1993, modified the flow processes, trapped most sediment in the reservoir, and greatly reduced the amount of sediment transporting downstream. Such reduction of sediment together with excessive sand mining in the lower Minjiang River resulted in the severe downcutting of the riverbed and continuous river channel evolution of the lower Minjiang River. Due to such kind of river bed downward scouring and river bed evolution, the salty tide limit kept moving upward, river embankments and river-related structures are damaged, the navigation capacity of the Minjiang River are reduced. At present, the flow-sediment situation in the lower reaches of the Minjiang River has not reached equilibrium, and the river channel is still in the process of evolving dominated by downward cutting.

Data availability. The internal data of the project is not convenient to be published.

Author contributions. YY, WC, and WW designed research, performed research, analyzed data. YY wrote the paper and WC and WW modified the paper. FMM and BYC participated in project research and data processing.

Competing interests. The authors declare that they have no conflict of interest.

Special issue statement. This article is part of the special issue "Hydrological processes and water security in a changing world". It is a result of the 8th Global FRIEND–Water Conference: Hydrological Processes and Water Security in a Changing World, Beijing, China, 6–9 November 2018.

References

Basson, G. R.: Sedimentation and sustainable use of reservoirs and river systems, Kyoto, ICOLD (International Committee on Large Dams), Bulletin 147, 2012.

Chen, Z., Chen, D., Xu, K., Zhao, Y., Wei, T., Chen, J., Li, L., and Watanabe, M.: Acoustic Doppler current profiler surveys along the Yangtze River, J. Geomorphology, 85, 155–165, https://doi.org/10.1016/j.geomorph.2006.03.018, 2007.

China Ministry of Water Resources: Bulletin of Chinese River Sediment, Beijing, China Water and Power Press, 2017 (in Chinese).

Chu, Z. X., Zhai, S. K., Zhang, L., and Dong, M.: Effects of Three Gorges Reservoir water storages in 2003 on the median grain size of suspended sediment of the Yangtze River, J. T. Ocean. Limnol., 23–28, https://doi.org/10.1007/978-3-540-72108-6_3, 2007 (in Chinese).

Huang, Y. F.: Study on bed evolution and its influence in the lower Minjiang River, J. Water Sci. Technol., 15–17, 2010.

Kondolf, G.: Hungry water: Effects of dams and gravel mining on river channels, J. Environ Manage., 21, 533–555, https://doi.org/10.1007/s002679900048, 1997.

Lai, X., Yin, D., and Finlayson, B. L.: Will river erosion below the Three Gorges Dam stop in the middle Yangtze, J. Hydrol., 554, 24–31, https://doi.org/10.1016/j.jhydrol.2017.08.057, 2017.

Luo, X. X., Yang, S. L., Wang, R. S., Zhang, C. Y., and Li, P.: New evidence of Yangtze delta recession after closing of the Three Gorges Dam, J. Sci. Rep., 7, 41735, https://doi.org/10.1038/srep41735, 2017.

Magilligan, F. J. and Nislow, K. H.: Changes in hydrologic regime by dams, J. Geomorphology, 71, 61–78, https://doi.org/10.1016/j.geomorph.2004.08.017, 2004.

Qu, Y. Z., Liu, J. W., and Sun, G. S.: Change and analysis of sediment particles in lower reach of Yellow River before and after construction of Xiaolangdi Reservoir, J. Henan Water Conservancy and South-North Water Transfer, 66–67, 2010 (in Chinese).

Wang, W., Wang, X. G., and Zhou, X.: Impacts of Californian dams on flow regime and maximum/minimum flow probability distribution, J. Hydrol. Res., 42, 275–289,https://doi.org/10.2166/nh.2011.137, 2011.

Xia, J., Deng, S., Lu, J., Xu, Q., Zong, Q., and Tan, G.: Dynamic channel adjustments in the Jingjiang Reach of the Middle Yangtze River, J. Sci. Rep., 6, 22802, https://doi.org/10.1038/srep22802, 2016.

Yang, S. L., Zhang, J., Dai, S. B., Li, M., and Xu, X. J.: Effect of deposition and erosion within the main river channel and large lakes on sediment delivery to the estuary of the Yangtze River, J. Geophys. Res., 112, F02005, https://doi.org/10.1029/2006JF000484, 2007.

Yang, S. L., Xu, K. H., Milliman, J. D., Yang, H. F., and Wu, C. S.: Decline of Yangtze River water and sediment discharge: Impact from natural and anthropogenic changes, J. Sci. Rep., 5, 12581, https://doi.org/10.1038/srep12581, 2015.

Zheng, S. W., Xu, Y. J., Cheng, H. Q., Wang, B., Xu, W., and Wu, S. H.: Riverbed erosion of the final 565 kilometers of the Yangtze River (Changjiang) following construction of the Three Gorges Dam, J. Sci. Rep., 8, 11917, https://doi.org/10.1038/s41598-018-30441-6, 2018.

Zhou, Y. Q., Jeppesen, E., Li, J. B., Zhang, Y. L., Zhang, X. P., and Li, X. C.: Impacts of Three Gorges Reservoir on the sedimentation regimes in the downstream-linked two largest Chinese freshwater lakes, J. Sci. Rep., 6, 35396, https://doi.org/10.1038/srep35396, 2016.

Proc. IAHS, 383, 347–353, 2020
https://doi.org/10.5194/piahs-383-347-2020

Open Access
PIAHS

Optimization of water quality evaluation method in Three Gorges Reservoir area under different hydrological conditions

Jing Yuan[1], **Jin-cheng Wang**[2], **and Bao Qian**[2]

[1]Mid Changjiang River Bureau of Hydrological and Water Reasource Survey, Bureau of Hydrology, Changjiang Water Resources Commission, Wuhan, 430012, China
[2]Bureau of Hydrology, Changjiang Water Resources Commission, Wuhan, 430012, China

Correspondence: Jing Yuan (jingyuan647@126.com)

Published: 16 September 2020

Abstract. In order to solve the problem of lack of unified evaluation criteria for water quality evaluation in the Three Gorges Reservoir under different hydrologic condition, hydraulic retention time and flow rate were calculated wih the hydrological data of 6 intact hydrological years after the 175 m impoundment in the Three Gorges Reservoir using the concept of segment capacity. Results indicate that the hydrological conditions of the Three Gorges Reservoir have obvious spatial and temporal differences. The whole river should be evaluated according to the river standard during the wet season. In the dry season, the upstream of Wanxian should be evaluated according to the river standards and the downstream of Wanxian should be evaluated according to lake standard. The tributary bay should be evaluated according to the lake standard. The conclusion obtained by this evaluation method is consistent with actual situation of the reservoir, providing a theoretical basis and technical support for the targeted protection of water environment in river-type reservoirs.

1 Introduction

The operation of the reservoir changes the reservoir area periodically. In the "Environmental Quality Standards for Surface Water (GB3838-2002)", different quality assessment standards are given for lakes and rivers (Table 1). The lake is directly applied to the reservoir and no distinction is made between different types of reservoirs. This has caused many confusions and difficulties in the actual water quality assessment of the reservoir area. Taking the Three Gorges Project as an example, the total storage capacity of the reservoir reached 393 billion m^3 after 175 m of water storage in the reservoir. The reservoir area includes the main stream reservoir area and 29 tributaries. When assessing the water quality of the reservoir area, the water conservancy system shall evaluate the rivers according to the requirements of the "River reservoirs evaluated by rivers and lake-type reservoirs by lakes" in the Technical Regulations for Surface Water Resources Quality Evaluation (SL395-2007). Some monitoring units evaluate according to lake standards. As the total phosphorus (TP) and total nitrogen (TN) indexes are important indicators for assessing the nutritional status of water bodies, the differences in the selection of standards often lead to large differences in the conclusions of various parties.

Table 1. Part water quality standard limit of environmental quality standards for surface water (unit: mg L^{-1}.

		Class I	Class II	Class III	Class IV	Class V
TP (in P) ≤	river	0.02	0.1	0.2	0.3	0.4
	Lakes/ reservoirs	0.01	0.025	0.05	0.1	0.2
TN (in N) ≤	river	–	–	–	–	–
	Lakes/ reservoirs	0.2	0.5	1.0	1.5	2.0

The tributaries of the Three Gorges Reservoir have always adopted the river evaluation standard and the evaluation results have not changed much before and after the impoundment of the reservoir. However, the water bloom phenomenon rarely occurs in the tributary bays before the impoundment of the reservoir. After the impoundment of the reservoir, 26 of the 29 tributaries had blooms (Calf. Limnology, 2011). The conclusions drawn from the evaluation of rivers are inconsistent with the actual situation. The applicability of the existing evaluation standards for the Three Gorges Reservoir after impounding remains to be further verified.

At present, there is a lack of available research cases and few studies systematically considering the main reservoir area and tributaries evaluation criteria for the Three Gorges Reservoir after impoundment. This paper mainly analyzes the water quality evaluation criteria selection and influencing factors in the reservoir area under the heterogeneous hydrology through the data of 6 complete hydrological years after the impoundment of the Three Gorges Reservoir. Therefore, it provides the basis for the water quality evaluation of the Three Gorges Reservoir and also provides a reference for other people to evaluation the water quality of manual controlling water bodies (such as artificial lakes, artificially controlled rivers).

2 Research method

The main influencing factors of eutrophication in the Three Gorges Reservoir area can be divided into three categories: nutrition factors (nitrogen, phosphorus, etc.); environmental factors (water temperature, light, transparency, etc.) and hydrological factors (velocity, flow rate, hydraulic retention time). Changes in hydrological factors such as hydraulic retention time and flow rate is the key factor leading to eutrophication of water bodies and aggravating water bloom (Wang, 2012).

2.1 Analysis of the effect of hydraulic retention time

The hydraulic retention time is directly related to the sedimentation of suspended solids in reservoirs, the self-purification of nutrients, and the maintenance and reproduction of plankton populations (Wang et al., 2004). The increase in hydraulic retention time means the probability of algal overgrowth is greatly increased (Milan et al., 1995). The experience of some reservoirs shows that even if eutrophication conditions is reached, as long as the hydraulic retention time does not exceed 2 weeks, blooms rarely occur (Kawara et al., 1998; Lin et al., 2009), and the hydrological characteristics of the reservoir are similar to those of rivers. When blooms occur, the guarantee rate of maximum hydraulic retention time is usually greater than 80 %. Controlling the hydraulic retention time of the water can effectively control algal overgrowth. At 25 to 15 °C, Qingcaosha Reservoir reached 15 000 per mL of "bloom" alert value corresponding to the maximum hydraulic retention time is 15–40 d (Lin et al., 2009). The United Nations Environment Program (UNEP) Integrated River Basin Management Guidelines recommend that the hydraulic retention time of reservoirs generally not exceed 30 d and there are 4 weeks as the boundary (Gu et al., 2011). Some scholars also use the 20 and 300 d from the perspective of ecological indicators as the demarcation standard for river-type, transitional, and lake-type reservoirs (Zheng et al., 2006). Combined with practical experience, this paper uses the hydraulic retention time of 30 and 300 d as the boundary, as the distinguishing standard for river-type, transitional, and lake-type reservoirs.

2.2 Analysis of flow rate

The flow rate conditions have an important effect on the ecosystem in the water body. With the increase of the flow rate, the chlorophyll content of the algae increases and then decreases. Too high flow rate or too low flow rate is not suitable for the growth of algae. The critical flow rates of different algae ranged from 0.01, 0.03 to 0.04, and 0.05 m s^{-1} (Jiao, 2007).

Flow rate is also an important basis for distinguishing between rivers and lakes. Some scholars use 0.1 and 0.001–0.01 m s^{-1} as the boundary of river and lake velocity (Technological regulations for surface Water resources quality assessment, 2007). Some scholars also use 0.03 and 0.01 m s^{-1} as the boundary values of river-type, transition-type and lake-type reservoirs (Zheng et al., 2006). Considering the principle of "the most unfavorable" and the actual conditions of the Three Gorges Reservoir area, this paper proposes a rough classification of the speed of the water body type: the lake flow rate is < 0.01 m s^{-1}, the transitional flow rate is 0.01–0.05 m s^{-1}, suitable for Lake evaluation standard; river type flow rate is > 0.05 m s^{-1}, suitable for river evaluation standard.

When the criteria for dividing hydraulic retention time and flow velocity are inconsistent, the "least unfavorable" principle should be used to apply the lake evaluation standard.

2.3 Statistical methods

The total designed storage capacity of the Three Gorges reservoir area is 39.3 billion m^3. The scope of the reservoir area is shown in Fig. 1. The reservoir water level is not same everyday and the seasonal variation is between 145–175 m. This will have a great impact on the storage capacity. Based on the storage capacity curve of storage capacity-water level-flow, this paper derives the storage capacity of the reservoir under the dam water level and then estimates the total hydraulic retention time based on the inflow or outflow. In order to compare whether the change trend of storage capacity and hydraulic retention time is consistent, the storage capacity change trend and the hydraulic retention time trend are compared. The result is shown in Fig. 2.

Applying segment storage, river channel large section data, the telemetry data of the water level station and the real-time operation data of the reservoir, segment storage capac-

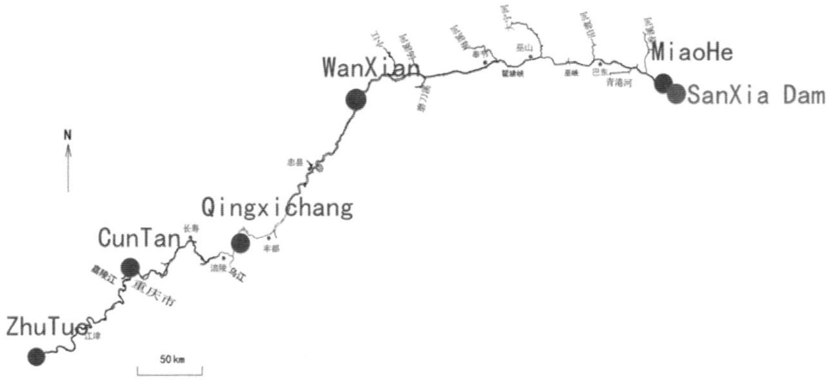

Figure 1. Study scope of the Three Gorges reservoir area.

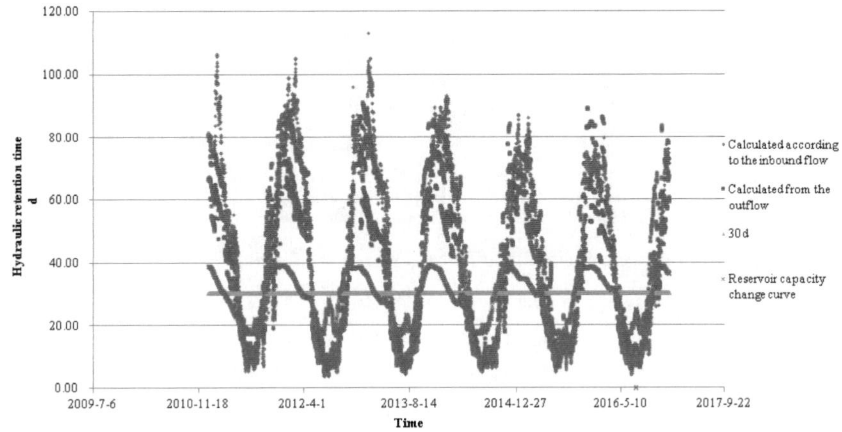

Figure 2. Hydraulic retention time of the Three Gorges Reservoir during 2011–2016.

ity, segment hydraulic retention time and the average cross-section velocity were calculated. Because the Three Gorges Reservoir was filled in water at 175 m in October of 2010, this paper has calculated the hydraulic retention time and flow rate information from 2011 to 2016. The typical daily statistics are shown in Table 3.

Segment hydraulic retention time = segment storage capacity / storage flow

Average cross-section flow rate = cross-sectional flow / cross-section area

3 Statistical analysis

3.1 Analysis of spatial and temporal changes in main stream

From Fig. 2, the hydraulic retention time of the reservoirs in different water periods varies significantly and differs by 2 orders of magnitude. The hydraulic retention time agrees well in the flood season and the flat water period. But in the dry season, the hydraulic retention time calculated from the outflow is less than the hydraulic retention time calculated from the inbound flow. Because the outbound flow reflects the actual operation of the Three Gorges Reservoir, such high dams should pay attention to the impact of the outflow on the hydraulic retention time.

During the period of statistics, all hydraulic retention time is less than 300 d, which is a transitional or river-type reservoir; The time shorter than 30 d is 955 d, about 43.26 % of the total observation time. From May to October of each year, the hydraulic retention time from the tail to the head of the reservoir is short. The flow velocity near the reservoir head has a downward trend, but difference is small and its hydrological characteristics are close to the river. The shortest and longest hydraulic retention time of the Three Gorges Reservoir is stable, which is in June to July and January to February every year. Therefore, the statistical data of the segment storage capacity and the segment hydraulic retention time collect the data of 1 February and 1 July of each year (the data of nearest date are selected if there is no data on these days).

From Table 3 and Fig. 3, the hydraulic retention time from the end of the reservoir to the head of the reservoir gradually

Table 2. Mileages of upstream and downstream from the Dam of the Three Gorges.

	1 Miaohe	2 Wanxian	3 Qingxichang	4 Cuntan	5 Zhutuo
Mileage from the dam site (km)	15	286.5	476	604	758
Interval length (km)	15	271.5	189.5	128	154

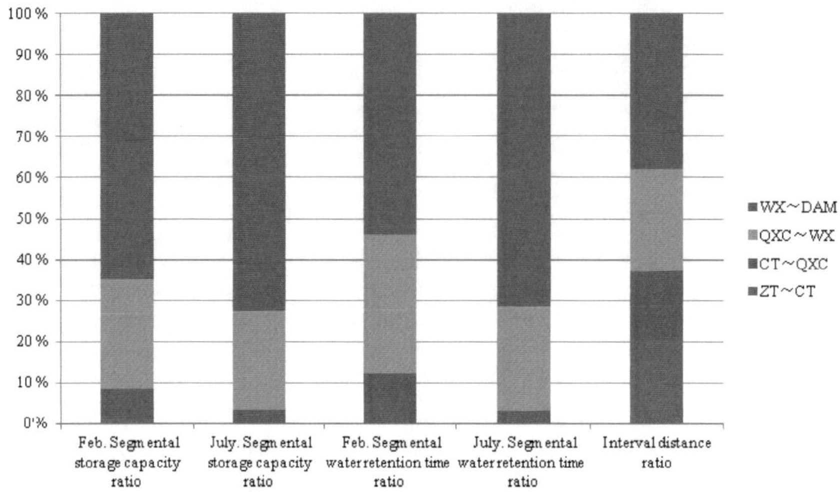

Figure 3. Statistics of the ratio of section storage capacity of the Three Gorges reservoir area from 2011 to 2016.

increases, and the flow rate gradually decreases. The water inventory of the three Gorges Reservoir presents obviously spatiotemporal heterogeneity. From November to next April each year, the reservoir has a high-water level, the water flow slows down, the storage capacity is large and the water environment is stable, which makes a lot of sediment deposition, high light transmittance and conducive to the occurrence of water bloom. The closer to the head of the library, the closer it is to the characteristics of the lake.

Due to the obvious longitudinal gradient distribution in the reservoir, the evaluation conclusions obtained by using single evaluation methods such as lakes or rivers do not conform to the actual situation. Therefore, new assessment methods need to be developed. However, there is no quantitative basis for the vertical division of the reservoir.

3.2 Evaluation of vertical zoning of main reservoir area during dry season

The Three Gorges Reservoir is a typical river-type reservoir. The relationship between the upper and lower reaches of the Three Gorges Reservoir area and the distance from the dam are detailed in Table 2.

The Zhuxi-Wanxian section accounts for 62.2 % of the total distance, but the storage capacity only accounts for about 30 %–40 % of the main stream. The fractional storage capacity accounts for a small proportion and the hydraulic retention time is shorter (the hydraulic detention time in the dry season from Zhuxi to Wanxian is approximately 30 d). It is applicable to river evaluation standards (including Zhuxi, Jiangjin, Tongguan, Cuntan, Changshou, Qing Xichang, Wanxian and other sections).

Wanxian to the dam is a transitional area (Fengjie, Guandukou, Badong, Miaohe, etc.). The distance is about 40 % of the distance, the hydraulic retention time is short in the wet period, using river evaluation standards. But in the dry season, the storage capacity of about 60 %, hydraulic retention time is long (see in Table 3), the flow rate is significantly lower than the river area, so it should be evaluated according to the lake standard.

3.3 Tributary evaluation

Before and after dam construction in the Three Gorges Reservoir, the total nutrient concentrations in the mainstream and tributaries of the Yangtze River did not change significantly. The TP concentration and TN concentration were 0.1–0.5 and 0.8–1.5 $mg\,L^{-1}$, which were close to or more than the accepted eutrophication conditions (TP and TN concentrations of 0.025 and 0.2 $mg\,L^{-1}$, respectively) (Wang, 2012). However, serious water blooms occurred in many tributaries after impoundment, especially at the late spring, early summer, late summer and early autumn. The river section where water bloom occurred was developed from 3 tributaries in 2003 to 26 tributaries (Wang, 2012). The water bloom was mostly found in the backwater section, estuary

Table 3. Statistical table of characteristics of the Three Gorges Reservoir in 2011–2016.

Time	Dam front water level (m)	Inbound traffic ($m^3 s^{-1}$)	Total capacity (billion m^3)	Outbound traffic ($m^3 s^{-1}$)	Hydraulic retention time with in/out flow meter (d)		ZT	CT	QXC ($m s^{-1}$)	WX	MH	ZT~CT	CT ~ QXC Main stream section storage/billon m^3	QXC~ WX	ZT ~ CT Mainstream hydraulic retention time (d)	ZT ~ QXC	ZT ~ WX
1 Feb 2011	171.12	5100	35.463	5510	80.48	74.49	0.92	0.49	0.22	0.11	0.08	0.363	2.74	9.361	5.37×10^{-6}	7.18	27.57
1 Jul 2011	146.96	16300	18.123	18900	12.87	11.10	2.23	2.69	1.05	0.49	0.30	0.015	0.657	4.337	2.52×10^{-6}	0.35	3.82
1 Feb 2012	172.17	5400	36.48	6200	78.19	68.10	0.96	0.34	0.16	0.10	0.09	0.406	2.858	9.651	5.33×10^{-6}	10.02	34.57
1 Jul 2012	145.47	25 000	17.382	20840	8.05	9.65	2.75	2.81	1.84	0.72	0.63	0.011	0.587	4.07	2.17×10^{-6}	0.24	2.19
1 Feb 2013	170.43	4000	34.805	7008	100.71	57.48	1.01	0.34	0.19	0.10	0.11	0.337	2.663	9.203	5.41×10^{-6}	8.94	32.25
1 Jul 2013	145.47	13 000	17.382	17438	15.48	11.54	2.48	4.03	1.40	0.38	0.48	0.011	0.587	4.07	2.24×10^{-6}	0.26	3.40
1 Feb 2014	168.15	4400	32.707	5075	86.03	74.59	1.11	0.48	0.14	0.10	0.11	0.259	2.414	8.771	5.62×10^{-6}	8.64	34.97
1 Jul 2014	146.3	20 000	17.793	19145	10.30	10.76	2.00	2.42	1.25	0.60	0.34	0.014	0.625	4.239	2.90×10^{-6}	0.40	3.19
1 Feb 2015	170.59	5800	34.957	5975	69.76	67.71	1.08	0.51	0.23	0.12	0.10	0.343	2.681	9.24	5.38×10^{-6}	6.78	25.91
1 Jul 2015	145.24	39 000	17.269	31 706	5.12	6.30	1.98	2.62	2.08	1.01	0.56	0.011	0.577	4.029	2.25×10^{-6}	0.22	1.67
1 Feb 2016	170.86	6300	35.214	7928	64.69	51.41	1.14	0.53	0.21	0.12	0.11	0.353	2.711	9.302	5.35×10^{-6}	6.86	26.38
1 Jul 2016	147.8	48 000	18.546	29754	4.47	7.21	2.51	2.62	1.93	1.02	0.52	0.017	0.701	4.49	2.53×10^{-6}	0.29	1.89
Feb	170.55	5166.7	34.938	6282.67	79.98	65.63	1.04	0.45	0.19	0.11	0.10	0.343	2.678	9.255	5.41×10^{-6}	8.07	30.27
Jul	146.21	26 883	17.749	22 963.83	9.38	9.43	2.33	2.80	1.59	0.70	0.47	0.013	0.623	4.206	2.44×10^{-6}	0.29	2.69

Note: Sectional section statistics are calculated based on actual measured flow rates, measured water levels, storage capacity, and average flow rates on 1 February and 1 July of each year. Abbreviation: ZT – ZhuTuo, CT– CunTan, QXC– QingXiChang, WX – WanXian, MH –MiaoHe.

and bay area of tributaries. The highest outbreak probability was Xiangxi River, followed by Meixi River, Daning River and Xiaojiang River.

The phytoplankton stock is highest in the backwater area, followed by the uninundated area and the lowest in the main stream. The impact of the impoundment of the Three Gorges Project on the phytoplankton and water quality in the tributary backwater area is greater than that in the mainstream of the Yangtze River (Wang, 2012). The declining capacity of pollutant degradation in the backwater area further exacerbate the eutrophication level of the tributaries (Hu et al., 2012). According to the existing research results, the velocity of the bay tributaries in the backwater area is generally lower than $0.05\,m\,s^{-1}$ (generally between 0.01 and $0.05\,m\,s^{-1}$). The lake standard should be considered for evaluation.

4 Results and discussion

According to hydraulic retention time, flow rate and other factors, Four sections of water quality such as Cuntan, Qingxichang, Wanxian, Badong, Miaohe, Daninghe, and Xiangxi estuaries were evaluated according to river or lake evaluation standards respectively. The results are shown in Table 5.

According to the results, in February, the cross section of Tongguanyi and Wanxian was classified into Class III. The main stream near the head of the reservoir is Class IV or V. The assessment criteria for tributary waters have changed from rivers to lake and the results have changed from Class II and III to non-compliance water bodies. It indicates that the risk of water bloom in the reservoir area is high. The main stream in July was river state, and the evaluation result was II and III.

Because the flow rate of the tributary is in the lake type, it is evaluated in accordance with the lake standard according to the most unfavourable principle. Most of the water quality is IV to inferior V, indicating a higher risk of blooms. This is consistent with the fact that tributaries have a high probability of occurrence of blooms. The improved evaluation results are more in line with the actual occurrence of water blooms and can better play an early warning effect.

5 Conclusion

This paper statistically analyzes the characteristics of water level, flow rate, reservoir capacity, and residence time of reservoirs in the Xiangxi and Daning rivers of the Three Gorges Reservoir (see in Table 4). It can be seen that the hydraulic retention time of the tributaries is much longer than the hydraulic retention time of the main stream. The average hydraulic retention time of the Xiangxi River and the Daning River in February and July were all longer than 30 d. The Xiangxi River near the head of the reservoir has a longer hydraulic retention time than the Daning River. This may be the

Table 5. Comparison of some parameter evaluation grades in February and July of the Three Gorges Reservoir in 2016.

		CunTan		QingXiChang		WanXian		BaDong		MiaoHe		Daning River estuary		Xiangxi estuary	
		TP	TN	TP	TN	TP	TN	TP	TN	TP	TN	TP	TN	TP	TN
February	concentration ($mg\,L^{-1}$)	0.11	1.77	0.11	2.33	0.1	2.11	0.1	1.89	0.11	1.97	0.1	2.14	0.13	1.84
	environmental protection system evaluation	V	V	V	Poor V	IV	Poor V	IV	V	V	V	II	–	III	–
	Water system evaluation	III	–	III	–	III	–	II	–	III	–	II	–	III	–
	Evaluation by this method	River type III	River type –	River type III	River type –	River type III	River type –	Lake type IV	Lake type V	Lake type V	Lake type V	Lake type IV	Lake type Poor V	Lake type V	Lake type V
July	concentration ($mg\,L^{-1}$)	0.13	1.6	0.12	1.9	0.09	2.1	0.13	2.02	0.12	2.03	0.06	2.04	0.07	2.29
	environmental protection system evaluation	V	V	V	V	IV	Poor V	V	Poor V	V	Poor V	II	–	II	–
	Water system evaluation	III	–	III	–	II	–	III	–	III	–	II	–	II	–
	Evaluation by this method	River type III	River type –	River type III	River type –	River type II	River type –	River type III	River type –	River type III	River type –	Lake type IV	Lake type Poor V	Lake type V	Lake type Poor V

Table 4. Statistical tables of typical tributary characteristics of the Three Gorges Reservoir in 2011–2016.

	Time	1 Feb 2011	7 Jul 2011	1 Feb 2012	1 Jul 2012	1 Feb 2013	1 Jul 2013	1 Feb 2014	1 Jul 2014	1 Feb 2015	1 Jul 2015	1 Feb 2016	1 Jul 2016	1 Feb 2017	1 Jul 2017
Dam front water level	(m)	171.12	146.96	172.17	145.47	170.43	145.47	168.15	146.3	170.59	145.24	170.86	147.8	170.55	146.21
Xiangxi River	Xingshan water level (m)	169.05	155.88	172.04	156.3	168.42	155.74	165.7	155.74	168.66	157.06	168.86	155.7	168.79	156.07
	flow ($m^3\,s^{-1}$)	6.37	37.8	6.37	110	9.1	51.4	20.5	31.4	NULL	130	20	140	12.47	83.43
	Hydraulic retention time (d)	1177.17	100.47	1208.41	32.93	809.84	70.48	339.11	245.94	–	27.66	372.49	27.85	781.4	84.22
Daning River	Dachang water level (m)	203.65	204.33	203.6	204.2	203.69	203.88	203.65	204.82	203.43	204.82	203.43	205.8	203.58	204.64
	flow ($m^3\,s^{-1}$)	–	–	–	62.3	17.1	29.4	16.3	181	16.3	181	10.2	567	14.98	204.14
	Hydraulic retention time (d)	–	–	–	57.58	578.36	122.02	562.93	20.66	609.89	19.59	983.09	7.1	683.57	45.39

https://doi.org/10.5194/piahs-383-347-2020

reason for the high probability of outbreak of water blooms in the Xiangxi River tributaries.

At present, rivers entering warehouses generally adopt river water quality assessment standards. Water quality category is still II–III. This conclusion may be reasonable before the impoundment of the Three Gorges. However, the evaluation results after the impoundment often do not match the nutritional level. It fails to reach the warning effect and do not match the occurrence frequency of the tributaries of the tributaries. Therefore, this paper proposes that the tributaries for sampling in lakes should be used to evaluate the evaluation criteria of lake banks.

China also has many manually controlled bodies of water, such as gated lakes, canals, and rivers that run on a cascade. The hydrological and ecological characteristics of these bodies of water have changed and there are significant differences between the natural rivers and lakes. This assessment method can be used to statistically analyse the specific hydrological situation, and then select suitable water quality assessment standards to provide technical support for targeted water conservation work.

Data availability. This article is about the operation and scheduling method of the Three Gorges Reservoir. The data is not publicly accessible and therefore cannot be made public.

Author contributions. JY and JCW contributed to the conception of the study. JY contributed significantly to analysis and manuscript preparation; JY performed the data analyses and wrote the manuscript; JY and BQ helped perform the analysis with constructive discussions.

Competing interests. The authors declare that they have no conflict of interest.

Special issue statement. This article is part of the special issue "Hydrological processes and water security in a changing world". It is a result of the 8th Global FRIEND–Water Conference: Hydrological Processes and Water Security in a Changing World, Beijing, China, 6–9 November 2018.

Acknowledgements. Thanks to Wang Jun (Former Director of Hydrology Bureau of Yangtze River Water Resources Commission) for the question of water quality evaluation method in Three Gorges Reservoir. His question promoted my thinking.

References

Gu, B.-H., Liu, Z.-W., and Li, K.-Y.: Limnology: Inland Water Ecosystems, Higher Education Press, Beijing, 2011.

Hu, N.-S., Liu, D.-F., Ji, D.-B., and Yang, Z.-J.: Effects of backflow of the main stream of the Three Gorges Reservoir on nutrient distribution in tributary bay, Environ. Sci. Technol., 35, 12–17, 2012.

Jiao, S.-J.: Effects of Velocity on Algae Growth in Low Velocity Rivers in the Three Gorges Reservoir Region, Southwest University, Chongqing, 2007.

Kawara, O., Yura, E., Fujii, S., and Matsumoto, T.: A study on the role of hydraulic retention time in eutrophication of the Asahi River Dam reservoir, Water Sci. Technol., 37, 245–252, 1998.

Lin, W.-Q., Gu, Y.-L., Lu, S.-Q., and Wang, X.-P.: Preliminary study on reasonable hydraulic retention time for Qingcaosha Reservoir to prevent overgrowth of algae, Water Wastewater Eng., 35, 60–62, 2009.

Milan, S., Dostálková, I., Hejzlar, J., and Vyhnálek, V.: The Effect of Reservoirs on Phosphorus Concentration, Int. Rev. Hydrobiol., 80, 403–413, 1995.

Technological regulations for surface Water resources quality assessment: SL395-2007, China Water Conservancy and Hydropower Press, Beijing, 2007.

Wang, X.-Q.: Coupling model of eutrophication and hydrodynamic water quality of the Pengxi River (Xiaojiang) in the Three Gorges reservoir area, Chongqing University, Chongqing, 2012.

Wu, H.: Achieving Water Conservation Safety of Surveillance and Surveillance – 2016 State Technological Invention Award, Second Class Project "Sensing Technique and Practice of Complex Hydraulic Concrete Structure Service Diagnosis", China Science and Technology Awards, 66, 2017.

Zheng, B.-H., Zhang, Y., Fu, G., and Liu, H.-L.: Study on the assessment standards for nutrition status in the Three Gorges Reservoir, Acta Scientiae Circumstantiae, 26, 1022–1030, 2006.

Proc. IAHS, 383, 355–365, 2020
https://doi.org/10.5194/piahs-383-355-2020

Open Access
PIAHS

Climate change impact on water resources in Klong Yai River Basin, Thailand

Chaiwat Ekkawatpanit[1], Weerayuth Pratoomchai[2], Chatchapol Khemngoen[1], and Patchanok Srivihok[3]

[1]Department of Civil Engineering, King Mongkut's University of Technology Thonburi, Bangkok 10140, Thailand
[2]Department of Civil Engineering, King Mongkut's University of Technology North Bangkok, Bangkok 10800, Thailand
[3]Environmental Technology Research Department, Innovation Institute, PTT Public Company Limited, Ayuttaya 13170, Thailand

Correspondence: Chaiwat Ekkawatpanit (chaiwat.ekk@kmutt.ac.th)

Published: 16 September 2020

Abstract. There is a 95 % scientific consensus that human activities contributed to climate change and altered hydrological processes. This study focused on a regional scale climate change impact assessment on water resources in the Klong Yai River basin in Thailand using multiple Global Climate Models (GCMs) under 3 scenarios of greenhouse gas emission called the Representative Concentration Pathway (RCPs), i.e., low (RCP2.6), intermediate (RCP4.5), and high (RCP8.5) scenarios. According to the projections, surface air temperature will increase relative to the reference period (1991–2000). Maximum surface air temperature is projected to increase around 0.3–1.35 °C. Precipitation in the near future (2017–2026 and 2027–2036) shows an increasing trend. The projected precipitation and temperature were used in the SWAT model for the Klong Yai River basin in order to assess the amount of inflow into reservoirs under the climate change conditions. Despite the future projections still have an uncertainty, but it is useful as a guideline for managing of our water resources and planning.

1 Introduction

Greenhouse gas concentration change in the atmosphere is a major cause of climate change and climate variation (Stocker et al., 2013; Hanson and Cramer, 2015). Since the industrial revolution, human added amount of greenhouse gases into the atmosphere by burning fuel fossil, deforestation and other activities (Alcamo, 2007). The average global temperature has increased 0.6 °C in the past three decades and 0.8 °C in the past century due to greenhouse gas emissions (Hansen et al., 2006). In Thailand, Limjirakan and Limsakul (2012a), A collected the observed data and analysed the trend of temperature change. The results showed the increase in minimum and maximum temperature of 0.26 °C per decade and 0.23 °C per decade respectively. Moreover, Limjirakan and Limsakul (2012b) collected temperature and evaporation data over the whole country. The statistical analysis result showed that the mean annual temperature is increasing with the 95 % of confidence level.

This research aims at evaluating the effect of climate change towards water resources assessment in Klong Yai River basin in the near future (2017 to 2036). 3 GCMs out of the total 10 GCMs that showed good performance to represent/mimic past climate patterns in the study were selected for assessing climate change impacts. This study was conducted under 3 climate change sceanrios, i.e., RCP2.6, RCP4.5, and RCP8.5. For the estimate of inflow into the reservoirs in the basin, SWAT model, which is a commonly used hydrological model, was applied to simulate the hydrological components in the future.

2 Study Area

The Klong Yai River basin covers about 1800 square kilometers of watershed area as shown in Fig. 1. It is the sub-

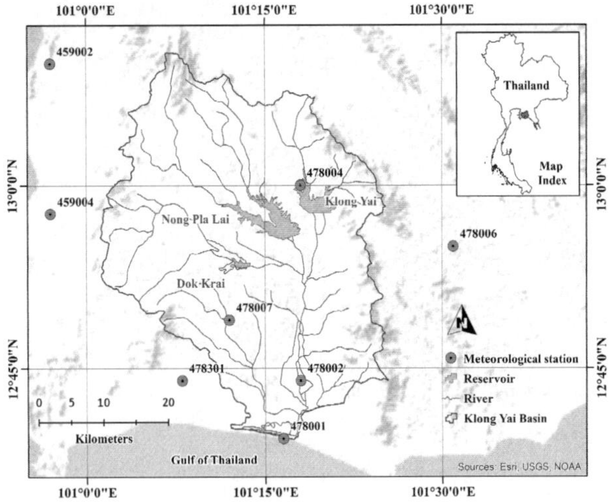

Figure 1. Location of the Klong Yai River basin in Thailand.

basin of the East coast watershed. The Northern part of the basin is the mountain range whereas the south is adjacent to the gulf of Thailand. River delineation lie along the north-south direction based upon the topography. There are 3 reservoirs in the Klong Yai River basin, namely Dok Krai reservoir (storage capacity of 72.5 million cubic meter), Nong Pla Lai reservoir (storage capacity of 164.7 million cubic meter) and Klong Yai reservoir (storage capacity of 8.5 million cubic meter). The annual precipitation was varied from 1300 to 2000 mm yr^{-1} and the annual stream flow was approximately 577 Mm^3.

3 Selection of the Global Climate Models (GCMs)

Performance based concept is used for the GCMs selection. It was a comparison between 1991–2000 GCM results and 1991–2000 observed data for rainfall and temperature in mean monthly basis. A GCMs that shows good performance to mimic observed climate patterns will be chosen (Lee and Wang, 2012; Watanabe, 2014).

In this study, rainfall is the most sensitive patterns and considered as a critical criterion for the GCMs selection. Table 1 shows the list of GCMs that use in the study. Since the improper of horizontal resolution of original GCMs and hydrological aspects of the study area; therefore, downscaling and bias correction are mandatory.

A simple 4-point linear interpolation and a so called shifting and scaling technique were applied for downscaling and bias correction (e.g. Lehner et al., 2006; Hanasaki et al., 2013). This is one of the simplest and most popular techniques for correcting biased GCMs data. Hanasaki et al (2013) also expressed this method in mathematical terms for surface air temperature and precipitation (rainfall) as shown in Eqs. (1) and (2):

$$T^{\mathrm{cor}}_{y,m,d} = T^{\mathrm{obs}}_{y,m,d} + \left(\overline{T}^{\mathrm{org}}_{\mathrm{future},m} - \overline{T}^{\mathrm{org}}_{\mathrm{baseline},m}\right) \tag{1}$$

$$P^{\mathrm{cor}}_{y,m,d} = P^{\mathrm{obs}}_{y,m,d} \times \left(\overline{P}^{\mathrm{org}}_{\mathrm{future},m} \div \overline{P}^{\mathrm{org}}_{\mathrm{baseline},m}\right) \tag{2}$$

where T is temperature, P is precipitation, obs is 10 year actual record data in the study area (1991–2000), baseline is data obtained from GCMs in 1991–2000, and future is data obtained from GCMs in 2017–2026 and 2027–2036.

4 SWAT Model

This study aims at evaluating the future inflow into the reservoirs changes using the application of Soil and Water Assessment Tool or SWAT model. It is a physically semi-distributed hydrological model which was developed by Arnold et al. (1998). This model requires input data from physical features of a watershed such as Digital Elevation Model (DEM) or topography, weather data, soil data, land use land cover, and model parameters. SWAT model can simulate many processes related to soil and water, for example, surface and sub-surface runoff, soil erosion and sedimentation, impact of soil and land use changes on water resources. However, streamflow simulation was the main focus in this study. Surface runoff was calculated using the SCS curve number method. The potential evapotranspiration was based on Penman-Monteith method (Monteith, 1965). The hydrological cycle that simulated by the SWAT model was based on the water balance concept as expressed in Eq. (3):

$$\mathrm{SW}_t = \mathrm{SW}_0 + \sum\nolimits_{t=1}^{t}\left(R_{\mathrm{day}} - Q_{\mathrm{surf}} - E_{\mathrm{a}} - w_{\mathrm{seep}} - Q_{\mathrm{gw}}\right) \tag{3}$$

where SW_t is the final soil water content (mm), SW_0 is the initial soil water content on day i (mm), t is the time (d), R_{day} is the amount of precipitation on day i (mm), Q_{surf} is the amount of surface runoff on day i (mm), E_{a} is the amount of evapotranspiration on day i (mm), w_{seep} is the amount of water entering the vadose zone from the soil profile on day i (mm), and Q_{gw} is the amount of return flow on day i (mm).

The streamflow or river-runoff was simulated and calibrated with observed data to find out the appropriate model parameters that able to represent the basin characteristics. This study was simulated during 2007–2016 and location of gauging stations is shown in Fig. 1. The criteria to evaluate model performance were coefficient of determination (R^2) (Polanco et al., 2017) and graphical comparison of observed and simulated stream flow.

$$R^2 = \left[\frac{\sum_{i=1}^{n}\left(Q_{\mathrm{obs}} - Q_{\mathrm{avg}}\right)\left(Q_{\mathrm{sim}} - \overline{Q_{\mathrm{sim}}}\right)}{\sum_{i=1}^{n}\left(Q_{\mathrm{obs}} - Q_{\mathrm{avg}}\right)\sum_{i=1}^{n}\left(Q_{\mathrm{sim}} - \overline{Q_{\mathrm{sim}}}\right)}\right]^2 \tag{4}$$

where Q_{obs}, Q_{avg}, Q_{sim} and $\overline{Q_{\mathrm{sim}}}$ are the observed, averaged observed, simulated, and average simulated streamflow and n is the length of the time series.

Table 1. List of the 10 GCMs used in this study.

Model	Climate Modeling Organization
CNRM-CM5	Center National de Recherches Meteorologiques/Centre Europeen de Recherche et Formation Avancees en Calcul Scientifique (CNRMCERFACS)
CSIRO-Mk3-6-0	Commonwealth Scientific and Industrial Research Organization and the Queensland Climate Change Centre of Excellence (CSIRO-QCCCE)
GFDL-ESM2M	Geophysical Fluid Dynamics Laborator (NOAA GFDL)
HadGEM2-ES	Met Office Hadley Centre (MOHC)
INM-CM4	Institute for Numerical Mathematics (INM)
GISS-E2-R	NASA Goddard Institute for Space Studies (NASA GISS)
MIROC5	Atmosphere and Ocean Research Institute University of Tokyo, National Institute for Environmental Studies, and Japan Agency for Marine-Earth Science and Technology (MIROC)
MPI-ESM-LR	Max Planck Institute for Meteorology (MPI-M)
MRI-CGCM3	Meteorological Research Institute (MRI)
NorESM1-M	Norwegian Climate Centre (NCC)

Source: Coupled Model Intercomparison Project 5 (CMIP5).

Table 2. Calibration results of SWAT model for reservoirs over Klong Yai River basin.

Reservoir Name	Mean Annual Inflow Observed (Mm^3)	Mean Annual Inflow Simulated (Mm^3)	R^2
Dok Krai	165.0	144.7	0.61
Nong Pla Lai	191.0	217.9	0.77
Klong Yai	112.6	121.7	0.71

Source: Royal Irrigation Department (RID).

5 Results and Discussion

5.1 SWAT model calibration

The SWAT model was simulated in 10 years (2007–2016) and compared the simulate result with the observed inflow into the Dok Krai, Nong Pla Lai and Klong Yai reservoirs as shown in Fig. 2 and Table 2. The simulation results were closed to the observation for each reservoir and the summarized of mean annual inflow of the model simulation and observation were also shown in Table 2. The result showed the similarity of annual inflow between simulation and observation.

5.2 GCMs selection

For the GCMs selection, 10 CMIP5 GCMs of historical monthly rainfall and temperature data during period 1991–2000 are compared with observed data for selecting the suitable GCMs in Klong Yai River Basin. The standard deviation of the monthly observed data is represented the similar pattern of the rainfall and temperature in this river basin. 3 GCMs were selected to evaluate the impact of climate change based on the period 2017–2026 and 2027–2036 relative to 1991–2000 using bias correction in the next step. The comparison of selected GCM precipitation output and observed data from 1991–2000 was shown in Fig. 3. The blue band represented the standard deviation range of observation (mean monthly over the period 1986–2000). Three suitable GCMs namely GISS, MIROC and MPI showed similar precipitation patterns with the observation.

Moreover, the analysis result showed that there were 2 tropical monsoons in this Klong Yai catchment. Two periods are May–July and September–October.

5.3 Impacts of climate change on water resources

The study of precipitation changes under climate change scenario was analyzed based on the selected 3 GCMs. The results of monthly precipitation were shown in Figs. 4 (2017–2026) and 5 (2027–2036). The findings showed an increasing trend of annual rainfall in both 2 future periods. The increase rainfall occurs mainly during March–May and August–October. The future annual rainfall projects change were varied around 0.51 % to 19.05 % (6.24 to 232.31 mm),

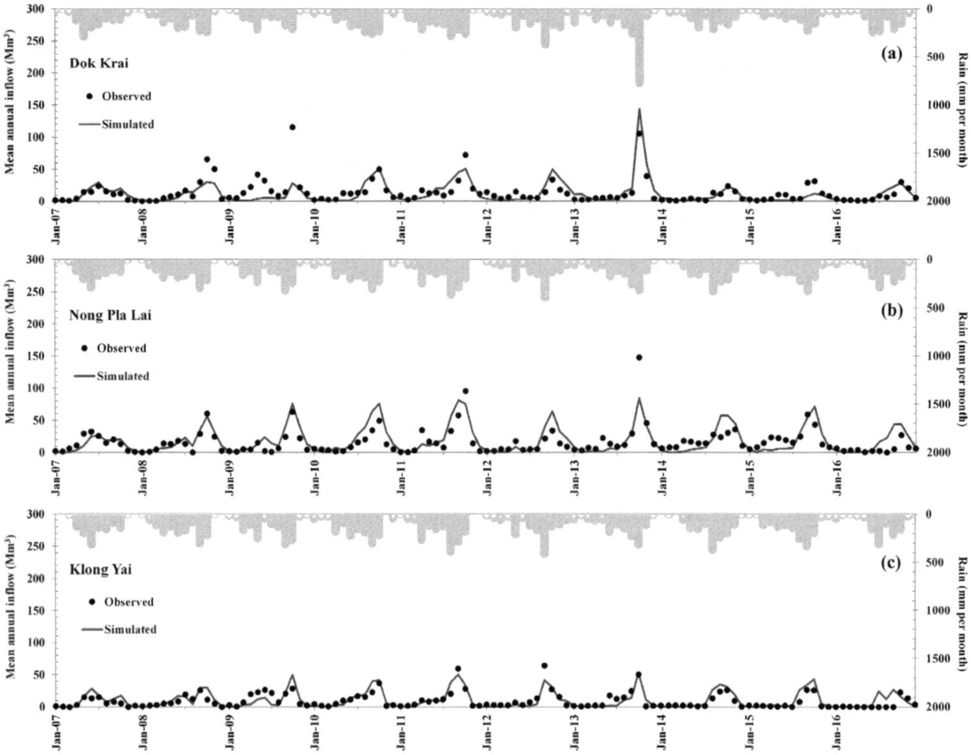

Figure 2. Calibrated inflow in 3 reserviors from SWAT model for the Klong Yai River basin.

and −7.22 % to 17.30 % (−88.04 to 211 mm) for periods 2017–2026, 2027–2036, respectively

For the future projection of temperature, the monthly temperature trend will be increased around 0.3–1.35 °C or increasing of a 1.25 % to 5.00 % by comparing to the average temperature during 1991–2000 as shown in Figs. 9 and 10.

April was the hottest month while December was the winter season in the past but temperature will increase about 1.0 °C. These projected results showed good agreement with the IPCC's report (IPCC, 2013) on 1.0–2.4 °C of surface temperature rising at the end of 21st century.

The selected GCMs (MIROC5, GISS and MPI GCMs) were applied to forecast the effect of climate change impact on reservoir inflow. There are 2 assessment periods which are 2017–2026 and 2027–2036. As shown in Figs. 8 to 13, there are the future simulations of the monthly reservoir inflow into the Dok Krai, Nong Pla Lai, and Klong Yai reservoirs in 2 periods, respectively. The comparison of monthly reservoirs inflow in the past (2007–2016) and under climate change conditions showed a decreasing trend due to the rising temperature and evaporation

Moreover, an increase in temperature will increase the agricultural water demand or crop water requirement in Klong Yai River basin. The climate change in the Klong Yai River basin is the negative impact on water resources. It can be developed adaptation options for address the climate change impact especially drought management for example increase water efficiency, reduce the water use or promote a water saving in the future.

6 Conclusions

3 GCMs, namely, MIROC5, GISS and MPI GCM models were applied to assess the potential impact of climate on precipitation and temperature in the Klong Yai River basin in Thailand. The future precipitation trend under climate change scenarios were slightly increase compared to the historical period (1991–2000). Temperature projection in the future, we found that the temperature in Klong Yai River basin trend were increase apparently, and the increasing trend of temperature varied between 0.3 to 1.35 °C relative to the past observation.

The impacts of climate change on water resources assessment in the Klong Yai River basin during 2017–2036 clearly are the reduction of streamflow volume. Relatively low amount of water storage for 3 reservoirs in the basin is expected as the direct impact of changing climate. An increasing rate of evapotranspiration is also expected due to surface temperature rise.

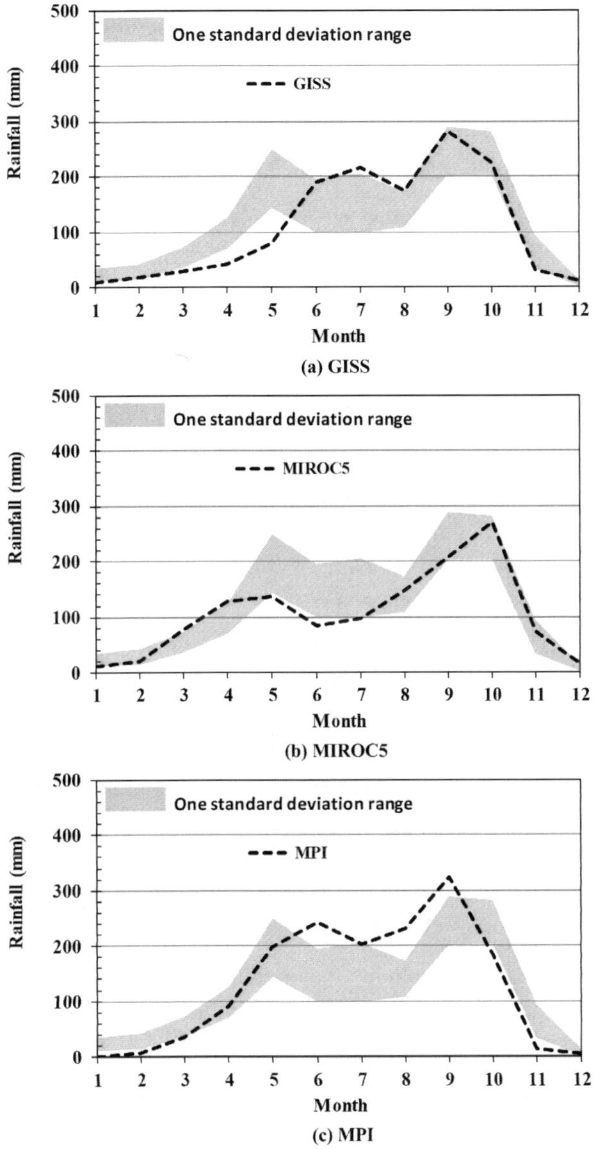

Figure 3. Comparison of Average Monthly Precipitation of the selected GCMs and observation (1991–2000) in Klong Yai River basin.

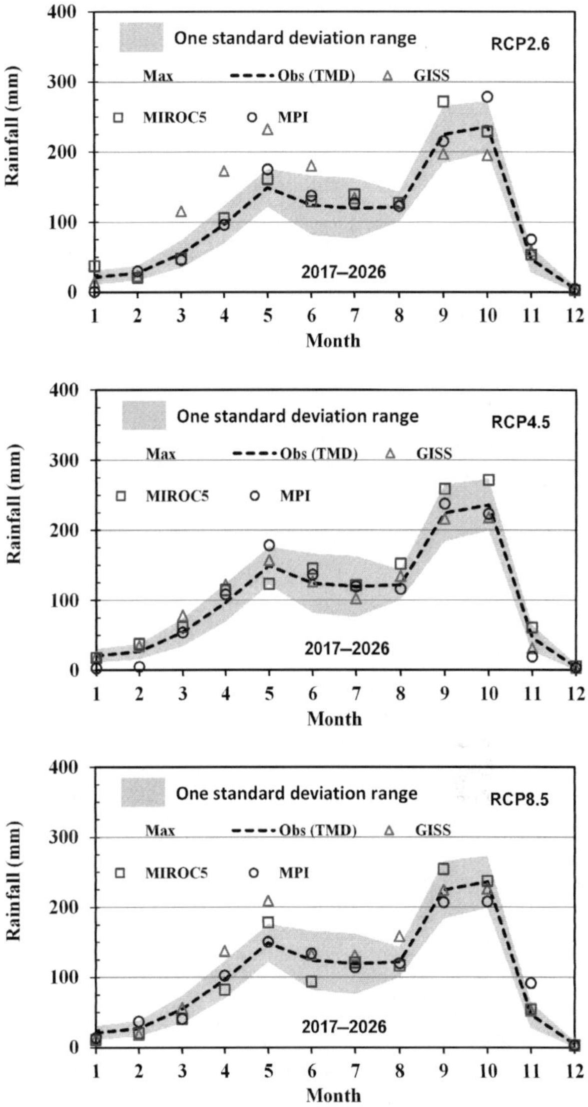

Figure 4. Comparison of mean monthly precipitation variation in 2017–2026 (under the climate change scenarios) and 1991–2000.

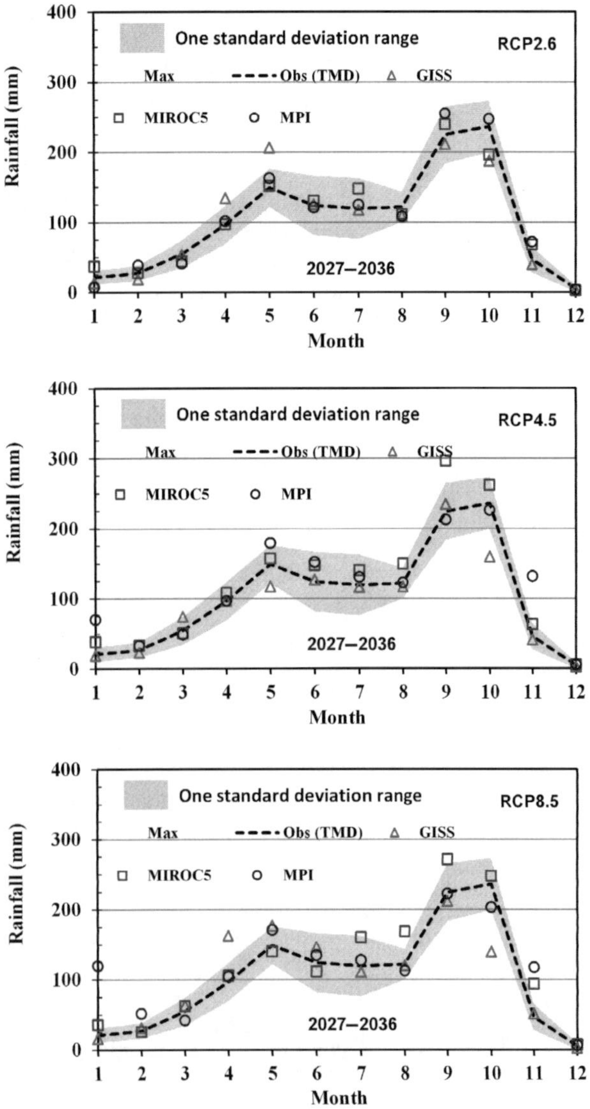

Figure 5. Comparison of mean monthly precipitation variation in 2027–2036 (under the climate change scenarios) and 1991–2000.

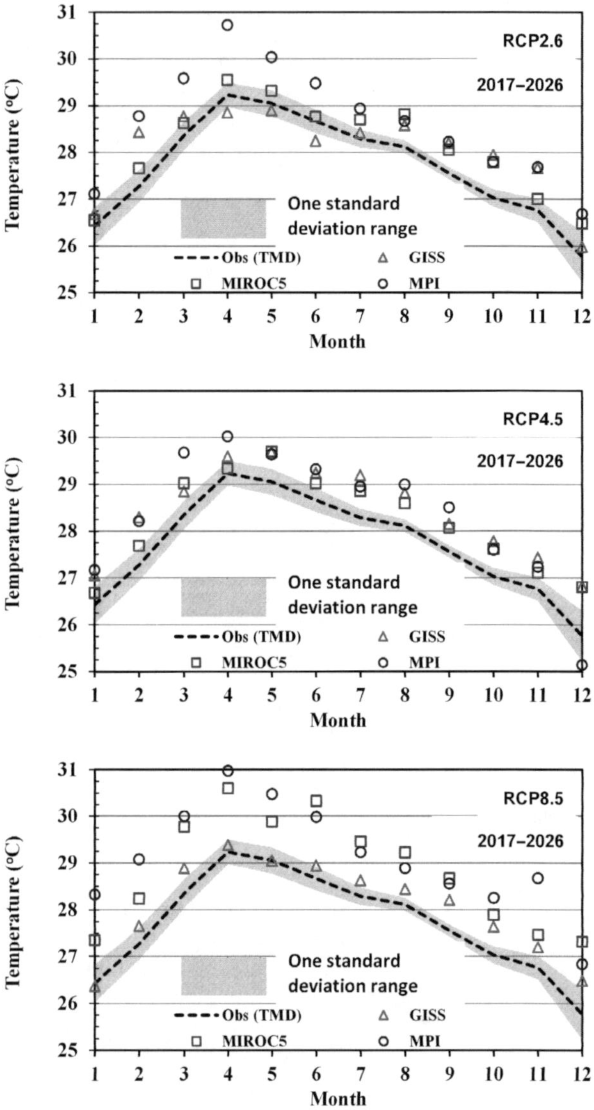

Figure 6. Comparison of mean monthly temperature variation during the periods of 2017–2026 (under the climate change scenarios) and 1991–2000.

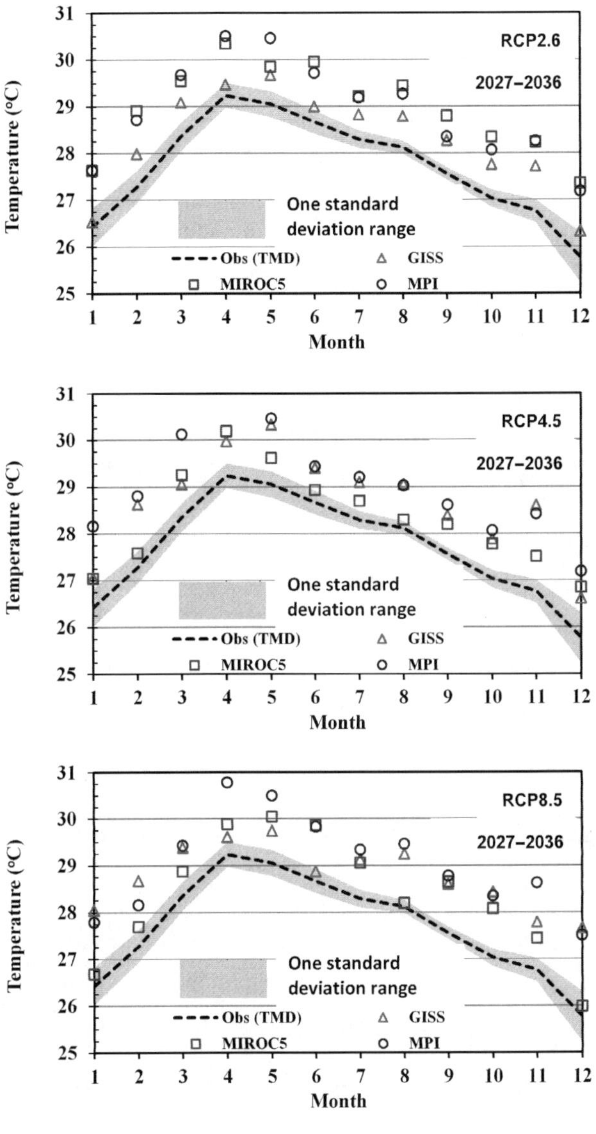

Figure 7. Comparison of mean temperature variation in during the periods of 2027–2036 (under the climate change scenarios) and 1991–2000.

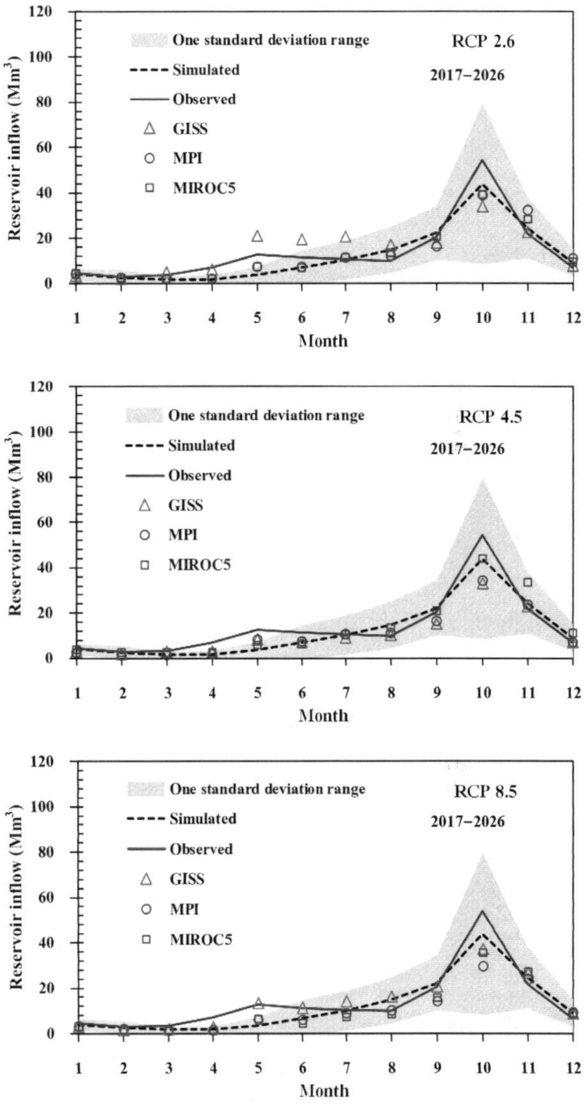

Figure 8. Projected mean monthly reservoir inflows under 3 scenarios for the years 2017–2026 at Dok Krai Resevoir.

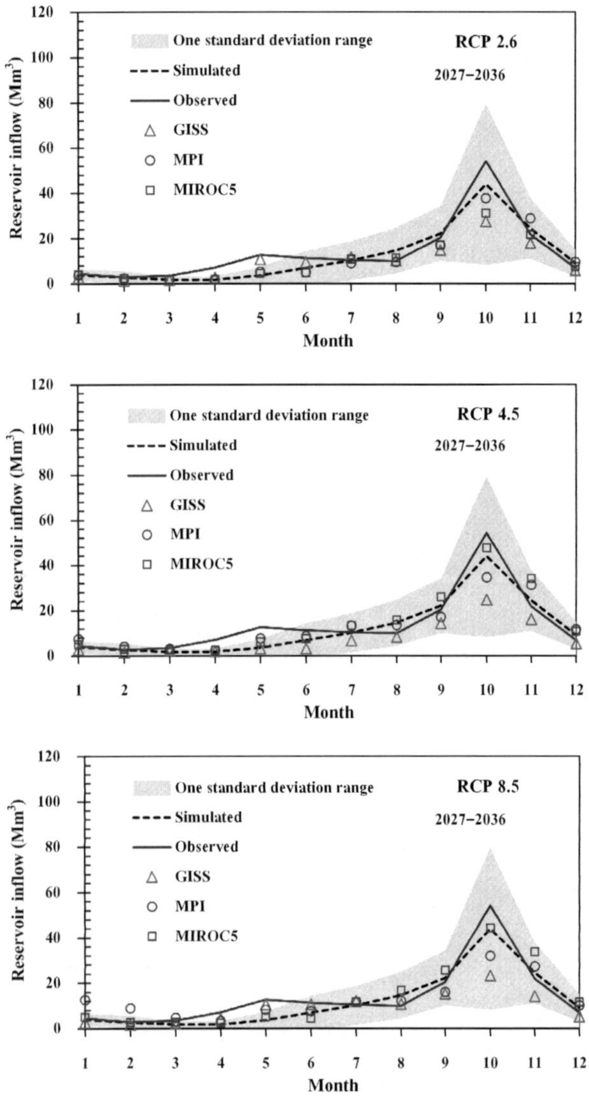

Figure 9. Projected mean monthly reservoir inflows under 3 scenarios for the years 2027–2036 at Dok Krai Resevoir.

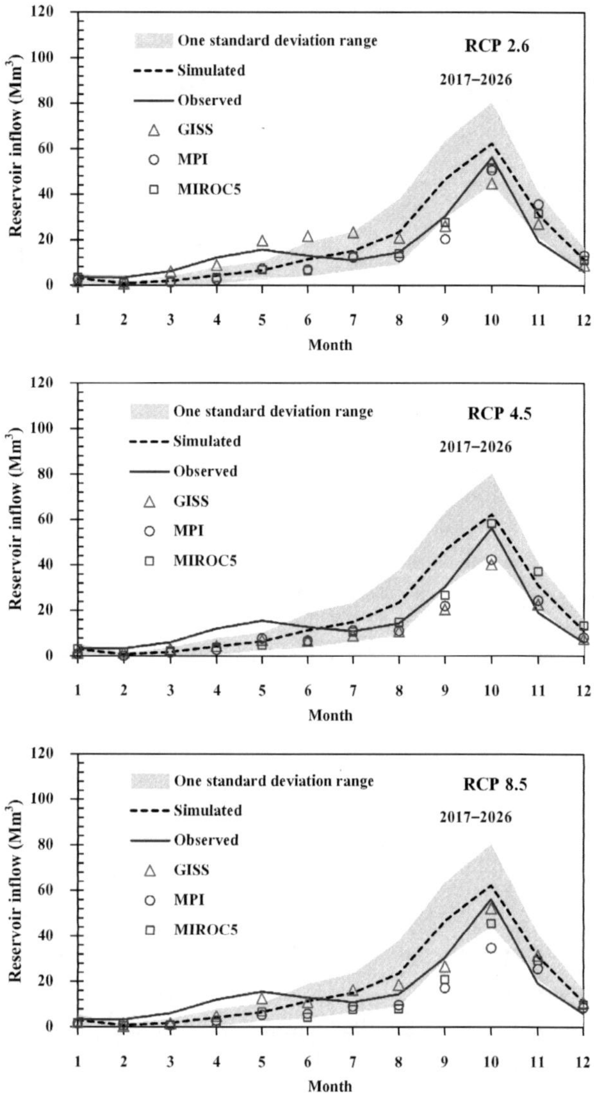

Figure 10. Projected mean monthly reservoir inflows under 3 scenarios for the years 2017–2026 at Nong Pla Lai Resevoir.

 https://doi.org/10.5194/piahs-383-355-2020

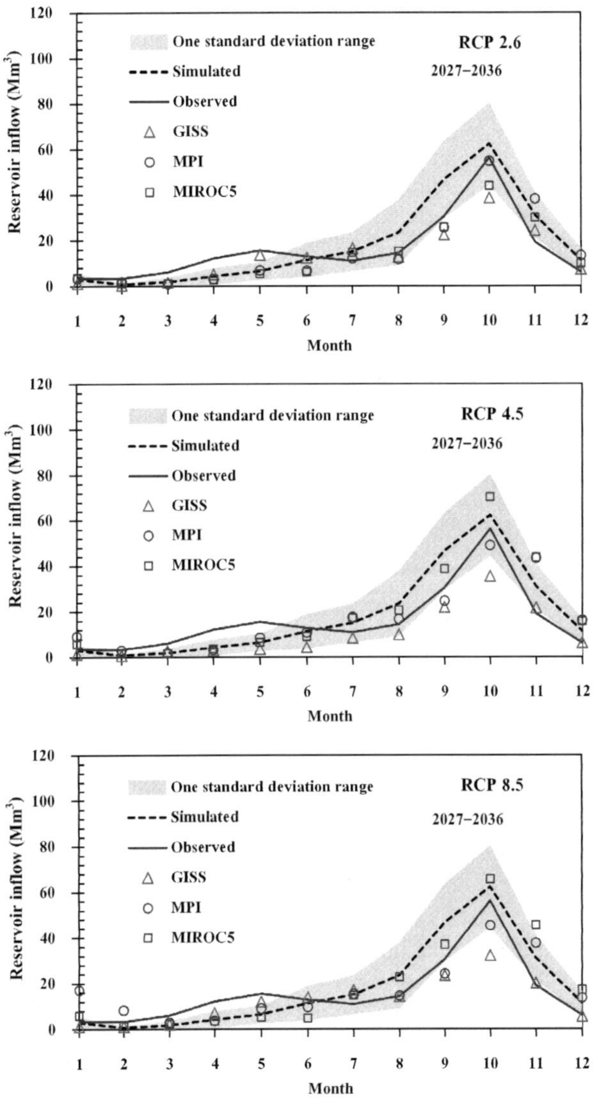

Figure 11. Projected mean monthly reservoir inflows under 3 scenarios for the years 2027–2036 at Nong Pla Lai Resevoir.

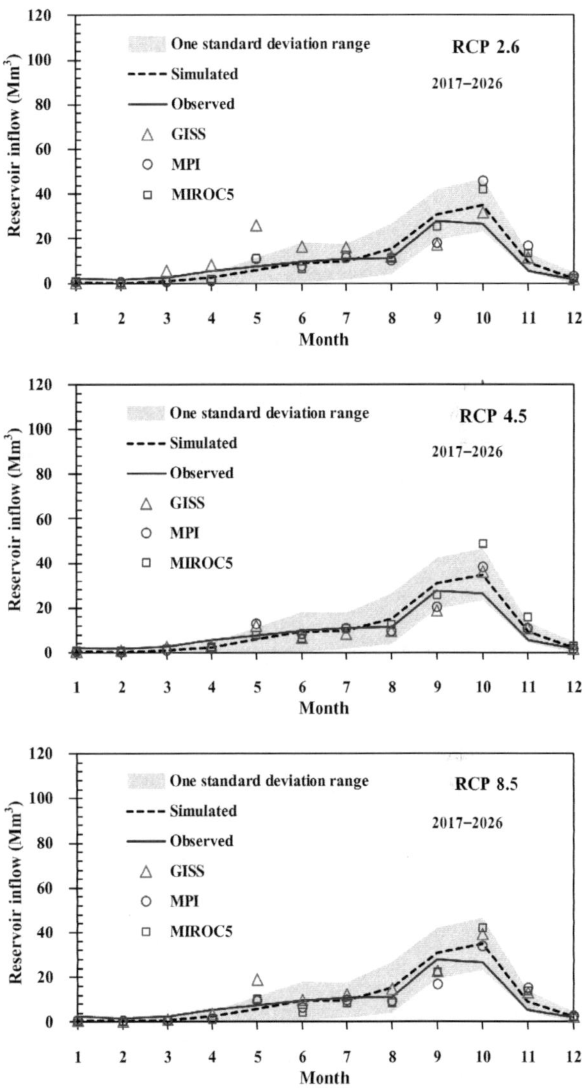

Figure 12. Projected mean monthly reservoir inflows under 3 scenarios for the years 2017–2026 at Klong Yai Resevoir.

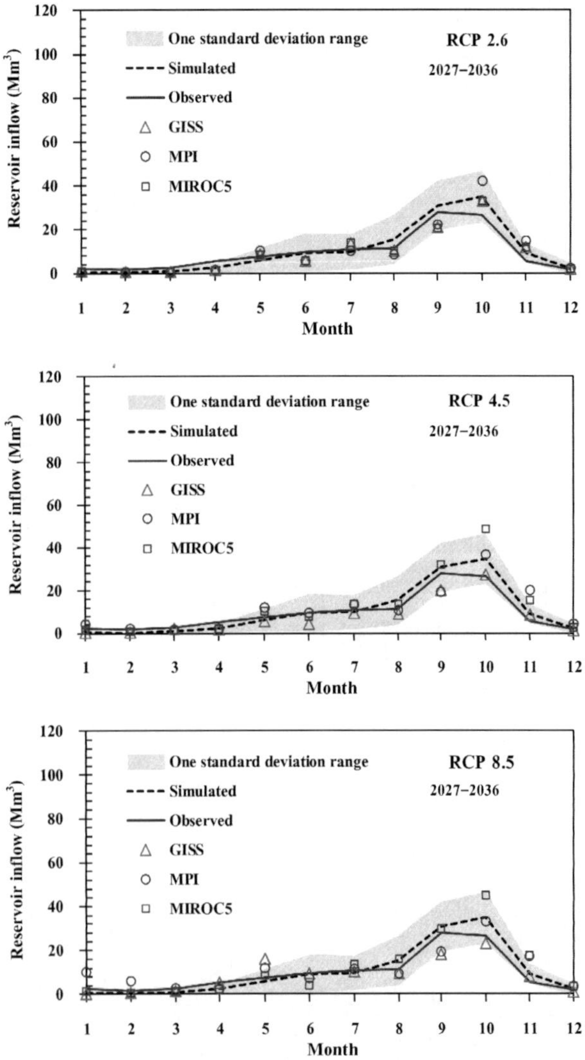

Figure 13. Projected mean monthly reservoir inflows under 3 scenarios for the years 2027–2036 at Klong Yai Resevoir.

Data availability. River flow data were obtained from Royal Irrigation Department (RID), Thailand and are not available to the public as far as the authors know.

Author contributions. CE and CK developed the SWAT model. CE, WP and PS have contributed to the data processing and analysis. CE overlooked this work as the project leader. CE prepared the manuscript with contributions from all co-authors.

Competing interests. The authors declare that they have no conflict of interest.

Special issue statement. This article is part of the special issue "Hydrological processes and water security in a changing world". It is a result of the 8th Global FRIEND–Water Conference: Hydrological Processes and Water Security in a Changing World, Beijing, China, 6–9 November 2018.

Acknowledgements. This work was supported by the Petroleum Authority of Thailand (PTT) Research and Technology Institute. The authors wish to thank the Thai Meteorological Department (TMD)and Royal Irrigation Department (RID) for providing the climate and hydrological data used in this work.

Financial support. This research was supported by the Petroleum Authority of Thailand (PTT) Research and Technology Institute (grant no. PR1110010318).

References

Arnold, J. G., Srinivasan, R., Muttiah, R. S., and Williams, J.: Large area hydrologic modelling and assessment, Part I: Model development, J. Am. Water Resour. As., 34, 73–89, https://doi.org/10.1111/j.1752-1688.1998.tb05961.x, 1998.

Alcamo, J., Flörke, M., and Märker, M.: Future long-term changes in global water resources driven by socio-economic and climatic changes, Hydrol. Sci. J., 52, 247–275, https://doi.org/10.1623/hysj.52.2.247, 2007.

Monteith, J. L.: Evaporation and environment, Sym. Soc. Exp. Biol., 19, 205–234, https://doi.org/10.1002/qj.49710745102, 1965.

Hansen, J., Sato, M., Ruedy, R., Lo, K., Lea, D. W., and MedinaElizade, M.: Global temperature change, P. Natl. Acad. Sci. USA, 103, 14288–14293, 2006.

Hanasaki, N., Fujimori, S., Yamamoto, T., Yoshikawa, S., Masaki, Y., Hijioka, Y., Kainuma, M., Kanamori, Y., Masui, T., Takahashi, K., and Kanae, S.: A global water scarcity assessment under Shared Socio-economic Pathways – Part 2: Water availability and scarcity, Hydrol. Earth Syst. Sci., 17, 2393–2413, https://doi.org/10.5194/hess-17-2393-2013, 2013.

Jones, N.: Climate assessments: 25 years of the IPCC, Nature, 501, 298–299, https://doi.org/10.1038/501298a, 2013.

Lee, J. and Wang, B.: Future change of global monsoon in the CMIP5, Clim. Dynam., 42, 101–119, https://doi.org/10.1007/s00382-012-1564-0, 2012.

Lehner, B., Döll, P., Alcamo, J., Henrichs, T., and Kaspar, F.: Estimating the Impact of Global Change on Flood and Drought Risks in Europe: A Continental, Integrated Analysis, Climatic Change, 75, 273–299, https://doi.org/10.1007/s10584-006-6338-4, 2006.

Limjirakan, S. and Limsakul, A.: Observed Trends in Surface Air Temperatures and Their Extremes in Thailand from 1970 to 2009, J. Meteorol. Soc. Jpn., 90, 647–662, https://doi.org/10.2151/jmsj.2012-505, 2012a.

Limjirakan, S. and Limsakul, A.: Trends in Thailand pan evaporation from 1970 to 2007, Atmos. Res., 108, 122–127, https://doi.org/10.1016/j.atmosres.2012.01.010, 2012a.

Polanco, E. I., Fleifle, A., Ludwig, R., and Disse, M.: Improving SWAT model performance in the upper Blue Nile Basin using meteorological data integration and subcatchment discretization, Hydrol. Earth Syst. Sci., 21, 4907–4926, https://doi.org/10.5194/hess-21-4907-2017, 2017.

Stocker, T. F., Qin, D. G., Plattner, K.,Tignor, M., Allen, S., Boschung, J., Nauels, A., Xia, Y., Bex, V., and Midgley, P.: Climate change 2013: the physical science basis., Cambridge University Press, Cambridge, 159–254, https://doi.org/10.1017/CBO9781107415324, 2013.

Watanabe, S., Hirabayashi, Y., Kotsuki, S., Hanasaki, N., Tanaka, K., Mateo, C., Kiguchi, M., Ikoma, E., Kanae, S., and Oki, T.: Application of performance metrics to climate models for projecting future river discharge in the Chao Phraya River basin, Hydrol. Res. Lett., 8, 33–38, https://doi.org/10.3178/hrl.8.33, 2014.

Proc. IAHS, 383, 367–374, 2020
https://doi.org/10.5194/piahs-383-367-2020

Impact of hydropower dam on total suspended sediment and total organic nitrogen fluxes of the Red River (Vietnam)

Nhu Da Le[1,2], **Thi Phuong Quynh Le**[1,2], **Thi Xuan Binh Phung**[3], **Thi Thuy Duong**[4], **and Orange Didier**[5]

[1]Graduate University of Science and Technology, 18 Hoang Quoc Viet road, Cau Giay district, Hanoi city, Vietnam
[2]Institute of Natural Product Chemistry, Vietnam Academy of Science and Technology, 18 Hoang Quoc Viet road, Cau Giay district, Hanoi city, Vietnam
[3]Electric Power University, 235 Hoang Quoc Viet, Bac Tu Liem, Hanoi, Vietnam
[4]Institute of Environmental Technology, Vietnam Academy of Science and Technology, 18 Hoang Quoc Viet road, Cau Giay district, Hanoi city, Vietnam
[5]IRD, ECO&SOLS, University of Montpellier, CIRAD, INRA, Supagro Montpellier, Place Viala, 34060 Montpellier CEDEX 2, France

Correspondence: Le Thi Phuong Quynh (quynhltp@gmail.com, quynhltp@yahoo.com) and Le Nhu Da (dalenhu@gmail.com)

Published: 16 September 2020

Abstract. The Red River is a typical example of the Southeast Asian rivers, which has been strongly affected by human activities. This paper analyses the change of total suspended sediment (TSS) load of the Red River from 1960 to 2015 in which numerous new dams in both China and Vietnam have been constructed. A strong decrease of TSS load of the whole Red River (from $79 \pm 26 \times 10^6\,\mathrm{t\,yr^{-1}}$ in 1960s to $6 \pm 1 \times 10^6\,\mathrm{t\,yr^{-1}}$ in 2010s) allocated to the dam impoundments in spite of population and deforestation increase. Base on the experimental equation describing the relationship between TSS and total organic nitrogen (TON) concentrations, and on the available data of TSS concentration and river discharge, the longterm TON concentrations and fluxes were calculated for the three tributaries and the whole Red River. The annual average of TON concentrations spatially varied from 0.41 to $3.19\,\mathrm{mg\,L^{-1}}$, averaging $0.98\,\mathrm{mg\,L^{-1}}$ for the whole period; the lowest was found for the Da River where the new dams have been impounded. The highest TON concentrations and fluxes occurred in the wet season in relationship with the highest sediment loads and river discharges. The riverine TON fluxes transferred to estuary significantly decreased from $141 \times 10^3 \pm 38 \times 10^3\,\mathrm{t\,yr^{-1}}$ (equivalent to $902 \pm 247\,\mathrm{kg\,km^{-2}\,yr^{-1}}$) in 1960s to $32 \times 10^3 \pm 5 \times 10^3\,\mathrm{t\,yr^{-1}}$ (equivalent to $207 \pm 35\,\mathrm{kg\,km^{-2}\,yr^{-1}}$) in 2010s. The TSS flux decrease has driven a clear reduction of associated elements like nitrogen, which let to hypothesis a change in biogeochemical processes in the coastal zone.

1 Introduction

Total suspended solids (TSS) transported by the Asian rivers to the coastal ocean play an important role in the global delivery of TSS and associated elements (C, N, P) due to the significant quantities of suspended solids. However, in recent decades, TSS and material fluxes of these rivers have been altered drastically as a result of reservoir impoundment, land use, population, and climate change (Walling and Fang, 2003). Within the factors affected to the riverine TSS load, dams and reservoirs are known to be the most important. The impact of dam impoundment was clearly observed for numerous Asian Rivers such as the Yangtze River (Changjiang) (Wu et al., 2007), the Yellow River (Huanghe) (in China) (Wang et al., 2007; Hu et al., 2015), the Mekong River (Kummu and Varis, 2007)... Dam impoundment has reduced TSS fluxes of these rivers by more than half. The loss of sediment and associated nutrient fluxes from the continent has dramatically affected deltas (Syvitski and Saito, 2007). The habitats of many aquatic species in coastal zone have significantly impacted the recovery of the mangroves and coastal aquaculture area. Thus, Beusen et al. (2005) emphasized that it is important to understand the response of riverine TSS and associated nutrient fluxes to regional and global changes, and the main factors that control the fluvial sediment transport to the oceans.

The Red River (Vietnam and China) is a typical Asian river which is strongly impacted by both climate and human activities. Indeed, human activities have altered the hydrology, suspended sediment and associated element (C, N, P) loads of the Red River though dam impoundment, land use change and population increased (Le et al., 2015). Two large dams (Hoa Binh, Thac Ba) have been impounded in the Red River systems, in the Vietnamese territory since 1970s. And recently, since the year 2010, new series of dams have been set up in the Red River upstream in period 2000–2015 in both China and Vietnam sectors. This study firstly aims to estimate the recent change of TSS flux due to new series of dams constructed in both China and Vietnam in the period 1960–2015 basing a data set of daily TSS concentrations and daily river discharge. And then, basing on the experimental equation describing the relationship between the TSS and total organic nitrogen (TON) concentrations, and on the available data of TSS concentrations and river discharges, the longterm TON concentrations and fluxes were calculated for the three tributaries and the whole Red River.

2 Site description and methodology

2.1 Site description

2.1.1 The Red River basin

The Red River situates in South-East Asia, drains a watershed area of 156 451 km^2, 50.3 % of which is situated in Vietnam, 48.8 % in China and 0.9 % in Laos. The Red River includes three tributaries Thao, Da, Lo Rivers which join at Viet Tri city, and then forms a large delta before flowing into the South-East Asian Sea through four distributaries called, Ba Lat, Lach Gia, Tra Ly and Day (Dang, 2000) (Fig. 1).

There are two distinct seasons in the Red River basin: rainy season (May–October) and dry season (November–April). The annual rainfall strongly varies from 700 to 4800 mm yr^{-1} across the basin with about 80 % of rainfall occurring during the rainy season. Noted that highest incidence of typhoons occur in July and August in the basin area.

The three upstream sub-basins of Thao, Da and Lo rivers have differences in their lithology. The Red main channel basin is dominated by metamorphic rocks, except in the upper reaches where the Mesozoic sedimentary deposits are found. The Da drainage basin is composed of sedimentary rocks from the Mesozoic and Paleozoic period with minor felsic intrusions. The Lo drainage has low-grade metamorphic rocks and Proterozoic to Paleozoic sedimentary rocks with some granitic intrusions. In the central Red River Delta, silt and clay are the most abundant sediments while fine sand presents in small amounts from the Holocene succession (Moon et al., 2007).

Land use and population density are different for the four sub-basins of the Red River. Low population (< 100 inhabitant km^{-2}) and high proportion of forestland are found for the Da river basin. The moderate population and industrial-agricultural and forest land is observed in the Lo and Thao river basins. Very high population (> 1000 inhabitant km^{-2}) and high proportion of agricultural land are characterized for the Delta area. Change in landuse (deforestation and intensive agricultural land) and population increased for longterm period from 1960s–2010s have been revealed in a recent study (Le et al., 2015, 2018).

2.1.2 Hydrology and Reservoir impoundment

Hydrology

Daily river discharges at the outlets of the three tributaries Thao (at Yen Bai station), Da (at Hoa Binh station) and Lo (at Vu Quang station) and of the downstream main axe of the Red River (at Hanoi station) were available for the period 1960–2015 (MONRE, 1960–2015) (Fig. 1). The river discharge follows the variation of rainfall during the year with clear higher river discharges observed in the rainy season than in the dry one. The average river discharge of the Thao (at Yen Bai), Da (at Hoa Binh), Lo (at Vu Quang) and in the main axe at Hanoi stations were 720; 1670; 1020 and 2500 $m^3 s^{-1}$ respectively for the period 1960–2015 (MONRE, 1960–2015). Of the three upstream tributaries, the Da River accounts for a half of the total Red River discharge. Long-term data on the decade average of water discharge showed a clear decrease for the Thao, Da tributaries and for the main downstream Red River in the 1960–2015 period (Table 1). During the whole study period, extreme low river discharges of the main axe Red River at Son Tay (368 $m^3 s^{-1}$ in May 1960 or 380 $m^3 s^{-1}$ in March 2002) hydrological stations were observed (Quach, 2011; Pham et al., 2015) and extreme high river discharge (37 800 $m^3 s^{-1}$) at Son Tay station was observed in August 1971.

Dam impoundment

In the Red River basin in both China and Vietnam territory, a series of reservoirs and dams have been constructed and operated (Ha and Vu, 2012; Le et al., 2018) for multipurpose including generating hydropower, controlling floods, providing water for the agricultural land, regulating water resource during dry season for downstream. In the upstream Thao River, two main intercepting dams (namely Namsha and Madushan, 140 km from the Vietnamese border to

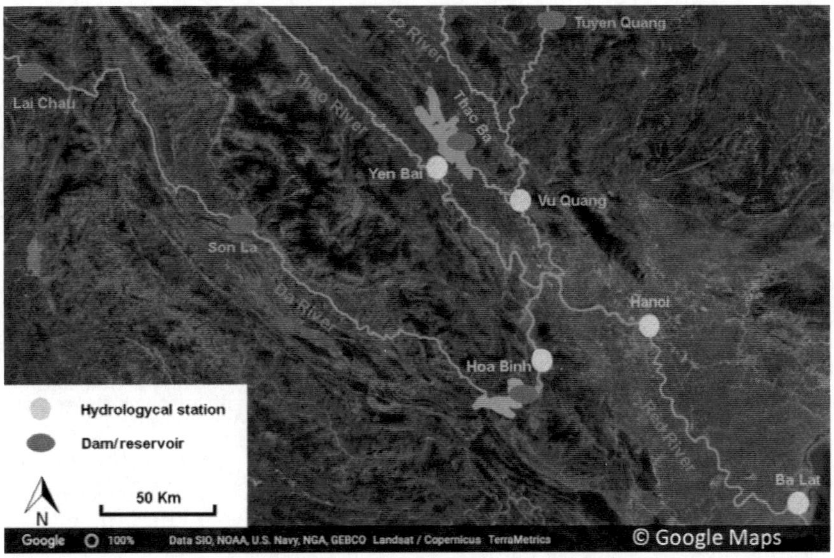

Figure 1. The Red River and its tributaries: location of the large reservoirs and hydrological stations.

Table 1. Decade average and standard deviation (SD) of river discharge, TSS and TON concentrations ($mg\,L^{-1}$) of the Thao, Da and Lo tributaries and of the main axis of the Red River at Hanoi station in different periods from 1960 to 2015.

Hydrological station	River	Variables	1960–1969	1970–1979	1980–1989	1990–1999	2000–2009	2010–2015
Yen Bai	Thao	River discharge, $m^3\,s^{-1}$	774 ± 126	818 ± 208	674 ± 103	752 ± 137	708 ± 140	517 ± 34
		TSS concentration, $mg\,L^{-1}$	883 ± 274	965 ± 79	1054 ± 454	1479 ± 344	928 ± 297	278 ± 62
		TON concentration, $mg\,L^{-1}$	1.5 ± 0.4	1.7 ± 0.1	1.8 ± 0.6	2.3 ± 0.4	1.6 ± 0.4	0.7 ± 0.1
Hoa Binh	Da	River discharge, $m^3\,s^{-1}$	1720 ± 219	1706 ± 213	1544 ± 236	1840 ± 301	1798 ± 230	1276 ± 216
		TSS concentration, $mg\,L^{-1}$	475 ± 144	477 ± 75	354 ± 169	74 ± 15	56 ± 26	15 ± 6
		TON concentration, $mg\,L^{-1}$	1.0 ± 0.2	1.0 ± 0.1	0.9 ± 0.2	0.5 ± 0.0	0.5 ± 0.0	0.4 ± 0.0
Vu Quang	Lo	River discharge, $m^3\,s^{-1}$	955 ± 144	1077 ± 227	1085 ± 165	1098 ± 158	897 ± 213	990 ± 439
		TSS concentration, $mg\,L^{-1}$	158 ± 28	167 ± 54	163 ± 56	207 ± 51	140 ± 53	56 ± 22
		TON concentration, $mg\,L^{-1}$	0.6 ± 0.0	0.6 ± 0.1	0.6 ± 0.1	0.7 ± 0.1	0.6 ± 0.1	0.5 ± 0.0
Hanoi	Main down-	River discharge, $m^3\,s^{-1}$	2610 ± 317	2821 ± 380	2607 ± 439	2660 ± 390	2310 ± 310	1673 ± 262
	stream Red	TSS concentration, $mg\,L^{-1}$	505 ± 112	554 ± 105	434 ± 116	389 ± 74	343 ± 172	77 ± 8
	River branch	TON concentration, $mg\,L^{-1}$	1.1 ± 0.1	1.1 ± 0.1	1.0 ± 0.2	0.9 ± 0.1	0.8 ± 0.2	0.5 ± 0.0

China) have been constructed in the period 2010s (IMRR, 2010). In the Lo River, the Thac Ba was constructed the earliest, in 1972 originally designed to supply water and generate hydropower (440 MW) but did not control floods. Recently, the Tuyen Quang is impounded in the Vietnamese part and at least 8 hydropower reservoirs have been built on the upstream Lo River in the Chinese part (Ha and Vu, 2012). In the Da River, a series of dams/reservoirs have been constructed since 1989. Besides the 3 large reservoirs (Hoa Binh, Son La and Lai Chau), small hydroelectric power stations have been set up (Ban Chat, Huoi Quang) on the Nam Mu river, a branch of the Da River in Vietnam and about 11 small hydrological dams constructed in the Chinese part since 2007 (Ha and Vu, 2012).

2.2 Methodology

2.2.1 Data collection

Daily TSS concentrations and the daily river discharge data of the 3 main tributaries: Yen Bai station (outlet of the Thao river), Hoa Binh station (outlet of the Da river), Vu Quang station (outlet of the Lo river), and in the main axe (at Hanoi station) of the Red River (Fig. 1, Table 1) from 1960 to 2015 were obtained from the MONRE (1960–2015). Daily TSS concentrations were determined at each gauging station by the Vietnamese standard TCVN 6625:2000 method (MOSTE, 2000).

2.2.2 Calculation of TSS fluxes

Daily TSS flux at the outlet of each tributary (Thao, Da and Lo) and at Hanoi station (main downstream axe of the Red River) was calculated based on the daily TSS concentrations with the daily river discharge. Annual TSS flux export ($10^6\,\mathrm{t\,yr^{-1}}$) at the outlet of the Thao, Da and Lo and at Hanoi was calculated as the sum of daily export at each tributary and at Hanoi over the course of a year. The total TSS fluxes of the whole Red River were calculated by extrapolating the flux measured at the Hanoi station taking into account the respective Delta area in the watershed.

2.2.3 Calculation of TON concentrations and fluxes

Previous studies revealed that riverine particulate nitrogen loads are strongly correlated with TSS loads (Ittekkot and Zhang, 1989; Ludwig and Probst, 1996). Our previous study (Le et al., 2005) demonstrated a relationship between TSS concentrations and Total organic nitrogen (TON) concentrations for the Red River water, basing on field measurements, as shown in Eq. (1):

$$\mathrm{TON}\ (\mathrm{mg\,L^{-1}}) = 0.4 + 0.0013\mathrm{TSS}\ (\mathrm{mg\,L^{-1}}) \quad (r^2 = 0.91) \tag{1}$$

Recently, we have analysed about 20 Red River samples taken in Hanoi hydrological station for testing this equation and found the possibility for application of this equation. Thus, we use this equation for calculating daily TON concentrations at different stations Yen Bai (Thao River), Hoa Binh (Da River), Vu Quang (Lo River) and Hanoi in the main branch Red River for the period 1960–2015 basing on the available daily TSS concentrations at the respective stations in this period. After obtaining daily TON concentrations, daily and annual TON fluxes were calculated by the same method for TSS flux calculation.

3 Results and discussion

3.1 Total suspended solids (TSS) and total organic nitrogen (TON) concentrations

3.1.1 Spatial and temporal variation

TSS concentrations which varied 8 to $2144\,\mathrm{mg\,L^{-1}}$ showed spatially different variation for the three tributaries and of the main branch of the Red River (at Hanoi stations). The highest TSS was observed at Yen Bai station (Thao River) in 1986 when the big flood occurred in this river whereas the lowest value was found at Hoa Binh station (Da River) in 2015 after the Son La dam has been in operation. For the whole period 1960–2015, within the three upstream tributaries, the mean TSS concentrations were the highest at the Thao station ($966 \pm 432\,\mathrm{mg\,L^{-1}}$) where less dams were impounded in its main course. The TSS concentrations in the Lo and Da Rivers were much lower and averaged 157 ± 59 and $297 \pm 554\,\mathrm{mg\,L^{-1}}$, respectively. The TSS concentration of main axe Red River at Hanoi station, averaging $411 \pm 170\,\mathrm{mg\,L^{-1}}$ over the period 1960–2015 was lower than the one of the Thao River due to the dilution process because of higher discharges and lower TSS concentrations of the Da and Lo Rivers than those of the Thao River (Table 1).

Regarding temporal variation, the TSS concentrations at all sites observed showed a clear decrease over the 1960–2015 period. Within the three tributaries, a significant decrease of TSS concentration was found for the Da River (from $475 \pm 144\,\mathrm{mg\,L^{-1}}$ in 1960s to $15 \pm 6\,\mathrm{mg\,L^{-1}}$ in 2010s) where a series of dams were operated in its main course (Table 1). Presently (2010–2015 period), within three main tributaries, the TSS concentration of the Da river is lowest (Table 1). The Thao and the Lo showed a clear reduction of TSS concentrations (from $883 \pm 274\,\mathrm{mg\,L^{-1}}$ in 1960s to $278 \pm 62\,\mathrm{mg\,L^{-1}}$ in 2010s) and (from $158 \pm 28\,\mathrm{mg\,L^{-1}}$ in 1960s to $56 \pm 22\,\mathrm{mg\,L^{-1}}$ in 2010s, respectively) (Table 1, Fig. 2), especially in the last decade where dams have been impounded in the Thao and Lo rivers.

TON concentration followed the same spatial variation of the TSS, being highest for the Thao River ($1.7 \pm 0.6\,\mathrm{mg\,L^{-1}}$) and lowest for the Lo River ($0.6 \pm 0.1\,\mathrm{mg\,L^{-1}}$) over the 1960–2015 period. A reduction of TON concentrations was also observed at different stations, however, much lower in comparing with TSS concentration decrease. Indeed, TON concentrations were reduced by a haft at all sites observed (Table 1). For the whole period, the annual average of TON concentrations spatially fluctuated from 0.41 to $3.19\,\mathrm{mg\,L^{-1}}$, averaging $0.98\,\mathrm{mg\,L^{-1}}$. Presently (2010–2015 period), within three main tributaries, the TON concentration of the Da river was lowest (Table 1).

3.1.2 Seasonal variation

The TSS concentrations were higher in the rainy season than in dry season at all four stations during all periods observed. Indeed, over the whole period, the average TSS concentrations in the Thao, Lo and Da Rivers during the rainy season averaging 1495 ± 744, 271 ± 112 and $83 \pm 262\,\mathrm{mg\,L^{-1}}$ respectively were 5 times higher than the ones in dry season averaging 301 ± 159, 35 ± 18 and $24 \pm 48\,\mathrm{mg\,L^{-1}}$ respectively. In the main branch of the Red River, at Hanoi station, TSS concentrations were 3 times higher in the rainy season ($655 \pm 276\,\mathrm{mg\,L^{-1}}$) than in the dry season ($171 \pm 90\,\mathrm{mg\,L^{-1}}$). As known, most of riverine TSS fluxes of the tropical river systems have transported during the rainy season, mainly at high river discharge (Van Maren and Hoekstra, 2004). Higher rainfall in rainy season accelerates soil erosion and leaching,

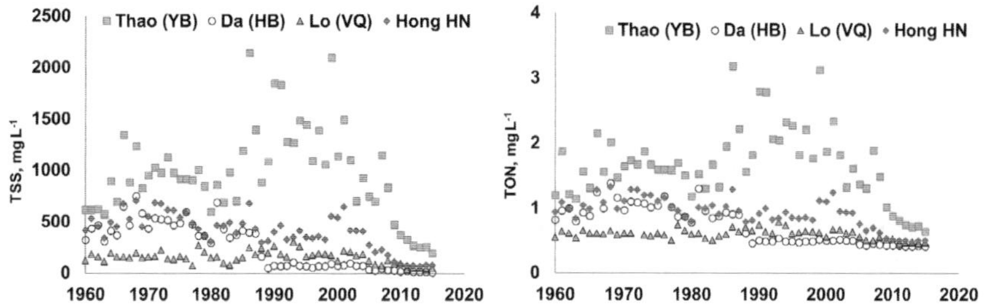

Figure 2. Annual average of TSS, TON concentrations of the Thao, Da and Lo Rivers and of the main axe at Hanoi station during different periods from 1960 to 2015.

Table 2. Annual average and standard deviation of TSS fluxes (10^6 t yr^{-1}) and TON flux (10^3 t yr^{-1}) of the Thao, Da and Lo tributaries and of the whole Red River system during different periods from 1960 to 2015.

Period		1960–1969	1970–1979	1980–1989	1990–1999	2000–2009	2010–2015
Thao (YB)	TSS	34 ± 18	38 ± 12	36 ± 25	59 ± 23	41 ± 21	7 ± 1
	TON	58 ± 25	59 ± 17	55 ± 33	86 ± 30	62 ± 28	16 ± 2
Da (HB)	TSS	57 ± 24	53 ± 16	32 ± 19	7 ± 2	5 ± 3	1 ± 0
	TON	95 ± 33	90 ± 23	61 ± 27	32 ± 6	29 ± 6	17 ± 3
Lo (VQ)	TSS	8 ± 3	10 ± 4	9 ± 4	12 ± 3	8 ± 5	2 ± 1
	TON	23 ± 5	27 ± 8	25 ± 6	30 ± 5	22 ± 9	15 ± 7
Whole RR	TSS	79 ± 26	86 ± 22	63 ± 24	58 ± 17	44 ± 25	6 ± 1
	TON	141 ± 38	154 ± 33	121 ± 36	114 ± 26	92 ± 36	32 ± 5

leading to higher riverine TSS concentrations in tropical region.

Associated with TSS, TON concentrations at all sites were higher in the rainy season than in the dry season. Indeed, over the whole period, the average TSS concentrations during the rainy season in the three tributaries Thao (2.5 ± 1.0 mg L^{-1}), Lo (0.7 ± 0.1 mg L^{-1}) and Da (0.6 ± 0.3 mg L^{-1}) Rivers and the main axe at Hanoi station (1.2 ± 0.4 mg L^{-1}) were about 1.5–3.0 times higher than the ones in dry season (0.8 ± 0.2, 0.5 ± 0.1; 0.5 ± 0.1 and 0.6 ± 0.1 mg L^{-1} respectively.

3.1.3 Total suspended solids (TSS) and total organic nitrogen (TON) fluxes

Our calculations showed that the TSS fluxes of the Da River at the Hoa Binh hydrological station decreased significantly from $57 \times 10^6 \pm 22 \times 10^6$ t yr^{-1} for the period 1960–1969 to $1.0 \times 10^6 \pm 0 \times 10^6$ t yr^{-1} for the period 2000–2015 (Table 2).

Comparing with the period 1960–1969 (before the Hoa Binh reservoir presence), the TSS fluxes of the Da River in the period 1990s decreased about 87.8 %. Since the additional presence of the Son La, Lai Chau, Huoi Quang reservoirs in Vietnam and 11 small hydroelectric powers in China, the TSS fluxes at the outlet of the Da River further decreased at the total rate of 98.6 %. It clearly shows the important role of the reservoirs in TSS storage, leading to a clearly decrease of TSS fluxes at the outlet of the Da River (Table 2). For the Lo and the Thao Rivers, similarly, in the period 2000–2015, the TSS fluxes at the outlets of these rivers decreased by 72 % and 80 % respectively, comparing to the pre-construction the hydropower plants in both Vietnam and China (Table 2).

Previous study found that about 70 % reduction of sediment flux was observed for the Red River after the impoundment of the Hoa Binh and Thac Ba reservoirs on the Da and Lo rivers (Le et al., 2007; Dang et al., 2010). Our study revealed that actually, at Hanoi site, the TSS fluxes in the period 2010–2015 decreased by 92 % in comparing with the period without reservoir impoundment in the whole Red River system (1960s) (Table 2). In addition, at Hanoi station (downstream of the Red River), we observed that the sum of the annual TSS fluxes of the three tributaries Thao, Da and Lo (input to the Delta area) was higher than the mean annual fluxes of the Hanoi site (output). This is due to the complexity of the hydrological network in the Delta area where some distributaries flow out from the main branch of the Red River (Duong, Day and Nhue Rivers) (Luu et al., 2010) and the important deposition process in the section from the Viet Tri – Son Tay – Hanoi, the floodplains (Lu et al., 2015).

Overall, the TSS fluxes of the whole Red River transferred to the estuary significantly decreased from $79 \times 10^6 \pm 26 \times 10^6$ t yr^{-1} (equivalent to a specific sediment yield of 505 ± 166 t km^{-2} yr^{-1}) in the period 1960s to $6 \times 10^6 \pm 1 \times$

$10^6\,t\,yr^{-1}$ (corresponding to a specific sediment yield of $38 \pm 6\,t\,km^{-2}\,yr^{-1}$) in 2010–2015 period.

The TSS fluxes of the Red River at present situation (2010s) were much lower than that of the large Asian Rivers such as the Ayeyarwady (1966 to 1996): $325 \times 10^6\,t\,yr^{-1}$ (Furuichi et al., 2009).

In Asia, reservoirs have been constructed on large rivers for multi-purposes, including hydropower, flood control/flow management, irrigation/water supply... The dam/reservoir impoundment resulted in dramatically reducing of sediment flux discharged to the sea which was observed for numerous Asian rivers. For example, the sediment discharge of the Yangtze River (Changjiang) decreased from $480 \times 10^6\,t\,yr^{-1}$ to approximately $150 \times 10^6\,t\,yr^{-1}$ over a 20-year period (Wang et al., 2008) whereas the Yellow River (Huanghe) (in China) reduced from 1080×10^6 to $150 \times 10^6\,t\,yr^{-1}$ over a 40-year period (Wang et al., 2007). For some cases of Asian river, e.g the Minjiang River (Xu and Yan, 2010) and the Upper Mekong River (Kummu and Varis, 2007), different reservoirs were built in different times, and thus, the sediment trapping induced by different reservoirs has various phases, driven to the complicated effects caused by a single reservoir.

In addition, associated with riverine TSS, the reduction of TSS flux could decrease a large particulate nutrients (N, P) and carbon load to the estuary. In this study, TON load of the whole Red River reduced from $141 \times 10^3 \pm 38 \times 10^3\,t\,yr^{-1}$ (equivalent to a specific yield of $902 \pm 247\,kg\,km^{-2}\,yr^{-1}$) in the period 1960s to $32 \times 10^3 \pm 5 \times 10^3\,t\,yr^{-1}$ (corresponding to a specific sediment yield of $207 \pm 35\,kg\,km^{-2}\,yr^{-1}$) in 2010–2015 period.

The construction of reservoirs has caused sudden loss of huge amounts of fresh water, sediment and nutrients from the rivers into the estuaries and coastal areas. Some major river systems in Asia, such as the Yellow River, the Changjiang River, the Mekong River, ... have significant decreases in water flow and suspended solid load due to environmental changes, especially the human impacts in recent decades (Lu and Siew, 2006; Kummu and Varis, 2007; Wang et al., 2012; Lu et al., 2015). The decreases of suspended solids and associated substances (C, N, P and Si) loads can have serious consequences, such as increasing coastal erosion, reducing nutrient elements which provide foods for phytoplankton and aquaculture, decreasing aquaculture production, loss of shelter and breeding grounds... in coastal zone, as observed for numerous Asian rivers (Chen, 2000).

The same is likely to be true for the Red River where the impoundment of the large dams resulted in a clear decrease of both TSS concentration and fluxes in this study. Noted that a clear population and deforestation increases were observed for the Red River basin in the period 1960s–2010s (Le et al., 2015). This may approve the important role of dam impoundment in reducing suspended solids and associated nutrients transfer of the in the Red River to the coastal zone. The decrease of TSS and TON fluxes of the Red River may give different sequences (increased erosion, salinization of cultivated land or damaged ecosystems) in the coastal zone.

3.2 Bias in our calculation

Our calculations of TON concentration and flux of the Red River are subject to a number of possible biases when the TON concentrations were extrapolated from the relationship between TON and TSS concentrations given in (Le et al., 2005) with daily TSS concentrations obtained over the period 1960–2015. The use of this equation may not be suitable for the entire longterm period (about 55 years for the Red River), however, this bias is difficult for evaluating when no contemporaneous observation data exist. In addition, nitrogen transformation and lost towards the atmosphere as N_2 was not considered in our calculation of the Red River TON fluxes transported to the coastal zone. However, despite these potential errors, the results presented here show the importance of the change in the Red River TON flux over the longterm period 1960–2015.

4 Conclusion

Longterm variation of TSS and TON concentrations and fluxes of the Red River were calculated for the period 1960–2015. The results showed that a clear decrease of both TSS and TON concentrations and fluxes of the three main tributaries and the whole Red River in this period due to a series of dams constructed in both China and Vietnam, especially for a recent period 2010–2015, in spite of population and deforestation increase in the whole basin.

For detail, spatial and seasonal variations of TSS and TON concentrations and fluxes of the Red River were clearly observed. The annual average of TSS concentrations spatially varied from 15 ± 5 to 1479 ± 326 whereas the annual average of TON concentrations spatially fluctuated from 0.41 to $3.19\,mg\,L^{-1}$, averaging $0.98\,mg\,L^{-1}$ for the whole period. Presently, the lowest values of both TON and TSS were detected for the Da River where several new dams (including the Son La and Lai Chau) appeared in the last decade 2010s. The higher TSS and TON concentrations were found in wet than in dry season for all sites during different periods.

Overall, the TSS fluxes of the Red River transferred to the estuary significantly decreased from $79 \times 10^6 \pm 26 \times 10^6\,t\,yr^{-1}$ (equivalent to a specific sediment yield of $505 \pm 166\,t\,km^{-2}\,yr^{-1}$) in the period 1960s to $6 \times 10^6 \pm 1 \times 10^6\,t\,yr^{-1}$ (corresponding to a specific sediment yield of $38 \pm 6\,t\,km^{-2}\,yr^{-1}$) in 2010s. Similarly, the riverine TON fluxes transferred to estuary significantly decreased from $141 \times 10^3 \pm 38 \times 10^3\,t\,yr^{-1}$ (equivalent to $902 \pm 247\,kg\,km^{-2}\,yr^{-1}$) in 1960s to $32 \times 10^3 \pm 5 \times 10^3\,t\,yr^{-1}$ (equivalent to $207 \pm 35\,kg\,km^{-2}\,yr^{-1}$) in 2010s. A strong decrease of TSS and TON fluxes of the whole Red River may give different sequences (increased erosion, salinization of cultivated land or damaged ecosystems) in the coastal zone of the Red River, as observed for numerous Asian rivers.

Data availability. Data is not publicly accessible.

Author contributions. LND and PTXB calculated TON fluxes. LND and LTPQ prepared the manuscript with contributions from all co-authors.

Competing interests. The authors declare that they have no conflict of interest.

Special issue statement. This article is part of the special issue "Hydrological processes and water security in a changing world". It is a result of the 8th Global FRIEND–Water Conference: Hydrological Processes and Water Security in a Changing World, Beijing, China, 6–9 November 2018.

Financial support. This research has been supported by the Graduate University of Science and Technology /Ministry of Industry and Trade of the Socialist Republic of Vietnam (GUST.STS.ĐT 2020-MT01 /DTKHCN.008/19 project).

References

Beusen, A. H. W., Dekkers, A. L. M, Bouwman, A. F., Ludwig, W., and Harrison, J.: Estimation of global river transport of sediments and associated particulate C, N, and P, Global Biogeochem. Cy., 19, GB4S05, https://doi.org/10.1029/2005GB002453, 2005.

Chen, C. T. A.: The Three Gorges Dam: Reducing the upwelling and thus productivity in the East China Sea, Geophys. Res. Lett., 27, 381–383, 2000.

Dang, A. T.: The Red River Delta – The Cradle of the Nation, Hanoi National University Publisher, Hanoi, 53 pp., 2000.

Dang, T. H., Coynel, A., Orange, D., Blanc, G., Etcheber, H., and Le, L. A.: Long-term monitoring (1960–2008) of the river-sediment transport in the Red River Watershed (Vietnam): Temporal variability and dam-reservoir impact, Sci. Total Environ., 408, 4654–4664, 2010.

Furuichi, T., Win, Z., and Robert, J. W.: Discharge and suspended sediment transport in the Ayeyarwady River, Myanmar: Centennial and decadal changes, Hydrol. Process. 23, 1631–1641, 2009.

Ha, V. K. and Vu, T. M. H.: Analysis of the effects of the reservoirs in the upstream Chinese section to the lower section flow of the Da and Thao Rivers, J. Water Resour. Environ. Eng., 38, 3–8, 2012.

Hu, B., Li, J., Bi, N., Wang, H., Wei, H., Zhao, J., Xie, L., Zou, L., Cui, R., Li, S., Liu, M., and Li, G.: Effect of human-controlled hydrological regime on the source, transport, and flux of particulate organic carbon from the lower Huanghe (Yellow River), Earth Surf. Proc. Land., 40, 1029–1042, https://doi.org/10.1002/esp.3702, 2015.

IMRR: WP3 report, Integrated and sustainable water Management of Red-Thai Binh Rivers System in changing climate (IMRR Project), 8 pp., 2010.

Ittekkot, V. and Zhang, S.: Pattern of particulate nitrogen transport in world rivers, Global Biogeochem. Cy., 3, 383–391, 1989.

Kummu, M. and Varis, O.: Sediment-related impacts due to upstream reservoir trapping, the Lower Mekong River, Geomorphology, 85, 275–293, 2007.

Le, T. P. Q., Billen, G., Garnier, J., Thery, S., Fezard, C., and Chau, V. M.: Nutrient (N, P) budgets for the Red River basin (Vietnam and China), Global Biogeochem. Cy., 19, 1–16, https://doi.org/10.1029/2004GB002405, 2005.

Le, T. P. Q., Garnier, J., Billen, G., Théry, S., and Chau, V. M.: The changing flow regime and sediment load of the Red River, Viet Nam, J. Hydrol., 334, 199–214, https://doi.org/10.1016/j.jhydrol.2006.10.020, 2007.

Le, T. P. Q., Billen, G., Garnier, J., Thery, S., and Chau, V.M.: Long-term biogeochemical functioning of the Red River (Vietnam): past and present situations, Reg. Environ. Change, 15, 329–339, https://doi.org/10.1007/s10113-014-0646-4, 2015.

Le, T. P. Q., Le, N. D., Dao, V. N., Rochelle-Newall, E., Nguyen T. M. H., Marchand, C., Duong, T. T., and Phung, T. X. B.: Change in carbon flux (1960–2015) of the Red River (Vietnam), Environ Earth Sci., 77, 658, https://doi.org/10.1007/s12665-018-7851-2, 2018.

Lu, X. X. and Siew, R. Y.: Water discharge and sediment flux changes over the past decades in the Lower Mekong River: possible impacts of the Chinese dams, Hydrol. Earth Syst. Sci., 10, 181–195, https://doi.org/10.5194/hess-10-181-2006, 2006.

Lu, X. X., Oeurng, C., Le, T. P. Q., and Duong, T. T.: Sediment budget of the lower Red River as affected by dam construction, Geomorphology, 248, 125–133, https://doi.org/10.1016/j.geomorph.2015.06.044, 2015.

Ludwig, W., Probst, J. L., and Kempe, S.: Predicting the oceanic input of organic carbon by continental erosion, Global Biogeochem. Cy., 10, 23–41, https://doi.org/10.1029/95GB02925, 1996.

Luu, T. N. M., Garnier, J., Billen, G., Orange D., Némery J., Le, T. P. Q., Tran, H. T., and Le, L. A.: Hydrological regime and water budget of the Red River Delta (Northern Vietnam), J. Asian Earth Sci., 37, 219–228, https://doi.org/10.1016/j.jseaes.2009.08.004, 2010.

MONRE: Report Annual on Hydrological Observation in Vietnam, Ministry of Environment and Natural Resources, Vietnam, 1960–2015.

MOSTE: Water quality – Determination suspended solids by filtration through glass-fibre filters, Ministry of Science, Technology and Environment, Vietnam, TCVN 6625:2000, 11 pp., 2000.

Moon, S., Huh, Y., Qin, J., and Nguyen, V. P.: Chemical weathering in the Hong (Red) River basin: Rates of silicate weathering and their controlling factors, Geochim. Cosmochim. Ac., 71, 1411–1430, https://doi.org/10.1016/j.gca.2006.12.004, 2007.

Pham, H. V.: Using ENSO information to improve the operation of the Hoa Binh reservoir, Vietnam, Master of Science in Environmental and Geomatic Engineering, Politecnico Di Milano, Italy, 70 pp., 2015.

Quach, X.: Assessing and optimizing the operation of the HoaBinh reservoir, Vietnam, by multi-objective optimal control techniques, PhD thesis, Politecnico di Milano, Italy, 138 pp., 2011.

Syvitski, J. P. M. and Saito, Y.: Morphodynamics of Deltas under the Influence of Humans, Glob. Planet. Change, 57, 261–282, 2007.

Van Maren, D. S. and Hoekstra, P.: Seasonal variation of hydrodynamics and sediment dynamics in a shallows subtropical estuary: the Ba Lat River, Vietnam, Estuar. Coast Shelf Sci., 60, 529–540, 2004.

Wang, F. S., Wang, F. C., Zhang, J., Hu, H., and Wei, X. G.: Human impact on the historical change of CO_2 degassing flux in River Changjiang, Geochem. Trans., 8, 7, https://doi.org/10.1186/1467-4866-8-7, 2007.

Wang, H., Yang, Z., Wang, Y., Saito, Y., and Liu, J. P.: Reconstruction of sediment flux from the Chanjiang (Yangtze River) to the sea since the 1860, J. Hydrol., 349, 318–332, https://doi.org/10.1016/j.jhydrol.2007.11.005, 2008.

Wang, X., Ma, H., Li, R., Song, Z., and Wu, J.: Seasonal fluxes and source variation of organic carbon transported by two major Chinese Rivers – The Yellow River and Changjiang (Yangtze) River, Global Biogeochem. Cy., 26, GB2025, https://doi.org/10.1029/2011GB004130, 2012.

Walling, D. E. and Fang, D.: Recent trends in the suspended sediment loads of the world's rivers, Glob. Planet. Change, 39, 111–126, https://doi.org/10.1016/S0921-8181(03)00020-1, 2003.

Wu, Y., Zhang, J., Liu, S. M., Zhang, Z. F., Yao, Q. Z., Hong, G. H., and Cooper, L.: Sources and distribution of carbon within the Yangtze River system, Estuar. Coast Shelf Sci., 71, 13–25, https://doi.org/10.1016/j.ecss.2006.08.016, 2007.

Xu, J. and Yan, Y.: Effect of reservoir construction on suspended sediment load in a large river system: thresholds and complex response, Earth Surf. Proc. Land, 35, 1666–1673, https://doi.org/10.1002/esp.2006, 2010.

Proc. IAHS, 383, 375–379, 2020
https://doi.org/10.5194/piahs-383-375-2020

A new approach to aquatic ecosystem monitoring of the Ob river

Aleksandr Andreevich Tskhai[1,2], **Aleksandr Vasilievich Puzanov**[1], **Nelley Mikhailovna Kovalevskaya**[1], **and Vladimir Viktorovich Kirillov**[1]

[1]Institute for Water and Environmental Problems, Siberian Branch, Russian Academy of Sciences, Barnaul 656038, Russia
[2]Polzunov Altai State Technical University, Barnaul 656038 Russia

Correspondence: Aleksandr Andreevich Tskhai (taa1956@mail.ru)

Published: 16 September 2020

Abstract. An integrated approach to aquatic ecosystem monitoring is being developed for the Ob river. The novelty of the approach is in transition to high-performance computing (HPC) under processing of remote sensing data and modeling of aquatic ecosystem degradation. Classes of optically complex surface waters of the Gulf of the Kara Sea – the Ob Bay are studied on the basis of the fuzzy cluster analysis of water objects emission. A ten-year dynamics of concentrations of optically active substances was estimated using the neural network simulation. The application of the structural-dynamic model allowed explaining the long-term trend in the development of the ecosystem of the Novosibirsk reservoir, the largest one in the Ob river basin.

1 Introduction

It includes (1) effective neural models for processing satellite data on ecosystem component characteristics; (2) advanced algorithms for big remote sensing data management; (3) original structural-dynamic adaptive models of biogeochemical cycles of transformation of limiting elements able to reconstruct the development of hydrobionts communities when the data are incomplete and uncertain. Only such components ensure the implementation of the new approach to real-time environmental monitoring.

2 Methods and materials

Generalization, standardization, serialization of data, etc. are an inevitable in the technological development of remote sensing. For subsequent application of remote sensing data from various sources additional tools are required. To study the distribution of chlorophyll *a* (CHL) and the absorption of dissolved organic matter (DOM), we integrate satellite data from RapidEye (5 m: 4 August 2015), Terra/ASTER (15 m: 7 July 2015, 16 July 2015, 4 August 2015) and Landsat-8 (30 m: 31 July 2015).

Based on bands of RapidEye (RedEdge and Green) satellites, heuristic models demonstrate the highest determination coefficient for defining of CHL concentrations using the in-situ data (Figs. 1, 2).

The models for CHL concentration determination with the use of Terra / ASTER data suggest concurrent application of Green and NIR data (Figs. 1, 3). The use of Landsat-8 data gives similar results for CHL estimates (Figs. 1, 4). Landsat-8 data-based models for evaluation of DOM absorption is distinguished by the efficiency of the CoastalBlue range corresponding to the DOM absorption maximum (Bukanova, 2014).

The obtained regression relationships are evidence of the necessity to consider simultaneously the data on the following ranges: for CHL concentrations – Green and RedEdge (or Red), for DOM absorption coefficient – CoastalBlue and Green.

Classification of the brightness space of features on the basis of the fuzzy cluster analysis of water objects emissivity is an important tool for the preliminary analysis of images of optically complex waters (Moore et al., 2014; Erlov, 1976). A characteristic curve in the brightness characteristic space corresponds to each cluster. In this case, the same raster element

Published by Copernicus Publications on behalf of the International Association of Hydrological Sciences.

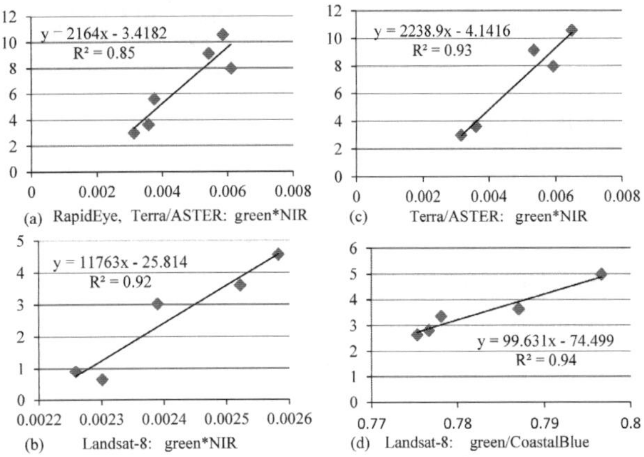

Figure 1. Regression dependences of in-situ data and satellite observations: **(a–c)** CHL concentrations, mg m^{-3}, **(d)** DOM absorption coefficient, 1 m^{-1}.

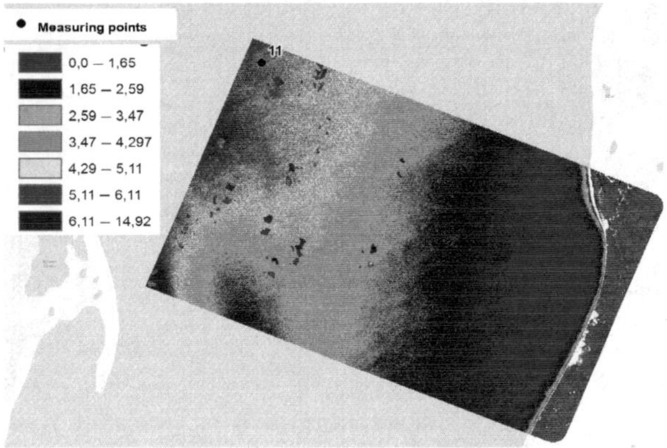

Figure 2. CHL distribution according to RapidEye data, mg m^{-3} (4 August 2015).

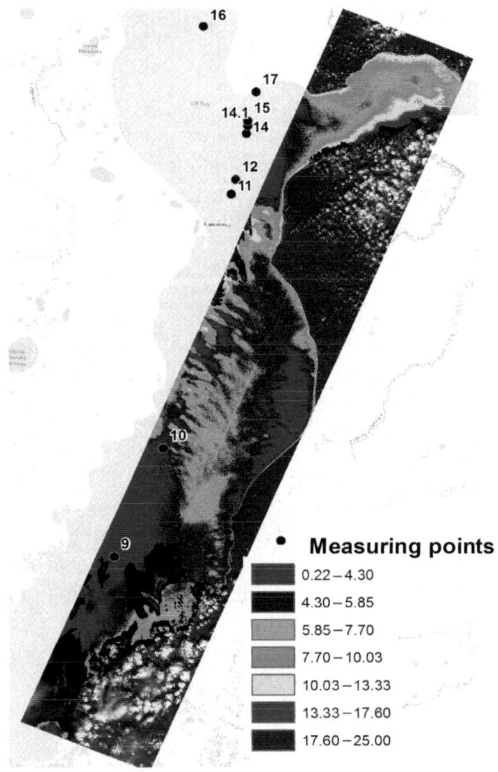

Figure 3. CHL distribution according to Terra/ASTER data, mg m^{-3} (16 July 2015).

$i(i = 1, N)$ can belong to several (or all) clusters $j(j = 1, C)$. Such a fuzzy clustering is formalized through the property matrix $U = (u_{ij})$ consisting of property levels u_{ij}:

$$\sum_{j=1}^{?} u_{ij} = 1,\ i = 1, \ldots, N.$$

High turbidity of the most water area under study corresponds to the 6th, 7th and 8th classes of optical classification.

The Novosibirsk reservoir- the largest in Western Siberia, was used for calibration of the aquatic ecosystem model (Tskhai and Ageikov, 2020). This reservoir was constructed on the Ob river in 1957 (the reservoir site is located upstream of the Novosibirsk). It is used mainly for energy production, navigation, and water supply. The average depth is 8.2 m; water storage and water-surface area at the normal pool level are 8.8 and 1070 km^2, respectively; and the active storage is 4.4 km^3.

The lake-like part of the reservoir is where the quality of water flowing into the lower pool is determined. At normal pool level, the lake-like part of the reservoir contains 73 % of the total amount of the reservoir water. With the reservoir drawdown, the share of the lake-like part increases to 90 %.

3 Results and discussion

In different parts of the water area, CHL concentrations vary greatly (Fig. 5). The noted earlier tendency (Gaevsky et al., 2009) in phytoplankton decrease is observed as it moves from the Ob delta towards the Kara Sea (Fig. 6). This is used for the neural model validation as a confirmation of the zonal CHL distribution depending on a salinity gradient and a biogenic element concentration (Sukhanova et al., 2010) in case of desalinated and frontal zones as well as a shallow shelf occurrence.

A common feature of optical properties of these waters is the influence of river water propagation, biological processes and gravitational deposition of solid matter (SM). The CHL content in the frontal zone is much lower than in the river. However, the structure of CHL concentration fields in this zone is characterized by extreme complexity at dynamic changes in the frontal zone's size (Flint and Poyarkov, 2015).

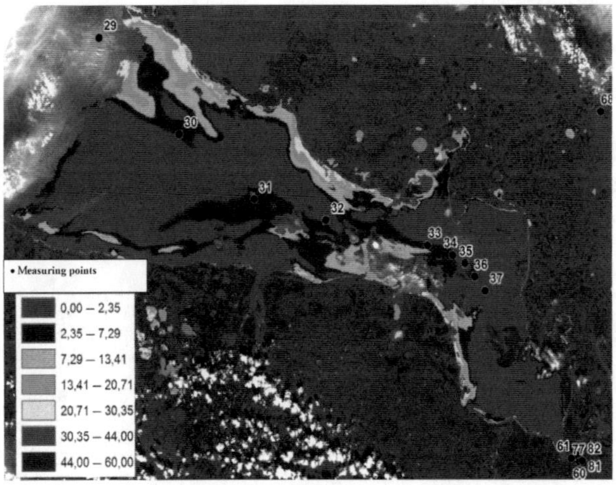

Figure 4. CHL distribution according to Landsat-8 data, $\mathrm{mg\,m^{-3}}$ (31 July 2015).

Maximal values of SM model concentrations are observed in the mouth of the Ob, Taz and Gydan Bays.

Maximum of DOM absorption falls on the main water stream from the Ob, Taz and Gydan catchments. The Taz water area is characterized by the least absorption coefficient. These waters extend for several kilometers to the north. The cooler and denser waters of the Taz Bay flowing along the right bank meet relatively warm regional stream generating a perturbance zone.

However, large gradients of SM and DOM in the studied water area lead to significant contrasts of transparency associated with a turbidity level in the mouth of the Ob, Taz and Gydan Bays. Huge inflow of SM and DOM with river waters brings to a large contrast between the first optical depth (0.1–11 m) and turbidity (0.4–89.4 FNU).

The dynamics study of model values (2003–2011) shows that SM concentrations and DOM absorption increase in the mouth and middle part of the Ob and Gydan Bays,

Calculations with the use of the aquatic ecosystem model (Fig. 7) of the 5th generation (Tskhai and Ageikov, 2020). demonstrate a clear increase in phytoplankton content during summer-autumn in case of several succeeding low-water years. Interestingly, it occurs starting from the 3rd year; curves of the first two years do not reflect such a difference (Fig. 8). The calculations confirm the revealed long-term trends in increasing bloom intensity and allow us to expect a probable shift of its peak to the earlier time in case of several succeeding low-water years. Moreover, all this ensures the assessment of changes in water body eutrophication depending on hydrological conditions.

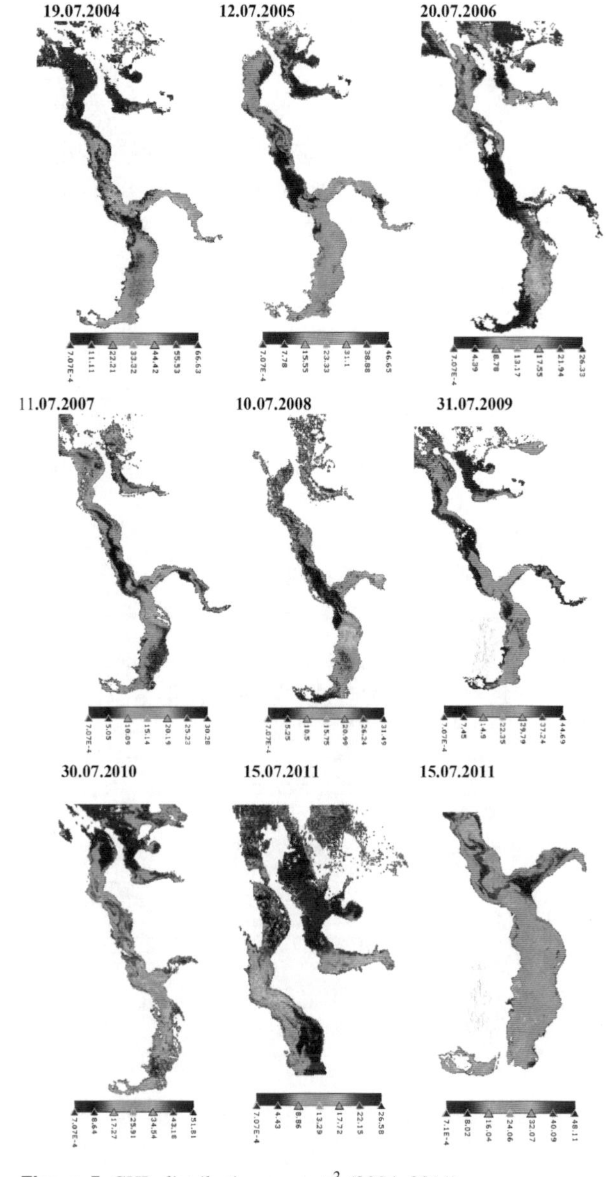

Figure 5. CHL distribution, $\mathrm{mg\,m^{-3}}$ (2004–2011).

4 Summary

High-performance processing of satellite information made it possible to obtain regular time series of MERIS data (2003–2011) processed by the open software environment BEAM-VIASAT/ESA. The method for fuzzy features clustering in the brightness space is applied to define the distribution of main optical classes on the basis of NS-algorithms for optically complex surface waters.

The use of efficient information technologies in satellite data processing allowed:

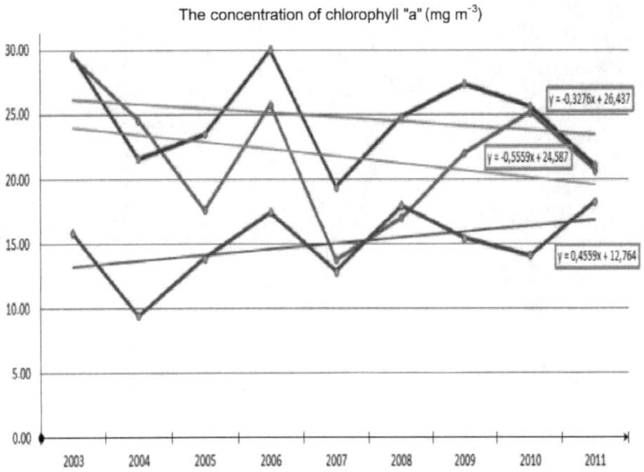

Figure 6. Averaged CHL concentrations in the Ob Bay (green – river, red – middle, blue – sea parts).

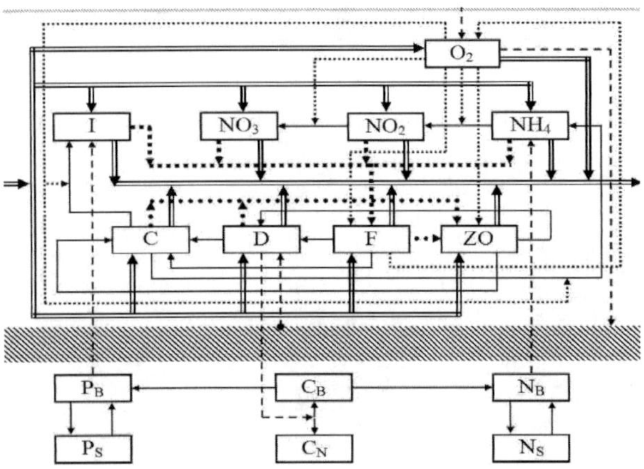

Figure 7. A scheme of biochemical transformation of aquatic ecosystem components.

- to specify optical classes of water and to present changes in optically active substances as well as in primary hydro-optical characteristics for almost a ten-year period;
- to present spatial details of the structure of the estuary frontal zone in the Ob Bay site of sea and river waters mixing;
- to reveal inter-annual variability of CHL, SM and DOM in the mouth, middle part, estuary, adjacent shelf of the Ob and Gydan Bays.
- to approximate the distribution of CHL and SM concentrations by a logonormal function, whereas DOM - by a bimodal function; to obtain the features of their relative stability; to develop the aquatic ecosystem model of the 5th generation for establishment of long-term trends;

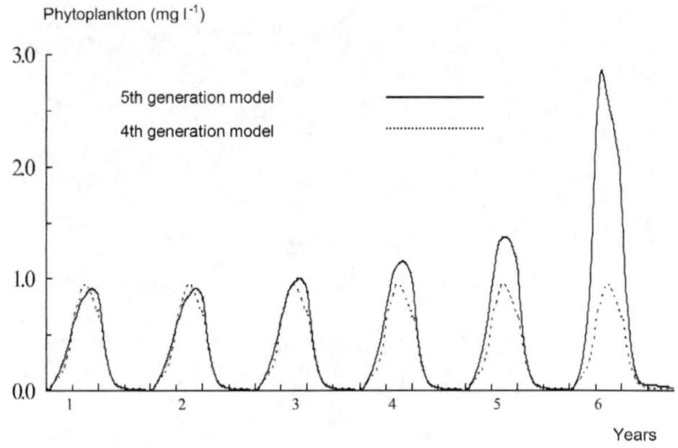

Figure 8. Modeling of inter-annual variability of phytoplankton content in the Novosibirsk reservoir (1981).

- to prove the preference of using these ranges (Green, RedEdge) CHL, (CoastalBlue, Green) DOM.
- to confirm proposals that the CHL content in the frontal zone of the Ob Bay is much lower than in its river part.

Data availability. The satellite data used in research is derived from the results of the ESA CoastColour Project – MERIS Case2 Water Processing – http://www.coastcolour.org (last access: 16 July 2020). The modeling data availability is via reference to the publications on these studies upon on reasonable request.

Author contributions. AAT developed the model of development for aquatic ecosystem and wrote the article. AVP and VVK selected the research place and provided planning and administrative support. NMK executed the satellite data processing and analysis.

Competing interests. The authors declare that they have no conflict of interest.

Special issue statement. This article is part of the special issue "Hydrological processes and water security in a changing world". It is a result of the 8th Global FRIEND–Water Conference: Hydrological Processes and Water Security in a Changing World, Beijing, China, 6–9 November 2018.

Financial support. The study was carried out within the Research Program of IWEP SB RAS and funded by the Russian Foundation for Basic Research and the government of Altai Krai, of the Russian Federation (grant no. 18-41-220002).

References

Bukanova, T. V.: Trends in eutrophication of the South-Eastern Baltic Sea according to satellite data: Diss. PhD, AB IO RAS, Kaliningrad, Russia, 142 p., 2014 (in Russian).

Erlov, N. G.: Marine optics, 2nd ed., Elsevier Sci. Ltd, New York, USA, 231 p., 1976.

Flint, M. V. and Poyarkov, S. G.: Comprehensive research on the Kara Sea ecosystem (128th cruise of Research Vessel "Professor Shtokman"), Oceanology, 55, 657–659, https://doi.org/10.1134/S0001437015040074, 2015.

Gaevsky, N. A., Semenova, L. A., and Matkovsky, A. K.: Trophic status of waters of the ecosystem of the Ob-Taz estuary region in terms of phytoplankton, Bull. Ecol. Forest. Landsc. Stud., 10, 170–179, 2009 (in Russian).

Moore, T. S., Dowell, M. D., Bradt, S., and Verdu, A. R.: An optical water type framework for selecting and blending retrievals from bio-optical algorithms in lakes and coastal waters, Remote Sens. Environ., 143, 97–111, https://doi.org/10.1016/j.rse.2013.11.021, 2014.

Sukhanova, I. N., Flint, M. V., Mosharov, S. A., and Sergeeva, V. M.: Structure of the phytoplankton communities and primary production in the Ob river estuary and over the adjacent Kara sea shelf, Oceanology, 50, 743–758, 2010.

Tskhai, A. A. and Ageikov, V. Yu.: Simulating Variations in Reservoir Eutrophication Level by Reproducing Biogeochemical Cycles, Water Resour., 47, 147–155, https://doi.org/10.31857/S0321059620010149, 2020.

Proc. IAHS, 383, 381–389, 2020
https://doi.org/10.5194/piahs-383-381-2020

PREMHYCE: An operational tool for low-flow forecasting

Pierre Nicolle[1], **François Besson**[2], **Olivier Delaigue**[1], **Pierre Etchevers**[2], **Didier François**[3], **Matthieu Le Lay**[4], **Charles Perrin**[1], **Fabienne Rousset**[2], **Dominique Thiéry**[5], **François Tilmant**[1], **Claire Magand**[6], **Timothée Leurent**[6], **and Élise Jacob**[7]

[1]Paris-Saclay University, Inrae, HYCAR Research Unit, Antony, France
[2]Météo-France, Direction of Climatology, Toulouse, France
[3]LOTERR, Lorraine University, Metz, France
[4]EDF-DTG, Grenoble, France
[5]BRGM, Orléans, France
[6]French Office for Biodiversity (OFB), Vincennes, France
[7]Ministry for the ecological transition, Water and biodiversity direction, La Défense, France

Correspondence: Pierre Nicolle (pierre.nicolle@univ-eiffel.fr)

Published: 16 September 2020

Abstract. In many countries, rivers are the primary supply of water. A number of uses are concerned (drinking water, irrigation, hydropower, etc.) and they can be strongly affected by water shortages. Therefore, there is a need for the early anticipation of low-flow periods to improve water management. This is strengthened by the perspective of having more severe summer low flows in the context of climate change. Several French institutions (Inrae, BRGM, Météo-France, EDF and Lorraine University) have been collaborating over the last years to develop an operational tool for low-flow forecasting, called PREMHYCE. It was tested in real time on 70 catchments in continental France in 2017, and on 48 additional catchments in 2018. PREMHYCE includes five hydrological models: one uncalibrated physically-based model and four storage-type models of various complexity, which are calibrated on gauged catchments. The models assimilate flow observations or implement post-processing techniques. Low-flow forecasts can be issued up to 90 d ahead, based on ensemble streamflow prediction (ESP) using historical climatic data as ensembles of future input scenarios. These climatic data (precipitation, potential evapotranspiration and temperature) are provided by Météo-France with the daily gridded SAFRAN reanalysis over the 1958–2017 period, which includes a wide range of conditions. The tool provides numerical and graphical outputs, including the forecasted ranges of low flows, and the probability to be under low-flow warning thresholds provided by the users. Outputs from the different hydrological models can be combined through a simple multi-model approach to improve the robustness of forecasts. Results are illustrated for the Ill River at Didenheim (northeastern France) where the 2017 low-flow period was particularly severe and for which PREMHYCE provided useful forecasts.

1 Introduction

1.1 Why anticipating low flows?

In many countries, rivers are the primary supply of water. In France in 2013, 73 % of total withdrawals (38 km^3) came from rivers (Banque National des Prélèvements en Eau, Chataigner and Michon, 2017). A number of uses are concerned (thermal power plant cooling, hydropower, drinking water, irrigation, industry, navigation) and can be strongly affected by water shortages in rivers (Bousquet et al., 2003). Furthermore, uses should be compatible with maintaining the quality of aquatic life, through environmental constraints like minimum environmental flows (Acreman and Dunbar, 2004).

There is a need for the early anticipation of low-flow periods to improve water management and to take more timely measures to mitigate the socio-economic and ecological impact of water shortages (Chiew and McMahon, 2002; Karamouz and Araghinejad, 2008). Extreme droughts which

Published by Copernicus Publications on behalf of the International Association of Hydrological Sciences.

occurred in France in 1976, 2003, and more recently in 2011, 2015 and 2017 underline the need for forecasting systems, which is strengthened by the perspective of having more frequent and severe low flows in summer in the context of climate change.

In 2011, the French Agency for Biodiversity (formerly ONEMA) and the Ministry for the environment launched a research project to compare and evaluate the ability of various hydrological models to produce low-flow forecasts useful for real-time decision making. This project led to the development of a low-flow forecasting tool that includes the tested hydrological models.

1.2 Hydrological tools for low-flow forecasting

There are a few existing approaches and tools for low-flow forecasting. A detailed review can be found in Nicolle et al. (2014). More recently, a few works have been carried out on this topic. Some examples are commented here:

- conditioning methods for input scenarios for seasonal streamflow forecasting, as tested by Crochemore et al. (2017). These authors investigated the impact of conditioning methods on the performance of seasonal streamflow forecasts, to identify forecast attributes leading to improvement or deterioration using these methods.

- investigation of the skill of seasonal ensemble low-flow forecasts in the Moselle River (Demirel et al., 2015). The authors compared three data-driven and conceptual hydrological models for low-flow forecasting, and assessed the effect of ensemble seasonal forecasts on low-flow forecasts quality.

- proposing a framework for low-flow forecasting in Mediterranean streams (Risva et al., 2018). The authors provided a simple and effective tool for low-flow forecasting up to six month ahead, which needs limited data, based on the improvement of the linear reservoir concept that represents streamflow recession.

In France, a few operational tools have been recently developed, indirectly or directly linked to low-flow forecasting. One can mention:

- the Aqui-FR project (Habets et al., 2015) that aims integrating hydrogeological models to monitor and forecast groundwater resource at medium range to seasonal scale, on the main aquifers of France.

- *E*-tiage, which is an online service to help end-users for water management. It allows monitoring river streamflow over past days and forecast streamflows on the Adour-Garonne and Charentes catchment (south-west part of France).

Table 1. Summary of the main characteristics of the 118 catchments.

	Min	25 %	Median	75 %	Max
Area (km^2)	9	145	275	732	110 188
Median elevation (m)	52	137	188	433	1794
Flow availability (yr)	4	34	45	51	60
Gap rate (%)	0	0	1	3	45

1.3 Objectives of the study

The objectives of the article are to present the main characteristics of the low-flow forecasting tool and the results obtained in operational conditions over the 2017 summer period on a case study.

2 Material and methods

2.1 Catchment set and data

2.1.1 Selection of catchments

The catchment set was built in cooperation with a variety of institutions involved in low-flow management, on the regulatory or operational sides (environment directions at regional and department levels, regional irrigation managers, etc.). They provided lists of target catchments, where human influences were requested to be limited, given the current version of the tool does not account for upstream influences. This resulted in the selection of 118 catchments, mainly located in north-east, north-west, south-west, and centre of France (see Fig. 1). The catchments show various hydrological regimes ranging from oceanic to Mediterranean or mountainous. Table 1 shows the main characteristics of the catchment set, with catchment sizes ranging from 9 to 111 000 km^2, median elevation ranging from 52 to 1794 m and historical streamflow data covering periods from 4 to 60 years.

2.1.2 Data

Daily streamflow records were retrieved from the French national discharge archive (HYDRO database, available at http://www.hydro.eaufrance.fr, last access: 20 November 2018). Daily precipitation and temperature data originate from the gridded (8 km × 8 km) SAFRAN climate reanalysis developed by Météo-France (Vidal et al., 2010). Potential evapotranspiration (PE) was computed using the formula proposed by Oudin et al. (2005). The climatic series are continuously available on the 1959–2018 period over France. This period includes severe droughts conditions (e.g. in summers 1976, 1989, 2003, 2005, 2011, 2015 and 2017).

Table 2 displays the ranges of climate and flow characteristics of the catchment set. Hydroclimatic conditions in France are quite variable in terms of mean annual precipitation, PE and streamflow. There is also a strong interannual variabil-

Table 2. Percentiles of the distribution of a few climatic and hydrological characteristics of the 118 catchments. Interannual variability values correspond to coefficients of variation calculated on the 1958–2009 period for P and PE, and on streamflow available period for Q. $Q50$, $Q80$ and $Q90$ are respectively the 50th, 80th and 90th exceedance percentiles of the flow duration curve.

	Min	25 %	Median	75 %	Max
Mean annual precipitation P_A (mm)	623	746	843	956	1757
Interannual variability of P_A	0.13	0.16	0.16	0.17	0.28
Mean annual potential evapotranspiration PE_A (mm)	488	657	684	718	852
Interannual variability of PE_A	0.07	0.10	0.10	0.10	0.11
Mean annual streamflow Q_A (mm yr^{-1})	57	182	266	341	1277
Interannual variability of Q_A	0.16	0.31	0.39	0.47	4.29
Runoff ratio Q_A/P_A (%)	6	24	31	37	75
Base-flow index (BFI) (%)	22.6	48.1	58.3	73.0	96.3
Q^*_{90}/Q^*_{50} (%)	0.00	0.19	0.35	0.49	0.73
Q^*_{80} (mm d^{-1})	0.01	0.11	0.19	0.33	1.53

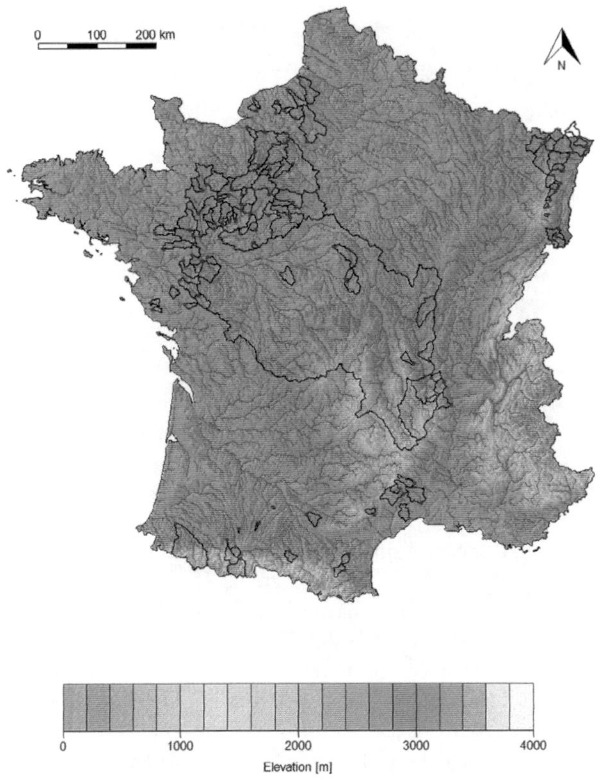

Figure 1. Location of the 118 catchments in France. Each outlet is shown by a red dot.

ity, especially for streamflow. On average, 31 % of rainfall become runoff for the catchment set, but this ratio varies between 6 % and 75 %.

2.2 Ensemble low-flow forecasting using hydrological models

Models are expected to forecast streamflow from time steps $t+1$ to $t+L$ (with L the lead time), knowing both observed meteorological inputs and streamflow until time step t and making assumptions (i.e. choosing scenarios) for the future meteorological inputs from $t+1$ to $t+L$. Streamflow observations can be used within an assimilation scheme and/or a statistical correction procedure.

Figure 2 presents the successive steps of the forecasting method for low-flow forecasting with hydrological models:

1. internal states of hydrological models are initialized using climatic observations of past conditions until the day of forecast;

2. last streamflow observation(s) can be assimilated, typically by correcting model internal states or by applying streamflow post-processing (e.g. model error correction);

3. several meteorological scenarios are used as model input to provide an ensemble of streamflow forecasts from $t+1$ to $t+L$;

4. streamflow forecast are statistically analyzed to provide confidence intervals of possible future streamflows over the time horizon.

This approach is quite classical. The originality in the case of PREMHYCE is that it is applied in a multi-model framework. This has two potential advantages: to improve the resulting forecasts and to better account for structural uncertainty.

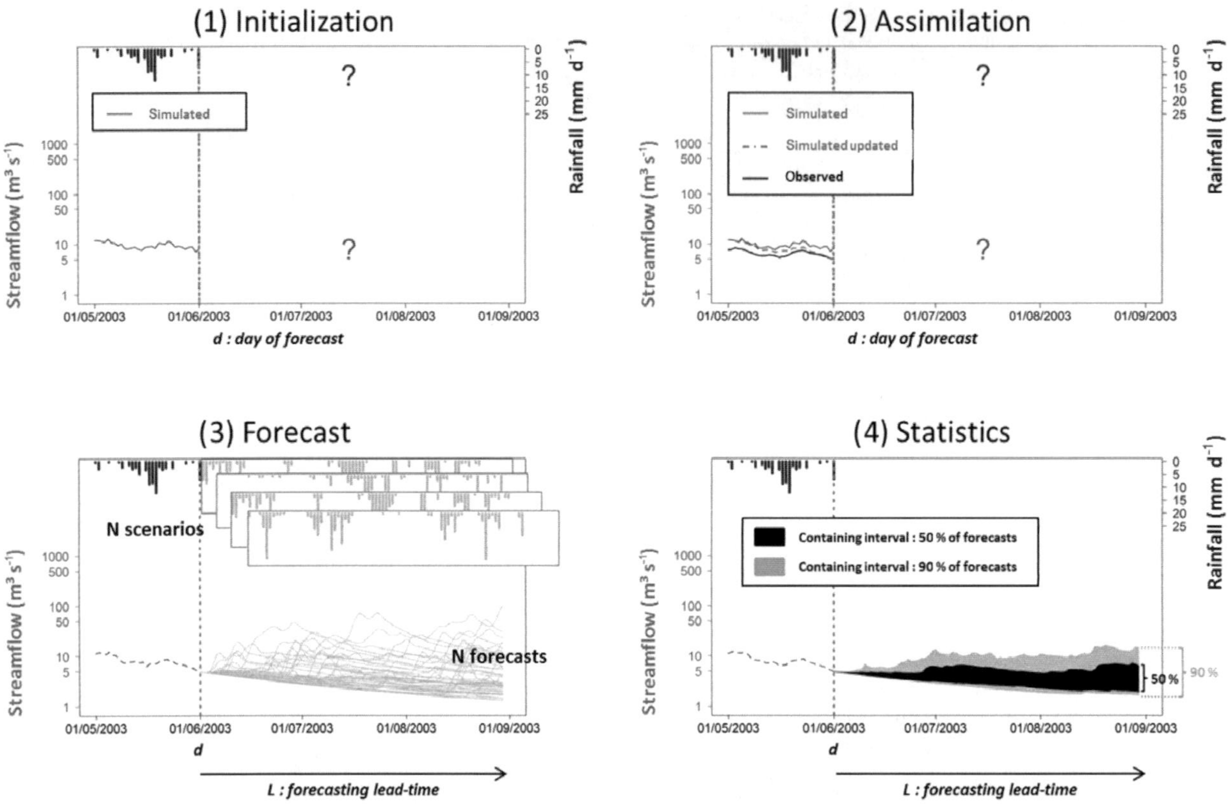

Figure 2. Forecasting method for low-flow forecasting with hydrological models.

2.3 Input forecasting scenarios

The PREMHYCE operational tool adopts the classical ESP approach (Day, 1985) in terms of future meteorological inputs. For a given catchment, let us consider that N years of past historical climatic observations are available. In real time, one wishes to make a forecast on a calendar day t of a year Y within the test period, i.e. to forecast flows between calendar days $t+1$ and $t+L$. The observed meteorological data available between days $t+1$ and $t+L$ in the years 1 to N (i.e. N scenarios) are used as input scenarios to the model, considering that they are likely meteorological conditions for this period of the year. Here, 57 years (1959–2016) of daily climate data from the SAFRAN reanalysis were used for the tests during the 2017 year and 58 scenarios (1959–2017) for the 2018 year.

A zero-precipitation scenario (i.e. precipitation equal to 0 for the L next days) is also used to provide the worst-case streamflow forecast. It is associated to a daily interannual average of potential evapotranspiration.

3 The low-flow forecasting tool PREMHYCE

3.1 Platform presentation

PREMHYCE is a low-flow forecasting platform that aims at providing low-flow forecasts at a daily time step for the next 90 d to end-users. It is based on the airGR package (Coron et al., 2017a, b) and includes five hydrological models: Gardenia (BRGM), GR6J (Inrae), Mordor (EDF, currently implemented under a simplified form for technical reasons, here called IrMo), Presages (Lorraine University) and the Safran-Isba-Modcou (SIM) modelling suite (Météo-France). More detailed information on these models and their practical implementation for forecasting are given by Nicolle et al. (2014). Note that the SIM model is run by Météo-France independently from the platform and only the SIM outputs are uploaded into the PREMHYCE tool. This platform is currently hosted by a server at Inrae and allows data exchange via FTP protocol.

The tool includes two main modules:

- An off-line module for the calibration of hydrological models, to estimate models parameters and evaluate models reliability. The module functions allow to:

Catchment : Ill at Didenheim		HYDRO code	A1080330	
Hydrological model	GR6J	Date of forecast	11/04/2017	Forecasting lead-time (days) 90
Threshold	Vigilance (m³ s⁻¹) : 1.10	Alert (m³ s⁻¹) : 0.80	Alert renf (m³ s⁻¹) : 0.73	Crisis (m³ s⁻¹) : 0.65

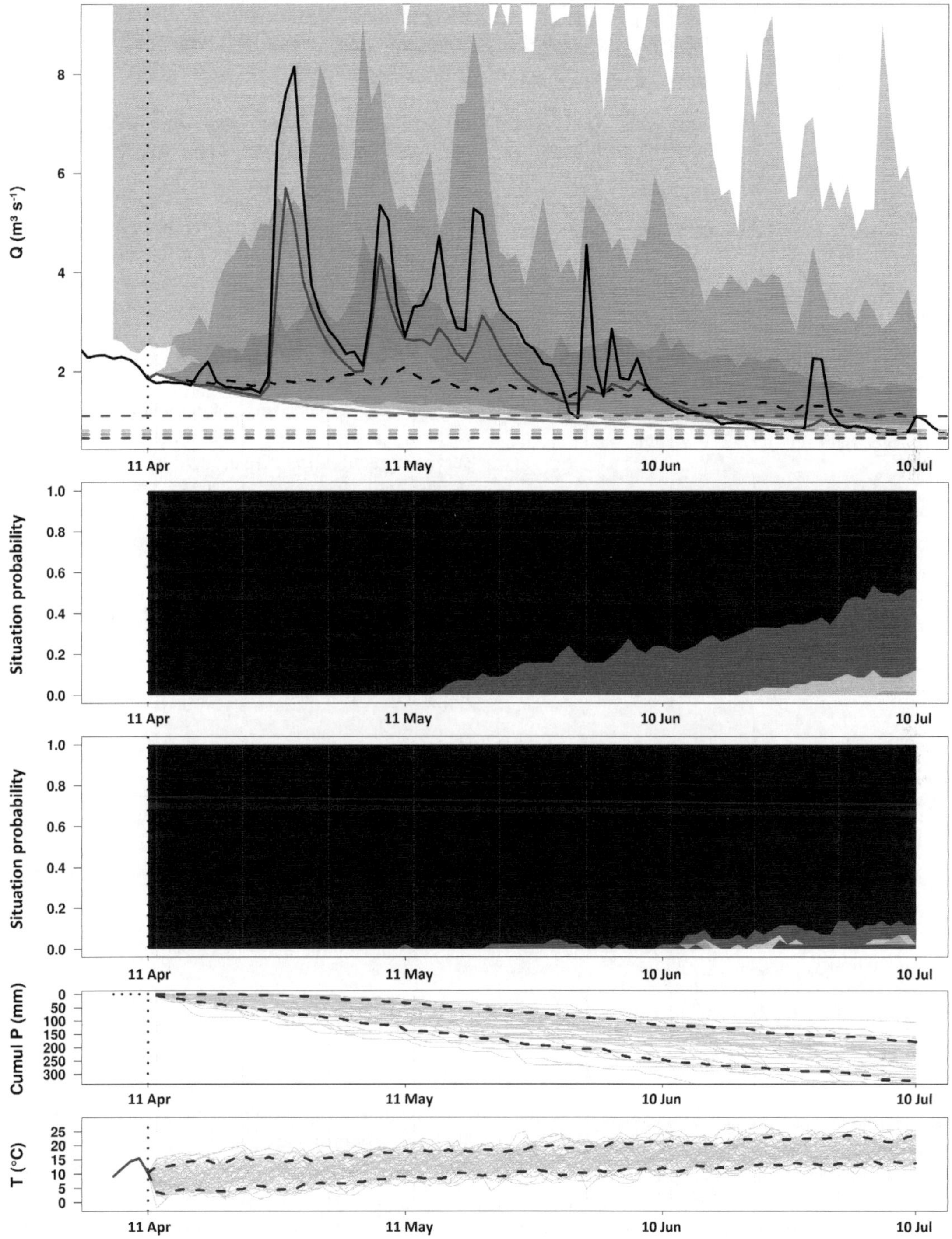

Figure 3. Example of streamflow forecast synthesis plot provided by PREMHYCE on 11 April 2017 for the Ill River at Didenheim with GR6J.

Table 3. Probability of detection to be under the vigilance and reinforced alert threshold for each model and for the 7 and 30 d lead-times for the Ill River at Didenheim.

Lead-time (d)	Threshold	Hydrological model			
		PRES	IrMo	GR6J	GARD
7	Vigilance	0.92	0.83	0.85	0.84
30	Vigilance	0.76	0.68	0.74	0.76
7	Reinf alert	0.75	0.58	0.53	0.58
30	Reinf alert	0.68	0.24	0.41	0.71

  - import the catchment database (observed time series of streamflows, precipitation, PE and temperature);
  - calibrate the hydrological models using these time series;
  - copy the calibration database to a real-time database.
- An online real-time low-flow forecasting module, to produce forecasts every day. The module functions allow to:
  - update the real time database using last observed data until the day of forecast;
  - update hydrological models internal states until the day of forecast;
  - compute streamflow forecasts up to 90 d ahead using historical scenarios;
  - provide synthetic graphical results.

3.2 Model calibration

The model parameters can be either calibrated within the platform or specified by the user. As a physically-based model, SIM has been set up all over France and its parameters are not further tuned for the PREMHYCE objectives. Note that SIM is the only model for which no calibration against observed flow data at the catchment outlet is performed. The spatially distributed parameters used in this model are estimated regionally.

Users provide a list of catchments that have to be in the French HYDRO database. For each catchment, up to four operational streamflow thresholds can be also provided, which will be used to interpret the severity of future forecasted low flows.

Models are calibrated using a gradient-type method for the GR6J, IrMo and Presages models. Two objective functions (KGE, Gupta et al., 2009; NSE, Nash and Sutcliffe, 1970) can be used with three prior transformations of streamflow (Q, $\sqrt{(Q)}$, $\ln(Q)$). A compromise can be sought between several parameter sets obtained by these various functions. Gardenia is calibrated using the Rosenbrock method with the NSE objective function calculated on $\ln(Q)$.

3.3 Real-time daily operation for low-flow forecasting

The tool can be run every day to provide low-flow forecasts at the daily time step. Given the computing constraints associated with the SIM model, the SIM streamflow forecasts are provided only once a month.

Several operations are made every day to provide low-flow forecasting.

The last observed data until the day of forecast are first used to update model internal states. A data import process has been implemented to retrieve meteorological and hydrological data.

Every day, Météo-France provides real-time gridded SAFRAN data (precipitation, temperature) for the day before forecasting, as daily data for the day of forecast is not yet available. Every month, Météo-France provides the gridded SAFRAN reanalysis of these meteorological data for the month before, to consolidate real-time SAFRAN data. These data are averaged at the catchment scale for each catchment.

Users can provide the last observed streamflows up to the day before forecasting, as daily streamflow for a catchment at the day of forecast is not yet available. These observed streamflow data are used within assimilation schemes or post-correction procedures for low-flow forecasting. If observed data is not available during the seven last days, models do not use assimilation schemes or post-correction procedure.

The database containing the streamflow and meteorological data is updated on a daily basis and initial models states are computed with these data at the day of forecast.

The forecasts are then computed by applying the ESP method described above.

3.4 Results

For each model, the PREMHYCE tool provides streamflow forecasts for each scenario and for the no precipitation scenario, for the next 90 d, under numerical or graphical formats. Figure 3 shows an example of synthesis plot provided by the tool for a forecast issued on 11 April 2017 for the Ill River at Didenheim with the GR6J model. It represents:

- Quantiles (0.1, 0.25, 0.5, 0.75 and 0.9) of the distribution of streamflow forecasts for the next 90 d (blue envelop curve and blue dashed line).
- Streamflow forecasted with the no precipitation scenario (orange line).
- Quantile 0.1 and 0.9 of the natural variability of observed streamflow (grey envelop curve), defined for a given calendar day d as the distribution of available streamflows in the historical years for this day. It is used as a benchmark to compare the streamflow forecasts.

- Probability for streamflow forecasts to be under each threshold provided by user for the next 90 d (i.e. number of scenarios under each threshold).
- Probability for the natural variability of observed streamflow to be under each threshold provided by user for the next 90 d.
- Cumulative precipitation of each meteorological scenario for the next 90 d.
- Temperature of each meteorological scenario for the next 90 d.

Here are also represented observed streamflow (black line), and simulated streamflow by GR6J using observed meteorological P and PE (brown line), that are normally not available in real-time.

4 Case study: 2017 low-flow forecasts for the Ill River at Didenheim

In 2017, the PREMHYCE tool was launched every day from the 1 March to the 1 October. Here, results will not include the SIM model, which was only supplied once a month.

We chose to present the results of low-flow forecasts for the Ill River at Didenheim. The Ill River is located in the north-east part of France. Catchment area is 660 km^2 and mean annual streamflow is 6700 $L\,s^{-1}$. Daily streamflow data are available over the 1974–2018 period. Operational threshold provided by users are 1100 $L\,s^{-1}$ for vigilance, 800 $L\,s^{-1}$ for alert, 730 $L\,s^{-1}$ for reinforced alert and 650 $L\,s^{-1}$ for crisis. They correspond respectively to the percentiles 8 %, 4 %, 2 % and 1 % of the daily flow duration curve. The year 2017 has been particularly severe in terms of duration and severity of low flows: streamflow remained under the vigilance threshold during 71 d over the March-October period, and 23 d under the crisis threshold.

The quality of low-flow forecasts is evaluated using the probability of detection (POD). It is based on the contingency table for low flows considering a threshold (Schaefer, 1990), and is computed considering the number of Hits and Correct misses as follow:

$$\mathrm{POD} = \frac{\mathrm{Hits}}{\mathrm{Hits} + \mathrm{Correct\,Misses}} \quad (1)$$

Here, an event is considered well forecasted if more than 50 % of members are below the low-flow threshold.

Table 3 presents the probability of detection (POD) for the vigilance and reinforced alert threshold for each model and for the 7 and 30 d ahead lead-times. All the models show a good ability to detect vigilance threshold at 7 d ahead, between 0.83 for IrMo and 0.92 for PRESAGES. This ability decreases when lead-time increases, whatever the model and the threshold. Models have more difficulties to anticipate cross of thresholds early, which is an expected result. Surprisingly, Gardenia presents a better POD for 30 d lead-time than for 7 d lead-time. This may come from the post-correction method used, which is different from the three other models. More investigations are needed to better understand this result.

POD also decreases for the lower threshold (i.e. reinforced alert), whatever the model or lead-time. Models have more difficulties to detect extreme low-flows. This could be due to the use of meteorological input scenarios, bearing in mind that 2017 is among the driest year on record and that the use of ESP always tends to provide meteorological scenarios that are statistically wetter for severe low-flow periods, leading to an overestimation of low-flows.

Hydrological models present the same trends, but significant differences can be observed for the reinforced alert threshold where PRESAGES seems to be slightly better.

Models reliability has been evaluated using the containing ratio (see Nicolle et al., 2014). For this catchment, models appear to be quite reliable, especially Presages and IrMo for the 7 d lead-time. For Gardenia and GR6J, reliability is improved for the 30 d lead-time.

Models are able to forecast the cross of threshold 35 d in advance on average (at least 50 % of ensemble members below 80th percentile of the streamflow distribution), even if there are some differences between models. Overall, comparing hydrological models to the natural variability of streamflow shows the interest of using hydrological models: streamflow ensemble forecast from hydrological models provide sharper and more accurate ensemble. Moreover, using natural variability of streamflow as an ensemble forecast does not allow detecting the crossing of thresholds: most of the observed streamflows in the past years are superior to the thresholds. The representation of soil humidity conditions by hydrological models at the day of forecast is essential to improve low-flow forecasts, although this representation remains less important than having reliable meteorological input scenarios for longer lead-time.

5 Discussion and conclusion

Improving forecasting input scenarios seems to be the most promising line of action, in order to increase the efficiency of hydrological forecasts at longer lead-times. Low-flow forecasting is less efficient when users need it the most, i.e. when the conditions are particularly dry, because the ESP method tends to overestimate real conditions by construction. Using conditioning methods to constrain input scenarios could be an interesting way to improve low-flow forecasting, as well as using ensemble forecasts from meteorological models. The use of seamless inputs scenarios combining both ESP and ensemble forecasts (typically from Météo-France or the European Centre for Medium-Range Weather Forecasts – ECMWF) could also be implemented.

The results shown on the Ill River are only an example. The relative merits of the models are different among catchments. A more thorough analysis is needed to get a more general evaluation on all the catchments where the PREMHYCE was run.

The PREMHYCE project has implemented several hydrological models for low-flow forecasting in a common structure. Results on the 2017 low-flow periods showed the interest of using such a tool to help end-users decisions.

There is ongoing work to improve low-flow forecasting by integrating short-term or mid-term meteorological forecasts as inputs, and by taking into account human influences such as dam or irrigation. The tool will also give the possibility to combine streamflow forecasts in a multi-model approach. The operational prototype currently tested by operational users will be more widely spread to practitioners in the coming months.

Data availability. Readers can access streamflow observations used in this study at the HYDRO database website (http://www.hydro.eaufrance.fr/) and climatic data from the Météo-France portal (https://publitheque.meteo.fr/). Hydrological and climatic data were processed into a joint database by Delaigue et al. (2020), with synthesis files available at https://doi.org/10.15454/UV01P1.

Author contributions. PN wrote the first draft of the article. PN and FT performed the computations. PN, FB, OD, PE, DF, MLL, CP, FR, DT and FT contributed the methodological developments, modelling experiments and results analysis. All authors discussed the results and contributed to the final manuscript.

Competing interests. The authors declare that they have no conflict of interest.

Special issue statement. This article is part of the special issue "Hydrological processes and water security in a changing world". It is a result of the 8th Global FRIEND–Water Conference: Hydrological Processes and Water Security in a Changing World, Beijing, China, 6–9 November 2018.

Acknowledgements. The authors thank Météo-France for providing meteorological data and the national hydrometeorological forecasting centre (SCHAPI) for providing streamflow data. The Regional Directions for the Environment (DREAL) are also thanked for providing streamflow data and their feedback on the project. The comments of an anonymous reviewer helped to improve the manuscript.

Financial support. The PREMHYCE project was funded by the French Office for Biodiversity (OFB) and the Direction for Fresh-Water and Biodiversity of the French Ministry for the Ecological Transition (MTES).

References

Acreman, M. C. and Dunbar, M. J.: Defining environmental river flow requirements – a review, Hydrol. Earth Syst. Sci., 8, 861–876, https://doi.org/10.5194/hess-8-861-2004, 2004.

Bousquet, S., Gaume, E., and Lancelot, B.: Évaluation des enjeux socio-économiques liés aux étiages de la Seine, Houille Blanche, 145–149, https://doi.org/10.1051/lhb/2003066, 2003.

Chataigner, J. and Michon, J.: Prélèvements quantitatifs sur la ressource en eau (données 2013) (Note de synthèse Bulletin no. 2), BNPE, Agence française pour la biodiversité, 2017.

Chiew, F. H. S. and McMahon, T. A.: Global ENSO-streamflow teleconnection, streamflow forecasting and interannual variability, Hydrol. Sci. J., 47, 505–522, https://doi.org/10.1080/02626660209492950, 2002.

Coron, L., Perrin, C., and Michel, C.: airGR: Suite of GR Hydrological Models for Precipitation-Runoff Modelling, available at: https://webgr.Inrae.fr/en/airGR/ (last access: 27 July 2020), 2017a.

Coron, L., Thirel, G., Delaigue, O., Perrin, C., and Andréassian, V.: The Suite of Lumped GR Hydrological Models in an R package, Environ. Model. Softw., 94, 166–171, https://doi.org/10.1016/j.envsoft.2017.05.002, 2017b.

Crochemore, L., Ramos, M.-H., Pappenberger, F., and Perrin, C.: Seasonal streamflow forecasting by conditioning climatology with precipitation indices, Hydrol. Earth Syst. Sci., 21, 1573–1591, https://doi.org/10.5194/hess-21-1573-2017, 2017.

Day, G.: Extended Streamflow Forecasting Using NWSRFS, J. Water Resour. Plan. Manag., 111, 157–170, 1985.

Delaigue, O., Génot, B., Lebecherel, L., Brigode, P., and Bourgin, P. Y.: Database of watershed-scale hydroclimatic observations in France, Université Paris-Saclay, INRAE, HYCAR Research Unit, Hydrology group, Antony, https://doi.org/10.15454/UV01P1, 2020.

Demirel, M. C., Booij, M. J., and Hoekstra, A. Y.: The skill of seasonal ensemble low-flow forecasts in the Moselle River for three different hydrological models, Hydrol. Earth Syst. Sci., 19, 275–291, https://doi.org/10.5194/hess-19-275-2015, 2015.

Gupta, H. V., Kling, H., Yilmaz, K. K., and Martinez, G. F.: Decomposition of the mean squared error and NSE performance criteria: Implications for improving hydrological modelling, J. Hydrol., 377, 80–91, https://doi.org/10.1016/j.jhydrol.2009.08.003, 2009.

Habets, F., Ackerer, P., Amraoui, N., Augeard, B., Besson, F., Caballero, Y., Dreuzy, J.-R. D., Moigne, P. L., Longuevergne, L., Martin, E., Morel, T., Regimbeau, F., Rousseau, M., Rousseau-Gueutin, P., Thiery, D., Soubeyroux, J.-M., Vergnes, J.-P., and Viennot, P.: Aqui-FR, un système multi-modèle hydrogéologique à l'échelle nationale, Géologues, 187, 105–109, 2015.

Karamouz M. and Araghinejad S.: Drought Mitigation through Long-Term Operation of Reservoirs: Case Study, J. Irrig. Drain. Eng., 134, 471–478, https://doi.org/10.1061/(ASCE)0733-9437(2008)134:4(471), 2008.

Nash, J. E. and Sutcliffe, J. V.: River flow forecasting through conceptual models, part I. A discussion of principles, J. Hydrol., 10, 282–290, https://doi.org/10.1016/0022-1694(70)90255-6, 1970.

Nicolle, P., Pushpalatha, R., Perrin, C., François, D., Thiéry, D., Mathevet, T., Le Lay, M., Besson, F., Soubeyroux, J.-M., Viel, C., Regimbeau, F., Andréassian, V., Maugis, P., Augeard, B., and Morice, E.: Benchmarking hydrological models for low-flow simulation and forecasting on French catchments, Hydrol. Earth Syst. Sci., 18, 2829–2857, https://doi.org/10.5194/hess-18-2829-2014, 2014.

Oudin, L., Hervieu, F., Michel, C., Perrin, C., Andréassian, V., Anctil, F., and Loumagne, C.: Which potential evapotranspiration input for a lumped rainfall–runoff model?: Part 2-Towards a simple and efficient potential evapotranspiration model for rainfall–runoff modelling, J. Hydrol., 303, 290–306, https://doi.org/10.1016/j.jhydrol.2004.08.026, 2005.

Risva, K., Nikolopoulos, D., Efstratiadis, A., and Nalbantis, I.: A Framework for Dry Period Low Flow Forecasting in Mediterranean Streams, Water Resour. Manag., 32, 4911–4932, https://doi.org/10.1007/s11269-018-2060-z, 2018.

Schaefer, J. T.: The Critical Success Index as an Indicator of Warning Skill, Weather Forecast., 5, 570–575, https://doi.org/10.1175/1520-0434(1990)005<0570:TCSIAA>2.0.CO;2, 1990.

Vidal, J.-P., Martin, E., Franchistéguy, L., Baillon, M., and Soubeyroux, J.-M.: A 50-year high-resolution atmospheric reanalysis over France with the Safran system, Int. J. Climatol., 30, 1627–1644, https://doi.org/10.1002/joc.2003, https://doi.org/10.1002/joc.2003, 2010.

Proc. IAHS, 383, 391–399, 2020
https://doi.org/10.5194/piahs-383-391-2020

Climatic variability in the Sine-Saloum basin and its impacts on water resources: case of the Sob and Diohine watersheds in the region of Niakhar

Waly Faye[1,2], **Awa Niang Fall**[1], **Didier Orange**[2,3], **Frédéric Do**[2], **Olivier Roupsard**[2,4], **and Alioune Kane**[1]

[1]UCAD, Université Cheikh Anta Diop, Faculté de Géographie, UMI RESILIENCES, Dakar, Sénégal
[2]IRD, Université de Montpellier, CIRAD, INRA, SupAgro Montpellier, UMR Eco&Sols, Montpellier, France
[3]LMI IESOL, Centre ISRA/IRD, Bel-Air, Dakar, Sénégal
[4]CIRAD, Université de Montpellier, CIRAD, INRA, SupAgro Montpellier, UMR Eco&Sols, Montpellier, France

Correspondence: Didier Orange (didier.orange@ird.fr)

Published: 16 September 2020

Abstract. Local peoples from Niakhar in the Senegalese peanut basin highlight a dramatic increase of water access problems due to marked rainfall deficits and salinization of surface and ground water resources. The chemical quality of groundwaters is often critical because of the salinization process, whereas water surfaces, which should be used in such situations, are up early. More and more, lowlands and rivers beds are pervaded by salt crusts. Then the salinization of wells is increasing, leading to the extension of tans (salty of acidified soils). To study the impacts of climatic pejoration on the agroecosystems and on the living conditions of the populations, we carried out the analysis of the time series of the precipitations with daily and annual time steps from 1950 to 2015 on 6 meteorological stations, in situ measurements on 78 wells for an area of 311 km^2, as well as local population interviews and field observation. The results confirm an important climatic break in the region in 1970. The long dry period, from 1970 to 2009, has increased the annual rain variability, decreased the number of rainy days per year. We confirm a real and large extension of well salinization, and salt crusting in the lowlands and the riverbeds. From the local people, it seems the process of degradation of the aquifers continues to progress from a large tidal event in 1984. The rainfall increase noted in the last decade does not seem to be enough to reverse the trend and to ensure both the rise of the piezometric level of the aquifers and the desalinization of surface and ground waters.

1 Introduction

Agricultural production, river discharge and groundwater productivity are related to the abundance and the regularity of rainfalls, mainly under semi-arid zones. In the Sahel, the environmental sustainability is very fragile, the rainfall is the most immediate climatic element for water harvesting to meet the natural needs from the ecosystems and for human demands. The drastic rainfall deficit that was almost continuous in the Sahel from 1968 to the end of the 1980s (Servat et al, 1999) led to a depletion of the surface waters of the Sudano-Sahelian zone (Descroix et al., 2009). The severity of the rainfall deficit has brought the Senegal river to its lowest level in 1984–1985 (Gac et Kane, 1986; Orange, 1992). In the Sine-Saloum basin, the rainfall deficit from 1971 to 2000 is more than 300 mm compared to the period 1941 to 1970, as a reduction of the annual rainfall of more than 30 % in the estuarine part and 15 % in the upper basin (DGPRE, 2014). In this region, this altered climate condition has led to a decrease of river flows, a drop of groundwater table, a drying up of the wells, a desiccation of the pools which are very important for the breeders, a significant degradation of the abiotic environment in particular with the salinization of soils, waters and wells, and by a considerable degradation of the vegetation cover (Sagna et al., 2015).

But in despite the drought, the population density has largely increased, and the region still continues to be an important agricultural production area for Senegal. The density which was 100 inhabitants km^{-2} in the years 1960–1970 (Lericollais, 1970) rose to more than 200 inhabitants km^{-2} in the years 2010 (Delaunay et al., 2009), increasing the pressure on the natural resources. This recent demographic increase is more accentuated by the fact that the region of Sine-Saloum (Fatick and Kaolack) was already the most populated region of Senegal until the 1980s, combining a very large rural population and the densest urban network after Dakar (Becker et al., 1987). According to Pontié et al. (1999), the analysis of the land management methods takes a particular relief in Sine, because of the high density of population, which entails as a corollary an exploitation of all the lands and a strong land pressure.

To understand why this environment is particularly resilient, we have focused our research on two small agricultural watersheds in the Sine-Saloum basin, with the villages of Sob and Diohine. These two villages are in the center of the region studied by IRD and its partners for the Niakhar Population, Health and Environment Observatory (OPSE) in the administrative region of Fatick (Delaunay et al., 2009). Already Lericolais (1989) indicated that the water level of the central well of Sob which was at 14 m depth in 1989, was at 6 m depth in 1965. Moreover, this well is no longer used since 1989 due to its high salt content. Then the local population is convinced that the impact of the pluviometric decrease on water resources is reflected not only in a continuous decline of piezometric levels but with an important salinization process.

This study aims to provide a first understanding of the impact of the long recent drought on the rainfall variability, and secondly its consequences on the groundwater quality, highlighted through the local population perception on the water demand stress of the populations of Sob and Diohine watersheds. We propose to analyze the long serial rainfall records on 6 stations to characterize the effective climatic evolution in this part of the agricultural region of the Senegalese peanut basin. Then, because there is no hydrological data, we will discuss the impact on the water resources through the actual salinity measurements and interviews of local people.

2 Location of Sob and Diohine watersheds

Sob and Diohine are two studied villages of the Niakhar Population, Health and Environmental Observatory (OPSE) implemented by the Research Institute for Development (IRD) and its partners in Senegal since the 1960s. For this purpose, we have described two watersheds centered on these two villages respectively. The both watersheds are oriented NNE-SSW and the respective areas are 115 and 96 km^2 (Fig. 1). Their outlets, chosen at the national road RN1 on the left and right of the village of Diouroup, are about 7 km from the entrance to the Sine-Saloum estuar. The topography of the region is relatively flat, with just some circle lowlands, the elevation does not exceed 20 m, and the Sob watershed is the highest. The climatic zone is sudanian, reputed with an annual rainfall of 600 mm. The soils consist mainly of Dior soil, the most widespread, Deck and Deck-Dior soils, with Tann soil often in the lowlands. The Dior soil has a coarse sandy texture to sandy-silty with less than 40 % clay in the first 40 cm; Deck soil corresponds to soil with sandy-silty to silty-sandy texture on the surface and more clayey at the bottom; Deck-Dior soil is with a texture very sandy-silty (with more than 15 % of clay) to silty; and Tann soil is salted or acidified soils entirely non-fertile. The structure of the deep horizon is massive with however a polyhedral substructure. The main activities of the population are agriculture, breeding, small gardening market, and more and more cattle fattening.

3 Data and method

3.1 Data used in the study

To investigate the rainfall pattern and climate variability in the watersheds of Sob and Diohine, two types of data were used. The first data base concerns the range of monthly rainfall data from 1950 to 2015 for 6 stations Fatick, Diourbel, Gossas, Bambey, Thiadiaye and Mbour. The daily rainfall concerned only Fatick, Diourbel, Bambey and Gossas stations. The choice of these stations whose data are provided by the National Agency of Civil Aviation and Meteorology (ANACIM), is focus on the quality of data (continuous and homogeneous) and proximity to the two studied basins.

The second database is dealing with in situ water quality measurements done in June 2018 on a sample of 78 wells equally distributed on Sob and Diohine watersheds. On each well, the static level (NS), the pH, the Conductivity and the Salinity were measured in situ and a GPS survey carried out systematically for the spatialization of the data. The NS of a well or borehole is the depth to water from the ground to the surface of the water before pumping. Conductivity, pH and salinity were measured in situ by the following equipment respectively: LAQUAtwin conductivity EC-22 measured in mS, LAQUAtwin-pH-22 pH meter and LAQUAtwin-Salt-22 salt-meter measured in percent. All these measurements are associated with field observations through an exploration of the two basins up to the outlet and interviews with rural populations to collect primary data on their perceptions on the trends and variability of rainfall and water resources of the area.

3.2 Method implemented

3.2.1 Standardized Precipitation Index (SPI)

SPI index is used to examine the nature of the trends and enables the determination of the dry and wet years in the

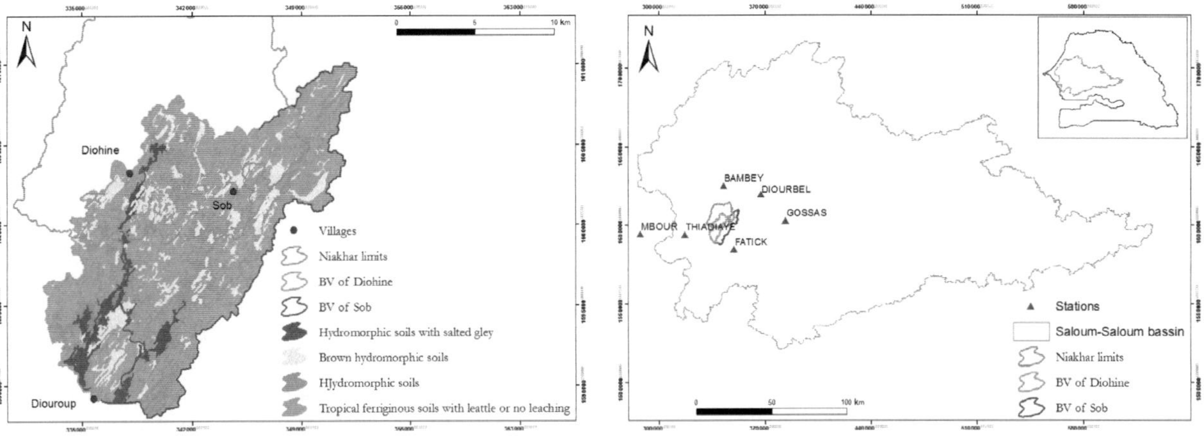

Figure 1. Location of the Sob and Diohine watersheds in the Niakhar Population Health and Environment Observatory (OPSE), and location of the 6 studied rainfall stations.

Table 1. The intervals of the SPI index values for identifying rainfall anomalies (from Shah et al., 2015).

Values of SPI	Characteristic of the year
$Z > +2.0$	Extremely wet
$1.5 < Z < 1.99$	Wet
$1 < Z < 1.49$	Moderately wet
$-0.99 < Z < 0.99$	Normal
$-1.49 < Z < -1$	Moderately dry
$-1.99 < Z < -1.5$	Dry
$Z < -2$	Extremely dry

record (Hadgu, 2013). Standardized precipitation is defined as the difference in precipitation from the mean for a specified period divided by the standard deviation. This index has the advantage of being able to determine the water deficit at both the scale of the seasons and the year (Elbouqdaoui et al., 2006). It is given by the following formula:

$$Z = \frac{P_i - \overline{P}}{\sigma} \quad (1)$$

Where P_i is the rainfall totals of a given year, $\overline{P}$ the mean of the range, σ is the standard deviation of annual precipitation of the observed period. This index measures meteorological drought and is negative for droughts and positive for wet conditions. The intervals of the index values for identifying rainfall anomalies are indicated in Table 1 (Elbouqdaoui et al., 2006; Shah et al., 2015).

3.2.2 Coefficient of variation

The annual coefficient of variation (CV) is calculated to assess the variability of each year. The annual coefficient of variation allows the comparison between weakly or abundantly watered stations. CV is the ratio between annual standard deviation σ and annual mean rainfall $\overline{P}$.

$$\mathrm{CV} = \frac{\sigma}{\overline{P}} \quad (2)$$

The CV is used to classify the degree of variability of rainfall events as less ($\mathrm{CV} < 20$), moderate ($20 < \mathrm{CV} < 30$) and high ($\mathrm{CV} > 30$) according to Asfaw et al. (2018).

4 Results

4.1 A rainfall break in 1970

The rainfall evolution at the studied stations highlights the rainfall anomaly that has been observed in the region since the rainfall break in 1970. For all the stations, the meteorological drought is persistent and the rainfall pattern from 1950 to nowadays is very similar for each one (Table 2, Fig. 2). Just few humid years occurred at Fatick and Mbour since 2009. However, we could not say anything for Gossas and Thiadiaye since the serial data ended before 2009 for these two stations.

The intervals of the SPI index values allow to identify rainfall anomalies, that is finely very similar for the six stations. At last, the serial data shows a resumption of rainfall over the last decade but with rainfall that is still far from normal. To conclude, we identify two homogeneous rain periods: one wet period until 1969, and one dry period from 1970 until 2008. More data are needed to qualify the recent period from 2009, which seems wetter than previously.

4.2 Increase of annual variability and reduction of rainy days per year

The evolution of the annual coefficient of variation for all the studied stations shows a strong variability of the average annual rainfall, mainly during the last dry homogenous period (Table 3): less it rains, more it is yearly variable. Indeed,

Table 2. Rainfall anomalies for the studied stations from 1950 to 2009 for Gossas, from 1950 to 2011 for Thiadiaye, and from 1950 to 2016 for Bambey, Fatick, Diourbel and Mbour.

Stations	Extremely wet	Wet	Moderately wet	Normal	Moderately dry	Dry	Extremely dry
Gossas	2	3	6	39	10	0	0
Fatick	2	4	7	43	8	3	0
Diourbel	4	0	5	45	11	2	0
Bambey	2	2	7	43	12	0	0
Mbour	3	1	6	44	8	3	1
Thiadiaye	3	2	3	34	9	6	0

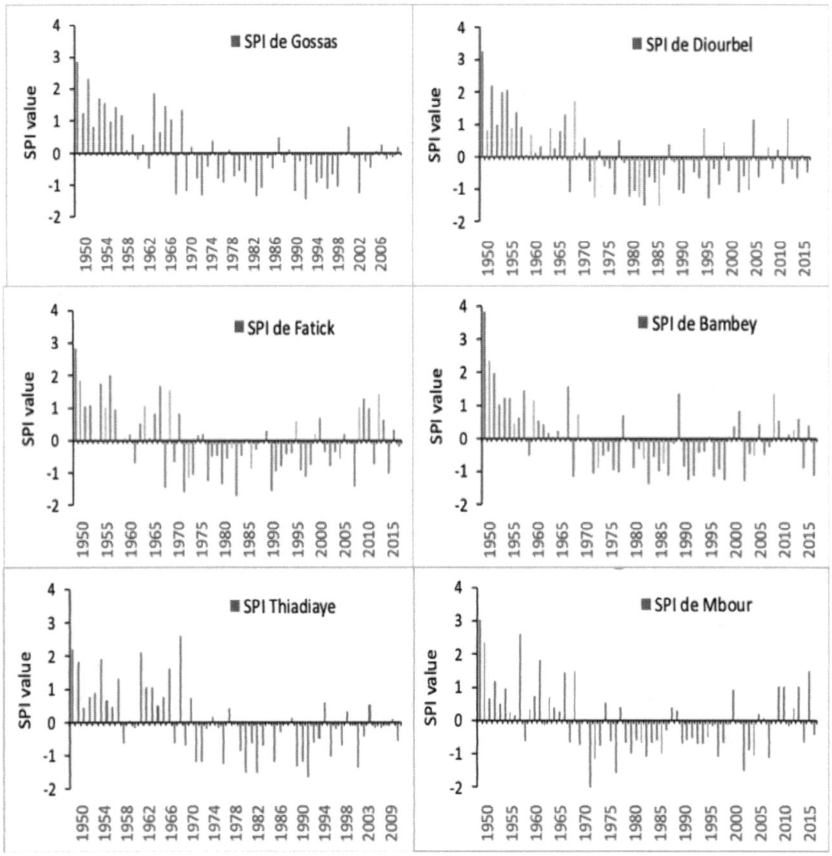

Figure 2. Standard precipitation index (SPI) of the 6 studied rainfall stations, from 1950 to 2009 for Gossas, from 1950 to 2011 for Thiadiaye, and from 1950 to 2016 for Bambey, Fatick, Diourbel and Mbour.

the variability is more pronounced over the 1970–2009 sequence, as shown in Fig. 3. The wettest period, 1950–1969, generally has the lowest coefficients of variation, whereas the period 1970–2009 shows high coefficients of variation. The period 1970–2009 in addition to be the dry sequence is very variable, which demonstrates particularly drastic climatic conditions that characterized this period.

Moreover, this variability increases along an east-west positive gradient, and a smoothly positive gradient from the North to the South. In resume, more the station is at the West (near the coast such as Mbour) and more in the South, more the annual variability is important (Table 2).

Moreover the annual variability, the number of rainy days per year has also changed between the two homogeneous climatic periods. From 1950 to 1968, the number of rainy days has constantly decreased during the whole dry period until a certain threshold relatively constant from 1970 to 2009 (Fig. 4). Gossas has the lowest average number of rainy days

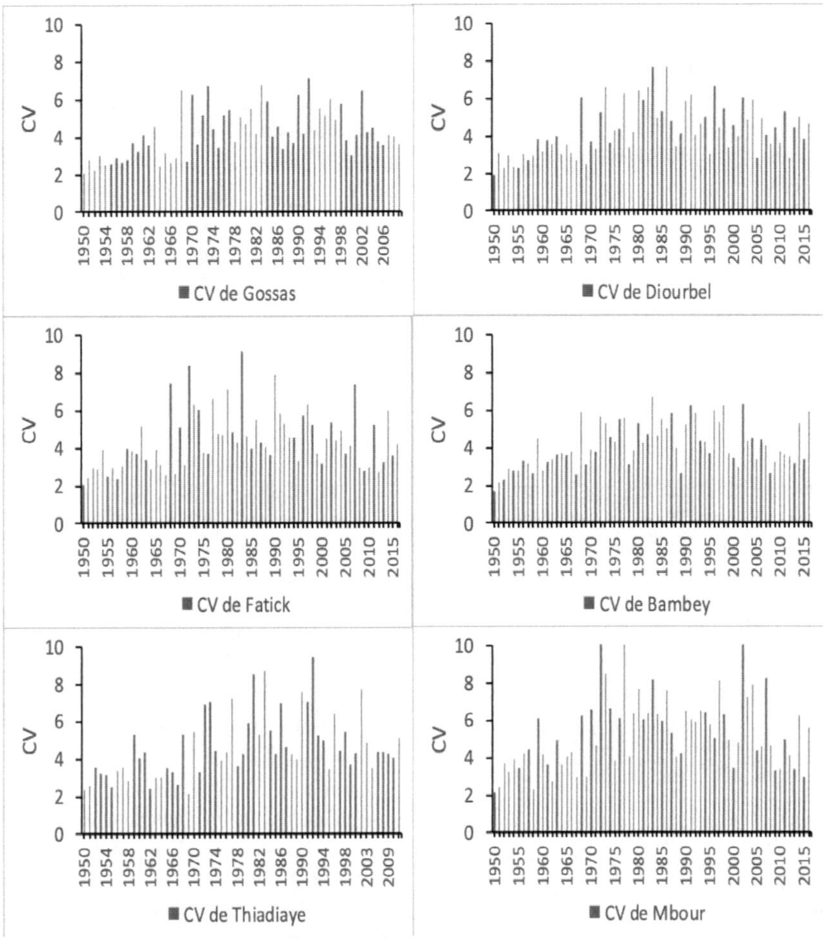

Figure 3. Evolution of the coefficient of variation for the 6 studied rainfall stations, from 1950 to 2009 for Gossas, from 1950 to 2011 for Thiadiaye, and from 1950 to 2016 for Bambey, Fatick, Diourbel and Mbour.

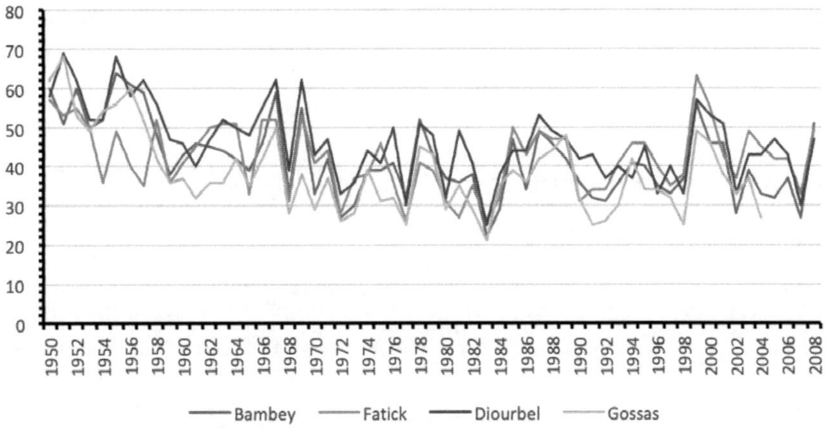

Figure 4. The evolution of the number of rainy days per year at Gossas, Fatick, Bambey and Diourbel stations from 1950 to 2008.

Table 3. Evolution of the coefficient of variation of the 6 studied rainfall stations for the two homogeneous climatic period of 1950–1969 and 1970–2009 (the geographical location is qualified by North, South, East and West, related to the 2 studied watersheds of Sob and Diohine; example: Gossa station with EE-N is at the extreme East and the North of the studied watershed).

Rainfall stations	Geographical location	1950–1969	1970–2009
Gossas	EE-N	3.1	4.8
Djourbel	E-N	3.1	4.9
Fatick	E-S	3.4	5.0
Bambey	N	3.2	4.6
Thiadiaye	W-S	3.3	5.4
Mbour	WW-S	3.8	6.6

with $34\,\mathrm{d\,yr^{-1}}$, and Diourbel has the highest average with $43\,\mathrm{d\,yr^{-1}}$, while Fatick and Bambey have respectively an average of 43 and $38\,\mathrm{d\,yr^{-1}}$. This can have dramatic consequences on agricultural crop cycles and on the water quality of rivers and groundwater recharge.

4.3 Rainfall amount per period and geographical trend

The first important result of the local evolution of this long serial rainfall data leads to affirm that in the region we should consider at least two homogeneous climatic period with a break in 1970. For the recent period, our data until 2016 seems to announce on a new positive break which could be in 2009. Then we could consider a wet period from 1950 to 1969 and a dry period from 1970 to 2009 (Fig. 5). However a need of at least 6 years more of rainfall records is requested to be sure to conclude on that issue.

Due to the small geographical trend from East to West and from North to South, we can consider the mathematical average of the 4 nearest rainfall stations from the OPSE (Diourbel, Bambey, Fatick and Thiadiaye) to estimate the average annual rainfall amount occurring on this area, since our studied area is well-centered between these 4 rainfall stations (Fig. 1).

Then the annual rainfall amount has changed in average from 753 to $495\,\mathrm{mm\,yr^{-1}}$, which represents a drastic loss of rainfall water by 34 %. From 2010, the annual rainfall in the region would be better with an average around $590\,\mathrm{mm\,yr^{-1}}$. But more than 10 years of measurements will be needed to confirm.

4.4 Drought impact on the quality of groundwater resources

Salinization has become a major constraint in the sylvopastoral area of Niakhar. And it concerns both water resources and soils. In situ measurements, observations and interviews with the rural populations allow to tackle the extent of the groundwater degradation in Sob and Diohine watersheds.

The current situation is highlighted by the high rate of abandonment of wells in most villages. This water problem has led to the specialization of wells in terms of uses: some wells are reserved for the population, others for domestic use and livestock, and others for the brick building in banco. In addition, there is an early drying of some wells, or a drop in the static level of wells that can no longer meet the water needs of populations in the middle of the dry season. Our survey of the static level in 78 wells indicated an average of static level around 4 to 5 m depth (Fig. 6). The water level in the wells is less deep in the Diohine watershed than in the Sob watershed. Here, we have to notice that the elevation of Sob watershed is smoothly upper. Globally, it seems that the static level of the groundwater is following the topographic slope. However, in details, the highest static levels are out of the lowland areas.

The measurements of water salinity in the wells are not correlated to the static levels. The salt concentration ranges from 0 % to 3 % (Fig. 7). The highest salt concentrations have been measured in the Sob watershed at Yenguele II, a village near Niakhar with 3.1 %. However, we notice a high rate of abandonment of the wells in the two basins because of the aquifers' salinization so that the cattle no longer drink them. Wells, which have the softest waters, are in downstream of the two basins.

There are no old measurements on wells to determine some changes in groundwater quality. But in the opinion of the local people, it is a long time process, specially from 1984 within both climatic drought and big tidal events. From this date, the wells have begun to be salty until they are abandoned because even the animals do not drink the water anymore. Nowadays, many new constructed wells directly capture salt water and are immediately abandoned. In addition to this degradation of the aquifers, the local people underlined an early drying up of the lowlands, river beds and pools which nevertheless drain large volumes of water each year. Moreover, crystalline salt is exploited in the Diohine valley downstream to Diouroup in the dry season, while towards the outlet of the Sob Basin salt crusts on lowland soils were visible in June. This situation shows the intensity of lowlands salinization mainly in downstream of the both basins. And it should be noted that these places are far from the Yenguele II area where the saltiest well is measured.

5 Discussion

The analysis of the standardized precipitation index (SPI) combined to the rain annual coefficient of variation and to the number of rainy days per year allow to conclude that in the whole studied area, the annual evolution of the rainfall is homogeneous on the studied period (1950–2016). We confirm a large and fundamental climatic break around 1969. In-

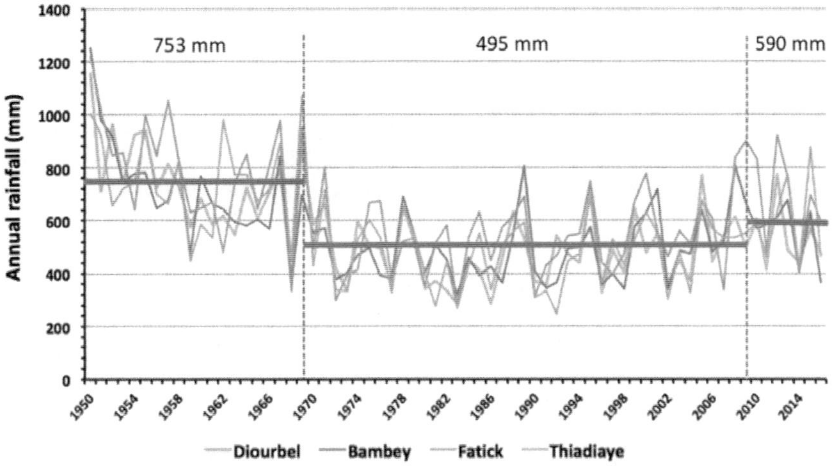

Figure 5. Evolution of the annual rainfall of the 4 climatic stations around the Diohine and Sob watersheds (from the East to West: Diourbel and Bambey in the North, Fatick and Thiadiaye in the South) to identify three homogeneous climatic period (1950–1969, 1970–2009 and 2010–2016), with the mean annual rainfall (in mm) for each considered period.

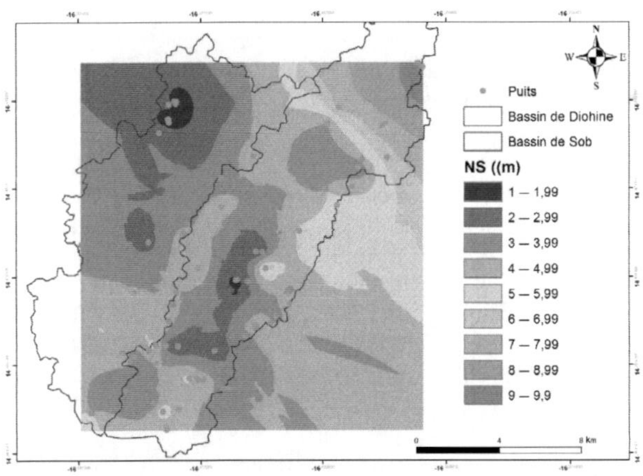

Figure 6. Distribution of static levels (NS) of wells in the Sob and Diohine watersheds.

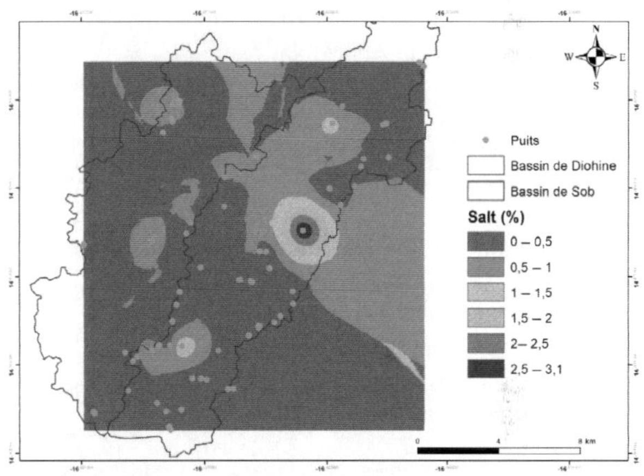

Figure 7. Percentage of water salinity distribution in wells distributed in Sob and Diohine watersheds, one-shot measurements campaign in June 2018.

deed, 1970 marked the beginning of a long dry period within the rainfall data series. This drought has increased the annual variability of the rainfall and decreased the number of rainy days per year. From 2010, it seems that the annual rainfall amount would have increased a little bit. But this positive trend should be confirmed in the next years.

The interviews of local people done in June 2018 within 100 persons, make us confirm that this long drought period has impacted the hydrological pattern in the region. The particularly random nature of the rainfall highlights the vulnerability of water resources in this area. In Sob and Diohine watersheds, the long and pronounced drought period and population growth have exacerbated the strong pressure on water resources and the salinization became one of the limiting factors for the water use. We hypothesize the pronounced groundwater recharge deficits and the increased of water use for agricultural development in the region have primed and accelerated the water salinization in wells and probably the lowering of the piezometric level. Nowadays, we don't have any old hydrological measurements to confirm this statement. From the local people, it seems the process of degradation of the aquifers continues to progress from a large tidal event of 1984. And the small increase of annual rainfall amount noted in the last decades does not seem to turn off. More and more, the recent dug wells capture salt water while the old wells are abandoned because of the strong salt water content. Many villages in these two basins now only have salty and brackish tablecloths for their domestic, pastoral and market garden uses. We noticed that the few wells that produce fresh water dry up because supply is far below demand. Our measurements showed also a non-uniform salt concentration between wells inside a same concession. The spatial variability is significant and seems to indicate several causes of the saline water pollution in the wells.

On the other hand, the long drought has also led to wind revitalization that has led to a silting up of the lowlands by the eolian dust and probably contributed to the salty pollution of soil surfaces. The salt crusts are also visible into the riverbeds and in the lowlands leading to hypothesize a saline pollution by hydraulic lift through capillary process. It is clear that the long drought enhanced also the evaporation process, facilitating the capillary process and the salt soil crusting. The current long dry period, from 1970 to 2009, brought many favorable factors for soil degradation largely linked to the salt dynamics. Indeed, the soil and water salinization are effective. But currently we don't have any clue to discuss about the origin of this salt and about the salinization process in the region.

Nowadays, people harvest salt in the Diohine valley before the outlet, which is very far from the estuary while salt crusts are visible in the Sob valley around Diouroup. This situation is indicative of the degradation of the hydrological conditions of the Sob and Diohine watersheds. The potential of the lowlands is strongly affected by evaporative salt concentrations in the dry season; it is a constraint for rice cultivation, which has disappeared due to soil salinization. In this context, it is essential to know the volumes of water flowing through these two rivers and to identify the origins of the salt deposited in the lowlands and its relationship with salinization of aquifers by future advanced studies.

6 Conclusion

Our study confirmed that the agricultural region of Senegalese peanut basin has been strongly marked by the climate break in 1970 in the rainfall distribution. The dry period is effective in the whole Niakhar region from 1970 to 2009, characterized by an increase of the annual rain variability and a decrease of the number of rainy days per year. Then the climatic water pressure on the water demand has been drastic. But unfortunately, there are no hydrological measurements and data in the region to evaluate the climatic impact of the surface and groundwater resources.

Basically, from the local people interview, the degradation of water resources is nowadays one of the major constraints of the watersheds of Sob and Diohine in the sylvopastoral area of Niakhar due mainly to the soil and water salinization we confirmed by our one-shot field measurements in June 2018. From local people, it seems that the salinization of sols and surface waters has been accelerated by a large tidal event in 1984.

Moreover we hypothesize that the pressure on the hydrosystems exerted by the increase of water demand largely driven by the population growth has breaking off the balance between the needs of the population and available water resources, increasing the salinization process while water needs continue to increase. The water scarcity is now a fundamental concern in this area, a limiting and constraining factor for the activities of the population. The rainfall increase noted in the last decade, does not seem to be enough to reverse the trend and to ensure both the rise of the piezometric level of the aquifers and the desalinization of surface and ground waters.

If we consider the future growing needs of water because of the population increase, because of the agricultural pressure for both livestock breeding and local farming market, how to match with a sustainable water supply for the rural population in this region?

Data availability. All the rainfall data are provided by the National Agency of Civil Aviation and Meteorology (ANACIM), these data are not publicly accessible. The original data for the water quality have been provided as Supplement.

Supplement. The supplement related to this article is available online at: https://doi.org/10.5194/piahs-383-391-2020-supplement.

Author contributions. Conceptualization was done by WF, ANF and DO; methodology has been done by WF and DO, data curation by WF and the formal analysis by WF and DO; writing of the original draft preparation has been done by WF; and the writing for review and editing was done by WF, ANF, DO, FD and OR; visualization was done by AK. All authors have read and agreed to the published version of the manuscript.

Competing interests. The authors declare that they have no conflict of interest.

Special issue statement. This article is part of the special issue "Hydrological processes and water security in a changing world". It is a result of the 8th Global FRIEND–Water Conference: Hydrological Processes and Water Security in a Changing World, Beijing, China, 6–9 November 2018.

Acknowledgements. This research was carried out under the umbrella of the PhD Scholarship from the French Embassy in Senegal with the support of Campus-France, and the funding support from IRD, LMI IESOL, the research unit Eco&Sols and the international research unit UMI0236 RESILIENCES. The study got technical support from the Faidherbia-Flux Program, "a new highly instrumented collaborative Observatory in a semi-arid agro-silvo-pastoral system of Western Africa (Niakhar-Senegal)", and the People-Health-Environment Observatory of Niakhar (OPSE, from IRD).

Authors wish to thank researchers and technicians from Faidherbia-Flux Programme, specially Ibou Diouf, and from the OPSE, specially Laurence Fleury and Robert Diatte for their help on field and discussion. Special thanks for Luc Descroix for his scientific recommendations. And the reviewers are also thanked for the time dedicated and their comments.

Financial support. This research has been supported by the French Embassy in Senegal, the IRD, the LMI IESOL, the research unit Eco&Sols, and the international research unit UMI0236 RESILIENCES.

References

Asfaw, A., Simane, B., Hassen, A., and Bantider, A.: Variability and time series trend analysis of rainfall and temperature in north-central Ethiopia: A case study in Woleka sub-basin, Weather and Climate Extremes, 19, 29–41, 2018.

Becker, C., Diouf, M., and Mbodj, M.: L'Evolution démographique régionale du Sénégal et du bassin arachidier (Sine-Saloum) au vingtième siècle, 1904–1976, in: African population and capitalism: historical perspectives, edited by: Cordell, D. D. and Gregory, J. W., The University of Wiscousin Press, Madison, USA, 76–94, 1987.

Daulenay, V., Deschamps-Cottin, M., Bertaudière, V., Vila, B., Oliveau, S., Dos Santos, S., Soumare, A., and Lalou, R.: Dynamique démographique et dynamique du parc agroforestier à Faidherbia albida (Del.) A Chev. en pays Serer (Sob, Sénégal), Proceeding in XXVIth International Population Conf., 2 October 2009, Marrakech, Morocco, available at: https://www.researchgate.net/publication/242201196 (last access: 28 July 2020), 2009.

Descroix, L., Mahé, G., Lebel, T., Favreau, G., Galle, S., Gautier, E., Olivry, J.-C., Albergel, J., Amogu, O., Cappelaere, B., Dessouassi, R., Diedhiou, A., Le Breton, E., Mamadou, I., and Sighomnou, D.: Spatio-temporal variability of hydrological regimes around the boundaries between Sahelian and Sudanian areas of West Africa: A synthesis, J. Hydrol., 375, 90–102, 2009.

DGPRE: Plan d'action de gestion intégrée des ressources en eau (PAGIRE) projet de mise en œuvre du PAGIRE dans le bassin arachidier (PAGIRE-BA) étude d'évaluation et de suivi des ressources en eau de surface, R3-rapport sur la modélisation des ressources en eau et les normes de conception des ouvrages, Ministère Ressources en Eau, Dakar, Sénégal, 2014.

Elbouqdaoui, K. Haida, S., Aboulabbes, O., Benaabidat, L., Zahraoui, M., and Ozer, A.: Evolution du régime pluviométrique et hydrométrique du bassin versant du Srou (Moyen Atlas, Maroc), Geo-Eco-Trop, 30, 41–56, 2006.

Gac, J. and Kane, A.: Le fleuve Sénégal: I. Bilan hydrologique et flux continentaux de matières particulaires à l'embouchure. Senegal river. Part I. Water budget and flux of carried load through the estuary, Bull. Sciences Géologiques, ULP, Université de Strasbourg, Strasbourg, France, 99–130, 1986.

Hadgu, G.: Trend and variability of rainfall in Tigray, Northern Ethiopia: Analysis of meteorological data and farmers perception, J. Agric. Res., 1, 88–100, 2013.

Lericolais, A.: La détérioration d'un terroir Sob, en pays Sérèr (Sénégal), Etudes rurales, no. 37/39, Terroirs Africains et Malgaches, 113–128, EHESS, Paris, France, 1970.

Lericolais, A.: La mort de l'arbre à Sob, en pays Sereer (Sénégal), in: Tropiques Lieux et Liens, ORSTOM, Paris, France, 187–197, 1989.

Orange, D.: Hydroclimatologie du Fouta Djalon et dynamique actuelle d'un vieux paysage latéritique (Afrique de l'Ouest), Mém. Sci. Géol., 93, 206 pp., Université de Strasbourg, Strasbourg, France, 1992.

Pontié, G., Guigou, B., and Lericollais, A.: La gestion des terres dans le Sine, in: Paysans Sereer: dynamique agraire et mobilité au Sénégal, edited by: Lericollais, A., IRD, Paris, France, 146–194, 1999.

Sagna, P., Ndiaye, O., Diop, C., Diongue Niang, A., and Sambou, P. C.: Les variations récentes du climat constatées au Sénégal sont-elles en phase avec les descriptions données par les scenarios du GIEC? Pollution Atmosphérique, 227, https://doi.org/10.4267/pollution-atmospherique.5320, 2015.

Servat, E., Paturel, E., Lubès-Niel, H., Kouamé, B., Masson, J. M., Travaglio, M., and Marieuet, B.: De différents aspects de la variabilité de la pluviométrie en Afrique de l'Ouest et Centrale non sahélienne, Revue des Sciences de l'Eau, 12, 363–387, 1999.

Shah, R., Bharadiya, N., and Manekar, V.: Drought Index Computation Using Standardized Precipitation Index (SPI) Method For Surat District, Gujarat, Aquat. Pr., 4, 1243–1249, https://doi.org/10.1016/j.aqpro.2015.02.162, 2015.

AUTHOR INDEX